Fundamental Physical Constants (to six significant digits)

acceleration due to gravity (*g*)	9.806 65 m/s^2
Avogadro constant (N_A)	6.022 14 × 10^{23}/mol
mass of electron (m_e)	9.109 38 × 10^{-31} kg
mass of neutron (m_n)	1.674 93 × 10^{-27} kg
mass of proton (m_p)	1.672 62 × 10^{-27} kg
molar gas constant (*R*)	8.314 47 J/mol·K
molar volume of gas at STP	22.414 0 L/mol
speed of light *in vacuo* (*c*)	2.997 92 × 10^8 m/s^2
unified atomic mass (*u*)	1.660 54 × 10^{-27} kg

Common SI Prefixes

tera (T)	10^{12}	centi (c)	10^{-2}
giga (G)	10^9	milli (m)	10^{-3}
mega (M)	10^6	micro (μ)	10^{-6}
kilo (k)	10^3	nano (n)	10^{-9}
deci (d)	10^{-1}	pico (p)	10^{-12}

Quick Guidelines for Significant Digits

Multiplication & Division: keep the least number of digits of any number in the question.

(e.g. 3.06 × **4.0** = 12.24 → **12**)

Addition & Subtraction: keep the least significant (left most) placeholder.

(e.g. 4.1**7** + 3.206 = 7.376 → 7.3**8**)

(For more on significant digits, see Tables 1.4 and 1.5 on pages 17 and 20)

Useful Math Relationships

$D = \frac{m}{V}$ $\quad P = \frac{F}{A}$ $\quad \pi = 3.1416$

volume of sphere $= \frac{4}{3\pi^3}$

volume of cylinder $= \frac{4}{3}\pi r^3$

Conversion Factors

Quantity	SI base unit	In SI base units	Other units commonly used	Relationships between units
length	metre (m)		kilometre (km) centimetre (cm) millimetre (mm) picometre (pm)	1 m = 10^{-3} km = 10^2 cm = 10^3 mm 1 pm = 10^{-12} m
mass	kilogram (kg)		gram (g) metric ton (t) atomic mass unit (u)	1 kg = 10^3 g = 10^{-3} t 1 u = 1.66 × 10^{-27} kg
temperature	kelvin (K)		degrees Celsius (°C)	0 K = −273.15°C *T* (K) = *T* (°C) + 273.15 *T* (°C) = *T* (K) − 273.15 mp of H_2O = 273.15 K (0°C) bp of H_2O = 373.15 K (100°C)
volume		cubic metre (m^3)	litre (L) millitre (mL) cubic decimetre (dm^3) cubic centimetre (cm^3)	1 L = 1 dm^3 = 10^{-3} m^3 = 10^3 mL 1 mL = 1 cm^3
pressure		pascal (Pa) (kg/m·s^2, often as N/m^2)	kilopascal (kPa) millimetres of mercury (mm Hg) torr atmosphere (atm)	101 325 Pa = 101.325 kPa = 760 mm Hg = 760 torr = 1 atm
density		kilograms per cubic metre (kg/m^3)	grams per cubic centimetre (g/cm^3) grams per millilitre (g/mL) grams per litre (g/L)	1 kg/m^3 = 10^3 g/m^3 = 10^{-3} g/mL = 1 g/L
energy		joule (J) (kg·m^2/s^2, often as N·m)	electron volt (eV)	1 J = 6.24 × 10^{18} eV

Index to Useful Tables and Figures

McGraw-Hill Ryerson

Chemistry 11

Author Team

Frank Mustoe
University of Toronto Schools
Toronto, Ontario

Michael P. Jansen
Crescent School
Toronto, Ontario

Ted Doram
Bowness High School
Calgary, Alberta

John Ivanco
Former Head of Science
Anderson Collegiate and Vocational Institute
Whitby, Ontario

Christina Clancy
Loyola Catholic Secondary School
Mississauga, Ontario

Anita Ghazariansteja
Birchmount Collegiate
Scarborough, Ontario

Contributing Authors

Christa Bedry
Professional Writer
Kananaskis, Alberta

Christy Hayhoe
Professional Writer
Toronto, Ontario

Consultants

Greg Wisnicki
Anderson Collegiate and Vocational Institute
Whitby, Ontario

Dr. Penny McLeod
Former Head of Science
Thornhill Secondary School
Thornhill, Ontario

Dr. Audrey Chastko
Science Curriculum Leader
Springbank Community High School
Calgary, Alberta

Probeware Specialist

Kelly Choy
Minnedosa, Manitoba

Technology Consultants

Alex Annab
Head of Science
Iona Catholic Secondary School
Mississauga, Ontario

Wilf Kazmaier
Multimedia Specialist
Calgary, Alberta

Toronto Montréal New York Burr Ridge Bangkok Beijing Bogotá Caracas Dubuque
Kuala Lumpur Lisbon London Madison Madrid Mexico City Milan New Delhi
San Francisco Santiago St. Louis Seoul Singapore Sydney Taipei

McGraw-Hill Ryerson Limited

A Subsidiary of The ***McGraw-Hill*** *Companies*

COPIES OF THIS BOOK MAY BE OBTAINED BY CONTACTING:

McGraw-Hill Ryerson Ltd.

WEB SITE:

http://www.mcgrawhill.ca

E-MAIL:

orders@mcgrawhill.ca

TOLL FREE FAX:

1-800-463-5885

TOLL FREE CALL:

1-800-565-5758

OR BY MAILING YOUR ORDER TO:

McGraw-Hill Ryerson
Order Department
300 Water Street
Whitby ON, L1N 9B6

Please quote the ISBN and title when placing your order.

STUDENT TEXT ISBN

0-07-088681-4

E-BOOK CD-ROM ISBN

0-07-089356-X

McGraw-Hill Ryerson Chemistry 11

0-07-088681-4

http://www.mcgrawhill.ca

2 3 4 5 6 7 8 9 0 TRI 0 9 8 7 6 5 4 3 2 1

Printed and bound in Canada

National Library of Canada Cataloguing in Publication Data

Main entry under title:

McGraw-Hill Ryerson chemistry 11

Includes index.

ISBN 0-07-088681-4

1. Chemisty. I. Clancy, Christina. II. Title: Chemistry 11. III. Title: McGraw-Hill Ryerson chemistry eleven.

QD33. M33 2001 540 C2001-930333-5

The Chemistry 11 Development Team

SCIENCE PUBLISHERS: Jane McNulty, Trudy Rising
PROJECT MANAGER: Jane McNulty
SENIOR DEVELOPMENTAL EDITOR: Jonathan Bocknek
DEVELOPMENTAL EDITORS: Christa Bedry, Sara Goodchild, Christy Hayhoe, Keith Owen Richards
SENIOR SUPERVISING EDITOR: Linda Allison
PROJECT CO-ORDINATORS: Valerie Janicki, Shannon Leahy
COPY EDITORS: Paula Pettitt-Townsend, May Look
PROOFREADER: Carol Ann Freeman
PERMISSIONS EDITOR: Pronk&Associates Inc.
SPECIAL FEATURES CO-ORDINATOR: Jill Bryant
PRODUCTION CO-ORDINATOR: Jennifer Vassiliou
COVER DESIGN, INTERIOR DESIGN, AND ART DIRECTION: Pronk&Associates Inc.
ELECTRONIC PAGE MAKE-UP: Pronk&Associates Inc.
TECHNICAL ILLUSTRATIONS: Pronk&Associates Inc., Jun Park, Theresa Sakno, Bill Sakno
ILLUSTRATORS: Steve Attoe, Scott Cameron
SET-UP PHOTOGRAPHY: Ian Crysler
SET-UP PHOTOGRAPHY CO-ORDINATOR: Shannon O'Rourke
COVER IMAGE: James Bell/Science Photo Library

Acknowledgements

We extend sincere thanks to the following people: Lois Edwards, for the gift of her insights early in the project; Dr. Michael J. Webb, who generously and tirelessly shared his chemistry expertise with the development team; Otto Pike, for his assistance in revising Unit 3; Catherine Little, who authored the Unit 1 Project; Doug Hayhoe, who provided guidance regarding assessment and the Achievement Chart; and Andrew Cherkas, who wrote material for Appendix E. We are deeply grateful to Frank Mustoe, Malisa Mezenberg, Greg Dick, and Greg Wisnicki for their help with the set-up photography sessions, and we thank the students of University of Toronto Schools and Loyalist Catholic Secondary School who participated in these sessions. We also wish to thank the following professional writers who authored the Special Features in *Chemistry 11*: Linda Cornies, Meaghan Craven, Paul Halpern, Marian Hughes, Carol Johnstone, Jill Lazenby, Natasha Marko, Andrea Rutty, Christopher Rutty, Elma Schemenauer, Erik Spigel, and Jean-Louis Trudel. We benefited greatly from the many thoughtful suggestions provided by our excellent team of reviewers from across the country, as well as from the recommendations supplied by our safety reviewer. The authors, publishers, consultants, and editors convey their profound thanks to these talented and dedicated educators.

Pedagogical and Academic Reviewers

Bruce Beamer
Formerly of John G. Diefenbaker High School
Calgary, Alberta

David Bocknek
King Secondary School
King City, Ontario

Andrew Cherkas
Stouffville District High School
Stouffville, Ontario

David Clayton
NorQuest College
Edmonton, Alberta

André Dumais
Hearst High School
Hearst, Ontario

Spencer R. Fosbury
Former Head of Science
Aurora High School
Aurora, Ontario

Christopher Freure
South Lincoln High School
Smithville, Ontario

Theresa George
St. Paul Secondary School
Mississauga, Ontario

Keith Gibbons
Science Department Head
Catholic Central High School
London, Ontario

Gail Gislason
Cresent Heights High School
Calgary, Alberta

Dana Griffiths
Science Department Head
Bishops College
St. John's, Newfoundland

Stephen Houlden
Formerly of Birchmount Park Collegiate Institute
Toronto, Ontario

Sarah Houlden
Twin Lakes Secondary School
Orillia, Ontario

Terry Kilroy
Widdifield Secondary School
North Bay, Ontario

Lucy Kisway
Glendale Secondary School
Hamilton, Ontario

John Kozelko
Sisler High School
Winnipeg, Manitoba

Peter MacDonald
Science Co-ordinator
Charles P. Allen High School
Bedford, Nova Scotia

Cheryl Madeira
Marshall McLuhan Catholic Secondary School
Toronto, Ontario

Dermot O'Hara
Head of Science
Mother Theresa Secondary School
Scarborough, Ontario

Henry Pasma
Cawthra Park Secondary School
Mississauga, Ontario

Otto Pike
Formerly of College of the North Atlantic
St. John's, Newfoundland

John Purificati
St. Patrick's High School
Ottawa, Ontario

Chris Schramek
John Paul II Catholic Secondary School
London, Ontario

Mario Simon
Holy Heart of Mary Regional School
St. John's, Newfoundland

Donna Stack-Durward
St. Mary's High School
Hamilton, Ontario

James Stewart
Formerly of Bayside Secondary School
Hastings County, Ontario

Dr. Michael J. Webb
Michael J. Webb Consulting Inc.
Toronto, Ontario

Safety Reviewer

Dr. Margaret-Ann Armour
Assistant Chair, Department of Chemistry
University of Alberta
Edmonton, Alberta

Our cover: A polarized light micrograph of a liquid crystal, a kind of substance you are familiar with from digital displays on products such as pocket calculators and laptop computers. Although liquid crystal flows like a fluid, its molecular arrangement exhibits some order, as in a solid. You will learn more about the movement of solid, liquid, and gas molecules in Unit 3.

Contents

Safety in Your Chemistry Laboratory and Classroom

The following *Safety Precautions* symbols appear throughout *Chemistry 11*, whenever an investigation or ExpressLab presents possible hazards.

appears when there is a danger to the eyes, and safety goggles, safety glasses, or a face shield should be worn

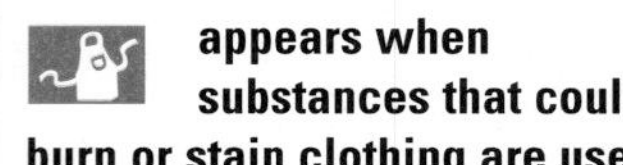

appears when substances that could burn or stain clothing are used

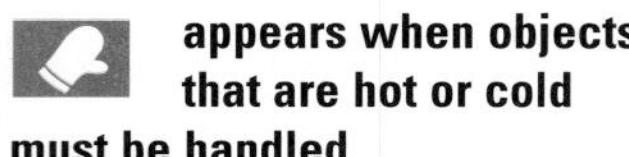

appears when objects that are hot or cold must be handled

appears when sharp objects are used, to warn of the danger of cuts and punctures

appears when toxic substances that can cause harm through ingestion, inhalation, or skin absorption are used

appears when corrosive substances, such as acids and bases, that can damage tissue are used

warns of caustic substances that could irritate the skin

appears when chemicals or chemical reactions that could cause dangerous fumes are used and ventilation is required

appears as a reminder to be careful when you are around open flames and when you are using easily flammable or combustible materials

warns of danger of electrical shock or burns from live electrical equipment

Actively engaging in laboratory investigations is essential to gaining a hands-on understanding of chemistry. Following safe laboratory procedures should not be seen as an inconvenience in your investigations. Instead, it should be seen as a positive way to ensure your safety and the safety of others who share a common working environment. Familiarize yourself with the following general safety rules and procedures. It is your responsibility to follow them when completing any of the investigations or ExpressLabs in this textbook, or when performing other laboratory procedures.

General Precautions

- Always wear safety glasses and a lab coat or apron in the laboratory. Wear other protective equipment, such as gloves, as directed by your teacher or by the Safety Precautions at the beginning of each investigation.
- If you wear contact lenses, always wear safety goggles or a face shield in the laboratory. Inform your teacher that you wear contact lenses. Generally, contact lenses should not be worn in the laboratory. If possible, wear eyeglasses instead of contact lenses, but remember that eyeglasses are not a substitute for proper eye protection.
- Know the location and proper use of the nearest fire extinguisher, fire blanket, fire alarm, first aid kit, and eyewash station (if available). Find out from your teacher what type of fire-fighting equipment should be used on particular types of fires. (See "Fire Safety" on page xiii.)
- Do not wear loose clothing in the laboratory. Do not wear open-toed shoes or sandals. Accessories may get caught on equipment or present a hazard when working with a Bunsen burner. Ties, scarves, long necklaces, and dangling earrings should be removed before starting an investigation.
- Tie back long hair and any loose clothing before starting an investigation.
- Lighters and matches must not be brought into the laboratory.
- Food, drinks, and gum must not be brought into the laboratory.
- Inform your teacher if you have any allergies, medical conditions, or physical problems (including hearing impairment) that could affect your work in the laboratory.

Before Beginning Laboratory Investigations

- Listen carefully to the instructions that your teacher gives you. Do not begin work until your teacher has finished giving instructions.
- Obtain your teacher's approval before beginning any investigation that you have designed yourself.
- Read through all of the steps in the investigation before beginning. If there are any steps that you do not understand, ask your teacher for help.
- Be sure to read and understand the Safety Precautions at the start of each investigation or Express Lab.

- Always wear appropriate protective clothing and equipment, as directed by your teacher and the Safety Precautions.
- Be sure that you understand all safety labels on materials and equipment. Familiarize yourself with the WHMIS symbols on this page.
- Make sure that your work area is clean and dry.

During Laboratory Investigations

- Make sure that you understand and follow the safety procedures for different types of laboratory equipment. Do not hesitate to ask your teacher for clarification if necessary.
- Never work alone in the laboratory.
- Remember that gestures or movements that may seem harmless could have dangerous consequences in the laboratory. For example, tapping people lightly on the shoulders to get their attention could startle them. If they are holding a beaker that contains an acid, for example, the results could be very serious.
- Make an effort to work slowly and steadily in the laboratory. Be sure to make room for other students.
- Organize materials and equipment neatly and logically. For example, do not place materials that you will need during an investigation on the other side of a Bunsen burner from you. Keep your bags and books off your work surface and out of the way.
- Never taste any substances in the laboratory.
- Never touch a chemical with your bare hands.
- Never draw liquids or any other substances into a pipette or a tube with your mouth.
- If you are asked to smell a substance, do not hold it directly under your nose. Keep the object at least 20 cm away, and waft the fumes toward your nostrils with your hand.
- Label all containers holding chemicals. Do not use chemicals from unlabelled containers.
- Hold containers away from your face when pouring liquids or mixing reactants.
- If any part of your body comes in contact with a potentially dangerous substance, wash the area immediately and thoroughly with water.
- If you get any material in your eyes, do not touch them. Wash your eyes immediately and continuously for 15 min, and make sure that your teacher is informed. A doctor should examine any eye injury. If you wear contact lenses, take them out immediately. Failing to do so may result in material becoming trapped behind the contact lenses. Flush your eyes with water for 15 min, as above.
- Do not touch your face or eyes while in the laboratory unless you have first washed your hands.
- Do not look directly into a test tube, flask, or the barrel of a Bunsen burner.
- If your clothing catches fire, smother it with the fire blanket or with a coat, or get under the safety shower.
- If you see any of your classmates jeopardizing their safety or the safety of others, let your teacher know.

WHMIS (Workplace Hazardous Materials Information System) symbols are used in Canadian schools and workplaces to identify dangerous materials. Familiarize yourself with the symbols below.

Poisonous and Infectious Material Causing Immediate and Serious Toxic Effects

Poisonous and Infectious Material Causing Other Toxic Effects

Flammable and Combustible Material

Compressed Gas

Corrosive Material

Oxidizing Material

Dangerously Reactive Material

Biohazardous Infectious Material

Heat Source Safety

- When heating any item, wear safety glasses, heat-resistant safety gloves, and any other safety equipment that your teacher or the Safety Precautions suggests.
- Always use heat-proof, intact containers. Check that there are no large or small cracks in beakers or flasks.
- Never point the open end of a container that is being heated at yourself or others.
- Do not allow a container to boil dry unless specifically instructed to do so.
- Handle hot objects carefully. Be especially careful with a hot plate that may look as though it has cooled down, or glassware that has recently been heated.
- Before using a Bunsen burner, make sure that you understand how to light and operate it safely. Always pick it up by the base. Never leave a Bunsen burner unattended.
- Before lighting a Bunsen burner, make sure there are no flammable solvents nearby.
- If you do receive a burn, run cold water over the burned area immediately. Make sure that your teacher is notified.
- When you are heating a test tube, always slant it. The mouth of the test tube should point away from you and from others.
- Remember that cold objects can also harm you. Wear appropriate gloves when handling an extremely cold object.

Electrical Equipment Safety

- Ensure that the work area, and the area of the socket, is dry.
- Make sure that your hands are dry when touching electrical cords, plugs, sockets, or equipment.
- When unplugging electrical equipment, do not pull the cord. Grasp the plug firmly at the socket and pull gently.
- Place electrical cords in places where people will not trip over them.
- Use an appropriate length of cord for your needs. Cords that are too short may be stretched in unsafe ways. Cords that are too long may tangle or trip people.
- Never use water to fight an electrical equipment fire. Severe electrical shock may result. Use a carbon dioxide or dry chemical fire extinguisher. (See "Fire Safety" on the next page.)
- Report any damaged equipment or frayed cords to your teacher.

Glassware and Sharp Objects Safety

- Cuts or scratches in the chemistry laboratory should receive immediate medical attention, no matter how minor they seem. Alert your teacher immediately.
- Never use your hands to pick up broken glass. Use a broom and dustpan. Dispose of broken glass as directed by your teacher. Do not put broken glassware into the garbage can.

- Cut away from yourself and others when using a knife or another sharp object.
- Always keep the pointed end of scissors and other sharp objects pointed away from yourself and others when walking.
- Do not use broken or chipped glassware. Report damaged equipment to your teacher.

Fire Safety

- Know the location and proper use of the nearest fire extinguisher, fire blanket, and fire alarm.
- Understand what type of fire extinguisher you have in the laboratory, and what type of fires it can be used on. (See below.) Most fire extinguishers are the ABC type.
- Notify your teacher immediately about any fires or combustible hazards.
- Water should only be used on Class A fires. Class A fires involve ordinary flammable materials, such as paper and clothing. Never use water to fight an electrical fire, a fire that involves flammable liquids (such as gasoline), or a fire that involves burning metals (such as potassium or magnesium).
- Fires that involve a flammable liquid, such as gasoline or alcohol (Class B fires) must be extinguished with a dry chemical or carbon dioxide fire extinguisher.
- Live electrical equipment fires (Class C) must be extinguished with a dry chemical or carbon dioxide fire extinguisher. Fighting electrical equipment fires with water can cause severe electric shock.
- Class D fires involve burning metals, such as potassium and magnesium. A Class D fire should be extinguished by smothering it with sand or salt. Adding water to a metal fire can cause a violent chemical reaction.
- If someone's hair or clothes catch on fire, smother the flames with a fire blanket. Do not discharge a fire extinguisher at someone's head.

Clean-Up and Disposal in the Laboratory

- Clean up all spills immediately. Always inform your teacher about spills.
- If you spill acid or base on your skin or clothing, wash the area immediately with a lot of cool water.
- You can neutralize small spills of acid solutions with sodium hydrogen carbonate (baking soda). You can neutralize small spills of basic solutions with sodium hydrogen sulfate or citric acid.
- Clean equipment before putting it away, as directed by your teacher.
- Dispose of materials as directed by your teacher, in accordance with your local School Board's policies. Do not dispose of materials in a sink or a drain unless your teacher directs you to do so.
- Wash your hands thoroughly after all laboratory investigations.

INTRODUCING *Chemistry 11*

Here is a quick glimpse at the learning that lies before you in this course. Expand your knowledge and skills from earlier courses and experience chemistry in action.

In Unit 1, you will take a close look at the periodic table. How does the arrangement of electrons affect the behaviour of an atom? How do different elements combine to form compounds? You will be able to predict the products of reactions, then test out the reactions in the lab. Also, you will take a look at common chemical substances you use every day. This unit concludes with a project in which you develop a chemistry newsletter.

How can you measure what happens in a reaction? How can you use mass to find out the vast number of atoms and molecules involved? In Unit 2, you will discover how to describe and calculate quantitative measurements in chemical reactions. At the end of Unit 2, you will design an investigation to determine the quantitative composition of a mixture.

Why does water dissolve other substances so well? How is this important—and when does it become a problem? Solutions are essential to your life, and you use them every day. In Unit 3, you will investigate why and how things dissolve. Your understanding of solutions will prepare you for an end-of-unit simulation on a societal and environmental issue.

A rocket takes off in a cloud of vapour, and a volcano erupts, shooting molten rock into the air. What do these events have to do with gases? In Unit 4, you will find out how gases behave. You will also see how gases are used in medical and industrial situations. At the end of Unit 4, you will debate an issue related to air pollution.

Imagine your life without plastics, gasoline, or natural gas. All these products come from a common source, rooted far back in time. In Unit 5, you will look at hydrocarbons, and discover their importance in your life. Later in the unit, you will see how society obtains energy from hydrocarbons. At the end of Unit 5, you will carry out a project to investigate the chemistry of a product of your choice.

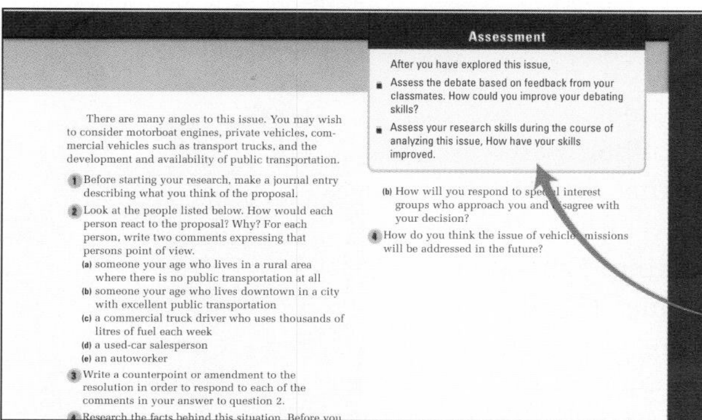
There are many angles to this issue. You may wish to consider motorboat engines, private vehicles, commercial vehicles such as transport trucks, and the development and availability of public transportation.

1. Before starting your research, make a journal entry describing what you think of the proposal.
2. Look at the people listed below. How would each person react to the proposal? Why? For each person, write two comments expressing that persons point of view.
   (a) someone your age who lives in a rural area where there is no public transportation at all
   (b) someone your age who lives downtown in a city with excellent public transportation
   (c) a commercial truck driver who uses thousands of litres of fuel each week
   (d) a used-car salesperson
   (e) an autoworker
3. Write a counterpoint or amendment to the resolution in order to respond to each of the comments in your answer to question 2.

(b) How will you respond to special interest groups who approach you and disagree with your decision?

4. How do you think the issue of vehicle emissions will be addressed in the future?

Assessment

After you have explored this issue,

- Assess the debate based on feedback from your classmates. How could you improve your debating skills?
- Assess your research skills during the course of analyzing this issue. How have your skills improved.

You will probably be designing rubrics to assess your end-of-unit project, issue, or investigation. Remember to include criteria that address all of the Achievement Chart categories, as shown in Chapter 1, page 10. As you work on these tasks, refer back to these rubrics.

Following the five units of the textbook is a Chemistry Course Challenge. This is your opportunity to demonstrate an understanding of the concepts covered in the course. You will apply your skills of inquiry to explore the possibility of settling an unknown planet. At the end of the challenge, you will communicate your ideas to a government task force. By applying practical chemistry skills to a new, "real-life" situation, you will see how science and technology connect with society and the environment.

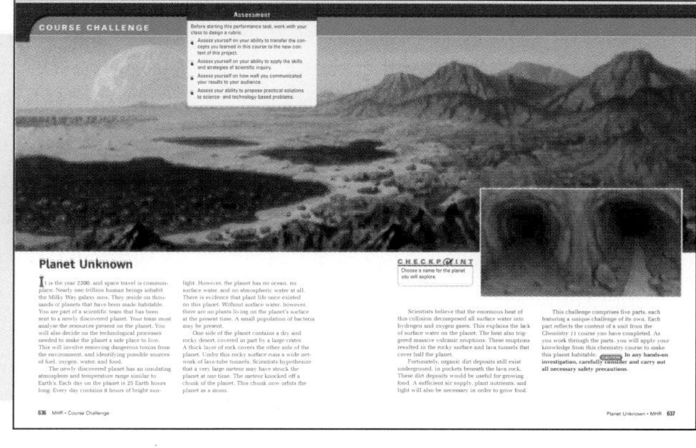

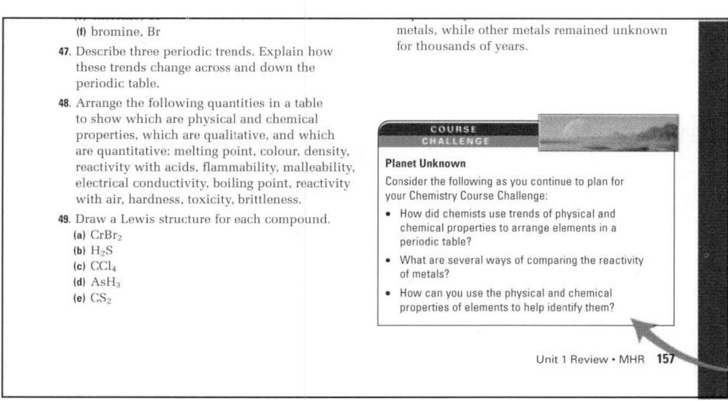
(f) bromine, Br

47. Describe three periodic trends. Explain how these trends change across and down the periodic table.
48. Arrange the following quantities in a table to show which are physical and chemical properties, which are qualitative, and which are quantitative: melting point, colour, density, reactivity with acids, flammability, malleability, electrical conductivity, boiling point, reactivity with air, hardness, toxicity, brittleness.
49. Draw a Lewis structure for each compound.
   (a) $CrBr_2$
   (b) H_2S
   (c) CCl_4
   (d) AsH_3
   (e) CS_2

metals, while other metals remained unknown for thousands of years.

COURSE CHALLENGE

Planet Unknown

Consider the following as you continue to plan for your Chemistry Course Challenge:

- How did chemists use trends of physical and chemical properties to arrange elements in a periodic table?
- What are several ways of comparing the reactivity of metals?
- How can you use the physical and chemical properties of elements to help identify them?

Unit 1 Review • MHR 157

Watch for this feature in text margins throughout the textbook and in the Unit Reviews to help you begin planning for your Course Challenge. The cues are designed to trigger your thought processes, and point you to a line of research.

UNIT 1

Matter and Chemical Bonding

UNIT 1 CONTENTS

UNIT 1 OVERALL EXPECTATIONS

- What are the relationships among periodic trends, types of chemical bonds, and properties of compounds?
- How can laboratory investigations help you represent the structures and interactions of chemicals in chemical reactions, and classify these reactions?
- How can understanding the properties and behaviour of matter lead to the development of useful substances and new technologies?

Unit Project Prep

Begin collecting ideas and resources for the project at the end of Unit 1.

Name ten things in your life that do not, in some way, involve the products and processes of chemistry. Take your time.

Are you having trouble? Can you name five things that do not involve chemistry?

Are you still thinking? Consider each room in your home. Think about the bathroom, for example. Does soap involve chemistry? Do toothpaste, cosmetics, and shampoo involve chemistry? Think about the light in the bathroom. Without chemistry, there is no glass to make lightbulbs.

Move to another room. Walk quickly. The floor is disappearing beneath your feet. Pause briefly to watch the paint fade away from the walls. In a moment, the walls will be gone, too.

The story is the same if you step outdoors. There are no sidewalks, vehicles, people, trees, or animals.

A world without chemistry is a world without anything! Everything in the world, including you, is made up of matter. Chemistry is the study of matter: its composition, its properties, and the changes it undergoes when it interacts with other matter. In this unit, you will explore matter. You will learn how to predict the kinds of bonds (the chemical combinations) and the reactions that occur during these interactions.

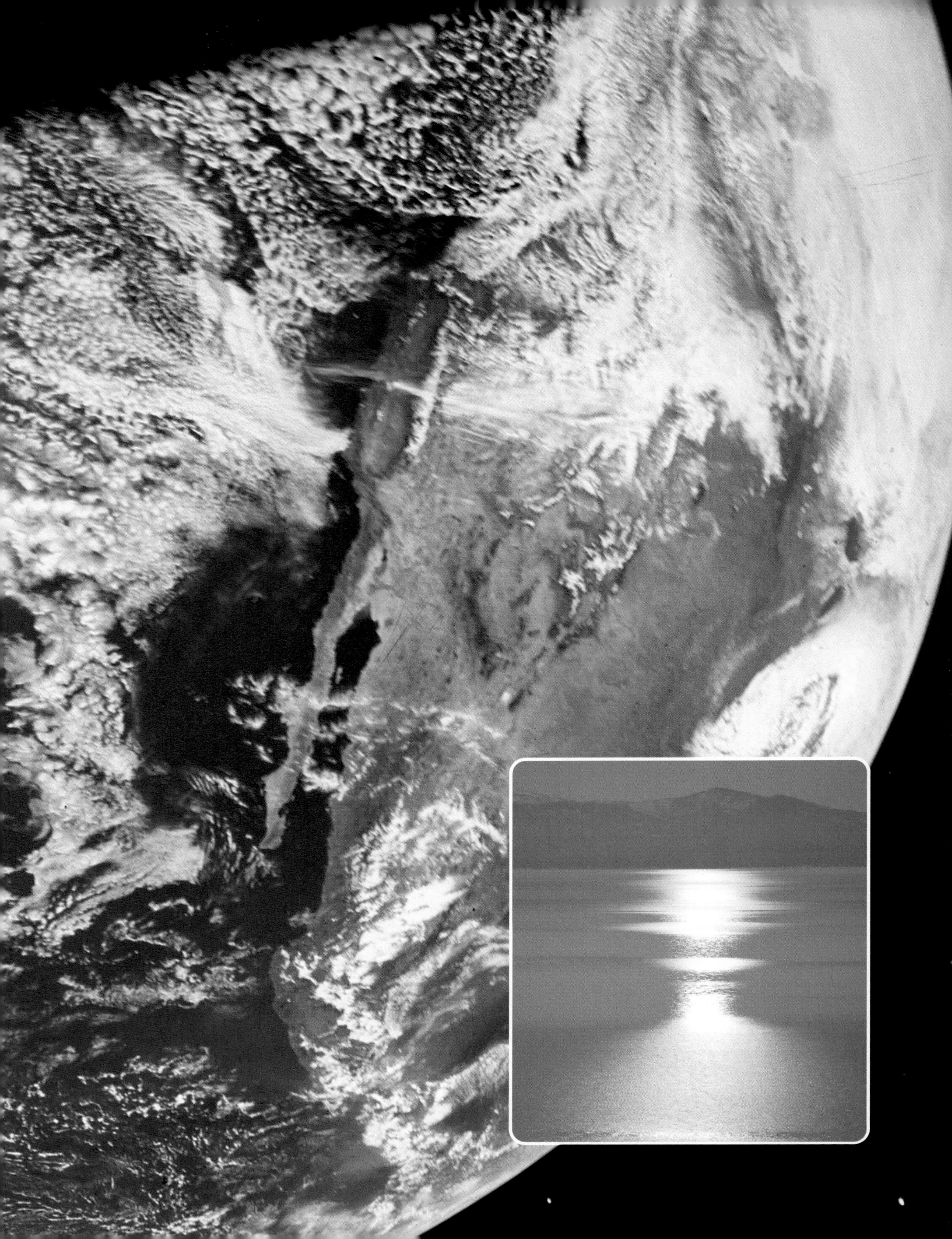

CHAPTER 1 Observing Matter

Imagine a chemical that
- is a key ingredient in most pesticides
- contributes to environmental hazards, such as acid rain, the greenhouse effect, and soil erosion
- helps to spread pollutants that are present in all contaminated rivers, lakes, and oceans
- is used in vast quantities by every industry on Earth
- can produce painful burns to exposed skin
- causes severe illness or death in either very low or very high concentrations in the body
- is legally discarded as waste by individuals, businesses, and industries
- has been studied extensively by scientists throughout the world

In 1996, a high school student wrote a report about this chemical, dihydrogen monoxide, for a science fair project. The information in the student's report was completely factual. As a result, 86% of those who read the report—43 out of 50 students—voted in favour of banning the chemical. What they did not realize was that "dihydrogen monoxide" is simply another name for water.

What if you did not know that water and dihydrogen monoxide are the same thing? What knowledge and skills can help you distinguish genuine environmental issues from pranks like this one? What other strategies can help you interpret all the facts, opinions, half-truths, and falsehoods that you encounter every day?

This chapter will reacquaint you with the science of chemistry. You will revisit important concepts and skills from previous grades. You will also prepare to extend your knowledge and skills in new directions.

Chapter Preview

What mistake in measuring matter nearly resulted in an airplane disaster in 1983? Read on to find the answer to this question later in this chapter.

1.1 The Study of Chemistry

Section Preview/ Specific Expectations

In this section, you will

- **identify** examples of chemistry and chemical processes in everyday use
- **communicate** ideas related to chemistry and its relationship to technology, society, and the environment, using appropriate scientific vocabulary
- **communicate** your understanding of the following terms: *chemistry, STSE*

Many people, when they hear the word "chemistry," think of scientists in white lab coats. They picture bubbling liquids, frothing and churning inside mazes of laboratory glassware.

Is this a fair portrayal of chemistry and chemists? Certainly, chemistry happens in laboratories. Laboratory chemists often do wear white lab coats, and they do use lots of glassware! Chemistry also happens everywhere around you, however. It happens in your home, your school, your community, and the environment. Chemistry is happening right now, inside every cell in your body. You are alive because of chemical changes and processes.

Chemistry is the study of matter and its composition. Chemistry is also the study of what happens when matter interacts with other matter. When you mix ingredients for a cake and put the batter in the oven, that is chemistry. When you pour soda water on a stain to remove it from your favourite T-shirt, that is chemistry. When a scientist puts a chunk of an ice-like solid into a beaker, causing white mist to ooze over the rim, that is chemistry, too. Figure 1.1 illustrates this interaction, as well as several other examples of chemistry in everyday life.

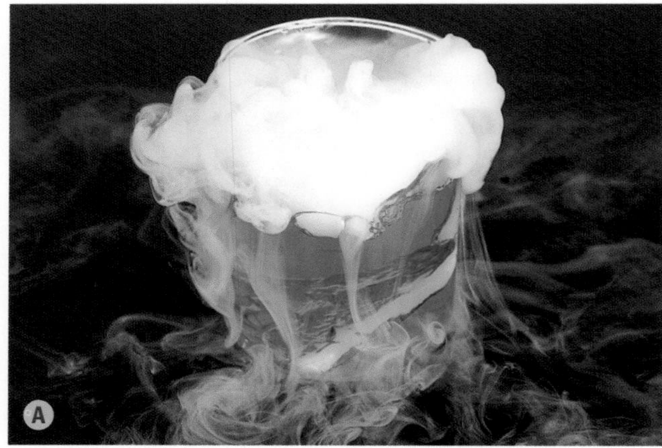

Figure 1.1

A Frozen (solid) carbon dioxide is also known as "dry ice." It changes to a gas at temperatures higher than −78°C. In this photograph, warm water has been used to speed up the process, and food colouring has been added.

B Dry ice is also used to create special effects for rock concerts, stage plays, and movies.

C Nitrogen gas becomes a liquid at −196°C. Liquid nitrogen is used to freeze delicate materials, such as food, instantly.

Chemistry: A Blend of Science and Technology

Like all scientists, chemists try to describe and explain the world. Chemists start by asking questions such as these:

- Why is natural gas such an effective fuel?
- How can we separate a mixture of crude oil and water?
- Which materials dissolve in water?
- What is rust and why does it form?

To answer these questions, chemists develop models, conduct experiments, and seek patterns. They observe various types of chemical reactions, and they perform calculations based on known data. They build continuously on the work and the discoveries of other scientists.

Long before humans developed a scientific understanding of the world, they invented chemical techniques and processes. These techniques and processes included smelting and shaping metals, growing crops, and making medicines. Early chemists invented technological instruments, such as glassware and distillation equipment.

Present-day chemical technologists continue to invent new equipment. They also invent new or better ways to provide products and services that people want. Chemical technologists ask questions such as the following:

- How can we redesign this motor to run on natural gas?
- How can we contain and clean up an oil spill?
- What methods can we use, or develop, to make water safe to drink?
- How can we prevent iron objects from rusting?

D Green plants use a chemical process, called photosynthesis, to convert water and carbon dioxide into the food substances they need to survive. All the foods that you eat depend on this process.

E Your body uses chemical processes to break down food and to release energy.

F Your home is full of products that are manufactured by chemical industries. The products that are shown here are often used for cleaning. Some of these products, such as bleach and drain cleaner, can be dangerous if handled improperly.

Chemistry, Technology, Society, and the Environment

Today we benefit in many ways from chemical understanding and technologies. Each benefit, however, has risks associated with it. The risks and benefits of chemical processes and technologies affect us either directly or indirectly. Many people—either on their own, in groups, or through their elected government officials—assess these risks and benefits. They ask questions such as the following:

- Is it dangerous to use natural gas to heat my home?
- Why is the cost of gasoline so high?
- Is my water really clean enough to drink and use safely?
- How does rust degrade machinery over time?

During your chemistry course this year, you will study the interactions among science, technology, society, and the environment. These interactions are abbreviated as **STSE**. Throughout the textbook—in examples, practice problems, activities, investigations, and features—STSE interactions are discussed. The issues that appear at the end of some units are especially rich sources for considering STSE interactions. In these simulations, you are encouraged to assess and make decisions about important issues that affect society and the environment.

STSE Issue: Are Phosphates Helpful or Harmful?

Phosphorus is an essential nutrient for life on Earth. Plants need phosphorus, along with other nutrients, in order to grow. Phosphorus is a component of bones and teeth. In addition, phosphorus is excreted as waste from the body. Thus, it is present in human sewage.

Since phosphorus promotes plant growth, phosphates are excellent fertilizers for crops. (*Phosphates* are chemicals containing phosphorus. You will learn more about phosphates later in this unit.) Phosphates are also used as food additives, and as components in some medicines. In addition, they are an important part of dishwasher and laundry detergents. For example, sodium tripolyphosphate (STPP) acts to soften water, and keep dirt suspended in the water. Before the 1970s, STPP was a major ingredient in most detergents.

Phosphates Causing Trouble

In the 1960s, residents around Lake Erie began to notice problems. Thick growths of algae carpeted the surface of the water. Large amounts of the algae washed onto beaches, making the beaches unfit for swimming. The water in the lake looked green, and had an unpleasant odour. As time passed, certain fish species in Lake Erie began to decrease.

In 1969, a joint Canadian and American task force pinpointed the source of the problem. Phosphates and other nutrients were entering the lake, causing algae to grow rapidly. The algae used up dissolved oxygen in the water. As a result, fish and other water species that needed high levels of oxygen were dying off.

The phosphate pollution arrived in the lake from three main sources: wastewater containing detergents, sewage, and run-off from farms carrying phosphate fertilizers. The task force recommended reducing the amount of phosphate in detergents. They also suggested removing phosphorus at wastewater treatment plants before the treated water entered the lake.

Detergent manufacturers were upset by the proposed reduction in phosphates. Without this chemical, their detergents would be less effec-

Eutrophication is the process in which excess nutrients in a lake or river cause algae to grow rapidly. Look up this term in a reference book or on the Internet. Is eutrophication always caused by human action?

tive. Also, it would be expensive to develop other chemicals to do the same job. After pressure from the government, detergent companies reduced the amount of phosphate in their products by about 90%. Cities on Lake Erie spent millions of dollars adding phosphorus removal to their waste treatment. Today, Lake Erie has almost completely recovered.

The connection between technology (human-made chemical products) and the environment (Lake Erie) is an obvious STSE connection in this issue. What other connections do you see?

Canadians  in Chemistry

John Charles Polanyi was born in Berlin, Germany, into a family of Hungarian origin. Polanyi was born on the eve of the Great Depression, shortly before the Nazi takeover. His father moved to England to become a chemistry professor at Manchester University. Polanyi was sent to Canada for safety during the darkest years of World War II.

John Polanyi went back to England to earn a doctorate in chemistry at Manchester University in 1952. He returned to Canada a few years later. Soon after, he took up a position at the University of Toronto. There Dr. Polanyi pursued the research that earned him a share of the Nobel Prize for chemistry in 1986. He pioneered the field of reaction dynamics, which addresses one of the most basic questions in chemistry: What happens when two substances interact to produce another substance? Polanyi's father had once investigated the same question.

Dr. Polanyi tried to provide some answers by studying the very faint light that is given off by molecules as they undergo chemical changes. This light is invisible to the unaided eye, because it is emitted in the infrared range of energy. It can be detected, however, with the right instruments. Dr. Polanyi's work led to the invention of the laser. As well, his research helped to explain what happens to energy during a chemical reaction.

Dr. Polanyi believes that people must accept the responsibility that comes with scientific understanding and technological progress. He believes, as well, that a vital element of hope lies at the heart of modern science. To Dr. Polanyi, human rights are integral to scientific success. "Science must breathe the oxygen of freedom," he stated in 1999.

This is why Dr. Polanyi says that scientists must take part in the debate on technological, social, and political affairs. Dr. Polanyi points to the political role played by scientists such as Andrei Sakharov in the former Soviet Union, Linus Pauling in the United States, and Fang Lizhi in China.

Make Connections

1. Research the scientists whom Dr. Polanyi mentioned: Andrei Sakharov, Linus Pauling, and Fang Lizhi. What work distinguished them as scientists? What work distinguished them as members of society?
2. Throughout history, chemists have laboured to present the truth as they know it to their fellow scientists and to society. Some of them, such as Linus Pauling, have been scorned and ridiculed by the scientific community. Do further research to discover two other chemists who have struggled to communicate their ideas, and have succeeded.

COURSE CHALLENGE

At the end of this course, you will have a chance to use what you have learned to help you in the Course Challenge: Planet Unknown. In this challenge, you are a member of a science team sent to a new planet. It is your task to analyze the planet's resources. You will design and carry out hands-on investigations and analyze your results. Then you will prepare a presentation to persuade the Canadian government to invest in the establishment of a community on the planet. As you work through this book, keep a research portfolio of notes and ideas that may help you in the Course Challenge.

Section Wrap-up

During this chemistry course, your skills of scientific inquiry will be assessed using the same specific set of criteria (Table 1.1). You will notice that all review questions are coded according to this chart.

Table 1.1 Achievement Chart Criteria, Ontario Science Curriculum

Knowledge and Understanding (K/U)	Inquiry (I)	Communication (C)	Making Connections (MC)
• understanding of concepts, principles, laws, and theories • knowledge of facts and terms • transfer of concepts to new contexts • understanding of relationships between concepts	• application of the skills and strategies of scientific inquiry • application of technical skills and procedures • use of tools, equipment, and materials	• communication of information and ideas • use of scientific terminology, symbols, conventions, and standard (SI) units • communication for different audiences and purposes • use of various forms of communication • use of information technology for scientific purposes	• understanding of connections among science, technology, society, and the environment • analysis of social and economic issues involving science and technology • assessment of impacts of science and technology on the environment • proposing of courses of practical action in relation to science-and technology-based problems

Section Review

1. K/U Based on your current understanding of chemistry, list five ways in which chemistry and chemical processes affect your life.
2. I Earlier in this section, you learned that fertilizers containing phosphorus can cause algae to grow faster. Design an investigation on paper to determine the effect of phosphorus-containing detergents on algae growth.
3. C Design a graphic organizer that clearly shows the connections among science, technology, society, and the environment.
4. MC For each situation, identify which STSE interaction is most important.
   (a) Research leads to the development of agricultural pesticides.
   (b) The pesticides prevent insects and weeds from destroying crops.
   (c) Rain soaks the excess pesticides on farm land into the ground. It ends up in groundwater systems.
   (d) Wells obtain water from groundwater systems. Well-water in the area is polluted by the pesticides. It is no longer safe to drink.

Describing and Measuring Matter

1.2

As you can see in the photograph at the beginning of this chapter, water is the most striking feature of our planet. It is visible from space, giving Earth a vivid blue colour. You can observe water above, below, and at Earth's surface. Water is a component of every living thing, from the smallest bacterium to the largest mammal and the oldest tree. You drink it, cook with it, wash with it, skate on it, and swim in it. Legends and stories involving water have been a part of every culture in human history. No other kind of matter is as essential to life as water.

As refreshing as it may be, water straight from the tap seems rather ordinary. Try this: Describe a glass of water to someone who has never seen or experienced water before. Be as detailed as possible. See how well you can distinguish water from other kinds of matter.

In addition to water, there are millions of different kinds of matter in the universe. The dust specks suspended in the air, the air itself, your chair, this textbook, your pen, your classmates, your teacher, and you—all these are examples of matter. In the language of science, **matter** is anything that has mass and volume (takes up space). In the rest of this chapter, you will examine some key concepts related to matter. You have encountered these concepts in previous studies. Before you continue, complete the Checkpoint activity to see what you recall and how well you recall it. As you proceed through this chapter, assess and modify your answers.

Section Preview/ Specific Expectations

In this section, you will

- **select** and **use** measuring instruments to collect and record data
- **express** the results of calculations to the appropriate number of decimal places and significant digits
- **select** and **use** appropriate SI units
- **communicate** your understanding of the following terms: *matter, properties, physical property, chemical property, significant digits, accuracy, precision*

Describing Matter

You must observe matter carefully to describe it well. When describing water, for example, you may have used statements like these:

- Water is a liquid.
- It has no smell.
- Water is clear and colourless.
- It changes to ice when it freezes.
- Water freezes at 0°C.
- Sugar dissolves in water.
- Oil floats on water.

Characteristics that help you describe and identify matter are called **properties**. Figure 1.2 on the next page shows some properties of water and hydrogen peroxide. Examples of properties include physical state, colour, odour, texture, boiling temperature, density, and flammability (combustibility). Table 1.2 on the next page lists some common properties of matter. You will have direct experience with most of these properties during this chemistry course.

CHECKPOINT

From memory, explain and define each of the following concepts. Use descriptions, examples, labelled sketches, graphic organizers, a computer FAQs file or Help file, or any combination of these. Return to your answers frequently during this chapter. Modify them as necessary.

- states of matter
- properties of matter
- physical properties
- chemical properties
- physical change
- chemical change
- mixture
- pure substance
- element
- compound

Figure 1.2 Liquid water is clear, colourless, odourless, and transparent. Hydrogen peroxide (an antiseptic liquid that many people use to clean wounds) has the same properties. It differs from water, however, in other properties, such as boiling point, density, and reactivity with acids.

Table 1.2 Common Properties of Matter

Physical Properties		Chemical Properties
Qualitative	Quantitative	
physical state	melting point	reactivity with water
colour	boiling point	reactivity with air
odour	density	reactivity with pure oxygen
crystal shape	solubility	reactivity with acids
malleability	electrical conductivity	reactivity with pure substances
ductility	thermal conductivity	combustibility (flammability)
hardness		toxicity
brittleness		decomposition

Properties may be physical or chemical. A **physical property** is a property that you can observe without changing one kind of matter into something new. For example, iron is a strong metal with a shiny surface. It is solid at room temperature, but it can be heated and formed into different shapes. These properties can all be observed without changing iron into something new.

A **chemical property** is a property that you can observe when one kind of matter is converted into a different kind of matter. For example, a chemical property of iron is that it reacts with oxygen to form a different kind of matter: rust. Rust and iron have completely different physical and chemical properties.

Figure 1.3 shows another example of a chemical property. Glucose test paper changes colour in the presence of glucose. Thus, a chemical property of glucose test paper is that it changes colour in response to glucose. Similarly, a chemical property of glucose is that it changes the colour of glucose test paper.

Recall that some properties of matter, such as colour, and flammability, are *qualitative*. You can describe them in words, but you cannot measure them or express them numerically. Other properties, such as density and boiling point, can be measured and expressed numerically. Such properties are *quantitative*. In Investigation 1-A you will use both qualitative and quantitative properties to examine a familiar item.

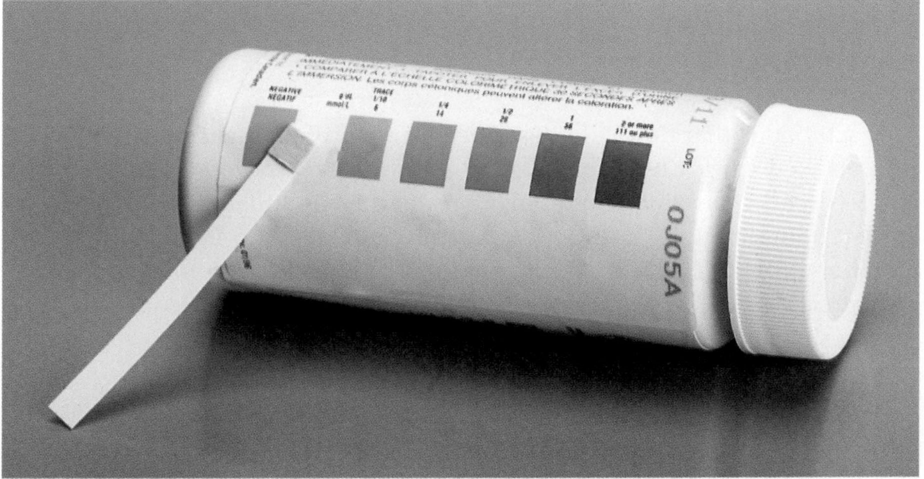

Figure 1.3 People with diabetes rely on a chemical property to help them monitor the amount of glucose (a simple sugar) in their blood.

Investigation 1-A

SKILL FOCUS

Initiating and Planning

Performing and recording

Analyzing and interpreting

Observing Aluminum Foil

You can easily determine the length and width of a piece of aluminum foil. You can use a ruler to measure these values directly. What about its thickness? In this investigation, you will design a method for calculating the thickness of aluminum foil.

Problem

How can you determine the thickness of a piece of aluminum foil, in centimetres?

Safety Precautions

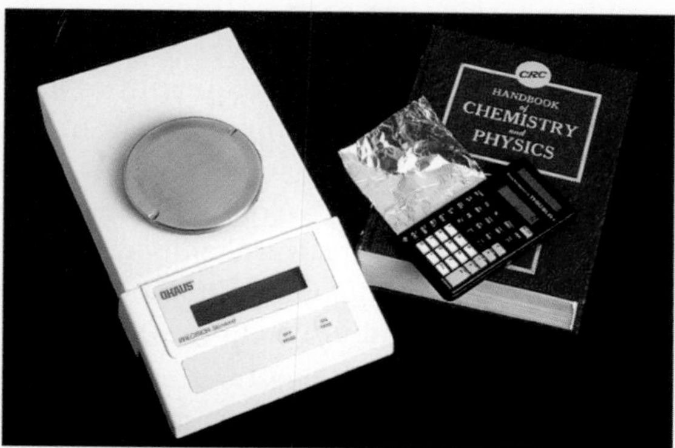

Materials

10 cm × 10 cm square of aluminum foil
ruler
electronic balance
calculator
chemical reference handbook

Procedure

1. Work together in small groups. Brainstorm possible methods for calculating the thickness of aluminum foil.
2. Observe and record as many physical properties of aluminum foil as you can. CAUTION Do not use the property of taste. Never taste anything in a laboratory.
3. As a group, review the properties you have recorded. Reflect on the possible methods you brainstormed. Decide on one method, and try it. (If you are stuck, ask your teacher for a clue.)

Analysis

1. Consider your value for the thickness of the aluminum foil. Is it reasonable? Why or why not?
2. Compare your value with the values obtained by other groups.
   (a) In what ways are the values similar?
   (b) In what ways are the values different?

Conclusion

3. (a) Explain how you decided on the method you used.
   (b) How much confidence do you have in your method? Explain why you have this level of confidence.
   (c) How much confidence do you have in the value you calculated? Give reasons to justify your answer.

Applications

4. Pure aluminum has a chemical property in common with copper and iron. It reacts with oxygen in air to form a different substance with different properties. This substance is called aluminum oxide. Copper has the same chemical property. The substance that results when copper reacts with oxygen is called a patina. Similarly, iron reacts with oxygen to form rust. Do research to compare the properties and uses (if any) of aluminum oxide, copper patina, and rust. What technologies are available to prevent their formation? What technologies make use of their formation?

Using Measurements to Describe Matter

In the investigation, you measured the size and mass of a piece of aluminum foil. You have probably performed these types of measurement many times before. Measurements are so much a part of your daily life that you can easily take them for granted. The clothes you wear come in different sizes. Much of the food you eat is sold by the gram, kilogram, millilitre, or litre. When you follow a recipe, you measure amounts. The dimensions of paper and coins are made to exact specifications. The value of money is itself a measurement.

Measurements such as clothing size, amounts of food, and currency are not standard, however. Clothing sizes in Europe are different from those in North America. European chefs tend to measure liquids and powdered solids by mass, rather than by volume. Currencies, of course, differ widely from country to country.

To communicate effectively, scientists rely on a standard system of measurement. As you have learned in previous studies, this system is called the *International System of Units* (Le système international d'unités, *SI*). It allows scientists anywhere in the world to describe matter in the same quantitative language. There are seven base SI units, and many more units that are derived from them. The metre (m), the kilogram (kg), and the second (s) are three of the base SI units. You will learn about two more base units, the mole (mol) and the kelvin (K), later in this book.

When you describe matter, you use terms such as mass, volume, and temperature. When you measure matter, you use units such as grams, cubic centimetres, and degrees Celsius. Table 1.3 lists some quantities and units that you will use often in this course. You are familiar with all of them except, perhaps, for the mole and the kelvin. The mole is one of the most important units for describing amounts of matter. You will be introduced to the mole in Unit 2. The kelvin is used to measure temperature. You will learn more about the kelvin scale in Unit 5. Consult Appendix E if you would like to review other SI quantities and units.

Table 1.3 Important SI Quantities and Their Units

Quantity	Definition	SI units or their derived equivalents	Equipment use to measure the quantity
mass	the amount of matter in an object	kilogram (kg) gram (g) milligram (mg)	balance
length	the distance between two points	metre (m) centimetre (cm) millimetre (mm)	ruler
temperature	the hotness or coldness of a substance	kelvin (K) degrees Celsius (°C)	thermometer
volume	the amount of space that an object occupies	cubic metre (m^3) cubic centimetre (cm^3) litre (L) millilitre (mL)	beaker, graduated cylinder, or pipette; may also be calculated
mole	the amount of a substance	mole (mol)	calculated not measured
density	the mass per unit of volume of a substance	kilograms per cubic metre (kg/m^3) grams per cubic centimetre (g/cm^3)	calculated or measured
energy	the capacity to do work (to move matter)	joule (J)	calculated not measured

Measurement and Uncertainty

Before you look more closely at matter, you need to know how much you can depend on measurements. How can you recognize when a measurement is trustworthy? How can you tell if it is only an approximation? For example, there are five Great Lakes. Are you sure there are five? Is there any uncertainty associated with the value "five" in this case? What about the number of millilitres in 1 L, or the number of seconds in 1 min? Numbers such as these—numbers that you can count or numbers that are true by definition—are called *exact numbers*. You are certain that there are five Great Lakes (or nine books on the shelf, or ten students in the classroom) because you can count them. Likewise, you are certain that there are 1000 mL in 1 L, and 60 s in 1 min. These relationships are true by definition.

Now consider the numbers you used and the calculations you did in Investigation 1-A. They are listed in Figure 1.4.

CHECKPOINT

Give five examples of exact numbers that you have personally experienced today or over the past few days.

- The area of the aluminum square measured 100 cm^2 (10 cm × 10 cm).

  Did you verify these dimensions? Are you certain that each side measured exactly 10 cm? Could it have been 9.9 cm or 10.1 cm?

- The mass of the aluminum square, as measured by an electronic balance, may have been about 3.26 g.

  If you used an electronic balance, are you certain that the digital read-out was accurate? Did the last digit fluctuate at all? If you used a triple-beam balance, are you certain that you read the correct value? Could it have been 3.27 g or 3.25 g?

- The density of aluminum is 2.70 g/cm^3.

  What reference did you use to find the density? Did you consult more than one reference? Suppose that the density was actually 2.699 g/cm^3. Would this make a difference in your calculations? Would this make a difference in the certainty of your answer?

- The thickness of the aluminum square, calculated using a calculator, may have been about 0.012 074 cm.

  Are you certain that this value is fair, given the other values that you worked with? Is it fair to have such a precise value, with so many digits, when there are so few digits (just two: the 1 and the 0) in your dimensions of the aluminum square?

Figure 1.4 **Numbers and calculations from Investigation 1-A**

During the investigations in this textbook, you will use equipment such as rulers, balances, graduated cylinders, and thermometers to measure matter. You will calculate values with a calculator or with specially programmed software. How exact can your measurements and calculations be? How exact *should* they be?

Two main factors affect your ability to record and communicate measurements and calculations. One factor is the instruments you use. The other factor is your ability to read and interpret what the instruments tell you. Examine Figures 1.5 and 1.6. They will help you understand which digits you can know with certainty, and which digits are uncertain.

What is the length measured by ruler A? Is it 4.2 cm, or is it 4.3 cm? You cannot be certain. The 2 of 4.2 is an estimate. The 3 of 4.3 is also an estimate. In both cases, therefore, you are uncertain about the last (farthest right) digit.

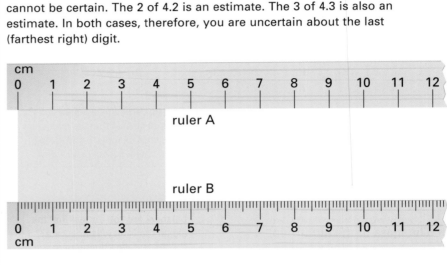

cm
0 1 2 3 4 5 6 7 8 9 10 11 12
ruler A

ruler B
0 1 2 3 4 5 6 7 8 9 10 11 12
cm

What is the length measured by ruler B? Is it 4.27 cm or 4.28 cm? Again, you cannot be certain. Ruler B lets you make more precise measurements than ruler A. Despite ruler B's higher precision, however, you must still estimate the last digit. The 7 of 4.27 is an estimate. The 8 of 4.28 is also an estimate.

Figure 1.5 **These two rulers measure the same length of the blue square. Ruler A is calibrated into divisions of 1 cm. Ruler B is calibrated into divisions of 0.1 cm. Which ruler can help you make more precise measurements?**

Figure 1.6 **These two thermometers measure the same temperature. Thermometer A is calibrated into divisions of 0.1°C. Thermometer B is calibrated into divisions of 1°C. Which thermometer lets you make more precise measurements? Which digits in each thermometer reading are you certain about? Which digits are you uncertain about?**

Significant Digits, Certainty, and Measurements

All measurements involve uncertainty. One source of this uncertainty is the measuring device itself. Another source is your ability to perceive and interpret a reading. In fact, you cannot measure anything with complete certainty. The last (farthest right) digit in any measurement is always an estimate.

The digits that you record when you measure something are called **significant digits**. Significant digits include the digits that you are certain about *and* a final, uncertain digit that you estimate. For example, 4.28 g has three significant digits. The first two digits, the 4 and the 2, are certain. The last digit, the 8, is an estimate. Therefore, it is uncertain. The value 4.3 has two significant digits. The 4 is certain, and the 3 is uncertain.

How Can You Tell Which Digits Are Significant?

You can identify the number of significant digits in any value. Table 1.4 lists some rules to help you do this.

Table 1.4 Rules for Determining Significant Digits

Rules	Examples
1. All non-zero numbers are significant.	7.886 has four significant digits. 19.4 has three significant digits. 527.266 992 has nine significant digits.
2. All zeros that are located between two non-zero numbers are significant.	408 has three significant digits. 25 074 has five significant digits.
3. Zeros that are located to the left of a value are *not* significant.	0.0907 has three significant digits. They are the 9, the third 0 to the right, and the 7. The function of the 0.0 at the begining is only to locate the decimal. 0.000 000 000 06 has one significant digit.
4. Zeros that are located to the right of a value may or may not be significant.	22 700 may have three significant digits, *or* it may have five significant digits. See the box below to find out why.

Explaining Three Significant Digits

The Great Lakes contain 22 700 km^3 of water. Is there exactly that amount of water in the Great Lakes? No, 22 700 km^3 is an approximate value. The actual volume could be anywhere from 22 651 km^3 to 22 749 km^3. You can use scientific notation to rewrite 22 700 km^3 as 2.27×10^4 km. This shows that only three digits are significant. (See Appendix E at the back of the book, if you would like to review scientific notation.)

Explaining Five Significant Digits

What if you were able to measure the volume of water in the Great Lakes? You could verify the value of 22 700 km^3. Then all five digits (including the zeros) would be significant. Here again, scientific notation lets you show clearly the five significant digits: 2.2700×10^4 km^3.

Practice Problems

1. Write the following quantities in your notebook. Beside each quantity, record the number of significant digits.

(a) 24.7 kg	**(e)** 8.930×10^5 km
(b) 247.7 mL	**(f)** 2.5 g
(c) 247.701 mg	**(g)** 0.0003 mL
(d) 0.247 01 L	**(h)** 923.2 g

2. Consider the quantity 2400 g.
   (a) Assume that you measured this quantity. How many significant digits does it have?
   (b) Now assume that you have no knowledge of how it was obtained. How many significant digits does it have?

Accuracy and Precision

In everyday speech, you might use the terms "accuracy" and "precision" to mean the same thing. In science, however, these terms are related to certainty. Each, then, has a specific meaning.

Accuracy refers to how close a given quantity is to an accepted or expected value. (See Figure 1.7.) **Precision** may refer to the exactness of a measurement. For example, ruler B in Figure 1.5 lets you measure length with greater precision than ruler A. Precision may also refer to the closeness of a series of data points. Data that are very close to one another are said to be precise. Examine Figure 1.8. Notice that a set of data can be precise but not accurate.

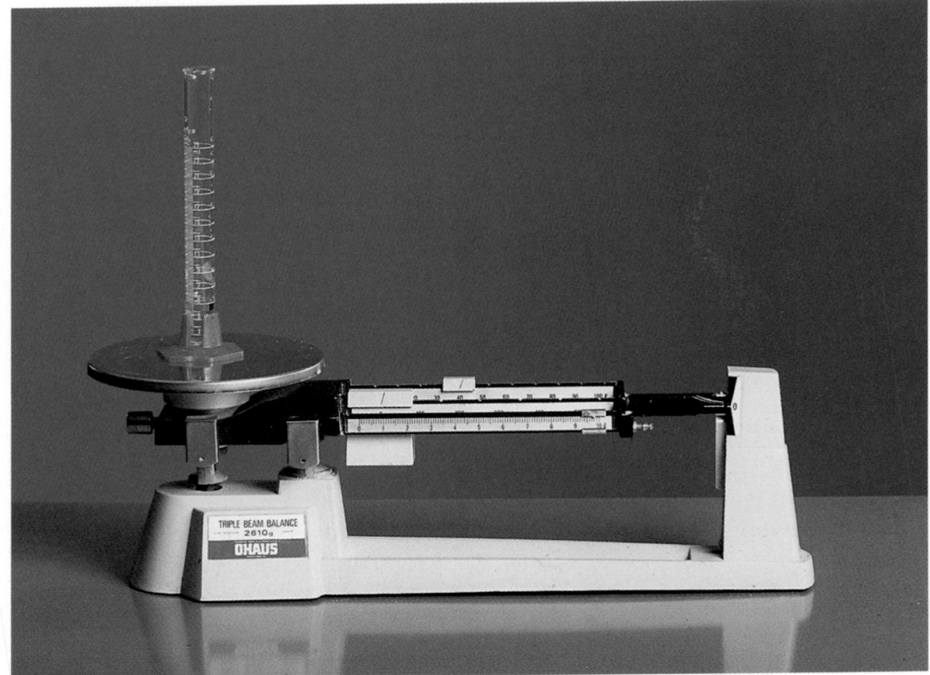

Figure 1.7 Under standard conditions of temperature and pressure, 5 mL of water has a mass of 5 g. Why does the reading on this balance show a different value?

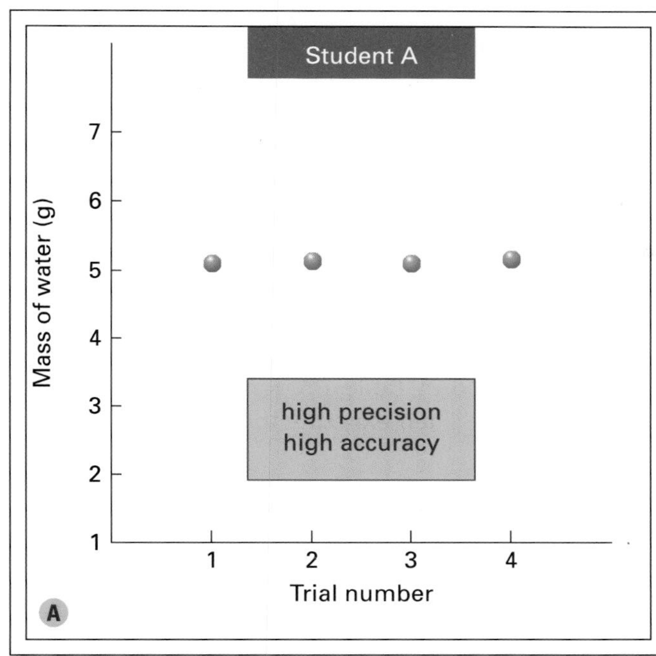

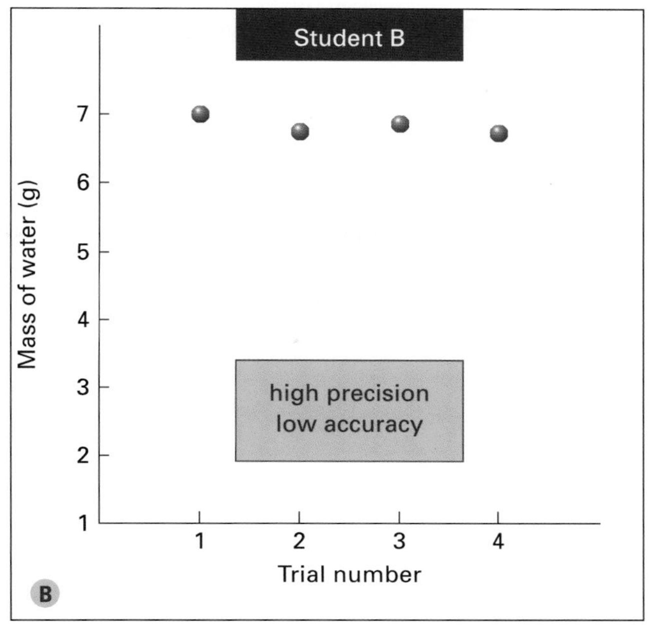

Figure 1.8 **Compare student A's results with results obtained by student B.**

Two students conducted four trials each to measure the volumes and masses of 5 mL of water. The graphs in Figure 1.8 show their results. The expected value for the mass of water is 5 g. Student A's results show high precision and high accuracy. Student B's results show high precision but low accuracy.

In the following Express Lab, you will see how the equipment you use affects the precision of your measurements.

ExpressLab  Significant Digits

You know that the precision of a measuring device affects the number of significant digits that you should report. In this activity, each group will use different glassware and a different balance to collect data.

Materials

glassware for measuring volume: for example, graduated cylinders, Erlenmeyer flasks, pipettes or beakers

balance

water

Procedure

1. Obtain the glassware and balance assigned to your group.
2. Determine the mass and volume of an amount of water. (The amount you use is up to you to decide.)
3. From the data you collect, calculate the density of water.
4. Enter your values for mass, volume, and density in the class table.

Analysis

1. Examine each group's data and calculated value for density. Note how the number of significant digits in each value for density compares with the number of significant digits in the measured quantities.
2. Propose a rule or guideline for properly handling significant digits when you multiply and divide measured quantities.

Calculating with Significant Digits

In this course, you will often take measurements and use them to calculate other quantities. You must be careful to keep track of which digits in your calculations and results are significant. Why? Your results should not imply more certainty than your measured quantities justify. This is especially important when you use a calculator. Calculators usually report results with far more significant figures—greater certainty—than your data warrant. Always remember that calculators do not make decisions about certainty. You do.

There are three rules for reporting significant digits in calculated answers. These rules are summarized in Table 1.5. Reflect on how they apply to your previous experiences. Then examine the Sample Problems that follow.

Table 1.5 Rules for Reporting Significant Digits in Calculations

Rule 1: Multiplying and Dividing The value with the fewest number of significant digits, going into the calculation, determines the number of significant digits that you should report in your answer.
Rule 2: Adding and Subtracting The value with the fewest number of decimal places, going into the calculation, determines the number of decimal places that you should report in your answer.
Rule 3: Rounding To get the appropriate number of significant digits (rule 1) or decimal places (rule 2), you may need to round your answer. If your answer ends in a number that is greater than 5, increase the preceding digit by 1. For example, 2.346 can be rounded to 2.35. If your answer ends with a number that is less than 5, leave the preceding number unchanged. For example, 5.73 can be rounded to 5.7. If your answer ends with 5, increase the preceding number by 1 if it is odd. Leave the preceding number unchanged if it is even. For example, 18.35 can be rounded to 18.4, but 18.25 is rounded to 18.2.

Sample Problem

Reporting Volume Using Significant Digits

Problem

A student measured a regularly shaped sample of iron and found it to be 6.78 cm long, 3.906 cm wide, and 11 cm tall. Determine its volume to the correct number of significant digits.

What Is Required?

You need to calculate the volume of the iron sample. Then you need to write this volume using the correct number of significant digits.

Continued ...

Continued ...
FROM PAGE 20

What Is Given?

You know the three dimensions of the iron sample.

Length = 6.78 cm (three significant digits)

Width = 3.906 cm (four significant digits)

Height = 11 cm (two significant digits)

Plan Your Strategy

To calculate the volume, use the formula

$$\text{Volume} = \text{Length} \times \text{Width} \times \text{Height}$$
$$V = l \times w \times h$$

Find the value with the smallest number of significant digits. Your answer can have only this number of significant digits.

Act on Your Strategy

$$\begin{aligned} V &= l \times w \times h \\ &= 6.78 \text{ cm} \times 3.906 \text{ cm} \times 11 \text{ cm} \\ &= 291.309\ 48 \text{ cm}^3 \end{aligned}$$

The value 11 cm has the smallest number of significant digits: two. Thus, your answer can have only two significant digits. In order to have only two significant digits, you need to put your answer into scientific notation.

$$V = 2.9 \times 10^2 \text{ cm}^3$$

Therefore, the volume is 2.9×10^2 cm^3, to two significant digits.

Check Your Solution

- Your answer is in cm^3. This is a unit of volume.
- Your answer has two significant digits. The least number of significant digits in the question is also two.

Sample Problem

Reporting Mass Using Significant Digits

Problem

Suppose that you measure the masses of four objects as 12.5 g, 145.67 g, 79.0 g, and 38.438 g. What is the total mass of the objects?

What Is Required?

You need to calculate the total mass of the objects.

What Is Given?

You know the mass of each object.

Continued ...

Continued ...
FROM PAGE 21

Plan Your Strategy

- Add the masses together, aligning them at the decimal point.
- Underline the estimated (farthest right) digit in each value. This is a technique you can use to help you keep track of the number of estimated digits in your final answer.
- In the question, two values have the fewest decimal places: 12.5 and 79.0. You need to round your answer so that it has only one decimal place.

PROBLEM TIP

Notice that adding the values results in an answer that has three decimal places. Using the underlining technique mentioned in "Plan Your Strategy" helps you count them quickly.

Act on Your Strategy

$$\begin{array}{r} 12.\underline{5} \\ 145.6\underline{7} \\ 79.\underline{0} \\ +\ 38.43\underline{8} \\ \hline 275.\underline{608} \end{array}$$

Total mass = 275.608 g

Therefore, the total mass of the objects is 275.6 g.

Check Your Solution

- Your answer is in grams. This is a unit of mass.
- Your answer has one decimal place. This is the same as the values in the question with the fewest decimal places.

Practice Problems

3. Do the following calculations. Express each answer using the correct number of significant digits.

(a) 55.671 g + 45.78 g

(b) 1.9 mm + 0.62 mm

(c) 87.9478 L − 86.25 L

(d) 0.350 mL + 1.70 mL + 1.019 mL

(e) 5.841 g × 6.03 g

(f) $\frac{0.6 \text{ kg}}{15 \text{ L}}$

(g) $\frac{17.51 \text{ g}}{2.2 \text{ cm}^3}$

Chemistry, Calculations, and Communication

Mathematical calculations are an important part of chemistry. You will need your calculation skills to help you investigate many of the topics in this textbook. You will also need calculation skills to communicate your measurements and results clearly when you do activities and investigations. Chemistry, however, is more than measurements and calculations. Chemistry also involves finding and interpreting patterns. This is the focus of the next section.

Chemistry Bulletin

Science | Technology | Society | Environment

Air Canada Flight 143

Air Canada Flight 143 was en route from Montréal to Edmonton on July 23, 1983. The airplane was one of Air Canada's first Boeing 767s, and its systems were almost completely computerized.

While on the ground in Montréal, Captain Robert Pearson found that the airplane's fuel processor was malfunctioning. As well, all three fuel gauges were not operating. Pearson believed, however, that it was safe to fly the airplane using manual fuel measurements.

Partway into the flight, as the airplane passed over Red Lake, Ontario, one of two fuel pumps in the left wing failed. Soon the other fuel pump failed and the left engine flamed out. Pearson decided to head to the closest major airport, in Winnipeg. He began the airplane's descent. At 8400 m, and more than 160 km from the Winnipeg Airport, the right engine also failed. The airplane had run out of fuel.

In Montréal, the ground crew had determined that the airplane had 7682 L of fuel in its fuel tank. Captain Pearson had calculated that the mass of fuel needed for the trip from Montréal to Edmonton was 22 300 kg. Since fuel is measured in litres, Pearson asked a mechanic how to convert litres into kilograms. He was told to multiply the amount in litres by 1.77.

By multiplying 7682 L by 1.77, Pearson calculated that the airplane had 13 597 kg of fuel on board. He subtracted this value from the total amount of fuel for the trip, 22 300 kg, and found that 8703 kg more fuel was needed.

To convert kilograms back into litres, Pearson divided the mass, 8703 kg, by 1.77. The result was 4916 L. The crew added 4916 L of fuel to the airplane's tanks.

This conversion number, 1.77, had been used in the past because the density of jet fuel is 1.77 *pounds* per litre. Unfortunately, the number that should have been used to convert litres into kilograms was 0.803. The crew should have added 21 163 L of fuel, not 4916 L.

First officer Maurice Quintal calculated their rate of descent. He determined that they would never make Winnipeg. Pearson turned north and headed toward Gimli, an abandoned Air Force base. Gimli's left runway was being used for drag-car and go-kart races. Surrounding the runway were families and campers. It was into this situation that Pearson and Quintal landed the airplane.

Tires blew upon impact. The airplane skidded down the runway as racers and spectators scrambled to get out of the way. Flight 143 finally came to rest 1200 m later, a mere 30 m from the dazed onlookers.

Miraculously no one was seriously injured. As news spread around the world, the airplane became known as "The Gimli Glider."

Making Connections

1. You read that the airplane should have received 21 163 L of fuel. Show how this amount was calculated.
2. Use print or electronic resources to find out what caused the loss of the *Mars Climate Orbiter* spacecraft in September 1999. How is this incident related to the "Gimli Glider" story? Could a similar incident happen again? Why or why not?

Section Wrap-up

In this section, you learned how to judge the accuracy and precision of your measurement. You learned how to recognize significant digits. You also learned how to give answers to calculations using the correct number of significant digits.

In the next section, you will learn about the properties and classification of matter.

Section Review

1. K/U Explain the difference between accuracy and precision in your own words.
2. C What SI-derived unit of measurement would you use to describe:
   - **(a)** the mass of a person
   - **(b)** the mass of a mouse
   - **(c)** the volume of a glass of juice
   - **(d)** the length of your desk
   - **(e)** the length of your classroom
3. K/U Record the number of significant digits in each of the following values:
   - **(a)** 3.545
   - **(b)** 308
   - **(c)** 0.000876
4. K/U Complete the following calculations and give your answer to the correct number of significant digits.
   - **(a)** 5.672 g + 92.21 g
   - **(b)** 32.34 km × 93.1 km
   - **(c)** 66.0 mL × 0.031 mL
   - **(d)** 11.2 g ÷ 92 mL
5. I What lab equipment would you use in each situation? Why?
   - **(a)** You need 2.00 mL of hydrogen peroxide for a chemical reaction.
   - **(b)** You want approximately 1 L of water to wash your equipment.
   - **(c)** You are measuring 250 mL of water to heat on a hot plate.
   - **(d)** You need 10.2 mL of alcohol to make up a solution.
6. I Review the graphs in Figure 1.8. Draw two more graphs to show
   - **(a)** data that have high accuracy but low precision
   - **(b)** data that have low accuracy and low precision

Classifying Matter and Its Changes

1.3

Matter is constantly changing. Plants grow by converting matter from the soil and air into matter they can use. Water falls from the sky, evaporates, and condenses again to form liquid water in a never-ending cycle. You can probably suggest many more examples of matter changing.

Matter changes in response to changes in energy. Adding energy to matter or removing energy from matter results in a change. Figure 1.9 shows a familiar example of a change involving matter and energy.

Section Preview/ Specific Expectations

In this section, you will

- **identify** chemical substances and chemical changes in everyday life
- **demonstrate** an understanding of the need to use chemicals safely in everyday life
- **communicate** your understanding of the following terms: *physical changes, chemical changes, mixture, pure substance, element, compound*

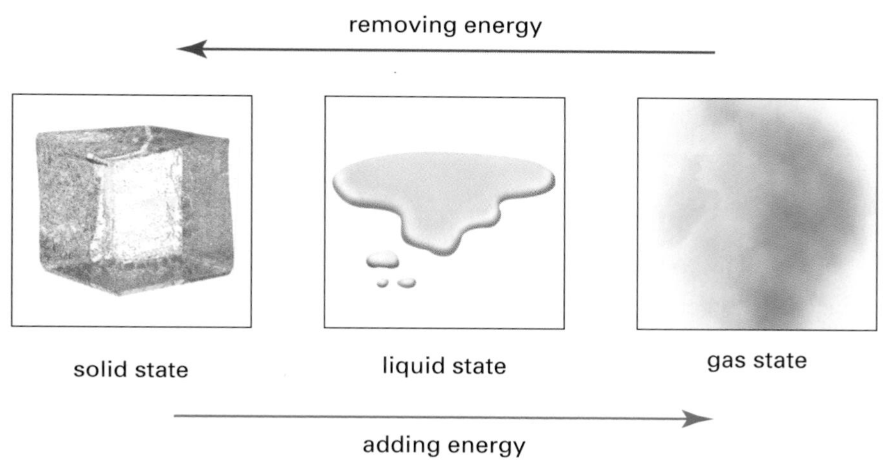

Figure 1.9 Like all matter, water can change its state when energy is added or removed.

Physical and Chemical Changes in Matter

A change of state alters the appearance of matter. The composition of matter remains the same, however, regardless of its state. For example, ice, liquid water, and water vapour are all the same kind of matter: water. Melting and boiling other kinds of matter have the same result. The appearance and some other physical properties change, but the matter retains its identity—its composition. Changes that affect the physical appearance of matter, but *not* its composition, are **physical changes**.

Figure 1.10 shows a different kind of change involving water. Electrical energy is passed through water, causing it to decompose. Two completely different kinds of matter result from this process: hydrogen gas and oxygen gas. These gases have physical and chemical properties that are different from the properties of water and from each other's properties. Therefore, decomposing water is a change that affects the composition of water. Changes that alter the composition of matter are called **chemical changes**. Iron rusting, wood burning, and bread baking are three examples of chemical changes.

You learned about physical and chemical properties earlier in this chapter. A physical change results in a change of physical properties only. A chemical change results in a change of both physical and chemical properties.

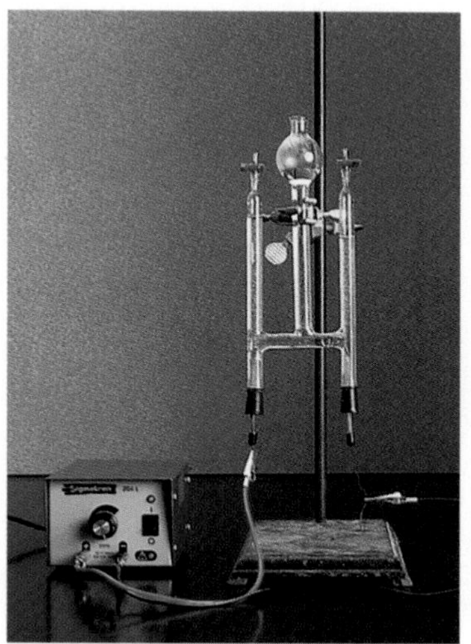

Figure 1.10 An electrical current is used to decompose water. This process is known as electrolysis.

mind STRETCH

Before adopting the metric system, Canadians measured temperature in units called Fahrenheit degrees (°F). Based on the Fahrenheit scale, water boils at 212°F and freezes at 32°F. A few countries, including the United States, still use the Fahrenheit scale. Without checking any reference materials, design a method for converting Fahrenheit temperatures to Celsius temperatures, and back again. Show your work, and explain your reasoning.

Practice Problems

4. Classify each situation as either a physical change or a chemical change. Explain your reasoning.
   - **(a)** A rose bush grows from a seed that you have planted and nourished.
   - **(b)** A green coating forms on a copper statue when the statue is exposed to air.
   - **(c)** Your sweat evaporates to help balance your body temperature.
   - **(d)** Frost forms on the inside of a freezer.
   - **(e)** Salt is added to clear chicken broth.
   - **(f)** Your body breaks down the food you eat to provide energy for your body's cells.
   - **(g)** Juice crystals dissolve in water.
   - **(h)** An ice-cream cone melts on a hot day.

Figure 1.11 **To see the components of soil, add some soil to a glass of water. What property is responsible for separating the components?**

Classifying Matter

All matter can be classified into two groups: mixtures and pure substances. A **mixture** is a physical combination of two or more kinds of matter. For example, soil is a mixture of sand, clay, silt, and decomposed leaves and animal bodies. If you look at soil under a magnifying glass, you can see these different components. Figure 1.11 shows another way to see the components of soil.

The components in a mixture can occur in different proportions (relative quantities). Each individual component retains its identity. Mixtures in which the different components are clearly visible are called *heterogeneous mixtures*. The prefix "hetero-" comes from the Greek word *heteros*, meaning "different."

Mixtures in which the components are blended together so well that the mixture looks like just one substance are called *homogeneous mixtures*. The prefix "homo- " comes from the Greek word *homos*, meaning "the same." Saltwater, clean air, and grape juice are common examples. Homogeneous mixtures are also called solutions. You will investigate solutions in Unit 3.

A **pure substance** has a definite composition, which stays the same in response to physical changes. A lump of copper is a pure substance. Water (with nothing dissolved in it) is also a pure substance. Diamond, carbon dioxide, gold, oxygen, and aluminum are pure substances, too.

Pure substances are further classified into elements and compounds. An **element** is a pure substance that cannot be separated chemically into any simpler substances. Copper, zinc, hydrogen, oxygen, and carbon are examples of elements.

A **compound** is a pure substance that results when two or more elements combine chemically to form a different substance. Compounds can be broken down into elements using chemical processes. For example, carbon dioxide is a compound. It can be separated into the elements carbon and oxygen. The Concept Organizer on the next page outlines the classification of matter at a glance. The ThoughtLab reinforces your understanding of properties, mixtures, and separation of substances.

Word LINK

The word "pure" can be used to mean different things. In ordinary conversation, you might say that orange juice is "pure" if no other materials have been added to it. How is this meaning of pure different from the scientific meaning in the term "pure substance?"

Concept Organizer The Classification System for Matter.

Matter
- anything that has mass and volume
- found in three physical states: solid, liquid, gas

Mixtures
- physical combinations of matter in which each component retains its identity

Heterogenous Mixtures (Mechanical Mixtures)
- all components are visible

Homogeneous Mixtures (Solutions)
- components are blended so that it looks like a single substance.

Physical Changes

Pure Substances
- matter that has a definite composition

Elements
- matter that cannot be decomposed into simpler substances

Chemical Changes

Compounds
- matter in which two or more elements are chemically combined

ThoughtLab  Mixtures, Pure Substances, and Changes

You frequently use your knowledge of properties to make and separate mixtures and substances. You probably do this most often in the kitchen. Even the act of sorting clean laundry, however, depends on your ability to recognize and make use of physical properties. This activity is a "thought experiment." You will use your understanding of properties to mix and separate a variety of chemicals, all on paper. Afterward, your teacher may ask you to test your ideas, either in the laboratory or at home in the kitchen.

Procedure

1. Consider the following chemicals: table salt, water, baking soda, sugar, iron filings, sand, vegetable oil, milk, and vinegar. Identify each chemical as a mixture or a pure substance.
2. Which of these chemicals can you mix together *without* producing a chemical change? In your notebook, record as many of these physical combinations as you can.
3. Which of these chemicals can you mix together to produce a chemical change? Record as many of these chemical combinations as you can.
4. Record a mixture that is made with four of the chemicals. Then suggest one or more techniques that you can use to separate the four chemicals from one another. Write notes and sketch labelled diagrams to show your techniques. Identify the properties that your techniques depend on.

Analysis

1. In step 2, what properties of the chemicals did you use to determine your combinations?
2. In step 3, what properties did you use to determine your combinations?

Application

3. Exchange your four-chemical mixture with a partner. Do not include your notes and diagrams. Challenge your partner to suggest techniques to separate the four chemicals. Then assess each other's techniques. What modifications, if any, would you make to your original techniques?

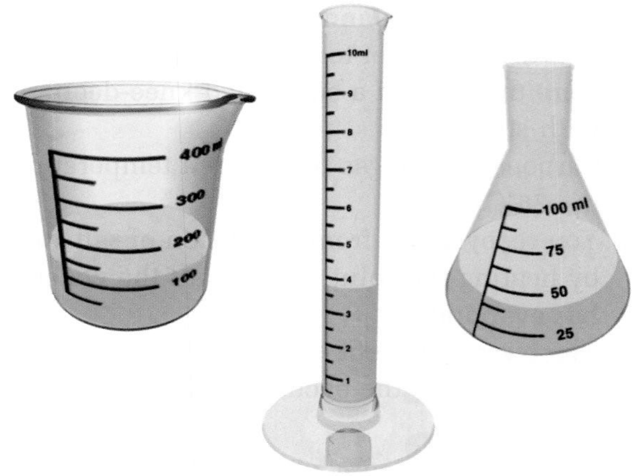

6. Make each conversion below.
(a) 10 kg to grams (g)
(b) 22.3 cm to metres (m)
(c) 52 mL to centimetres cubed (cm^3)
(d) 1.0 L to centimetres cubed (cm^3)

7. Identify the number of significant digits in each value.
(a) 0.002 cm
(b) 3107 km
(c) 5 g
(d) 8.6×10^{10} m^3
(e) 4.0003 mL
(f) 5.432×10^2 km^2
(g) 91 511 L

8. (a) Explain why the value 5700 km could have two, three, or four significant digits.
(b) Write 5700 km with two significant digits.
(c) Write 5700 km with four significant digits.

9. Complete each calculation. Express your answer to the correct number of significant digits.
(a) 4.02 mL + 3.76 mL + 0.95 mL
(b) $(2.7 \times 10^2$ m$) \times (4.23 \times 10^2$ m$)$
(c) 5 092 kg ÷ 23 L
(d) $2 - 0.3 + 6 - 7$
(e) $(6.853 \times 10^3$ L$) + (5.40 \times 10^3$ L$)$
(f) (572.3 g + 794.1 g) ÷ (24 mL + 52 mL)

10. Round each value to the given number of significant digits.
(a) 62 091 to three significant digits
(b) 27 to one significant digit
(c) 583 to one significant digit
(d) 17.25 to three significant digits

11. A plumber installs a pipe that has a diameter of 10 cm and a length of 2.4 m. Calculate the volume of water (in cm^3) that the pipe will hold. Express your answer to the correct number of significant digits. **Note:** The formula for the volume of a cylinder is $V = \pi r^2 h$, where r is the radius and h is the height or length.

12. During an investigation, a student monitors the temperature of water in a beaker. The data from the investigation are shown in the table below.
(a) What was the average temperature of the water? Express your answer to the appropriate number of significant digits.
(b) The thermometer that the student used has a scale marked at 1° intervals. Which digits in the table below are estimated?

Time (min)	Temperature (°C)
0.0	25
1	24.3
2	24
3	23.7
4	23.6

13. Identify each change as either physical or chemical.
(a) Over time, an iron swing set becomes covered with rust.
(b) Juice crystals "disappear" when they are stirred into a glass of water.
(c) Litmus paper turns pink when exposed to acid.
(d) Butter melts when you spread it on hot toast.

Inquiry

14. Your teacher asks the class to measure the mass of a sample of aluminum. You measure the mass three times, and obtain the following data: 6.74, 6.70, and 6.71 g. The actual value is 6.70 g. Here are the results of three other students:

Student A 6.50, 6.49, and 6.52 g

Student B 6.57, 6.82, and 6.71 g

Student C 6.61, 6.70, and 6.87 g

(a) Graph the four sets of data. (Call yourself "student D.")

(b) Which results are most precise?
(c) Which results are most accurate?
(d) Which results have the highest accuracy *and* precision?

15. (a) Design an investigation to discover some of the physical and chemical properties of hydrogen peroxide, H_2O_2.
(b) List the materials you need to carry out your investigation.
(c) What specific physical and/or chemical properties does your investigation test for?
(d) What variables are held constant during your investigation? What variables are changed? What variables are measured?
(e) If you have time, obtain some hydrogen peroxide from a drugstore. Perform your investigation, and record your observations.

Communication

16. Choose one of the common chemicals listed below. In your notebook, draw a concept web that shows some of the physical properties, chemical properties, and uses of this chemical.
    - table salt (sodium chloride)
    - water
    - baking soda (sodium hydrogen carbonate)
    - sugar (sucrose)
17. In your notebook, draw a flowchart or concept web that illustrates the connections between the following words:
    - mixture
    - pure substance
    - homogeneous
    - heterogeneous
    - solution
    - matter
    - water
    - cereal
    - aluminum
    - apple juice
18. Is salad dressing a homogenous mixture or a heterogeneous mixture? Use diagrams to explain.

Making Connections

19. Locate 3 cleaning products in your home. For each product, record the following information:
    - the chemical(s) most responsible for its cleaning action
    - any safety symbols or warnings on the packaging or container
    - any hazards associated with the chemical(s) that the product contains
    - suggestions for using the product safely

Back in class, share and analyze the chemicals that everyone found.
(a) Prepare a database that includes all the different chemicals, the products in which they are found, their hazards, and instructions for their safe use. Add to the database throughout the year. Make sure that you have an updated copy at all times.
(b) Identify the cleaning products that depend mainly on chemical changes for their cleaning action. How can you tell?

20. At the beginning of this chapter, you saw how water, a very safe chemical compound, can be misrepresented to appear dangerous. Issues about toxic and polluting chemicals are sometimes reported in newspapers or on television. List some questions you might ask to help you determine whether or not an issue was being misrepresented.
21. Describe the most important STSE connections for each situation.
(a) Car exhaust releases gases such as sulfur dioxide, $SO_{2(g)}$, and nitrogen oxide, $NO_{(g)}$. These gases lead to smog in cities. As well, they are a cause of acid rain.
(b) In the past, people used dyes from plants and animals to colour fabrics. These natural dyes produced a limited range of colours, and they faded quickly. Today long-lasting artificial dyes are available in almost every possible colour. These dyes were invented by chemists. They are made in large quantities for the fabric and clothing industries.

Answers to Practice Problems and Short Answers to Section Review Questions

Practice Problems: 1.(a) 3 (b) 4 (c) 6 (d) 5 (e) 4 (f) 2 (g) 1 (h) 4 2.(a) 4 (b) 2 3.(a) 101.45 g (b) 2.5 mm (c) 1.70 L (d) 3.07 mL (e) 35.2 g^2 (f) 4×10^{-2} kg/L (g) 8.0 g/cm^3 4.(a) chemical (b) chemical (c) physical (d) physical (e) physical (f) chemical (g) physical (h) physical
Section Review: 1.2: 2.(a) kg (b) g (c) mL (d) cm (e) m 3.(a) 4 (b) 3 (c) 3 4.(a) 97.88 g (b) 3.01×10^3 km^2 (c) 2.0 mL^2 (d) 0.12 g/mL 5.(a) pipette (b) Erlenmeyer flask or large beaker (c) 250 mL beaker (d) graduated cylinder

BEACHEY AIRSHIP

CHAPTER 2

Elements and the Periodic Table

Today, if you want to travel by air across the country or overseas, you take an airplane. During the first three decades of the twentieth century, you would have boarded a hydrogen-filled balloon such as the one shown in the black-and-white photograph. Large and small airships such as these, called dirigibles, were common sights in the skies above many North American and European cities. Unfortunately, during a landing in Lakehurst, New Jersey in 1937, the hydrogen in one of these airships, the *Hindenburg,* ignited. The resulting explosion killed 36 people, and marked the end of the use of hydrogen for dirigibles.

Gas-filled airships and balloons like the one shown in the colour photograph now use helium gas instead of hydrogen. Helium, unlike hydrogen, does not burn. In fact, helium is a highly unreactive gas. What is it about hydrogen that makes it so reactive? Why is helium so unreactive? The answer lies in the structure of the atoms of these elements. In previous science courses, you traced the history of our understanding of atoms and their structure. You also learned how chemists use properties to arrange elements, and the atoms of which they are made, into a remarkable tool called the periodic table. This chapter highlights and expands on key ideas from your earlier studies. By the end of the chapter, you will have a greater understanding of the properties of elements at an atomic level. This understanding is a crucial foundation for concepts that you will explore in your chemistry course this year.

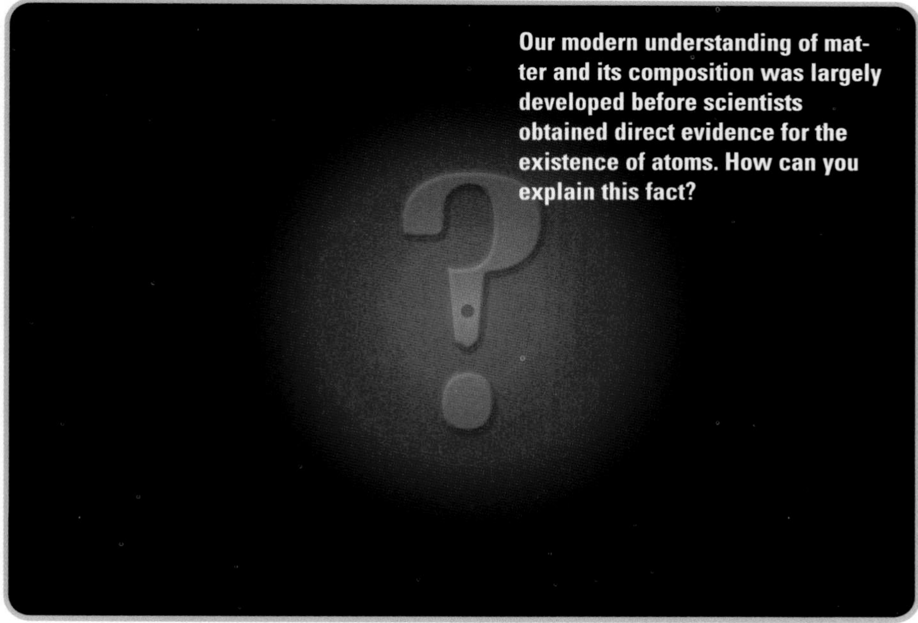

Our modern understanding of matter and its composition was largely developed before scientists obtained direct evidence for the existence of atoms. How can you explain this fact?

Chapter Preview

Concepts and Skills You Will Need

Before you begin this chapter, review the following concepts and skills:

- expressing the results of calculations to the appropriate number of decimal places and significant digits (Chapter 1, section 1.2)

2.1 Atoms and Their Composition

Section Preview/ Specific Expectations

In this section, you will

- **define** and **describe** the relationships among atomic number, mass number, atomic mass, isotope, and radioisotope
- **communicate** your understanding of the following terms: *atom, atomic mass unit (u), atomic number (Z), mass number (A), atomic symbol, isotopes, radioactivity, radioisotopes*

Technology LINK

How scientists visualize the atom has changed greatly since Dalton proposed his atomic theory in the early nineteenth century. Technology has played an essential role in these changes. At a library or on the Internet, research the key modifications to the model of the atom. Create a summary chart to show your findings. Include the scientists involved, the technologies they used, the discoveries they made, and the impact of their discoveries on the model of the atom. If you wish, use a suitable graphics program to set up your chart.

Elements are the basic substances that make up all matter. About 90 elements exist naturally in the universe. The two smallest and least dense of these elements are hydrogen and helium. Yet hydrogen and helium account for nearly 98% of the mass of the entire universe!

Here on Earth, there is very little hydrogen in its pure elemental form. There is even less helium. In fact, there is such a small amount of helium on Earth that it escaped scientists' notice until 1895.

Regardless of abundance, any two samples of hydrogen—from anywhere on Earth or far beyond in outer space—are identical to each other. For example, a sample of hydrogen from Earth's atmosphere is identical to a sample of hydrogen from the Sun. The same is true for helium. This is because *each element is made up of only a single kind of atom.* For example, the element hydrogen contains only hydrogen atoms. The element helium contains only helium atoms. What, however, is an atom?

The Atomic Theory of Matter

John Dalton was a British teacher and self-taught scientist. In 1809, he described atoms as solid, indestructible particles that make up all matter. (See Figure 2.1.) Dalton's concept of the atom is one of several ideas in his atomic theory of matter, which is outlined on the next page. Keep in mind that scientists have modified several of Dalton's ideas, based on later discoveries. You will learn about these modifications at the end of this section. See if you can infer what some of them are as you study the structure of the atom on the next few pages.

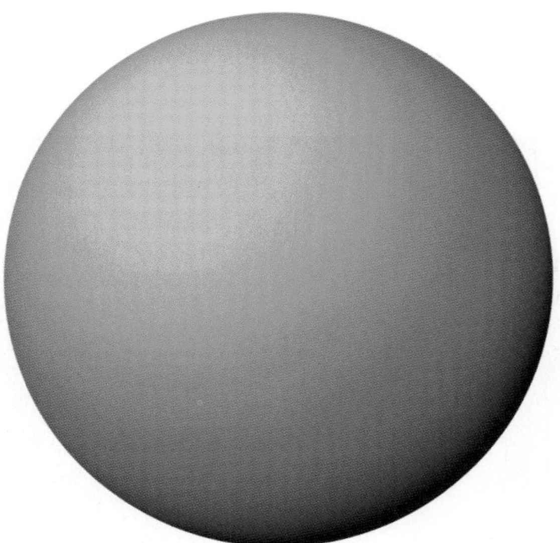

Figure 2.1 This illustration shows an atom as John Dalton (1766–1844) imagined it. Many reference materials refer to Dalton's concept of the atom as the "billiard ball model." Dalton, however, was an avid lawn bowler. His concept of the atom was almost certainly influenced by the smooth, solid bowling balls used in the game.

Dalton's Atomic Theory (1809)

- All matter is made up of tiny particles called atoms. An atom cannot be created, destroyed, or divided into smaller particles.
- The atoms of one element cannot be converted into the atoms of any another element.
- All the atoms of one element have the same properties, such as mass and size. These properties are different from the properties of the atoms of any other element.
- Atoms of different elements combine in specific proportions to form compounds.

CHECKPOINT

The atomic theory was a convincing explanation of the behaviour of matter. It explained two established scientific laws: the law of conservation of mass and the law of definite composition.

- *Law of conservation of mass*: During a chemical reaction, the total mass of the substances involved does not change.
- *Law of definite proportion*: Elements always combine to form compounds in fixed proportions by mass. (For example, pure water always contains the elements hydrogen and oxygen, combined in the following proportions: 11% hydrogen and 89% oxygen.)

How does the atomic theory explain these two laws?

The Modern View of the Atom

An **atom** is the smallest particle of an element that still retains the identity and properties of the element. For example, the smallest particle of the writing material in your pencil is a carbon atom. (Pencil "lead" is actually a substance called graphite. Graphite is a form of the element carbon.)

An average atom is about 10^{-10} m in diameter. Such a tiny size is difficult to visualize. If an average atom were the size of a grain of sand, a strand of your hair would be about 60 m in diameter!

Atoms themselves are made up of even smaller particles. These *subatomic particles* are protons, neutrons, and electrons. Protons and neutrons cluster together to form the central core, or *nucleus*, of an atom. Fast-moving electrons occupy the space that surrounds the nucleus of the atom. As their names imply, subatomic particles are associated with electrical charges. Table 2.1 and Figure 2.2 summarize the general features and properties of an atom and its three subatomic particles.

Table 2.1 Properties of Protons, Neutrons, and Electrons

Subatomic particle	Charge	Symbol	Mass (in g)	Radius (in m)
electron	1–	e^-	9.02×10^{-28}	smaller than 10^{-18}
proton	1+	p^+	1.67×10^{-24}	10^{-15}
neutron	0	n^0	1.67×10^{-24}	10^{-15}

Expressing the Mass of Subatomic Particles

As you can see in Table 2.1, subatomic particles are incredibly small. Suppose that you could count out protons or neutrons equal to

602 000 000 000 000 000 000 000 (or 6.02×10^{23})

and put them on a scale. They would have a mass of about 1 g. This means that one proton or neutron has a mass of

$$\frac{1\text{ g}}{6.02 \times 10^{23}} = 0.000\ 000\ 000\ 000\ 000\ 000\ 000\ 001\ 66\text{ g}$$
$$= 1.66 \times 10^{-24}\text{ g}$$

It is inconvenient to measure the mass of subatomic particles using units such as grams. Instead, chemists use a unit called an **atomic mass unit** (symbol **u**). A proton has a mass of about 1 u, which is equal to 1.66×10^{-24} g.

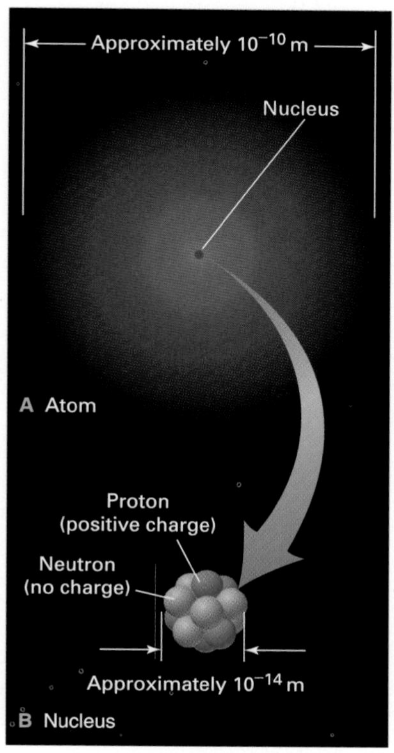

Figure 2.2 This illustration shows the modern view of an atom. Notice that a fuzzy, cloud-like region surrounds the atomic nucleus. Electrons move rapidly throughout this region, which represents most of the atom's volume.

The number 6.02×10^{23} is called the Avogadro constant. In Chapter 5, you will learn more about the Avogadro constant.

A proton is about 1837 times more massive than an electron. According to Table 2.1, the mass of an electron is 9.02×10^{-28} g. This value is so small that scientists consider the mass of an electron to be approximately equal to zero. Thus, electrons are not taken into account when calculating the mass of an atom.

The Nucleus of an Atom

All the atoms of a particular element have the same number of protons in their nucleus. For example, all hydrogen atoms—anywhere in the universe—have one proton. All helium atoms have two protons. All oxygen atoms have eight protons. Chemists use the term **atomic number** (symbol ***Z***) to refer to the number of protons in the nucleus of each atom of an element.

As you know, the nucleus of an atom also contains neutrons. In fact, the mass of an atom is due to the combined masses of its protons and neutrons. Therefore, an element's **mass number** (symbol ***A***) is the total number of protons and neutrons in the nucleus of one of its atoms. Each proton or neutron is counted as one unit of the mass number. For example, an oxygen atom, which has 8 protons and 8 neutrons in its nucleus, has a mass number of 16. A uranium atom, which has 92 protons and 146 neutrons, has a mass number of 238.

Information about an element's protons and neutrons is often summarized using the chemical notation shown in Figure 2.3. The letter X represents the **atomic symbol** for an element. (The atomic symbol is also called the *element symbol*.) Each element has a different atomic symbol. All chemists, throughout the world, use the same atomic symbols. Over the coming months, you will probably learn to recognize many of these symbols instantly. Appendix G, at the back of this book, lists the elements in alphabetical order, along with their symbols. You can also find the elements and their symbols in the periodic table on the inside back cover of this textbook, and in Appendix C. (You will review and extend your understanding of the periodic table, in section 2.2.)

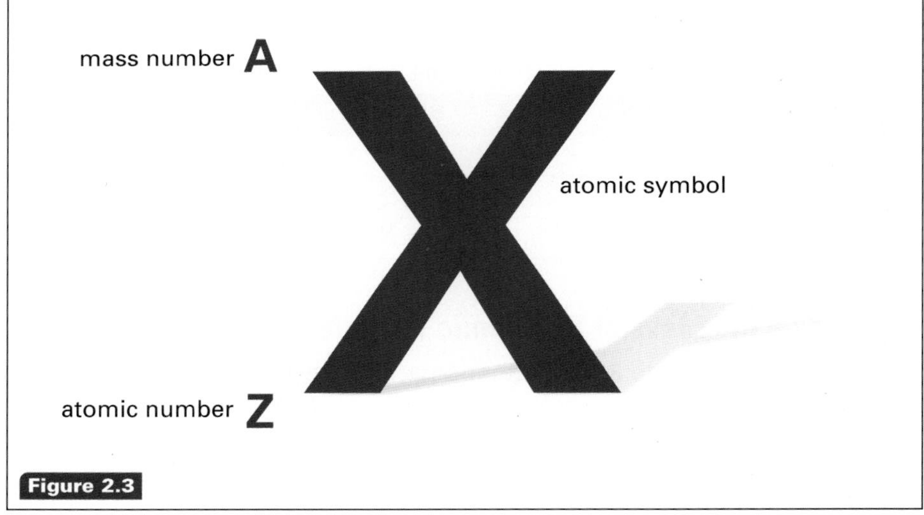

Figure 2.3

Notice what the chemical notation in Figure 2.3 does, and does not, tell you about the structure of an element's atoms. For example, consider the element fluorine: $^{19}_{9}F$. The mass number (the superscript 19) indicates that fluorine has a total of 19 protons and neutrons. The atomic number (subscript 9) indicates that fluorine has 9 protons. Neither the mass number nor the atomic number tells you how many neutrons fluorine has. You can calculate this value, however, by subtracting the atomic number from the mass number.

$$\text{Number of neutrons} = \text{Mass number} - \text{Atomic number}$$
$$= A - Z$$

Thus, for fluorine,

$$\text{Number of neutrons} = A - Z = 19 - 9 = 10$$

Now try a few similar calculations in the Practice Problem below.

Practice Problems

1. Copy the table below into your notebook. Fill in the missing information. Use a periodic table, if you need help identifying the atomic symbol.

Chemical notation	Element	Number of protons	Number of neutrons
$^{11}_{5}B$	(a)	(b)	(c)
$^{208}_{82}Pb$	(d)	(e)	(f)
(g)	tungsten	(h)	110
(i)	helium	(j)	2
$^{239}_{94}Pu$	(k)	(l)	(m)
$^{56}_{26}$(n)	(o)	26	(p)
(q)	bismuth	(r)	126
(s)	(t)	47	107
$^{20}_{10}$(u)	(v)	(w)	(x)

Math LINK

Expressing numerical data about atoms in units such as metres is like using a bulldozer to move a grain of sand. Atoms and subatomic particles are so small that they are not measured using familiar units. Instead, chemists often measure atoms in nanometres ($1 \text{ nm} = 1 \times 10^{-9} \text{ m}$) and picometres ($1 \text{ pm} = 1 \times 10^{-12} \text{ m}$).

- Convert the diameter of a proton and a neutron into nanometres and then picometres.
- Atomic and subatomic sizes are hard to imagine. Create an analogy to help people visualize the size of an atom and its sub-atomic particles. (The first sentence of this feature is an example of an analogy.)

Using the Atomic Number to Infer the Number of Electrons

As just mentioned, the atomic number and mass number do not give you direct information about the number of neutrons in an element. They do not give you the number of electrons, either. You can infer the number of electrons, however, from the atomic number. The atoms of each element are electrically neutral. This means that their positive charges (protons) and negative charges (electrons) must balance one another. In other words, *in the neutral atom of any element, the number of protons is equal to the number of electrons*. For example, a neutral hydrogen atom contains one proton, so it must also contain one electron. A neutral oxygen atom contains eight protons, so it must contain eight electrons.

Isotopes and Atomic Mass

All neutral atoms of the same element contain the same number of protons and, therefore, the same number of electrons. The number of neutrons can vary, however. For example, most of the oxygen atoms in nature have eight neutrons in their atomic nuclei. In other words, most oxygen atoms have a mass number of 16 (8 protons + 8 neutrons). As you can see in Figure 2.4 on the next page, there are also two other naturally occurring forms of oxygen. One of these has nine neutrons, so $A = 17$. The other has ten neutrons, so $A = 18$. These three forms of oxygen are called isotopes. **Isotopes** are atoms of an element that have the same number of protons but different numbers of neutrons.

Web LINK

www.school.mcgrawhill.ca/resources/

The atomic symbols are linked to the names of the elements. The links are not always obvious, however. Many atomic symbols are derived from the names of the elements in a language other than English, such as Latin, Greek, German, or Arabic. With your classmates, research the origin and significance of the name of each element. Go to the web site above. Go to **Science Resources**, then to **Chemistry 11** to find out where to go next. See if you can infer the rules that are used to create the atomic symbols from the names of the elements.

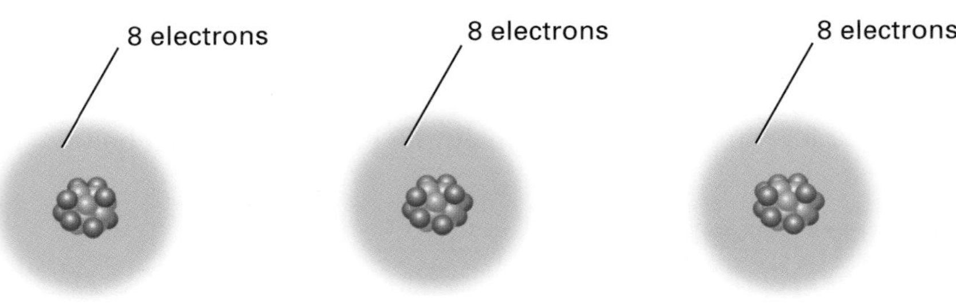

Figure 2.4 Oxygen has three naturally occurring isotopes. Notice that oxygen-16 has the same meaning as $^{16}_{8}O$. Similarly, oxygen-17 has the same meaning as $^{17}_{8}O$ and oxygen-18 has the same meaning as $^{18}_{8}O$.

The isotopes of an element have very similar chemical properties because they have the same number of protons and electrons. They differ in mass, however, because they have different numbers of neutrons.

Some isotopes are more unstable than others. Their *nuclei* (plural of *nucleus*) are more likely to decay, releasing energy and subatomic particles. This process, called **radioactivity**, happens spontaneously. All uranium isotopes, for example, have unstable nuclei. They are called radioactive isotopes, or **radioisotopes** for short. Many isotopes are not radioisotopes. Oxygen's three naturally occurring isotopes, for example, are stable. In contrast, chemists have successfully synthesized ten other isotopes of oxygen, all of which are unstable radioisotopes. (What products result when radioisotopes decay? You will find out in Chapter 4.)

CHEM FACT

Radioisotopes decay because their nuclei are unstable. The time it takes for a nucleus to decay varies greatly. For example, it takes billions of years for only half of the nucleus of naturally occurring uranium-238 to decay. The nuclei of other radioisotopes — mainly those that scientists have synthesized — decay much more rapidly. The nuclei of some isotopes, such as sodium-22, take about 20 years to decay. For calcium-47, this decay occurs in a matter of days. The nuclei of most synthetic radioisotopes decay so quickly, however, that the radioisotopes exist for mere fractions of a second.

Electrons in Atoms

So far, much of the discussion about the atom has concentrated on the nucleus and its protons and neutrons. What about electrons? What is their importance to the atom? Recall that electrons occupy the space surrounding the nucleus. Therefore, they are the first subatomic particles that are likely to interact when atoms come near one another. In a way, electrons are on the "front lines" of atomic interactions. The number and arrangement of the electrons in an atom determine how the atom will react, if at all, with other atoms. As you will learn in section 2.2, and throughout the rest of this unit, electrons are responsible for the chemical properties of the elements.

CHECKPOINT

When chemists refer symbolically to oxgyen-16 atoms, they often leave out the atomic number. They write ^{16}O. You can write other isotopes of oxygen, and all other elements, the same way. Why is it acceptable to leave out the atomic number?

Revisiting the Atomic Theory

John Dalton did not know about subatomic particles when he developed his atomic theory. Even so, the modern atomic theory (shown on the next page) retains many of Dalton's ideas, with only a few modifications. Examine the comments to the right of each point. They explain how the modern theory differs from Dalton's.

The atomic theory is a landmark achievement in the history of chemistry. It has shaped the way that all scientists, especially chemists, think about matter. In the next section, you will investigate another landmark achievement in chemistry: the periodic table.

The Modern Atomic Theory

- All matter is made up of tiny particles called atoms. Each atom is made up of smaller subatomic particles: protons, neutrons, and electrons.

  Although an atom is divisible, it is still the smallest particle of an element that has the properties and identity of the element.

- The atoms of one element cannot be converted into the atoms of any another element by a chemical reaction.

  Nuclear reactions (changes that alter the composition of the atomic nucleus) may, in fact, convert atoms of one element into atoms of another.

- Atoms of one element have the same properties, such as average mass and size. These properties are different from the properties of the atoms of any other element.

  Different isotopes of an element have different numbers of neutrons and thus different masses. As you will learn in Chapter 5, scientists treat elements as if their atoms have an average mass.

- Atoms of different elements combine in specific proportions to form compounds.

  This idea has remained basically unchanged.

Section Review

1 (C) Copy the table below into your notebook. Use a graphic organizer to show the relationship among the titles of each column. Then fill in the blanks with the appropriate information. (Assume that the atoms of each element are neutral.)

Element	Atomic number	Mass number	Number of protons	Number of electrons	Number of neutrons
(a)	(b)	108	(c)	47	(d)
(e)	(f)	(g)	33	(h)	42
(i)	35	(j)	(k)	(l)	45
(m)	79	179	(n)	(o)	(p)
(q)	(r)	(s)	(t)	50	69

2 (K/U) Explain the difference between a stable isotope and a radioisotope. Provide an example other than oxygen to support your answer.

3 (K/U) Examine the information represented by the following pairs: $^{3}_{1}H$ and $^{3}_{2}He$; $^{14}_{6}C$ and $^{16}_{7}N$; $^{19}_{9}F$ and $^{18}_{9}F$.

(a) For each pair, do both members have the same number of protons? electrons? neutrons?

(b) Which pair or pairs consist of atoms that have the same value for Z? Which consists of atoms that have the same value for A?

4 (C) Compare Dalton's atomic theory with the modern atomic theory. Explain why scientists modified Dalton's theory.

5 (C) In your opinion, should chemistry students learn about Dalton's theory if scientists no longer agree with it completely? Justify your answer.

Not all chemists believed that Dalton's atoms existed. In 1877, one skeptical scientist called Dalton's atoms "stupid hallucinations." Other scientists considered atoms to be a valuable *idea* for understanding matter and its behaviour. They did not, however, believe that atoms had any physical reality. The discovery of electrons (and, later, the other subatomic particles) finally convinced scientists that atoms are more than simply an idea. Atoms, they realized, must be matter.

2.2 Atoms, Elements, and the Periodic Table

Section Preview/ Specific Expectations

In this section, you will

- **state**, in your own words, the periodic law
- **describe** elements in the periodic table in terms of energy levels and the electron arrangements
- **use** Lewis structures to represent valence electrons
- **communicate** your understanding of the following terms: *energy levels, periodic trends, valence electrons, Lewis structures, stable octet, octet*

Language LINK

The term *periodic* means "repeating in an identifiable pattern." For example, a calendar is periodic. It organizes the days of the months into a repeating series of weeks. What other examples of periodicity can you think of?

By the mid 1800's, there were 65 known elements. Chemists studied these elements intensively and recorded detailed information about their reactivity and the masses of their atoms. Some chemists began to recognize patterns in the properties and behaviour of many of these elements. (See Figure 2.5.)

Other sets of elements display similar trends in their properties and behaviour. For example, oxygen (O), sulfur (S), selenium (Se), and tellurium (Te) share similar properties. The same is true of fluorine (F), chlorine (Cl), bromine (Br), and iodine (I). These similarities prompted chemists to search for a fundamental property that could be used to organize all the elements. One chemist, Dmitri Mendeleev (1834–1907), sequenced the known elements in order of increasing atomic mass. The result was a table of the elements, organized so that elements with similar properties were arranged in the same column. Because Mendeleev's arrangement highlighted periodic (repeating) patterns of properties, it was called a *periodic table.*

The modern periodic table is a modification of the arrangement first proposed by Mendeleev. Instead of organizing elements according to atomic mass, the modern periodic table organizes elements according to atomic number. According to the **periodic law**, *the chemical and physical properties of the elements repeat in a regular, periodic pattern when they are arranged according to their atomic number.*

Figures 2.6 and 2.7 outline the key features of the modern periodic table. Take some time to review these features. Another version of the periodic table, containing additional data, appears on the inside back cover of this textbook, as well as in Appendix C.

Figure 2.5 These five elements share many physical and chemical properties. However, they have widely differing atomic masses.

lithium, Li

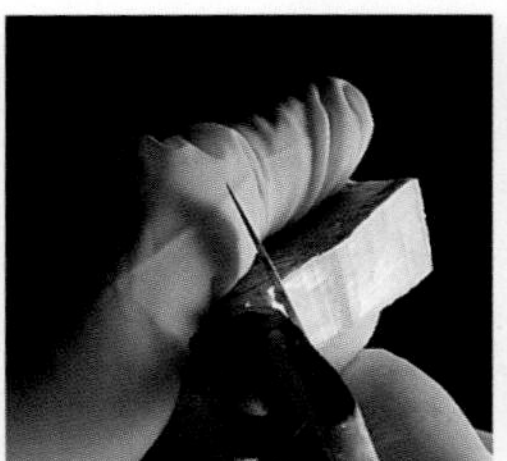
Sodium, Na

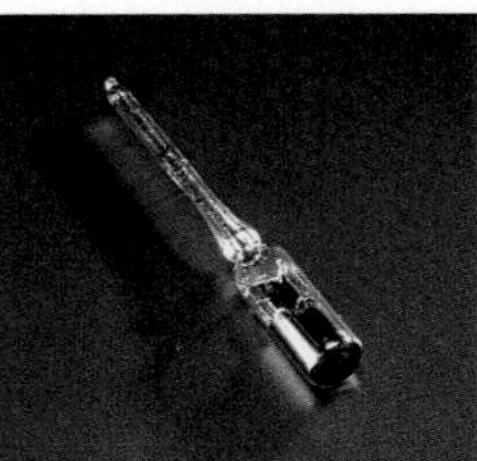
Potassium, K

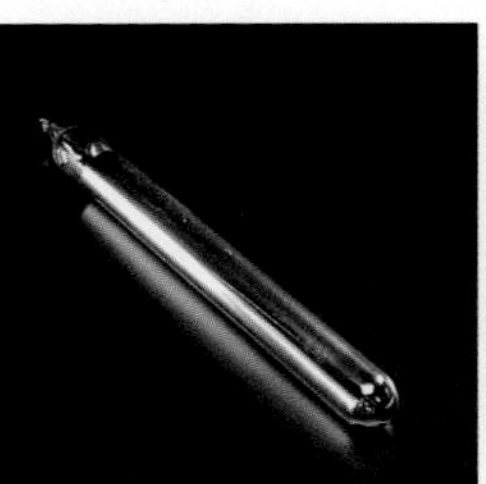
Rubidium, Rb

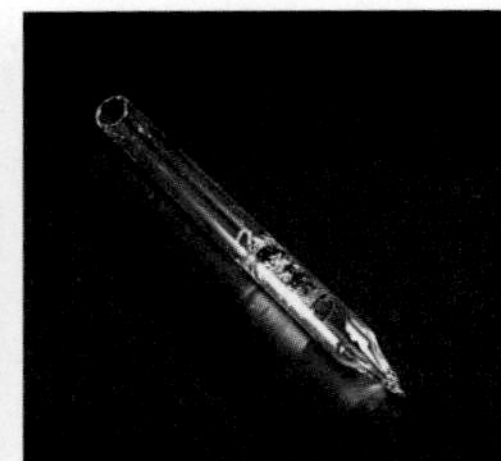
Cesium, Cs

Shared Physical Properties

- soft
- metallic (therefore malleable, ductile, and good conductors of electricity)

Shared Chemical Properties

- are very reactive
- react vigorously (and explosively) with water
- combine with chlorine to form a white solid that dissolves easily in water

MAIN GROUP ELEMENTS — TRANSITION ELEMENTS — MAIN-GROUP ELEMENTS

- metals (main group)
- metals (transition)
- metals (inner transition)
- metalloids
- nonmetals

Period	1 (IA)	2 (IIA)	3 (IIIB)	4 (IVB)	5 (VB)	6 (VIB)	7 (VIIB)	8	9 (VIIIB)	10	11 (IB)	12 (IIB)	13 (IIIA)	14 (IVA)	15 (VA)	16 (VIA)	17 (VIIA)	18 (VIIIA)
1	1 H 1.01																	2 He 4.003
2	3 Li 6.941	4 Be 9.012											5 B 10.81	6 C 12.01	7 N 14.01	8 O 16.00	9 F 19.00	10 Ne 20.18
3	11 Na 22.99	12 Mg 24.13											13 Al 26.98	14 Si 28.09	15 P 30.97	16 S 32.07	17 Cl 35.45	18 Ar 39.95
4	19 K 39.10	20 Ca 40.08	21 Sc 44.96	22 Ti 47.88	23 V 50.94	24 Cr 52.00	25 Mn 54.94	26 Fe 55.85	27 Co 58.93	28 Ni 58.69	29 Cu 63.55	30 Zn 65.39	31 Ga 69.72	32 Ge 72.61	33 As 74.92	34 Se 78.96	35 Br 79.90	36 Kr 83.80
5	37 Rb 85.47	38 Sr 87.62	39 Y 88.91	40 Zr 91.22	41 Nb 92.91	42 Mo 95.94	43 Tc (98)	44 Ru 101.1	45 Rh 102.9	46 Pd 106.4	47 Ag 107.9	48 Cd 112.4	49 In 114.8	50 Sn 118.7	51 Sb 121.8	52 Te 127.6	53 I 126.9	54 Xe 131.3
6	55 Cs 132.9	56 Ba 137.3	57 La 138.9	72 Hf 178.5	73 Ta 180.9	74 W 183.9	75 Re 186.2	76 Os 190.2	77 Ir 192.2	78 Pt 195.1	79 Au 197.0	80 Hg 200.6	81 Tl 204.4	82 Pb 207.2	83 Bi 209.0	84 Po (209)	85 At (210)	86 Rn (222)
7	87 Fr (223)	88 Ra (226)	89 Ac (227)	104 Rf (261)	105 Db (262)	106 Sg (266)	107 Bh (262)	108 Hs (265)	109 Mt (266)	110 Uun (269)	111 Uuu (272)	112 Uub (277)		114 Uuq (285)		116 Uuh (289)		118 Uuo (293)

INNER TRANSITION ELEMENTS

Period														
6	58 Ce 140.1	59 Pr 140.9	60 Nd 144.2	61 Pm (145)	62 Sm 150.4	63 Eu 152.0	64 Gd 157.3	65 Tb 158.9	66 Dy 162.5	67 Ho 164.9	68 Er 167.3	69 Tm 168.9	70 Yb 173.0	57 Lu 175.0
7	90 Th 232.0	91 Pa (231)	92 U 238.0	93 Np (237)	94 Pu (242)	95 Am (243)	96 Cm (247)	97 Bk (247)	98 Cf (251)	99 Es (252)	100 Fm (257)	101 Md (258)	102 No (259)	89 Lr (260)

- Each element is in a separate box, with its atomic number, atomic symbol, and atomic mass. (Different versions of the periodic table provide additional data and details.)
- Elements are arranged in seven numbered periods (horizontal rows) and 18 numbered groups (vertical columns).
- Groups are numbered according to two different systems. The current system numbers the groups from 1 to 18. An older system numbers the groups from I to VIII, and separates them into two categories labelled A and B. Both of these systems are included in this textbook.
- The elements in the eight A groups are the main-group elements. They are also called the representative elements. The elements in the ten B groups are known as the transition elements. (In older periodic tables, Roman numerals are used to number the A and B groups.)
- Within the B group transition elements are two horizontal series of elements called inner transition elements. They usually appear below the main periodic table. Notice, however, that they fit between the elements in Group 3 (IIIB) and Group 4 (IVB).
- A bold "staircase" line runs from the top of Group 13 (IIIA) to the bottom of Group 16 (VIA). This line separates the elements into three broad classes: metals, metalloids (or semi-metals), and non-metals. (See Figure 2.7 on the next page for more information.)
- Group 1 (IA) elements are known as *alkali metals*. They react with water to form alkaline, or basic, solutions.
- Group 2 (IIA) elements are known as *alkaline earth metals*. They react with oxygen to form compounds called oxides, which react with water to form alkaline solutions. Early chemists called all metal oxides "earths."
- Group 17 (VIIA) elements are known as *halogens*, from the Greek word *hals*, meaning "salt." Elements in this group combine with other elements to form compounds called salts.
- Group 18 (VIIIA) elements are known as *noble gases*. Noble gases do not combine naturally with any other elements.

Figure 2.6 The basic features of the periodic table are summarized here. Most of your work in this course will focus on the representative elements.

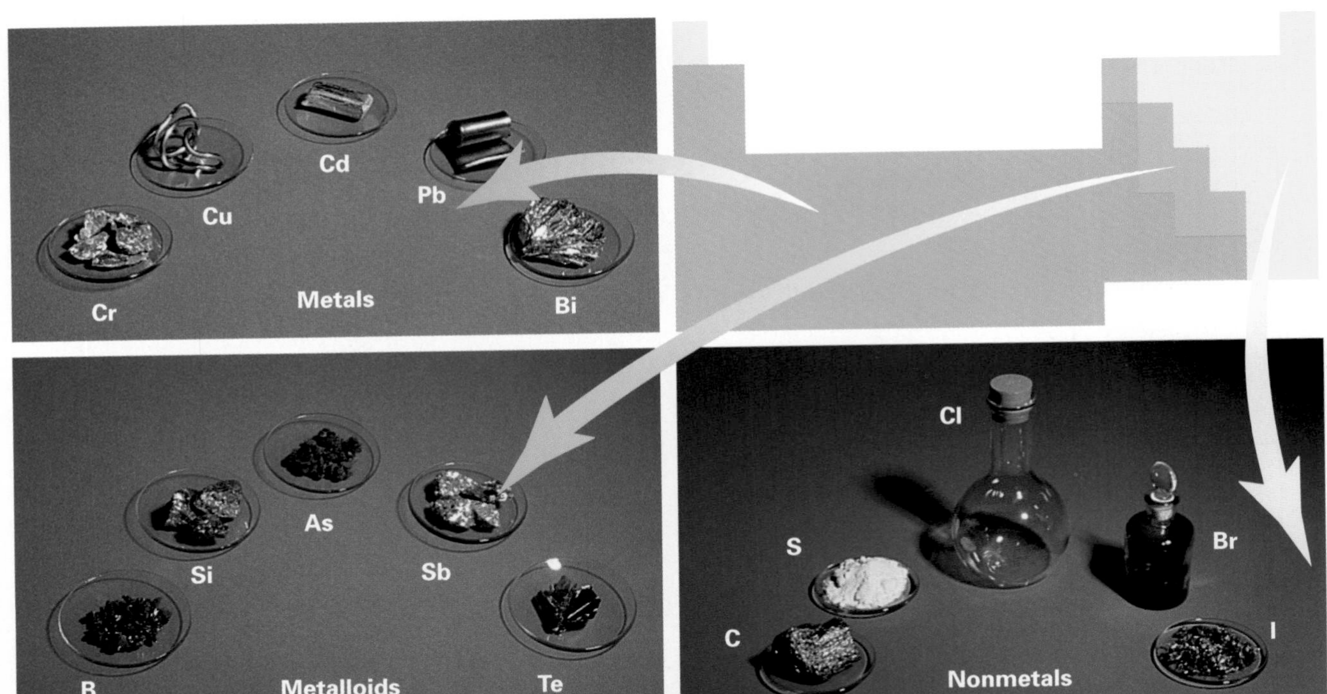

Figure 2.7 Several examples from each of the three main classes of elements are shown here. Find where they appear in the periodic table in Figure 2.6.

Practice Problems

2. Identify the name and symbol of the elements in the following locations of the periodic table:
   (a) Group 14 (IV A), Period 2
   (b) Group 11 (I B), Period 4
   (c) Group 18 (VIII A), Period 6
   (d) Group 1 (I A), Period 1
   (e) Group 12 (II B), Period 5
   (f) Group 2 (II A), Period 4
   (g) Group 17 (VII A), Period 5
   (h) Group 13 (III A), Period 3

Electrons and the Periodic Table

You have seen how the periodic table organizes elements so that those with similar properties are in the same group. You have also seen how the periodic table shows a clear distinction among metals, non-metals, and metalloids. Other details of the organization of the periodic table may seem baffling, however. Why, for example, are there different numbers of elements in the periods?

The reason for this, and other details of the periodic table's organization, involves the number and arrangement of electrons in the atoms of each element. To appreciate the importance of electrons to the periodic table, it is necessary to revisit the structure of the atom.

In the following ExpressLab, you will observe elements in much the same way that scientists did in the early twentieth century. In doing so, these scientists set the stage for a new understanding of matter and the electrical structure of its atoms.

History LINK

Mendeleev did not develop his periodic table in isolation. He built upon work that had been done by other chemists, in other parts of the world, over several decades. Research other ideas that were proposed for organizing the elements. Include Mendeleev's work in your research. What was it about his arrangement that convinced chemists to adopt it?

ExpressLab Observing the Spectra of Elements

In this activity, you will use a device called a *diffraction grating*. It separates light into banded patterns of colour (a spectrum). Different colours of light have different frequencies and wavelengths, so they have different amounts of energy. Red light is less energetic, for example, than blue light.

Safety Precautions

- Gas discharge tubes operate at a voltage that is high enough to cause serious injury. Observe them only from a safe distance, as determined by your teacher.

Materials

diffraction grating
incandescent light source
gas discharge tubes containing different elements

Procedure

1. Use the diffraction grating to observe the light that is emitted from an ordinary incandescent light bulb. Make a quick sketch to record your observations.
2. Observe the light that is emitted from the hydrogen gas discharge tube. CAUTION You should be about 1 m from the discharge tube. Come no farther than your teacher directs. Sketch your observations.
3. Observe the light that is emitted from the discharge tubes of other elements. Sketch your observations for each element.

Analysis

1. If the electrons in a discharge tube are moving everywhere in the space around the nucleus, their spectrum should look like the spectrum of an ordinary light bulb. What does hydrogen's spectrum look like? How do the spectra of the other elements compare with the spectrum of a light bulb and the spectrum of hydrogen?
2. Hydrogen has only one electron. Why, then, does its spectrum have four coloured lines?
3. Why is the light that is emitted by hydrogen different from the light that is emitted by the other elements? Explain the difference in terms of electrons.

Application

4. What do gas discharge tubes have in common with street lights? Do research to find out which gases are used in street lamps, and why certain gases are chosen for certain locations.

Electrons and Energy Levels

Electrons cannot move haphazardly. Their movement around an atomic nucleus is restricted to fixed regions of space. These regions are three-dimensional, similar to the layers of an onion.

Figure 2.8 shows a representation of these regions. Keep in mind that they are *not* solid. They are volumes of space in which electrons may be found. You may have heard these regions called *energy shells* or *shells*. In this textbook, they are called **energy levels**. An electron that is moving in a lower energy level is close to the nucleus. It has less energy than it would if it were moving in a higher energy level.

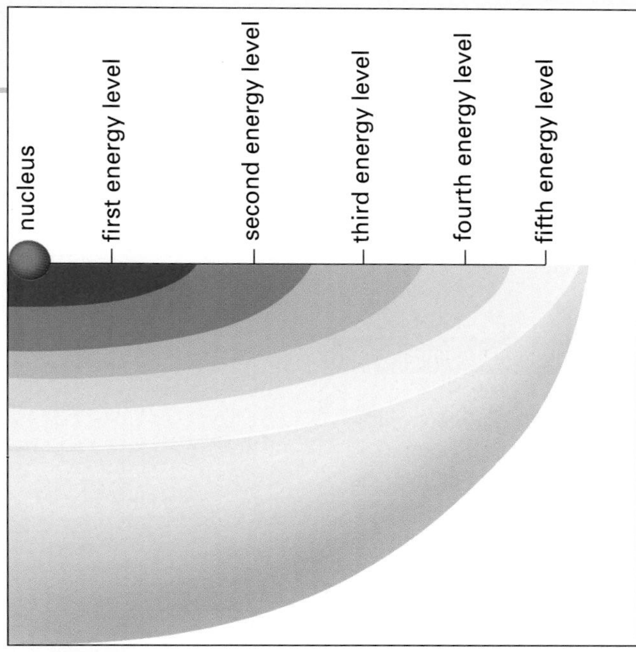

Figure 2.8 Energy levels of an atom from the fifth period

mind STRETCH

Examine the following illustration. Then answer these questions.

- Which book possesses more potential energy? Why?
- Can a book sit between shelves instead of on a shelf as shown?
- How does the potential energy of a book on a higher shelf change if it is moved to a lower shelf?
- How do you think this situation is related to electrons and the potential energy they possess when they move in different energy levels?

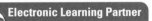

Electronic Learning Partner

Your Chemistry 11 Electronic Learning Partner has an interactive activity to help you assess your understanding of the relationship among elements, their atomic number, and their position in the periodic table.

There is a limit to the number of electrons that can occupy each energy level. For example, a maximum of two electrons can occupy the first energy level. A maximum of eight electrons can occupy the second energy level. The **periodic trends** (repeating patterns) that result from organizing the elements by their atomic number are linked to the way in which electrons occupy and fill energy levels. (See Figure 2.9.)

As shown in Figure 2.9A, a common way to show the arrangement of electrons in an atom is to draw circles around the atomic symbol. Each circle represents an energy level. Dots represent electrons that occupy each energy level. This kind of diagram is called a Bohr-Rutherford diagram. It is named after two scientists who contributed their insights to the atomic theory.

Figure 2.9B shows that the first energy level is full when two electrons occupy it. Only two elements have two or fewer electrons: hydrogen and helium. Hydrogen has one electron, and helium has two. These elements, with their electrons in the first energy level, make up Period 1 of the periodic table.

As you can see in Figure 2.9C, Period 2 elements have two occupied energy levels. The second energy level is full when eight electrons occupy it. Neon, with a total of ten electrons, has its first and second energy levels filled. Notice how the second energy level fills with electrons as you move across the period from lithium to fluorine.

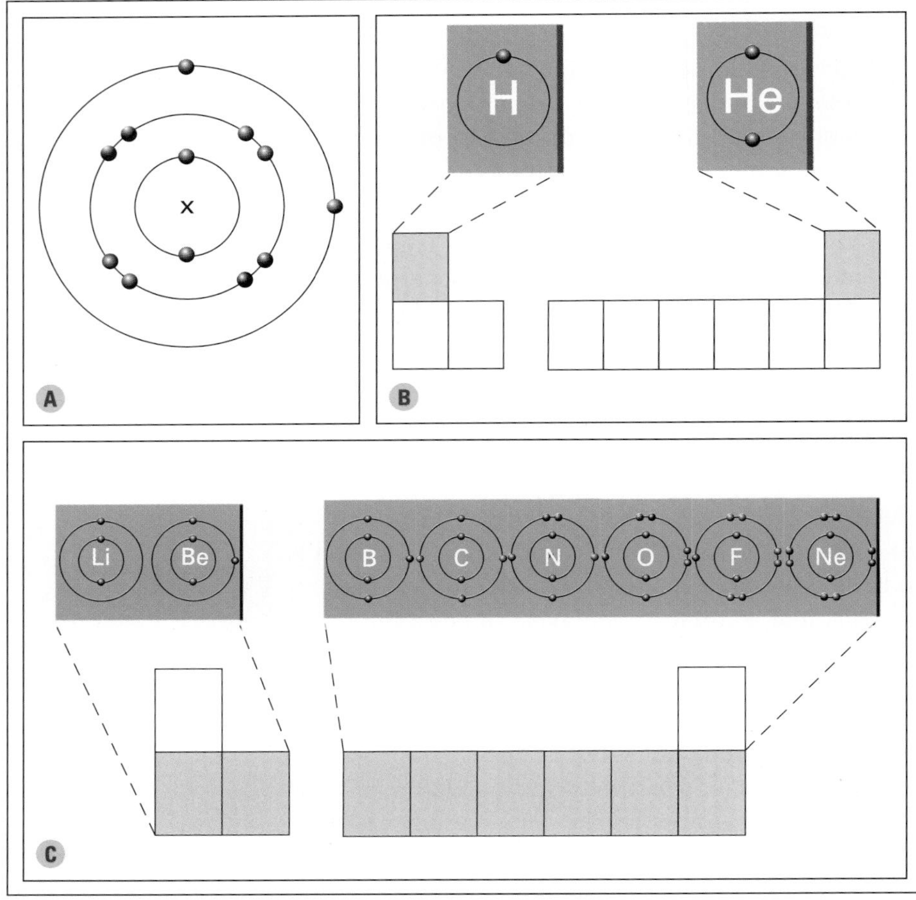

Figure 2.9 (A) A Bohr-Rutherford diagram (B) Hydrogen and helium have a single energy level. (C) The eight Period 2 elements have two energy levels.

Patterns Based on Energy Levels and Electron Arrangements

The structure of the periodic table is closely related to energy levels and the arrangement of electrons. Two important patterns result from this relationship. One involves periods, and the other involves groups.

The Period-Related Pattern

As you can see in Figure 2.9, elements in Period 1 have electrons in one energy level. Elements in Period 2 have electrons in two energy levels. This pattern applies to all seven periods. *An element's period number is the same as the number of energy levels that the electrons of its atoms occupy.* Thus, you could predict that Period 5 elements have electrons that occupy five energy levels. This is, in fact, true.

What about the inner transition elements—the elements that are below the periodic table? Figure 2.10 shows how this pattern applies to them. Elements 58 through 71 belong in Period 6, so their electrons occupy six energy levels. Elements 90 through 103 belong in Period 7, so their electrons occupy seven energy levels. Chemists and chemical technologists tend to use only a few of the inner transition elements (notably uranium and plutonium) on a regular basis. Thus, it is more convenient to place all the inner transition elements below the periodic table.

CHEM FACT

Energy levels and the arrangement of electrons involve ideas from theoretical physics. These ideas are beyond the scope of this course. Appendix D at the back of this book provides a brief introduction to these ideas. If you pursue your studies in chemistry next year and beyond, you will learn a more complete theory of electron arrangement.

Period	1	2															3	4	5	6	7	8	9	10	11	12	13	14	15	16	17	18
	IA	IIA															IIIB	IVB	VB	VIB	VIIB	VIIIB			IB	IIB	IIIA	IVA	VA	VIA	VIIA	VIIIA
1	1 H																															2 He
2	3 Li	4 Be																									5 B	6 C	7 N	8 O	9 F	10 Ne
3	11 NA	12 Mg																									13 Al	14 Si	15 P	16 S	17 Cl	18 Ar
4	19 K	20 Ca															21 Sc	22 Ti	23 V	24 Cr	25 Mn	26 Fe	27 Co	28 Ni	29 Cu	30 Zn	31 Ga	32 Ge	33 As	34 Se	35 Br	36 Kr
5	37 Rb	38 Sr															39 Y	40 Zr	41 Nb	42 Mo	43 Tc	44 Ru	45 Rh	46 Pd	47 Ag	48 Cd	49 In	50 Sn	51 Sb	52 Te	53 I	54 Xe
6	55 Cs	56 Ba	57 La	58 Ce	59 Pr	60 Nd	61 Pm	62 Sm	63 Eu	64 Gd	65 Tb	66 Dy	67 Ho	68 Er	69 Tm	70 Yb	71 Lu	72 Hf	73 Ta	74 W	75 Re	76 Os	77 Ir	78 Pt	79 Au	80 Hg	81 Tl	82 Pb	83 Bi	84 Po	85 At	86 Rn
7	87 Fr	88 Ra	89 Ac	90 Th	91 Pa	92 U	93 Np	94 Pu	95 Am	96 Cm	97 BK	98 Cf	99 Es	100 Fm	101 Md	102 No	103 Lr	104 Rf	105 Db	106 Sg	107 Bh	108 Hs	109 Mt	110 Uun	111 Uuu	112 Uub		114 Uuq		116 Unh		118 Uuo

Figure 2.10 The "long form" of the periodic table includes the inner transition metals in their proper place.

The Group-Related Pattern

The second pattern emerges when you consider the electron arrangements in the main-group elements: the elements in Groups 1 (1A), 2 (2A), and 13 (3A) to 18 (8A). All the elements in each main group have the same number of electrons in their highest (outer) energy level. The electrons that occupy the outer energy level are called **valence electrons**. The term "valence" comes from a Latin word that means "to be strong." "Valence electrons" is a suitable name because the outer energy level electrons are the electrons involved when atoms form compounds. In other words, valence electrons are responsible for the chemical behaviour of elements.

You can infer the number of valence electrons in any main-group element from its group number. For example, Group 1 (1A) elements have one valence electron. Group 2 (2A) elements have two valence electrons. For elements in Groups 13 (3A) to 18 (8A), the number of valence electrons is the same as the second digit in the current numbering system. It is the same as the only digit in the older numbering system. For example, elements in Group 15 (5A) have 5 valence electrons. The elements in Group 17 (7A) have 7 valence electrons.

Using Lewis Structures to Represent Valence Electrons

It is time-consuming to draw electron arrangements using Bohr-Rutherford diagrams. It is much simpler to use **Lewis structures** to represent elements and the valence electrons of their atoms. To draw a Lewis structure, you replace the nucleus and inner energy levels of an atom with its atomic symbol. Then you place dots around the atomic symbol to represent the valence electrons. The order in which you place the first four dots is up to you. You may find it simplest to start at the top and proceed clockwise: right, then bottom, then left.

Examine Figure 2.11, and then complete the Practice Problems that follow. In Chapter 3, you will use Lewis structures to help you visualize what happens when atoms combine to form compounds.

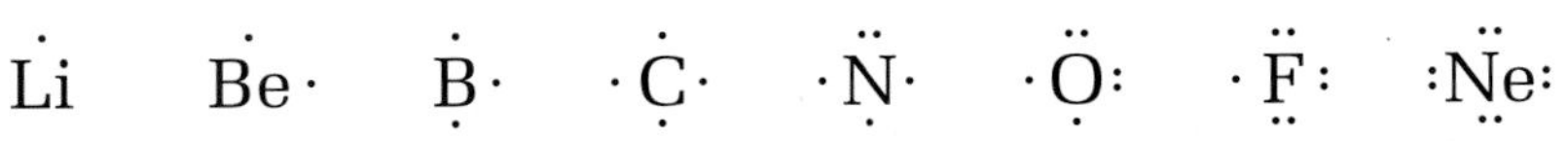

Li Be· B· ·C· ·N· ·O: ·F: :Ne:

Figure 2.11 Examine these Lewis structures for the Period 2 elements. Place a dot on each side of the element—one dot for each valence electron. Then start pairing dots when you reach five or more valence electrons.

Practice Problems

3. Draw boxes to represent the first 20 elements in the periodic table. Using Figure 2.9 as a guide, sketch the electron arrangements for these elements.

4. Redraw the 20 elements from Practice Problem 2 using Lewis structures.

5. Identify the number of valence electrons in the outer energy levels of the following elements:

(a) chlorine
(b) helium
(c) indium
(d) strontium
(e) rubidium
(f) lead
(g) antimony
(h) selenium
(i) arsenic
(j) xenon

6. Use the periodic table to draw Lewis structures for the following elements: barium (Ba), gallium (Ga), tin (Sn), bismuth (Bi), iodine (I), cesium (Cs), krypton (Kr), xenon (Xe).

The Significance of a Full Outer Energy Level

The noble gases in Group 18 (VIII A) are the only elements that exist as individual atoms in nature. They are extremely *un*reactive. They do not naturally form compounds with other atoms. (Scientists *have* manipulated several of these elements in the laboratory to make them react, however.) What is it about the noble gases that explains this behaviour?

Recall that chemical reactivity is determined by valence electrons. Thus, there must be something about the arrangement of the electrons in the noble gases that explains their *un*reactivity. All the noble gases have outer energy levels that are completely filled with the maximum number of electrons. Helium has a full outer energy level of two valence electrons. The other noble gases have eight valence electrons in the outer energy level. Chemists reason that having a full outer energy level must be a very stable electron arrangement.

What does this stability mean? It means that a full outer energy level is unlikely to change. Scientists have observed that, in nature, situations or systems of lower energy are favoured over situations or systems of higher energy. For example, a book on a high shelf has more potential energy (is less stable) than a book on a lower shelf. If you move a book from a high shelf to a lower shelf, it has less potential energy (is more stable). If you move a book to the floor, it has low potential energy (is much more stable).

When atoms have eight electrons in the outer energy level (or two electrons for hydrogen and helium), chemists say that they have a **stable octet**. Often this term is shortened to just **octet**. An octet is a very stable electron arrangement. As you will see in Chapter 3, an octet is often the result of changes in which atoms combine to form compounds.

Section Wrap-up

You have seen that the structure of the periodic table is directly related to energy levels and arrangements of electrons. The patterns that emerge from this relationship enable you to predict the number of valence electrons for any main group element. They also enable you to predict the number of energy levels that an element's electrons occupy. The relationship between electrons and the position of elements in the periodic table leads to other patterns, as well. You will examine several of these patterns in the next section.

Section Review

1. K/U State the periodic law, and provide at least two examples to illustrate its meaning.
2. K/U Identify the group number for each of these sets of elements. Then choose two of these groups and write the symbols for the elements within it.
   - alkali metals
   - noble gases
   - halogens
   - alkaline earth metals

3 (a) K/U Identify the element that is described by the following information. Refer to a periodic table as necessary.
- It is a Group 14 (III A) metalloid in the third period.
- It is a Group 15 (V A) metalloid in the fifth period.
- It is the other metalloid in Group 15 (V A).
- It is a halogen that exists in the liquid state at room temperature.

(b) C Develop four more element descriptions like those in part (a). Exchange them with a classmate and identify each other's elements.

4 K/U What is the relationship between electron arrangement and the organization of elements in the periodic table?

5 C In writing, sketches, or both, explain to someone who has never seen the periodic table how it can be used to tell at a glance the number of valence electrons in the atoms of an element.

6 (a) K/U How many valence electrons are there in an atom of each of these elements?

neon	sodium	magnesium
bromine	chlorine	silicon
sulfur	helium	
strontium	tin	

(b) Present your answers from part (a) in the form of Lewis structures.

(c) Without consulting a periodic table, classify each element from part (a) as a metal, non-metal, or metalloid.

7 K/U How many elements are liquids at room temperature? Name them.

8 K/U Compare and contrast the noble gases with the other elements.

9 I An early attempt to organize the elements placed them in groups of three called triads. Examine the three triads shown below.

Triad 1	Triad 2	Triad 3
Mn	Li	S
Cr	Na	Se
Fe	K	Te

(a) Infer the reasoning for grouping the elements in this way.

(b) Which of the elements in these three triads still appear together in the same group of the modern periodic table?

10 MC Using print or electronic resources, or both, find at least one common technological application for each of the following elements:

(a) europium	(f) mercury
(b) neodymium	(g) ytterbium
(c) carbon	(h) bromine
(d) nitrogen	(i) chromium
(e) silicon	(j) krypton

11 (a) C Draw Lewis structures for each of these elements: lithium, sodium, potassium, magnesium, aluminum, carbon.

(b) Which of these elements have the same number of occupied energy levels?

(c) Which have the same number of valence electrons?

Periodic Trends Involving the Sizes and Energy Levels of Atoms

2.3

Section Preview/ Specific Expectations

In this section, you will

- **use** your understanding of electron arrangement and forces in atoms to explain the following periodic trends: atomic radius, ionization energy, electron affinity
- **analyze** data involving atomic radius, ionization energy, and electron affinity to identify and describe general periodic trends
- **communicate** your understanding of the following terms: *ion, anion, cation ionization energy, electron affinity*

In section 2.1, you learned that the size of a typical atom is about 10^{-10} m. You know, however, that the atoms of each element are distinctly different. For example, the atoms of different elements have different numbers of protons. This means, of course, that they also have different numbers of electrons. You might predict that the size of an atom is related to the number of protons and electrons it has. Is there evidence to support this prediction? If so, is there a pattern that can help you predict the relative size of an atom for any element in the periodic table?

In Investigation 2-A, you will look for a pattern involving the size of atoms. Chemists define, and measure, an atom's size in terms of its radius. The radius of an atom is the distance from its nucleus to the *approximate* outer boundary of the cloud-like region of its electrons. This boundary is approximate because atoms are not solid spheres. They do not have a fixed outer boundary.

Figure 2.12 represents how the radius of an atom extends from its nucleus to the approximate outer boundary of its electron cloud. Notice that the radius line in this diagram is just inside the outer boundary of the electron cloud. An electron may also spend time beyond the end of the radius line.

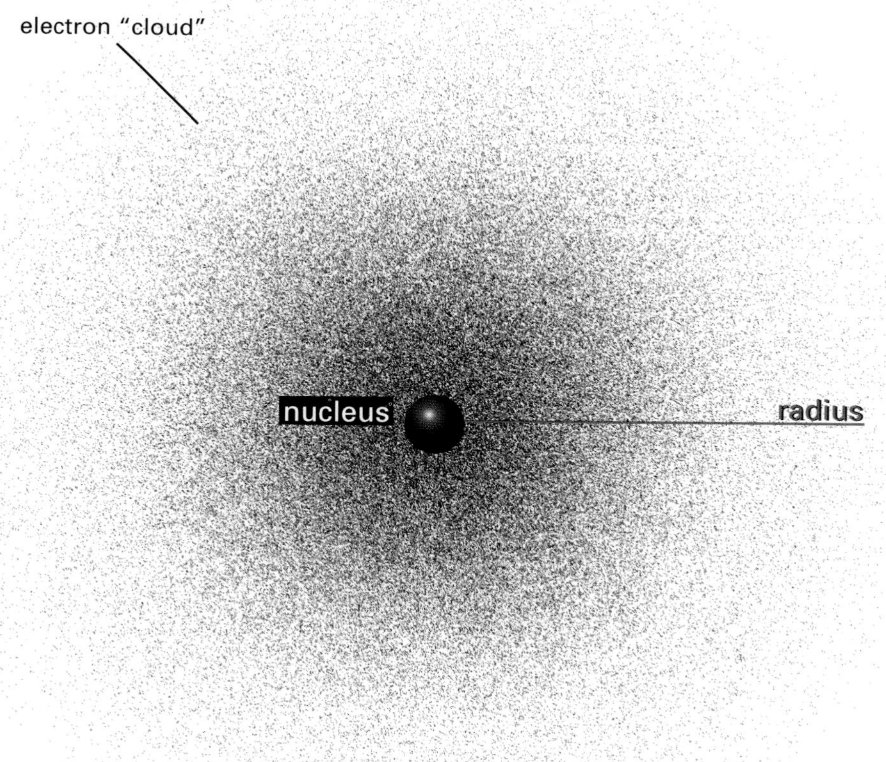

Figure 2.12 A representation of the radius of an atom

SKILL FOCUS

Predicting

Performing and recording

Analyzing and interpreting

Investigation 2-A

Analyzing Atomic Radius Data

Examine the main-group elements in the periodic table. Imagine how their size might change as you move down a group or across a period. What knowledge and reasoning can you use to infer the sizes of the atoms?

Question

How do the sizes of main-group atoms compare within a group and across a period?

Prediction

Predict a trend (pattern) that describes how the sizes of main-group atoms change down a group and across a period. Include a brief explanation to justify your prediction.

Safety Precautions

Be careful when handling any sharp instruments or materials that you choose to use.

Materials

to be decided in class

Procedure

1. The table below lists the atomic *radii* (plural of *radius*) for the main-group elements. Design different scale models that could help you visualize and compare the sizes of the atoms. Your models can be two-dimensional or three-dimensional, large or small.
2. Discuss your designs as a group. Choose a design that you think will best show the information you require.
3. Build your models. Arrange them according to their positions in the periodic table.

Atomic Radii of Main-Group Elements

Name of element	Atomic radius in picometres (pm)	Name of element	Atomic radius in picometres (pm)	Name of element	Atomic radius in picometres (pm)
aluminum	143	gallium	141	polonium	167
antimony	159	germanium	137	potassium	235
argon	88	helium	49	radon	134
astatine	145	hydrogen*	79	rubidium	248
barium	222	indium	166	selenium	140
beryllium	112	iodine	132	silicon	132
bismuth	170	krypton	103	sodium	190
boron	98	lead	175	strontium	215
bromine	112	lithium	155	sulfur	127
calcium	197	magnesium	160	tellurium	142
carbon	91	neon	51	thallium	171
cesium	267	nitrogen	92	tin	162
chlorine	97	oxygen	65	xenon	124
fluorine	57	phosphorus	128		

*Quantum mechanical value for a free hydrogen atom

Analysis

1. How do atomic radii change as you look from top to bottom within a group?
2. How do atomic radii change as you look from left to right across a period?
3. Compare your observations with your prediction. Explain why your results did, or did not, agree with your prediction.

Conclusion

4. State whether or not atomic radius is a periodic property of atoms. Give evidence to support your answer.

Application

5. Would you expect atoms of the transition elements to follow the same trend you observed for the main-group elements? Locate atomic radius data for the transition elements (not including the inner transition elements). Make additional models, or draw line or bar graphs, to verify your expectations.

Trends for Atomic Size (Radius)

There are two general trends for atomic size:

- *As you go down each group in the periodic table, the size of an atom increases.* This makes sense if you consider energy levels. As you go down a group, the valence electrons occupy an energy level that is farther and farther from the nucleus. Thus, the valence electrons experience less attraction for the nucleus. In addition, electrons in the inner energy levels block, or *shield,* the valence electrons from the attraction of the nucleus. As a result, the total volume of the atom, and thus the size, increases with each additional energy level.
- *As you go across a period, the size of an atom decreases.* This trend might surprise you at first, since the number of electrons increases as you go across a period. You might think that more electrons would occupy more space, making the atom larger. You might also think that repulsion from their like charges would force the electrons farther apart. The size of an atom decreases, however, because the positive charge on the nucleus also increases across a period. As well, without additional energy the electrons are restricted to their outer energy level. For example, the outer energy level for Period 2 elements is the second energy level. Electrons cannot move beyond this energy level. As a result, the positive force exerted by the nucleus pulls the outer electrons closer, reducing the atom's total size.

Figure 2.13 summarizes the trends for atomic size. The Practice Problems that follow give you a chance to apply your understanding of these trends.

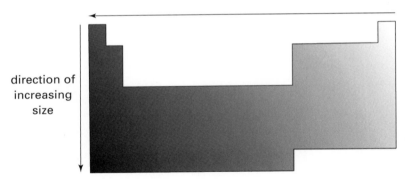

Figure 2.13 **Atomic size increases down a group and decreases across a row in the periodic table.**

Practice Problems

7. Using only their location in the periodic table, rank the atoms in each set by decreasing atomic size. Explain your answers.
   - **(a)** Mg, Be, Ba
   - **(b)** Ca, Se, Ga
   - **(c)** Br, Rb, Kr
   - **(d)** Se, Br, Ca
   - **(e)** Ba, Sr, Cs
   - **(f)** Se, Br, Cl
   - **(g)** Mg, Ca, Li
   - **(h)** Sr, Te, Se
   - **(i)** In, Br, I
   - **(j)** S, Se, O

Trends for Ionization Energy

A neutral atom contains equal numbers of positive charges (protons) and negative charges (electrons). The particle that results when a neutral atom gains electrons or gives up electrons is called an **ion**. Thus, an ion is a charged particle. *An atom that gains electrons becomes a negatively charged* **anion**. *An atom that gives up electrons becomes a positively charged* **cation**. Figure 2.14 shows the formation of ions for several elements. As you examine the diagrams, pay special attention to

- the energy level from which electrons are gained or given up
- the charge on the ion that is formed when an atom gains or gives up electrons
- the arrangement of the electrons that remain after electrons are gained or given up

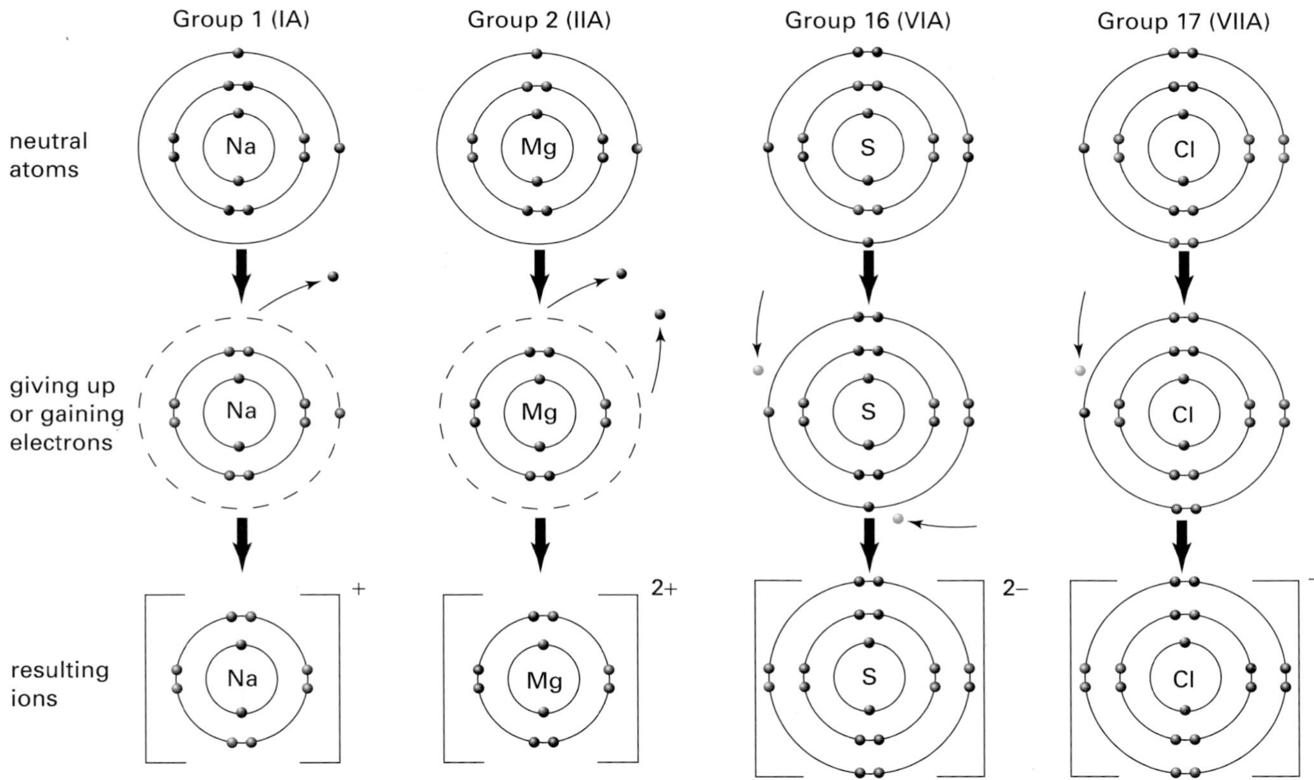

Figure 2.14 **These diagrams show the ions that are formed from neutral atoms of sodium, magnesium, sulfur, and chlorine. What other element has the same electron arrangement that sodium, magnesium, sulfide, and chloride ions have?**

Try to visualize the periodic table as a cylinder, rather than a flat plane. Can you see a relationship between ion formation and the electron arrangement of noble gases? Examine Figure 2.14 as well as 2.15 on the next page. The metals that are main-group elements tend to *give up* electrons and form ions that have the same number of electrons as the nearest noble gases. Non-metals tend to *gain* electrons and form ions that have the same number of electrons as the nearest noble gases. For example, when a sodium atom gives up its single valence electron, it becomes a positively charged sodium ion. Its outer electron arrangement is like neon's outer electron arrangement. When a fluorine atom gains an electron, it becomes a negatively charged ion with an outer electron arrangement like that of neon.

Figure 2.15 can help you determine the charge on an ion. Count the number of groups an ion is from the nearest noble gas. That number is the charge on the ion. For example, aluminum is three groups away from neon. Thus, an aluminum ion has a charge of 3+. Sulfur is two groups away from argon. Thus, a sulfide ion has a charge of 2–. **Remember:** Metals form positive ions (cations) and non-metals form negative ions (anions).

Figure 2.15 **Examine the relationship between ion charge and noble gas electron arrangement.**

It takes energy to overcome the attractive force of a nucleus and pull an electron away from a neutral atom. The energy that is required to remove an electron from an atom is called **ionization energy**. The bar graph in Figure 2.16 shows the ionization energy that is needed to remove one electron from the outer energy level of the atoms of the main-group elements. This energy is called the *first ionization energy.* It is measured in units of kJ/mol. A kilojoule (kJ) is a unit of energy. A mole (mol) is an amount of a substance. (You will learn about the mole in Unit 2.)

As you can see, atoms that give up electrons easily have low ionization energies. You would probably predict that the alkali metals of Group 1 (1A) would have low ionization energies. These elements are, in fact, extremely reactive because it takes so little energy to remove their single valence electron.

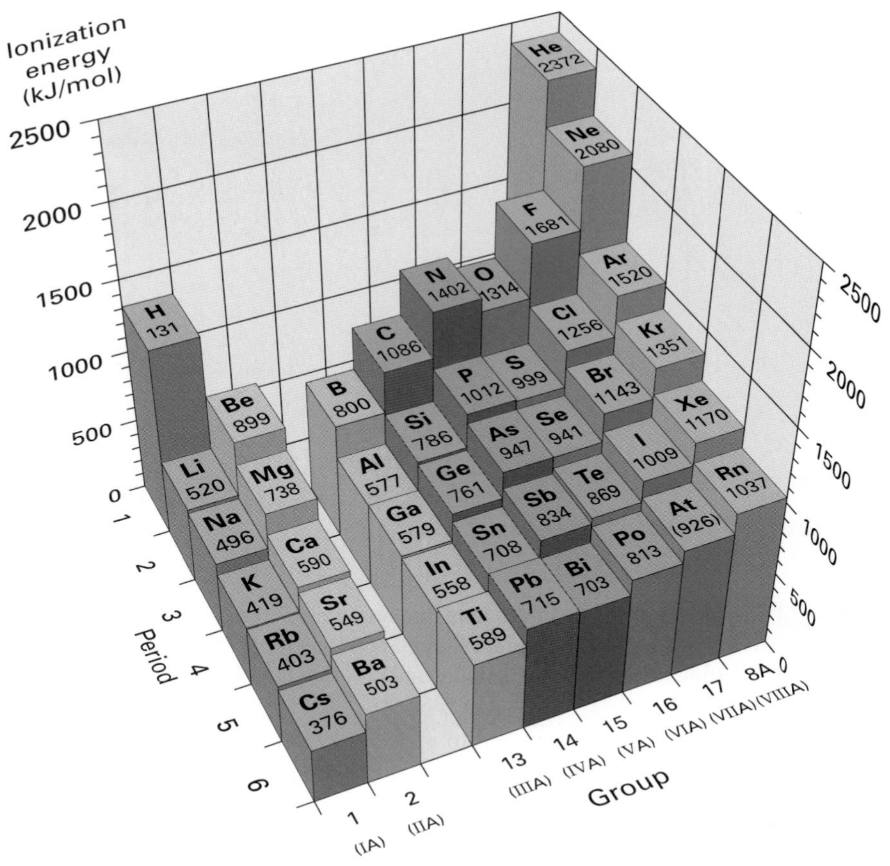

Figure 2.16 **This graph represents the first ionization energy for the main-group elements.**

CHEM FACT

All elements, except hydrogen, have more than one electron that can be removed. Therefore, they have more than one ionization energy. The energy that is needed to remove a second electron is called the *second ionization energy.* The energy that is needed to remove a third electron is the *third ionization energy,* and so on. What trend would you expect to see in the values of the first, second, and third ionization energies for main-group elements? What is your reasoning?

Summarizing Trends for Ionization Energy

Although you can see a few exceptions in Figure 2.16, there are two general trends for ionization energy:

- *Ionization energy tends to decrease down a group.* This makes sense in terms of the energy level that the valence electrons occupy. Electrons in the outer energy level are farther from the positive force of the nucleus. Thus, they are easier to remove than electrons in lower energy levels.
- *Ionization energy tends to increase across a period.* As you go across a period, the attraction between the nucleus and the electrons in the outer energy level increases. Thus, more energy is needed to pull an electron away from its atom. For this trend to be true, you would expect a noble gas to have the highest ionization energy of all the elements in the same period. As you can see in Figure 2.16, they do.

Figure 2.17 summarizes these general trends for ionization energy. The Practice Problems below give you a chance to apply your understanding of these trends.

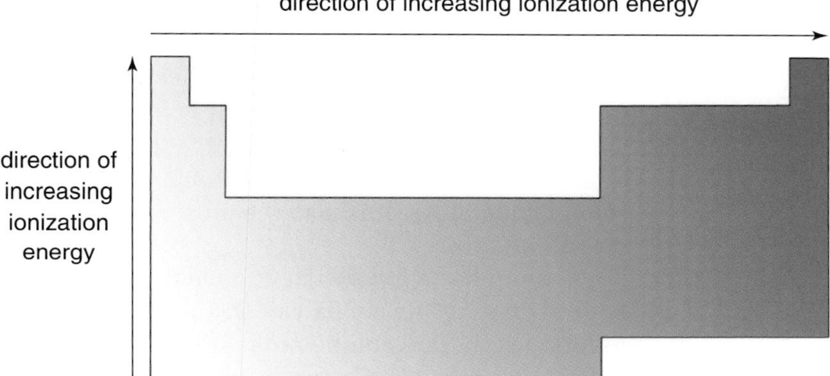

Figure 2.17 **Ionization energy tends to decrease down a group and increase across a period.**

Practice Problems

8. Using only a periodic table, rank the elements in each set by increasing ionization energy. Explain your answers.
   (a) Xe, He, Ar
   (b) Sn, In, Sb
   (c) Sr, Ca, Ba
   (d) Kr, Br, K
   (e) K, Ca, Rb
   (f) Kr, Br, Rb
9. Using only a periodic table, identify the atom in each of the following pairs with the *lower* first ionization energy.
   (a) B, O
   (b) B, In
   (c) I, F
   (d) F, N
   (e) Ca, K
   (f) B, Tl

COURSE CHALLENGE

Your understanding of periodic trends such as atomic radius and ionization energy will help you identify some unknown elements in the Chemistry Course Challenge at the end of this book.

Chemistry Bulletin

Science | Technology | Society | Environment

Manitoba Mine Specializes in Rare Metals

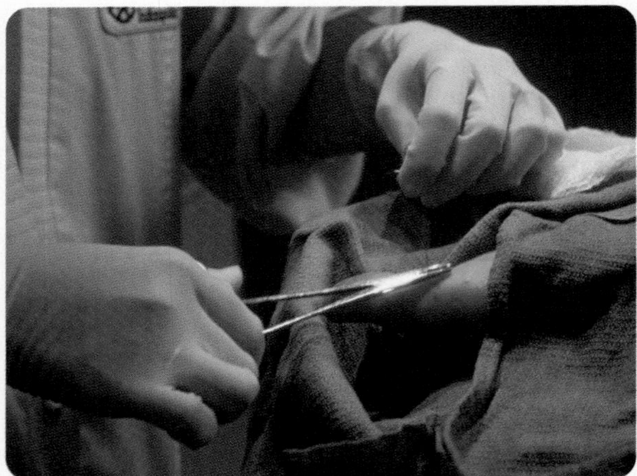

At TANCO in Bernic Lake, Manitoba, miners are busy finding and processing two rare and very different metallic elements: tantalum and cesium. Both of these metals are important parts of "high-tech" applications around the world. They are used in nuclear reactors and as parts of aircraft, missiles, camera lenses, and surgical instruments like the one shown above.

Tantalum is found only in Canada, Australia, Brazil, Zaire, and China. TANCO (the Tantalum Mining Corporation of Canada) is the only mine in North America that produces tantalum. TANCO is also the world's main producer of cesium. Other than the fact that tantalum and cesium are both found at TANCO and both are used in high-tech applications, they share little in common.

Tantalum is a heavy, hard, and brittle grey metal. In its pure form, it is extremely ductile and can be made into a fine wire. This has proved useful for making surgical sutures. Another property that makes tantalum useful is its resistance to corrosion by most acids, due to its very limited reactivity. At normal temperatures, tantalum is virtually non-reactive. In fact, tantalum has about the same resistance to corrosion as glass. Tantalum can withstand higher temperatures than glass, however. It has a melting point of 3290 K — higher than the melting points of all other elements, except tungsten and rhenium. Tantalum's resistance to corrosion and high melting point make it suitable for use in surgical equipment and implants. For example, some of the pins that are used by surgeons to hold a patient's broken bones together are made of tantalum.

Tantalum is resistant to corrosion because a thin film of tantalum oxide forms when tantalum is exposed to oxygen. The metal oxide acts as a protective layer. The oxide also has special refractive properties that make it ideal for use in camera lenses.

Cesium is quite different from tantalum, but it, too, has many high-tech applications. Cesium is a silvery-white metal. It is found in a mineral called pollucite. Cesium is the softest of all the metals and is a liquid at just above room temperature. It is also the most reactive metal on Earth.

Cesium has a low ionization energy. It readily gives up its single valence electron to form crystalline compounds with all the halogen non-metals. Cesium is also very photoelectric. This means that it easily gives up its lone outer electron when it is exposed to light. Thus, cesium is used in television cameras and traffic signals. As well, it has the potential to be used in ion propulsion engines for travel into deep space.

Making Connections

1. Make a table to show the differences and similarities between tantalum and cesium. For each metal, add a column to describe how its different properties make it useful for specific applications.

2. Bernic Lake is one of the few locations where tantulum can be found. As well, it is the most important cesium source in the world. Research and describe what geographical conditions led to the presence of two such rare metals in one location.

Trends for Electron Affinity

In everyday conversation, if you like something, you may say that you have an affinity for it. For example, what if you enjoy pizza and detest asparagus? You may say that you have a high affinity for pizza and a low affinity for asparagus. If you prefer asparagus to pizza, your affinities are reversed.

Atoms are not living things, so they do not like or dislike anything. You know, however, that some atoms have a low attraction for electrons. Other atoms have a greater attraction for electrons. **Electron affinity** is a measure of the change in energy that occurs when an electron is added to the outer energy level of an atom to form a negative ion.

Figure 2.18 identifies the electron affinities of the main-group elements. If energy is released when an atom of an element gains an electron, the electron affinity is expressed as a negative integer. When energy is absorbed when an electron is added, electron affinity is low, and is expressed as a positive integer. Notice, for example, that fluorine has the highest electron affinity (indicated by a large, negative integer). This indicates that fluorine is very likely to be involved in chemical reactions. In fact, fluorine is the most reactive of all the elements.

Metals have very low electron affinities. This is especially true for the Group 1 (1A) and 2 (2A) elements. Atoms of these elements form stable positive ions. A negative ion that is formed by the elements of these groups is unstable. It breaks apart into a neutral atom and a free electron.

Examine Figure 2.18. What trends can you observe? How regular are these trends?

1 (IA)	2 (IIA)	13 (IIIA)	14 (IVA)	15 (VA)	16 (VIA)	17 (VIIA)	18 (VIIIA)
H −72.8							He (+21)
Li −59.6	Be (+241)	B −26.7	C −122	N 0	O −141	F −328	Ne (+29)
Na −52.9	Mg (+230)	Al −42.5	Si −134	P −72.0	S −200	Cl −349	Ar (+34)
K −48.4	Ca (+156)	Ga −28.9	Ge −119	As −78.2	Se −195	Br −325	Kr (+39)
Rb −46.9	Sr (+167)	In −28.9	Sn −107	Sb −103	Te −190	I −295	Xe (+40)
Cs −45.5	Ba (+52)	Tl −19.3	Pb −35.1	Bi −91.3	Po −183	At −270	Rn (+41)

Figure 2.18 **The units for electron affinity are the same as the units for ionization energy: kJ/mol. High negative numbers mean a high electron affinity. Low negative numbers and any positive numbers mean a low electron affinity.**

Tools & Techniques

Analyzing the Ice Man's Axe

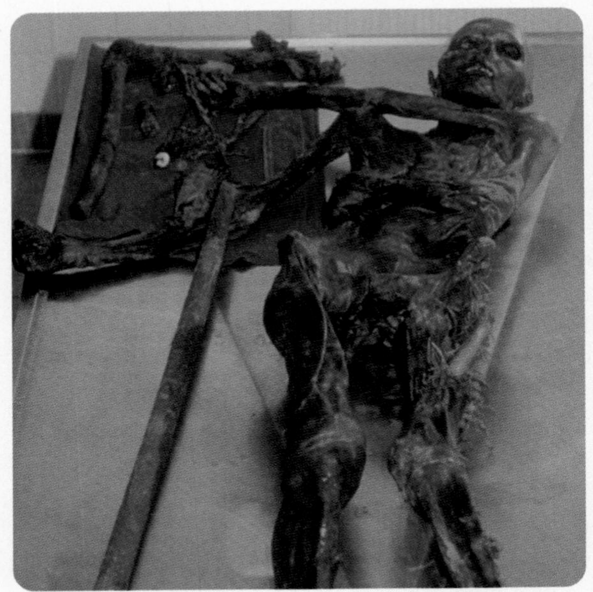

In September 1991, hikers in the Alps Mountains near the Austrian-Italian border discovered the body of a man who had been trapped in a glacier. He was almost perfectly preserved. With him was an assortment of tools, including an axe with a metal blade.

Scientists were particularly interested in the axe. At first, they believed that it was bronze, which is an alloy of copper and tin. There was a complication, however. Dating techniques that were used for the clothing and body suggested that the "Ice Man" was about 5300 years old. Bronze implements do not appear in Europe's fossil record until about 4000 years ago. Either Europeans were using bronze earlier than originally thought, or the axe was made of a different material. Copper was consistent with the Ice Man's age, since it has been used for at least the past 6000 years.

One technique to determine a metal's identity is to dissolve it in acid. The resulting solution is examined for evidence of ions. Scientists did not want to damage the precious artifact in any way, though.

The solution was an analytical technique called *X-ray fluorescence*. The object is irradiated with high-energy X-ray radiation. Its atoms absorb the radiation, causing electrons from a lower energy level to be ejected from the atom. This causes electrons from an outer energy level to "move in," to occupy the vacated space. As the electrons fall to a less energetic state, they emit X-rays. The electrons of each atom emit X-rays of a particular wavelength. Scientists use this energy "signature" to identify the atom.

Analysis by X-ray fluorescence revealed that the metal in the blade of the axe was almost pure copper.

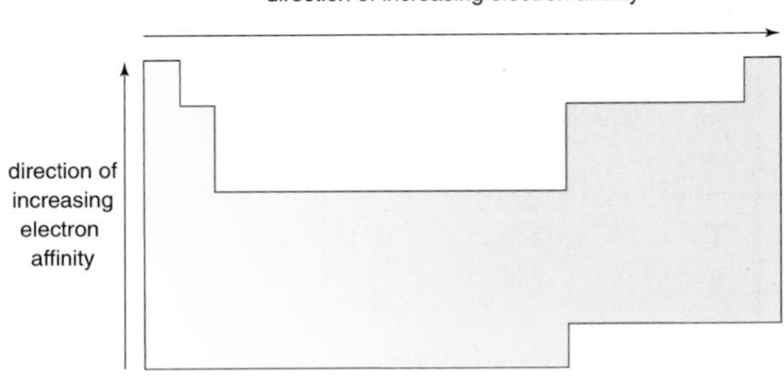

Figure 2.19 **Electron affinity tends to decrease down a group and increase across a period.**

The trends for electron affinity, shown in Figure 2.19, are more irregular than the trends for ionization energy and atomic radius. Nevertheless, the following *general* trends can be observed:

- *Electron affinity tends to decrease down a group.* For example, fluorine has a higher electron affinity than iodine.
- *Electron affinity tends to increase across a period.* For example, calcium has a lower electron affinity than sulfur.

ThoughtLab  Design an Annotated Periodic Table

You have learned a great deal about the properties of the elements. In the following chapters, you will learn more. With your classmates, develop your own large-scale periodic table to record the properties and common uses of the elements.

Procedure

1. Use print and electronic resources (including this textbook) to find information about one element. Consult with your classmates to make sure that everyone chooses a different element.
2. Find the following information about your element:
   - atomic number
   - atomic mass
   - atomic symbol
   - melting point
   - boiling point
   - density
   - atomic radius
   - ionization energy
   - electron affinity
   - place and date discovered, and the name of the scientist who discovered it
   - uses, both common and unusual
   - hazards and methods for safe handling

   If possible, find a photograph of the element in its natural form. If this is not possible, find a photograph that shows one or more compounds in which the element is commonly found.
3. Record your findings on a sheet of notepaper or blank paper. Arrange all the sheets of paper, for all the elements, in the form of a periodic table on a wall in the classroom. Make sure that you leave space to insert additional properties and uses of your element as you learn about them during this course.

Analysis

1. What uses of your element did you know about? Which uses surprised you? Why?
2. Examine the dates on which the elements were discovered. What pattern do you notice? How can you explain this pattern?
3. Do you think that scientists have discovered all the naturally occurring elements? Do you think they have discovered all the synthetically produced elements? Give reasons to justify your opinions.

Section Wrap-up

Despite some irregularities and exceptions, the following periodic trends summarize the relationships among atomic size, ionization energy, and electron affinity:

- Trends for atomic size are the reverse of trends for ionization energy and electron affinity. Larger atoms tend to have lower ionization energies and lower electron affinities.
- Group 16 (VI A) and 17 (VII A) elements attract electrons strongly. They do not give up electrons readily. In other words, they have a strong tendency to form negative ions. Thus, they have high ionization energies and high electron affinities.
- Group 1 (I A) and 2 (II A) elements give up electrons readily. They have low or no attraction for electrons. In other words, they have a strong tendency to form positive ions. Thus, they have low ionization energies and low electron affinities.
- Group 18 (VIII A) elements do not attract electrons and do not give up electrons. In other words, they do not naturally form ions. (They are very stable.) Thus, they have very high ionization energies and very low electron affinities.

13. Arrange the following elements into groups that share similar properties: Ca, K, Ga, P, Si, Rb, B, Sr, Sn, Cl, Bi, Br. How much confidence do you have in your groupings, and why?

14. Use a drawing of your choice to show clearly the relationship among the following terms: valence, stable octet, electron, energy level.

15. In what ways are periodic trends related to the arrangement of electrons in atoms?

Inquiry

16. Imagine hearing on the news that somebody has discovered a new element. The scientist who discovered this element claims that it fits between tin and antimony on the periodic table.
   (a) How likely is it for this claim to be true? Justify your answer.
   (b) Write at least three questions that you could ask this scientist. What is your reasoning for asking these questions? (In other words, what do you expect to hear that could help convince you that the scientist is right or that you are?)

17. Technetium, with an atomic number of 43, was discovered after Mendeleev's death. Nevertheless, he used the properties of manganese, rhenium, molybdenum, and ruthenium to predict technetium's properties.
   (a) Use a chemical database to find the following properties for the above-mentioned elemental "neighbours" of technetium:
   - atomic mass
   - appearance
   - melting point
   - density

   If you would like to truly follow in Mendeleev's footsteps, you could also look for the chemical formulas of the compounds that these elements form with oxygen and chlorine. (Such compounds are called oxides and chlorides.)
   (b) Use this data to predict the properties for technetium.
   (c) Consult a chemical data base to assess your predictions against the observed properties for technetium.

Communication

18. Explain how you would design a data base to display information about the atomic numbers, atomic masses, the number of subatomic particles, and the number of electrons in the outer energy levels of the main-group elements. If you have access to spreadsheet software, construct this table.

19. **(a)** Decide on a way to compare, in as much detail as you can, the elements sodium and helium. The following terms should appear in your answer. Use any other terms that you think are necessary to complete your answer fully.

atom	element
nucleus	proton
neutron	electron
energy level	valence
periodic table	periodic trend
group	period
atomic radius	electron affinity
ionization energy	

   (b) Modify your answer to part (a) so that a class of grade 4 students can understand it.

20. Element A, with three electrons in its outer energy level, is in Period 4 of the periodic table. How does the number of its valence electrons compare with that of Element B, which is in Group 13 (IIIA) and Period 6? Use Lewis structures to help you express your answer.

21. Which elements would be affected if the elements in Periods 1, 2, 3, and 4 were arranged based on their atomic mass, rather than their atomic number? Based on what you have learned in this chapter, how can you be reasonably sure that arranging elements by their atomic number is accurate?

Making Connections

22. "When she blew her nose, her handkerchief glowed in the dark." The woman who made this statement in the early 1900s was one of several factory workers who were hired to paint clock and watch dials with luminous paint. This paint glowed in the dark, because it contained radium (atomic number 88), which is

highly radioactive and toxic. Marie and Pierre Curie discovered radium in 1898. Chemists knew as early as 1906 that the element was dangerous. Nevertheless, it was used not only for its "glowing effects," but also as a medicine. In fact, several companies produced drinks, skin applications, and foods containing radium.

Choose either one of the topics below for research.

- the uses and health-related claims made for radium during the early 1900s
- the story of the so-called "radium girls"—the factory workers who painted clock and watch faces with radium paint

How does the early history of radium and its uses illustrate the need for people to understand the connections among science, technology, society, and the environment?

23. Have you ever heard someone refer to aluminum foil as "tin foil"? At one time, the foil was, in fact, made from elemental tin. Find out why manufacturers phased out tin in favour of aluminum. Compare their chemical and physical properties. Identify and classify the products made from or with aluminum. What are the technological costs and benefits of using aluminum? What health-related and environment-related issues have surfaced as a result of its widespread use in society? Write a brief report to assess the economic, social, and environment impact of our use of aluminum.

Answers to Practice Problems and Short Answers to Section Review Questions:

Practice Problems: 1. (a) boron **(b)** 5 **(c)** 6 **(d)** lead **(e)** 82 **(f)** 126 **(g)** ${}^{184}_{74}W$ **(h)** 74 **(i)** ${}^{4}_{2}He$ **(j)** 2 **(k)** plutonium **(l)** 94 **(m)** 145 **(n)** Fe **(o)** iron **(p)** 30 **(q)** ${}^{209}_{83}Bi$ **(r)** 83 **(s)** ${}^{154}_{47}Ag$ **(t)** silver **(u)** Ne **(v)** neon **(w)** 10 **(x)** 10 **2. (a)** carbon **(b)** copper **(c)** radon **(d)** hydrogen **(e)** cadmium **(f)** calcium **(g)** iodine **(h)** aluminum **3.** H and He have one energy level; H has 1 electron, He has 2. Li, Be, B, C, N, O, F, and Ne have two energy levels. First energy level is filled with two electrons. Second energy of Li has 1 electron, and electrons increase by one, totalling 8 in outer energy level for Ne. Na, Mg, Al, Si, P, S, Cl, Ar have three energy levels. First two energy levels are full. Third energy level of Na has 1 electron, and electrons increase by one, totalling 8 in outer energy level for Ar. K and Ca have four energy levels. First three energy levels are full. K has 1 electron and Ca has two in outer energy level. **4.** The pattern of outer energy level electrons from Practice Problem 3 is repeated by placing dots around the atomic symbol for each element. **5. (a)** 7 **(b)** 2 **(c)** 3 **(d)** 2 **(e)** 1 **(f)** 4 **(g)** 5 **(h)** 6 **(i)** 5 **(j)** 8 **6.** Ba has two dots; Ga has 3 dots; Sn has 4 dots; Bi has 5 dots; I has 7 dots; Cs has 1 dot; Kr has 8 dots; Xe has 8 dots. **7. (a)** Ba, Mg, Be **(b)** Ca, Ga, Se **(c)** Rb, Br, Kr **(d)** Ca, Se, Br **(e)** Cs, Ba, Sr **(f)** Se, Br, Cl **(g)** Ca, Mg, Li **(h)** Sr, Te, Se **(i)** In, I, Br **(j)** Se, S, O **8. (a)** Xe, Ar, He **(b)** In, Sn, Sb **(c)** Ba, Sr, Ca **(d)** K, Br, Kr **(e)** Rb, K, Ca **(f)** Rb, Br, Kr **9. (a)** B **(b)** In **(c)** I **(d)** N **(e)** K **(f)** Tl

Section Review: 2.1: 1. (a) silver **(b)** 47 **(c)** 47 **(d)** 61 **(e)** arsenic **(f)** 33 **(g)** 75 **(h)** 33 **(i)** bromine **(j)** 80 **(k)** 35 **(l)** 35 **(m)** gold **(n)** 79 **(o)** 79 **(p)** 100 **(q)** tin **(r)** 50 **(s)** 119 **(t)** 50 **3. (a)** last pair has same protons and electrons, but different neutrons **(b)** last pair has same value for Z; first pair has same value for A **2.2: 2.** 1 (1A), 18 (8A), 17 (7A), 2 (2A) **3. (a)** Si; Sb; Te; Br **6. (a)** 8; 7; 6; 2; 1; 7; 2; 5; 2; 4. **(c)** non-metal; non-metal; non-metal; metal; metal; non-metal; non-metal; metal, metal; metalloid **7.** two: mercury and bromine **9. (b)** Triads 2 and 3 **11. (b)** Na, Mg, and Al have same number of energy levels, as do Li and C **(c)** Li, Na, and K have same number of valence electrons **2.3: 2. (a)** Cl, S, Mg **(b)** B, Al, In **(c)** Ne, Ar, Xe **(d)** Xe, Te, Rb **(e)** F, P, Na **(f)** O, N, S **3. (a)** Cl, Br, I **(b)** Se, Ge, Ga **(c)** Kr, Ca, K **(d)** Li, Na, Cs **(e)** Cl, Br, S **(f)** Ar, Cl, K **4. (a)** Ca, **(b)** Li **(c)** Se **(d)** Cs

CHAPTER 3

Chemical Compounds and Bonding

The year was 1896. A chance discovery sent a message echoing from Yukon's Far North to the southern reaches of the United States: "Gold!" People migrated in great numbers to the Yukon Territory, hoping to make their fortunes. Within two years, these migrants transformed a small fishing village into bustling Dawson City—one of Canada's largest cities at the time. They also launched the country's first metal-mining industry.

Gold, like all metals, is shiny, malleable, ductile, and a good conductor of electricity and heat. Unlike most metals and other elements, however, gold is found in nature in its pure form, as an element. Most elements are chemically combined in the form of compounds. Why is this so? Why do atoms of some elements join together as compounds, while others do not? In this chapter, you will use the periodic trends you examined in Chapter 2 to help you answer these questions. You will learn about the bonds that hold elements together in compounds. At the same time, you will learn how to write chemical formulas and how to name compounds.

Chapter Preview

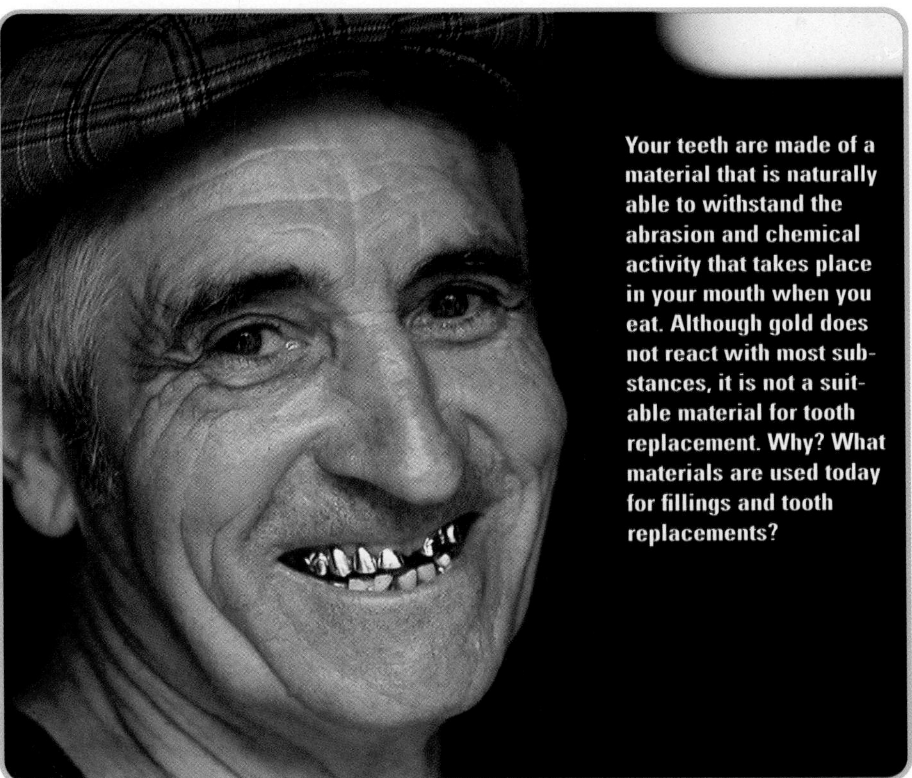

Your teeth are made of a material that is naturally able to withstand the abrasion and chemical activity that takes place in your mouth when you eat. Although gold does not react with most substances, it is not a suitable material for tooth replacement. Why? What materials are used today for fillings and tooth replacements?

Concepts and Skills You Will Need

Before you begin this chapter, review the following concepts and skills:

- drawing Lewis structures to represent valence electrons in the outer energy levels of atoms (Chapter 2, section 2.1)
- identifying and explaining periodic trends (Chapter 2, section 2.2)
- identifying elements by name and by symbol (Chapter 2, section 2.2)

3.1 Classifying Chemical Compounds

Section Preview/ Specific Expectations

In this section, you will

- **describe** how electron arrangement and forces in atoms can explain the periodic trend associated with electronegativity
- **perform** a Thought Lab to classify compounds as ionic or covalent according to their properties
- **predict** the ionic character of a given bond using electronegativity values
- **communicate** your understanding of the following terms: *chemical bonds, ionic bond, covalent bond, electronegativity*

As you learned in the chapter opener, most elements do not exist in nature in their pure form, as elements. Gold, silver, and platinum are three metals that can be found in Earth's crust as elements. They are called "precious metals" because this occurrence is so rare. Most other metals, and most other elements, are found in nature only as compounds.

As the prospectors in the Yukon gold rush were searching for the element gold, they were surrounded by compounds. The streams they panned for gold ran with water, H_2O, a compound that is essential to the survival of nearly every organism on this planet. To sustain their energy, the prospectors ate food that contained, among other things, starch. Starch is a complex compound that consists of carbon, hydrogen, and oxygen. To flavour their food, they added sodium chloride, NaCl, which is commonly called table salt. Sometimes a compound called pyrite, also known as "fool's gold," tricked a prospector. Pyrite (iron disulfide, FeS_2) looks almost exactly like gold, as you can see in Figure 3.1. Pyrite, however, will corrode, and it is not composed of rare elements. Thus, it was not valuable to a prospector.

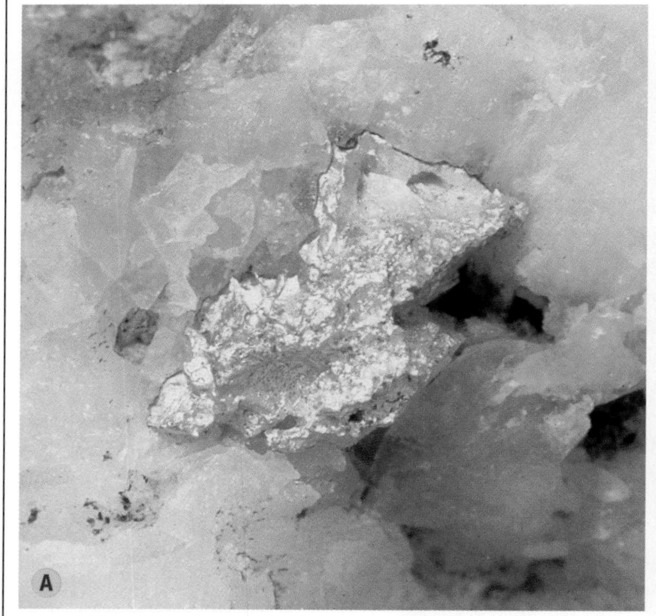

Figure 3.1 Prospectors used the physical properties of gold and pyrite to distinguish between them. Can you tell which of these photos shows gold and which shows pyrite?

There are only about 90 naturally occurring elements. In comparison, there are thousands upon thousands of different compounds in nature, and more are constantly being discovered. Elements combine in many different ways to form the astonishing variety of natural and synthetic compounds that you see and use every day.

Because there are so many compounds, chemists have developed a classification system to organize them according to their properties, such as melting point, boiling point, hardness, conductivity, and solubility. In the following Express Lab, you will use the property of magnetism to show that an element has formed a compound.

ExpressLab  A Metal and a Compound

Humans have invented ways to extract iron from its compounds in order to take advantage of its properties. Does iron remain in its uncombined elemental form once it has been extracted? No, it doesn't. Instead, it forms rust, or iron(III) oxide, Fe_2O_3. How do we know that rust and iron are different substances? One way to check is to test a physical property, such as magnetism. In this activity, you will use magnetism to compare the properties of iron and rust.

Safety Precautions

Procedure

1. Obtain a new iron nail and a rusted iron nail from your teacher.
2. Obtain a thin, white piece of cardboard and a magnet. Wrap your magnet in plastic to keep it clean.
3. Test the iron nail with the magnet. Record your observations.
4. Gently rub the rusted nail with the other nail over the cardboard. Some rust powder will collect on the cardboard.
5. Hold up the cardboard horizontally. Move the magnet back and forth under the cardboard. Record your observations.

Analysis

1. How did the magnet affect the new iron nail? Based on your observations, is iron magnetic?
2. What did you observe when you moved the magnet under the rust powder?
3. What evidence do you have to show that iron and rust are different substances?
4. Consider what you know about iron and rust from your everyday experiences. Is it more likely that rust will form from iron, or iron from rust?

Properties of Ionic and Covalent Compounds

Based on their physical properties, compounds can be classified into two groups: ionic compounds and covalent compounds. Some of the properties of ionic and covalent compounds are summarized in Table 3.1.

Table 3.1 Comparing Ionic and Covalent Compounds

Property	Ionic compound	Covalent compound
state at room temperature	crystalline solid	liquid, gas, solid
melting point	high	low
electrical conductivity as a liquid	yes	no
solubility in water	most have high solubility	most have low solubility
conducts electricity when dissolved in water	yes	not usually

In the following Thought Lab, you will use the properties of various compounds to classify them as covalent or ionic.

ThoughtLab Ionic or Covalent?

Imagine that you are a chemist. A colleague has just carried out a series of tests on the following compounds:
ethanol
carbon tetrachloride
glucose
table salt (sodium chloride)
water
potassium permanganate

You take the results home to organize and analyze them. Unfortunately your colleague labelled the tests by sample number and forgot to write down which compound corresponded to each sample number. You realize, however, that you can use the properties of the compounds to identify them. Then you can use the compounds' properties to decide whether they are ionic or covalent.

Procedure

1. Copy the following table into your notebook.

Sample	Compound name	Dissolves in water?	Conductivity as a liquid or when dissolved in water	Melting point	Appearance	Covalent or ionic?
1		yes	high	801°C	clear, white crystalline solid	
2		yes	low	0.0°C	clear, colourless liquid	
3		yes	high	240°C	purple, crystalline solid	
4		yes	low	146°C	white powder	
5		no	low	−23°C	clear, colourless liquid	
6		yes	low	−114°C	clear, colourless liquid	

2. Based on what you know about the properties of compounds, decide which compound corresponds to each set of properties. Write your decisions in your table. Once you have identified the samples, share your results as a class and come to a consensus. **Hint:** Carbon tetrachloride is not soluble in water.
3. Examine the properties associated with each compound. Decide whether each compound is ionic or covalent. If you are unsure, leave the space blank. Discuss your results as a class, and come to a consensus.

Analysis

1. Write down the reasoning you used to identify each compound, based on the properties given.
2. Write down the reasoning you used to decide whether each compound was ionic or covalent.
3. Were you unsure how to classify any of the compounds? Which ones, and why?
4. Think about the properties in the table you filled in, as well as your answers to questions 1 to 3. Which property is most useful for deciding whether a compound is ionic or covalent?
5. Suppose that you could further subdivide the covalent compounds into two groups, based on their properties. Which compounds would you group together? Explain your answer.

Applications

6. Use a chemistry reference book or the Internet to find an MSDS for ethanol, carbon tetrachloride, and potassium permanganate.
   (a) Write down the health hazards associated with each compound.
   (b) What precautions would a chemist who was performing tests on ethanol and carbon tetrachloride need to take?

Table Salt: An Ionic Compound

Figure 3.2 Sodium chloride is used to melt ice because salt water has a lower melting point than pure water.

Sodium chloride, NaCl, is a familiar compound. You know it as table salt. The sodium in sodium chloride plays a vital role in body functions. We need to ingest about 500 mg of sodium a day. Too much sodium chloride, however, may contribute to high blood pressure. In the winter, sodium chloride is put on roads and sidewalks to melt the ice, as shown in Figure 3.2. Although this use of sodium chloride increases the safety of pedestrians and drivers, there are several drawbacks. For example, the saltwater discolours and damages footwear, and it corrodes the metal bodies of cars and trucks. Also, as shown in Figure 3.3, deer and moose that are attracted to the salt on the roads can be struck by vehicles.

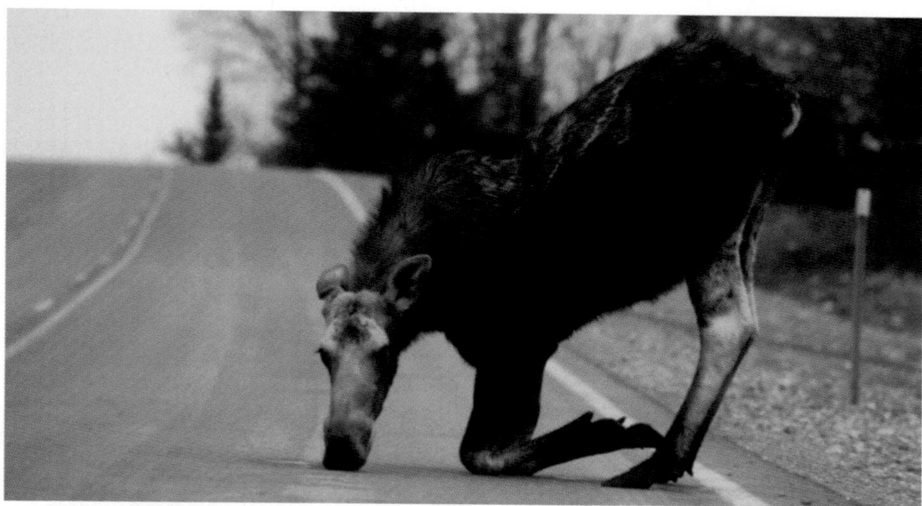

Figure 3.3 This moose was attracted to the sodium chloride that was put on the road to melt snow and ice. Humans, like most organisms, need sodium to maintain normal body functions.

Sodium chloride is a typical ionic compound. Like most ionic compounds, it is a crystalline solid at room temperature. It melts at a very high temperature, at 801°C. As well, it dissolves easily in water. A solution of sodium chloride in water is a good conductor of electricity. Liquid sodium chloride is also a good electrical conductor.

Carbon Dioxide: A Covalent Compound

The cells of most organisms produce carbon dioxide, CO_2, during cellular respiration: the process that releases energy from food. Plants, like the ones shown in Figure 3.4, synthesize their own food from carbon dioxide and water using the Sun's energy.

Carbon dioxide has most of the properties of a typical covalent compound. It has a low melting point (–79°C). At certain pressures and temperatures, carbon dioxide is a liquid. Liquid carbon dioxide is a weak conductor of electricity.

Figure 3.4 Plants use carbon dioxide and water to produce their own food, using the Sun's energy.

Figure 3.5 **The bubbles fizzing out of the soft drink contain carbon dioxide.**

Carbon dioxide is somewhat soluble in water, especially at high pressures. This is why soft drinks are bottled under pressure. When you open a bottle of pop, some of the carbon dioxide comes out of solution. Often, this happens too quickly, as you can see in Figure 3.5. A solution of carbon dioxide in water is a weak conductor of electricity.

What Is Bonding?

Why are carbon dioxide and sodium chloride so different? Why can we divide compounds into two categories that display distinct physical properties? The answers come from an understanding of **chemical bonds**: the forces that attract atoms to each other in compounds. *Bonding involves the interaction between the valence electrons of atoms.* Usually the formation of a bond between two atoms creates a compound that is more stable than either of the two atoms on their own.

The different properties of ionic and covalent compounds result from the manner in which chemical bonds form between atoms in these compounds. Atoms can either exchange or share electrons.

When two atoms exchange electrons, one atom loses its valence electron(s) and the other atom gains the electron(s). This kind of bonding usually occurs between a metal and a non-metal. Recall, from Chapter 2, that metals have low ionization energies and non-metals have high electron affinities. That is, metals tend to lose electrons and non-metals tend to gain them. When atoms exchange electrons, they form an **ionic bond**.

Atoms can also share electrons. This kind of bond forms between two non-metals. It can also form between a metal and a non-metal when the metal has a fairly high ionization energy. *When atoms share electrons, they form a* **covalent bond**.

How can you determine whether the bonds that hold a compound together are ionic or covalent? Examining the physical properties of the compound is one method. This method is not always satisfactory, however. Often a compound has some ionic characteristics and some covalent characteristics. You saw this in the previous Thought Lab.

For example, hydrogen chloride, also known as hydrochloric acid, has a low melting point and a low boiling point. (It is a gas at room temperature.) These properties might lead you to believe that hydrogen chloride is a covalent compound. Hydrogen chloride, however, is extremely soluble in water, and the water solution conducts electricity. These properties are characteristic of an ionic compound. Is there a clear, theoretical way to decide whether the bond between hydrogen and chlorine is ionic or covalent? The answer lies in a periodic trend.

mind STRETCH

Do you think that water is a covalent compound or an ionic compound? List water's physical properties. Can you tell whether water is a covalent compound or an ionic compound based only on its physical properties? Why or why not?

Electronegativity: Attracting Electrons

When two atoms form a bond, each atom attracts the other atom's electrons in addition to its own. The **electronegativity** of an atom is a measure of an atom's ability to attract electrons in a chemical bond. *EN* is used to symbolize electronegativity. There is a specific electronegativity associated with each element.

As you can see in Figure 3.6, electronegativity is a periodic property, just as atomic size, ionization energy, and electron affinity are. Atomic size, ionization energy, and electron affinity, however, are properties of single atoms. In contrast, electronegativity is a property of atoms that are involved in chemical bonding.

Electronegativities

1 H 2.20																	2 He -
3 Li 0.98	4 Be 1.57											5 B 2.04	6 C 2.55	7 N 3.04	8 O 3.44	9 F 3.98	10 Ne -
11 Na 0.93	12 Mg 1.31											13 Al 1.61	14 SI 1.90	15 P 2.19	16 S 2.58	17 Cl 3.16	18 Ar -
19 K 0.82	20 Ca 1.00	21 Sc 1.36	22 Ti 1.54	23 V 1.63	24 Cr 1.66	25 Mn 1.55	26 Fe 1.83	27 Co 1.88	28 Ni 1.91	29 Cu 1.90	30 Zn 1.65	31 Ga 1.81	32 Ge 2.01	33 As 2.18	34 Se 2.55	35 Br 2.96	36 Kr -
37 Rb 0.82	38 Sr 0.95	39 Y 1.22	40 Zr 1.33	41 Nb 1.6	42 Mo 2.16	43 Tc 2.10	44 Ru 2.2	45 Rh 2.28	46 Pd 2.20	47 Ag 1.93	48 Cd 1.69	49 In 1.78	50 Sn 1.96	51 Sb 2.05	52 Te 2.1	53 I 2.66	54 Xe -
55 Cs 0.79	56 Ba 0.89	71 Lu 1.0	72 Hf 1.3	73 Ta 1.5	74 W 1.7	75 Re 1.9	76 Os 2.2	77 Ir 2.2	78 Pt 2.2	79 Au 2.4	80 Hg 1.9	81 Tl 1.8	82 Pb 1.8	83 Bi 1.9	84 Po 2.0	85 At 2.2	86 Rn -
87 Fr 0.7	88 Ra 0.9	103 Lr -	104 Rf -	105 Db -	106 Sg -	107 Bh -	108 Hs -	109 Mt -	110 Uun -	111 Uuu -	112 Uub -	113 - 	114 Uuq -	115 - 	116 Uuh -	117 - 	118 Uuo -

57 La 1.10	58 Ce 1.12	59 Pr 1.13	60 Nd 1.14	61 Pm -	62 Sm 1.17	63 Eu -	64 Gd 1.20	65 Tb -	66 Dy 1.22	67 Ho 1.23	68 Er 1.24	69 Tm 1.25	70 Yb -
89 Ac 1.1	90 Th 1.3	91 Pa 1.5	92 U 1.7	93 Np 1.3	94 Pu 1.3	95 Am -	96 Cm -	97 BK -	98 Cf -	99 Es -	100 Fm -	101 Md -	102 No -

Figure 3.6 Electronegativity is a periodic trend. It increases up a group and across a period.

The trend for electronegativity is the reverse of the trend for atomic size. Examine Figure 3.7, on the next page, to see what this means. In general, as atomic size decreases from left to right across a period, electronegativity increases. Why? The number of protons (which attract electrons) in the nucleus increases. At the same time, the number of filled, inner electron energy levels (which shield the protons from valence electrons) remains the same. Thus the electrons are pulled more tightly to the nucleus, resulting in a smaller atomic size. The atom attracts a bonding pair of electrons more strongly, because the bonding pair can move closer to the nucleus.

In the second period, for example, lithium has the largest atomic size and the lowest electronegativity. As atomic size decreases across the second period, the electronegativity increases. Fluorine has the smallest atomic size in the third period (except for neon) and the highest electronegativity. Because noble gases do not usually participate in bonding, their electronegativities are not given.

Similarly, as atomic size increases down a group, electronegativity decreases. As you move down a group, valence electrons are less strongly attracted to the nucleus because the number of filled electron energy levels between the nucleus and the valence electrons increases. In a compound, increasing energy levels between valence electrons and the nucleus mean that the nucleus attracts bonding pairs less strongly.

For example, in Group 2 (IIA), beryllium has the smallest atomic radius and the largest electronegativity. As atomic size increases down the group, electronegativity decreases.

Figure 3.7 shows the relationship between atomic size and electronegativity for the main-group elements in periods 2 to 6.

CHECKPOINT

Which element is the most electronegative? Not including the noble gases, which element is the least electronegative?

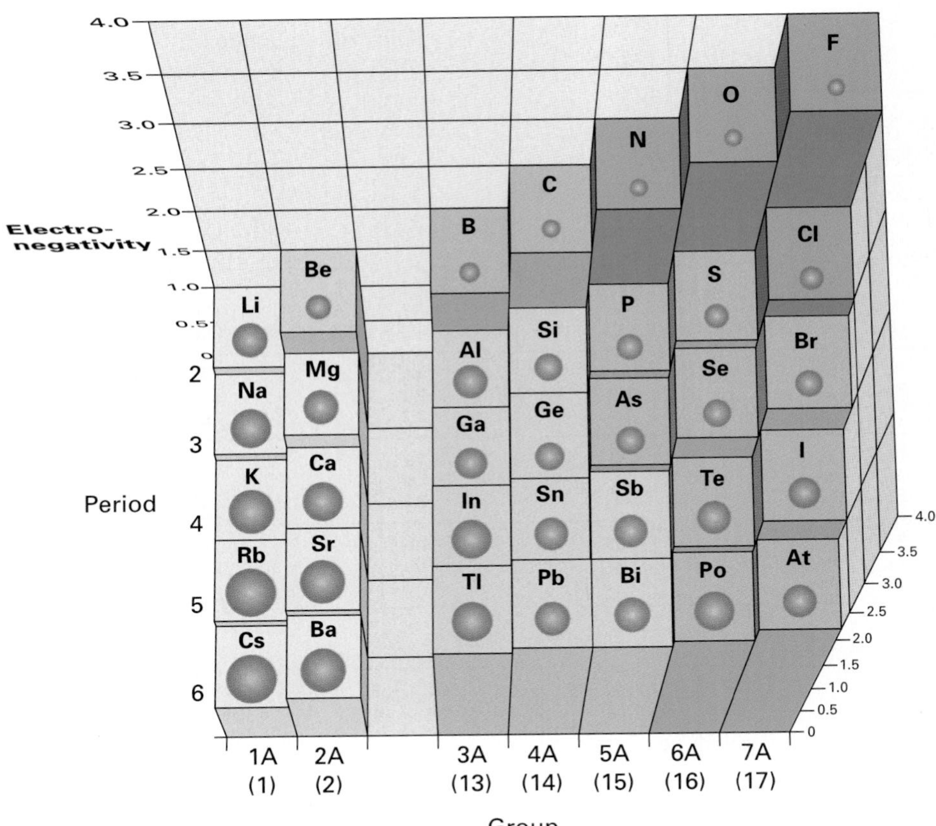

Figure 3.7 Periodic trends for electronegativity (bars) and atomic size (spheres) are inversely related.

Predicting Bond Type Using Electronegativity

You can use the differences between electronegativities to decide whether the bond between two atoms is ionic or covalent. The symbol ΔEN stands for the difference between two electronegativity values. When calculating the electronegativity difference, the smaller electronegativity is always subtracted from the larger electronegativity, so that the electronegativity difference is always positive.

How can the electronegativity difference help you predict the type of bond? By the end of this section, you will understand the aswer to this question. Consider three different substances: potassium fluoride, KF, oxygen, O_2, and hydrochloric acid, HCl. Potassium fluoride is an ionic compound made up of a metal and a non-metal that have very different electronegativities. Potassium's electronegativity is 0.82. Fluorine's electronegativity is 3.98. Therefore, ΔEN for the bond between potassium and fluorine is 3.16.

Now consider oxygen. This element exists as units of two atoms held together by covalent bonds. Each oxygen atom has an electronegativity of 3.44. The bond that holds the oxygen atoms together has an electronegativity difference of 0.00 because each atom in an oxygen molecule has an equal attraction for the bonding pair of electrons.

Finally, consider hydrogen chloride, or hydrochloric acid. Hydrogen has an electronegativity of 2.20, and chlorine has an electronegativity of 3.16. Therefore, the electronegativity difference for the chemical bond in hydrochloric acid, HCl, is 0.96. Hydrogen chloride is a gas at room temperature, but its water solution conducts electricity. Is hydrogen chloride a covalent compound or an ionic compound? Its ΔEN can help you decide, as you will see below.

The Range of Electronegativity Differences

When two atoms have electronegativities that are identical, as in oxygen, they share their bonding pair of electrons equally between them in a covalent bond. When two atoms have electronegativities that are very different, as in potassium fluoride, the atom with the lower electronegativity loses an electron to the atom with the higher electronegativity. In potassium fluoride, potassium gives up its valence electron to fluorine. Therefore, the bond is ionic.

It is not always clear whether atoms share electrons or transfer them. Atoms with different electronegativities can share electrons unequally without exchanging them. How unequal does the sharing have to be before the bond is considered ionic?

Figure 3.8 shows the range of electronegativity differences. These values go from mostly covalent at 0.0 to mostly ionic at 3.3. Chemists consider bonds with an electronegativity difference that is greater than 1.7 to be ionic, and bonds with an electronegativity difference that is less than 1.7 to be covalent.

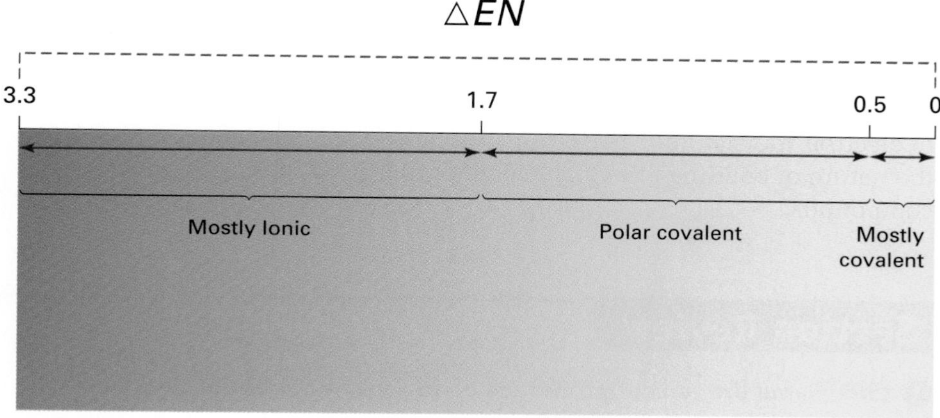

Figure 3.8 **Chemical bonds range in character from mostly ionic to mostly covalent.**

Table 3.2 shows how you can think of bonds as having a percent ionic character or percent covalent character, based on their electronegativity differences. When bonds have nearly 50% ionic or covalent character, they have characteristics of both types of bonding.

Table 3.2 Character of Bonds

Electronegativity difference	0.00	0.65	0.94	1.19	1.43	1.67	1.91	2.19	2.54	3.03
Percent ionic character	0%	10%	20%	30%	40%	50%	60%	70%	80%	90%
Percent covalent character	100%	90%	80%	70%	60%	50%	40%	30%	20%	10%

Based on Table 3.2, what kind of bond forms between hydrogen and chlorine? ΔEN for the bond in hydrogen chloride, HCl, is 0.96. This is lower than 1.7. Therefore, the bond in hydrogen chloride is a covalent bond.

Calculate ΔEN and predict bond character in the following Practice Problem.

Practice Problems

1. Determine ΔEN for each bond shown. Indicate whether each bond is ionic or covalent.
   - **(a)** O—H
   - **(b)** C—H
   - **(c)** Mg—Cl
   - **(d)** B—F
   - **(e)** Cr—O
   - **(f)** C—N
   - **(g)** Na—I
   - **(h)** Na—Br

Section Wrap-up

In this section, you learned that most elements do not exist in their pure form in nature. Rather, they exist as different compounds. You reviewed the characteristic properties of ionic and covalent compounds. You considered the periodic nature of electronegativity, and you learned how to use the electronegativity difference to predict the type of bond. You learned, for example, that ionic bonds form between two atoms with very different electronegativities.

In section 3.2, you will explore ionic and covalent bonding in terms of electron transfer and sharing. You will use your understanding of the nature of bonding to explain some properties of ionic and covalent compounds.

Section Review

1. K/U Name the typical properties of an ionic compound. Give two examples of ionic compounds.

2. K/U Name the typical properties of covalent compounds. Give two examples of covalent compounds.

3. C In your own words, describe and explain the periodic trend for electronegativity.

4. K/U Based only on their position in the periodic table, arrange the elements in each set in order of increasing attraction for electrons in a bond.
   - **(a)** Li, Br, Zn, La, Si
   - **(b)** P, Ga, Cl, Y, Cs

5. K/U Determine ΔEN for each bond. Indicate whether the bond is ionic or covalent.
   - **(a)** N—O
   - **(b)** Mn—O
   - **(c)** H—Cl
   - **(d)** Ca—Cl

6. I A chemist analyzes a white, solid compound and finds that it does not dissolve in water. When the compound is melted, it does not conduct electricity.
   - **(a)** What would you expect to be true about this compound's melting point?
   - **(b)** Are the atoms that make up this compound joined with covalent or ionic bonds? Explain.

Ionic and Covalent Bonding: The Octet Rule

3.2

In section 3.1, you reviewed your understanding of the physical properties of covalent and ionic compounds. You learned how to distinguish between an ionic bond and a covalent bond based on the difference between the electronegativities of the atoms. By considering what happens to electrons when atoms form bonds, you will be able to explain some of the characteristic properties of ionic and covalent compounds.

Section Preview/ Specific Expectations

In this section, you will

- **demonstrate** an understanding of the formation of ionic and covalent bonds, and **explain** the properties of the products
- **explain** how different elements combine to form covalent and ionic bonds, using the octet rule
- **represent** the formation of ionic and covalent bonds using diagrams
- **communicate** your understanding of the following terms: *octet rule, isoelectronic, pure covalent bond, diatomic elements, double bond, triple bond, molecular compounds, intramolecular forces, intermolecular forces, metallic bond, alloy*

The Octet Rule

Why do atoms form bonds? When atoms are bonded together, they are often more stable. We know that noble gases are the most stable elements in the periodic table. What evidence do we have? The noble gases are extremely unreactive. They do not tend to form compounds. What do the noble gases have in common? They have a filled outer electron energy level. When an atom loses, gains, or shares electrons through bonding to achieve a filled outer electron energy level, the resulting compound is often very stable.

According to the **octet rule**, atoms bond in order to achieve an electron configuration that is the same as the electron configuration of a noble gas. When two atoms or ions have the same electron configuration, they are said to be **isoelectronic** with one another. For example, Cl^- is isoelectronic with Ar because both have 18 electrons and a filled outer energy level. This rule is called the octet rule because all the noble gases (except helium) have eight electrons in their filled outer energy level. (Recall that helium's outer electron energy level contains only two electrons.)

Ionic Bonding

In Section 5.1 you learned that the electronegativity difference for the bond between sodium and chlorine is 2.1. Thus, the bond is an ionic bond. Sodium has a very low electronegativity, and chlorine has a very high electronegativity. Therefore, when sodium and chlorine interact, sodium transfers its valence electron to chlorine. As shown in Figure 3.9, sodium becomes Na^+ and chlorine becomes Cl^-.

How does the formation of an ionic bond between sodium and chlorine reflect the octet rule? Neutral sodium has one valence electron. When it loses this electron to chlorine, the resulting Na^+ cation has an electron energy level that contains eight electrons. It is isoelectronic with the noble gas neon. On the other hand, chlorine has an outer electron energy level that contains seven electrons. When chlorine gains sodium's

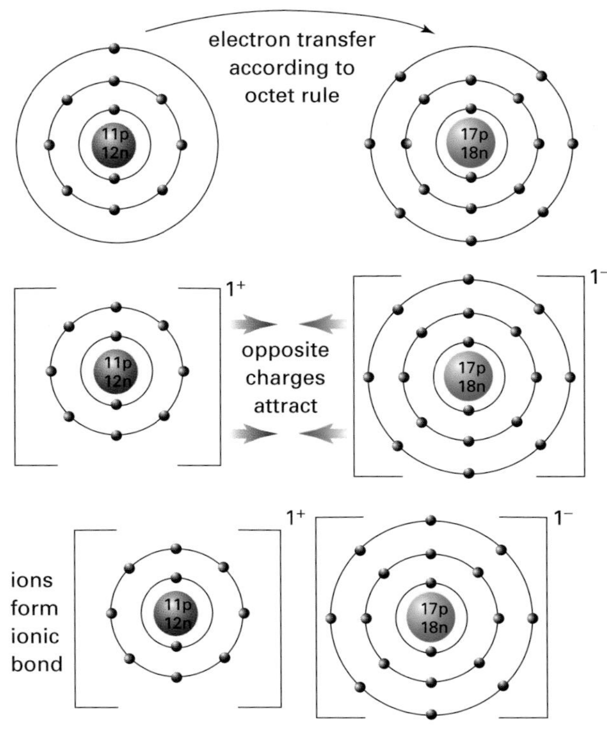

Figure 3.9 Sodium's electron is transferred to chlorine. The atoms become oppositely charged ions with stable octets. Because they are oppositely charged, they are strongly attracted to one another.

electron, it becomes an anion that is isoelectronic with the noble gas argon. As you can see in Figure 3.10, you can represent the formation of an ionic bond using Lewis structures.

Thus, in an ionic bond, electrons are transferred from one atom to another so that they form oppositely charged ions. The strong force of attraction between the oppositely charged ions is what holds them together.

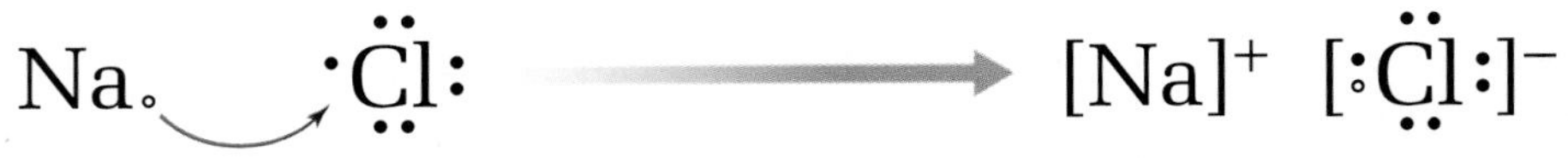

Na∘ ·C̤̈l: → $[Na]^+$ $[:Cl:]^-$

Figure 3.10 These Lewis structures show the formation of a bond between a sodium atom and a chlorine atom.

Transferring Multiple Electrons

In sodium chloride, NaCl, one electron is transferred from sodium to chlorine. In order to satisfy the octet rule, two or three electrons may be transferred from one atom to another. For example, consider what happens when magnesium and oxygen combine.

The electronegativity difference for magnesium oxide is $3.4 - 1.3 = 2.1$. Therefore, magnesium oxide is an ionic compound. Magnesium contains two electrons in its outer shell. Oxygen contains six electrons in its outer shell. In order to become isoelectronic with a noble gas, magnesium needs to lose two electrons and oxygen needs to gain two electrons. Hence, magnesium transfers its two valence electrons to oxygen, as shown in Figure 3.11. Magnesium becomes Mg^{2+}, and oxygen becomes O^{2-}.

PROBLEM TIP

When you draw Lewis structures to show the formation of a bond, you can use different colours or symbols to represent the electrons from different atoms. For example, use an "x" for an electron from sodium, and an "o" for an electron from chlorine. Or, use open and closed cirlces as is shown here. This will make it easier to see how the electrons have been transferred.

Mg ·O: → $[Mg]^{2+}$ $[:O:]^{2-}$

Figure 3.11 These Lewis structures show the formation of a bond between a magnesium atom and an oxygen atom.

Try the following problems to practise representing the formation of ionic bonds between two atoms.

Practice Problems

2. For each bond below, determine ΔEN. Is the bond ionic or covalent?
   (a) Ca—O
   (b) K—Cl
   (c) K—F
   (d) Li—F
   (e) Li—Br
   (f) Ba—O
3. Draw Lewis structures to represent the formation of each bond in question 2.

Careers in Chemistry

Metallurgist

Alison Dickson

Alison Dickson is a metallurgist at Polaris, the world's northernmost mine. Polaris is located on Little Cornwallis Island in Nunavut. It is a lead and zinc mine, operated by Cominco Ltd., the world's largest producer of zinc concentrate.

After ore is mined at Polaris, metallurgists must separate the valuable lead- and zinc-bearing compounds from the waste or "slag." First the ore is crushed and ground with water to produce flour-like particles. Next a process called *flotation* is used to separate the minerals from the slag. In flotation, chemicals are added to the metal-containing compounds. The chemicals react with the lead and zinc to make them very insoluble in water, or *hydrophobic*. Air is then bubbled through the mineral and water mixture. The hydrophobic particles attach to the bubbles and float to the surface. They form a stable froth, or concentrate, which is collected. The concentrate is filtered and dried, and then stored for shipment.

Dickson says that she decided on metallurgy as a career because she wanted to do something that was "hands-on." After completing her secondary education in Malaysia, where she grew up, Dickson moved to Canada. She studied mining and mineral process engineering at the University of British Columbia.

Dickson says that she also wanted to do something adventurous. She wanted to travel and live in different cultures. As a summer student, Dickson worked at a Chilean copper mine. Her current job with Cominco involves frequent travel to various mines. "Every day provides a new challenge," Dickson says. When she is at Polaris, Dickson enjoys polar bear sightings on the tundra.

Making Career Connections

Are you interested in a career in mining and metallurgy? Here are two ways that you can get information:

1. Explore the web site of The Canadian Institute of Mining, Metallurgy and Petroleum. Go to **www.school.mcgrawhill.ca/resources/**, to Science Resources, then to Chemistry 11 to know where to go next. It has a special section for students who are interested in mining and metallurgy careers. This section lists education in the field, scholarships and bursaries, and student societies for mining and metallurgy.
2. To discover the variety of jobs that are available for metallurgists, search for careers at Infomine. Go to **www.school.mcgrawhill.ca/resources/**, to Science Resources, then to Chemistry 11 to know where to go next. Many of the postings are for jobs overseas.

Ionic Bonding That Involves More Than Two Ions

Sometimes ionic compounds contain more than one atom of each element. For example, consider the compound that is formed from calcium and fluorine. Because the electronegativity difference between calcium and fluorine is 3.0, you know that a bond between calcium and fluorine is ionic. Calcium has two electrons in its outer energy level, so it needs to lose two electrons according to the octet rule. Fluorine has seven electrons in its outer energy level, so it needs to gain one electron, again according to the octet rule. How do the electrons of these elements interact so that each element achieves a filled outer energy level?

Figure 3.12 These Lewis structures show the formation of bonds between one atom of calcium and two atoms of fluorine.

Examine Figure 3.11. Because calcium tends to lose two electrons and fluorine tends to gain one electron, one calcium atom bonds with two fluorine atoms. Calcium loses one of each of its valence electrons to each fluorine atom. Calcium becomes Ca^{2+}, and fluorine becomes F^-. They form the compound calcium fluoride, CaF_2.

In the following Practice Problems, you will predict the kind of ionic compound that will form from two elements.

Electronic Learning Partner

Your Chemistry 11 Electronic Learning Partner has an interactive simulation on forming ionic compounds.

Practice Problems

4. For each pair of elements, determine ΔEN.
   (a) magnesium and chlorine
   (b) calcium and chlorine
   (c) lithium and oxygen
   (d) sodium and oxygen
   (e) potassium and sulfur
   (f) calcium and bromine
5. Draw Lewis structures to show how each pair of elements in question 4 forms bonds to achieve a stable octet.

Explaining the Conductivity of Ionic Compounds

Now that you understand the nature of the bonds in ionic compounds, can you explain some of their properties? Consider electrical conductivity. Ionic compounds do not conduct electricity in their solid state. They are very good conductors in their liquid state, however, or when they are dissolved in water. To explain these properties, ask yourself two questions:

1. What is required for electrical conductivity?
2. What is the structure of ionic compounds in the liquid, solid, and dissolved states?

An electrical current can flow only if charged particles are available to move and carry the current. Consider sodium chloride as an example. Is there a mobile charge in solid sodium chloride? No, there is not. In the solid state, sodium and chlorine ions are bonded to each other by strong ionic bonds. Like all solid-state ionic compounds, the ions are arranged in a rigid lattice formation, as shown in Figure 3.13.

In the solid state, the ions cannot move very much. Thus, there is no mobile charge. Solid sodium chloride does not conduct electricity.

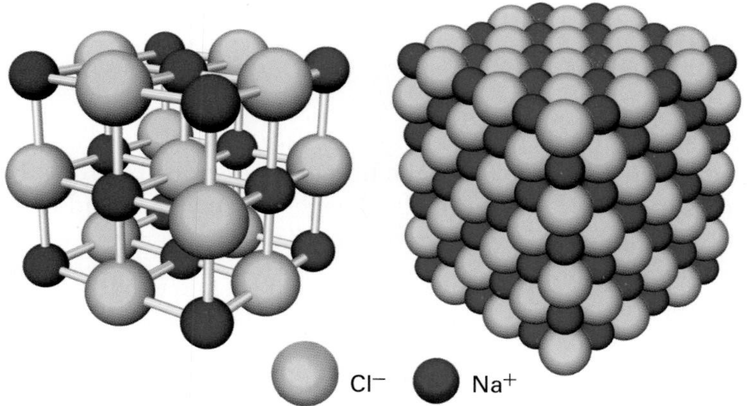

Figure 3.13 **In solid sodium chloride, NaCl, sodium and chlorine are arranged in a rigid lattice pattern.**

In molten sodium chloride, the rigid lattice structure is broken. The ions that make up the compound are free to move, and they easily conduct electricity. Similarly, when sodium chloride is dissolved, the sodium and chlorine ions are free to move. The solution is a good conductor of electricity, as shown in Figure 3.14. You will learn more about ionic compounds in solution in Chapter 9.

mind STRETCH

Go back to Table 3.1. What other properties of ionic compounds can you now explain with your new understanding of ionic bonding?

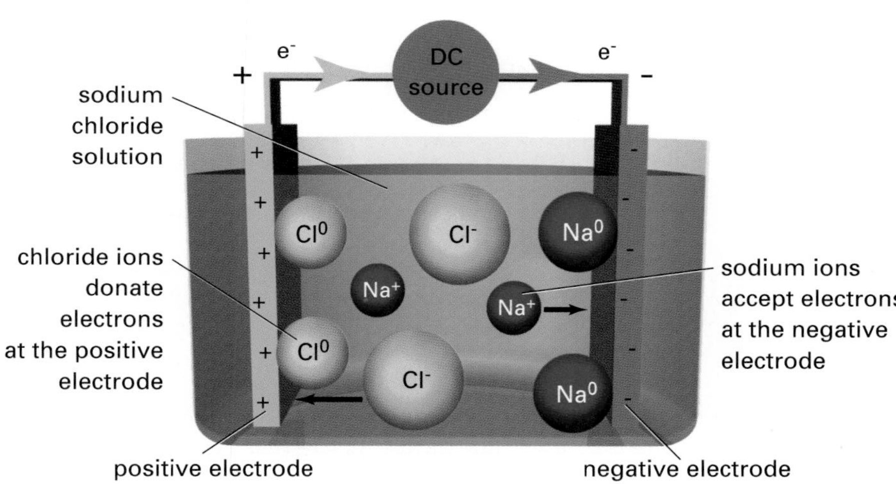

Figure 3.14 **Aqueous sodium chloride is a good conductor of electricity.**

You are probably familiar with the ionic crystals in caves. Stalagmites and stalactites are crystal columns that form when water, containing dissolved lime, drips very slowly from the ceiling of a cave onto the floor below. How do these ionic crystals grow?

When a clear solution of an ionic compound is poured over a seed crystal of the same compound, the ions align themselves according to the geometric arrangement in the seed crystal. Your will observe this for yourself in Investigation 3-A.

Investigation 3-A

SKILL FOCUS
- Performing and recording
- Analyzing and interpreting

Crystalline Columns

In this investigation, you will prepare a supersaturated solution of sodium acetate. (A supersaturated solution contains more dissolved solute at a specific temperature than is normally possible.) Then you will use the solution to prepare your own ionic crystal.

Question

How can you build a crystal column on your laboratory bench?

Prediction

If the solution drips slowly enough, a tall crystal column will form.

Safety Precautions

Materials

water
sodium acetate trihydrate crystals
balance
10 mL graduated cylinder
100 mL Erlenmeyer flask
hot plate
squirt bottle
burette
burette stand
forceps

Procedure

1. Place 50 g of sodium acetate trihydrate in a clean 100 mL Erlenmeyer flask.
2. Add 5 mL of water. Heat the solution slowly.
3. Swirl the flask until the solid completely dissolves. If any crystals remain inside the flask or on the neck, wash them down with a small amount of water.
4. Remove the flask from the heat. Pour the solution into a clean, dry buret.
5. Raise the burette as high as it will go. Place it on the lab bench where you intend to grow your crystal column.
6. Pour some sodium acetate trihydrate crystals onto the lab bench. Using clean and dry forceps, choose a relatively large crystal (the seed crystal). Place it directly underneath the burette spout.
7. Turn the buret stopcock slightly so that the solution drips out slowly. Adjust the position of the seed crystal so that the drops fall on it. (You can drip the solution right onto the bench, or onto a glass plate if you prefer.)
8. Observe the crystal for 10 min. Record your observations about the crystal column or your apparatus. Continue to make observations every 10 min.

Analysis

1. Describe your observations.
2. Why does the column form upward?
3. What was the purpose of the seed crystal?
4. What improvements would you suggest for better results in the future?

Conclusions

5. What kind of reaction is taking place when a crystal forms? Is the reaction chemical or physical? Explain.

Application

6. Repeat steps 1 to 3 in the Procedure. Once you remove the solution from the heat, seal the flask with a clean, dry rubber stopper. Allow the flask to cool to room temperature. Next, remove the stopper and carefully add only one crystal of sodium acetate trihydrate to the flask. Record your observations, and explain what is happening.

Covalent Bonding

You have learned what happens when the electronegativity difference between two atoms is greater than 1.7. The atom with the lower electronegativity transfers its valence electron(s) to the atom with the higher electronegativity. The resulting ions have opposite charges. They are held together by a strong ionic bond.

What happens when the electronegativity difference is very small? What happens when the electronegativity difference is zero? As an example, consider chlorine. Chlorine is a yellowish, noxious gas. What is it like at the atomic level? Each chlorine atom has seven electrons in its outer energy level. In order for chlorine to achieve the electron configuration of a noble gas according to the octet rule, it needs to gain one electron. When two chlorine atoms bond together, their electronegativity difference is zero. The electrons are equally attracted to each atom.

Therefore, instead of transferring electrons, the two atoms each share one electron with each other. In other words, each atom contributes one electron to a covalent bond. *A covalent bond consists of a pair of shared electrons.* Thus, each chlorine atom achieves a filled outer electron energy level, satisfying the octet rule. Examine Figure 3.15 to see how to represent a covalent bond with a Lewis structure.

When two atoms of the same element form a bond, they share their electrons equally in a **pure covalent bond**. Elements that bond to each other in this way are known as **diatomic elements**.

When atoms such as carbon and hydrogen bond to each other, their electronegativities are so close that they share their electrons almost equally. Carbon and hydrogen have an electronegativity difference of only $2.6 - 2.2 = 0.4$. In Figure 3.16, you can see how one atom of carbon forms a covalent bond with four atoms of hydrogen. The compound methane, CH_4, is formed.

Each hydrogen atom shares one of its electrons with the carbon. The carbon shares one of its four valence electrons with each hydrogen. Thus, each hydrogen atom achieves a filled outer energy level, and so does carbon. (Recall that elements in the first period need only two electrons to fill their outer energy level.) *When analyzing Lewis structures that show covalent bonds, count the shared electrons as if they belong to each of the bonding atoms.* In the following Practice Problems, you will represent covalent bonding using Lewis structures.

:Cl· ·Cl:

:Cl:Cl:

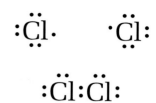

Figure 3.15 These Lewis structures show the formation of a bond between two atoms of chlorine.

CHEM FACT

Some examples of diatomic elements are chlorine, Cl_2, bromine, Br_2, iodine, I_2, nitrogen, N_2, and hydrogen, H_2.

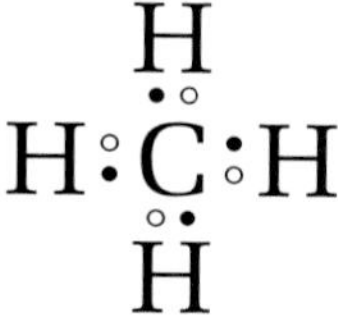

Figure 3.16 This Lewis structure shows a molecule of methane, CH_4.

Practice Problems

6. Show the formation of a covalent bond between two atoms of each diatomic element.
   (a) iodine
   (b) bromine
   (c) hydrogen
   (d) fluorine

7. Use Lewis structures to show the simplest way in which each pair of elements forms a covalent bond, according to the octet rule.
   (a) hydrogen and oxygen
   (b) chlorine and oxygen
   (c) carbon and hydrogen
   (d) iodine and hydrogen
   (e) nitrogen and hydrogen
   (f) hydrogen and rubidium

Electronic Learning Partner

Your Chemistry 11 Electronic Learning Partner has several animations that show ionic and covalent bonding.

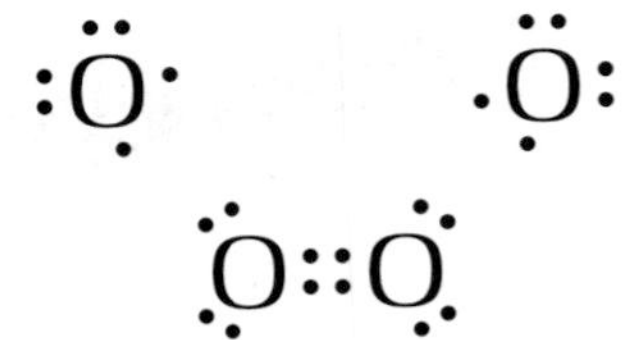

Figure 3.17 These Lewis structures show the formation of a double bond between two atoms of oxygen.

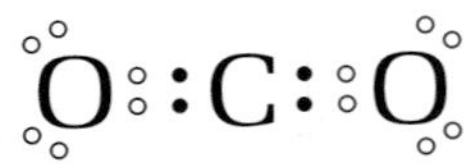

Multiple Covalent Bonds

Atoms sometimes transfer more than one electron in ionic bonding. Similarly, in covalent bonding, atoms sometimes need to share two or three pairs of electrons, according to the octet rule. For example, consider the familiar diatomic element oxygen. Each oxygen atom has six electrons in its outer energy level. Therefore, each atom requires two additional electrons to achieve a stable octet. When two oxygen atoms form a bond, they share two pairs of electrons, as shown in Figure 3.17. This kind of covalent bond is called a **double bond**.

Double bonds can form between different elements, as well. For example, consider what happens when carbon bonds to oxygen in carbon dioxide. To achieve a stable octet, carbon requires four electrons, and oxygen requires two electrons. Hence, two atoms of oxygen bond to one atom of carbon. Each oxygen forms a double bond with the carbon, as shown in Figure 3.18.

When atoms share three pairs of electrons, they form a **triple bond**. Diatomic nitrogen contains a triple bond, as you can see in Figure 3.19. Try the following problems to practise representing covalent bonding using Lewis structures. Watch for multiple bonding!

O::C::O

Figure 3.18 This Lewis structure shows the double bond in a molecule of carbon dioxide, CO_2.

Practice Problems

8. One carbon atom is bonded to two sulfur atoms. Use a Lewis structure to represent the bonds.
9. A molecule contains one hydrogen atom bonded to a carbon atom, which is bonded to a nitrogen atom. Use a Lewis structure to represent the bonds.
10. Two carbon atoms and two hydrogen atoms bond together, forming a molecule. Each atom achieves a full outer electron level. Use a Lewis structure to represent the bonds.

:N:::N:

Figure 3.19 This Lewis structure shows the triple bond in a molecule of nitrogen, N_2.

Explaining the Low Conductivity of Covalent Compounds

Covalent compounds have a wider variety of properties than ionic compounds. Some dissolve in water, and some do not. Some conduct electricity when molten or dissolved in water, and some do not. If you consider only covalent compounds that contain bonds with an electronegativity difference that is less than 0.5, you will notice greater consistency. For example, consider the compounds carbon disulfide, CS_2, dichlorine monoxide, Cl_2O, and carbon tetrachloride, CCl_4. What are some of the properties of these compounds? They all have low boiling points. None of them conducts electricity in the solid, liquid, or gaseous state.

How do we explain the low conductivity of these pure covalent compounds? The atoms in each compound are held together by strong covalent bonds. Whether the compound is in the liquid, solid, or gaseous state, these bonds do not break. Thus, covalent compounds (unlike ionic compounds) do not break up into ions when they melt or boil. Instead, their atoms remain bonded together as molecules. For this reason, covalent compounds are also called **molecular compounds**. The molecules that make up a pure covalent compound cannot carry a current, even if the compound is in its liquid state or in solution.

PROBLEM TIP

When drawing Lewis structures to show covalent bonding, you can use lines between atoms to show the bonding pairs of electrons. One line (–) signifies a single bond. Two lines (=) signify a double bond. Three lines (≡) signify a triple bond. Non-bonding pairs are shown as dots in the usual way.

Evidence for Intermolecular Forces

You have learned that pure covalent compounds are not held together by ionic bonds in lattice structures. They do form liquids and solids at low temperatures, however. Something must hold the molecules together when a covalent compound is in its liquid or solid state. The forces that bond the *atoms* to each other within a molecule are called **intramolecular forces**. Covalent bonds are intramolecular forces. In comparison, the forces that bond *molecules* to each other are called **intermolecular forces**.

You can see the difference between intermolecular forces and intramolecular forces in Figure 3.20. Because pure covalent compounds have low melting and boiling points, you know that the intermolecular forces must be very weak compared with the intramolecular forces. It does not take very much energy to break the bonds that hold the molecules to each other.

There are several different types of intermolecular forces. You will learn more about them in section 3.3, as well as in Chapters 8 and 11.

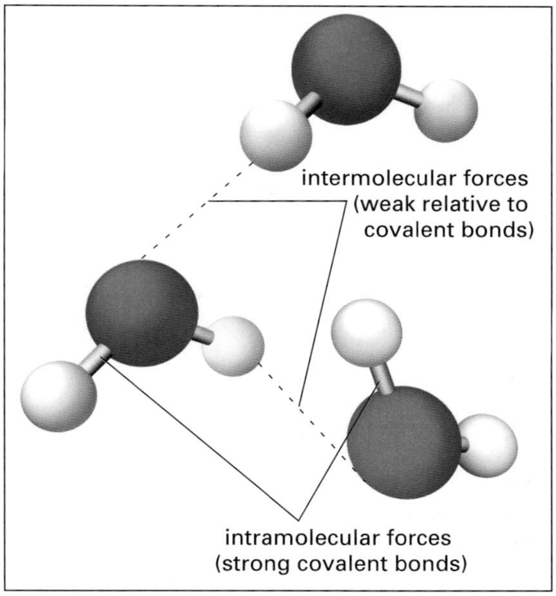

Figure 3.20 **Strong intramolecular forces (covalent bonds) hold the atoms in molecules together. Relatively weak intermolecular forces act between molecules.**

Metallic Bonding

In this chapter, you have seen that non-metals tend to form ionic bonds with metals. Non-metals tend to form covalent bonds with other non-metals and with themselves. How do metals bond to each other?

We know that elements that tend to form ionic bonds have very different electronegativities. Metals bonding to themselves or to other metals do not have electronegativity differences that are greater than 1.7. Therefore, metals probably do not form ionic bonds with each other.

Evidence bears this out. A pure metal, such as sodium, is soft enough to be cut with a butter knife. Other pure metals, such as copper or gold, can be drawn into wires or hammered into sheets. Ionic compounds, by contrast, are hard and brittle.

Do metals form covalent bonds with each other? No. They do not have enough valence electrons to achieve stable octets by sharing electrons. Although metals do not form covalent bonds, however, they do share their electrons.

In metallic bonding, atoms release their electrons to a shared pool of electrons. You can think of a metal as a non-rigid arrangement of metal ions in a sea of free electrons, as shown in Figure 3.21. The force that holds metal atoms together is called a **metallic bond**. Unlike ionic or covalent bonding, metallic bonding does not have a particular orientation in space. Because the electrons are free to move, the metal ions are not rigidly held in a lattice formation. Therefore, when a hammer pounds metal, the atoms can slide past one another. This explains why metals can be easily hammered into sheets.

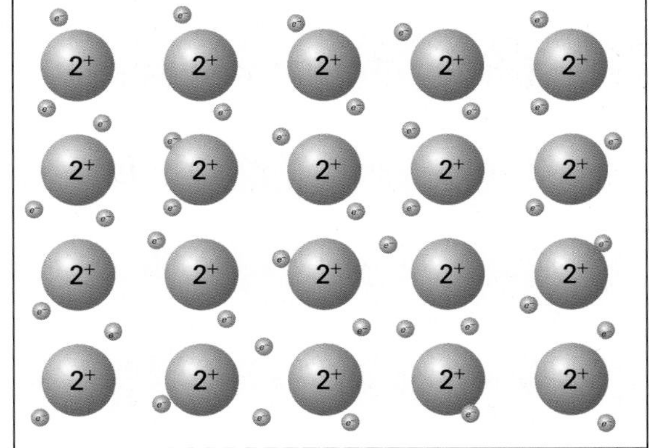

Figure 3.21 **In magnesium metal, the two valence electrons from each atom are free to move in an "electron sea." The valence electrons are shared by all the metal ions.**

Pure metals contain metallic bonds, as do alloys. An **alloy** is a homogeneous mixture of two or more metals. Different alloys can have different amounts of elements. Each alloy, however, has a uniform composition throughout. One example of an alloy is bronze. Bronze contains copper, tin, and lead, joined together with metallic bonds. You will learn more about alloys in Chapter 4 and Chapter 8.

Section Wrap-up

In this section, you learned how to distinguish between an ionic bond and a pure covalent bond. You learned how to represent these bonds using Lewis structures. You were also introduced to metallic bonding.

In section 3.3, you will learn about "in between" covalent bonds with ΔEN greater than 0.5 but less than 1.7. You will learn how the nature of these bonds influences the properties of the compounds that contain them. As well, you will examine molecules in greater depth. You will explore ways to visualize them in three dimensions, which will help you further understand the properties of covalent compounds.

Section Review

1 K/U Use Lewis structures to show how each pair of elements forms an ionic bond.

- (a) magnesium and fluorine
- (b) potassium and bromine
- (c) rubidium and chlorine
- (d) calcium and oxygen

2 K/U Use Lewis structures to show how the following elements form covalent bonds.

- (a) one silicon atom and two oxygen atoms
- (b) one carbon atom, one hydrogen atom, and three chlorine atoms
- (c) two nitrogen atoms
- (d) two carbon atoms bonded together with three hydrogen atoms bonded to one carbon, and one hydrogen atom and one oxygen atom bonded to the other carbon

3 K/U Use what you know about electronegativity differences to decide what kind of bond would form between each pair of elements.

- (a) palladium and oxygen
- (b) carbon and bromine
- (c) silver and sulfur
- (d) sodium and iodine
- (e) beryllium and fluorine
- (f) phosphorus and calcium

4 C "In general, the farther away two elements are from each other in the periodic table, the more likely they are to participate in ionic bonding." Do you agree with this statement? Explain why or why not.

5 C Covalent bonding and metallic bonding both involve electron sharing. Explain how covalent bonding is different from metallic bonding.

6 MC Ionic compounds are extremely hard. They hold their shape extremely well.

- (a) Based on what you know about ionic bonding within an ionic crystal, explain these properties.
- (b) Give two reasons to explain why, in spite of these properties, it is not practical to make tools out of ionic compounds.

Polar Covalent Bonds and Polar Molecules

3.3

In section 3.2, you learned what kind of bond forms when the electronegativity difference between two atoms is very small or very large. You now understand how electrons are shared or transferred in bonds. Thus, you can explain the properties of ionic compounds, and some of the properties of covalent compounds.

How can you explain the wide variety of properties that covalent compounds have? Covalent compounds may be solids, liquids, or gases at different temperatures. Some covalent compounds dissolve in water, and some do not. In fact, water itself is a covalent compound! Examine Figures 3.22 and 3.23. Why are the bonds in water different from the bonds in dinitrogen oxide? Both of these compounds are made up of two elements, and each molecule contains three atoms. The differences in the properties of these compounds are explained in part by the ΔEN of their bonds.

Section Preview/ Specific Expectations

In this section, you will

- **construct** molecular models
- **predict** the polarity of a given bond, using electronegativity values
- **predict** the overall polarity of molecules, using electronegativity values and molecular models
- **communicate** your understanding of the following terms: *polar covalent bond, lone pairs, bonding pairs, polar molecule, dipolar molecules, non-polar molecule*

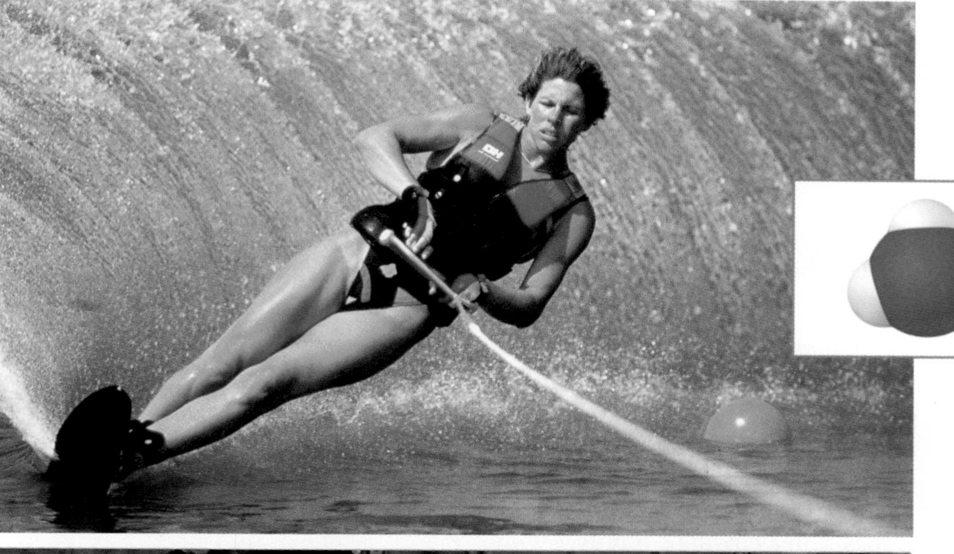

Figure 3.22 Water may be liquid, solid, or gas in nature. Why does the water that is sprayed up by this skier form a sheet?

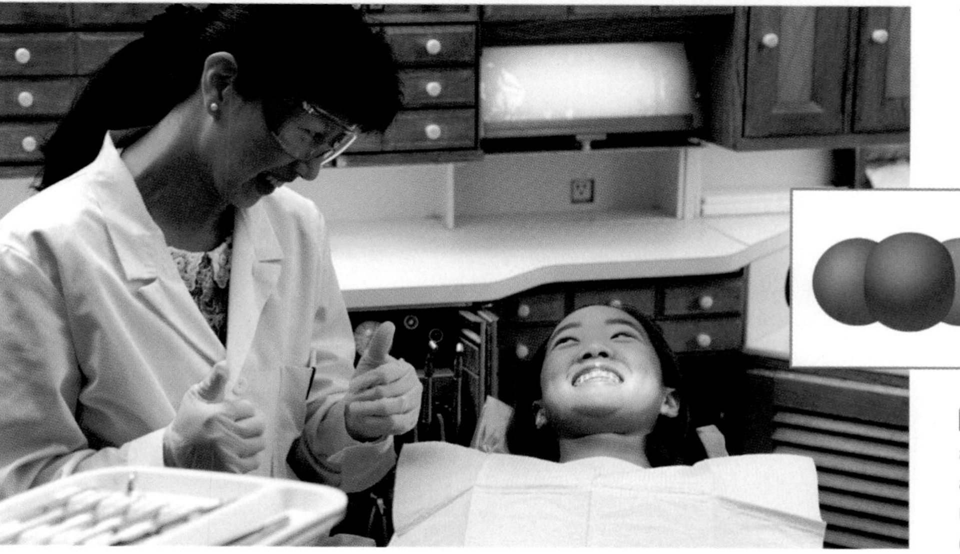

Figure 3.23 Dinitrogen oxide, also known as laughing gas, boils at about –89°C. laughing gas is used as an anaesthetic for dental work.

Polar Covalent Bonds: The "In-Between" Bonds

When two bonding atoms have an electronegativity difference that is greater than 0.5 but less than 1.7, they are considered to be a particular type of covalent bond called a **polar covalent bond**. In a polar covalent bond, the atoms have significantly different electronegativities. The electronegativity difference is not great enough, however, for the less electronegative atom to transfer its valence electrons to the other, more electronegative atom. The difference *is* great enough for the bonding electron pair to spend more time near the more electronegative atom than the less electronegative atom.

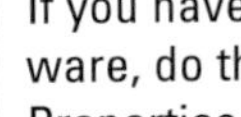

PROBEWARE

If you have access to probeware, do the Chemistry 11 lab, Properties of Bonds, now.

For example, the bond between oxygen and hydrogen in water has an electronegativity difference of 1.24. Because this value falls between 0.5 and 1.7, the bond is a polar covalent bond. The oxygen attracts the electrons more strongly than the hydrogen. Therefore, the oxygen has a slightly negative charge and the hydrogen has a slightly positive charge. Since the hydrogen does not completely transfer its electron to the oxygen, their respective charges are not +1 and –1, but rather δ^+ and δ^-. The symbol δ^+ (delta plus) stands for a partial positive charge. The symbol δ^- (delta minus) stands for a partial negative charge. Figure 3.24 illustrates the partial negative and positive charges across an oxygen-hydrogen bond. Figure 3.25 shows the polar covalent bond between hydrogen and chlorine.

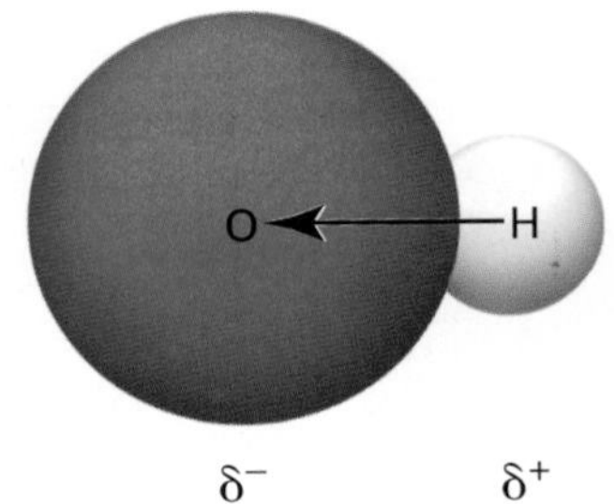

Figure 3.24 **The O end of an O—H bond has a partial negative charge. The H end has a partial positive charge.**

δ^+ δ^-

H:Cl:

Figure 3.25 **The Cl end of a H—Cl bond has a partial negative charge. The H end has a partial positive charge.**

Try the following problems to practise identifying the partial charges across polar covalent bonds.

Practice Problems

11. Predict whether each bond will be covalent, polar covalent, or ionic.

(a) C—F **(b)** O—N **(c)** Cl—Cl **(d)** Cu—O **(e)** Si—H **(f)** Na—F **(g)** Fe—O **(h)** Mn—O

12. For each polar covalent bond in problem 11, indicate the locations of the partial charges.

13. Arrange the bonds in each set in order of increasing polarity. (A completely polarized bond is an ionic bond.)

(a) H—Cl, O—O, N—O, Na—Cl

(b) C—Cl, Mg—Cl, P—O, N—N

Electronic Learning Partner

Your Chemistry 11 Electronic Learning Partner has a film clip that shows the formation of bonds in water, hydrogen gas, and sodium chloride.

Comparing Molecular Models

Throughout this chapter, you have seen several different types of diagrams representing molecules. These diagrams, or models, are useful for highlighting various aspects of molecules and bonding. Examine Figure 3.26 to see the various strengths of the different models.

A A *Lewis structure* shows you exactly how many electrons are involved in each bond in a compound. Some Lewis structures show bonding pairs as lines between atoms.

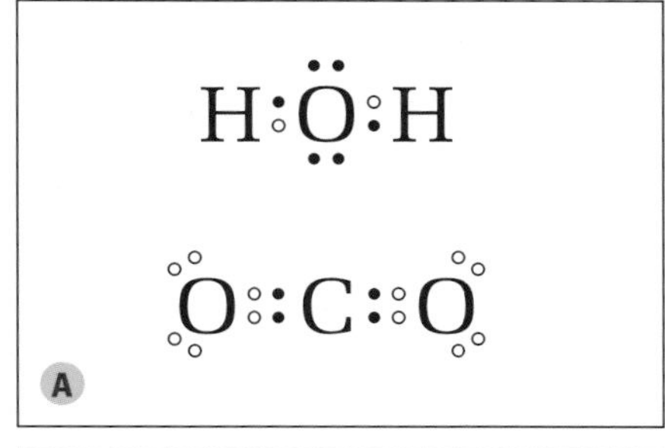

B A *structural diagram* shows single bonds as single lines and multiple bonds as multiple lines. It does not show non-bonding pairs. It is less cluttered than a Lewis structure. It clearly shows whether the bonds involved are single, double, or triple bonds.

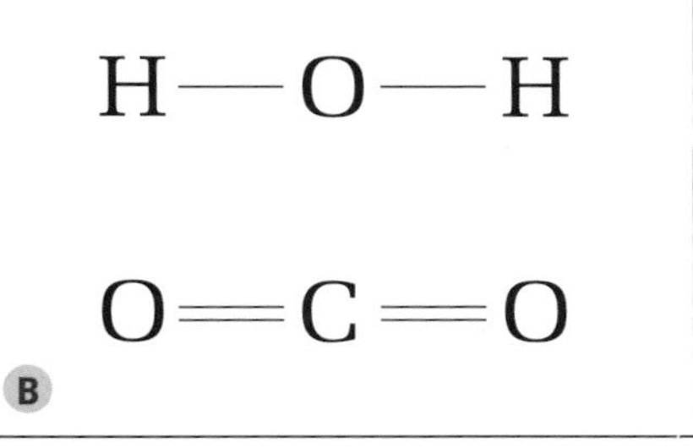

C A *ball-and-stick model* shows atoms as spheres and bonds as sticks. It accurately shows how the bonds within a molecule are oriented in three-dimensional space. The distances between the atoms are exaggerated, however. In this model, you can see the differences in the shapes of carbon dioxide and water.

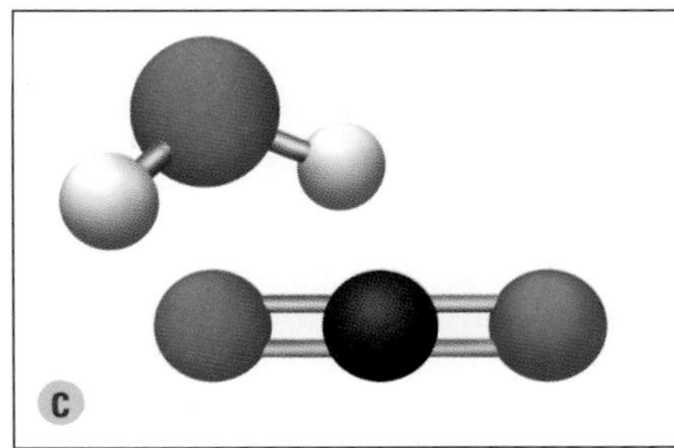

D A *space-filling model* shows atoms as spheres. It is the most accurate representation of the shape of a real molecule.

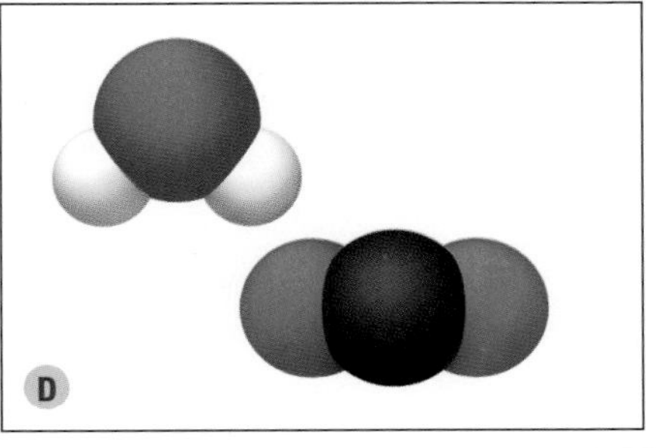

Figure 3.26 You can compare a molecule of water with a molecule of carbon dioxide using a variety of different models.

Electronic Learning Partner

Your Chemistry 11 Electronic Learning Partner has a video clip that explains how to draw Lewis structures.

Web LINK

There are ways of representing molecules in addition to the ones shown in Figure 3.26. Seach for some examples on the Internet. Go to **www.school.mcgrawhill.ca/resources/** for some ideas on where to start.

Consider a molecule of water and a molecule of carbon dioxide. Both water and carbon dioxide contain two atoms of the same element bonded to a third atom of another element. According to Figure 3.26, however, water and carbon dioxide molecules are different shapes. Why does carbon dioxide have a linear shape while water is bent?

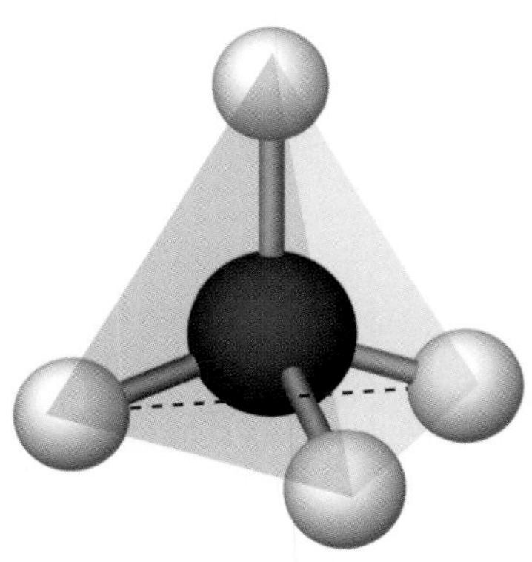

Figure 3.27 A tetrahedron has four equal sides.

To understand why molecules have different shapes, consider how electron arrangement affects shape. The Lewis structure for water, for example, shows that the oxygen is surrounded by four electron pairs. As shown in figure 3.26, two of the pairs are involved in bonding with the hydrogen atoms and two of the pairs are not. Electron pairs that are not involved in bonding are called **lone pairs**. Electron pairs that are involved in bonding are called **bonding pairs**.

Electron pairs are arranged around molecules so that they are a maximum distance from each other. This makes sense, because electrons are negatively charged and they repel each other. The shape that allows four electron pairs to be a maximum distance from each other around an atom is a tetrahedron. Figure 3.27 shows a tetrahedron.

The Shape of a Water Molecule

In a water molecule, there are four electron pairs around the oxygen atom. Two of these pairs bond with the hydrogen. The electron pairs are arranged in a shape that is nearly tetrahedral. When you draw the molecule, however, you draw only the oxygen atom and the two hydrogen atoms. This is where the bent shape comes from, as you can see in Figure 3.28.

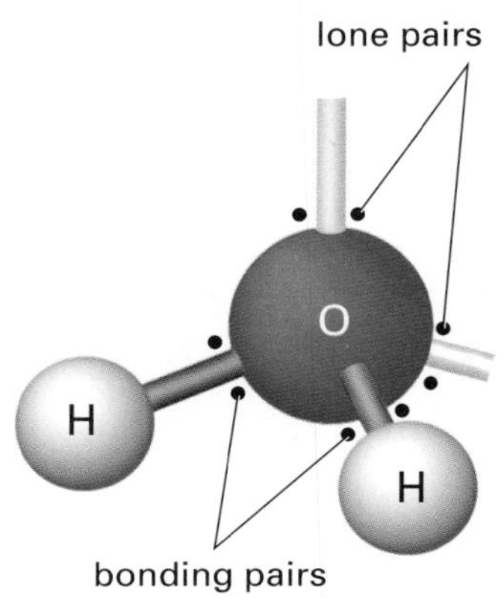

Figure 3.28 Two non-bending pairs account for water's bent shape.

The Shape of a Carbon Dioxide Molecule

Now consider carbon dioxide, CO_2. Why does a carbon dioxide molecule have a linear shape? Examine the Lewis structure for carbon dioxide. The central carbon atom is surrounded by eight electrons (four pairs), like the oxygen atom in a water molecule. In a carbon dioxide molecule, though, all the electrons are involved in bonding. There are no lone pairs. Because the bonding electrons spend most of their time between the carbon and oxygen atoms, they are arranged in a straight line. This allows them to be as far away from each other as possible, as you can see in Figure 3.29.

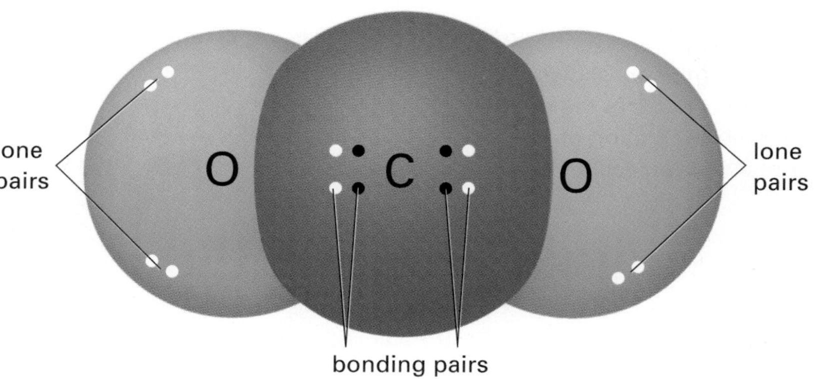

Figure 3.29 Carbon dioxide is linear in shape.

The shapes of the water molecule and the carbon dioxide molecule, as shown in the diagram you have seen, make sense based on what we know about electron pairs. These shapes have also been supported by experiment. You will learn more about experimental evidence for the structure of carbon dioxide and water later in this chapter.

Drawing the Lewis structure of a molecule can help you determine the molecule's shape. In Figure 3.30, you can see the shape of the ammonia, NH_3, molecule. The ammonia molecule has three bonding electron pairs and one lone pair on its central atom, all arranged in a nearly tetrahedral shape. Because there is one lone pair, the molecule's shape is pyramidal. The molecule methane, CH_4, is shown in Figure 3.31. This molecule has four bonding pairs on its central atom and no lone pairs. It is shaped like a perfectly symmetrical tetrahedron.

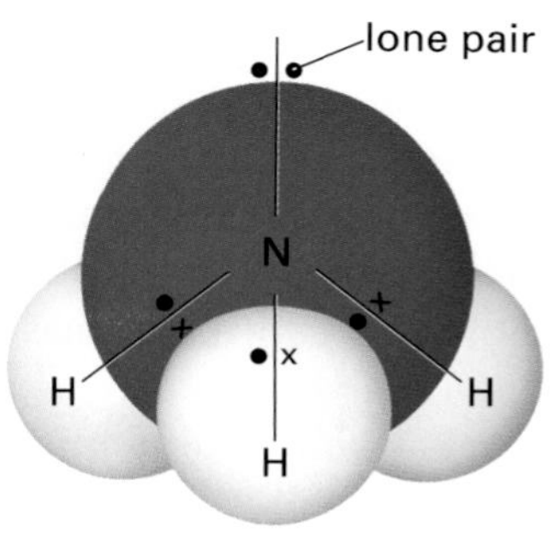

Figure 3.30 An ammonia molecule is shaped like a pyramid.

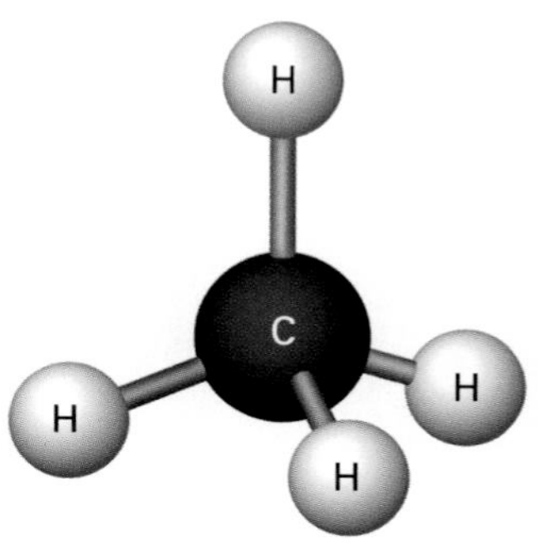

Figure 3.31 A methane molecule is shaped as though its hydrogen atoms were on the corners of a tetrahedron.

Canadians in Chemistry

Dr. Geoffrey Ozin

His work has flown on a Space Shuttle, and it has been hailed as art. It may well be part of the next computing revolution.

What does Dr. Geoffrey Ozin do? As little as possible, for he believes in letting the atoms do most of the work. This approach has made him one of the more celebrated chemists in Canada. Time and again, he has brought together organic and inorganic molecules, polymers, and metals in order to create materials with just the right structure for a specific purpose.

Self-assembly is the key. Atoms and molecules are driven into pre-designed shapes by intermolecular forces and geometrical constraints. At the University of Toronto, Dr. Ozin teaches his students the new science of intentional design, instead of the old trial-and-error methods.

Born in London, England, in 1943, Geoffrey Ozin earned a doctorate in chemistry at Oxford University. He joined the University of Toronto in 1969. Ozin's father was a tailor. In a way, Ozin is continuing the family tradition. Ozin, however, uses ionic and covalent bonds, atoms and molecules, acids, gases, and solutions to fashion his creations.

In 1996, Dr. Ozin demonstrated the self-assembly of crystals with a porous structure in space, under the conditions (such as microgravity) found aboard a Space Shuttle. Since then, he has shown how the self-assembly of many materials can be controlled to produce their structure.

Dr. Ozin's latest achievement involves structure. Ozin was part of an international research team that created regular microscopic cavities inside a piece of silicon. This material can transmit light photons in precisely regulated ways. In the future, this material might be used to build incredibly fast computers that function by means of photons instead of electrons!

Polar Bonds and Molecular Shapes

Water molecules are attracted to one another. Because we are surrounded by water, we are surrounded by evidence of this attraction. Re-examine the water skier in Figure 3.22. If water molecules did not attract one another, do you think the spray from the ski would form a "sheet" as shown? Try filling a glass with water. As you near the rim, add water very slowly. If you are careful, you can fill the glass so that the water bulges over the rim. After a rainfall, you have probably seen beads of water on the surface of vehicles. In Figure 3.32, you can see further evidence of the attraction of water molecules to one another.

Why do water molecules "stick together"? To answer this question, you need to consider both the nature of the bonds within a water molecule and its shape.

Figure 3.32 The shape of water droplets are evidence that water molecules are attracted to one another. This property of water can be explained by the polarity of its O—H bonds.

The Polar Water Molecule

First consider the shape of a water molecule. You have discovered that a water molecule has a bent shape. Each oxygen-hydrogen bond is polar. The hydrogen atom has a partial positive charge and the oxygen atom has a partial negative charge. You know that the bonds are polar, but what about the molecule as a whole? Because the molecule is bent, there is a partial negative charge on the oxygen end and a partial positive charge on the hydrogen end, as shown in Figure 3.33.

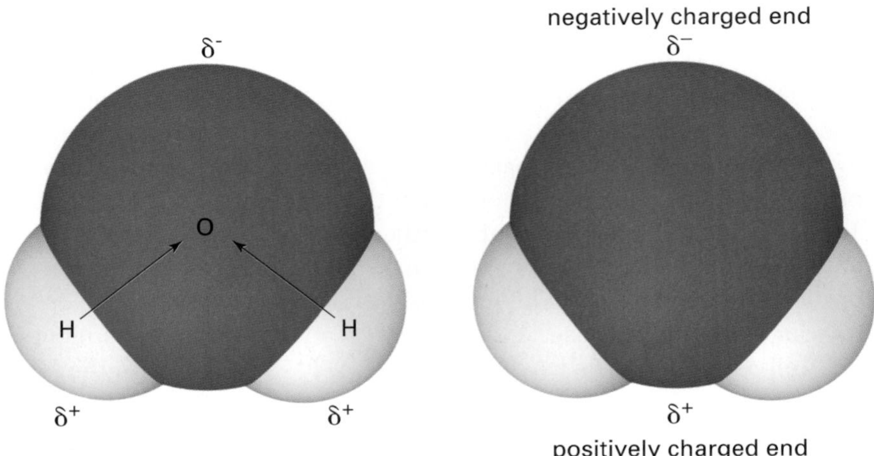

Figure 3.33 Water is a polar molecule because of its shape and the polarity of its bonds.

Because the water molecule *as a whole* has a partial negative charge on one end and a partial positive charge on the other end, it is called a **polar molecule**. Because water is polar, its negative and positive ends attract each other. This explains why liquid water "sticks" to itself. Figure 3.34 shows how water molecules attract each other in the liquid state.

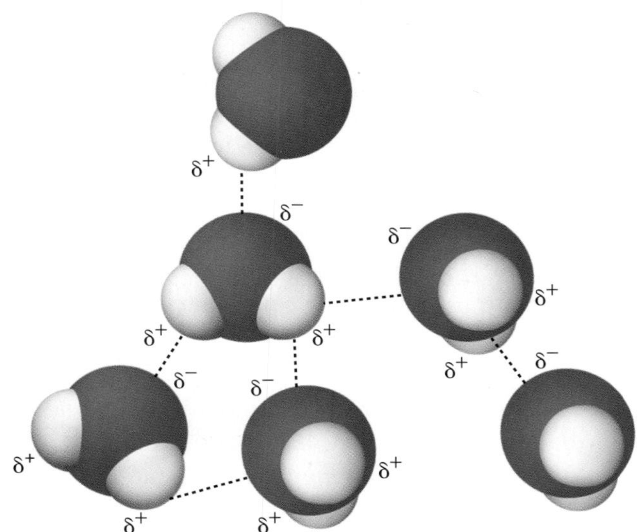

Figure 3.34 **The negative ends of water molecules attracts the positive ends. Some of the resulting intermolecular forces are shown here.**

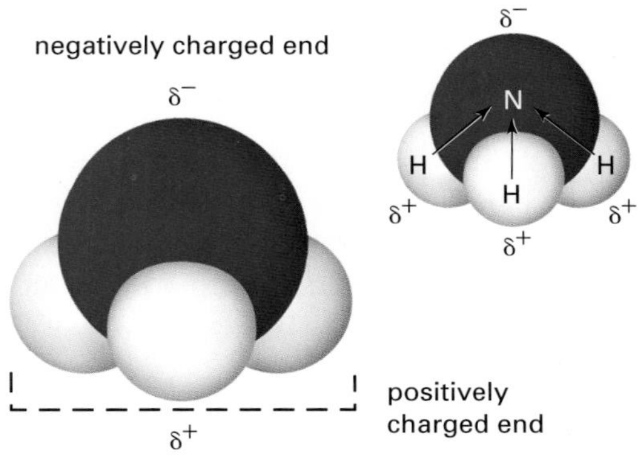

negatively charged end

δ−

δ+

positively charged end

N

H H H

Figure 3.35 **Because an ammonia molecule contains polar bonds and is asymmetrical, it is a polar molecule.**

Two other examples of polar molecules are ammonia and hydrogen chloride, shown in Figures 3.35 and 3.36. Polar molecules are also called **dipolar molecules** because they have a negative pole and a positive pole.

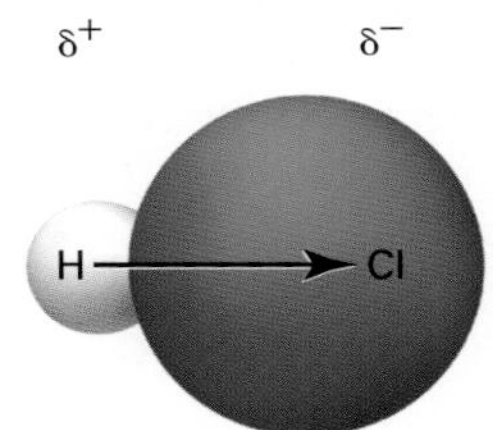

Figure 3.36 **Hydrogen chloride contains one polar bond. Therefore, the molecule is polar.**

The Non-Polar Carbon Dioxide Molecule

The bond between carbon and oxygen is polar. It has an electronegativity difference of 1.0. Does this mean that carbon dioxide, a molecule that contains two carbon-oxygen double bonds, is a polar molecule? No, it does not. The oxygen atoms have partial negative charges, and the carbon atom has a partial positive charge. The molecule, however, is straight and symmetrical. As you can see in Figure 3.37, the effects of the polar bonds cancel each other out. Therefore, while carbon dioxide contains polar bonds, it is a **non-polar molecule**. It has neither a positive pole nor a negative pole.

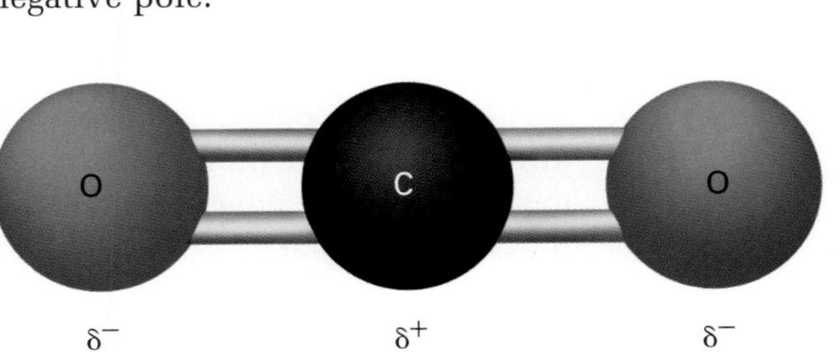

Figure 3.37 **Carbon dioxide is a non-polar molecule because it is symmetrical.**

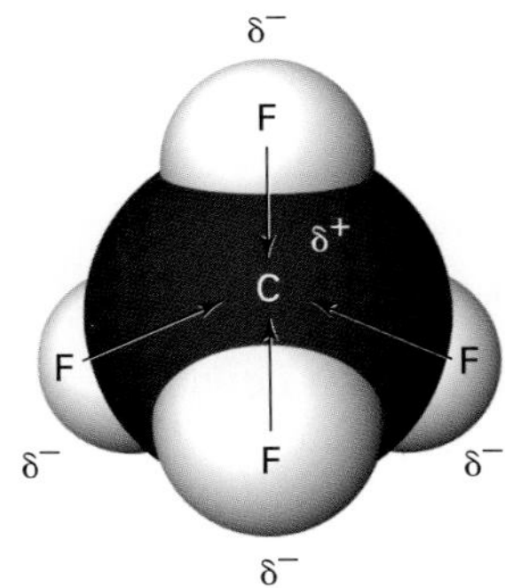

Figure 3.38 **Carbon tetrafluoride, CF_4, contains four polar bonds. Because of its symmetry, however, it is a non-polar molecule.**

Carbon tetrafluoride, CF_4, shown in Figure 3.38, is another example of a non-polar molecule that contains polar bonds.

Investigation 3-B

SKILL FOCUS

Modelling concepts

Analyzing and interpreting

Communicating results

Modelling Molecules

We cannot see molecules with our eyes or with a light microscope. We can predict their shapes, however, based on what we know about their electron configurations. In this investigation, you will practise working with a kit to build models of molecules.

Question

How can you build models of molecules to help you predict their shape and polarity?

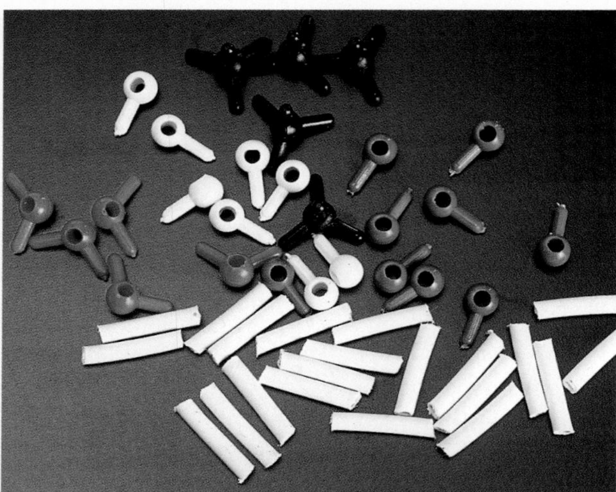

Materials

molecular model kit
pen
paper

Procedure

1. Obtain a model kit from your teacher.
2. Draw a Lewis structure for each molecule below.
   - **(a)** hydrogen bonded to a hydrogen: H_2
   - **(b)** chlorine bonded to a chlorine: Cl_2
   - **(c)** oxygen bonded to two hydrogens: H_2O
   - **(d)** carbon bonded to two oxygens: CO_2
   - **(e)** nitrogen bonded to three hydrogens: NH_3
   - **(f)** carbon bonded to four chlorines: CCl_4
   - **(g)** boron bonded to three fluorines: BF_3
3. Build a three-dimensional model of each molecule using your model kit.
4. Sketch the molecular models you have built.
5. In your notebook, make a table like the one below. Give it a title, fill in your data, and exchange your table with a classmate.

Compound	Lewis structure for compound	Sketch of predicted shape of molecule

Analysis

1. Compare your models with the models that your classmates built. Discuss any differences.
2. How did your Lewis structures help you predict the shape of each molecule?

Conclusion

3. Summarize the strengths and limitations of creating molecular models using molecular model kits.

Applications

4. Calculate the electronegativity difference for each bond in the molecules you built. Show partial charges. Based on the electronegativity difference and the predicted shape of each molecule, decide whether the molecule is polar or non-polar.
5. Look back through Chapter 3, and locate some different simple molecules. Build models of these molecules. Predict whether they are polar or non-polar.

Properties of Polar and Non-Polar Molecules

Because water is made up of polar molecules with positive and negative ends that attract one another, water tends to "stick" to itself. This means that it has a high melting point and boiling point, relative to other covalent compounds. For example, carbon dioxide is made up of non-polar molecules. These molecules do not attract each other as much as polar molecules do, because they do not have positive and negative poles. Compounds that are made up of non-polar molecules generally have lower melting points and boiling points than compounds that are made up of polar molecules. In fact, compounds with non-polar molecules, like carbon dioxide, are often gases at room temperature.

CHEM FACT

Differences in melting points and boiling points are due in part to the polarity of the molecules that make up the compounds. They are also due to the different masses of the individual molecules.

Section Wrap-up

In section 3.2, you learned about the strong bonds that hold ions in clearly-defined lattice patterns. You learned that these bonds are responsible for the properties of ionic compounds. You also learned how to describe the properties of compounds that are made up of molecules with covalent bonds. In this section, you discovered that the properties of compounds with polar covalent bonds depend on their shape. The following Concept Organizer summarizes some of the properties of covalent compounds that are made up of polar and non-polar molecules.

In sections 3.2 and 3.3, you learned how to represent compounds using Lewis structures and molecular models. In the next section, you will learn how chemists name compounds and represent them using symbols.

Concept Organizer Melting Point and Bonding Concepts

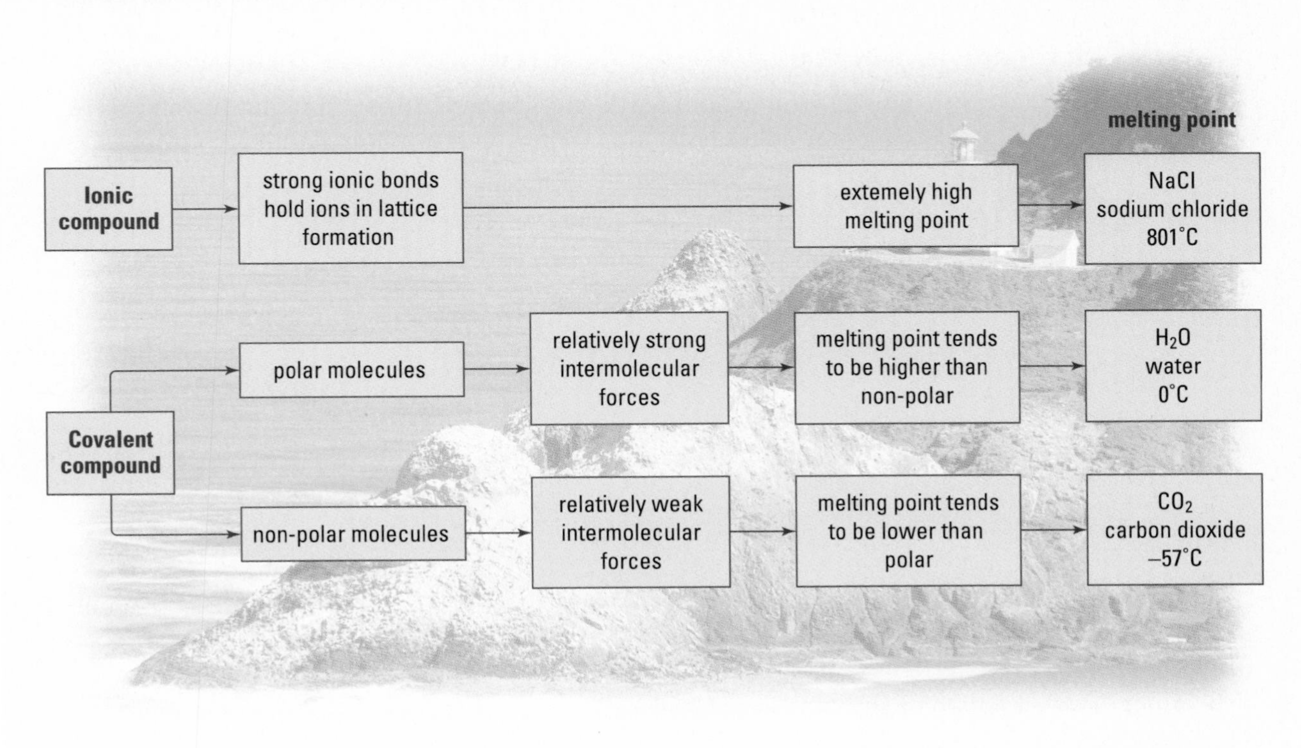

Section Review

Unit Project Prep

Before beginning your Unit Project, think about properties of compounds that would be useful in common chemical products. What kinds of properties would an abrasire or a window-cleaning fluid need to have? What kinds of compounds exhibit these properties?

1 K/U Determine ΔEN for each bond. Is the bond ionic, covalent, or polar covalent?

(a) B—F
(b) C—H
(c) Na—Cl
(d) Si—O
(e) S—O
(f) C—Cl

2 K/U For each polar covalent bond in question 1, label the partial negative and partial positive charges on each end.

3 C Explain how a non-polar molecule can contain polar bonds.

4 K/U Arrange each set of bonds in order of increasing polarity, using only their position in the periodic table.

(a) H—Cl, H—Br, O—F, K—Br
(b) C—O, C—F, C—H, C—Br

5 K/U Check your arrangements in question 4 by determining the ΔEN for each bond. Explain any discrepancies between your two sets of predictions.

6 I A molecule of chloroform, $CHCl_3$, has the same shape as a molecule of methane, CH_4. However, methane's boiling point is −164°C and chloroform's boiling point is 62°C. Explain the difference between the two boiling points.

7 K/U Determine the shape of each molecule by drawing a Lewis structure and considering the distribution of electron pairs around the atoms. Determine ΔEN for the bonds. Use the ΔEN and the shape to predict whether the molecule is polar or non-polar.

(a) $SiCl_4$
(b) PCl_3

8 MC How would Earth and life on Earth be different if water were a non-polar molecule? Write a paragarph explaining your ideas.

Writing Chemical Formulas and Naming Chemical Compounds

3.4

In this section, you will

- **write** the formulas of binary and tertiary compounds, including compounds that contain elements with multiple valences
- **communicate** formulas using IUPAC and traditional systems
- **recognize** the formulas of compounds in various contexts
- **communicate** your understanding of the following terms: *chemical formula, valence, polyatomic ions, zero sum rule, chemical nomenclature, binary compound, Stock system, tertiary compounds*

You have used Lewis structures to demonstrate how ionic and covalent bonds form between atoms. When given two elements, you determined how many atoms of each element bond together to form a compound, according to the octet rule. For example, you used the periodic table and your understanding of the octet rule to determine how calcium and bromine bond to form an ionic compound. Using a Lewis structure, you determined that calcium and bromine form a compound that contains two bromine atoms for every calcium atom, as shown in Figure 3.39.

Chemical Formulas

Lewis structures are helpful for keeping track of electron transfers in bonding and for making sure that the octet rule is obeyed. As well, Lewis structures can be used to help determine the ratio of the atoms in a compound. To communicate this ratio, chemists use a special kind of shorthand called a **chemical formula**. A chemical formula provides two important pieces of information:

1. the elements that make up the compound
2. the number of atoms of each element that are present in a compound

The order in which the elements are written also communicates important information. The less electronegative element or ion is usually listed first in the formula, and the more electronegative element or ion comes second. For example, the ionic compound that is formed from calcium and bromine is written $CaBr_2$. Calcium, a metal with low electronegativity, is written first. The subscript 2 after the bromine indicates that there are two bromine atoms for every calcium atom.

$$Ca\cdot\cdot \quad \cdot\ddot{\underset{..}{Br}}: \quad \cdot\ddot{\underset{..}{Br}}:$$

$$[Ca]^{2+}\ [:\ddot{\underset{..}{Br}}:]^- \ [:\ddot{\underset{..}{Br}}:]^-$$

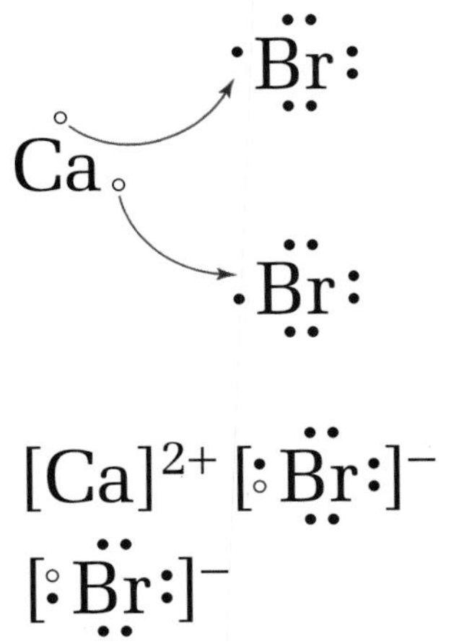

Figure 3.39 These Lewis structures show the formation of calcium bromide.

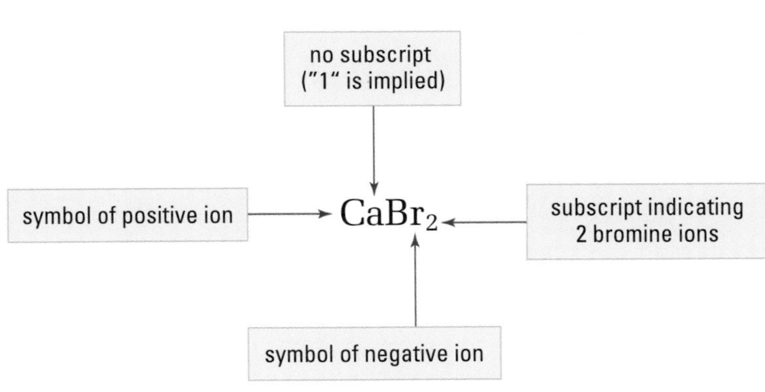

Figure 3.40 $CaBr_2$ is the chemical formula of the compound formed by calcium and bromine. When a subscript is omitted, only one atom is present per formula unit.

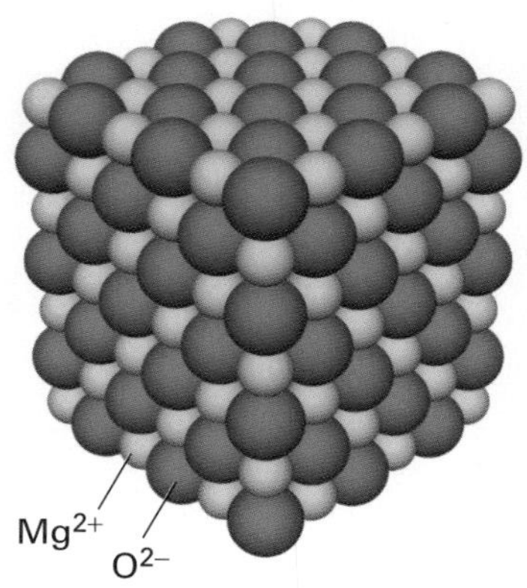

Figure 3.41 In a crystal of magnesium oxide, MgO, magnesium and oxygen atoms exist in a 1:1 ratio.

CHECKPOINT

What does the formula of calcium bromide represent?

What a Chemical Formula Represents

For covalent compounds, the chemical formula represents how many of each type of atom are in each molecule. For example, the formula NH_3 signifies that a molecule of ammonia contains one nitrogen atom and three hydrogen atoms. The formula C_2H_6 tells you that a molecule of propane contains two atoms of carbon and six atoms of hydrogen.

For ionic compounds, the formula represents *a ratio rather than a discrete particle*. For example, the formula for magnesium oxide, MgO, signifies that magnesium and oxygen exist in a one-to-one atomic ratio. Recall that MgO exists in a lattice structure held together by ionic bonds, as shown in Figure 3.41. The formula MgO represents the ratio in which ions are present in the compound.

Using Valence Numbers to Describe Bonding Capacity

You have seen how Lewis structures can help you draw models of ionic, covalent, and polar covalent compounds. When you draw a Lewis structure, you can count how many electrons are needed by each atom to achieve a stable octet. Thus, you can find out the ratio in which the atoms combine. Once you know the ratio of the atoms, you can write the chemical formula of the compound. Drawing Lewis structures can become overwhelming, however, when you are dealing with large molecules. Is there a faster and easier method for writing chemical formulas?

Every element has a certain capacity to combine with other atoms. An atom of a Group 1 (IA) element, for example, has the capacity to lose one electron from its valence level in order to bond with another atom. A number is assigned to each element to describe the element's bonding capacity. This number is called the **valence**. Thus, Group 1 (IA) elements, such as sodium and lithium, have a valence of +1. The 1 indicates that these elements tend to have one electron involved in bonding. This makes sense, because Group 1 elements have only one electron in their outer electron energy level. The + indicates that these elements tend to give up their electrons, becoming positively charged ions. They may transfer their electrons, or they may attract the electron relatively weakly in a polar covalent bond.

On the other hand, Group 17 (VIIA) elements (the halogens) have a valence of −1. Again, the 1 indicates that these elements tend to have one electron involved in bonding. However, they need to *gain* an electron to achieve a stable octet. In general, halogens become more negatively charged when they participate in bonding.

As a general rule, if two atoms form an ionic bond, the valence tells you the charges on the ions that are formed. If a covalent bond is formed, the valence tells you how many electrons the atoms contribute to the covalent bond.

You can use the periodic table to predict valence numbers. For example, Group 2 (IIA) elements have two electrons in their outer energy level. To achieve a stable octet, they need to lose these two electrons. Therefore, the valence for all Group 2 elements is +2.

Practice Problems

14. Use the periodic table to predict the most common valences of the atoms in Groups 16 (VIA) and 17 (VIIA).
15. If you had to assign a valence to the noble gases, what would it be? Explain your answer.

The smaller atoms of elements in the first two periods usually have only one common valence, which is easily determined from the periodic table. Many larger elements, however, have more than one valence because the electron distribution in these elements is much more complex. Therefore, you will have to memorize the valences of the elements that are commonly used in this course. Some useful valences are listed in Table 3.3, with the most common valences listed first.

Table 3.3 Common Valences of Selected Elements

1	2	3	4	5	6	7	8	9	10	11	12	13	14	15	16	17
H(+1)																
Li(+1)	Be(+2)													N(±3)	O(−2)	F(−1)
Na(+1)	Mg(+2)													P(+3)	S(−2)	Cl(−1)
K(+1)	Ca(+2)				Cr(+3) Cr(+2) Cr(+6)		Fe(+3) Fe(+2)	Co(+2) Co(+3)	Ni(+2) Ni(+3)	Cu(+2) Cu(+1)	Zn(+2)	Ga(+3)				Br(−1)
Rb(+1)	Sr(+2)									Ag(+1)	Cd(+2)		Sn(+4) Sn(+2)			I(−1)
Cs(+1)	Ba(+2)									Au(+3) Au(+1)	Hg(+2) Hg(+1)		Pb(+2) Pb(+4)			

Polyatomic Ions

Some compounds contain ions that are made from more than one atom. These ions are called **polyatomic ions**. (The prefix *poly* means "many.") Calcium carbonate, $CaCO_3$, which is found in chalk, contains one calcium cation and one polyatomic anion called carbonate, CO_3^{2-}.

Polyatomic ions are in fact *charged molecules*. For example, the carbonate ion consists of one carbon atom covalently bonded to three oxygen atoms. The entire molecule has a charge of −2. Therefore, its valence is −2, as well. Polyatomic ions remain unchanged in simple chemical reactions because of the strong bonds that hold the component atoms together. They behave as a single unit and should be treated as a single ion.

Table 3.4 on the next page gives the valences, formulas, and names of many common polyatomic ions.

Language LINK

It is important to learn the names and valences of the five most common polyatomic ions: nitrate, carbonate, chlorate, sulfate, and phosphate. These ions form many of the chemicals in nature and in common use. While the task seems overwhelming, it may help to learn the "big five" using a mnemonic, or memory aid. You can use the following mnemonic to remember their names, valences, and number of oxygen atoms:

NICK the CAMEL had a CLAM for SUPPER in PHOENIX.

The first letter identifies the polyatomic ion. The number of vowels represents the valence. The number of consonants represents the number of oxygen atoms. For example, NICK (nitrate) has three consonants and one vowel. Therefore, nitrate contains three oxygen atoms and has a valence of –1. (All of these valences are negative.)

Try to come up with your own mnemonic.

The most common polyatomic cation is the ammonium ion, [NH_4^+]. The five atoms in NH_4^+ form a particle with a +1 charge. Because the atoms are bonded together strongly, the polyatomic ion is not altered in most chemical reactions. For example, when ammonium chloride is dissolved in water, the only ions in the solution are ammonium ions and chloride ions.

Table 3.4 Names and Valences of Some Common Ions

Valence = –1			
Ion	**Name**	**Ion**	**Name**
CN^-	cyanide	$H_2PO_3^-$	dihydrogen phosphite
CH_3COO^-	acetate	$H_2PO_4^-$	dihydrogen phosphate
ClO^-	hypochlorite	MnO_4^-	permangante
ClO_2^-	chlorite	NO_2^-	nitrite
ClO_3^-	chlorate	NO_3^-	nitrate
ClO_4^-	perchlorate	OCN^-	cyanate
HCO_3^-	hydrogen carbonate	HS^-	hydrogen sulfide
HSO_3^-	hydrogen sulfite	OH^-	hydroxide
HSO_4^-	hydrogen sulfate	SCN^-	thiocyanate

Valence = –2			
Ion	**Name**	**Ion**	**Name**
CO_3^{2-}	carbonate	O_2^{2-}	peroxide
$C_2O_4^{2-}$	oxalate	SiO_3^{2-}	silicate
CrO_4^{2-}	chromate	SO_3^{2-}	sulfite
$Cr_2O_7^{2-}$	dichromate	SO_4^{2-}	sulfate
HPO_3^{2-}	hydrogen phosphite	$S_2O_3^{2-}$	thiosulfate
HPO_4^{2-}	hydrogen phosphate		

Valence = –3			
Ion	**Name**	**Ion**	**Name**
AsO_3^{3-}	arsenite	PO_3^{3-}	phosphite
AsO_4^{3-}	arsenate	PO_4^{3-}	phosphate

Writing Chemical Formulas Using Valences

You can use valences to write chemical formulas. This method is faster than using Lewis structures to determine chemical formulas. As well, you can use this method for both ionic and covalent compounds. In order to write a chemical formula using valences, you need to know which elements (or polyatomic ions) are in the compound, and their valences. You also need to know how to use the **zero sum rule**: *For neutral chemical formulas containing ions, the sum of positive valences plus negative valences of the atoms in a compound must equal zero.*

In the compound potassium fluoride, KF, each potassium ion has a charge of +1. Each fluoride ion has a charge of –1. Because there is one of each ion in the formula, the sum of the valences is zero.

What is the formula of a compound that consists of magnesium and chlorine? You know that the valence of magnesium, Mg, is +2. The valence of chlorine, Cl, is −1. The formula MgCl is not balanced, however, because it does not yet obey the zero sum rule. How can you balance this formula? You might be able to see, at a glance, that two chlorine atoms are needed for every magnesium atom. If it is not obvious how to balance a formula, you can follow these steps:

1. Write the unbalanced formula. Remember that the metal is first and the non-metal is second.

$$\text{Mg Cl}$$

2. Place the valence of each element on top of the appropriate symbol.

$$\overset{+2}{\text{Mg}} \quad \overset{-1}{\text{Cl}}$$

3. Using arrows, bring the numbers (without the signs) down to the subscript positions *by crossing over.*

$$\overset{+2}{\text{Mg}} \times \overset{-1}{\text{Cl}}$$

4. Check the subscripts. Any subscript of "1" can be removed.

$$MgCl_2$$

You can check your formula by drawing a Lewis structure, as shown in Figure 3.42.

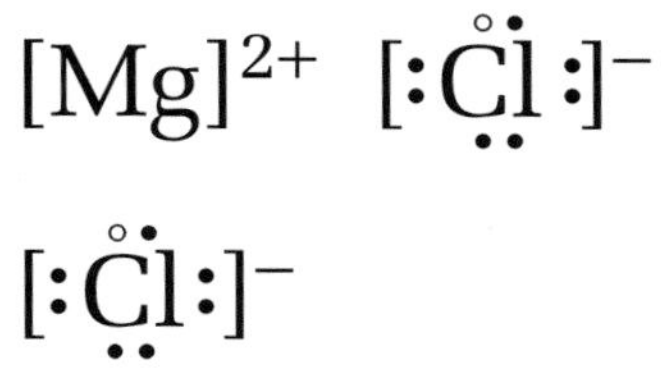

$[Mg]^{2+}$ $[:\ddot{Cl}:]^-$ $[:\ddot{Cl}:]^-$

Figure 3.42 This Lewis structure represents magnesium chloride, $MgCl_2$. Each atom has achieved a stable octet.

Practice Problems

16. Write a balanced formula for a compound that contains sulfur and each of the following elements. Use a valence of −2 for sulfur.
    - (a) sodium
    - (b) calcium
    - (c) barium
    - (d) aluminum
    - (e) rubidium
    - (f) hydrogen

17. Write a balanced formula for a compound that contains calcium and each of the following elements.
    - (a) oxygen
    - (b) sulfur
    - (c) chlorine
    - (d) bromine
    - (e) phosphorus
    - (f) fluorine

PROBLEM TIP

After the crossing over step, you may need to reduce the subscripts to their lowest terms. For example, Mg_2O_2 becomes MgO. Be_2O_2 becomes BeO. Remember, formulas for ionic compounds represent ratios of ions.

How do you write and balance formulas that contain polyatomic ions? The same steps can be used, as long as you keep the atoms that belong to a polyatomic ion together. The easiest way to do this is to place brackets around the polyatomic ion at the beginning.

For example, suppose that you want to write a balanced formula for a compound that contains potassium and the phosphate ion. Use the following steps as a guide.

1. Write the unbalanced formula. Place brackets around any polyatomic ions that are present.

$K \quad PO_4$

$K \quad (PO_4)$

2. Write the valence of each ion above it. (Refer to Table 3.4.)

$+1 \quad -3$

$K \quad (PO_4)$

3. Cross over, and write the subscripts.

$K_3(PO_4)_1$

4. Tidy up the formula. Remember that you omit the subscript if there is only one particle in the ionic compound or molecule. Here the brackets are no longer needed, so they can be removed.

K_3PO_4

Pay close attention to the brackets when you are writing formulas that contain polyatomic ions. For example, how would you write the formula for a compound that contains ammonium and phosphate ions?

PROBLEM TIP

In this formula, the brackets must remain around the ammonium ion to distinguish the subscripts. The subscript 4 refers to how many hydrogen atoms are in each ammonium ion. The subscript 3 refers to how many ammonium ions are needed to form an ionic compound with the phosphate ion.

1. Write the unbalanced formula. Place brackets around any polyatomic ions that are present.

$NH_4 \quad PO_4$

$(NH_4) \quad (PO_4)$

2. Write the valence above each ion.

$+1 \quad -3$

$(NH_4) \quad (PO_4)$

3. Cross over, and write in the subscripts.

$(NH_4)_3(PO_4)_1$

4. Tidy up the formula. Remove the brackets only when the polyatomic ion has a subscript of "1".

$(NH_4)_3PO_4$

Try the following problems to practise writing formulas for compounds containing polyatomic ions.

Practice Problems

18. Use the information in Table 3.4 to write a chemical formula for a compound that contains sodium and each of the following polyatomic ions.

(a) nitrate **(c)** sulfite **(e)** thiosulfate

(b) phosphate **(d)** acetate **(f)** carbonate

19. Repeat question 19 using magnesium instead of sodium.

Naming Chemical Compounds

When writing a chemical formula, you learned that you write the metal element first. Similarly, the metal comes first when naming a chemical compound. For example, sodium chloride is formed from the metal sodium and the non-metal chlorine. Think of other names you have seen in this chapter, such as beryllium chloride, calcium oxide, and aluminum oxide. In each case, the metal is first and the non-metal is second. In other words, the cation is first and the anion is second. This is just one of the rules in **chemical nomenclature**: the system that is used in chemistry for naming compounds.

A chemical formula identifies a specific chemical compound because it reveals the composition of the compound. Similarly, the name of a compound distinguishes the compound from all other compounds.

Figure 3.43 **Many common chemicals have trivial, or common, names.**

In the early days of chemistry, there were no rules for naming compounds. Often, compounds received the names of people or places. Some of the original names are still used today. They are called *trivial* or *common* names because they tell little or nothing about the chemistry of the compounds. For example, potassium nitrate, KNO_3, is commonly known as saltpetre. The Greek word for rock is petra, and saltpetre is a salt found crusted on rocks. The chemical name of a compound of ammonia and chlorine, NH_4Cl, is ammonium chloride. Long before NH_4Cl received this name, however, people commonly referred to it as sal ammoniac. They mined this ionic compound near the ancient Egyptian temple of Ammon in Libya. The name sal ammoniac literally means "salt of Ammon." Figure 3.43 shows other examples of common (trivial) names for familiar compounds.

Early chemists routinely gave trivial names to substances before understanding their chemical structure and behaviour. This situation changed during the mid- to late-1800s. By this time, chemistry was firmly established as a science. Chemists observed and discovered new patterns of chemical relationships (such as periodicity). As well, chemists discovered new chemical compounds with tremendous frequency. The rapidly increasing number of chemical compounds required a more organized method of nomenclature.

The International Union of Pure and Applied Chemistry (IUPAC) was formed in 1919 by a group of chemists. The main aim of IUPAC was to establish international standards for masses, measurement, names, and symbols used in the discipline of chemistry. To further that aim, IUPAC developed, and continues to develop, a consistent and thorough system of nomenclature for compounds.

Table 3.5 contains the IUPAC names of selected common compounds as well as their common names.

Table 3.5 Common Chemical Compounds

IUPAC name	Chemical formula	Common name	Use or propety
aluminum oxide	Al_2O_3	alumina	abrasive
calcium carbonate	$CaCO_3$	limestone, marble	building, sculpting
calcium oxide	CaO	lime	neutralizing acidified lakes
hydrochloric acid	HCl	muriatic acid	cleaning metal
magnesium hydroxide	$Mg(OH)_2$	milk of magnesia	antacid
dinitrogen monoxide	N_2O	laughing gas	used in dentistry as an anaesthetic
silicon dioxide	SiO_2	quartz sand	manufacturing glass
sodium carbonate	Na_2CO_3	washing soda	general cleaner
sodium chloride	$NaCl$	table salt	enhancing flavour
sodium hydrogen carbonate	$NaHCO_3$	baking soda	making baked goods rise
sodium hydroxide	$NaOH$	lye	neutralizing acids
sodium thiosulfate	NaS_2O_3	hypo	fixer in photography

Naming Binary Compounds Containing a Metal and a Non-metal

A **binary compound** is an inorganic compound that contains two elements. Binary compounds may contain a metal and a non-metal or two non-metals. Binary compounds are often ionic compounds. To name a binary ionic compound, name the cation first and the anion second. For example, the compound that contains sodium and chlorine is called sodium chloride.

In the subsections that follow, you will examine the rules for naming metals and non-metals in binary compounds.

Naming Metals in Chemical Compounds: The Stock System

The less electronegative element in a binary compound is always named first. Often this element is a metal. You use the same name as the element. For example, *sodium* chloride, NaCl, *calcium* oxide, CaO, and *zinc* sulfide, ZnS, contain the metals sodium, calcium, and zinc.

Many of the common metals are transition elements that have more than one possible valence. For example, tin is able to form the ions Sn^{2+} and Sn^{4+}, iron can form Fe^{2+} and Fe^{3+}, and copper can form Cu^{+} and Cu^{2+}. (The most common transition metals with more than one valence number are listed in Table 3.3.) The name of a compound must identify which ion is present in the compound. To do this, the element's name is used, followed by the valence in parentheses, written in Roman numerals. Therefore, Sn^{4+} is tin(IV), Fe^{3+} is iron(III), and Cu^{2+} is copper(II). This naming method is called the **Stock system** after Alfred Stock, a German chemist who first used it. Some examples of Stock system names are listed in Table 3.6.

Another Method for Naming Metals with Two Valences

In a method that predates the Stock system, two different endings are used to distinguish the valences of metals. The ending *-ic* is used to represent the *larger* valence number. The ending *-ous* is used to represent the *smaller* valence number. Thus, the ions Sn^{2+} and Sn^{4+} are named stann*ous* ion and stann*ic* ion. To use this system, you need to know the Latin name of an element. For example, the two ions of lead are the plumbous and plumbic ions. See Table 3.6 for more examples.

This naming method has several drawbacks. Many metals have more than two oxidation numbers. For example, chromium can form three different ions, and manganese can form five different ions. Another drawback is that the name does not tell you what the valence of the metal is. It only tells you that the valence is the smaller or larger of two.

Table 3.6 Two ways to Name Cations with Two Valences

element	ion	Stock system	Alternative system
copper	Cu^+	copper(I)	cuprous
	Cu^{2+}	copper(II)	cupric
mercury	Hg^+	mercury(I)	mercurous
	Hg^{2+}	mercury(II)	mercuric
lead	Pb^{2+}	lead(II)	plumbous
	Pb^{4+}	lead(IV)	plumbic

mind STRETCH

Some types of ionic compounds can absorb water so that each formula unit is attached to a specific number of water molecules. They are called *hydrates*. $BaOH_2 \bullet 8H_2O$ is called barium hydroxide octahydrate. $CaSO_4 \bullet 2H_2O$ is called calcium sulfate dihydrate. Can you see the pattern? Try naming $MgSO_4 \bullet 7H_2O$. Use Table 3.8 to help you. You will learn more about hydrates in Chapter 6.

Naming Non-Metals in Chemical Compounds

To distinguish the non-metal from the metal in the name of a chemical compound, the non-metal (or more electronegative element) is always written second. Its ending changes to *-ide*. For example, hydrogen changes to *hydride*, carbon changes to *carbide*, sulfur changes to *sulfide*, and iodine changes to *iodide*.

Putting It All Together

To name a binary compound containing metal and a non-metal, write the name of the metal first and the name of the non-metal second. For example, a compound that contains potassium as the cation and bromine as the anion is called *potassium bromide*. Be sure to indicate the valence if necessary, using the Stock system. For example, a compound that contains Pb^{2+} and oxygen is called *lead(II) oxide*.

Practice Problems

20. Write the IUPAC name for each compound.

(a) Al_2O_3
(b) $CaBr_2$
(c) Na_3P
(d) Cu_2S
(e) MgN_3
(f) HgI_2

21. Write the formula of each compound.

(a) iron(II) sulfide
(b) stannous oxide
(c) chromium(II) oxide
(d) cobaltous chloride
(e) manganese(II) iodide
(f) zinc oxide

Naming Compounds That Contain Hydrogen

If a binary compound contains hydrogen as the less electronegative element, "hydrogen" is used first in the name of the compound. For example, HCl is called *hydrogen chloride* and H_2S is called *hydrogen sulfide*. Sometimes hydrogen can be the anion, usually in a compound that contains a Group 1 metal. If hydrogen is the anion, its ending must be changed to *-ide*. For example, NaH is called *sodium hydride* and LiH is called *lithium hydride*. Hydrogen-containing compounds can also be formed with the Group 15 elements. These compounds are usually referred to by their common names as opposed to their IUPAC names. For example, NH_3 is called *ammonia*, PH_3 is called *phosphine*, AsH_3 is called *arsine*, and SbH_3 is called *stibine*.

Many compounds that contain hydrogen are also acids. For example, H_2SO_4, hydrogen sulfate, is also called sulfuric acid. You will learn about acid nomenclature in Chapter 10.

Practice Problems

22. Write the IUPAC name for each compound.

(a) H_2Se

(b) HCl

(c) HF

(d) LiH

(e) CaH_2

(f) PH_3

Naming Compounds That Contain Polyatomic Ions

Many compounds contain one or more polyatomic ions. Often these compounds contain three elements, in which case they are called **tertiary compounds**. Although they are not binary compounds, they still contain one type of anion and one type of cation. The same naming rules that apply to binary compounds apply to these compounds as well. For example, NH_4Cl is called *ammonium chloride*. Na_2SO_4 is called *sodium sulfate*. $NiSO_4$ is called *nickel(II) sulfate*. NH_4NO_3 is called *ammonium nitrate*.

The non-metals in the periodic table are greatly outnumbered by the metals. There are many negatively charged polyatomic ions, however, to make up for this. In fact, polyatomic anions are commonly found in everyday chemicals. Refer back to Table 3.4 for the names of the most common polyatomic anions.

When you are learning the names of polyatomic ions, you will notice a pattern. For example, consider the polyatomic ions that contain chlorine and oxygen:

ClO^-	hypochlorite
ClO_2^-	chlorite
ClO_3^-	chlorate
ClO_4^-	perchlorate

Can you see the pattern? Each ion has the same valence, but different numbers of oxygen atoms. The base ion is the one with the "ate" ending chlorate. It contains three oxygen atoms. When the ending is changed to "ite," subtract an oxygen atom from the chlorate ion. The resulting chlorite ion contains two oxygen atoms. Add "hypo" to "chlorite," and subtract one more oxygen atom. The resulting hypochlorite ion has one oxygen atom. Adding "per" to "chlorate," means that you should add an oxygen to the chlorate ion. The perchlorate ion has four oxygen atoms.

The base "ate" ions do not always have three oxygen atoms like chlorate does. Consider the polyatomic ions that contain sulfur and oxygen. In this case, the base ion, sulfate, SO_4^{2-}, contains four oxygen atoms. The sulfite ion, SO_3^{2-}, therefore contains three oxygen atoms. The hyposulfite ion, SO_2^{2-}, contains two oxygen atoms. Once you know the meanings of the prefixes and suffixes, you need only memorize the formulas of the "ate" ions. You can work out the formulas for the related ions using their prefixes and suffixes. The meanings of the prefixes and suffixes are summarized in Table 3.7. In this table, the "x" stands for the number of oxygen atoms in the "ate" ion.

Table 3.7 Meaning of prefixes and suffixes

Prefix and suffix		Number of oxygen atoms
hypo	ite	$x - 2$ oxygen atoms
	ite	$x - 1$ oxygen atoms
	ate	x oxygen atoms
per	ate	$x + 1$ oxygen atoms

Language LINK

The prefix "thio" in the name of a polyatomic ion means that an oxygen atom in the base "ate" ion has been replaced by a sulfur atom. For example, the sulfate ion is SO_4^{2-}, while the thiosulfate ion is $S_2O_3^{2-}$. Notice that the valence does not change.

Practice Problems

23. Write the IUPAC name for each compound.

(a) $(NH_4)_2SO_3$

(b) $Al(NO_2)_3$

(c) Li_2CO_3

(d) $Ni(OH)_2$

(e) Ag_3PO_4

(f) $Cu(CH_3COO)_2$

Naming Binary Compounds That Contain Two Non-Metals

To indicate that a binary compound is made up of two non-metals, a prefix is usually added to both non-metals in the compound. This prefix indicates the *number of atoms of each element* in one molecule or formula unit of the compound. For example, P_2O_5 is named *diphosphorus pentoxide*. Alternatively, the Stock System may be used, and P_2O_5 can be named *phosphorus (V) oxide*. $AsBr_3$ is named phosphorus tribromide. The prefix *mono-* is often left out when there is only one atom of the first element in the name. A list of numerical prefixes is found in Table 3.8.

Table 3.8 Numerical Prefixes for Binary Compounds That Contain Two Non-Metals

Number	Prefix
1	mono-
2	di-
3	tri-
4	tetra-
5	penta-
6	hexa-
7	hepta-
8	octa-
9	nona-
10	deca-

Practice Problems

24. Write the IUPAC name for each compound.

(a) SF_6

(b) N_2O_5

(c) PCl_5

(d) CF_4

14. Without calculating ΔEN, arrange each set of bonds from most polar to least polar. Then calculate ΔEN for each bond to check your arrangement.
   (a) Mn—O, Mn—N, Mn—F
   (b) Be—F, Be—Cl, Be—Br
   (c) Ti—Cl, Fe—Cl, Cu—Cl, Ag—Cl, Hg—Cl
15. What kind of diagram would you use (Lewis structure, structural diagram, ball-and-stick model, or space-filling model) to illustrate each idea?
   (a) When sodium and chlorine form an ionic bond, sodium loses an electron and chlorine gains an electron.
   (b) Water, H_2O, is a molecule that has a bent shape.
   (c) Each oxygen atom in carbon dioxide, CO_2, has two bonding pairs and two non-bonding pairs.
   (d) Silicon tetrachloride, $SiCl_4$, is a tetrahedral molecule.
16. Write the valences for the elements in each compound. If the compound is ionic, indicate the charge that is associated with each valence.
   (a) AgCl
   (b) Mn_3P_2
   (c) PCl_5
   (d) CH_4
   (e) TiO_2
   (f) HgF_2
   (g) CaO
   (h) FeS
17. Write the formula of each compound.
   (a) tin(II) fluoride
   (b) barium sulfate
   (c) hydrogen cyanide
   (d) cesium bromide
   (e) ammonium hydrogen phosphate
   (f) sodium periodate
   (g) potassium bromate
   (h) sodium cyanate
18. Write the name of each compound.
   (a) HIO_2
   (b) $KClO_4$
   (c) CsF
   (d) N_2Cl_2
   (e) $NaHSO_4$
   (f) $Al_2(SO_3)_3$
   (g) $K_2Cr_2O_7$
   (h) $Fe(IO_4)_3$
19. Name each compound in two different ways.
   (a) FeO
   (b) $SnCl_4$
   (c) $CuCl_2$
   (d) $CrBr_3$
   (e) PbO_2
   (f) HgO

Inquiry

20. Suppose that you have two colourless compounds. You know that one is an ionic compound and the other is a covalent compound. Design an experiment to determine which compound is which. Describe the tests you would perform and the results you would expect.
21. You have two liquids, A and B. You know that one liquid contains polar molecules, and the other liquid contains non-polar molecules. You do not know which is which, however. You pour each liquid so that it falls in a steady, narrow stream. As you pour, you hold a negatively charged ebonite rod to the stream. The stream of liquid A is deflected toward the rod. The rod does not affect the stream of liquid B. Which liquid is polar? Explain your answer.

Communication

22. Explain how you would predict the most common valences of the elements of the second period (Li, Be, B, C, N, O, F, and Ne) if you did not have access to the periodic table. Use Lewis structures to illustrate your explanation.
23. Create a concept map to summarize what you learned in this chapter about the nature of bonding and the ways in which bonding models help to explain physical and chemical properties.
24. Compare and contrast ionic bonding and metallic bonding. Include the following ideas:
   - Metals do not bond to other metals in definite ratios. Metals do bond to non-metals in definite ratios.
   - Solid ionic compounds do not conduct electricity, but solid metals do.
25. Explain why it was important for chemists worldwide to decide on a system for naming compounds.

Making Connections

26. Chemists do not always agree on names, not just for compounds but even for elements. As new elements are synthesized in laboratories, they must be named. Until 1997, there was a controversy over the names of elements 104 to

109 (called the *transfermium* elements). The periodic table at the back of this textbook gives the names that have now been accepted. Do some research to find out what other names were proposed for those elements. Find out what justification was given for the alternative names and the accepted names. Then write an essay in which you evaluate the choice that was made. Do you agree or disagree? Justify your opinion.

Answers to Practice Problems and Short Answers to Section Review Questions:

Practice Problems: 1.(a) 1.24, covalent **(b)** 0.50, covalent **(c)** 1.85, ionic **(d)** 1.94, ionic **(e)** 1.78, ionic **(f)** 0.49, covalent **(g)** 1.73, ionic **(h)** 2.03, ionic **2.(a)** 2.44, ionic **(b)** 2.34, ionic **(c)** 3.16, ionic **(d)** 3.00, ionic **(e)** 1.98, ionic **(f)** 2.55, ionic **3.(a)** one calcium atom gives up two electrons to one oxygen atom **(b)** one potassium atom gives up one electron to one chlorine atom **(c)** one potassium atom gives up one electron to one fluorine atom **(d)** one lithium atom gives up one electron to one fluorine atom **(e)** one lithium atom gives up one electron to one bromine atom **(f)** one barium atom gives up two electrons to one oxygen atom **4.(a)** 1.85 **(b)** 2.16 **(c)** 2.46 **(d)** 2.51 **(e)** 1.76 **(f)** 1.96 **5.(a)** one magnesium atom gives up one electron to each of two chlorine atoms **(b)** one calcium atom gives up one electron to each of two chlorine atoms **(c)** two lithium atoms each give up one electron to one oxygen atom **(d)** two sodium atoms each give up one electron to one oxygen atom **(e)** two potassium atoms each give up one electron to one sulfur atom **(f)** one calcium atom gives up one electron to each of two bromine atoms **6.(a)** Each iodine atom has seven electrons. Two iodine atoms bonded together share one pair of electrons so each has access to eight electrons. **(b)** Each bromine atom has seven electrons. Two bromine atoms bonded together share one pair of electrons so that each has access to eight electrons. **(c)** Each hydrogen atom has one electron. Two hydrogen atoms bonded together share one pair of electrons so that each has access to two electrons. **(d)** Each fluorine atom has seven electrons. Two fluorine atoms bonded together share one pair of electrons so that each has access to eight electrons. **7.(a)** One hydrogen atom bonds to one oxygen atom, sharing one electron pair. **(b)** Two chlorine atoms bond to one oxygen atom. Each chlorine atom shares one pair of electrons with the oxygen atom. **(c)** One carbon atom bonds to four hydrogen atoms. Each hydrogen atom shares one pair of electrons with the carbon atom. **(d)** One iodine atom bonds to one hydrogen atom. They share an electron pair. **(e)** One nitrogen atom bonds to three hydrogen atoms. Each hydrogen atom shares a pair of electrons with the nitrogen atom. **(f)** One hydrogen atom bonds to one rubidium atom. They share a pair of electrons. **8.** The carbon atom shares two pairs of electrons with each sulfur atom, so it has two double bonds. **9.** The carbon atom shares one pair of electrons with hydrogen, and three pairs of electrons with the nitrogen atom. **10.** The two carbon atoms share three pairs of electrons in a triple bond. Each carbon atom shares one pair of electrons with one hydrogen atom. **11.(a)** polar covalent **(b)** covalent **(c)** covalent **(d)** polar covalent **(e)** covalent **(f)** ionic **(g)** polar covalent **(h)** ionic **12.(a)** C δ^+, F δ^- **(d)** Cu δ^+, O δ^- **(g)** Fe δ^+, O δ^- **13.(a)** O—O, N—O, H—Cl, Na—Cl **(b)** N—N, P—O, C—Cl, Mg—C **14.** Group 13, 3, Group 16, −2, Group 17, −1 **15.** 0, do not tend to gain or lose electrons **16.(a)** Na_2S **(b)** CaS **(c)** BaS **(d)** Al_2S_3 **(e)** Rb_2S **(f)** H_2S **17.(a)** CaO **(b)** CaS **(c)** $CaCl_2$ **(d)** $CaBr_2$ **(e)** Ca_3P_2 **(f)** CaF_2 **18.(a)** $NaNO_3$ **(b)** Na_3PO_4 **(c)** Na_2SO_3 **(d)** $NaCH_3COO$ **(e)** $Na_2S_2O_3$ **(f)** Na_2CO_3 **19.** $Mg(NO_3)_2$ **(b)** $Mg_3(PO_4)_2$ **(c)** $MgSO_3$ **(d)** $Mg(CH_3COO)_2$ **(e)** MgS_2O_3 **20.(a)** aluminum oxide **(b)** calcium bromide **(c)** sodium phosphide **(d)** copper(I) sulfide **(e)** magnesium nitride **(f)** mercury(II) iodide **21.(a)** FeS **(b)** SnO (c) CrO **(d)** $CoCl_2$ **(e)** MnI_2 **(f)** ZnO **22.(a)** hydrogen selenide **(b)** hydrogen chloride **(c)** hydrogen fluoride **(d)** lithium hydride **(e)** calcium hydride **(f)** phosphorus(III) hydride **23.(a)** ammonium sulfite **(b)** aluminum nitrite **(c)** lithium carbonate **(d)** nickel(II) hydroxide **(e)** silver phosphate **(f)** copper(II) acetate **24.(a)** sulfur hexafluoride **(b)** dinitrogen pentoxide **(c)** phosphorus pentachloride **(d)** carbon tetrafluoride

Section Review: 3.1: 4.(a) Li, La, Zn, Si, Br **(b)** Cs, Y, Ga, P, Cl **5.(a)** 0.40, covalent **(b)** 1.89, ionic **(c)** 0.96, covalent **(d)** 2.16, ionic **6.(a)** low **(b)** covalent **3.2: 1.(a)** one magnesium atom gives up one electron to each of two fluorine atoms **(b)** one potassium electron gives up one electron to one bromine atom **(c)** one rubidium atom gives up one electron to one chlorine atom **(d)** one calcium atom gives up two electrons to one oxygen atom **2.(a)** The hydrogen atom and chlorine atoms all bond to the carbon atom. **(b)** Each hydrogen and chlorine atom shares one electron pair with the carbon atom. **(c)** The two nitrogen atoms share three pairs of electrons in a triple covalent bond. **3.(a)** covalent **(b)** covalent **(c)** covalent **(d)** ionic **3.3: 1.(a)** 1.94, ionic **(b)** 0.35, covalent **(c)** 2.23, ionic **(d)** 1.54, polar covalent **(e)** 0.86, polar covalent **(f)** 0.61, polar covalent **2.(d)** Si δ^+, O δ^- **(e)** S δ^+, O δ^- **(f)** C δ^+, Cl δ^- **4.(a)** O—F, H—Br, H—Cl, K—Br **(b)** C—H, C—Br, C—O, C—F **7.(a)** 1.26, non-polar molecule **(b)** 0.97, polar molecule **(c)** 1.55, non-polar molecule **3.4: 1.(a)** potassium chromate **(b)** ammonium nitrate **(c)** sodium sulfate **(d)** strontium phosphate **(e)** potassium nitrite **(f)** barium hypochlorite **2.(a)** magnesium chloride **(b)** sodium oxide **(c)** iron(III) chloride **(d)** copper(II) oxide **(e)** zinc sulfate **(f)** aluminum bromide **3.(a)** $NaHCO_3$ **(b)** $K_2Cr_2O_7$ **(c)** NaClO **(d)** LiOH **(e)** $KMnO_4$ **(f)** NH_4Cl **(g)** $Ca_3(PO_4)_2$ **(h)** $Na_2S_2O_3$ **4.(a)** any two of: vanadium(II) oxide, VO, vanadium(III) oxide, V_2O_3, vanadium(IV) oxide, VO_2, vanadium(V) oxide, V_2O_5 **(b)** iron(II) sulfide, FeS, iron(I) sulfide, Fe_2S **(c)** nickel(II) oxide, NiO, nickel(III) oxide, Ni_2O_3

CHAPTER 4

Classifying Reactions: Chemicals in Balance

Picture a starry night and tired hikers sitting around a campfire. It is hard to believe that everywhere in this peaceful setting, chemical reactions are taking place! The cellulose in the firewood reacts with the oxygen in the air, producing carbon dioxide and water. The light and heat of the campfire are evidence of the chemical reaction. If someone roasts a marshmallow its sugars react with oxygen. The soft, white marshmallow forms a brown and brittle crust. When someone eats a marshmallow, the chemicals in the stomach react with the sugar molecules to digest them. A person telling a story exhales carbon dioxide with every breath. Carbon dioxide is the product of respiration, another chemical reaction.

In each star in the night sky above, another type of reaction is taking place. This type of reaction is called a **nuclear reaction**, because it involves changes within the nucleus of the atom. Nuclear reactions are responsible for the enormous amounts of heat and light generated by all the stars, including our Sun.

Back on Earth, however, chemical reactions are everywhere in our daily lives. We rely on chemical reactions for everything from powering a car to making toast. In this chapter, you will learn how to write balanced chemical equations for these reactions. You will look for patterns and similarities between the chemical equations, and you will classify the reactions they represent. As well, you will learn how to balance and classify equations for nuclear reactions.

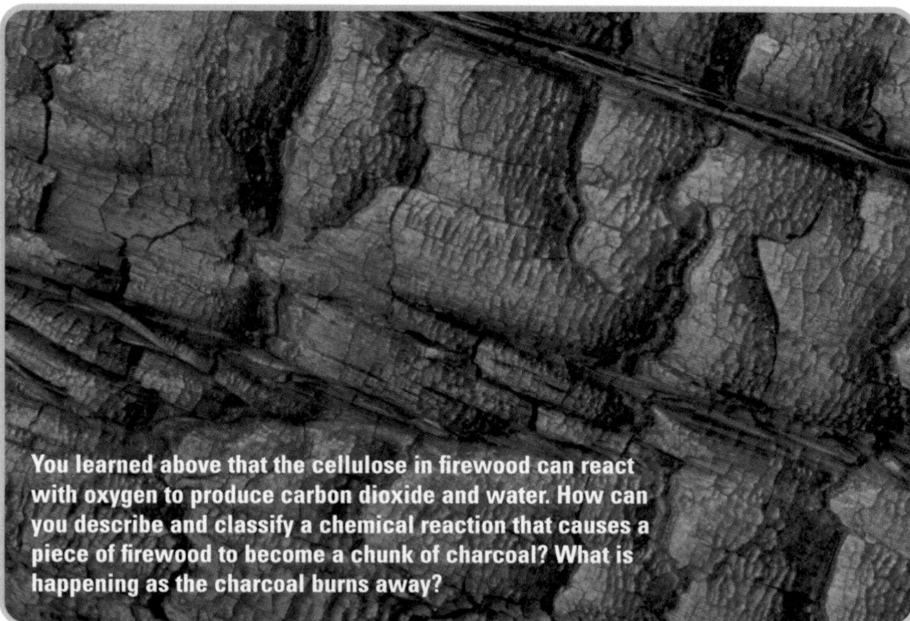

You learned above that the cellulose in firewood can react with oxygen to produce carbon dioxide and water. How can you describe and classify a chemical reaction that causes a piece of firewood to become a chunk of charcoal? What is happening as the charcoal burns away?

Chapter Preview

Concepts and Skills You Will Need

Before you begin this chapter, review the following concepts and skills:

- defining and describing the relationships among atomic number, mass number, atomic mass, isotope, and radio isotope (Chapter 2, section 2.1)
- naming chemical compounds (Chapter 3, section 3.4)
- writing chemical formulas (Chapter 3, section 3.4)
- explaining how different elements combine to form covalent and ionic bonds using the octet rule (Chapter 3, sections 3.2 and 3.3)

4.1 Chemical Equations

Section Preview/ Specific Expectations

In this section, you will

- **use** word equations and skeleton equations to describe chemical reactions
- **balance** chemical equations
- **communicate** your understanding of the following terms: *chemical reactions, reactant, product, chemical equations, word equation, skeleton equation, law of conservation of mass, balanced chemical equation*

In Chapter 3, you learned how and why elements combine to form different compounds. In this section, you will learn how to describe what happens when elements and compounds interact with one another to form new substances. These interactions are called **chemical reactions**. A substance that undergoes a chemical reaction is called a **reactant**. A substance that is formed in a chemical reaction is called a **product**.

For example, when the glucose in a marshmallow reacts with oxygen in the air to form water and carbon dioxide, the glucose and oxygen are the reactants. The carbon dioxide and water are the products. Chemists use **chemical equations** to communicate what is occurring in a chemical reaction. Chemical equations come in several forms. All of these forms condense a great deal of chemical information into a short statement.

Word Equations

A **word equation** identifies the reactants and products of a chemical reaction by name. In Chapter 3, you learned that chlorine and sodium combine to form the ionic compound sodium chloride. This reaction can be represented by the following word equation:

$$\text{sodium} + \text{chlorine} \rightarrow \text{sodium chloride}$$

In this equation, "+" means "reacts with" and "→" means "to form." Try writing some word equations in the following Practice Problems.

CHECKPOINT

Write the chemical formulas of the products in the reactions described in Practice Problem 1.

Practice Problems

1. Describe each reaction using a word equation. Label the reactant(s) and product(s).
   (a) Calcium and fluorine react to form calcium fluoride.
   (b) Barium chloride and hydrogen sulfate react to form hydrogen chloride and barium sulfate.
   (c) Calcium carbonate, carbon dioxide, and water react to form calcium hydrogen carbonate.
   (d) Hydrogen peroxide reacts to form water and oxygen.
   (e) Sulfur dioxide and oxygen react to form sulfur trioxide.
2. Yeast can facilitate a reaction in which the sugar in grapes reacts to form ethanol and carbon dioxide. Write a word equation to describe this reaction.

CHEM FACT

Caffeine is an ingredient in coffee, tea, chocolate, and cola drinks. Its chemical name is 1,3,7-trimethylxanthene. You can see how long names like this would become unwieldy in word equations.

Word equations are useful because they identify the products and reactants in a chemical reaction. They do not, however, provide any chemical information about the compounds and elements themselves. If you did not know the formula for sodium chloride, for example, this equation would not help you understand the reaction very well. Another shortcoming of a word equation is that the names for chemicals are often very long and cumbersome. Chemists have therefore devised a more convenient way of representing reactants and products.

Skeleton Equations

Using a chemical formula instead of a chemical name simplifies a chemical equation. It allows you to see at a glance what elements and compounds are involved in the reaction. A **skeleton equation** lists the chemical formula of each reactant on the left, separated by a + sign if more than one reactant is involved, followed by an arrow →. The chemical formula of each product is listed on the right, again separated by a + sign if more than one product is produced. A skeleton equation also shows the state of each reactant by using the appropriate subscript, as shown in Table 4.1.

The reaction of sodium metal with chlorine gas to form sodium chloride can be represented by the following skeleton equation:

$$Na_{(s)} + Cl_{2(g)} \rightarrow NaCl_{(s)}$$

A skeleton equation is more useful to a chemist than a word equation, because it shows the formulas of the compounds involved. It also shows the state of each substance.Try writing some skeleton equations in the following Practice Problems.

Table 4.1 Symbols Used in Chemical Equations

Symbol	Meaning
+	reacts with (reactant side)
+	and (product side)
→	to form
(s)	solid or precipitate
(ℓ)	liquid
(g)	gas
(aq)	in aqueous (water) solution

Practice Problems

3. Write a skeleton equation for each reaction.

(a) Solid zinc reacts with chlorine gas to form solid zinc chloride.

(b) Solid calcium and liquid water react to form solid calcium hydroxide and hydrogen gas.

(c) Solid barium reacts with solid sulfur to produce solid barium sulfide.

(d) Aqueous lead(II) nitrate and solid magnesium react to form aqueous magnesium nitrate and solid lead.

4. In each reaction below, a solid reacts with a gas to form a solid. Write a skeleton equation for each reaction.

(a) carbon dioxide + calcium oxide → calcium carbonate

(b) aluminum + oxygen → aluminum oxide

(c) magnesium + oxygen → magnesium oxide

Why Skeleton Equations Are Incomplete

Although skeleton equations are useful, they do not fully describe chemical reactions. To understand why, consider the skeleton equation showing the formation of sodium chloride (above). According to this equation, one sodium atom reacts with one chlorine molecule containing two chlorine atoms. The product is one formula unit of sodium chloride, containing one atom of sodium and one atom of chlorine. Where has the extra chlorine atom gone?

The Law of Conservation of Mass

All atoms must be accounted for, according to an important law. The **law of conservation of mass** states that *in any chemical reaction, the mass of the products is always equal to the mass of the reactants.* In other words, according to this law, matter can be neither created nor destroyed. Chemical reactions proceed according to the law of conservation of mass, which is based on experimental evidence.

Many chemical reactions can go in either direction, so an arrow pointing in the opposite direction is often added to the equation. This can look like $\rightleftarrows$ or $\rightleftharpoons$. To indicate which reaction is more likely to occur, one arrow can be drawn longer than the other: for example: $\stackrel{\longrightarrow}{\leftarrow}$ or $\stackrel{\rightarrow}{\longleftarrow}$.

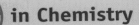

Careers in Chemistry

Food Chemist

How do you make a better tasting sports drink? How can you make a gravy mix that can be ready to serve in 5 min, yet maintain its consistency under a heat lamp for 8 h? Food chemists use their knowledge of chemical reactions to improve food quality and develop new products.

A good example of food chemistry in action is the red pimento stuffing in olives. Chopped pimentos (sweet red peppers) and sodium alginate are mixed. This mixture is then added to a solution of calcium chloride. The sodium alginate reacts with the calcium chloride. Solid calcium alginate forms, which causes the stream of pimento mixture to form a gel instantly. The gelled strip is then sliced thinly and stuffed into the olives.

Food chemists work in universities, government laboratories, and major food companies. To become a food chemist, most undergraduates take a food science degree with courses in chemistry. It is also possible to become a food chemist with an undergraduate chemistry degree plus experience in the food industry. Students can specialize in food chemistry at the graduate level.

Make Career Connections

1. If you are interested in becoming a food chemist, you can look for a summer or part-time job in the food industry.
2. To find out more about food science, search for Agriculture and Agri-Food Canada's web site and the Food Web web site. You can also contact the Food Science department of a university.

History LINK

Jan Baptista van Helmont (1577–1644) was a Flemish physician who left medical practice to devote himself to the study of chemistry. He used the mass balance in an important experiment that laid the foundations for the law of conservation of mass. He showed that a definite quantity of sand could be fused with excess alkali to form a kind of glass. He also showed that when this product was treated with acid, it regenerated the original amount of sand (silica). As well, Van Helmont is famous for demonstrating the existence of gases, which he described as "aerial fluids." Investigate on the Internet or in the library to find out how he did this.

Balanced Chemical Equations

A **balanced chemical equation** reflects the law of conservation of mass. This type of equation shows that there is the same number of each kind of atom on both sides of the equation. Some skeleton equations are, by coincidence, already balanced. For example, examine the reaction of carbon with oxygen to form carbon dioxide, shown in Figure 4.1. In the skeleton equation, one carbon atom and two oxygen atoms are on the left side of the equation, and one carbon atom and two oxygen atoms are on the right side of the equation.

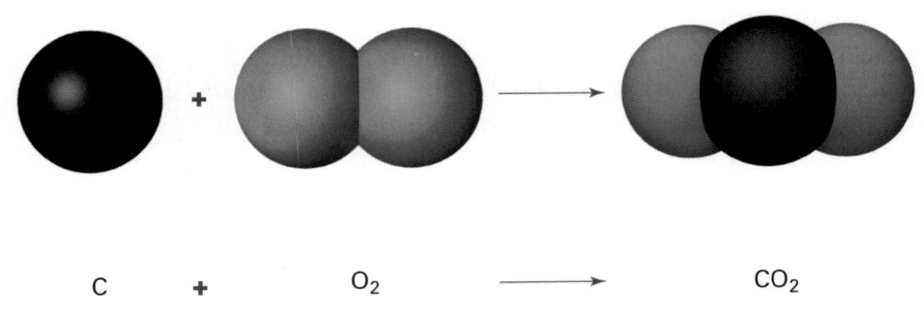

$$C + O_2 \longrightarrow CO_2$$

Figure 4.1 This skeleton equation is already balanced.

Most skeleton equations, however, are not balanced, such as the one showing the formation of sodium chloride. Examine Figure 4.2 to see why. There is one sodium atom on each side of the equation, but there are two chlorine atoms on the left side and only one chlorine atom on the right side.

To begin to balance an equation, you can add numbers in front of the appropriate formulas. The numbers that are placed in front of chemical formulas are called **coefficients**. They represent how many of each atom, molecule, or formula unit take part in each reaction. For example, if you add a coefficient of 2 to NaCl in the equation in Figure 4.2, you indicate that two formula units of NaCl are produced in the reaction. Is the equation balanced now? As you can see by examining Figure 4.3, it is not. The chlorine atoms are balanced, but now there is one sodium atom on the left side of the equation and two sodium atoms on the right side.

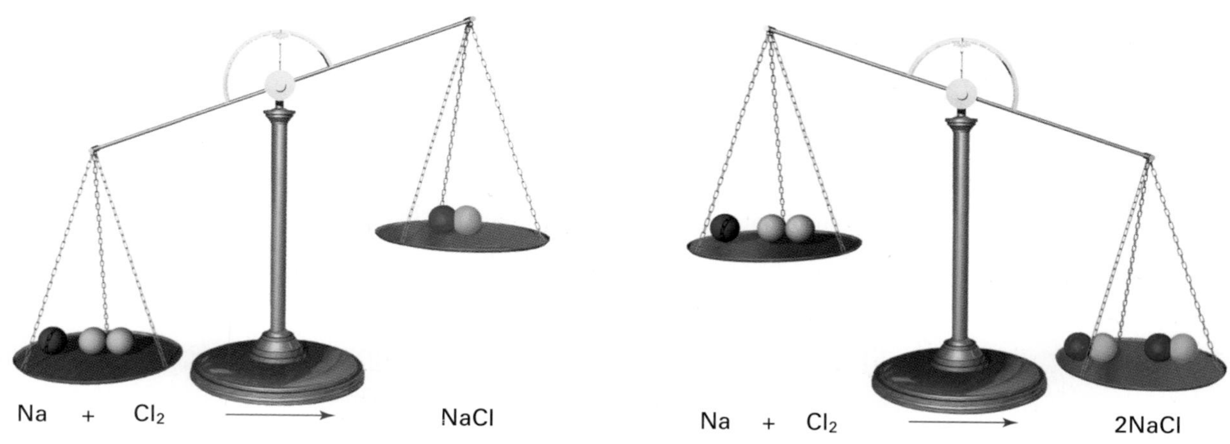

Figure 4.2 **This skeleton equation is unbalanced. The mass of the reactants is greater than the mass of the product.**

Figure 4.3 **The equation is still unbalanced. The mass of the product is now greater than the mass of the reactants.**

Add a coefficient of 2 to the sodium on the reactant side. As you can see in Figure 4.4, the equation is now balanced. The mass of the products is equal to the mass of the reactants. This balanced chemical equation satisfies the law of conservation of mass.

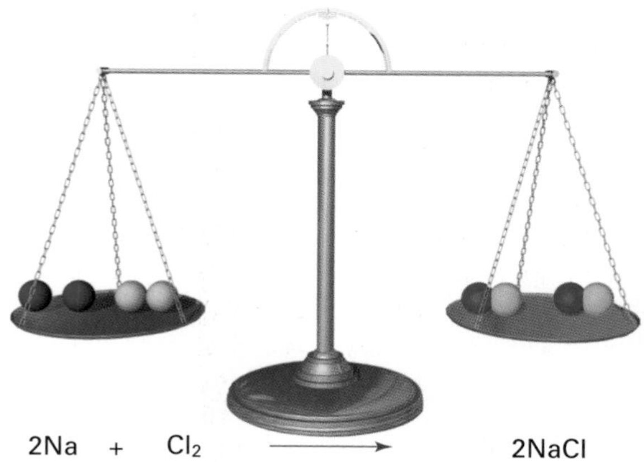

Figure 4.4 **The equation is now balanced according to the law of conservation of mass.**

mind STRETCH

Why is it *not* acceptable to balance the equation

$$Na + Cl_2 \rightarrow NaCl$$

by changing the formula of NaCl to $NaCl_2$? Would this not satisfy the law of conservation of mass? Write an explanation in your notebook.

You cannot balance an equation by changing any of the chemical formulas. The only way to balance a chemical equation is to put the appropriate numerical coefficient in front of each compound or element in the equation.

Many skeleton equations are simple enough to balance by a back-and-forth process of reasoning, as you just saw with the sodium chloride reaction. Try balancing the equations in the Practice Problems that follow.

Practice Problems

5. Copy each skeleton equation into your notebook, and balance it.

(a) $S_{(s)} + O_{2(g)} \rightarrow SO_{2(g)}$

(b) $P_{4(s)} + O_{2(g)} \rightarrow P_4O_{10(s)}$

(c) $H_{2(g)} + Cl_{2(g)} \rightarrow HCl_{(g)}$

(d) $SO_{2(g)} + H_2O_{(\ell)} \rightarrow H_2SO_{3(aq)}$

6. Indicate whether these equations are balanced. If they are not, balance them.

(a) $4Fe_{(s)} + 3O_{2(g)} \rightarrow 2Fe_2O_{3(s)}$

(b) $HgO_{(s)} \rightarrow Hg_{(\ell)} + O_{2(g)}$

(c) $H_2O_{2(aq)} \rightarrow 2H_2O_{(\ell)} + O_{2(g)}$

(d) $2HCl_{(aq)} + Na_2SO_{3(aq)} \rightarrow 2NaCl_{(aq)} + H_2O_{(\ell)} + SO_{2(g)}$

Steps for Balancing Chemical Equations

More complex chemical equations than the ones you have already tried can be balanced by using a combination of inspection and trial and error. Here, however, are some steps to follow.

Step 1 Write out the skeleton equation. Ensure that you have copied all the chemical formulas correctly.

Step 2 Begin by balancing the atoms that occur in the largest number on either side of the equation. Leave hydrogen, oxygen, and any other elements until later.

Step 3 Balance any polyatomic ions, such as sulfate, SO_4^{2-}, that occur on both sides of the chemical equation as an ion unit. That is, do not split a sulfate ion into 1 sulfur atom and 4 oxygen atoms. Balance this ion as one unit.

Step 4 Next, balance any hydrogen or oxygen atoms that occur in a combined and uncombined state. For example, combined oxygen might be in the form of CO_2, while uncombined oxygen occurs as O_2.

Step 5 Finally, balance any other element that occurs in its uncombined state: for example, Na or Cl_2.

Step 6 Check your answer. Count the number of each type of atom on each side of the equation. Make sure that the coefficients used are whole numbers in their lowest terms.

Examine the following Sample Problem to see how these steps work.

Go to the Chemistry 11 Electronic Learning Partner for some extra practice balancing chemical equations.

Sample Problem

Balancing Chemical Equations

Problem

Copper(II) nitrate reacts with potassium hydroxide to form potassium nitrate and solid copper(II) hydroxide.
Balance the equation.

$$Cu(NO_3)_{2(aq)} + KOH_{(aq)} \rightarrow Cu(OH)_{2(s)} + KNO_{3(aq)}$$

What Is Required?

The atoms of each element on the left side of the equation should equal the atoms of each element on the right side of the equation.

Plan Your Strategy

Balance the polyatomic ions first (NO_3^-, then OH^-). Check to see whether the equation is balanced. If not, balance the potassium and copper ions. Check your equation again.

Act on Your Strategy

There are two NO_3^- ions on the left, so put a 2 in front of KNO_3.

$$Cu(NO_3)_{2(aq)} + KOH_{(aq)} \rightarrow Cu(OH)_{2(s)} + 2KNO_{3(aq)}$$

To balance the two OH^- ions on the right, put a 2 in front of the KOH.

$$Cu(NO_3)_{2(aq)} + 2KOH_{(aq)} \rightarrow Cu(OH)_{2(s)} + 2KNO_{3(aq)}$$

Check to see that the copper and potassium ions are balanced. They are, so the equation above is balanced.

Check Your Solution

Tally the number of each type of atom on each side of the equation.

$$Cu(NO_3)_{2(aq)} + 2KOH_{(aq)} \rightarrow Cu(OH)_{2(s)} + 2KNO_{3(aq)}$$

Left Side		Right Side
Cu	1	1
NO_3^-	2	2
K	2	2
OH^-	2	2

Practice Problems

7. Copy each chemical equation into your notebook, and balance it.

(a) $SO_{2(g)} + O_{2(g)} \rightarrow SO_{3(g)}$

(b) $BaCl_{2(aq)} + Na_2SO_{4(aq)} \rightarrow NaCl_{(aq)} + BaSO_{4(s)}$

8. When solid white phosphorus, P_4, is burned in air, it reacts with oxygen to produce solid tetraphosphorus decoxide, P_4O_{10}. When water is added to the P_4O_{10}, it reacts to form aqueous phosphoric acid, H_3PO_4. Write and balance the chemical equations that represent these reactions.

Continued ...

Math LINK

What does it mean when a fraction is expressed in lowest terms? The fraction $\frac{5}{10}$, expressed in lowest terms, is $\frac{1}{2}$. Similarly, the equation

$4H_{2(g)} + 2O_{2(g)} \rightarrow 4H_2O_{(\ell)}$

is balanced, but it can be simplified by dividing all the coefficients by two.

$2H_{2(g)} + O_{2(g)} \rightarrow 2H_2O_{(\ell)}$

Write the balanced equation $6KClO_{3(s)} \rightarrow 6KCl_{(g)} + 9O_{2(s)}$ so that the coefficients are the lowest possible whole numbers. Check that the equation is still balanced.

Continued ...

FROM PAGE 117

9. Copy each chemical equation into your notebook, and balance it.
   (a) $As_4S_{6(s)} + O_{2(g)} \rightarrow As_4O_{6(s)} + SO_{2(g)}$
   (b) $Sc_2O_{3(s)} + H_2O_{(\ell)} \rightarrow Sc(OH)_{3(s)}$
   (c) $C_2H_5OH_{(\ell)} + O_{2(g)} \rightarrow CO_{2(g)} + H_2O_{(\ell)}$
   (d) $C_4H_{10(g)} + O_{2(g)} \rightarrow CO_{(g)} + H_2O_{(g)}$

Section Wrap-up

In this section, you learned how to represent chemical reactions using balanced chemical equations. Because there are so many different chemical reactions, chemists have devised different classifications for these reactions. In section 4.2, you will learn about five different types of chemical reactions.

Section Review

1. MC In your own words, explain what a chemical reaction is. Write descriptions of four chemical reactions that you encounter every day.

2. C Write a word equation, skeleton equation, and balanced equation for each reaction.
   (a) Sulfur dioxide gas reacts with oxygen gas to produce gaseous sulfur trioxide.
   (b) Metallic sodium reacts with liquid water to produce hydrogen gas and aqueous sodium hydroxide.
   (c) Copper metal reacts with an aqueous hydrogen nitrate solution to produce aqueous copper(II) nitrate, nitrogen dioxide gas, and liquid water.

3. K/U The equation for the decomposition of hydrogen peroxide, H_2O_2 is $H_2O_{2(aq)} \rightarrow H_2O_{(\ell)} + O_{2(g)}$. Explain why you cannot balance it by writing it as $H_2O_{2(aq)} \rightarrow H_2O_{(\ell)} + O_{(g)}$

4. K/U Balance the following chemical equations.
   (a) $Al_{(s)} + O_{2(g)} \rightarrow Al_2O_{3(s)}$
   (b) $Na_2S_2O_{3(aq)} + I_{2(aq)} \rightarrow NaI_{(aq)} + Na_2S_4O_{6(aq)}$
   (c) $Al_{(s)} + Fe_2O_{3(s)} \rightarrow Al_2O_{3(s)} + Fe_{(s)}$
   (d) $NH_{3(g)} + O_{2(g)} \rightarrow NO_{(g)} + H_2O_{(\ell)}$
   (e) $Na_2O_{(s)} + (NH_4)_2SO_{4(aq)} \rightarrow Na_2SO_{4(aq)} + H_2O_{(\ell)} + NH_{3(aq)}$
   (f) $C_5H_{12(\ell)} + O_{2(g)} \rightarrow CO_{2(g)} + H_2O_{(g)}$

5. I A student places 0.58 g of iron and 1.600 g of copper(II) sulfate in a reaction vessel. The reaction vessel has a mass of 40.32 g, and it contains 100.00 g of water. The aqueous copper sulfate and solid iron react to form solid copper and aqueous iron(II) sulfate. After the reaction, the reaction vessel plus the products have a mass of 142.5 g. Explain the results. Then write a balanced chemical equation to describe the reaction.

Synthesis and Decomposition Reactions

4.2

How are different kinds of compounds formed? In section 4.1, you learned that they are formed by chemical reactions that you can describe using balanced chemical equations. Just as there are different types of compounds, there are different types of chemical reactions. In this section, you will learn about five major classifications for chemical reactions. You will use your understanding of chemical formulas and chemical equations to predict products for each class of reaction.

Section Preview/ Specific Expectations

In this section, you will

- **distinguish** between synthesis, decomposition, and combustion reactions
- **write** balanced chemical equations to represent synthesis, decomposition, and combustion reactions
- **predict** the products of chemical reactions
- **demonstrate** an understanding of the relationship between the type of chemical reaction and the nature of the reactants
- **communicate** your understanding of the following terms: *synthesis reaction, decomposition reaction, combustion reaction, incomplete combustion*

Why Classify?

People use classifications all the time. For example, many types of wild mushrooms are edible, but many others are poisonous—even deadly! How can you tell which is which? Poisonous and deadly mushrooms have characteristics that distinguish them from edible ones, such as odour, colour, habitat, and shape of roots. It is not always easy to distinguish one type of mushroom from another; the only visible difference may be the colour of the mushroom's spores. Therefore, you should never try to eat any wild mushrooms without an expert's advice.

Examine Figure 4.5. Which mushroom looks more appetizing to you? An expert will always be able to distinguish an edible mushroom from a poisonous mushroom based on the characteristics that have been used to classify each type. By classifying, they can predict the effects of eating any wild mushroom.

Figure 4.5 **The mushroom on the left, called a chanterelle, is edible and very tasty. The mushroom on the right is called a death cap. It is extremely poisonous.**

In the same way, you can recognize similarities between chemical reactions and the types of reactants that tend to undergo different types of reactions. With this knowledge, you can predict what will happen when one, two, or more substances react. In this section, you will often see chemical reactions without the subscripts showing the states of matter. They are omitted deliberately because, in most cases, you are not yet in a position to predict the states of the products.

Synthesis Reactions

In a **synthesis reaction**, two or more elements or compounds combine to form a new substance. Synthesis reactions are also known as combination or formation reactions. A general equation for a synthesis reaction is

$$A + B \rightarrow C$$

In a simple synthesis reaction, one element reacts with one or more other elements to form a compound. Two, three, four, or more elements may react to form a single product, although synthesis reactions involving four or more reactants are extremely rare. Why do you think this is so? When two elements react together, the reaction is almost always a synthesis reaction because the product is almost always a single compound. There are several types of synthesis reactions. Recognizing the patterns of the various types of reactions will help you to predict whether substances will take part in a synthesis reaction.

When a metal or a non-metal element reacts with oxygen, the product is an oxide. Figure 4.6 shows a familiar example, in which iron reacts with oxygen according to the following equation:

$$3Fe_{(s)} + O_{2(g)} \rightarrow Fe_3O_{2(s)}$$

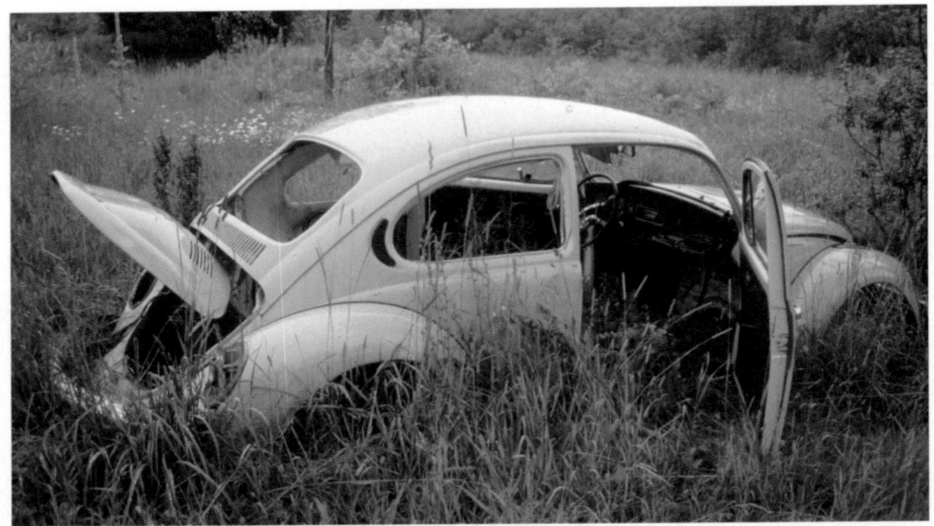

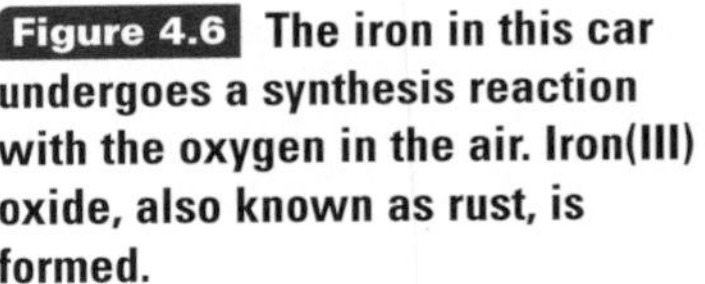

Figure 4.6 The iron in this car undergoes a synthesis reaction with the oxygen in the air. Iron(III) oxide, also known as rust, is formed.

Two other examples of this type of reaction are:

$$2H_{2(g)} + O_{2(g)} \rightarrow 2H_2O_{(g)}$$

$$2Mg_{(s)} + O_{2(g)} \rightarrow 2MgO_{(s)}$$

A second type of synthesis reaction involves the reaction of a metal and a non-metal to form a binary compound. One example is the reaction of potassium with chlorine.

$$2K_{(s)} + Cl_{2(g)} \rightarrow 2KCl_{(s)}$$

Synthesis Reactions Involving Compounds

In the previous two types of synthesis reactions, two elements reacted to form one product. There are many synthesis reactions in which one or more compounds are the reactants. For the purpose of this course, however, we will deal only with the two specific types of synthesis reactions involving compounds that you should recognize: oxides and water.

When a non-metallic oxide reacts with water, the product is an acid. You will learn more about acids and the rules for naming them in Chapter 10. The acids that form when non-metallic oxides and water react are composed of hydrogen cations and polyatomic anions containing oxygen and a non-metal. For example, one contributor to acid rain is hydrogen sulfate (sulfuric acid), H_2SO_4, which forms when sulfur trioxide reacts with water. The sulfur trioxide comes from sources such as industrial plants that emit the gas as a byproduct of burning fossil fuels, as shown in Figure 4.7.

$$SO_{3(g)} + H_2O_{(\ell)} \rightarrow H_2SO_{4(aq)}$$

Figure 4.7 **Sulfur trioxide, emitted by this factory, reacts with the water in the air. Sulfuric acid is formed in a synthesis reaction.**

Conversely, when a metallic oxide reacts with water, the product is a metal hydroxide. Metal hydroxides belong to a group of compounds called bases. You will learn more about bases in Chapter 10. For example, when calcium oxide reacts with water, it forms calcium hydroxide, $Ca(OH)_2$. Calcium oxide is also called lime. It can be added to lakes to counteract the effects of acid precipitation.

$$CaO_{(s)} + H_2O_{(\ell)} \rightarrow Ca(OH)_{2(aq)}$$

Sometimes it is difficult to predict the product of a synthesis reaction. The only way to really know the product of a reaction is to carry out the reaction and then isolate and identify the product. For example, carbon can react with oxygen to form either carbon monoxide or carbon dioxide. Therefore, if all you know is that your reactants are carbon and oxygen, you cannot predict with certainty which compound will form. You can only give options.

$$C_{(s)} + O_{2(g)} \rightarrow CO_{2(g)}$$

$$2C_{(s)} + O_{2(g)} \rightarrow 2CO_{(g)}$$

You would need to analyze the products of the reaction by experiment to determine which compound was formed.

History LINK

Today we have sophisticated lab equipment to help us analyze the products of reactions. In the past, when such equipment was not available, chemists sometimes jeopardized their safety and health to determine the products of the reactions they studied. Sir Humphry Davy (1778–1829), a contributor to many areas of chemistry, thought nothing of inhaling the gaseous products of the chemical reactions that he carried out. He tried to breathe pure CO_2, then known as *fixed air*. He nearly suffocated himself by breathing hydrogen. In 1800, Davy inhaled dinitrogen monoxide, N_2O, otherwise known as nitrous oxide, and discovered its anaesthetic properties. What is nitrous oxide used for today?

CHECKPOINT

As you begin learning about different types of chemical reactions, keep a separate list of each type of reaction. Add to the list as you encounter new reactions.

Try predicting the products of synthesis reactions in the following Practice Problems.

Practice Problems

10. Copy the following synthesis reactions into your notebook. Predict the product of each reaction. Then balance each chemical equation.
    - (a) $K + Br_2 \rightarrow$
    - (b) $H_2 + Cl_2 \rightarrow$
    - (c) $Ca + Cl_2 \rightarrow$
    - (d) $Li + O_2 \rightarrow$
11. Copy the following synthesis reactions into your notebook. For each set of reactants, write the equations that represent the possible products.
    - (a) $Fe + O_2 \rightarrow$
      (suggest two different synthesis reactions)
    - (b) $V + O_2 \rightarrow$
      (suggest four different synthesis reactions)
    - (c) $Co + Cl_2 \rightarrow$
      (suggest two different synthesis reactions)
    - (d) $Ti + O_2 \rightarrow$
      (suggest three different synthesis reactions)
12. Copy the following equations into your notebook. Write the product of each reaction. Then balance each chemical equation.
    - (a) $K_2O + H_2O \rightarrow$
    - (b) $MgO + H_2O \rightarrow$
    - (c) $SO_2 + H_2O \rightarrow$
13. Ammonia gas and hydrogen chloride gas react to form a solid compound. Predict what the solid compound is. Then write a balanced chemical equation.

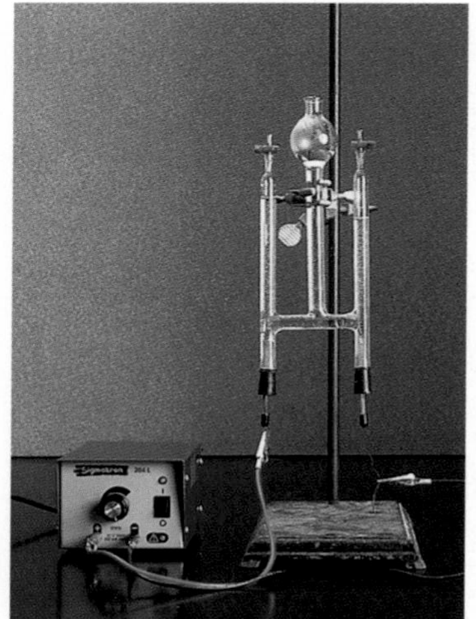

Figure 4.8 As electricity passes through the water, it decomposes to hydrogen and oxygen gas.

Decomposition Reactions

In a **decomposition reaction**, a compound breaks down into elements or other compounds. Therefore, *a decomposition reaction is the opposite of a synthesis reaction.* A general formula for a decomposition reaction is:

$$C \rightarrow A + B$$

The substances that are produced in a decomposition reaction can be elements or compounds. In the simplest type of decomposition reaction, a compound breaks down into its component elements. One example is the decomposition of water into hydrogen and oxygen. This reaction occurs when electricity is passed through water. Figure 4.8 shows an apparatus set up for the decomposition of water.

$$2H_2O \rightarrow 2H_2 + O_2$$

More complex decomposition reactions occur when compounds break down into other compounds. An example of this type of reaction is shown in Figure 4.9. The photograph shows the explosive decomposition of ammonium nitrate.

When ammonium nitrate is heated to a high temperature, it forms dinitrogen monoxide and water according to the following balanced equation:

$$NH_4NO_{3(s)} \rightarrow N_2O_{(g)} + 2H_2O_{(g)}$$

Try predicting the products of the decomposition reactions in the following Practice Problems.

Figure 4.9 At high temperatures, ammonium nitrate explodes, decomposing into dinitrogen monoxide and water.

Practice Problems

14. Mercury(II) oxide, or mercuric oxide, is a bright red powder. It decomposes on heating. What are the products of the decomposition of HgO?

15. What are the products of the following decomposition reactions? Predict the products. Then write a balanced equation for each reaction.

(a) $HI \rightarrow$

(b) $Ag_2O \rightarrow$

(c) $AlCl_3 \rightarrow$

(d) $MgO \rightarrow$

16. Calcium carbonate decomposes into calcium oxide and carbon dioxide when it is heated. Based on this information, predict the products of the following decomposition reactions.

(a) $MgCO_3 \rightarrow$

(b) $CuCO_3 \rightarrow$

Combustion Reactions

Combustion reactions form an important class of chemical reactions. The combustion of fuel—wood, fossil fuel, peat, or dung—has, throughout history, heated and lit our homes and cooked our food. The energy produced by combustion reactions moves our airplanes, trains, trucks, and cars.

A complete **combustion reaction** is the reaction of a compound or element with O_2 to form the most common oxides of the elements that make up the compound. For example, a carbon-containing compound undergoes combustion to form carbon dioxide, CO_2. A sulfur-containing compound reacts with oxygen to form sulfur dioxide, SO_2.

Combustion reactions are usually accompanied by the production of light and heat. In the case of carbon-containing compounds, complete combustion results in the formation of, among other things, carbon dioxide. For example, methane, CH_4, the primary constituent of natural gas, undergoes complete combustion to form carbon dioxide, (the most common oxide of carbon), as well as water. This combustion reaction is represented by the following equation:

$$CH_{4(g)} + 2O_{2(g)} \rightarrow CO_{2(g)} + 2H_2O_{(g)}$$

The combustion of methane, shown in Figure 4.10, leads to the formation of carbon dioxide and water.

The complete combustion of any compound that contains carbon, hydrogen, and oxygen (such as ethanol, C_2H_5OH) produces carbon dioxide and water.

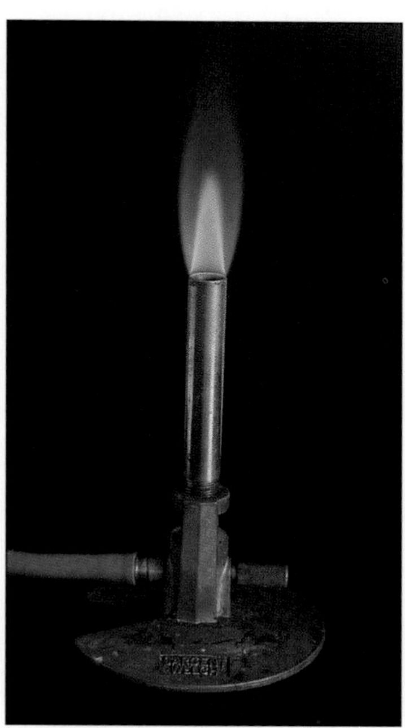

Figure 4.10 This photo shows the combustion of methane in a laboratory burner.

Compounds that contain elements other than carbon also undergo complete combustion to form stable oxides. For instance, sulfur-containing compounds undergo combustion to form sulfur dioxide, SO_2, a precursor to acid rain. Complete combustion reactions are often also synthesis reactions. Metals, such as magnesium, undergo combustion to form their most stable oxide, as shown in Figure 4.11.

$$2Mg_{(s)} + O_{2(g)} \rightarrow 2MgO_{(s)}$$

Figure 4.11 **Magnesium metal burns in oxygen. The smoke and ash that are produced in this combustion reaction are magnesium oxide.**

In the absence of sufficient oxygen, carbon-containing compounds undergo **incomplete combustion**, leading to the formation of carbon monoxide, CO, and water. Carbon monoxide is a deadly gas. You should always make sure that sufficient oxygen is present in your indoor environment for your gas furnace, gas stove, or fireplace.

Try the following problems to practise balancing combustion reactions.

CHECKPOINT

Copy the following skeleton equation for the combustion of pentane, C_5H_{12}, into your notebook and balance it.

$C_5H_{12} + O_2 \rightarrow CO_2 + H_2O$

If it took you a long time to balance this equation, chances are that you did not use the quickest method. Try balancing carbon first, hydrogen second, and finally oxygen. What is the advantage of leaving O_2 until the end? Would this method work for incomplete combustion reactions? Would this method work if the "fuel" contained oxygen in addition to C and H? Now try balancing the chemical equation for the combustion of heptane (C_7H_{16}) and the combustion of rubbing alcohol, isopropanol (C_3H_8O).

Practice Problems

17. The alcohol lamps that are used in some science labs are often fuelled with methanol, CH_3OH. Write the balanced chemical equation for the complete combustion of methanol.
18. Gasoline is a mixture of compounds containing hydrogen and carbon, such as octane, C_8H_{18}. Write the balanced chemical equation for the complete combustion of C_8H_{18}.
19. Acetone, C_3H_6O, is often contained in nail polish remover. Write the balanced chemical equation for the complete combustion of acetone.
20. Kerosene consists of a mixture of hydrocarbons. It has many uses including jet fuel and rocket fuel. It is also used as a fuel for hurricane lamps. If we represent kerosene as $C_{16}H_{34}$, write a balanced chemical equation for the complete combustion of kerosene.

Section Wrap-up

In this section, you learned about three major types of reactions: synthesis, decomposition, and combustion reactions. Using your knowledge about these types of reactions, you learned how to predict the products of various reactants. In section 5.3, you will increase your understanding of chemical reactions even further, learning about two major types of chemical reactions. As well, you will observe various chemical reactions in three investigations.

Section Review

1. K/U Write the product for each synthesis reaction. Balance the chemical equation.
   - **(a)** $Be + O_2 \rightarrow$
   - **(b)** $Li + Cl_2 \rightarrow$
   - **(c)** $Mg + N_2 \rightarrow$
   - **(d)** $Al + Br_2 \rightarrow$
   - **(e)** $K + O_2 \rightarrow$
2. K/U Write the products for each decomposition reaction. Balance the chemical equation.
   - **(a)** $K_2O \rightarrow$
   - **(b)** $CuO \rightarrow$
   - **(c)** $H_2O \rightarrow$
   - **(d)** $Ni_2O_3 \rightarrow$
   - **(e)** $Ag_2O \rightarrow$
3. C Write a balanced chemical equation for each of the following word equations. Classify each reaction.
   - **(a)** With heating, solid tin(IV) hydroxide produces solid tin(IV) oxide and water vapour.
   - **(b)** Chlorine gas reacts with crystals of iodine to form iodine trichloride.
4. K/U Write a balanced chemical equation for the combustion of butanol, C_4H_9OH.
5. I A red compound was heated, and the two products were collected. The gaseous product caused a glowing splint to burn brightly. The other product was a shiny pure metal, which was a liquid at room temperature. Write the most likely reaction that would explain these results. Classify the reaction. **Hint:** Remember that the periodic table identifies the most common valences.
6. MC Explain why gaseous nitrogen oxides emitted by automobiles and industries contribute to acid rain. Write balanced chemical reactions to back up your ideas. You may need to look up chemical formulas for your products.

Unit Project Prep

Before you design your Chemistry Newsletter at the end of Unit 1, consider that fuels are composed of compounds containing hydrogen and carbon (hydrocarbons). What kind of reaction have you seen in this section that involves those kinds of compounds? What type of warning would you expect to see on a container of lawnmower fuel? How is the warning related to the types of reaction that involve hydrocarbons?

4.3 Single Displacement and Double Displacement Reactions

Section Preview/ Specific Expectations

In this section, you will

- **distinguish** between synthesis, decomposition, combustion, single displacement, and double displacement reactions
- **write** balanced chemical equations to represent single displacement and double displacement reactions
- **predict** the products of chemical reactions and **test** your predictions through experimentation
- **demonstrate** an understanding of the relationship between the type of chemical reaction and the nature of the reactants
- **investigate**, through experimentation, the reactivity of different metals to produce an activity series
- **communicate** your understanding of the following terms: *single displacement reaction, activity series, double displacement reaction, precipitate, neutralization reactions*

In section 4.2, you learned about three different types of chemical reactions. In section 4.3, you will learn about two more types of reactions. You will learn how performing these reactions can help you make inferences about the properties of the elements and compounds involved.

Single Displacement Reactions

In a **single displacement reaction**, one element in a compound is displaced (or replaced) by another element. Two general reactions represent two different types of single displacement reactions. One type involves a metal replacing a metal cation in a compound, as follows:

$$A + BC \rightarrow AC + B$$

For example, $Zn_{(s)} + Fe(NO_3)_{2(aq)} \rightarrow Zn(NO_3)_{2(aq)} + Fe_{2(s)}$

The second type of single displacement reaction involves a non-metal (usually a halogen) replacing an anion in a compound, as follows:

$$DE + F \rightarrow DF + E$$

For example, $Cl_{2(g)} + CaBr_{2(aq)} \rightarrow CaCl_{2(aq)} + Br_{2(\ell)}$

Single Displacement Reactions and the Metal Activity Series

Most single displacement reactions involve one metal displacing another metal from a compound. In the following equation, magnesium metal replaces the zinc in $ZnCl_2$, thereby liberating zinc as the free metal.

$$Mg_{(s)} + ZnCl_{2(aq)} \rightarrow MgCl_{2(aq)} + Zn_{(s)}$$

The following three reactions illustrate the various types of single displacement reactions involving metals:

1. $Cu_{(s)} + 2AgNO_{3(aq)} \rightarrow Cu(NO_3)_{2(aq)} + 2Ag_{(s)}$
   In this reaction, *one metal replaces another metal in an ionic compound*. That is, copper replaces silver in $AgNO_3$. Because of the +2 charge on the copper ion, it requires two nitrate ions to balance its charge.

2. $Mg_{(s)} + 2HCl_{(aq)} \rightarrow MgCl_{2(aq)} + H_{2(g)}$
   In this reaction, magnesium metal replaces hydrogen from hydrochloric acid, $HCl_{(aq)}$. Since hydrogen is diatomic, it is "liberated" in the form of H_2. This reaction is similar to reaction 1 if
   - you treat hydrochloric acid as an ionic compound (which it technically is not), and if
   - you treat hydrogen as a metal (also, technically, not the case).

3. $2Na_{(s)} + 2H_2O_{(\ell)} \rightarrow 2NaOH_{(aq)} + H_{2(g)}$
   Sodium metal displaces hydrogen from water in this reaction. Again, since hydrogen is diatomic, it is produced as H_2. As above, you can understand this reaction better if
   - you treat hydrogen as a metal, and if
   - you treat water as an ionic compound, $H^+(OH^-)$.

All of the reactions just described follow the original general example of a single displacement reaction:

$$A + BC \rightarrow AC + B$$

Figures 4.12 and 4.13 show two examples of single displacement reactions. When analyzing single displacement reactions, use the following guidelines:

- Treat hydrogen as a metal.
- Treat acids, such as HCl, as ionic compounds of the form H^+Cl^-. (Treat sulfuric acid, H_2SO_4, as $H^+H^+SO_4^{2-}$).
- Treat water as ionic, with the formula $H^+(OH^-)$.

Figure 4.12 **Lithium metal reacts violently with water in a single displacement reaction. Lithium must be stored under kerosene or oil to avoid reaction with atmospheric moisture, or oxygen.**

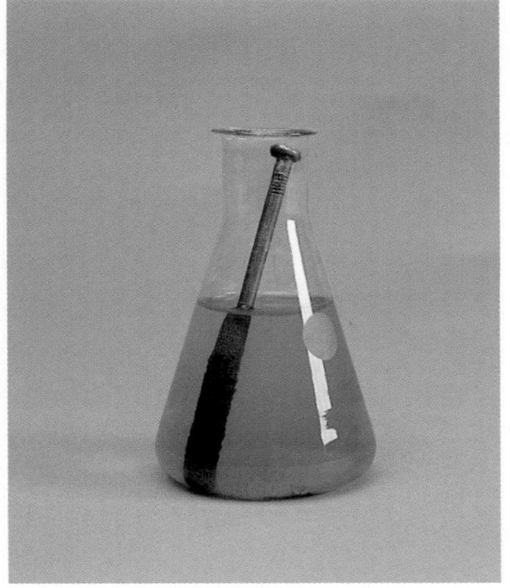

Figure 4.13 **When an iron nail is placed in a solution of copper(II) sulfate, a single displacement reaction takes place.**
$Fe_{(s)} + CuSO_{4(aq)} \rightarrow FeSO_{4(aq)} + Cu_{(s)}$
Notice the formation of copper metal on the nail.

Practice Problems

21. Each of the following incomplete equations represents a single displacement reaction. Copy each equation into your notebook, and write the products. Balance each chemical equation. When in doubt, use the most common valence.

(a) $Ca + H_2O \rightarrow$

(b) $Zn + Pb(NO_3)_2 \rightarrow$

(c) $Al + HCl \rightarrow$

(d) $Cu + AgNO_3 \rightarrow$

(e) $Pb + H_2SO_4 \rightarrow$

(f) $Mg + Pt(OH)_4 \rightarrow$

(g) $Ba + FeCl_2 \rightarrow$

(h) $Fe + Co(ClO_3)_2 \rightarrow$

Through experimentation, chemists have ranked the relative reactivity of the metals, including hydrogen (in acids and in water), in an **activity series**. The reactive metals, such as potassium, are at the top of the activity series. The unreactive metals, such as gold, are at the bottom. In Investigation 4-A, you will develop an activity series using single displacement reactions.

Investigation 4-A

MICROSCALE

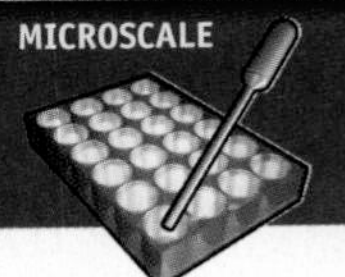

SKILL FOCUS

Predicting

Performing and recording

Analyzing and interpreting

Creating an Activity Series of Metals

Certain metals, such as silver and gold, are extremely unreactive, while sodium is so reactive that it will react with water. Zinc is unreactive with water. It *will*, however, react with acid. Why will magnesium metal react with copper sulfate solution, while copper metal will not react with aqueous magnesium sulfate? In Chapter 3, you learned that an alloy is a solution of two or more metals. Steel is an alloy that contains mostly iron. Is its reactivity different from iron's reactivity?

Question

How can you rank the metals, including hydrogen, in terms of their reactivity? Is the reactivity of an alloy very different from the reactivity of its major component?

Predictions

Based on what you learned in Chapter 3 about periodic trends, make predictions about the relative reactivity of copper, iron, magnesium, zinc, and tin. Explain your reasons for these predictions.

What do you know about alloys such as bronze, brass, and steel? Based on what you know, make a prediction about whether steel will be more or less reactive than iron, its main component.

Materials

well plate(s): at least a 6×8 matrix
wash bottle with distilled water
5 test tubes
test tube rack
dilute $HCl_{(aq)}$
6 small pieces each of copper, iron, magnesium, zinc, tin, steel, galvanized steel, stainless steel
dropper bottles of dilute solutions of $CuSO_4$, $FeSO_4$, $MgSO_4$, $ZnSO_4$, $SnCl_2$

Metal \ Cation or solution	HCl	H_2O	Cu^{2+}	Fe^{2+}	Mg^{2+}	Sn^{2+}
Cu						
Mg						
Sn						
Zn						
Fe						
steel						
galvanized steel						
stainless steel						

Safety Precautions

Handle the hydrochloric acid solution with care. It is corrosive. Wipe up any spills with copious amounts of water, and inform your teacher.

Procedure

1. Place your well plate(s) on a white sheet of paper. Label them according to the matrix on the previous page.
2. Place a rice-grain-sized piece of each metal in the appropriate well. Record the appearance of each metal.
3. Put enough drops of the appropriate solution to completely cover the piece of metal.
4. Record any changes in appearance due to a chemical reaction. In reactions of metal with acid, look carefully for the formation of bubbles. If you are unsure about any observation, repeat the experiment in a small test tube. This will allow you to better observe the reaction.
5. If you believe that a reaction has occurred, write "r" on the matrix. If you believe that no reaction has occurred, write "nr" on the matrix.
6. Dispose of the solutions in the waste beaker supplied by your teacher. Do not pour anything down the drain.

Analysis

1. For any reactions that occurred, write the corresponding single displacement reaction.
2. (a) What was the most reactive metal that you tested?
   (b) What was the least reactive metal that you tested?
3. Look at Figure 4.12. Lithium reacts violently with water to form aqueous lithium hydroxide and hydrogen gas. Do you expect lithium to react with hydrochloric acid?
   (a) Write the balanced chemical equation for this reaction.
   (b) Is lithium more or less reactive than magnesium?
4. What evidence do you have that hydrogen in hydrochloric acid is different from hydrogen in water?

Conclusion

5. (a) Write the activity series corresponding to your observations. Include hydrogen in the form of water and also as an ion (H^+). Do not include the alloys.
   (b) How did the reactivity of the iron compare with the reactivity of the various types of steel?
6. How do you think an activity series for metal would help you predict whether or not a single displacement reaction will occur? Use examples to help you explain your answer.
7. You have learned that an alloy is a homogeneous mixture (solution) of two or more metals. Steel consists of mostly iron.
   (a) Which type of steel appeared to be the most reactive? Which type was the least reactive? Did you notice any differences?
   (b) What other components make up steel, galvanized steel, and stainless steel?

Application

8. For what applications are the various types of steel used? Why would you not use iron for these applications?

SKILL FOCUS

Predicting

Performing and recording

Analyzing and interpreting

Investigation 4-B

MICROSCALE

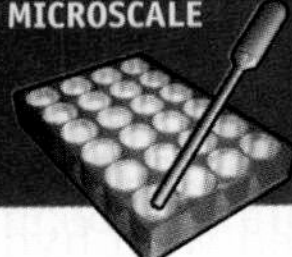

Observing Double Displacement Reactions

A double displacement reaction involves the exchange of cations between two ionic compounds, usually in aqueous solution. It can be represented with the general equation

$$AB + CD \rightarrow AD + CB$$

Most often, double displacement reactions result in the formation of a precipitate. However, some double displacement reactions result in the formation of an unstable compound which then decomposes to water and a gas.

The reaction of an acid and a base—a neutralization reaction—is also a type of double displacement reaction. It results in the formation of a salt and water.

Question

How can you tell if a double displacement reaction has occurred? How can you predict the products of a double displacement reaction?

Prediction

For each reaction in Tables A and B, write a balanced chemical equation. Use the following guidelines to predict precipitate formation in Table A.

- Hydrogen, ammonium, and Group I ions form soluble compounds with all negative ions.
- Chloride ions form compounds that are not very soluble when they bond to silver, lead(II), mercury(I), and copper(I) positive ions.
- All compounds that are formed from a nitrate and a positive ion are soluble.
- With the exception of the ions in the first bulleted point, as well as strontium, barium, radium, and thallium positive ions, hydroxide ions form compounds that do not dissolve.
- Iodide ions that are combined with silver, lead(II), mercury(I), and copper(I) are not very soluble.
- Chromate compounds are insoluble, except when they contain ions from the first bulleted point.

Materials

well plate
sheet of white paper
several test tubes
test tube rack
test tube holder
2 beakers (50 mL)
tongs
scoopula
laboratory burner
flint igniter
red litmus paper
wooden splint
wash bottle with distilled water
HCl solution
the following aqueous solutions in dropper bottles: $BaCl_2$, $CaCl_2$, $MgCl_2$, Na_2SO_4, NaOH, $AgNO_3$, $Pb(NO_3)_2$, KI, NH_4Cl, solid Na_2CO_3 and $NH_4C\ell$

Safety Precautions

- Hydrochloric acid is corrosive. Use care when handling it.
- Before lighting the laboratory burner, check that there are no flammable liquids nearby.
- If you accidentally spill a solution on your skin, immediately wash the area with copious amounts of water.
- Wash your hands thoroughly after the experiment.

Procedure

1. Copy Tables A and B into your notebook. Do not write in this textbook.
2. Place the well plate on top of the sheet of white paper.
3. Carry out each of the reactions in Table A by adding several drops of each solution to a well. Record your observations in Table A.

If you are unsure about the formation of a precipitate, repeat the reaction in a small test tube for improved visibility.

4. Place a scoopula tipful of Na_2CO_3 in a 50 mL beaker. Add 5 mL of HCl. Use a burning wooden splint to test the gas produced. Record your observations in Table B.

5. Place a scoopula tipful of NH_4Cl in a test tube. Add 2 mL NaOH. To detect any odour, gently waft your hand over the mouth of the test tube towards your nose. Warm the tube gently (do not boil) over a flame. Record your observations in Table B.

6. Dispose of all chemicals in the waste beaker supplied by your teacher. Do not pour anything down the drain.

Table A Double Displacement Reactions That May Form a Precipitate

Skeleton equation	Observations
$MgCl_2$ + NaOH	
$FeCl_3$ + NaOH	
$BaCl_2$ + Na_2SO_4	
$CaCl_2$ + $AgNO_3$	
$Pb(NO_3)_2$ + KI	

Table B Double Displacement Reactions That May Form a Gas

Reaction	Observations
Na_2CO_3 + HCI	
NH_4Cl + NaOH	

Analysis

1. Write the balanced chemical equation for each reaction in Table A.

2. For each reaction in Table B, write the appropriate balanced chemical equation for the double displacement reaction. Then write a balanced chemical equation for the decomposition reaction that leads to the formation of a gas and water.

Conclusion

3. How did you know when a double displacement reaction had occurred? How did your results compare with your predictions?

Application

4. Suppose that you did not have any information about the solubility of various compounds, but you did have access to a large variety of ionic compounds. What would you need to do before predicting the products of the displacement reactions above? Outline a brief procedure.

Investigation 4-C

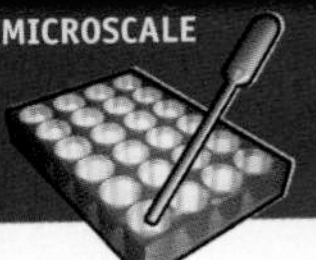

SKILL FOCUS

Predicting

Performing and Recording

Analyzing and Interpreting

From Copper to Copper

This experiment allows you to carry out the sequential conversion of copper metal to copper(II) nitrate to copper(II) hydroxide to copper(II) oxide to copper(II) sulfate and back to copper metal. This conversion is carried out using synthesis, decomposition, single displacement, and double displacement reactions.

Question

What type of chemical reaction is involved in each step of this investigation?

Prediction

Examine the five reactions outlined in the procedure. Predict what reactions will occur, and write equations to describe them.

Materials

hot plate
glass rod
wash bottle with distilled water
50 mL Erlenmeyer flask
beaker tongs
250 mL beaker containing ice and liquid water
red litmus paper
$Cu(NO_3)_2$ solution
*6 mol/L NaOH solution in dropping bottle
*3 mol/L H_2SO_4 solution in dropping bottle
0.8 g of flaked zinc

Safety Precautions

- Constantly stir, or swirl, any precipitate-containing solution that is being heated to avoid a sudden boiling over, or *bumping*.
- Unplug any hot plate not in use.
- Do not allow electrical cords to hang over the edge of the bench.
- NaOH and H_2SO_4 solutions are corrosive. Handle them with care. If you accidentally spill a solution on your skin, wash the area immediately with copious amounts of cool water. Inform your teacher.

Procedure

Reaction A: Reaction of Copper Metal with Copper(II) Nitrate

$$Cu_{(s)} + 4HNO_{3(aq)} \rightarrow Cu(NO_3)_{2(aq)} + 2NO_{2(g)} + 2H_2O_{(\ell)}$$

CAUTION Your teacher will carry out steps 1 to 4 in the fumehood before class. Concentrated nitric acid is required, and the $NO_{2(g)}$ produced is poisonous. Furthermore, this reaction is quite slow. Your teacher will perform a brief demonstration of this reaction so that you may record observations.

1. Place 0.100 g (100 mg) of Cu in a 50 mL Erlenmeyer flask.
2. Add 2 mL of 6 mol/L $HNO_{3(aq)}$ to the flask in the fumehood.
3. Warm the flask on a hot plate in the fumehood. The heating will continue until all the Cu dissolves and the evolution of brown $NO_{2(g)}$ ceases.
4. Cool the flask in a cool water bath.
5. Add about 2 mL of distilled water to the flask containing the $Cu(NO_3)_2$ solution.

*The unit mol/L refers to concentration. You will learn more about this in Unit 3. For now, you should know that 6 mol/L NaOH and 3 mol/L H_2SO_4 are highly corrosive solutions, and you should treat them with respect.

Reaction B: Preparation of Copper(II) Hydroxide

6. At room temperature, while stirring with a glass rod, add 6 mol/L NaOH, drop by drop, until the solution is basic to red litmus paper. (Red litmus paper turns blue in basic solution.) Do not put the red litmus paper in the solution. Dip the glass rod into the solution and touch it to the red litmus paper. Record your observations.

Reaction C: Preparation of Copper(II) Oxide

7. While constantly stirring the solution with a glass rod, heat the mixture from step 6 on a hot plate until a black precipitate is formed. If necessary, use the wash bottle to wash loose any unreacted light blue precipitate that is adhering to the side of the flask.

8. When all of the light blue precipitate has reacted to form the black precipitate, cool the flask in an ice bath or a cool water bath for several minutes.

Reaction D: Preparation of Copper(II) Sulfate Solution

9. Carefully add about 6 mL of 3 mol/L sulfuric acid to the flask. Stir it until all the black precipitate has dissolved. Record your observations. **CAUTION** The sulfuric acid is highly corrosive. If any comes in contact with your skin, rinse the area thoroughly and immediately with water.

Reaction E: Regeneration of Copper Metal

10. In the fumehood or in a well ventilated area, carefully add about 0.8 g of powdered zinc to the solution of copper(II) sulfate. Stir or swirl the solution until the blue colour disappears. Record your observations. **CAUTION** You should wear a mask for this step to avoid breathing in the powdered zinc.

11. When the reaction is complete, add 5 mL of 3 mol/L sulfuric acid while stirring or swirling the solution. This removes any unreacted zinc but does not affect the copper metal. Carefully decant the liquid into a clean waste container. Wash the copper metal carefully several times with water. Return the copper metal to your teacher. Wash your hands. **CAUTION** This sulfuric acid is highly corrosive. If any comes in contact with your skin, rinse the area thoroughly and immediately with water.

Analysis

1. What type of reaction is occurring in reactions A through E?
2. Write a balanced chemical equation for reactions B through E.
3. Explain why H_2SO_4 reacts with Zn but not with Cu. (See step 11 in the procedure.)
4. Could another metal have been used in place of Zn in step 10? Explain.
5. Why was powdered Zn used in step 10, rather than a single piece of Zn?
6. You used 0.100 g of Cu metal in reaction A. How much copper should theoretically be recovered at the end of reaction E?

Conclusion

7. Create a flowchart that shows each step of the reaction series. Include the balanced chemical equations.

Concept Organizer Predicting Chemical Reactions

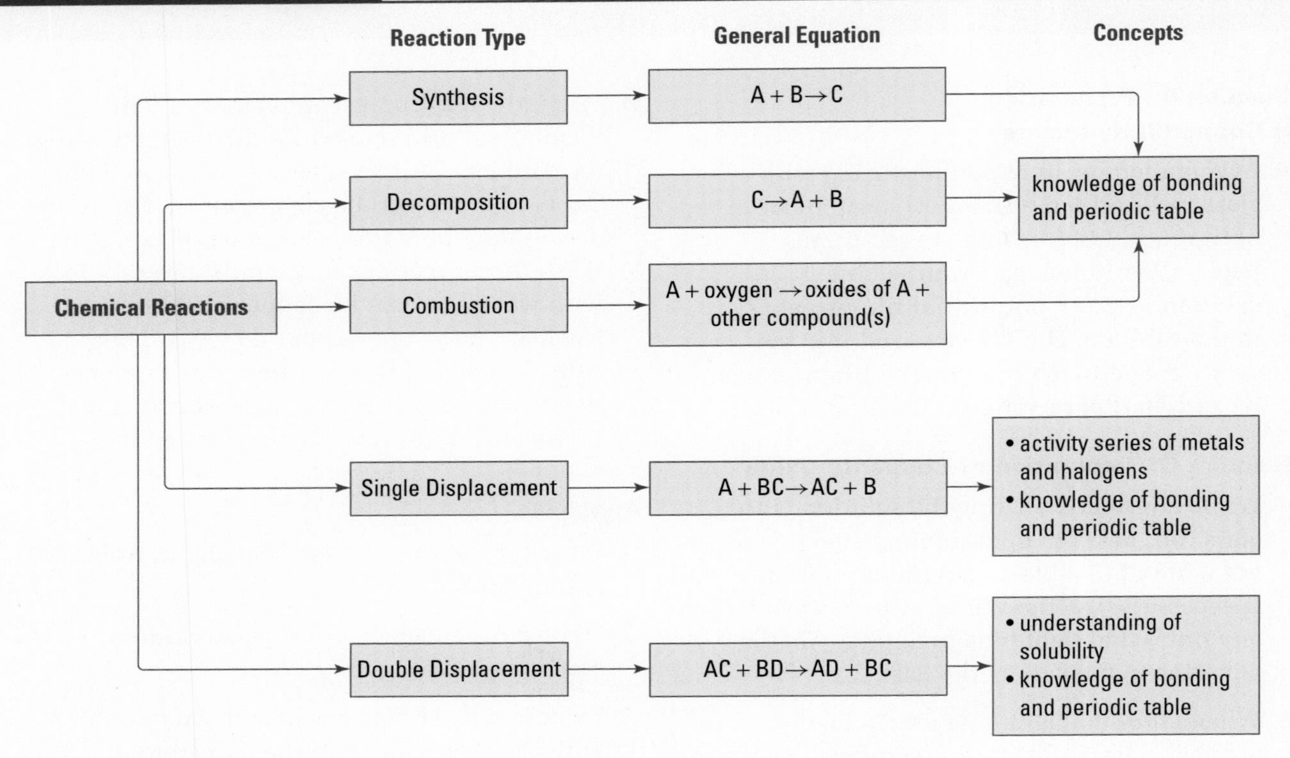

Section Wrap-up

In sections 4.2 and 4.3, you have examined five different types of chemical reactions: synthesis, decomposition, combustion, single displacement, and double displacement. Equipped with this knowledge, you can examine a set of reactants and predict what type of reaction will occur and what products will be formed. The Concept Organizer above provides a summary of the types of chemical reactions.

Section Review

Unit Project Prep

Before you design your Chemistry Newsletter at the end of Unit 1, take a look at some of the labels of chemical products in your home. Are there any warnings about mixing different products together? Use what you know about chemical reactions to explain why mixing some chemical products might be dangerous.

1 K/U Write the product(s) of each single displacement reaction. If you predict that there will be no reaction, write "NR." Balance each chemical equation.

(a) $Li + H_2O \rightarrow$

(b) $Sn + FeCl_2 \rightarrow$

(c) $F_2 + KI \rightarrow$

(d) $Al + MgSO_4 \rightarrow$

(e) $Zn + CuSO_4 \rightarrow$

(f) $K + H_2O \rightarrow$

2 K/U Complete each double displacement reaction. Be sure to indicate the physical state of each product. Then balance the equation.
Hint: Compounds containing alkali metal ions are soluble. Calcium chloride is soluble. Iron(III) hydroxide is insoluble.

(a) $NaOH + Fe(NO_3)_3 \rightarrow$

(b) $Ca(OH)_2 + HCl \rightarrow$

(c) $K_2CrO_4 + NaCl \rightarrow$

(d) $K_2CO_3 + H_2SO_4 \rightarrow$

3 **(a)** **C** Explain why the following chemical equation represents a double displacement reaction followed by a decomposition reaction.

$(NH_4)_2SO_{4(aq)} + KOH_{(aq)} \rightarrow NH_{3(g)} + H_2O_{(\ell)} + K_2SO_{4(aq)}$

(b) Balance the chemical equation in part (a).

4 **K/U** Identify each reaction as synthesis, combustion (complete or incomplete), decomposition, single displacement, or double displacement. Balance the equations, if necessary.

(a) $C_8H_{18(\ell)} + O_{2(g)} \rightarrow CO_{(g)} + H_2O_{(g)}$

(b) $Pb_{(s)} + H_2SO_{4(aq)} \rightarrow PbSO_{4(s)} + H_{2(g)}$

(c) $Al_2(SO_4)_{3(aq)} + K_2CrO_{4(aq)} \rightarrow K_2SO_{4(aq)} + Al_2(CrO_4)_{3(s)}$

(d) $C_3H_7OH_{(\ell)} + O_{2(g)} \rightarrow CO_{2(g)} + H_2O_{(g)}$

(e) $(NH_4)_2Cr_2O_{7(s)} \rightarrow N_{2(g)} + H_2O_{(g)} + Cr_2O_{3(s)}$

(f) $Mg_{(s)} + N_{2(g)} \rightarrow Mg_3N_{2(s)}$

(g) $N_2O_{4(g)} \rightarrow 2NO_{2(g)}$

5 **MC** Biosphere II was created in 1991 to test the idea that scientists could build a sealed, self-sustaining ecosystem. The carbon dioxide levels in Biosphere II were lower than scientists had predicted. Scientists discovered that the carbon dioxide was reacting with calcium hydroxide, a basic compound in the concrete.

(a) Write two balanced equations to show the reactions. Then classify the reactions. Hint: In the first reaction, carbon dioxide reacts with water in the concrete to form hydrogen carbonate. Hydrogen carbonate, an acid, reacts with calcium hydroxide, a base.

(b) Why do you think scientists failed to predict that this would happen?

(c) Suggest ways that scientists could have combatted the problem.

6 **K/U** What reaction is shown in the figure below? Write a balanced chemical equation to describe the reaction, then classify it.

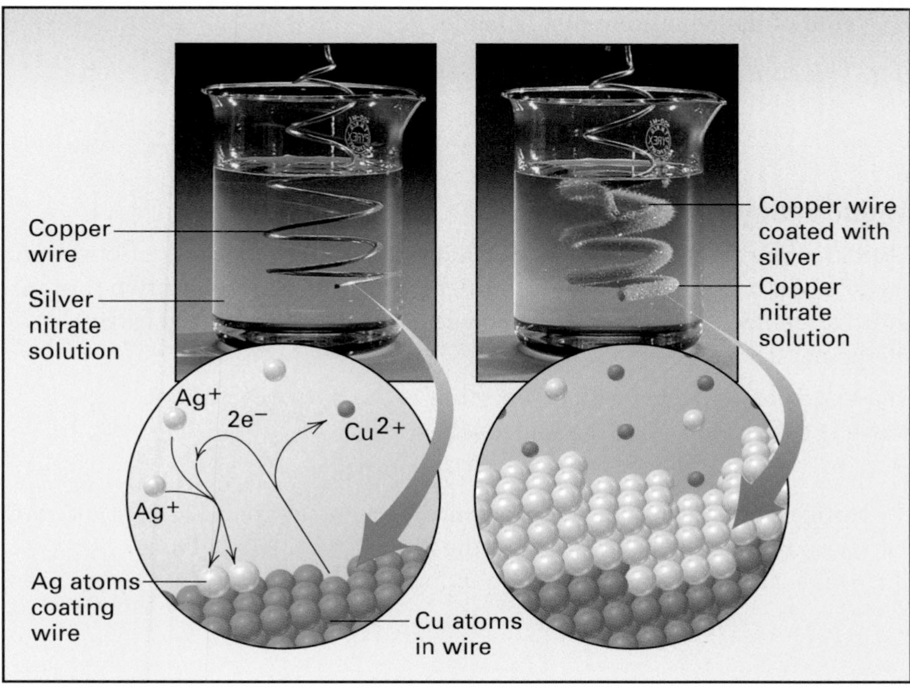

4.4 Simple Nuclear Reactions

Section Preview/ Specific Expectations

In this section, you will

- **balance** simple nuclear equations
- **communicate** your understanding of the following terms: *nuclear reactions, nuclear equation, alpha* (α) *particle emission, beta* (β) *decay, beta particle, gamma* (γ) *radiation, nuclear fission, nuclear fusion*

You have seen some chemical reactions that involve the formation and decomposition of different compounds. These reactions involve the rearrangement of atoms due to the breaking and formation of chemical bonds. Chemical bonds involve the interactions between the electrons of various atoms. There is another class of reactions, however, that are not *chemical*. These reactions involve changes that occur within the nucleus of atoms. These reactions are called **nuclear reactions**.

We know that nuclear weapons are capable of mass destruction, yet radiation therapy, shown in Figure 4.18, is a proven cancer fighter. Smoke detectors, required by law in all homes, rely on the radioactive decay of americium-241. The human body itself is radioactive, due to the presence of radioactive isotopes including carbon-14, phosphorus-32, and potassium-40. Most people view radioactivity and nuclear reactions with a mixture of fascination, awe, and fear. Since radioactivity is all around us, it is important to understand what it is, how it arises, and how we can deal with it safely.

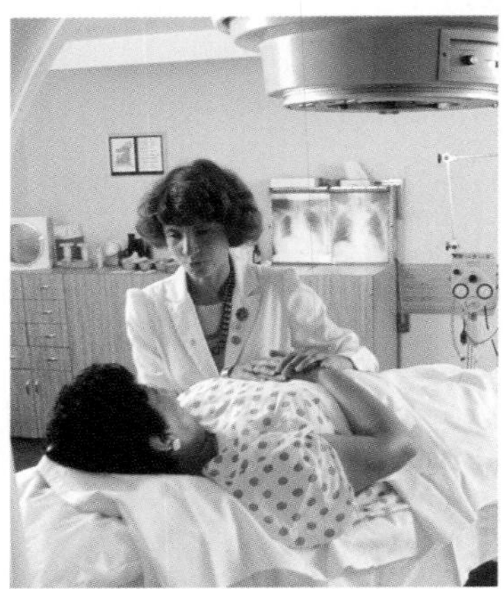

Figure 4.18 **This patient is about to undergo radiation therapy.**

Media LINK

The media are full of references to radioactivity. For example, the comic book hero Spiderman gained the abilities of a spider after being bitten by a radioactive spider. Bart Simpson reads *Radioactive Man* comics. Can you think of any other references to radioactivity in popular culture? What kind of "reputation" do radioactivity and nuclear reactions have? Do you think this reputation is deserved?

Types of Radioactive Decay and Balancing Nuclear Equations

There are three main types of radioactive decay: alpha particle emission, beta particle emission, and the emission of gamma radiation. When an unstable isotope undergoes radioactive decay, it produces one or more different isotopes. We represent radioactive decay using a **nuclear equation.** Two rules for balancing nuclear equations are given below.

Rules for Balancing Nuclear Equations

1. The sum of the mass numbers (written as superscripts) on each side of the equation must balance.
2. The sum of the atomic numbers (written as subscripts) on each side of the equation must balance.

Alpha Decay

Alpha (α) particle emission, or alpha decay, involves the loss of one alpha particle. An α particle is a helium nucleus, ${}^{4}_{2}He$, composed of two protons and two neutrons. Since it has no electrons, an alpha particle carries a charge of +2.

One example of alpha particle emission is the decay of radium. This decay is shown in the following equation:

$${}^{226}_{88}Ra \rightarrow {}^{222}_{86}Rn + {}^{4}_{2}He$$

Notice that the sum of the mass numbers on the right (222 + 4) equals the mass on the left (226). As well, the atomic numbers balance (88 = 86 + 2). Thus, this nuclear equation is balanced.

In another example of alpha particle emission, Berkelium-248 is formed by the decay of a certain radioisotope according to the balanced nuclear equation:

$$^{a}_{b}X \rightarrow \, ^{248}_{97}\text{Bk} + \, ^{4}_{2}\text{He}$$

Given this information, what is $^{a}_{b}X$? You can use your knowledge of how to balance a nuclear equation to determine the identity of a radioisotope undergoing alpha particle decay.

The total of the atomic masses on the right side is $(248 + 4) = 252$. The total of the atomic numbers on the right is $(97 + 2) = 99$. Therefore, $a = 252$ and $b = 99$. From the periodic table, you see that element number 99 is Es, einsteinium. The missing atom is $^{252}_{99}\text{Es}$, so the balanced nuclear equation is:

$$^{252}_{99}\text{Es} \rightarrow \, ^{248}_{97}\text{Bk} + \, ^{4}_{2}\text{He}$$

Try the following problems to practise balancing alpha emission nuclear reactions.

Practice Problems

29. Uranium was the first element shown to be radioactive. Complete the following reaction representing the alpha decay of uranium-238.

$$^{238}_{92}\text{U} \rightarrow \quad \_\_\_\_ \quad + \, ^{4}_{2}\text{He}$$

30. Radon-222, $^{222}_{86}\text{Rn}$, is known to decay by alpha particle emission. Write a balanced nuclear equation and name the element produced in this decay process.

31. Plutonium-242 decays by emitting an alpha particle. Write the balanced nuclear equation for this reaction.

32. Neodymium-144, $^{144}_{60}\text{Nd}$, decays by alpha particle emission. Write the balanced nuclear equation for this nuclear decay.

History LINK

Marie Curie discovered the element polonium, Po, in 1898. She named polonium after Poland, her homeland. Curie won two Nobel Prizes, one in Physics (1903) for sharing in the discovery of radioactivity, and one in Chemistry (1911) for the discovery of radium, which has been used to treat cancer. Radium-226 undergoes alpha decay to yield radon-222.

Beta Decay

Beta (β) decay occurs when an isotope emits an electron, called a **beta particle**. Because of its tiny mass and −1 charge, a beta particle, is represented as $^{0}_{-1}\text{e}$. For example, hydrogen-3, or tritium, emits a beta particle to form helium-3 as illustrated by the equation:

$$^{3}_{1}\text{H} \rightarrow \, ^{3}_{2}\text{He} + \, ^{0}_{-1}\text{e}$$

Notice that the total of the atomic masses and the total of the atomic numbers on each side of the nuclear equation balance. What is happening, however to the hydrogen-3 nucleus as this change occurs? In effect, the emission of a beta particle is accompanied by the conversion, inside the nucleus, of a neutron into a proton:

$^{1}_{0}\text{n}$	$\rightarrow$	$^{1}_{1}\text{H}$	+	$^{0}_{-1}\text{e}$
neutron		proton		electron (β particle)

CHEM FACT

One of the most harmful potential sources of radiation in the home is radon gas. Radon-222 is a product of the decay of uranium-containing rocks beneath Earth's surface. Since radon is denser than air, it can build up to dangerous levels in basements when it seeps through cracks in walls and floors. Simple radon detectors can be purchased at hardware stores.

CHEM FACT

The nucleus of the most common isotope of hydrogen consists of one proton. Therefore, a proton can be represented by H^+ or $^{1}_{1}H$.

Carbon-14 is a radioactive isotope of carbon. Its nucleus emits a beta, particle to form a nitrogen-14 nucleus, according to the balanced nuclear equation shown below in Figure 4.19.

$$^{14}_{6}C \rightarrow {}^{14}_{7}N + {}^{0}_{-1}e$$

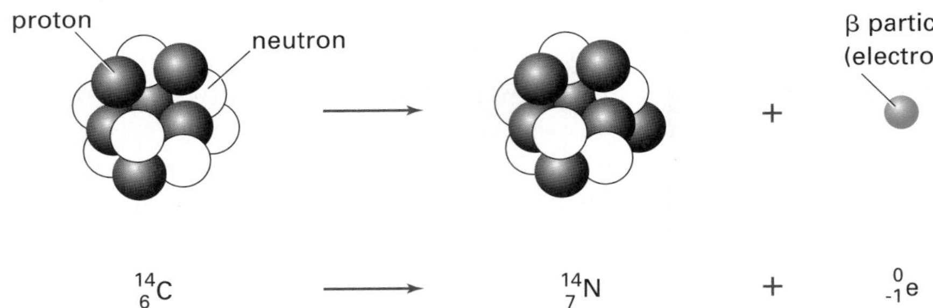

$$^{14}_{6}C \longrightarrow {}^{14}_{7}N + {}^{0}_{-1}e$$

Figure 4.19 Carbon-14 decays by emitting a beta particle and converting to nitrogen-14. Notice that a neutron in the nucleus of carbon is converted to a protron as the β particle is emitted.

CHEM FACT

The terms radiation and radioactivity are often confused. Radiation refers to electromagnetic radiation— everything from gamma rays, to X-rays, to visible light, to microwaves, to radio and television signals. Radioactivity, on the other hand, involves the emission of particles or energy from an unstable nucleus.

Radioactive waste from certain nuclear power plants and from weapons testing can lead to health problems. For example, ions of the radioactive isotope strontium-90, an alkali metal, exhibit chemical behaviour similar to calcium ions. This leads to incorporation of the ions in bone tissue, sending ionizing radiation into bone marrow, and possibly causing leukemia. Given the following equation for the decay of strontium-90, how would you complete it?

$$^{90}_{38}Sr \rightarrow \square + {}^{0}_{-1}e$$

Since both atomic numbers and mass numbers must balance, you can find the other product.

The mass number of the unknown element is equal to $90 + 0 = 90$. The atomic number of the unknown element is equal to $38 - (-1) = 39$. From the periodic table, you can see that element 39 is titanium, Ti. The balanced nuclear equation is therefore

$$^{90}_{38}Sr \rightarrow {}^{90}_{39}Ti + {}^{0}_{-1}e$$

You can check your answer by ensuring that the total mass number and the total atomic number on each side of the equation are the same.

Mass numbers balance: $90 = 90 + 0$

Atomic numbers balance: $38 = 39 + (-1)$

Try the following problems to practise balancing beta emission equations.

Practice Problems

33. Write the balanced nuclear equation for the radioactive decay of potassium-40 by emission of a β particle.

34. What radioisotope decays by β particle emission to form $^{47}_{21}Sc$?

35. Complete the following nuclear equation:

$$^{73}_{31}Ga \rightarrow {}^{0}_{-1}e + \square$$

Canadians in Chemistry

Harriet Brooks

It was the Nobel that just missed being Canadian. In 1907, Ernest Rutherford left Montréal's McGill University for a position in England. The following year, he received the Nobel Prize for Chemistry for his investigations of the chemistry of radioactive substances. Most of the work, however, had been done in Montréal. Moreover, one young Canadian woman had played an important role in putting it on the right track.

Harriet Brooks is nearly forgotten today, even though she helped to show that elements could be transformed. For over a century, chemists had rejected the dream of the ancient alchemists who thought that they might turn lead into gold with the help of the philosopher's stone. They believed that elements were forever fixed and unchangeable.

Then Harriet Brooks arrived on the scene. When she joined Rutherford's team, she was asked to measure the atomic mass of the isotopes that make up the mysterious vapour given off by radium. She determined that its atomic mass was between 40 and 100, whereas radium was known to have an atomic mass of over 140. Surely this was not just a gaseous form of radium. Somehow radium was turning into another element!

It turned out that Brooks' result was a mistake. Radon—as the mystery gas is now known—has almost the same atomic mass as radium. Brooks' result was a fruitful mistake, however. Her experiment led to a basic understanding of radioactivity and isotopes.

Why did Rutherford win a Nobel Prize for Chemistry? Both he and Brooks worked as physicists. By proving that elements transformed, Rutherford, Brooks, and their co-workers revolutionized traditional chemistry.

Gamma Radiation

Gamma (γ) radiation is high energy electromagnetic radiation. It often accompanies either alpha or beta particle emission. Since gamma radiation has neither mass nor charge, it is represented as ${}^{0}_{0}\gamma$, or simply γ. For example, cesium-137 is a radioactive isotope that is found in nuclear fall-out. It decays with the emission of a beta particle and gamma radiation, according to the equation

$${}^{137}_{55}\text{Cs} \rightarrow {}^{137}_{56}\text{Ba} + {}^{0}_{-1}\text{e} + {}^{0}_{0}\gamma$$

How is gamma radiation produced in a radioactive decay? When a radioactive nucleus emits an alpha or beta particle, the nucleus is often left in an unstable, high-energy state. The "relaxation" of the nucleus to a more stable state is accompanied by the emission of gamma radiation.

Nuclear Fission and Fusion

All cases of radioactive decay involve the atom's nucleus. Since these processes do not involve the atom's electrons, they occur regardless of the chemical environment of the nucleus. For example, radioactive hydrogen-3, or tritium, will decay by β particle emission whether it is contained in a water molecule or hydrogen gas, or in a complex protein.

Mode	Emission	Decay			Change in ...		
					Mass numbers	Atomic numbers	Number of neutrons
α Decay	α (^4_2He)	reactant →	product +	α expelled	−4	−2	−2
β Decay	$^0_{-1}\beta$	^1_0n in nucleus →	^1_1p in nucleus +	$^0_{-1}\beta$ β expelled	0	+1	−1
γ Emission	$^0_0\gamma$	excited nucleus →	stable nucleus +	$^0_0\gamma$ γ photon radiated	0	0	0

Figure 4.20 A summary of alpha decay, beta decay, and gamma emission

Many chemical reactions, once begun, can be stopped. For example, a combustion reaction, such as a fire, can be extinguished before it burns itself out. Nuclear decay processes, on the other hand, cannot be stopped.

The principles of balancing nuclear equations apply to all nuclear reactions. **Nuclear fission** occurs when a highly unstable isotope splits into smaller particles. Nuclear fission usually has to be induced in a particle accelerator. Here, an atom can absorb a stream of high-energy particles such as neutrons, ^1_0n. This will cause the atom to split into smaller fragments.

For example, when uranium-235 absorbs a high energy neutron, ^1_0n, it breaks up, or undergoes fission as follows:

$$^{235}_{92}U + ^1_0n \rightarrow ^{87}_{35}Br + \_\_\_\_ + 3^1_0n$$

How would you identify the missing particle? Notice that three neutrons, ^1_0n, have a mass number of 3 and a total atomic number of 0. The total atomic mass on the left side is $(235 + 1) = 236$. On the right we have $(87 + 3(1)) = 90$, and so $(236 - 90) = 146$ remains. The missing particle must have a mass number of 146.

The total atomic number on the left is 92. The total atomic number on the right is 35. This means that $(92 - 35) = 57$ is the atomic number of the missing particle. From the periodic table, atomic number 57 corresponds to La, lanthanum. The balanced nuclear equation is

$$^{235}_{92}U + ^1_0n \rightarrow ^{87}_{35}Br + ^{146}_{57}La + 3^1_0n$$

Check your answer by noting that the total mass number and the total atomic number are the same on both sides.

Nuclear fusion occurs when a target nucleus absorbs an accelerated particle. The reaction that takes place in a hydrogen bomb is a fusion reaction, as are the reactions that take place within the Sun, shown in Figure 4.21. Fusion reactions require very high temperatures to proceed but produce enormous amounts of energy. The fusion reaction that takes place in a hydrogen bomb is represented by the following equation:

$$^{6}_{3}Li + ^{1}_{0}n \rightarrow ^{3}_{1}H + ^{4}_{2}He$$

Notice that the total mass numbers and the total atomic numbers are the same on both sides.

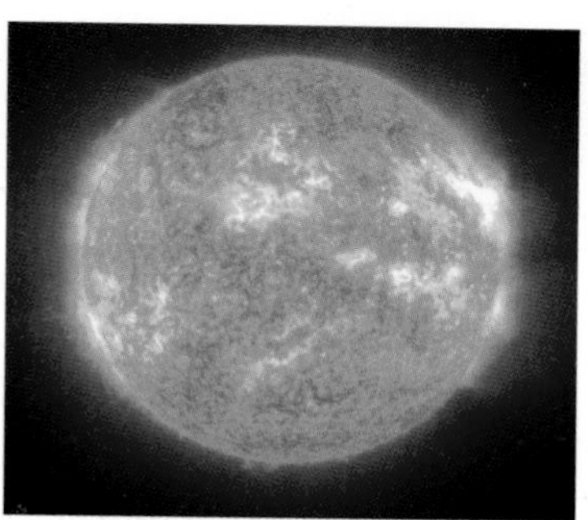

Figure 4.21 The Sun's interior has a temperature of about 15 000 000 °C, due to the energy provided by nuclear fusion reactions.

Practice Problems

36. Astatine can be produced by the bombardment of a certain atom with alpha particles, as follows:

$\_\_\_\_ + ^{4}_{2}He \rightarrow ^{211}_{85}At + 2^{1}_{0}n$

Identify the atom.

37. Balance the following equation by adding a coefficient.

$^{252}_{96}Cf + ^{10}_{5}B \rightarrow ^{257}_{101}Md + \_\_\_\_ ^{1}_{0}n$

38. How many neutrons are produced when U-238 is bombarded with C-12 nuclei in a particle accelerator? Balance the following equation.

$^{238}_{92}U + ^{12}_{6}C \rightarrow ^{246}_{98}Cf + \_\_\_\_ ^{1}_{0}n$

39. Aluminum-27, when it collides with a certain nucleus, transforms into phosphorus-30 along with a neutron. Write a balanced nuclear equation for this reaction.

Section Wrap-Up

In this chapter, you learned how atoms can interact with each other and how unstable isotopes behave. In the first three sections, you learned about chemical reactions. In section 4.4, you learned about different types of nuclear reactions: reactions in which atoms of one element change into atoms of another element. In Unit 2, you will learn how stable isotopes contribute to an indirect counting method for atoms and molecules.

Section Review

1 K/U Draw a chart in your notebook to show alpha decay, beta decay, gamma decay, nuclear fusion, and nuclear fission. Write a description and give an example of each type of reaction. Illustrate each example with a drawing.

2 K/U Complete each nuclear equation. Then state the type of nuclear reaction that each equation represents.

(a) $^{232}_{90}Th + \_\_\_\_ \rightarrow ^{233}_{90}Th$

(b) $^{233}_{91}Pa \rightarrow ^{233}_{92}U + \_\_\_\_$

(c) $^{226}_{88}Ra \rightarrow \_\_\_\_ + ^{4}_{2}He$

(d) $^{210}_{83}Bi \rightarrow ^{206}_{81}Tl + \_\_\_\_$

9. Write a balanced chemical equation corresponding to each word equation.
   (a) The reaction between aqueous sodium hydroxide and iron(III) nitrate produces a precipitate.
   (b) Powdered antimony reacts with chlorine gas to produce antimony trichloride.
   (c) Mercury(II) oxide is prepared from its elements.
   (d) Ammonium nitrite decomposes into nitrogen gas and water.
   (e) Aluminum metal reacts with a solution of zinc sulfate to produce aluminum sulfate and metallic zinc.
10. Consider the unbalanced chemical equation corresponding to the formation of solid lead(II) chromate, $PbCrO_4$:
   $Pb(NO_3)_{2(aq)} + K_2CrO_{4(aq)} \rightarrow PbCrO_{4(s)} + KNO_{3(aq)}$
   (a) What type of chemical reaction is this?
   (b) Balance the equation.
11. In general, what is formed when an oxide of a non-metal reacts with water? Give an example.
12. In general, what is formed when an oxide of a metal reacts with water? Give an example.
13. Complete and balance each nuclear equation. Then classify the reaction.
   (a) ${}^{2}_{1}H + {}^{3}_{1}H \rightarrow {}^{4}_{2}He +$ ______
   (b) ${}^{239}_{92}U \rightarrow$ ______ $+ {}^{0}_{-1}\beta$
   (c) ${}^{239}_{93}Np \rightarrow {}^{239}_{94}Pu +$ ______
   (d) ${}^{238}_{92}U \rightarrow {}^{234}_{90}Th +$ ______ $+ 2{}^{0}_{0}\gamma$
14. Write the product(s) for each reaction. If you predict that there will be no reaction, write "NR." Balance each chemical equation.
   (a) $BaCl_{2(aq)} + Na_2CO_{3(aq)} \rightarrow$
   (b) $Fe_{(s)} + CuSO_{4(aq)} \rightarrow$
   (c) $C_2H_{2(g)} + O_{2(g)} \rightarrow$
   (d) $PCl_{5(s)} \rightarrow$ ______ $+ Cl_{2(g)}$
   (e) $Mg_{(s)} + Fe_2O_3 \rightarrow$
   (f) $Ca_{(s)} + Cl_{2(g)} \rightarrow$
15. Iron often occurs as an oxide, such as Fe_2O_3. In the steel industry, Fe_2O_3 is reacted with carbon monoxide to produce iron metal and carbon dioxide. Write the balanced chemical equation for this reaction, and classify it.
16. Calcium chloride is often used to melt ice on roads and sidewalks, or to prevent it from forming. Calcium chloride can be made by reacting hydrochloric acid with calcium carbonate. Write the balanced chemical equation corresponding to this reaction, and classify it.

Inquiry

17. An American penny is composed of a zinc core clad in copper. Some of the copper is filed away, exposing the zinc, and placed in a solution of hydrochloric acid. Describe what will occur.
18. What will happen to a silver earring that is accidentally dropped into toilet bowl cleaner that contains hydrochloric acid?

Communication

19. Explain why it is advisable to store chemicals in tightly sealed bottles out of direct sunlight.
20. Why is smoking not allowed near an oxygen source? What would happen if a match were struck in an oxygen-rich atmosphere?
21. Even if a smoker is very careful not to let a lighted cigarette come in contact with liquid gasoline, why is it very dangerous to smoke when refuelling an automobile?
22. Solutions that have been used to process film contain silver ions, $Ag^{+}_{(aq)}$.
   (a) Explain how you could recover the silver, in the form of an ionic compound.
   (b) How could you recover the silver as silver metal?

Making Connections

23. Calcium oxide, CaO (lime), is used to make mortar and cement.
   (a) State two reactions that could be used to make lime. Classify each reaction, based on the types of reactions studied in this chapter.
   (b) In construction, cement is prepared by mixing the powdered cement with water. Write the chemical equation that represents the reaction of calcium oxide with water. Why are we cautioned not to expose skin to dry cement mix *and* wet cement? It may help you to know that bases are often corrosive. They can burn exposed skin.

Answers to Practice Problems and Short Answers to Section Review Questions

Practice Problems: 1.(a) calcium + fluorine (reactants) → calcium fluoride (product) **(b)** barium chloride + hydrogen sulfate → hydrogen chloride + barium sulfate **(c)** calcium carbonate + carbon dioxide + water → calcium hydrogen carbonate **(d)** hydrogen peroxide → water + oxygen **(e)** sulfur dioxide + oxygen → sulfur trioxide **2.** Sugar → ethanol + carbon dioxide **3.(a)** $Zn_{(s)} + Cl_{2(g)} \rightarrow ZnCl_{2(s)}$ **(b)** $Ca_{(s)} + H_2O_{(l)} \rightarrow Ca(OH)_{2(aq)} + H_{2(g)}$ **(c)** $Ba_{(s)} + S_{(s)} \rightarrow BaS_{(s)}$ **(d)** $Pb(NO_3)_{2(aq)} + Mg_{(s)} \rightarrow Mg(NO_3)_{2(aq)} + Pb_{(s)}$ **4.(a)** $CO2_{(g)} + CaO_{(s)} \rightarrow CaCO_{3(s)}$ **(b)** $Al_{(s)} + O_{2(g)} \rightarrow Al_2O_{3(s)}$ **(c)** $Mg_{(s)} + O_{2(g)} \rightarrow MgO_{(s)}$ **5.(a)** $S_{(s)} + O_{2(g)} \rightarrow SO_{2(g)}$ **(b)** $P_{4(s)} + 5O_{2(g)} \rightarrow P_4O_{10(s)}$ **(c)** $H_{2(g)} + Cl_{2(g)} \rightarrow 2HCl_{(g)}$ **(d)** $SO_{2(g)} + H_2O_{(l)} \rightarrow H_2SO_{3(aq)}$ **6.(a)** balanced **(b)** $2HgO_{(s)} \rightarrow 2Hg_{(l)} + O_{2(g)}$ **(c)** $H_2O_{2(aq)} \rightarrow 2H_2O_{(l)} + O_{2(g)}$ **(d)** balanced **7.(a)** $2SO_{2(g)} + O_{2(g)} \rightarrow 2SO_{3(g)}$ **(b)** $BaCl_{2(aq)} + Na_2SO_{4(aq)} \rightarrow NaCl_{(aq)} + BaSO_{4(s)}$ **8.** $P_{4(s)} + 5O_{2(g)} \rightarrow P_4O_{10(s)}; P_4O_{10(s)} + 6H_2O_{(l)} \rightarrow 4H_3PO_{4(aq)}$ **9.(a)** $As_4S_{6(s)} + 9O_{2(g)} \rightarrow As_4O_{6(s)} + 6SO_{2(g)}$ **(b)** $Sc_2O_{3(s)} + 3H_2O_{(l)} \rightarrow 2Sc(OH)_{3(s)}$ **(c)** $C_2H_5OH_{(l)} + 3O_{2(g)} \rightarrow 2CO_{2(g)} + 3H_2O_{(l)}$ **(d)** $2C_4H_{10(g)} + 9O_{2(g)} \rightarrow 8CO_{(g)} + 10H_2O_{(g)}$ **10.(a)** $2K + Br_2 \rightarrow 2KBr$ **(b)** $H_2 + Cl_2 \rightarrow 2HCl$ **(c)** $Ca + Cl_2 \rightarrow CaCl_2$ **(d)** $Li + O_2 \rightarrow LiO_2$ **11.(a)** products are Fe2O3, FeO **(b)** possible products: V_2O_5, VO, V_2O_3, VO_2 **(c)** possible products: TiO_2, TiO, Ti_2O_3 **12.(a)** $K_2O + H_2O \rightarrow 2KOH$ **(b)** $MgO + H_2O \rightarrow Mg(OH)_2$ **(c)** $SO_2 + H_2O \rightarrow H_2SO_3$ **13.** $NH_{3(g)} + HCl_{(g)} \rightarrow NH_4Cl_{(s)}$ **14.** Hg, O_2 **15.(a)** $2HI \rightarrow H_2 + I_2$ **(b)** $2Ag_2O \rightarrow 4Ag + O_2$ **(c)** $2AlCl_3 \rightarrow 2Al + 3Cl_2$ **(d)** $MgO \rightarrow Mg + O_2$ **16.(a)** $MgCO_3 \rightarrow MgO + CO_2$ **(b)** $CuCO_3 \rightarrow CuO + CO_2$ **17.** $2CH_3OH + 3O_2 \rightarrow 2CO_2 + 4H_2O$ **18.** $2C_8H_{18} + 25O_2 \rightarrow 16CO_2 + 18H_2O$ **19.** $C_3H_6O + O_2 \rightarrow CO_2 + H_2O$ **20.** $2C_{16}H_{34} + 49O_2 \rightarrow 32CO_2 + 34H_2O$ **21(a)** $Ca + H_2O \rightarrow CaO + H_2$ **(b)** $Zn + Pb(NO_3)_2 \rightarrow Zn(NO_3)_2 + Pb$ **(c)** $2Al + 6HCl \rightarrow 2AlCl_3 + 3H_2$ **(d)** $Li + AgNO_3 \rightarrow Ag + LiNO_3$ **(e)** $Pb + H_2SO_4 \rightarrow PbSO_4 + H_2$ **(f)** $2Mg + Pt(OH)_4 \rightarrow 2Mg(OH)_2 + Pt$ **(g)** $Ba + FeCl_2 \rightarrow BaCl_2 + Fe$ **(h)** $Fe + Co(ClO_3)_2 \rightarrow Fe(ClO_3)_3 + Co$ **22.(a)** NR **(b)** $Zn + FeCl_2 \rightarrow ZnCl_2 + Fe$ **(c)** $K + H_2O \rightarrow KOH + H_2$ **(d)** $2Al + 3H_2SO_4 \rightarrow Al_2(SO_4)_3 + 3H_2$ **(e)** NR **(f)** NR **(g)** $Zn + H_2SO_4 \rightarrow ZnSO_4 + H_2$ **(h)** $Mg + SnCl_2 \rightarrow MgCl_2 + Sn$ **23.(a)** NR **(b)** $Cl_2 + 2NaI \rightarrow 2NaCl + I_2$ **24.(a)** $2Pb + 2HCl \rightarrow 2PbCl + H_2$ **(b)** $KI + Br_2 \rightarrow KBr + I_2$ **(c)** NR **(d)** $Ca + H_2O \rightarrow Ca(OH)_2 + H_2$ **(e)** NR **(f)** $Ni + H_2SO_4 \rightarrow NiSO_4 + H_2$ **25.(a)** $Pb(NO_3)_{2(aq)} + 2KI_{(aq)} \rightarrow 2KNO_{3(aq)} + PbI_{2(s)}$ **(b)** NR **(c)** NR **(d)** $Ba(NO_3)_{2(aq)} + Mg(SO_4)_{(aq)} \rightarrow BaSO_{4(s)} + Mg(NO_3)_{2(aq)}$ **26.(a)** $Na_2SO_{3(aq)} + 2HCl_{(aq)} \rightarrow SO_{2(g)} + 2NaCl_{(aq)} + H_2O_{(l)}$ **(b)** $CaS_{(aq)} + H_2SO_{4(aq)} \rightarrow H_2S_{(g)} + CaSO_{4(l)}$ **27.(a)** $HCl_{(aq)} + LiOH_{(aq)} \rightarrow H_2O_{(l)} + LiCl_{(aq)}$ **(b)** $HclO_{4(aq)} + Ca(OH)_{2(aq)} \rightarrow H_2O_{(l)} + Ca(ClO_4)_{2(aq)}$ **(c)** $H_2SO_{4(aq)} + NaOH_{(aq)} \rightarrow Na_2SO_{4(aq)} + H_2O_{(l)}$ **28.(a)** $BaCl_{2(aq)} + Na_2CrO_{4(aq)} \rightarrow BaCrO_{4(s)} + 2NaCl_{(aq)}$ **(b)** $HNO_{3(aq)} + NaOH_{(aq)} \rightarrow H_2O_{(l)} + NaNO_{3(aq)}$ **(c)** $K_2CO_{3(aq)} + 2HNO_{3(aq)} \rightarrow H_2O_{(l)} + 2KNO_{3(aq)} + CO_{2(g)}$ **29.** [234/90]Th **30.** [222/86]Rn → [4/2]He + [218/82]Pb **31.** [242/94]Pu → [4/2]He + [238/92]U **32.** [144/60]Nd → [4/2]He + [140/58]Ce **33.** [40/19]K → [0/ – 1]e + [40/20]Ca **34.** [47/20]Ca **35.** [73/31]Ga → [0/ – 1]e + [73/32]Ge **36.** [208/83]Bi **37.** 5 **38.** 4 **39.** [27/13]Al+ [4/2]He → [30/15]P + [1/0]n

Section Review: 4.1: 2.(a) $2SO_{2(g)} + O_{2(g)} \rightarrow 2SO_{3(g)}$ **(b)** $Na_{(s)} + H_2O_{(l)} \rightarrow H_{2(g)} + NaOH_{(aq)}$ **(c)** $Cu_{(s)} + HNO_{3(aq)} \rightarrow Cu(NO_3)_{2(aq)} + NO_{2(g)} + H_2O_{(l)}$ **4.(a)** $4Al_{(s)} + 3O_{2(g)} \rightarrow 42Al_2O_{3(s)}$ **(b)** $2Na_2S_2O_{3(aq)} + I_{2(aq)} \rightarrow 2NaI_{(aq)} + Na_2S_4O_{6(aq)}$ **(c)** $2Al_{(s)} + Fe_2O_{3(s)} \rightarrow Al_2O_{3(s)} + 2Fe_{(s)}$ **(d)** $4NH_{3(g)} + 5O_{2(g)} \rightarrow 4NO_{(g)} + 6H_2O_{(l)}$ **(e)** $Na_2O_{(s)} + (NH_4)_2SO_{4(aq)} + H_2O_{(l)} + NH_{3(aq)}$ **(f)** $C_5H_{12(l)} + 8O_{2(g)} \rightarrow 5CO_{2(g)} + 6H_2O_{(g)}$ **5.** $Fe_{(s)} + CuSO_{4(aq)} \rightarrow Cu_{(s)} + FeSO_{4(aq)}$ **4.2: 1.(a)** $Be + O_2 \rightarrow BeO$ **(b)** $2Li + Cl_2 \rightarrow 2LiCl$ **(c)** $Mg + N_2 \rightarrow Mg_3N_2$ **(d)** $Ca + Br_2 \rightarrow CaBr_2$ **2.(a)** $2K_2O \rightarrow O_2 + 4K$ **(b)** $2CuO \rightarrow 2Cu + O_2$ **(c)** $2H_2O \rightarrow 2H_2 + O_2$ **(d)** $2Ni_2O_3 \rightarrow 4Ni + 3O_2$ **(e)** $2Ag_2O \rightarrow 4Ag + O_2$ **3.(a)** $Sn(OH)_{4(s)} \rightarrow SnO_{2(s)} + 2H_2O_{(g)}$, decomposition **(b)** $3Cl_2(g) + I_{2(s)} \rightarrow 2ICl_3$ synthesis **(c)** $C_4H_9OH + 6O_2 \rightarrow 5H_2O + 4CO_2$ **5.** $2HgO_{(s)} \rightarrow O_{2(g)} + 2Hg_{(s)}$ decomposition **4.3: 1. (a)** $Li + H_2O \rightarrow Li_2O + H_2$ **(b)** NR **(c)** $F_2 + 2KI \rightarrow 2KF + I_2$ **(d)** NR **(e)** $Zn + CuSO_4 \rightarrow Cu + ZnSO_4$ **(f)** $K + H_2O \rightarrow K_2O + H_2$ **2.(a)** $NaOH_{(aq)} + Fe(NO_3)_{3(aq)} \rightarrow NaNO_{3(aq)} + Fe(OH)_{3(s)}$ **(b)** $Ca(OH)_{2(aq)} + HCl_{(aq)} \rightarrow CaCl_{2(aq)} + H_2O)_{(l)}$ **(c)** NR **(d)** $K_2CO_{3(s)} + H_2SO_{4(aq)} \rightarrow K_2SO_{4(aq)} + CO_{2(g)} + H_2O_{(l)}$ **3.(b)** $(NH_4)_2SO_{4(aq)} + 2KOH_{(aq)} \rightarrow 2NH_{3(g)} + 2H_2O_{(l)} + K_2SO_{4(aq)}$ **4.(a)** incomplete combustion **(b)** single displacement **(c)** double displacement **(d)** complete combustion **(e)** decomposition **(f)** synthesis **(g)** decomposition **5.4: 2.(a)** [1/0]n **(b)** [0/ – 1]e **(c)** [222/26]Fe **(d)** [4/2]He **(e)** [236/92]U **(f)** [4/2]He

Developing a Chemistry Newsletter

cause dark spots (removable with silver polish).
• Do not spill on fabric – contains chlorine bleach.
For Tough Cleaning Jobs
• Pretreat by applying Cascade LiquiGel directly to dishes.
WARNING: DON'T LET YOUNG CHILDREN USE BOTTLE OR TOU
DISHWASHER. SCREW ON PROTECTIVE CAP TIGHTL
Contains chlorine bleach, carbonate and silicate salts. Not for hand dishwashing.
dishwashing liquids, other cleaning products or ammonia as irritating fumes may resu
EMERGENCY FIRST AID TREATMENT: •If swallowed or gets in mouth -
glass of water or milk, and call a poison control center or physician. •Eye
thoroughly with water. •Spilled on skin - rinse with water.
INGREDIENTS: Water softeners (potassium and/or sodium complex phosphates, a
cleaning agent (chlorine bleach), dishware, flatware, and dishwasher protection
potassium silicates), water, thickening agent, buffering agent, stabilizing agent, col
PHOSPHORUS CONTENT: This Liquid Cascade formula averages 4.4% ph
of phosphates, which is equivalent to 0.9 grams per tablespoon.
Questions? Comments? Call 1-800-765-5516.
TL
Bottle made from 25% or more post-consumer recycled plastic.

Background

Many of the chemicals in your school laboratory are hazardous. Some are corrosive, some are flammable, and some are poisonous. Many exhibit these properties when they are combined. You can work safely with these chemicals, as long as you treat them with care and respect, observe proper safety precautions, and follow the directions that are given by your teacher and this textbook.

Did you know that many of the chemical products in your home are hazardous, too? For example, common household bleach, when used as directed, is safe for disinfecting and whitening clothing. Hazard labels on bleaching products, however, warn against mixing bleach with acids, household ammonia, or products that contain these chemicals. Bleach, when combined with acids, produces toxic chlorine gas. The products of combining bleach with ammonia are explosive.

Most homes contain numerous chemical products, ranging from cleaners and disinfectants, to fertilizers and fuels. All potentially hazardous products have a warning on their containers or on paper inserts in their packaging. Many, but not all, have a list of the chemicals they contain. Some hazardous products advise users only to keep them away from children and pets.

How much do you know about the safe use of chemical products? Would you know what to do if an accidental spill occurred? Would the members of your family, or people in your community, know what to do?

Challenge

Design, produce, and distribute a newsletter to inform your community about the safe use of common chemical products. Include the potential hazards of these products to living things and the environment. Also include emergency procedures to follow if an accident occurred.

Materials

Select a medium for your newsletter, such as a traditional paper newsletter or an electronic version for the Internet. For a traditional newsletter, you will need to decide on methods of production and distribution. For an electronic version, you will need to use computer hardware and software.

Design Criteria

A As a class, develop a rubic listing criteria for assessing the newsletters. For example, one criterion may involve a newsletter's effectiveness in altering the behaviour of its readers. You may want to develop different criteria for traditional and electronic newsletters.

B Your newsletter must be factual, easy to read for a wide variety of audiences, and educational.

Action Plan

1 The following items must be part of your newsletter:
- examples of household chemical products and their uses
- hazards associated with each chemical product
- suggestions to encourage safe and responsible use
- environmental considerations for the disposal of the chemical products
- alternative products (if any), and hazards (if any) associated with these alternatives
- an interview with a professional who researches, develops, or works with household chemical products

2 Develop detailed steps to research, plan, and produce your newsletter. Include deadlines for completion and specific roles for the members of your group (for example, editor, writers, artists, and designers).

Evaluate

Present your completed newsletter to your class. Hold a focus group session to evaluate the content and impact of your newsletter. The focus group could include students from other classes, parents and relatives, and members of the community.

Assessment

After you complete this project:

- Assess the success of your project based on how similar the final project is to your action plan.
- Assess your project based on how clearly the chemistry concepts and safety recommendations are conveyed.
- Assess your project using the rubric designed in class.

(a) Be, Ca, Mg
(b) Kr, Se, Br
(c) Na, Cs, K

31. Write the chemical name for each compound.
(a) NH_4NO_3
(b) $Pb(C_2H_3O_2)_4$
(c) S_2Cl_2
(d) $Ba(NO_2)_2$
(e) P_4O_{10}
(f) Mn_2O_3

32. Write the chemical formula of each compound.
(a) strontium chloride
(b) lead(II) sulfite
(c) chromium(III) acetate
(d) hydrogen sulfide
(e) iodine heptafluoride

33. Explain why it is useful to classify reactions.

34. Balance each chemical equation, if necessary. State which class it belongs to.
(a) $Zn_{(s)} + AgNO_{3(aq)} \rightarrow Zn(NO_3)_{2(aq)} + Ag_{(s)}$
(b) $Fe_{(s)} + S_{(s)} \rightarrow FeS_{(s)}$
(c) $KClO_{3(s)} \rightarrow 2KCl_{(s)} + 3O_{2(g)}$
(d) $NaCO_{3(aq)} + MgSO_{4(aq)} \rightarrow MgCO_{3(s)} + Na_2SO_{4(aq)}$
(e) $C_2H_6O_{(\ell)} + O_{2(g)} \rightarrow CO_{2(g)} + H_2O_{(g)}$

35. Predict the products of each reaction. Then write a balanced chemical equation, and state which class the reaction belongs to.
(a) $Mg_{(s)} + HCl_{(aq)} \rightarrow$
(b) $HgO_{(s)} \rightarrow$
(c) $Al_{(s)} + O_{2(g)} \rightarrow$
(d) $C_6H_{12}O_{6(s)} + O_{2(g)} \rightarrow$
(e) $BaCl_{2(aq)} + Na_2SO_{4(aq)} \rightarrow$

Inquiry

36. Describe an experimental procedure to test three qualitative properties and two quantitative properties of lead.

37. Design an experiment that uses acids to test the reactivities of one metal from each of the following groups: alkali metals, alkaline earth metals, and transition metals.

38. Raja weighed calcium sulfate on filter papers for an experiment that he performed three times. His data are shown below.

	Mass of filter paper	Mass of paper + powder	Mass of calcium sulfate
Trial # 1	4.13 g	13.6 g	9.47 g
Trial # 2	4.2 g	12.81 g	8.51 g
Trial # 3	4.12 g	10.96 g	6.8 g

What errors did Raja make in his reporting and calculations?

39. What kinds of tests could be used to differentiate between unknown metal and non-metal samples in a laboratory? Design an experiment that includes these tests.

40. You are given a substance. You must decide whether it is an ionic compound or a covalent compound. The substance has roughly cube-shaped granules, which are translucent and colourless.
(a) Predict whether the compound is ionic or covalent.
(b) Explain your prediction.
(c) Design an experiment to collect data that will support your prediction.

41. A student drops a coil of metal wire, $X_{(s)}$, into a water solution of a metal sulfate, $ZSO_{4(aq)}$. The student observes that the colour of the solution changes, and that a metallic-looking substance appears to be forming on the metal wire. Based on these observations, answer the following questions.
(a) Has a reaction taken place? If so, what kind of reaction has taken place? Explain your answer.
(b) Which metal is more reactive, metal X or metal Z in compound ZSO_4? Explain your answer.
(c) Write the names of a real metal and a metal sulfate that you predict would behave this way in a laboratory.

Communication

42. Perform each calculation. Express the answer to the correct number of significant digits.
(a) 19.3 g + 2.22 g
(b) 14.2 cm × 1.1 cm × 3.69 cm
(c) 57.9 kg ÷ 3.000 dm^3
(d) 18.76 g – 1.3 g
(e) 25.2 + 273°C

43. Name four groups in the periodic table. Give characteristics of each group, and list three members of each group.

44. Copy the following table into your notebook, and fill in the missing information. If isotopic data are not given, use the atomic mass from the periodic table to find the number of neutrons.

Atom or ion with mass number	Number of protons	Number of neutrons	Number of electrons
$^{14}N^{3-}$			
	16		18
		2	2
$^{7}Li^{+}$	3		
		20	16

45. Draw a Lewis structure for each element.
(a) argon, Ar
(b) sodium, Na
(c) aluminum, Al
(d) boron, B

46. Draw a Lewis structure for each element. Explain the two patterns that appear.
(a) carbon, C
(b) neon, N
(c) oxygen, O
(d) fluorine, F
(e) chlorine, Cl
(f) bromine, Br

47. Describe three periodic trends. Explain how these trends change across and down the periodic table.

48. Arrange the following quantities in a table to show which are physical and chemical properties, which are qualitative, and which are quantitative: melting point, colour, density, reactivity with acids, flammability, malleability, electrical conductivity, boiling point, reactivity with air, hardness, toxicity, brittleness.

49. Draw a Lewis structure for each compound.
(a) $CrBr_2$
(b) H_2S
(c) CCl_4
(d) AsH_3
(e) CS_2

50. Draw diagrams to represent each class of reaction below. Use symbols or drawings to represent different kinds of atoms.
(a) synthesis
(b) decomposition
(c) combustion
(d) single displacement
(e) double displacement

51. Compare the boiling points of ammonia, NH_3, phosphorus trihydride, PH_3, and arsenic trihydride, AsH_3. Use the periodic table and the concept of molecular shape and polarity.

52. The molecule BF_3 contains polar covalent bonds, yet the molecule is not polar. Explain why. Include a diagram with your explanation.

Making Connections

53. What effects do accuracy, precision, and margin of error have in courts of law? Consider court cases that involve forensic analysis. What are the implications of inaccurate science in the courts?

54. Some metals (such as gold, lead, and silver) were known and widely used in ancient times. Other metals have only been discovered relatively recently. For example, both sodium and potassium were discovered in the early nineteenth century by Sir Humphrey Davy. Explain why ancient cultures knew about some metals, while other metals remained unknown for thousands of years.

COURSE CHALLENGE

Planet Unknown

Consider the following as you continue to plan for your Chemistry Course Challenge:

- How did chemists use trends of physical and chemical properties to arrange elements in a periodic table?
- What are several ways of comparing the reactivity of metals?
- How can you use the physical and chemical properties of elements to help identify them?

UNIT 2

Chemical Quantities

UNIT 2 CONTENTS

UNIT 2 OVERALL EXPECTATIONS

- What is the mole? Why is it important for analyzing chemical systems?
- How are the quantitative relationships in balanced chemical reactions used for experiments and calculations?
- Why are quantitative chemical relationships important in the home and in industries?

Unit Investigation Prep

Read pages 274–275 before beginning this unit. There, you will have the opportunity to determine the composition of a mixture. You can start planning your investigation well in advance by knowing the kind of skills and information you need to have as you progress through Unit 2.

In the 1939 film *The Wizard of Oz*, Dorothy and her companions collapsed in sleep in a field of poppies. This scene is not realistic, however. Simply walking in a field of poppies will not put you into a drugged sleep.

Poppy seeds do, however, contain a substance called opium. Opium contains the drugs morphine, codeine, and heroin, collectively known as opiates.

While you are unlikely to experience any physiological effects from eating the poppy seeds on a bagel, they could cost you your job! For some safety sensitive jobs, such as nursing and truck-driving, you may be required to take a drug test as part of the interview process.

Each gram of poppy seeds may contain 2 mg to 18 mg of morphine and 0.6 mg to 2.4 mg of codeine. Eating foods with large amounts of poppy seeds can cause chemists to detect opiates in urine. The opiates may be at levels above employers' specified limits.

Knowing about quantities in chemical reactions is crucial to interpreting the results of drug tests. Policy-makers and chemists need to understand how the proportions of codeine to morphine caused by eating poppy seeds differ from the proportions caused by taking opiates.

In this unit, you will carry out experiments and calculations based on the quantitative relationships in chemical formulas and reactions.

Robin Hood
NORMANDY-STYLE
CULTURED BUTTER
BEURRE DE
CULTURE
À LA NORMANDE
250 g
Quality
Baking Soda
Bicarbonate de soude
5 lb.
Acetic Acid, Glacial
'Baker Analyzed'

CHAPTER 5

Counting Atoms and Molecules: The Mole

A recipe for chocolate chip muffins tells you exactly what ingredients you will need. One recipe might call for flour, butter, eggs, milk, vinegar, baking soda, sugar, and chocolate chips. It also tells you how much of each ingredient you will need, using convenient units of measurement. Which of the ingredients do you measure by counting? Which do you measure by volume or by mass? The recipe does not tell you exactly how many chocolate chips or grains of sugar you will need. It would take far too long to count individual chocolate chips or grains of sugar. Instead, the amounts are given in millilitres or grams—the units of volume or mass.

In some ways, chemistry is similar to baking. To carry out a reaction successfully—in chemistry or in baking—you need to know how much of each reactant you will need. When you bake something with vinegar and baking soda, for example, the baking soda reacts with acetic acid in the vinegar to produce carbon dioxide gas. The carbon dioxide gas helps the batter rise. The chemical equation for this reaction is

$$NaHCO_{3(s)} + CH_3COOH_{(aq)} \rightarrow NaCH_3COO_{(aq)} + H_2O_{(\ell)} + CO_{2(g)}$$

According to the balanced equation, one molecule of baking soda reacts with one molecule of acetic acid to form a salt, water, and carbon dioxide. If you wanted to carry out the reaction, how would you know the amount of baking soda and vinegar to use? Their molecules are much too small and numerous to be counted like eggs.

In this chapter, you will learn how chemists count atoms by organizing large numbers of them into convenient, measurable groups. You will learn how these groups relate the number of atoms in a substance to its mass. Using your calculator and the periodic table, you will learn how to convert between the mass of a substance and the number of atoms it contains.

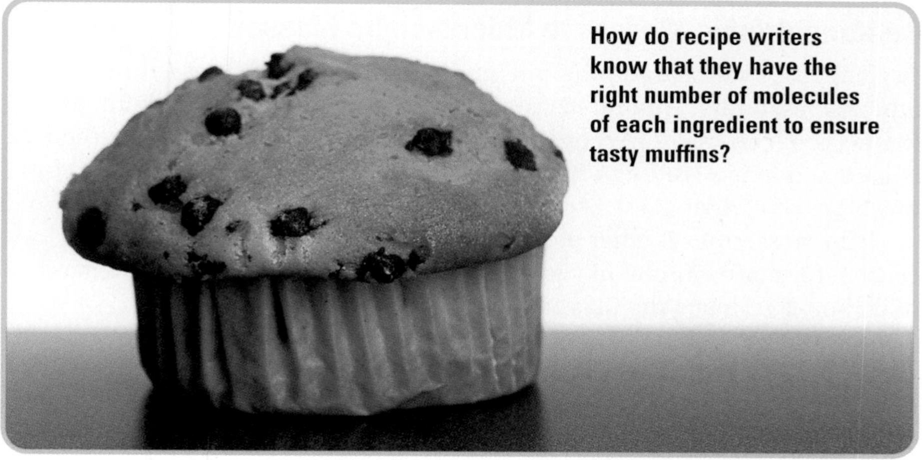

How do recipe writers know that they have the right number of molecules of each ingredient to ensure tasty muffins?

Chapter Preview

Concepts and Skills You Will Need

Before you begin this chapter, review the following concepts and skills:

- defining and describing the relationships among atomic number, mass number, atomic mass, and isotope (Chapter 2, section 2.1)
- writing chemical formulas and equations (Chapter 3, section 3.4)
- balancing chemical equations by inspection (Chapter 4, section 4.1)

5.1 Isotopes and Average Atomic Mass

Section Preview/ Specific Expectations

In this section, you will

- **describe** the relationship between isotopic abundance and average atomic mass
- **solve** problems involving percentage abundance of isotopes and relative atomic mass
- **explain** the significance of a weighted average
- **communicate** your understanding of the following terms: *isotopic abundance, average atomic mass, mass spectrometer, weighted average*

How does the mass of a substance relate to the number of atoms in the substance? To answer this question, you need to understand how the relative masses of individual atoms relate to the masses of substances that you can measure on a balance.

The head of a pin, like the one shown in Figure 5.1, is made primarily of iron. It has a mass of about 8×10^{-3} g, yet it contains about 8×10^{19} atoms. Even if you could measure the mass of a single atom on a balance, the mass would be so tiny (about 1×10^{-22} g for an iron atom) that it would be impractical to use in everyday situations. Therefore, you need to consider atoms in bulk, not individually.

How do you relate the mass of individual atoms to the mass of a large, easily measurable number of atoms? In the next two sections, you will find out.

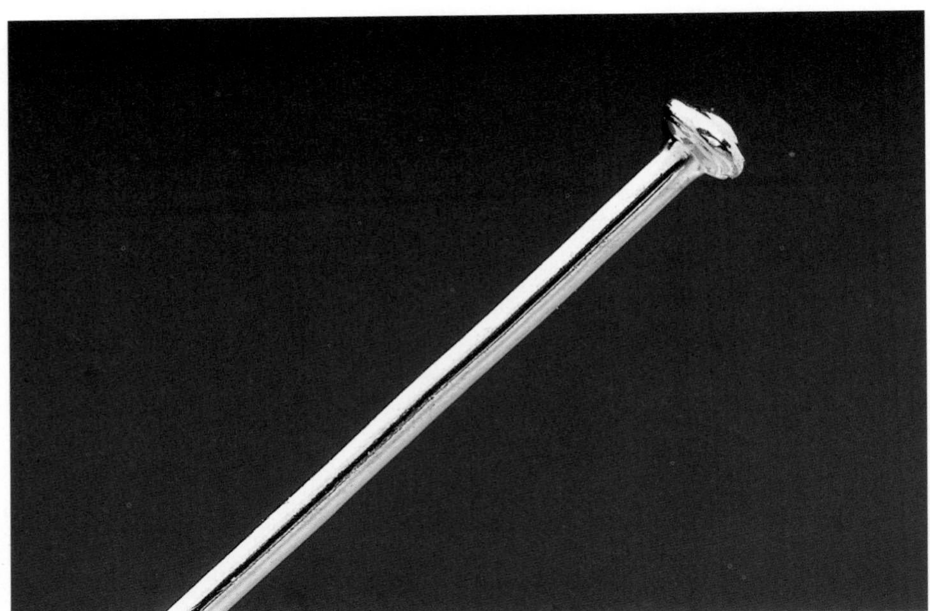

Figure 5.1 **The head of a typical pin contains about 80 quintillion atoms.**

Relating Atomic Masses to Macroscopic Masses

In Chapter 2, you learned that the mass of an atom is expressed in atomic mass units. Atomic mass units are a relative measure, defined by the mass of carbon-12. According to this definition, one atom of carbon-12 is assigned a mass of 12 u. Stated another way, 1 u = $\frac{1}{12}$ of the mass of one atom of carbon-12.

The masses of all other atoms are defined by their relationship to carbon-12. For example, oxygen-16 has a mass that is 133% of the mass of carbon-12. Hence the mass of an atom of oxygen-16 is $\frac{133}{100} \times 12.000 \text{ u} = 16.0 \text{ u}$.

Usually, not all the atoms in an element have the same mass. As you learned in Chapter 2, atoms of the same element that contain different numbers of neutrons are called isotopes. Most elements are made up of two or more isotopes. Chemists need to account for the presence of isotopes when finding the relationship between the mass of a large number of atoms and the mass of a single atom. To understand why this is important, consider the following analogy.

Imagine that you have the task of finding the total mass of 10 000 spoons. If you know the mass of a dessertspoon, can you assume that its mass represents the average mass of all the spoons? What if the 10 000 spoons include soupspoons, dessertspoons, and tablespoons? If you use the mass of a dessertspoon to calculate the total mass of all the spoons, you may obtain a reasonable estimate. Your answer will not be accurate, however, because each type of spoon has a different mass. You cannot calculate an accurate average mass for all the spoons based on knowing the mass of only one type. How can you improve the accuracy of your answer without determining the mass of all the spoons?

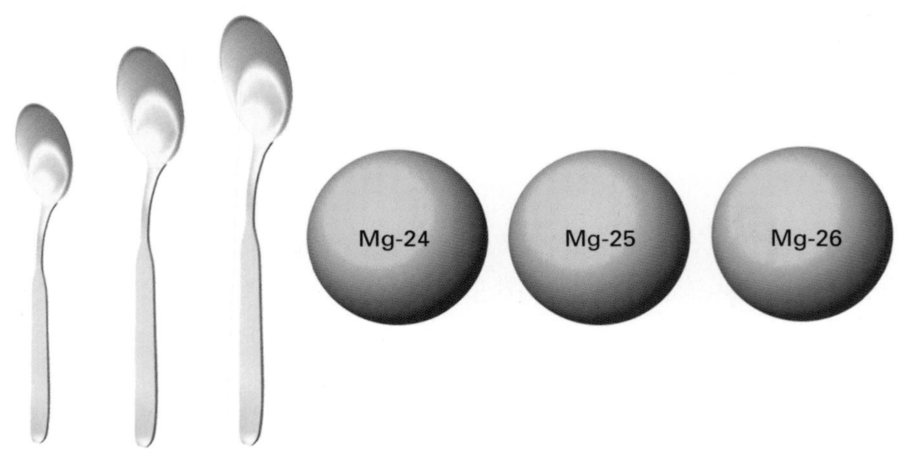

Figure 5.2 **Think about finding the average mass of a group of objects that have different masses. How is this similar to finding the average mass of an element that is composed of different isotopes?**

Isotopic Abundance

Chemists face a situation similar to the one described above. Because all the atoms in a given element do not have the same number of neutrons, they do not all have the same mass. For example, magnesium has three naturally occurring isotopes. It is made up of 79% magnesium-24, 10% magnesium-25, and 11% magnesium-26. Whether the magnesium is found in a supplement tablet (like the ones on the right) or in seawater as $Mg(OH)_2$, it is always made up of these three isotopes in the same proportion. The relative amount in which each isotope is present in an element is called the **isotopic abundance**. It can be expressed as a percent or as a decimal fraction. When chemists consider the mass of a sample containing billions of atoms, they must take the isotopic abundance into account.

Magnesium plays a variety of roles in the body. It is involved in energy production, nerve function, and muscle relaxation, to name just a few. The magnesium in these tablets, like all naturally occurring magnesium, is made up of three isotopes.

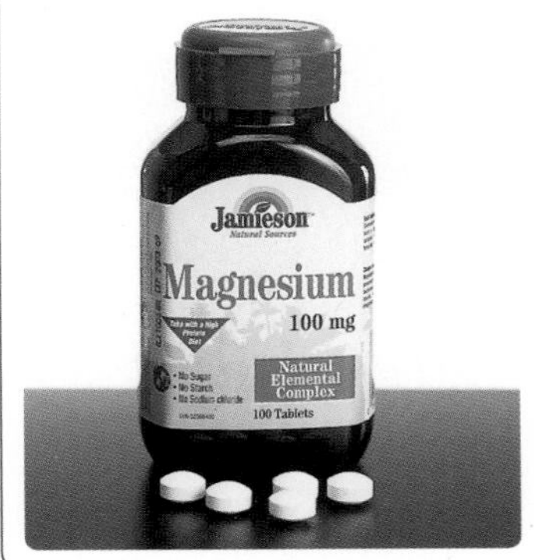

Average Atomic Mass and the Periodic Table

The **average atomic mass** of an element is the average of the masses of all the element's isotopes. It takes into account the abundance of each isotope within the element. The average atomic mass is the mass that is given for each element in the periodic table.

It is important to interpret averages carefully. For example, in 1996, the average size of a Canadian family was 3.1. Of course, no one family actually has 3.1 people. In the same way, while the average atomic mass of carbon is 12.01 u, no one atom of carbon has a mass of 12.01 u.

Examine Figure 5.3. Since the atomic mass unit is based on carbon-12, why does the periodic table show a value of 12.01 u, instead of exactly 12 u? Carbon is made up of several isotopes, not just carbon-12. Naturally occurring carbon contains carbon-12, carbon-13, and carbon-14. If all these isotopes were present in equal amounts, you could simply find the average of the masses of the isotopes. This average mass would be about 13 u, since the masses of carbon-13 and carbon-14 are about 13 u and 14 u respectively.

The isotopes, though, are not present in equal amounts. Carbon-12 comprises 98.9% of all carbon, while carbon-13 accounts for 1.1%. Carbon-14 is present in a very small amount—about 1×10^{-10}%. It makes sense that the average mass of all the isotopes of carbon is 12.01 u—very close to 12—since carbon-12 is by far the predominant isotope.

The only elements with only one naturally occurring isotope are beryllium, sodium, aluminum, and phosphorus.

Figure 5.3 The atomic mass that is given in the periodic table represents the average mass of all the naturally occurring isotopes of the element. It takes into account their isotopic abundances.

Thus chemists need to know an element's isotopic abundance and the mass of each isotope to calculate the average atomic mass. How do chemists determine the isotopic abundance associated with each element? How do they find the mass of each isotope? They use a **mass spectrometer**, a powerful instrument that generates a magnetic field to obtain data about the mass and abundance of atoms and molecules. You will learn more about the mass spectrometer in Tools & Techniques on page 166. You can use the data obtained with a mass spectrometer to calculate the average atomic mass given in the periodic table.

Working with Weighted Averages

If you obtain the isotopic abundance of an element from mass spectrometer data or a table, you can calculate the average atomic mass of the element. You do this by calculating the **weighted average** of each isotope's mass. A weighted average takes into account not only the values associated with a set of data, but also the abundance or importance of each value.

Normally, when you calculate the average of a set of data, you find the equally weighted average. You add the given values and divide the total by the number of values in the set. Each value in the average is given equal weight. For example, imagine that you have three objects: A, B, and C. A has a mass of 1.0 kg, B has a mass of 2.0 kg, and C has a mass of 3.0 kg. Their average mass is

$$\frac{\text{Mass of (A + B + C)}}{\text{Number of items}} = \frac{1.0\text{ kg} + 2.0\text{ kg} + 3.0\text{ kg}}{3}$$
$$= 2.0\text{ kg}$$

What if you have a set containing two of A, one of B, and three of C? Their average mass becomes

$$\frac{2(1.0\text{ kg}) + 2.0\text{ kg} + 3(3.0\text{ kg})}{6} = 2.2\text{ kg}$$

This is a weighted average.

Another way to calculate the same weighted average is to consider the relative abundance of each object. There are six objects in total. A is present as $\frac{2}{6}$ (33%) of the total, B is present as $\frac{1}{6}$ (17%) of the total, and C is present as $\frac{3}{6}$ (50%) of the total. Thus their average mass can be calculated in the following way:

$$(0.33)(1.0\text{ kg}) + (0.17)(2.0\text{ kg}) + (0.50)(3.0\text{ kg}) = 2.2\text{ kg}$$

mind STRETCH

We use weighted averages all the time! For example, course marks are often based on weighted averages. Suppose that the final mark in a chemistry course is determined as follows: laboratory 25%, tests 30%, homework and quizzes 5%, project 10%, and final exam 30%. A student obtains the following marks: laboratory $\frac{114}{130}$, tests $\frac{261}{300}$, homework and quizzes $\frac{90}{95}$, project $\frac{21}{25}$, and final exam $\frac{70}{80}$. What is the student's final mark in chemistry?

Calculating Average Atomic Mass

You can use a similar method to calculate average atomic mass. If you know the atomic mass of each isotope that makes up an element, as well as the isotopic abundance of each isotope, you can calculate the average atomic mass of the element.

For example, lithium exists as two isotopes: lithium-7 and lithium-6. As you can see in Figure 5.4, lithium-7 has a mass of 7.015 u and makes up 92.58% of lithium. Lithium-6 has a mass of 6.015 u and makes up the remaining 7.42%. To calculate the average atomic mass of lithium, multiply the mass of each isotope by its abundance.

$$\left(\frac{92.58}{100}\right)(7.015\text{ u}) + \left(\frac{7.42}{100}\right)(6.015\text{ u}) = 6.94\text{ u}$$

Looking at the periodic table confirms that the average atomic mass of lithium is 6.94 u. The upcoming Sample Problem gives another example of how to calculate average atomic mass.

CHECKPOINT

How is the average atomic mass of an element different from the mass number of the element? How are the average atomic mass and mass number similar?

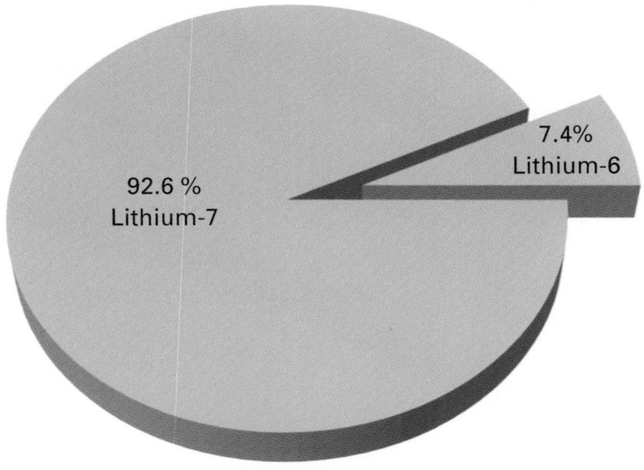

Figure 5.4 Naturally occurring lithium consists of two isotopes, $^{7}_{3}Li$ and $^{6}_{3}Li$.

Tools & Techniques

The Mass Spectrometer

Many chemists depend on instruments known as mass spectrometers. Mass spectrometers can detect trace pollutants in the atmosphere, provide information about the composition of large molecules, and help to determine the age of Earth's oldest rocks.

As well, mass spectrometers can find the relative abundance of each isotope in an element. In 1912, J.J. Thompson first detected neon-20 and neon-22 in a sample of neon gas by using a magnetic field to separate the isotopes.

Today's mass spectrometers also use a magnetic field to separate the isotopes of an element. Since a magnetic field can only affect the path of a charged particle, the atoms must first be charged, or "ionized." The magnetic field then deflects ions with the same charge, but different masses, onto separate paths. Imagine rolling a tennis ball at a right angle to the air current from a fan. The air current deflects the path of the ball. The same air current deflects a Ping-Pong™ ball more than a tennis ball. Similarly, a magnetic field deflects isotopes of smaller mass more than isotopes of larger mass.

In a mass spectrometer, elements that are not gases are vaporized by heating. Next, the gas atoms are ionized. In electron impact ionization, the gas atoms are bombarded with a stream of electrons from a heated filament. These electrons collide with the gas atoms, causing each atom to lose an electron and become a positive ion.

The ions are focussed and accelerated by electric fields, toward a magnetic field. The magnetic field deflects them and forces them to take a curved path. Lighter ions curve more than heavier ions. Therefore ions from different isotopes arrive at different destinations. Since ions have a charge, a detector can be used to register the current at each destination. The current is proportional to the number of ions that arrive at the destination. An isotope that has a larger relative abundance generates a larger current. From the currents at the different destinations, chemists can deduce the proportion of each isotope in the element.

Mass spectral data for neon, below, show the relative abundance of the isotopes of neon. Using mass spectrometers, chemists now detect a signal from neon-21, in addition to the signals from neon-20 and neon-22.

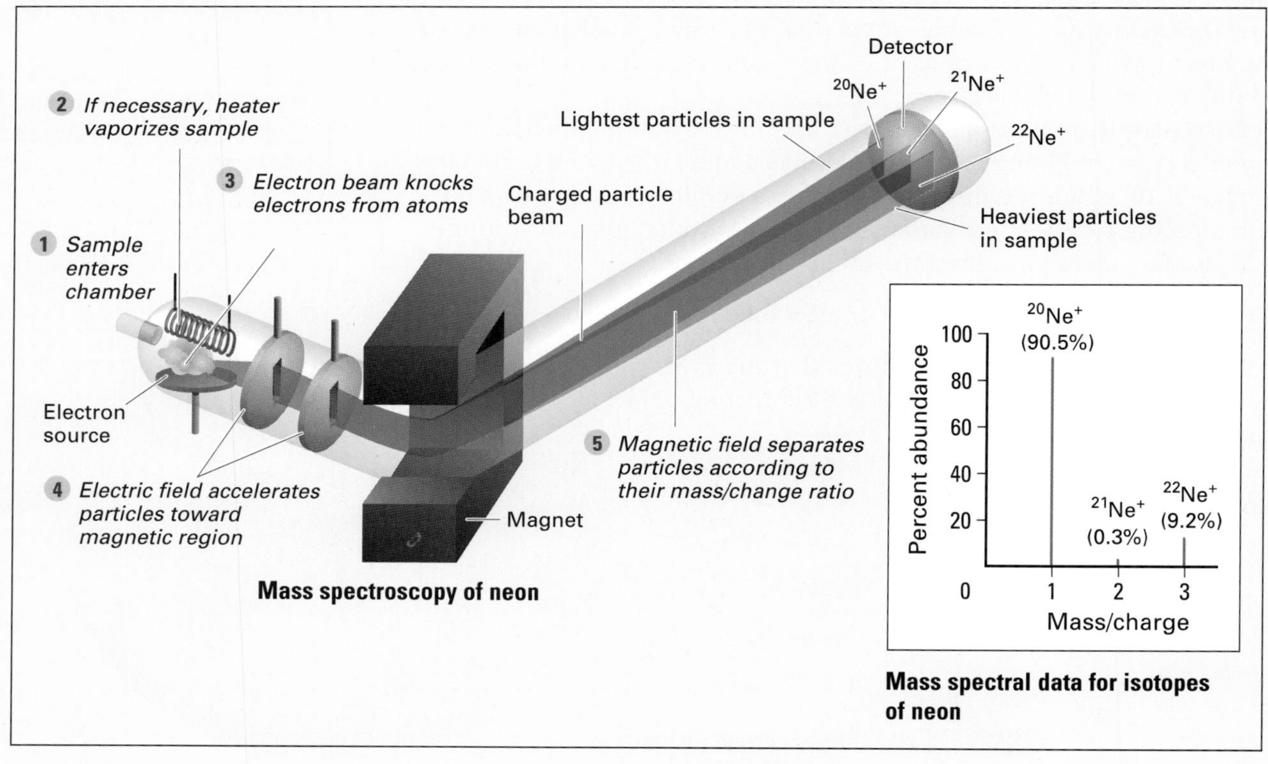

Mass spectroscopy of neon

Mass spectral data for isotopes of neon

Sample Problem

Average Atomic Mass

Problem

Naturally occurring silver exists as two isotopes. From the mass of each isotope and the isotopic abundance listed below, calculate the average atomic mass of silver.

Isotope	Atomic mass (u)	Relative abundance (%)
$^{107}_{47}Ag$	106.9	51.8
$^{109}_{47}Ag$	108.9	48.2

What Is Required?

You need to find the average atomic mass of silver.

What Is Given?

You are given the relative abundance and the atomic mass of each isotope.

Plan Your Strategy

Multiply the atomic mass of each isotope by its relative abundance, expressed as a decimal. That is, 51.8% expressed as a decimal is 0.518 and 48.2% is 0.482.

Act on Your Strategy

$$\text{Average atomic mass of Ag} = 106.9\text{ u }(0.518) + 108.9\text{ u }(0.482)$$
$$= 107.9\text{ u}$$

Check Your Solution

In this case, the abundance of each isotope is close to 50%. An average atomic mass of about 108 u seems right, because it is between 106.9 u and 108.9 u. Checking the periodic table reveals that the average atomic mass of silver is indeed 107.9 u.

Practice Problems

1. The two stable isotopes of boron exist in the following proportions: 19.78% $^{10}_{5}B$ (10.01 u) and 80.22% $^{11}_{5}B$ (11.01 u). Calculate the average atomic mass of boron.

2. In nature, silicon is composed of three isotopes. These isotopes (with their isotopic abundances and atomic masses) are $^{28}_{14}Si$ (92.23%, 27.98 u), $^{29}_{14}Si$ (4.67%, 28.97 u), and $^{30}_{14}Si$ (3.10%, 29.97 u). Calculate the average atomic mass of silicon.

Continued ...

CHEM FACT

Why are the atomic masses of individual isotopes not exact whole numbers? After all, ^{12}C has a mass of exactly 12 u. Since carbon has 6 neutrons and 6 protons, you might assume that protons and neutrons have masses of exactly 1 u each. In fact, protons and neutrons have masses that are close to, but slightly different from, 1 u. As well, the mass of electrons, while much smaller than the masses of protons and neutrons, must still be taken into account.

CHECKPOINT

Why is carbon-12 the only isotope with an atomic mass that is a whole number?

In some periodic tables, the average atomic mass is referred to as the atomic weight of an element. This terminology, while technically incorrect, is still in use and is generally accepted.

Continued ...
FROM PAGE 167

3. Copper is a corrosion-resistant metal that is used extensively in plumbing and wiring. Copper exists as two naturally occurring isotopes: $^{63}_{29}Cu$ (62.93 u) and $^{65}_{29}Cu$ (64.93 u). These isotopes have isotopic abundances of 69.1% and 30.9% respectively. Calculate the average atomic mass of copper.

4. Lead occurs naturally as four isotopes. These isotopes (with their isotopic abundances and atomic masses) are $^{204}_{82}Pb$ (1.37%, 204.0 u), $^{206}_{82}Pb$ (26.26%, 206.0 u), $^{207}_{82}Pb$ (20.82%, 207.0 u), and $^{208}_{82}Pb$ (51.55%, 208.0 u). Calculate the average atomic mass of lead.

ExpressLab  A Penny for your Isotopes

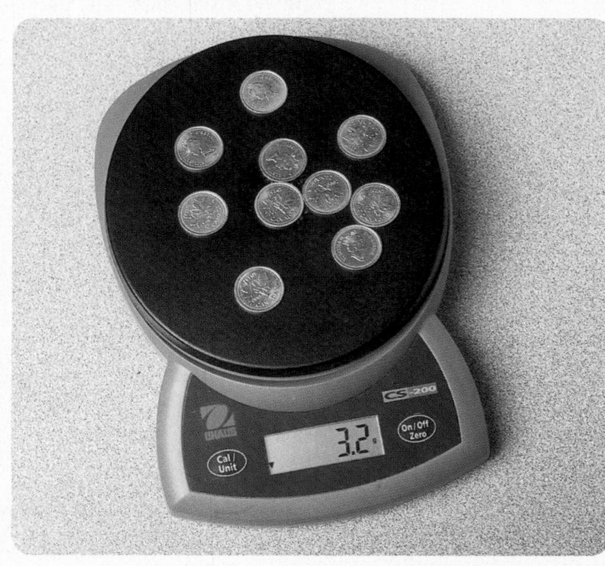

The mass of a Canadian penny has decreased several times over the years. Therefore you can use pennies to represent different "isotopes" of a fictitious element, *centium*. That is, each "atom" of *centium* reacts the same way—it is still worth 1¢—but the various isotopes have different characteristic masses.

Safety Precautions

Procedure

1. Obtain a bag of pennies from your teacher. Since the mass of a penny decreased in 1982 and 1997, your bag will contain pennies dated anywhere from 1982 to the present date.
2. Sort your pennies into groups of pre-1997 "isotopes" and post-1997 "isotopes" of *centium*.
3. Count the number of pennies in each group.
4. Find the mass of ten pennies from each group. Divide the total mass by 10 to get the mass of each *centium* "isotope."
5. Use the data you have just gathered to calculate the mass of the pennies, using a weighted average. This represents the "average atomic mass" of *centium*.

Analysis

1. In step 4, you used the average mass of ten pennies to represent the mass of one "isotope" of *centium*.
   (a) Why did you need to do this? Why did you not just find the mass of one penny from each group?
   (b) If you were able to find the mass of real isotopes for this experiment, would you need to do step 4? Explain.
2. Compare your "average atomic mass" for *centium* with the "average atomic mass" obtained by other groups.
   (a) Are all the masses the same? Explain any differences.
   (b) What if you were able to use real isotopes of an element, such as copper, for this experiment? Would you expect results to be consistent throughout the class? Explain.

Calculating Isotopic Abundance

Chemists use a mass spectrometer to determine accurate values for the isotopic abundance associated with each element. Knowing the average atomic mass of an element, you can use the masses of its isotopes to calculate the isotopic abundances.

Sample Problem

Isotopic Abundance

Problem

Boron exists as two naturally occurring isotopes: $^{10}_{5}B$ (10.01 u) and $^{11}_{5}B$ (11.01 u). Calculate the relative abundance of each isotope of boron.

What Is Required?

You need to find the isotopic abundance of boron.

What Is Given?

Atomic mass of $^{10}_{5}B$ = 10.01 u

Atomic mass of $^{11}_{5}B$ = 11.01 u

From the periodic table, the average atomic mass of boron is B = 10.81 u.

Plan Your Strategy

Express the abundance of each isotope as a decimal rather than a percent. The total abundance of both isotopes is therefore 1. Let the abundance of boron-10 be x. Let the abundance of boron-11 be $1 - x$. Set up an equation, and solve for x.

Act on Your Strategy

$$\text{Average atomic mass} = x(\text{atomic mass B-10}) + (1 - x)(\text{atomic mass B-11})$$
$$10.81 = x(10.01) + (1 - x)(11.01)$$
$$10.81 = 10.01x + 11.01 - 11.01x$$
$$11.01x - 10.01x = 11.01 - 10.81$$
$$x = 0.2000$$

The abundance of boron-10 is 0.2000.

The abundance of boron-11 is $1 - x$, or $1 - 0.2000 = 0.8000$.

The abundance of $^{10}_{5}B$ is therefore 20.00%. The abundance of $^{11}_{5}B$ is 80.00%.

Check Your Solution

The fact that boron-11 comprises 80% of naturally occurring boron makes sense, because the average atomic mass of boron is 10.81 u. This is closer to 11.01 u than to 10.01 u.

Continued ...

CHEM FACT

If you wear contact lenses, you may use boron every day. Boron is part of boric acid, H_3BO_3, which is contained in many cleaning solutions for contact lenses.

Continued ...
FROM PAGE 169

Practice Problems

5. Hydrogen is found primarily as two isotopes in nature: $^{1}_{1}H$ (1.0078 u) and $^{2}_{1}H$ (2.0140 u). Calculate the percentage abundance of each isotope based on hydrogen's average atomic mass.

6. Lanthanum is composed of two isotopes: $^{138}_{57}La$ (137.91 u) and $^{139}_{57}La$ (138.91 u). Look at the periodic table. What can you say about the abundance of $^{138}_{57}La$?

7. Rubidium ignites spontaneously when exposed to oxygen to form rubidium oxide, Rb_2O. Rubidium exists as two isotopes: $^{85}_{37}Rb$ (84.91 u) and $^{87}_{37}Rb$ (86.91 u). If the average atomic mass of rubidium is 85.47 u, determine the percentage abundance of $^{85}_{37}Rb$.

8. Oxygen is composed of three isotopes: $^{16}_{8}O$ (15.995 u), $^{17}_{8}O$ (16.999 u), and $^{18}_{8}O$ (17.999 u). One of these isotopes, $^{17}_{8}O$, comprises 0.037% of oxygen. Calculate the percentage abundance of the other two isotopes, using the average atomic mass of 15.9994 u.

Section Wrap-up

In this section, you learned how isotopic abundance relates to average atomic mass. Since you know the average mass of an atom in any given element, you can now begin to relate the mass of a single atom to the mass of a large number of atoms. First you need to establish how many atoms are in easily measurable samples. In section 5.2, you will learn how chemists group atoms into convenient amounts.

Section Review

1. K/U The average atomic mass of potassium is 39.1 u. Explain why no single atom of potassium has a mass of 39.1 u.

2. I Naturally occurring magnesium exists as a mixture of three isotopes. These isotopes (with their isotopic abundances and atomic masses) are Mg-24 (78.70%, 23.985 u), Mg-25 (10.13%, 24.985 u), and Mg-26 (11.17%, 25.983 u). Calculate the average atomic mass of magnesium.

3. C Assume that an unknown element, X, exists naturally as three different isotopes. The average atomic mass of element X is known, along with the atomic mass of each isotope. Is it possible to calculate the percentage abundance of each isotope? Why or why not?

4. C You know that silver exists as two isotopes: silver-107 and silver-109. However, radioisotopes of silver, such as silver-105, silver-106, silver-108, and silver-110 to silver-117 are known. Why do you not use the abundance and mass of these isotopes when you calculate the average atomic mass of silver? Suggest two reasons.

The Avogadro Constant and the Mole

5.2

In section 5.1, you learned how to use isotopic abundances and isotopic masses to find the average atomic mass of an element. You can use the average atomic mass, found in the periodic table, to describe the average mass of an atom in a large sample.

Why is relating average atomic mass to the mass of large samples important? In a laboratory, as in everyday life, we deal with macroscopic samples. These samples contain incredibly large numbers of atoms or molecules. Can you imagine a cookie recipe calling for six septillion molecules of baking soda? What if copper wire in a hardware store were priced by the atom instead of by the metre, as in Figure 5.5? What if we paid our water bill according to the number of water molecules that we used? The numbers involved would be ridiculously inconvenient. In this section, you will learn how chemists group large numbers of atoms into amounts that are easily measurable.

Section Preview/ Specific Expectations

In this section, you will

- **describe** the relationship between moles and number of particles
- **solve** problems involving number of moles and number of particles
- **explain** why chemists use the mole to group atoms
- **communicate** your understanding of the following terms: *mole, Avogadro constant*

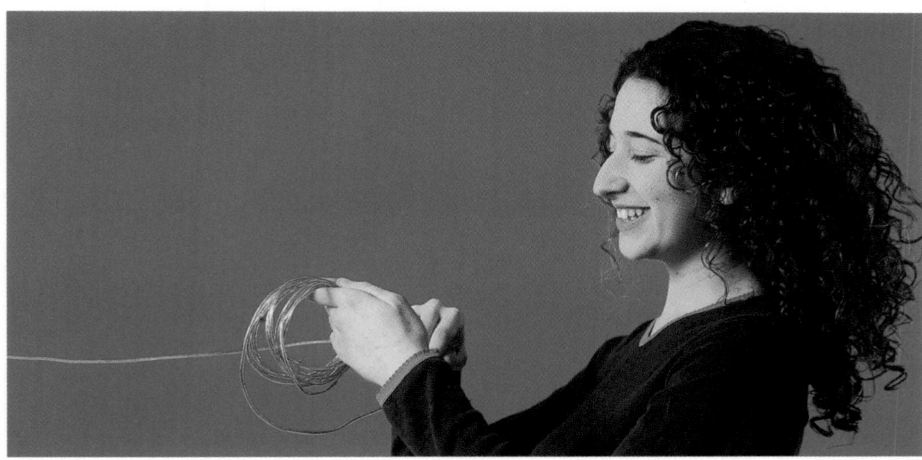

Figure 5.5 **Copper wire is often priced by the metre because the metre is a convenient unit. What unit do chemists use to work with large numbers of atoms?**

Grouping for Convenience

In a chemistry lab, as well as in other contexts, it is important to be able to measure amounts accurately and conveniently. When you purchase headache tablets from a drugstore, you are confident that each tablet contains the correct amount of the active ingredient. Years of testing and development have determined the optimum amount of the active ingredient that you should ingest. If there is too little of the active ingredient, the tablet may not be effective. If there is too much, the tablet may be harmful. When the tablets are manufactured, the active ingredient needs to be weighed in bulk. When the tablets are tested, however, to ensure that they contain the right amount of the active ingredient, chemists need to know how many molecules of the substance are present. How do chemists group particles so that they know how many are present in a given mass of substance?

On its own, the mass of a chemical is not very useful to a chemist. The chemical reactions that take place depend on the number of atoms present, not on their masses. Since atoms are far too small and numerous to count, you need a way to relate the numbers of atoms to masses that can be measured.

When many items in a large set need to be counted, it is often useful to work with groups of items rather than individual items. When you hear the word "dozen," you think of the number 12. It does not matter what the items are. A dozen refers to the quantity 12 whether the items are eggs or pencils or baseballs. Table 5.1 lists some common quantities that we use to deal with everyday items.

Table 5.1 Some Common Quantities

Item	Quantity	Amount
gloves	pair	2
soft drinks	six-pack	6
eggs	dozen	12
pens	gross (12 dozen)	144
paper	ream	500

You do not buy eggs one at a time. You purchase them in units of a dozen. Similarly, your school does not order photocopy paper by the sheet. The paper is purchased in bundles of 500 sheets, called a ream. It would be impractical to sell sheets of paper individually.

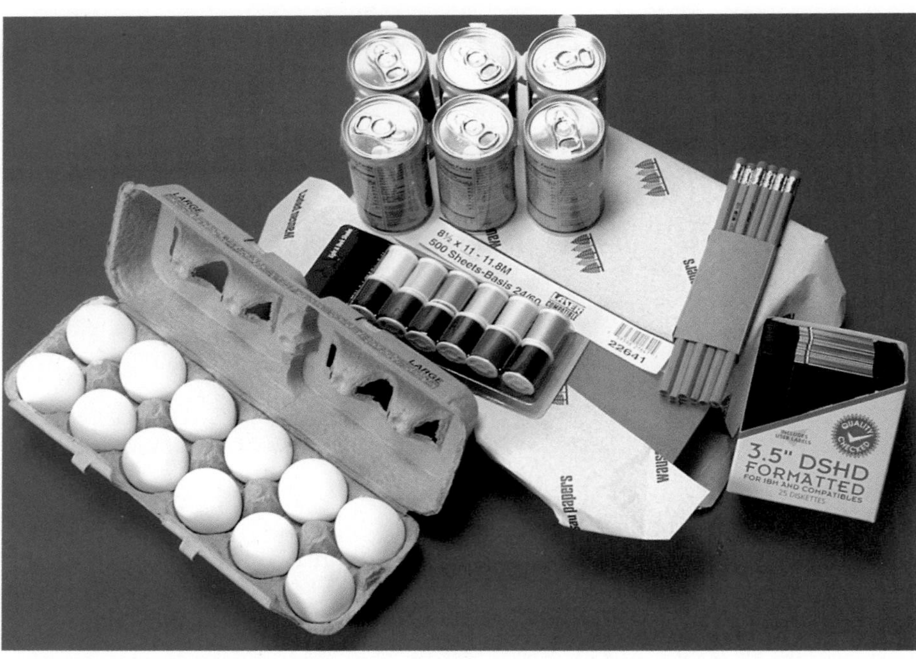

Figure 5.6 Certain items, because of their size, are often handled in bulk. Would you rather count reams of paper or individual sheets?

The Definition of the Mole

Convenient, or easily measurable, amounts of elements contain huge numbers of atoms. Therefore chemists use a quantity that is much larger than a dozen or a ream to group atoms and molecules together. This quantity is the **mole** (symbol **mol**).

- One mole (1 mol) of a substance contains $6.022\ 141\ 99 \times 10^{23}$ particles of the substance. This value is called the **Avogadro constant**. Its symbol is N_A.
- The mole is defined as the amount of substance that contains as many elementary entities (atoms, molecules, or formula units) as exactly 12 g of carbon-12.

The term *mole* is an abbreviation of another word. What do you think this other word is? Check your guess by consulting a dictionary.

For example, one mole of carbon contains 6.02×10^{23} atoms of C. One mole of sodium chloride contains 6.02×10^{23} formula units of NaCl. One mole of hydrofluoric acid contains 6.02×10^{23} molecules of HF.

The Avogadro constant is an experimentally determined quantity. Chemists continually devise more accurate methods to determine how many atoms are in exactly 12 g of carbon-12. This means that the accepted value has changed slightly over the years since it was first defined.

Web LINK

www.school.mcgrawhill.ca/resources/
Chemists have devised various ways to determine the Avogadro constant. To learn more about how this constant has been found in the past and how it is found today, go to the web site above. Go to **Science Resources**, then to **Chemistry 11** to find out where to go next. What are some methods that chemists have used to determine the number of particles in a mole? How has the accepted value of the Avogadro constant changed over the years?

The Chemist's Dozen

The mole is literally the chemist's dozen. Just as egg farmers and grocers use the dozen (a unit of 12) to count eggs, chemists use the mole (a much larger number) to count atoms, molecules, or formula units. When farmers think of two dozen eggs, they are also thinking of 24 eggs.

$$(2\ \cancel{\text{dozen}}) \times \left(\frac{12\ \text{eggs}}{\cancel{\text{dozen}}}\right) = 24\ \text{eggs}$$

Chemists work in a similar way. As you have learned above, 1 mol has 6.02×10^{23} particles. Thus 2 mol of aluminum atoms contain 12.0×10^{23} atoms of Al.

$$2\ \cancel{\text{mol}} \times \left(6.02 \times 10^{23}\ \frac{\text{atoms}}{\cancel{\text{mol}}}\right) = 1.20 \times 10^{24}\ \text{atoms of Al}$$

How Big Is the Avogadro Constant?

The Avogadro constant is a huge number. Its magnitude becomes easier to visualize if you imagine it in terms of ordinary items. For example, suppose that you created a stack of 6.02×10^{23} loonies, as in Figure 5.7. To determine the height of the stack, you could determine the height of one loonie and multiply by 6.02×10^{23}. The Avogadro constant needs to be this huge to group single atoms into convenient amounts. What does 1 mol of a substance look like? Figure 5.8 shows some samples of elements. Each sample contains 6.02×10^{23} atoms. Notice that each sample has a different mass. You will learn why in section 5.3. Examine the following Sample Problem to see how to work with the Avogadro constant.

Figure 5.7 Measure the height of a pile of five loonies. How tall, in kilometres, would a stack of 6.02×10^{23} loonies be?

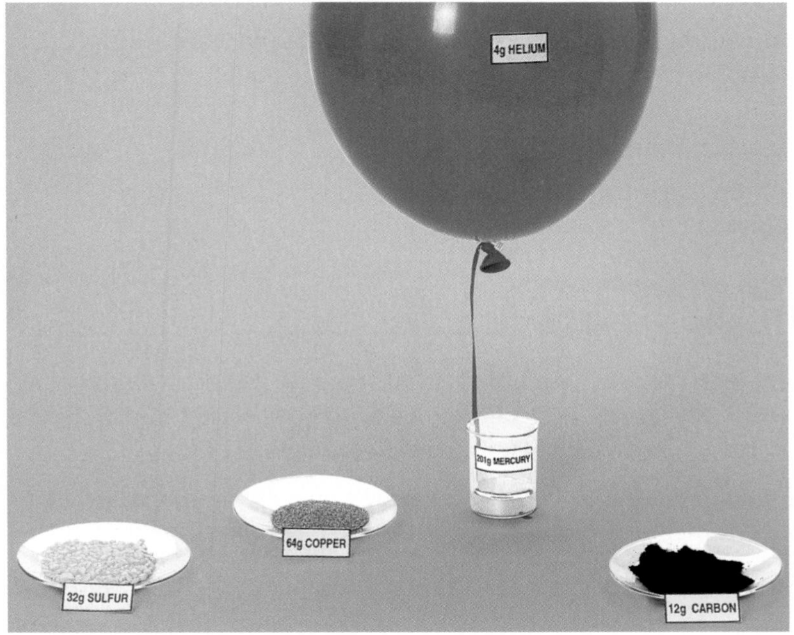

Figure 5.8 Each sample contains 1.00 mol, or 6.02×10^{23} atoms. Why do you think the mass of each sample is different?

Math LINK

Suppose that you invested $\$6.02 \times 10^{23}$ so that it earned 1% interest annually. How much money would you have at the end of ten years?

Figure 5.9 Toronto's SkyDome cost about $500 million to build. Spending $\$6.02 \times 10^{23}$ at the rate of one billion dollars per second is roughly equivalent to building two SkyDomes per second for over 19 million years.

Sample Problem

Using the Avogadro Constant

Problem

The distance "as the crow flies" from St. John's in Newfoundland to Vancouver in British Columbia is 5046 km. Suppose that you had 1 mol of peas, each of diameter 1 cm. How many round trips could be made between these cities, laying the peas from end to end?

What Is Required?

You need to find the number of round trips from St. John's to Vancouver (2×5046 km) that can be made by laying 6.02×10^{23} peas end to end.

What Is Given?

Each round trip is 2×5046 km or 10 092 km. A pea has a diameter of 1 cm.

Plan Your Strategy

First convert the round trip distance from kilometres to centimetres. Since each pea has a diameter of 1 cm, a line of 6.02×10^{23} peas is 6.02×10^{23} cm in length. Divide the length of the line of peas by the round-trip distance to find the number of round trips.

Act on Your Strategy

Converting the round-trip distance from kilometres to centimetres gives

$$(10\,092\ \cancel{\text{km}}) \times (10^5\ \text{cm}/\cancel{\text{km}}) = 1.01 \times 10^9\ \text{cm}$$

$$\text{Number of round trips} = \frac{6.02 \times 10^{23}\ \cancel{\text{cm}}}{(1.01 \times 10^9\ \cancel{\text{cm}}/\text{round trip})}$$

$$= 5.96 \times 10^{14}\ \text{round trips}$$

About 596 trillion round trips between St. John's and Vancouver could be made by laying one mole of peas end to end.

Check Your Solution

Looking at the magnitude of the numbers, you have $10^{23} \div 10^9$. This accounts for 10^{14} in the answer.

Practice Problems

9. The length of British Columbia's coastline is 17 856 km. If you laid 6.02×10^{23} metre sticks end to end along the coast of BC, how many rows of metre sticks would you have?

10. The area of Nunavut is 1 936 113 km^2. Suppose that you had 6.02×10^{23} sheets of pastry, each with the dimensions 30 cm $\times$ 30 cm. How many times could you cover Nunavut completely with pastry?

Continued ...

Continued ...
FROM PAGE 174

11. If you drove for 6.02×10^{23} days at a speed of 100 km/h, how far would you travel?
12. If you spent \$$6.02 \times 10^{23}$ at a rate of \$1.00/s, how long, in years, would the money last? Assume that every year has 365 days.

Technology LINK

Scientific calculators are made to accommodate scientific notation easily. To enter the Avogadro constant, for example, type 6.02, followed by the key labelled "EE" or "EXP." (The label on the key depends on the make of calculator you have.) Then enter 23. The 23 will appear at the far right of the display, *without* the exponential base of 10. Your calculator *understands* the number to be in scientific notation.

Converting Moles to Number of Particles

In the Thought Lab below, you can practise working with the mole by relating the Avogadro constant to familiar items. Normally the mole is used to group atoms and compounds. For example, chemists know that 1 mol of barium contains 6.02×10^{23} atoms of Ba. Similarly, 2 mol of barium sulfate contain $2 \times (6.02 \times 10^{23}) = 12.0 \times 10^{23}$ molecules of $BaSO_4$.

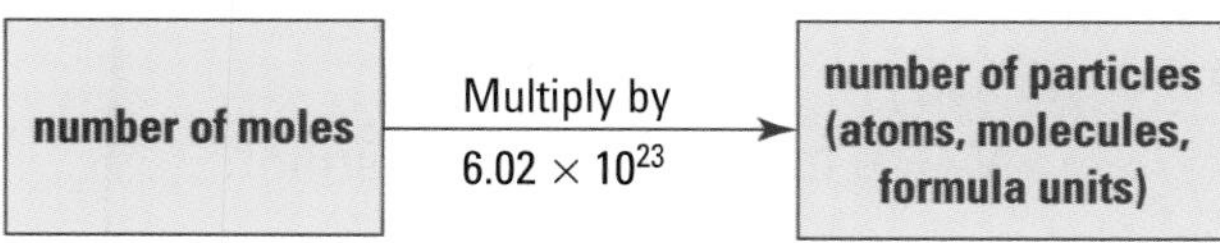

The mole is used to help us "count" atoms and molecules. The relationship between moles, number of particles, and the Avogadro constant is

$$N = \text{number of particles}$$
$$n = \text{number of moles}$$
$$N_A = \text{Avogadro constant}$$
$$N = n \times N_A$$

Try the next Sample Problem to learn how the number of moles of a substance relates to the number of particles in the substance.

ThoughtLab  The Magnitude of the Avogadro Constant

This activity presents some challenges related to the magnitude of the Avogadro constant. These questions are examples of *Fermi problems,* which involve large numbers (like the Avogadro constant) and give approximate answers. The Italian physicist, Enrico Fermi, liked to pose and solve these types of questions.

Procedure

Work in small groups. Use any reference materials, including materials supplied by your teacher and information on the Internet. For each question, brainstorm to determine the required information. Obtain this information, and answer the question. Be sure to include units throughout your calculation, along with a brief explanation.

Analysis

1. If you covered Canada's land mass with 1.00 mol of golf balls, how deep would the layer of golf balls be?
2. Suppose that you put one mole of five-dollar bills end to end. How many round trips from Earth to the Moon would they make?
3. If you could somehow remove 6.02×10^{23} teaspoons of water from the world's oceans, would you completely drain the oceans? Explain.
4. What is the mass of one mole of apples? How does this compare with the mass of Earth?
5. How many planets would we need for one mole of people, if each planet's population were limited to the current population of Earth?

Sample Problem

Moles to Atoms

Problem

A sample contains 1.25 mol of nitrogen dioxide, NO_2.

(a) How many molecules are in the sample?

(b) How many atoms are in the sample?

What Is Required?

You need to find the number of atoms and molecules in the sample.

What Is Given?

The sample consists of 1.25 mol of nitrogen dioxide molecules. Each nitrogen dioxide molecule is made up of three atoms: 1 N atom + 2 O atoms.

$$N_A = 6.02 \times 10^{23} \text{ molecules/mol}$$

Plan Your Strategy

(a) A molecule of NO_2 contains three atoms. Find the number of NO_2 *molecules* in 1.25 mol of nitrogen dioxide.

(b) Multiply the number of molecules by 3 to arrive at the total number of *atoms* in the sample.

Act on Your Strategy

(a) Number of molecules of NO_2

$$= (1.25 \cancel{\text{mol}}) \times \left(6.02 \times 10^{23} \frac{\text{molecules}}{\cancel{\text{mol}}}\right)$$

$$= 7.52 \times 10^{23} \text{ molecules}$$

Therefore there are 7.52×10^{23} molecules in 1.25 mol of NO_2.

(b) $$(7.52 \times 10^{23} \cancel{\text{molecules}}) \times \left(3 \frac{\text{atoms}}{\cancel{\text{molecule}}}\right) = 2.26 \times 10^{24} \text{ atoms}$$

Therefore there are 2.26×10^{24} atoms in 1.25 mol of NO_2.

Check Your Solution

Work backwards. One mol contains 6.02×10^{23} atoms. How many moles represent 2.2×10^{24} atoms?

$$2.2 \times 10^{24} \cancel{\text{atoms}} \times \frac{1 \text{ mol}}{6.02 \times 10^{23} \cancel{\text{atoms}}} = 3.7 \text{ mol}$$

There are 3 atoms in each molecule of NO_2

$$3.7 \cancel{\text{mol atoms}} \times \frac{1 \text{ mol molecule}}{3 \cancel{\text{mol atoms}}} = 1.2 \text{ molecules}$$

This is close to the value of 1.25 mol of molecules, given in the question.

Continued ...

Language LINK

The term *order of magnitude* refers to the size of a number—specifically to its exponent when in scientific notation. For example, a scientist will say that 50 000 (5×10^4) is two orders of magnitude (10^2 times) larger than 500 (5×10^2). How many orders of magnitude is the Avogadro constant greater than one billion?

Continued ...

FROM PAGE 176

Go to the Chemistry 11 Electronic Learning Partner for a video clip that describes the principles behind the Avogadro constant and the mole.

Practice Problems

13. A small pin contains 0.0178 mol of iron, Fe. How many atoms of iron are in the pin?

14. A sample contains 4.70×10^{-4} mol of gold, Au. How many atoms of gold are in the sample?

15. How many formula units are contained in 0.21 mol of magnesium nitrate, $Mg(NO_3)_2$?

16. A litre of water contains 55.6 mol of water. How many molecules of water are in this sample?

17. Ethyl acetate, $C_4H_8O_2$, is frequently used in nail polish remover. A typical bottle of nail polish remover contains about 2.5 mol of ethyl acetate.

(a) How many molecules are in the bottle of nail polish remover?

(b) How many atoms are in the bottle?

(c) How many carbon atoms are in the bottle?

18. Consider a 0.829 mol sample of sodium sulfate, Na_2SO_4.

(a) How many formula units are in the sample?

(b) How many sodium ions, Na^+, are in the sample?

Converting Number of Particles to Moles

Chemists very rarely express the amount of a substance in number of particles. As you have seen, there are far too many particles to work with conveniently. For example, you would never say that you had dissolved 3.21×10^{23} molecules of sodium chloride in water. You might say, however, that you had dissolved 0.533 mol of sodium chloride in water. When chemists communicate with each other about amounts of substances, they usually use units of moles (see Figure 5.10). To convert the number of particles in a substance to the number of moles, rearrange the equation you learned previously.

$$N = N_A \times n$$
$$n = \frac{N}{N_A}$$

To learn how many moles are in a substance when you know how many particles are present, find out how many times the Avogadro constant goes into the number of particles.

Try the next Sample Problem to practise converting the number of atoms, formula units, or molecules in a substance to the number of moles.

Figure 5.10 **Chemists rarely use the number of particles to communicate how much of a substance they have. Instead, they use moles.**

History LINK

The Avogadro constant is named to honour the Italian chemist Amedeo Avogadro. In 1811, Avogadro postulated what is now known as Avogadro's law: Equal volumes of gases, at equal temperatures and pressures, contain the same number of molecules. You will learn more about Avogadro's law in Unit 4.

Web LINK

www.school.mcgrawhill.ca/resources/
The International System of Units (SI) is based on seven base units for seven quantities. (One of these quantities is the mole.) The quantities are assumed to be all independent of each other. To learn about the seven base quantities, go to the web site above. Go to **Science Resources**, then to **Chemistry 11** to find out where to go next. All other SI units are derived from the seven base units.

Sample Problem

Molecules to Moles

Problem

How many moles are present in a sample of carbon dioxide, CO_2, made up of 5.83×10^{24} molecules?

What Is Required?

You need to find the number of moles in 5.83×10^{24} molecules of carbon dioxide.

What Is Given?

You are given the number of molecules in the sample.

$$N_A = 6.02 \times 10^{23} \text{ molecules } CO_2/\text{mol } CO_2$$

Plan Your Strategy

$$n = \frac{N}{N_A}$$

Act on Your Strategy

$$n = \frac{5.83 \times 10^{24} \cancel{\text{molecules } CO_2}}{(6.02 \times 10^{23} \cancel{\text{molecules } CO_2}/\text{mol } CO_2)}$$
$$= 9.68 \text{ mol } CO_2$$

There are 9.68 mol of CO_2 in the sample.

Check Your Solution

5.83×10^{24} molecules is approximately equal to 6×10^{24} molecules. Since the number of molecules is about ten times larger than the Avogadro constant, it makes sense that there are about 10 mol in the sample.

Practice Problems

19. A sample of bauxite ore contains 7.71×10^{24} molecules of aluminum oxide, Al_2O_3. How many moles of aluminum oxide are in the sample?

20. A vat of cleaning solution contains 8.03×10^{26} molecules of ammonia, NH_3. How many moles of ammonia are in the vat?

21. A sample of cyanic acid, HCN, contains 3.33×10^{22} atoms. How many moles of cyanic acid are in the sample? **Hint:** Find the number of molecules of HCN first.

22. A sample of pure acetic acid, CH_3COOH, contains 1.40×10^{23} carbon atoms. How many moles of acetic acid are in the sample?

Section Wrap-up

In section 5.1, you learned about the average atomic mass of an element. Then, in section 5.2, you learned how chemists group particles using the mole. In the final section of this chapter, you will learn how to use the average atomic masses of the elements to determine the mass of a mole of any substance. You will learn about a relationship that will allow you to relate the mass of a sample to the number of particles it contains.

Section Review

1 K/U In your own words, define the mole. Use three examples.

2 I Imagine that $\$6.02 \times 10^{23}$ were evenly distributed among six billion people. How much money would each person receive?

3 I A typical adult human heart beats an average of 60 times per minute. If you were allotted a mole of heartbeats, how long, in years, could you expect to live? You may assume each year has 365 days.

4 I Calculate the number of atoms in 3.45 mol of iron, Fe.

5 I A sample of carbon dioxide, CO_2, contains 2.56×10^{24} molecules.

(a) How many moles of carbon dioxide are present?

(b) How many moles of atoms are present?

6 I A balloon is filled with 0.50 mol of helium. How many atoms of helium are in the balloon?

7 I A sample of benzene, C_6H_6, contains 5.69 mol.

(a) How many molecules are in the sample?

(b) How many hydrogen atoms are in the sample?

8 I Aluminum oxide, Al_2O_3, forms a thin coating on aluminum when aluminum is exposed to the oxygen in the air. Consider a sample made up of 1.17 mol of aluminum oxide.

(a) How many molecules are in the sample?

(b) How many atoms are in the sample?

(c) How many oxygen atoms are in the sample?

9 C Why do you think chemists chose to define the mole the way they did?

10 I A sample of zinc oxide, ZnO, contains 3.28×10^{24} molecules of zinc oxide. A sample of zinc metal contains 2.78 mol of zinc atoms. Which sample contains more zinc: the compound or the element?

Unit Investigation Prep

In your Unit Investigation, you will need to determine the amount of several pure substances in a mixture. Do you think it will be more convenient for you to work with quantities expressed in moles or molecules?

5.3 Molar Mass

Section Preview/ Specific Expectations

In this section, you will

- **explain** the relationship between the average atomic mass of an element and its molar mass
- **solve** problems involving number of moles, number of particles, and mass
- **calculate** the molar mass of a compound
- **communicate** your understanding of the following term: *molar mass*

In section 5.2, you explored the relationship between the number of atoms or particles and the number of moles in a sample. Now you are ready to relate the number of moles to the mass, in grams. Then you will be able to determine the number of atoms, molecules, or formula units in a sample by finding the mass of the sample.

Mass and the Mole

You would never express the mass of a lump of gold, like the one in Figure 5.11, in atomic mass units. You would express its mass in grams. How does the mole relate the number of atoms to measurable quantities of a substance? The definition of the mole pertains to relative atomic mass, as you learned in section 5.1. One atom of carbon-12 has a mass of exactly 12 u. Also, by definition, one mole of carbon-12 atoms (6.02×10^{23} carbon-12 atoms) has a mass of exactly 12 g.

The Avogadro constant is the factor that converts the relative mass of individual atoms or molecules, expressed in atomic mass units, to mole quantities, expressed in grams.

How can you use this relationship to relate mass and moles? The periodic table tells us the average mass of a single atom in atomic mass units (u). For example, zinc has an average atomic mass of 65.39 u. *One mole of an element has a mass expressed in grams numerically equivalent to the element's average atomic mass expressed in atomic mass units.* One mole of zinc atoms has a mass of 65.39 g. This relationship allows chemists to use a balance to count atoms. You can use the periodic table to determine the mass of one mole of an element.

Table 5.2 Average Atomic Mass and Molar Mass of Four Elements

Element	Average atomic mass (u)	Molar mass (g)
hydrogen, H	1.01	1.01
oxygen, O	16.00	16.00
sodium, Na	22.99	22.99
argon, Ar	39.95	39.95

What is Molar Mass?

The mass of one mole of any element, expressed in grams, is numerically equivalent to the average atomic mass of the element, expressed in atomic mass units. The mass of one mole of a substance is called its **molar mass** (symbol ***M***). Molar mass is expressed in g/mol. For example, the average atomic mass of gold, as given in the periodic table, is 196.97 u. Thus the mass of one mole of gold atoms, gold's molar mass, is 196.97 g. Table 5.2 gives some additional examples of molar masses.

Express mass in atomic mass units.

Express mass in grams.

Figure 5.11 The Avogadro constant is a factor that converts from atomic mass to molar mass.

Finding the Molar Mass of Compounds

While you can find the molar mass of an element just by looking at the periodic table, you need to do some calculations to find the molar mass of a compound. For example, 1 mol of beryllium oxide, BeO, contains 1 mol of beryllium and 1 mol of oxygen. To find the molar mass of BeO, add the mass of each element that it contains.

$$\begin{aligned} M_{\text{BeO}} &= 9.01 \text{ g/mol} + 16.00 \text{ g/mol} \\ &= 25.01 \text{ g/mol} \end{aligned}$$

Examine the following Sample Problem to learn how to determine the molar mass of a compound. Following Investigation 5-A on the next page, there are some Practice Problems for you to try.

CHEM FACT

The National Institute of Standards and Technology (NIST) and most other standardization bodies use *M* to represent molar mass. You may see other symbols, such as *mm*, used to represent molar mass.

Sample Problem

Molar Mass of a Compound

Problem

What is the mass of one mole of calcium phosphate, $Ca_3(PO_4)_2$?

What Is Required?

You need to find the molar mass of calcium phosphate.

What Is Given?

You know the formula of calcium phosphate. You also know, from the periodic table, the average atomic mass of each atom that makes up calcium phosphate.

Plan Your Strategy

Find the total mass of each element to determine the molar mass of calcium phosphate. Find the mass of 3 mol of calcium, the mass of 2 mol of phosphorus, and the mass of 8 mol of oxygen. Then add these masses together.

Act on Your Strategy

$M_{\text{Ca}} \times 3 = (40.08 \text{ g/mol}) \times 3 = 120.24 \text{ g/mol}$

$M_{\text{P}} \times 2 = (30.97 \text{ g/mol}) \times 2 = 61.94 \text{ g/mol}$

$M_{\text{O}} \times 8 = (16.00 \text{ g/mol}) \times 8 = 128.00 \text{ g/mol}$

$$\begin{aligned} M_{\text{Ca}_3(\text{PO}_4)_2} &= 120.24 \text{ gmol} + 61.94 \text{ gmol} + 128.00 \text{ gmol} \\ &= 310.18 \text{ gmol} \end{aligned}$$

Therefore the molar mass of calcium phosphate is 310.18 g/mol.

Check Your Solution

Using round numbers for a quick check, you get

$$(40 \times 3) + (30 \times 2) + (15 \times 8) = 300$$

This estimate is close to the answer of 310.18 g/mol.

Continued ... ON PAGE 184

PROBLEM TIP

Once you are used to calculating molar masses, you will want to do the four calculations at left all at once. Try solving the Sample Problem using only one line of calculations.

Investigation 5-A

SKILL FOCUS
Modelling concepts
Analyzing and interpreting
Communicating results

Modelling Mole and Mass Relationships

Chemists use the mole to group large numbers of atoms and molecules into manageable, macroscopic quantities. In this way, they can tell how many atoms or molecules are in a given sample, even though the particles are too small to see. In this investigation, you will explore how to apply this idea to everyday objects such as grains of rice and nuts and bolts.

Question

How can you use what you know about the mole and molar mass to count large numbers of tiny objects using mass, and to relate numbers of objects you cannot see based on their masses?

Part 1 Counting Grains of Rice

Materials

electronic balance

40 mL dry rice

50 mL beaker

Procedure

1. Try to measure the mass of a grain of rice. Does the balance register this mass?
2. Count out 20 grains of rice. Measure and record their mass.
3. Find the mass of the empty beaker. Add the rice to the beaker. Find the mass of the beaker and the rice. Determine the mass of the rice.
4. Calculate the number of grains of rice in the 40 mL sample. Report your answer to the number of significant digits that reflects the precision of your calculation.

Part 2 Counting Objects Based on Their Relative Masses

Materials

electronic balance

10 small metal nuts (to represent the fictitious element *nutium*)

10 washers (to represent the fictitious element *washerium*)

2 opaque film canisters with lids

Procedure

1. Measure the mass of 10 nuts (*nutium* atoms). Then measure the mass of 10 washers (*washerium* atoms).
2. Calculate the average mass of a single "atom" of *nutium* and *washerium*.
3. Determine the mass ratio of *nutium* to *washerium*.
4. Obtain, from your teacher, a sealed film canister containing an unknown number of *nutium* atoms. Your teacher will tell you the mass of the empty film canister and lid.
5. Find the mass of the unknown number of *nutium* atoms.
6. You know that you need an equal number of *washerium* atoms to react with the unknown number of *nutium* atoms. What mass of *washerium* atoms do you need?

Analysis

1. Was it possible to get an accurate mass for an individual grain of rice? How did you solve this problem?

2. How did you avoid having to count every single grain of rice in order to determine how many there were in the sample?

3. Using your data, how many grains of rice would be in 6.5×10^3 g of rice?

4. A mole of helium atoms weighs 4.00 g.
   (a) How many atoms are in 23.8 g of helium?
   (b) What known relationship did you use to find your answer to part (a)?
   (c) What analogous relationship did you set up in order to calculate the number of grains of rice based on the mass of the rice?

5. You know the relative mass of nuts and washers. Suppose that you are given some washers in a sealed container. You know that you have the same number of nuts in another sealed container.
   (a) Can you determine how many washers are in the container without opening either container? Why or why not?
   (b) What, if any, additional information do you need?

6. The molar mass of carbon is 12.0 g. The molar mass of molecular oxygen is 32.0 g. Equal numbers of carbon atoms and oxygen molecules react to form carbon dioxide.
   (a) If you have 5.8 g of carbon, what mass of oxygen will react?
   (b) How does part (a) relate to step 6 in the Procedure for Part 2?

Conclusion

7. How do chemists use the mole and molar masses to count numbers and relative numbers of atoms and molecules? Relate your answer to the techniques you used to count rice, nuts, and washers.

Applications

8. Think about your answer to Analysis question 5(a). Did you need to use the Avogadro constant in your calculation? Explain why or why not.

9. Chemists rarely use the Avogadro constant directly in their calculations. What relationship do they use to avoid working with such a large number?

Continued ...
FROM PAGE 181

Practice Problems

23. State the molar mass of each element.
(a) xenon, Xe
(b) osmium, Os
(c) barium, Ba
(d) tellurium, Te

24. Find the molar mass of each compound.
(a) ammonia, NH_3
(b) glucose, $C_6H_{12}O_6$
(c) potassium dichromate, $K_2Cr_2O_7$
(d) iron(III) sulfate, $Fe_2(SO_4)_3$

25. Strontium may be found in nature as celestite, $SrSO_4$. Find the molar mass of celestite.

26. What is the molar mass of the ion $[Cu(NH_3)_4]^{2+}$?

Counting Particles Using Mass

Using the mole concept and the periodic table, you can determine the mass of one mole of a compound. You know, however, that one mole represents 6.02×10^{23} particles. Therefore you can use a balance to count atoms, molecules, or formula units!

For example, consider carbon dioxide, CO_2. One mole of carbon dioxide has a mass of 44.0 g and contains 6.02×10^{23} molecules. You can set up the following relationship:

$$6.02 \times 10^{23} \text{ molecules of } CO_2 \rightarrow 1 \text{ mol of } CO_2 \rightarrow 44.0 \text{ g of } CO_2$$

How can you use this relationship to find the number of molecules and the number of moles in 22.0 g of carbon dioxide?

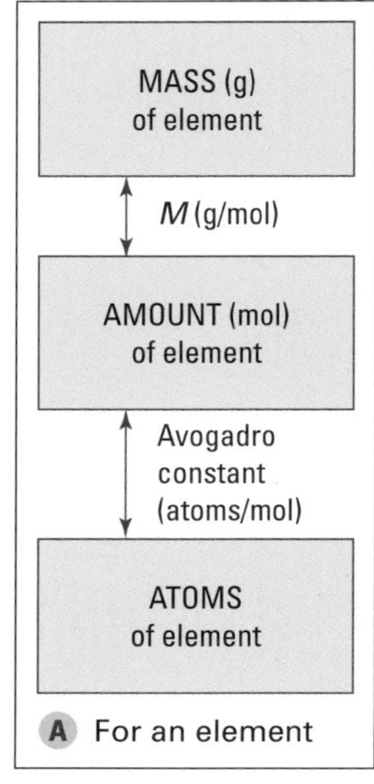

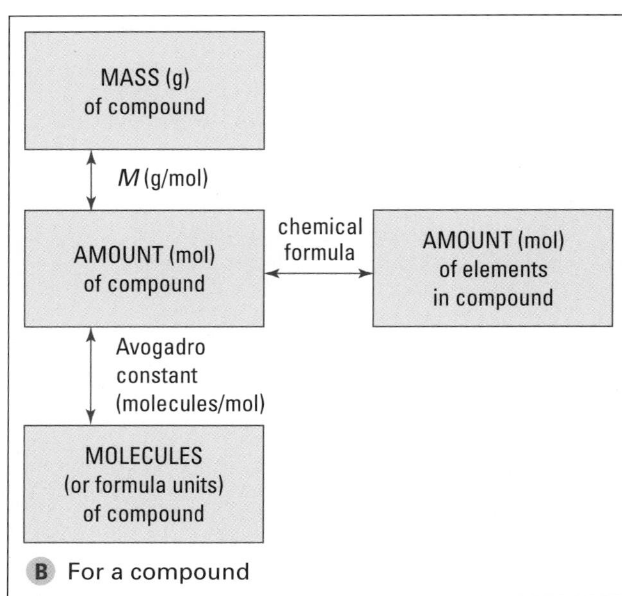

Figure 5.12 The molar mass relates the amount of an element or a compound, in moles, to its mass. Similarly, the Avogadro constant relates the number of particles to the molar amount.

Converting from Moles to Mass

Suppose that you want to carry out a reaction involving ammonium sulfate and calcium chloride. The first step is to obtain one mole of each chemical. How do you decide how much of each chemical you need? You convert the molar amount to mass. Then you use a balance to determine the mass of the proper amount of each chemical.

The following equation can be used to solve problems involving mass, molar mass, and number of moles:

$$\text{Mass} = \text{Number of moles} \times \text{Molar mass}$$
$$m = n \times M$$

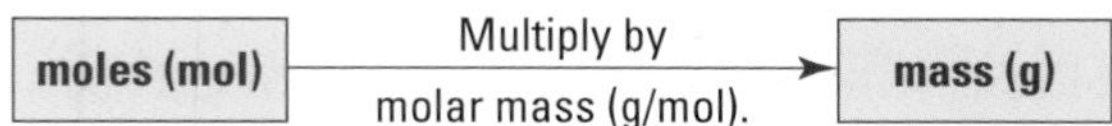

CHECKPOINT

Write the chemical formulas for calcium chloride and ammonium sulfate. Predict what kind of reaction will occur between them. Write a balanced chemical equation to show the reaction. Ionic compounds containing the ammonium ion are soluble. Ammonium sulfate is soluble, but barium chloride is not.

Sample Problem

Moles to Mass

Problem

A flask contains 0.750 mol of carbon dioxide gas, CO_2. What mass of carbon dioxide gas is in this sample?

What Is Required?

You need to find the mass of carbon dioxide.

What Is Given?

The sample contains 0.750 mol. You can determine the molar mass of carbon dioxide from the periodic table.

Plan Your Strategy

In order to convert moles to grams, you need to determine the molar mass of carbon dioxide from the periodic table.

Multiply the molar mass of carbon dioxide by the number of moles of carbon dioxide to determine the mass.

$$m = n \times M$$

Act on Your Strategy

$$M_{CO_2} = 2 \times (16.00 \text{ g/mol}) + 12.01 \text{ g/mol}$$
$$= 44.01 \text{ g/mol}$$

$$m = (0.750 \cancel{\text{mol}}) \times (44.01 \text{ g/}\cancel{\text{mol}})$$
$$= 33.0 \text{ g}$$

The mass of 0.75 mol of carbon dioxide is 33.0 g.

Check Your Solution

1 mol of carbon dioxide has a mass of 44 g. You need to determine the mass of 0.75 mol, or 75% of a mole. 33 g is equal to 75% of 44 g.

Continued ...

Continued ...

FROM PAGE 185

Practice Problems

27. Calculate the mass of each molar quantity.

(a) 3.90 mol of carbon, C

(b) 2.50 mol of ozone, O_3

(c) 1.75×10^7 mol of propanol, C_3H_8O

(d) 1.45×10^{-5} mol of ammonium dichromate, $(NH_4)_2Cr_2O_7$

28. For each group, which sample has the largest mass?

(a) 5.00 mol of C, 1.50 mol of Cl_2, 0.50 mol of $C_6H_{12}O_6$

(b) 7.31 mol of O_2, 5.64 mol of CH_3OH, 12.1 mol of H_2O

29. A litre, 1000 mL, of water contains 55.6 mol. What is the mass of a litre of water?

30. To carry out a particular reaction, a chemical engineer needs 255 mol of styrene, C_8H_8. How many kilograms of styrene does the engineer need?

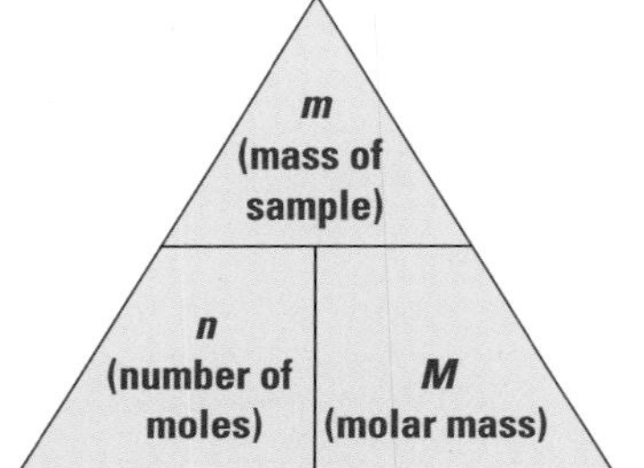

Figure 5.13 Use this triangle for problems involving number of moles, mass of sample, and molar mass. For what other scientific relationships might you use a triangle like this?

You might find the triangle shown in Figure 5.13 useful for problems involving number of moles, number of particles, and molar mass. To use it, cover the quantity that you need to find. The required operation—multiplication or division—will be obvious from the position of the remaining variables. For example, if you want to find the mass of a sample, cover the m in the triangle. You can now see that

$$\text{Mass} = \text{Number of moles} \times \text{Molar mass.}$$

Be sure to check that your units cancel.

Converting from Mass to Moles

In the previous Sample Problem, you saw how to convert moles to mass. Often, however, chemists know the mass of a substance but are more interested in knowing the number of moles. Suppose that a reaction produces 223 g of iron and 204 g of aluminum oxide. The masses of the substances do not tell you very much about the reaction. You know, however, that 223 g of iron is 4 mol of iron. You also know that 204 g of aluminum oxide is 2 mol of aluminum oxide. You may conclude that the reaction produces twice as many moles of iron as it does moles of aluminum oxide. You can perform the reaction many times to test your conclusion. If your conclusion is correct, the mole relationship between the products will hold. To calculate the number of moles in a sample, find out how many times the molar mass goes into the mass of the sample.

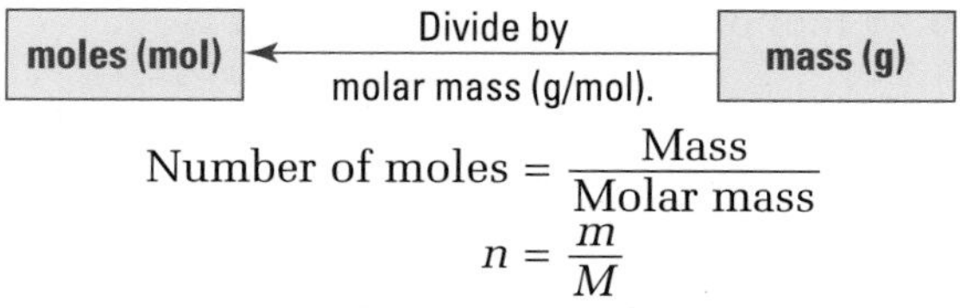

$$\text{Number of moles} = \frac{\text{Mass}}{\text{Molar mass}}$$

$$n = \frac{m}{M}$$

The following Sample Problem explains how to convert from the mass of a sample to the number of moles it contains.

Sample Problem

Mass to Moles

Problem

How many moles of acetic acid, CH_3COOH, are in a 23.6 g sample?

What Is Required?

You need to find the number of moles in 23.6 g of acetic acid.

What Is Given?

You are given the mass of the sample.

Plan Your Strategy

To obtain the number of moles of acetic acid, divide the mass of acetic acid by its molar mass.

Act on Your Strategy

The molar mass of CH_3COOH is
$(12.01 \times 2) + (16.00 \times 2) + (1.01 \times 4) = 60.06$ g.

$$n = \frac{m}{M_{CH_3COOH}}$$

$$n \text{ mol } CH_3COOH = \frac{23.6 \text{ } \cancel{g}}{60.06 \text{ } \cancel{g}/\text{mol}}$$

$$= 0.393 \text{ mol}$$

Therefore there are 0.393 mol of acetic acid in 23.6 g of acetic acid.

Check Your Solution

Work backwards. There are 60.06 g in each mol of acetic acid. So in 0.393 mol of acetic acid, you have 0.393 $\cancel{\text{mol}} \times 60.06$ g/$\cancel{\text{mol}}$ = 23.6 g of acetic acid. This value matches the question.

Practice Problems

31. Calculate the number of moles in each sample.

(a) 103 g of Mo

(b) 1.32×10^4 g of Pd

(c) 0.736 kg of Cr

(d) 56.3 mg of Ge

32. How many moles of compound are in each sample?

(a) 39.2 g of silicon dioxide, SiO_2

(b) 7.34 g of nitrous acid, HNO_2

(c) 1.55×10^5 kg of carbon tetrafluoride, CF_4

(d) 8.11×10^{-3} mg of 1-iodo-2,3-dimethylbenzene C_8H_9I

33. Sodium chloride, NaCl, can be used to melt snow. How many moles of sodium chloride are in a 10 kg bag?

34. Octane, C_8H_{18}, is a principal ingredient of gasoline. Calculate the number of moles in a 20.0 kg sample of octane.

Chemistry Bulletin

Science | Technology | Society | Environment

Chemical Amounts in Vitamin Supplements

Vitamins and minerals (micronutrients) help to regulate your metabolism. They are the building blocks of blood and bone, and they maintain muscles and nerves. In Canada, a standard called *Recommended Nutrient Intake* (*RNI*) outlines the amounts of micronutrients that people should ingest each day. Eating a balanced diet is the best way to achieve your RNI. Sometimes, however, you may need to take multivitamin supplements when you are unable to attain your RNI through diet alone.

The label on a bottle of supplements lists all the vitamins and minerals the supplements contain. It also lists the form and source of each vitamin and mineral, and the amount of each. The form of a mineral is especially important to know because it affects the quantity your body can use. For example, a supplement may claim to contain 650 mg of calcium carbonate, $CaCO_3$, per tablet. This does not mean that there is 650 mg of calcium. The amount of actual calcium, or *elemental calcium*, in calcium carbonate is only 260 mg. Calcium carbonate has more elemental calcium than the same amount of calcium gluconate, which only has 58 mg for every 650 mg of the compound. Calcium gluconate may be easier for your body to absorb, however.

Quality Control

Multivitamin manufacturers employ chemists, or analysts, to ensure that the products they make have the right balance of micronutrients. Manufacturers have departments devoted to *quality control* (*QC*). QC chemists analyze all the raw materials in the supplements, using standardized tests. Most manufacturers use tests approved by a "standardization body," such as the US Pharmacopoeia. Such standardization bodies have developed testing guidelines to help manufacturers ensure that their products contain what the labels claim, within strict limits.

To test for quality, QC chemists prepare samples of the raw materials from which they will make the supplements. They label the samples according to the "lot" of materials from which the samples were taken. They powder and weigh the samples. Then they extract the vitamins. At the same time, they prepare standard solutions containing a known amount of each vitamin.

Next the chemists compare the samples to the standards by subjecting both to the same tests. One test that is used is *high-performance liquid chromatography* (*HPLC*). HPLC produces a spectrum, or "fingerprint," that identifies each compound. Analysts compare the spectrum that is produced by the samples to the spectrum that is produced by the standard.

Analysts test tablets and capsules for dissolution and disintegration properties. The analyst may use solutions that simulate the contents of the human stomach or intestines for these tests. Only when the analysts are sure that the tablets pass all the necessary requirements are the tablets shipped to retail stores.

Making Connections

1. Why might consuming more of the daily RNI of a vitamin or mineral be harmful?
2. The daily RNI of calcium for adolescent females is 700 to 1100 mg. A supplement tablet contains 950 mg of calcium citrate. Each gram of calcium citrate contains 5.26×10^{-3} mol calcium. How many tablets would a 16-year-old female have to take to meet her daily RNI?

An analyst tests whether tablets will sufficiently dissolve within a given time limit.

Converting Between Moles, Mass, and Number of Particles

You can use what you now know about the mole to carry out calculations involving molar mass and the Avogadro constant. One mole of any compound or element contains 6.02×10^{23} particles. The compound or element has a mass, in grams, that is determined from the periodic table.

Now that you have learned how the number of particles, number of moles, and mass of a substance are related, you can convert from one value to another. Usually chemists convert from moles to mass and from mass to moles. Mass is a property that can be measured easily. The following graphic shows the factors used to convert between particles, moles, and mass. Moles are a convenient way to communicate the amount of a substance.

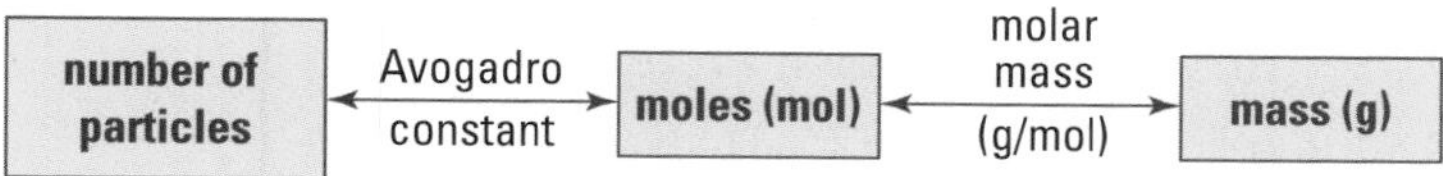

For example, suppose that you need 2.3 mol of potassium chloride to carry out a reaction. You need to convert the molar amount to mass so that you can measure the correct amount with a balance.

To be certain you understand the relationship among particles, moles, and mass, examine the following Sample Problem.

Sample Problem

Particles to Mass

Problem

What is the mass of 5.67×10^{24} molecules of cobalt(II) chloride, $CoCl_2$?

What Is Required?

You need to find the mass of 5.67×10^{24} molecules of cobalt(II) chloride.

What Is Given?

You are given the number of molecules.

Plan Your Strategy

Convert the number of molecules into moles by dividing by the Avogadro constant. Then convert the number of moles into grams by multiplying by the molar mass of cobalt(II) chloride.

Act on Your Strategy

$$\frac{\text{Number of molecules } CoCl_2}{\text{Number of molecules } CoCl_2/\text{mol } CoCl_2} \times \text{mass } CoCl_2/\text{mol } CoCl_2$$

$$= \frac{5.67 \times 10^{24} \cancel{\text{molecules } CoCl_2}}{6.02 \times 10^{23} \cancel{\text{molecules } CoCl_2}/\cancel{\text{mol } CoCl_2}} \times 129.84 \text{ g } CoCl_2/\cancel{\text{mol } CoCl_2}$$

$$= 1.22 \times 10^3 \text{ g } CoCl_2$$

Continued ...

Math LINK

Average atomic mass values in some periodic tables can have five or more significant digits. How do you know how many significant digits to use? When you enter values, such as average atomic mass, into your calculator, be sure that you use at least one more significant digit than is required in your answer.

Continued ...
FROM PAGE 189

Check Your Solution

5.67×10^{24} molecules is roughly 10 times the Avogadro constant. This means that you have about 10 mol of cobalt(II) chloride. The molar mass of cobalt(II) chloride is about 130 g, and 10 times 130 g is 1300 g.

Practice Problems

35. Determine the mass of each sample.

(a) 6.02×10^{24} formula units of $ZnCl_2$

(b) 7.38×10^{21} formula units of $Pb_3(PO_4)_2$

(c) 9.11×10^{23} molecules of $C_{15}H_{21}N_3O_{15}$

(d) 1.20×10^{29} molecules of N_2O_5

36. What is the mass of lithium in 254 formula units of lithium chloride, LiCl?

37. Express the mass of a single atom of titanium, Ti, in grams.

38. Vitamin B_2, $C_{17}H_{20}N_4O_6$, is also called riboflavin. What is the mass, in grams, of a single molecule of riboflavin?

What if you wanted to compare amounts of substances, and you only knew their masses? You would probably convert their masses to moles. The Avogadro constant relates the molar amount to the number of particles. Examine the next Sample Problem to learn how to convert mass to number of particles.

Sample Problem
Mass to Particles

Problem

Chlorine gas, Cl_2, can react with iodine, I_2, to form iodine chloride, ICl. How many molecules of iodine chloride are contained in a 2.74×10^{-1} g sample?

What Is Required?

You need to find the number of molecules in 2.74×10^{-1} g of iodine.

What Is Given?

You are given the mass of the sample.

Continued ...

Continued ...
FROM PAGE 190

Plan Your Strategy

First convert the mass to moles, using the molar mass of iodine. Multiplying the number of moles by the Avogadro constant will yield the number of molecules.

Act on Your Strategy

The molar mass of ICl is 162.36 g.
Dividing the given mass of ICl by the molar mass gives

$$n = \frac{2.74 \times 10^{-1}\ \cancel{g}}{162.36\ \cancel{g}/\text{mol}}$$
$$= 1.69 \times 10^{-3}\ \text{mol}$$

Now multiply the number of moles by the Avogadro constant. This gives the number of molecules in the sample.

$$(1.69 \times 10^{-3}\ \cancel{\text{mol}}) \times \frac{(6.02 \times 10^{23}\ \text{molecules})}{1\ \cancel{\text{mol}}} = 1.01 \times 10^{21}\ \text{molecules}$$

Therefore there are 1.01×10^{21} molecules in 2.74×10^{-1} g of iodine chloride.

Check Your Solution

Work backwards. Each mole of iodine chloride has a mass of 162.36 g/mol. Therefore 1.01×10^{21} molecules of iodine chloride have a mass of:

$$1.01 \times 10^{21}\ \cancel{\text{molecules}} \times \frac{1\ \cancel{\text{mol}}}{6.02 \times 10^{23}\ \cancel{\text{molecules}}} \times \frac{162.36\ \text{g}}{1\ \cancel{\text{mol}}}$$
$$= 2.72 \times 10^{-1}\ \text{g}$$

The answer is close to the value given in the question. Your answer is reasonable.

Practice Problems

39. Determine the number of molecules or formula units in each sample.

(a) 10.0 g of water, H_2O

(b) 52.4 g of methanol, CH_3OH

(c) 23.5 g of disulfur dichloride, S_2Cl_2

(d) 0.337 g of lead(II) phosphate, $Pb_3(PO_4)_2$

40. How many atoms of hydrogen are in 5.3×10^{4} molecules of sodium glutamate, $NaC_5H_8NO_4$?

41. How many molecules are in a 64.3 mg sample of tetraphosphorus decoxide, P_4O_{10}?

42. **(a)** How many formula units are in a 4.35×10^{-2} g sample of potassium chlorate, $KClO_3$?

(b) How many ions (chlorate and potassium) are in this sample?

Concept Organizer Relationships Among Particles, Moles, and Mass

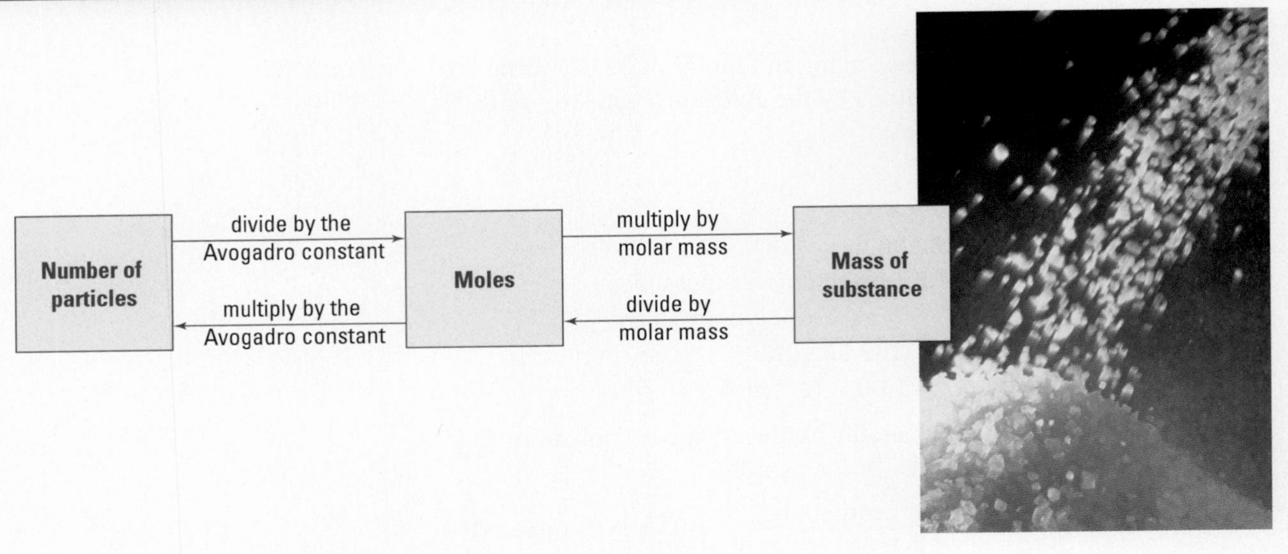

Section Wrap-up

In this chapter, you have learned about the relationships among the number of particles in a substance, the amount of a substance in moles, and the mass of a substance. Given the mass of any substance, you can now determine how many moles and particles make it up. In the next chapter, you will explore the mole concept further. You will learn how the mass proportions of elements in compounds relate to their formulas.

Section Review

Unit Investigation Prep

Before you design your experiment to determine the composition of a mixture, be sure you understand the relationship between moles and mass.

1. (C) Draw a diagram that shows the relationship between the atomic mass and molar mass of an element and the Avogadro constant.

2. (I) Consider a 78.6 g sample of ammonia, NH_3.
   (a) How many moles of ammonia are in the sample?
   (b) How many molecules of ammonia are in the sample?

3. (I) Use your understanding of the mole to answer the following questions.
   (a) What is the mass, in grams, of a single atom of silicon, Si?
   (b) What is the mass, in atomic mass units, of a mole of silicon atoms?

4. (I) Consider a 0.789 mol sample of sodium chloride, NaCl.
   (a) What is the mass of the sample?
   (b) How many formula units of sodium chloride are in the sample?
   (c) How many ions are in the sample?

5. (I) A 5.00 carat diamond has a mass of 1.00 g. How many carbon atoms are in a 5.00 carat diamond?

6. (I) A bottle of mineral supplement tablets contains 100 tablets and 200 mg of copper. The copper is found in the form of cupric oxide. What mass of cupric oxide is contained in each tablet?

CHAPTER 5 Review

Reflecting on Chapter 5

Summarize this chapter in the format of your choice. Here are a few ideas to use as guidelines:

- Describe the relationships among isotopic abundance, isotopic masses, and average atomic mass.
- Explain why you need to use a weighted average to calculate average atomic mass.
- Calculate isotopic abundance based on isotopic masses and average atomic mass.
- Explain how chemists use a mass spectrometer to determine isotopic abundance and the masses of different isotopes.
- Describe how and why chemists group atoms and molecules into molar amounts.
- Explain how chemists define the mole and why this definition is useful.
- Use the Avogadro constant to convert between moles and particles.
- Explain the relationship between average atomic mass and the mole.
- Find a compound's molar mass using the periodic table.
- Convert between particles, moles, and mass.

Reviewing Key Terms

For each of the following terms, write a sentence that shows your understanding of its meaning.

average atomic mass	Avogadro constant
isotopic abundance	mass spectrometer
mole	molar mass
weighted average	

Knowledge/Understanding

1. Distinguish between atomic mass and average atomic mass. Give examples to illustrate each term.
2. Explain why you use a weighted average, based on the masses and abundances of the isotopes, to calculate the average atomic mass of an element.
3. The periodic table lists the average atomic mass of chlorine as 35.45 u. Are there any chlorine atoms that have a mass of 35.45 u? Explain your answer.
4. Explain how the Avogadro constant, average atomic mass, and molar mass are related.
5. Explain how a balance allows a chemist to count atoms or molecules indirectly.
6. (a) Describe the relationship between the mole, the Avogadro constant, and carbon-12.
   (b) Why do chemists use the concept of the mole to deal with atoms and molecules?
7. How is the molar mass of an element related to average atomic mass?
8. Explain what the term molar mass means for each of the following, using examples.
   (a) a metallic element
   (b) a diatomic element
   (c) a compound

Inquiry

9. The isotopes of argon have the following relative abundances: Ar-36, 0.34%; Ar-38, 0.06%; and Ar-40, 99.66%. Estimate the average atomic mass of argon.
10. The isotopes of gallium have the following relative abundances: Ga-69, 60.0%; and Ga-71, 40.0%. Estimate the average atomic mass of gallium using the mass numbers of the isotopes.
11. Estimate the average atomic mass of germanium, given its isotopes with their relative abundances: Ge-70 (20.5%), Ge-72 (27.4%), Ge-73 (7.8%), Ge-74 (36.5%), and Ge-76 (7.8%).
12. Potassium exists as two naturally occurring isotopes: K-39 and K-41. These isotopes have atomic masses of 39.0 u and 41.0 u respectively. If the average atomic mass of potassium is 39.10 u, calculate the relative abundance of each isotope.
13. How many moles of the given substance are present in each sample below?
    (a) 0.453 g of Fe_2O_3
    (b) 50.7 g of H_2SO_4
    (c) 1.24×10^{-2} g of Cr_2O_3
    (d) 8.2×10^{2} g of $C_2Cl_3F_3$
    (e) 12.3 g of NH_4Br

14. Convert each quantity to an amount in moles.
   (a) 4.27×10^{21} atoms of He
   (b) 7.39×10^{23} molecules of ICl
   (c) 5.38×10^{22} molecules of NO_2
   (d) 2.91×10^{23} formula units of $Ba(OH)_2$
   (e) 1.62×10^{24} formula units of KI
   (f) 5.58×10^{20} molecules of C_3H_8

15. Copy the following table into your notebook and complete it.

Sample	Molar mass (g/mol)	Mass of sample (g)	Number of molecules in sample	Number of moles of molecules	Number of moles of atoms
NaCl	58.4	58.4	6.02×10^{23}	1.00	2.00
NH_3		24.8			
H_2O			5.28×10^{22}		
Mn_2O_3					0.332
K_2CrO_4		9.67×10^{-1}			
$C_8H_8O_3$			7.90×10^{24}		
$Al(OH)_3$				8.54×10^{2}	

16. Calculate the molar mass of each compound.
   (a) $PtBr_2$
   (b) $C_3H_5O_2H$
   (c) Na_2SO_4
   (d) $(NH_4)_2Cr_2O_7$
   (e) $Ca_3(PO_4)_2$
   (f) Cl_2O_7

17. Express each quantity as a mass in grams.
   (a) 3.70 mol of H_2O
   (b) 8.43×10^{23} molecules of PbO_2
   (c) 14.8 mol of $BaCrO_4$
   (d) 1.23×10^{22} molecules of Cl_2
   (e) 9.48×10^{23} molecules of HCl
   (f) 7.74×10^{19} molecules of Fe_2O_3

18. How many atoms of C are contained in 45.6 g of C_6H_6?

19. How many atoms of F are contained in 0.72 mol of BF_3?

20. Calculate the following.
   (a) the mass (u) of one atom of xenon
   (b) the mass (g) of one mole of xenon atoms
   (c) the mass (g) of one atom of xenon
   (d) the mass (u) of one mole of xenon atoms
   (e) the number of atomic mass units in one gram

21. How many atoms of C are in a mixture containing 0.237 mol of CO_2 and 2.38 mol of CaC_2?

22. How many atoms of H are in a mixture of 3.49×10^{23} molecules of H_2O and 78.1 g of CH_3OH?

23. How many nitrate ions are in a solution that contains 3.76×10^{-1} mol of calcium nitrate, $Ca(NO_3)_2$?

24. Ethanol, C_2H_5OH, is frequently used as the fuel in wick-type alcohol lamps. One molecule of C_2H_5OH requires three molecules of O_2 for complete combustion. What mass of O_2 is required to react completely with 92.0 g of C_2H_5OH?

25. Bromine exists as two isotopes: Br-79 and Br-81. Calculate the relative abundance of each isotope. You will need to use information from the periodic table.

26. Examine the following double displacement reaction.
   $NaCl_{(aq)} + AgNO_{3(aq)} \rightarrow AgCl_{(s)} + NaNO_{3(aq)}$
   In this reaction, one formula unit of NaCl reacts with one formula unit of $AgNO_3$.
   (a) How many moles of NaCl react with one mole of $AgNO_3$?
   (b) What mass of $AgNO_3$ reacts with 29.2 g of NaCl?

27. The planet Zoltan is located in a solar system in the Andromeda galaxy. On Zoltan, the standard unit for the amount of substance is the wog and the standard unit for mass is the wibble. The Zoltanians, like us, chose carbon-12 to define their standard unit for the amount of substance. By definition, one wog of C-12 atoms contains 2.50×10^{21} atoms. It has a mass of exactly 12 wibbles.
   (a) What is the mass, in wibbles, of 1 wog of nitrogen atoms?
   (b) What is the mass, in wibbles, of 5.00×10^{-1} wogs of O_2?
   (c) What is the mass, in grams, of 1 wog of hydrogen atoms?

Communication

28. Use the definition of the Avogadro constant to explain why its value must be determined by experiment.

29. Why is carbon-12 the only isotope with an atomic mass that is a whole number?

30. Draw a concept map for the conversion of mass (g) of a sample to amount (mol) of a sample to number of molecules in a sample to number of atoms in a sample. Be sure to include proper units.

31. Explain why 1 mol of carbon dioxide contains 6.02×10^{23} molecules and not 6.02×10^{23} atoms.

Making Connections

32. The RNI (Recommended Nutrient Intake) of iron for women is listed as 14.8 mg per day. Ferrous gluconate, $Fe(C_6H_{11}O_7)_2$ is often used as an iron supplement for those who do not get enough iron in their diet because it is relatively easy for the body to absorb. Some iron-fortified breakfast cereals contain elemental iron metal as their source of iron.
    - **(a)** Calculate the number of moles of elemental iron, Fe, required by a woman, according to the RNI.
    - **(b)** What mass, in milligrams, of ferrous gluconate, would satisfy the RNI for iron?
    - **(c)** The term *bioavailability* refers to the extent that the body can absorb a certain vitamin or mineral supplement. There is evidence to suggest that the elemental iron in these iron-fortified cereals is absorbed only to a small extent. If this is the case, should cereal manufacturers be allowed to add elemental iron at all? How could cereal manufacturers assure that the consumer absorbs an appropriate amount of iron? Would adding more elemental iron be a good solution? List the pros and cons of adding more elemental iron, then propose an alternative solution.

33. Vitamin B_3, also known as niacin, helps maintain the normal function of the skin, nerves, and digestive system. The disease pellagra results from a severe niacin deficiency. People with pellagra experience mouth sores, skin irritation, and mental deterioration. Niacin has the following formula: $C_6H_5NO_2$. Often vitamin tablets contain vitamin B_3 in the form of niacinamide, $C_6H_6N_2O$, which is easier for the body to absorb.
    - **(a)** A vitamin supplement tablet contains 100 mg of niacinamide. What mass of niacin contains an equivalent number of moles as 100 mg of niacinamide?
    - **(b)** Do some research to find out how much niacin an average adult should ingest each day.
    - **(c)** Do some research to find out what kinds of food contain niacin.
    - **(d)** What are the consequences of ingesting too much niacin?
    - **(e)** Choose another vitamin to research. Find out its chemical formula, its associated recommended nutrient intake, and where it is found in our diet. Prepare a poster to communicate your findings.

Answers to Practice Problems and Short Answers to Section Review Questions

Practice Problems: **1.** 10.81 u **2.** 28.09 u **3.** 63.55 u **4.** 207.2 **5.** 99.8%, 0.2% **6.** very low, **7.** 72% **8.** 99.8%, 0.2% **9.** 3.37×10^{16} rows **10.** 2.80×10^{14} **11.** 1.44×10^{27} km **12.** 1.91×10^{16} a **13.** 1.07×10^{22} **14.** 2.83×10^{20} **15.** 1.3×10^{23} **16.** 3.35×10^{25} **17.(a)** 1.5×10^{24} **(b)** 2.1×10^{25} **(c)** 6.0×10^{24} **18.(a)** 4.99×10^{23} **(b)** 9.98×10^{23} **19.** 12.8 mol **20.** 1.33×10^{3} mol **21.** 1.84×10^{-2} mol **22.** 1.16×10^{-1} mol **23.(a)** 131.29 g/mol **(b)** 190.23 g/mol **(c)** 137.33 g/mol **(d)** 127.60 g/mol **24.(a)** 17.04 g/mol **(b)** 180.2 g/mol **(c)** 294.2 g/mol **(d)** 399.9 g/mol **25.** 183.68 g/mol **26.** 131.7 g/mol **27.(a)** 46.8 g **(b)** 1.20×10^{2} g **(c)** 1.05×10^{9} **(d)** 3.66×10^{-3} **28.(a)** 1.5 mol Cl_2 **(b)** 7.31 mol O_2 **29.** 1.00×10^{3} g **30.** 26.6 kg **31.(a)** 1.07 mol **(b)** 1.24×10^{2} mol **(c)** 14.2 mol **(d)** 7.75×10^{-4} mol **32.(a)** 0.652 mol **(b)** 0.156 mol **(c)** 1.76×10^{6} **(d)** 3.49×10^{-5} **33.** 1.7×10^{2} mol **34.** 1.75×10^{2} mol **35.(a)** 1.36×10^{3} g **(b)** 9.95 g **(c)** 7.32×10^{2} g **(d)** 2.15×10^{7} **36.** 2.93×10^{-21} g **37.** 7.95×10^{-23} g **38.** 6.25×10^{-22} g **39.(a)** 3.34×10^{23} molecules **(b)** 9.84×10^{23} molecules **(c)** 1.05×10^{23} molecules **(d)** 2.50×10^{20} formula units **40.** 4.2×10^{5} atoms **41.** 1.36×10^{20} **42.(a)** 2.14×10^{20} **(b)** 4.27×10^{20}

Section Review: **5.1:** **2.** 24.31 u **3.** No **5.2:** **2.** $\$1.00 \times 10^{14}$ **3.** 1.9×10^{16} years **4.** 2.08×10^{24} atoms **5.(a)** 4.25 mol **(b)** 12.8 mol **6.** 3.0×10^{23} atoms **7.(a)** 3.43×10^{24} molecules **(b)** 2.06×10^{25} atoms **8.(a)** 7.04×10^{23} molecules **(b)** 3.52×10^{24} atoms **(c)** 2.11×10^{24} atoms **10.** Compound **5.3:** **2.(a)** 4.61 mol **(b)** 2.78×10^{24} **3.(a)** 4.67×10^{-23} g **(b)** 1.69×10^{25} u **4.(a)** 46.1 g **(b)** 4.75×10^{23} formula units **(c)** 9.50×10^{23} **5.** 5.01×10^{22} atoms **6.** 2.50×10^{-3} g

CHAPTER

6 Chemical Proportions in Compounds

How do chemists use what they know about molar masses? In Chapter 5, you learned how to use the periodic table and the mole to relate the mass of a compound to the number of particles in the compound. Chemists can use their understanding of molar mass to find out important information about compounds.

Sometimes chemists analyze a compound that is found in nature to learn how to produce it more cheaply in a laboratory. For example, consider the flavour used in vanilla ice cream, which may come from natural or artificial vanilla extract. Natural vanilla extract is made from vanilla seed pods, shown on the left. The seed pods must be harvested and processed before being sold as vanilla extract. The scent and flavour of synthetic vanilla come from a compound called vanillin, which can be produced chemically in bulk. Therefore its production is much cheaper. Similarly, many medicinal chemicals that are found in nature can be produced more cheaply and efficiently in a laboratory.

Suppose that you want to synthesize a compound such as vanillin in a laboratory. You must first determine the elements in the compound. Then you need to know the proportion of each element that is present. This information, along with your understanding of molar mass, will help you determine the chemical formula of the compound. Once you know the chemical formula, you are on your way to finding out how to produce the compound.

In this chapter, you will learn about the relationships between chemical formulas, molar masses, and the masses of elements in compounds.

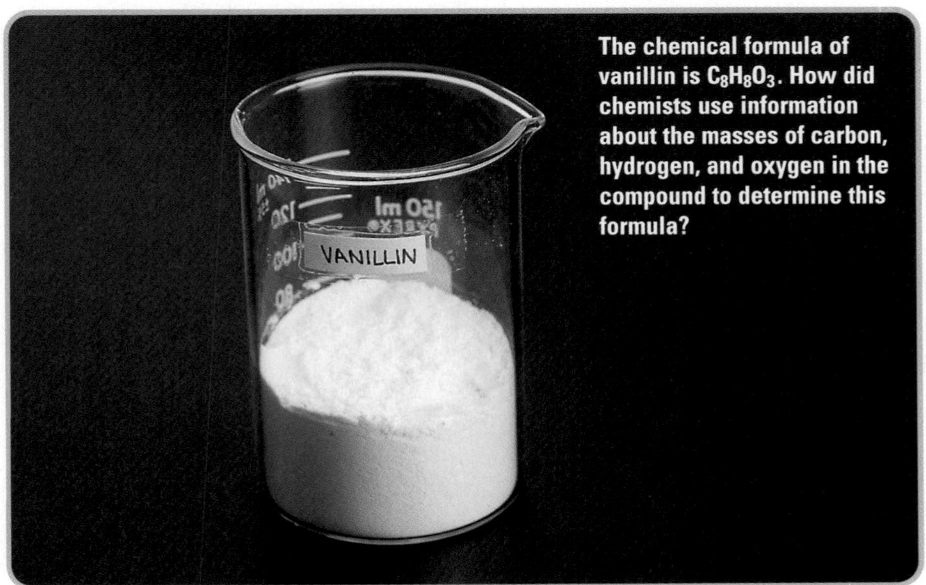

The chemical formula of vanillin is $C_8H_8O_3$. How did chemists use information about the masses of carbon, hydrogen, and oxygen in the compound to determine this formula?

Chapter Preview

Concepts and Skills You Will Need

Before you begin this chapter, review the following concepts and skills:

- naming chemical compounds (Chapter 3, section 3.5)
- understanding the mole (Chapter 5, section 5.2)
- explaining the relationship between the mole and molar mass (Chapter 5, section 5.3)
- solving problems involving number of moles, number of particles, and mass (Chapter 5, section 5.3)

6.1 Percentage Composition

Section Preview/ Specific Expectations

In this section, you will

- **explain** the law of definite proportions
- **calculate** the percentage composition of a compound using the formula and the relative atomic masses of the elements
- **communicate** your understanding of the following terms: *law of definite proportions, mass percent, percentage composition*

mind STRETCH

If a bicycle factory has 1000 wheels and 400 frames, how many bicycles can be made? How many wheels does each bicycle have? Is the number of wheels per bicycle affected by any extra wheels that the factory may have in stock? Relate these questions to the law of definite proportions.

When you calculate and use the molar mass of a compound, such as water, you are making an important assumption. You are assuming that every sample of water contains hydrogen and oxygen in the ratio of two hydrogen atoms to one oxygen atom. Thus you are also assuming that the masses of hydrogen and oxygen in pure water always exist in a ratio of 2 g:16 g. This may seem obvious to you, because you know that the molecular formula of water is always H_2O, regardless of whether it comes from any of the sources shown in Figure 6.1. When scientists first discovered that compounds contained elements in fixed mass proportions, they did not have the periodic table. In fact, the discovery of fixed mass proportions was an important step toward the development of atomic theory.

The Law of Definite Proportions

In the late eighteenth century, Joseph Louis Proust, a French chemist, analyzed many samples of copper(II) carbonate, $CuCO_3$. He found that the samples contained the same proportion of copper, carbon, and oxygen, regardless of the source of the copper(II) carbonate. This discovery led Proust to propose the **law of definite proportions**: *the elements in a chemical compound are always present in the same proportions by mass.*

Figure 6.1 Suppose that you distil pure water from each of these sources to purify it. Are the distilled water samples the same or different? What is the molar mass of the distilled water from each source?

The mass of an element in a compound, expressed as a percent of the total mass of the compound, is the element's **mass percent.** The mass percent of hydrogen in water from any of the sources shown in Figure 6.1 is 11.2%. Similarly, the mass percent of oxygen in water is always 88.8%. Whether the water sample is distilled from a lake, an ice floe, or a drinking fountain, the hydrogen and oxygen in pure water are always present in these proportions.

Chemical formulas such as CO and CO_2 reflect an important law called the *law of multiple proportions*. This law applies when two elements (such as carbon and oxygen) combine to form two or more different compounds. In these cases, the masses of the element (such as O_2 in CO and CO_2) that combine with a fixed amount of the second element are in ratios of small whole numbers. For example, two moles of carbon can combine with one mole of oxygen to form carbon monoxide, or with two moles of oxygen to form carbon dioxide. The ratio of the two different amounts of oxygen that combine with the fixed amount of carbon is 1:2.

Different Compounds from the Same Elements

The law of definite proportions does not imply that elements in compounds are always present in the same relative amounts. It is possible to have different compounds made up of different amounts of the same elements. For example, water, H_2O, and hydrogen peroxide, H_2O_2, are both made up of hydrogen and oxygen. Yet, as you can see in Figure 6.2, each compound has unique properties. Each compound has a different mass percent of oxygen and hydrogen. You may recognize hydrogen peroxide as a household chemical. It is an oxidizing agent that is used to bleach hair and treat minor cuts. It is also sold as an alternative to chlorine bleach.

Figure 6.3 shows a molecule of benzene, C_6H_6. Benzene contains 7.76% hydrogen and 92.2% carbon by mass. Octane, C_8H_{18}, is a major component of the fuel used for automobiles. It contains 84.1% carbon and 15.9% hydrogen.

Figure 6.2 Water contains hydrogen and oxygen, but it does not decompose in the presence of manganese dioxide. Hydrogen peroxide is also composed of hydrogen and oxygen. It is fairly unstable and decomposes vigorously in the presence of manganese(IV) oxide.

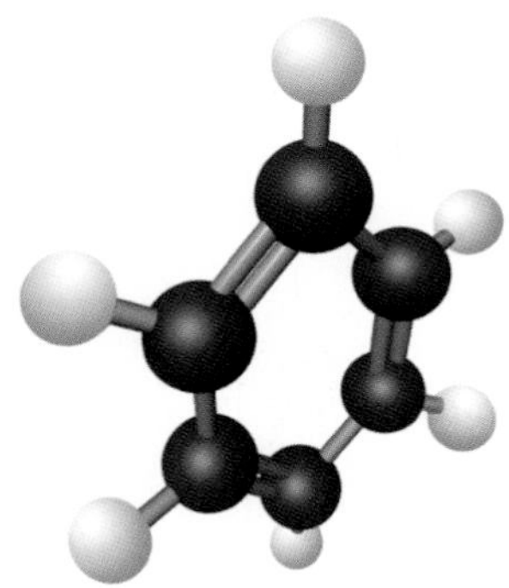

Figure 6.3 Benzene C_6H_6, is made up of six carbon atoms and six hydrogen atoms. Why does benzene not contain 50% of each element by mass?

Similarly, carbon monoxide, CO, and carbon dioxide, CO_2, are both made up of carbon and oxygen. Yet each compound is unique, with its own physical and chemical properties. Carbon dioxide is a product of cellular respiration and the complete combustion of fossil fuels. Carbon monoxide is a deadly gas formed when insufficient oxygen is present during the combustion of carbon-containing compounds. Carbon monoxide always contains 42.88% carbon by mass. Carbon dioxide always contains 27.29% carbon by mass.

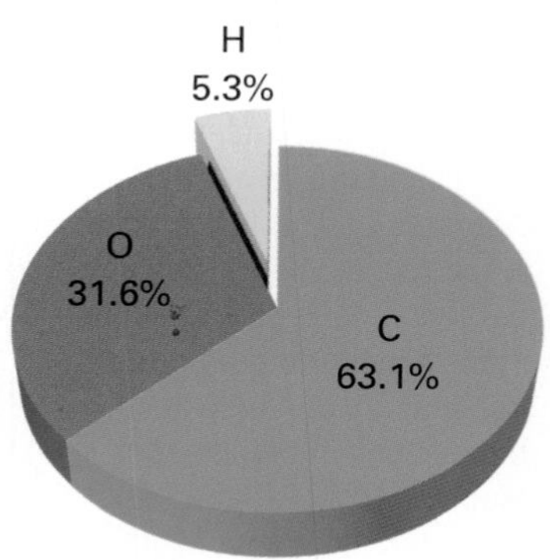

Figure 6.4 This pie graph shows the percentage composition of vanillin.

Percentage Composition

Carbon dioxide and carbon monoxide contain the same elements but have different proportions of these elements. In other words, they are composed differently. Chemists express the composition of compounds in various ways. One way is to describe how many moles of each element make up a mole of a compound. For example, one mole of carbon dioxide contains one mole of carbon and two moles of oxygen. Another way is to describe the percent mass of each element in a compound.

The **percentage composition** of a compound refers to the relative mass of each element in the compound. In other words, percentage composition is a statement of the values for mass percent of every element in the compound. For example, the compound vanillin, $C_8H_8O_3$, has a percentage composition of 63.1% carbon, 5.3% hydrogen, and 31.6% oxygen, as shown in Figure 6.4.

A compound's percentage composition is an important piece of information. For example, percentage composition can be determined experimentally, and then used to help identify the compound.

Examine the following Sample Problem to learn how to calculate the percentage composition of a compound from the mass of the compound and the mass of the elements that make up the compound. Then do the Practice Problems to try expressing the composition of substances as mass percents.

mind STRETCH

Does the unknown compound in the Sample Problem contain any elements other than zinc and sulfur? How do you know? Use the periodic table to predict the formula of the compound. Does the percentage composition support your prediction?

Sample Problem

Percentage Composition from Mass Data

Problem

A compound with a mass of 48.72 g is found to contain 32.69 g of zinc and 16.03 g of sulfur. What is the percentage composition of the compound?

What Is Required?

You need to find the mass percents of zinc and sulfur in the compound.

What Is Given?

You know the mass of the compound. You also know the mass of each element in the compound.

Mass of compound = 48.72 g
Mass of Zn = 32.69 g
Mass of S = 16.03 g

Plan Your Strategy

To find the percentage composition of the compound, find the mass percent of each element. To do this, divide the mass of each element by the mass of the compound and multiply by 100%.

Continued ...

Continued ...
FROM PAGE 200

Act on Your Strategy

$$\text{Mass percent of Zn} = \frac{\text{Mass of Zn}}{\text{Mass of compound}} \times 100\%$$
$$= \frac{32.69 \text{ g}}{48.72 \text{ g}} \times 100\%$$
$$= 67.10\%$$

$$\text{Mass percent of S} = \frac{\text{Mass of S}}{\text{Mass of compound}} \times 100\%$$
$$= \frac{16.03 \text{ g}}{48.72 \text{ g}} \times 100\%$$
$$= 32.90\%$$

The percentage composition of the compound is 67.10% zinc and 32.90% sulfur.

Check Your Solution

The mass of zinc is about 32 g per 50 g of the compound. This is roughly 65%, which is close to the calculated value.

Practice Problems

1. A sample of a compound is analyzed and found to contain 0.90 g of calcium and 1.60 g of chlorine. The sample has a mass of 2.50 g. Find the percentage composition of the compound.
2. Find the percentage composition of a pure substance that contains 7.22 g nickel, 2.53 g phosphorus, and 5.25 g oxygen only.
3. A sample of a compound is analyzed and found to contain carbon, hydrogen, and oxygen. The mass of the sample is 650 mg, and the sample contains 257 mg of carbon and 50.4 mg of hydrogen. What is the percentage composition of the compound?
4. A scientist analyzes a 50.0 g sample and finds that it contains 13.3 g of potassium, 17.7 g of chromium, and another element. Later the scientist learns that the sample is potassium dichromate, $K_2Cr_2O_7$. Potassium dichromate is a bright orange compound that is used in the production of safety matches. What is the percentage composition of potassium dichromate?

It is important to understand clearly the difference between percent by mass and percent by number. In the Thought Lab that follows, you will investigate the distinction between these two ways of describing composition.

mind STRETCH

Iron is commonly found as two oxides, with the general formula Fe_xO_y. One oxide is 77.7% iron. The other oxide is 69.9% iron. Use the periodic table to predict the formula of each oxide. Match the given values for the mass percent of iron to each compound. How can you use the molar mass of iron and the molar masses of the two iron oxides to check the given values for mass percent?

Web LINK

www.school.mcgrawhill.ca/resources/

Vitamin C is the common name for ascorbic acid, $C_6H_8O_6$. To learn about this vitamin, go to the web site above. Go to **Science Resources**, then to **Chemistry 11** to find out where to go next. Do you think there is a difference between natural and synthetic vitamin C? Both are ascorbic acid. Natural vitamin C comes from foods we eat, especially citrus fruits. Synthetic vitamin C is made in a laboratory. Why do the prices of natural and synthetic products often differ? Make a list to show the pros and cons of vitamins from natural and synthetic sources.

ThoughtLab  Percent by Mass and Percent by Number

A company manufactures gift boxes that contain two pillows and one gold brick. The gold brick has a mass of 20 kg. Each pillow has a mass of 1.0 kg.

Procedure

1. You are a quality control specialist at the gift box factory. You need to know the following information:
   - **(a)** What is the percent of pillows, in terms of the number of items, in the gift box?
   - **(b)** What is the percent of pillows, by mass, in the gift box?
   - **(c)** What is the percent of gold, by mass, in the gift box?
2. You have a truckload of gift boxes to inspect. You now need to know this information:
   - **(a)** What is the percent of pillows, in terms of the number of items, in the truckload of gift boxes?
   - **(b)** What is the percent of pillows, by mass, in the truckload of gift boxes?
   - **(c)** What is the percent of gold, by mass, in the truckload of gift boxes?

Analysis

1. The truckload of gift boxes, each containing 2 light pillows and 1 heavy gold brick, can be used to represent a pure substance, water, containing 2 mol of a "light" element, such as hydrogen, and 1 mol of a "heavy" element, such as oxygen.
   - **(a)** What is the percent of hydrogen, in terms of the number of atoms, in 1 mol of water?
   - **(b)** What is the mass percent of hydrogen in 1 mol of water?
   - **(c)** What is the mass percent of oxygen in 1 mol of water?
2. Would the mass percent of hydrogen or oxygen in question 1 change if you had 25 mol of water? Explain.
3. Why do you think chemists use mass percent rather than percent by number of atoms?

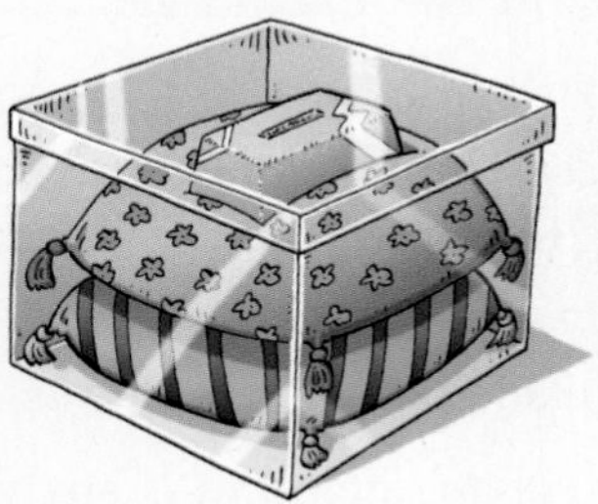

Calculating Percentage Composition from a Chemical Formula

In the previous Practice Problems, you used mass data to calculate percentage composition. This skill is useful for interpreting experimental data when the chemical formula is unknown. Often, however, the percentage composition is calculated from a known chemical formula. This is useful when you are interested in extracting a certain element from a compound. For example, many metals, such as iron and mercury, exist in mineral form. Mercury is most often found in nature as mercury(II) sulfide, HgS. Knowing the percentage composition of HgS helps a metallurgist predict the mass of mercury that can be extracted from a sample of HgS.

When determining the percentage composition by mass of a homogeneous sample, the size of the sample does not matter. According to the law of definite proportions, there is a fixed proportion of each element in the compound, no matter how much of the compound you have. This means that you can choose a convenient sample size when calculating percentage composition from a formula.

If you assume that you have one mole of a compound, you can use the molar mass of the compound, with its chemical formula, to calculate its percentage composition. For example, suppose that you want to find the percentage composition of HgS. You can assume that you have one mole of HgS and find the mass percents of mercury and sulfur in one mole of the compound.

$$\text{Mass percent of Hg in HgS} = \frac{\text{Mass of Hg in 1 mol of HgS}}{\text{Mass of 1 mol of HgS}} \times 100\%$$
$$= \frac{200.6\,\cancel{g}}{228.68\,\cancel{g}} \times 100\%$$
$$= 87.7\%$$

Mercury(II) sulfide is 87.7% mercury by mass. Since there are only two elements in HgS, you can subtract the mass percent of mercury from 100 percent to find the mass percent of sulfur.

$$\text{Mass percent of S in HgS} = 100\% - 87.7\% = 12.3\%$$

Therefore, the percentage composition of mercury(II) sulfide is 87.7% mercury and 12.3% sulfur.

Sometimes there are more than two elements in a compound, or more than one atom of each element. This makes determining percentage composition more complex than in the example above. Work through the Sample Problem below to learn how to calculate the percentage composition of a compound from its molecular formula.

History LINK

Before AD 1500, many alchemists thought that matter was composed of two "elements": mercury and sulfur. To impress their patrons, they performed an experiment with mercury sulfide, also called cinnabar, HgS. They heated the red cinnabar, which drove off the sulfur and left the shiny liquid mercury. On further heating, the mercury reacted to form a red compound again. Alchemists wrongly thought that the mercury had been converted back to cinnabar. What Hg(II) compound do you think was really formed when the mercury was heated in the air? What is the mass percent of mercury in this new compound? What is the mass percent of mercury in cinnabar?

Sample Problem

Finding Percentage Composition from a Chemical Formula

Problem

Cinnamaldehyde, C_9H_8O, is responsible for the characteristic odour of cinnamon. Determine the percentage composition of cinnamaldehyde by calculating the mass percents of carbon, hydrogen, and oxygen.

What Is Required?

You need to find the mass percents of carbon, hydrogen, and oxygen in cinnamaldehyde.

What Is Given?

The molecular formula of cinnamaldehyde is C_9H_8O.
Molar mass of C = 12.01 g/mol
Molar mass of H = 1.01 g/mol
Molar mass of O = 16.00 g/mol

Plan Your Strategy

From the molar masses of carbon, hydrogen, and oxygen, calculate the molar mass of cinnamaldehyde.

Then find the mass percent of each element. To do this, divide the mass of each element in 1 mol of cinnamaldehyde by the molar mass of cinnamaldehyde, and multiply by 100%. Remember that

Continued ...

Continued ...
FROM PAGE 203

there are 9 mol carbon, 8 mol hydrogen, and 1 mol oxygen in each mole of cinnamaldehyde.

Act on Your Strategy

$M_{C_9H_8O}$
$= (9 \times M_C) + (8 \times M_H) + (M_O)$
$= (9 \times 12.01\,g) + (8 \times 1.01\,g) + 16.00\,g$
$= 132\,g$

$$\text{Mass percent of C} = \frac{9 \times M_C}{M_{C_9H_8O}} \times 100\% = \frac{9 \times 12.01\,\cancel{g/mol}}{132\,\cancel{g/mol}} \times 100\% = 81.9\%$$

$$\text{Mass percent of H} = \frac{8 \times M_H}{M_{C_9H_8O}} \times 100\% = \frac{8 \times 1.01\,\cancel{g/mol}}{132\,\cancel{g/mol}} \times 100\% = 6.12\%$$

$$\text{Mass percent of O} = \frac{1 \times M_O}{M_{C_9H_8O}} \times 100\% = \frac{1 \times 16.00\,\cancel{g/mol}}{132\,\cancel{g/mol}} \times 100\% = 12.1\%$$

The percentage composition of cinnamaldehyde is 81.9% carbon, 6.12% hydrogen, and 12.1% oxygen.

Check Your Solution

The mass percents add up to 100%.

Practice Problems

5. Calculate the mass percent of nitrogen in each compound.
 (a) N_2O
 (b) $Sr(NO_3)_2$
 (c) NH_4NO_3
 (d) HNO_3

6. Sulfuric acid, H_2SO_4, is an important acid in laboratories and industries. Determine the percentage composition of sulfuric acid.

7. Potassium nitrate, KNO_3, is used to make fireworks. What is the mass percent of oxygen in potassium nitrate?

8. A mining company wishes to extract manganese metal from pyrolusite ore, MnO_2.
 (a) What is the percentage composition of pyrolusite ore?
 (b) Use your answer from part (a) to calculate the mass of pure manganese that can be extracted from 250 kg of pyrolusite ore.

CHECKPOINT

When it is heated, solid potassium nitrate reacts to form solid potassium oxide, gaseous nitrogen, and gaseous oxygen. Write a balanced chemical equation for this reaction. What type of reaction is it?

Section Wrap-up

In this section, you learned that you can calculate percentage composition using a chemical formula. Often, however, chemists do not know the chemical formula of the compound they are analyzing, as in Figure 6.5. Through experiment, they can determine the masses of the elements that make up the compound. Then they can use the masses to calculate the percentage composition. (You will learn about one example of this kind of experimental technique in section 6.4.) From the percentage composition, chemists can work backward to determine the formula of the unknown compound. In section 6.2, you will learn about the first step in using the percentage composition of a compound to determine its chemical formula.

mind STRETCH

You know that both elements and compounds are pure substances. Write a statement, using the term "percentage composition," to distinguish between elements and compounds.

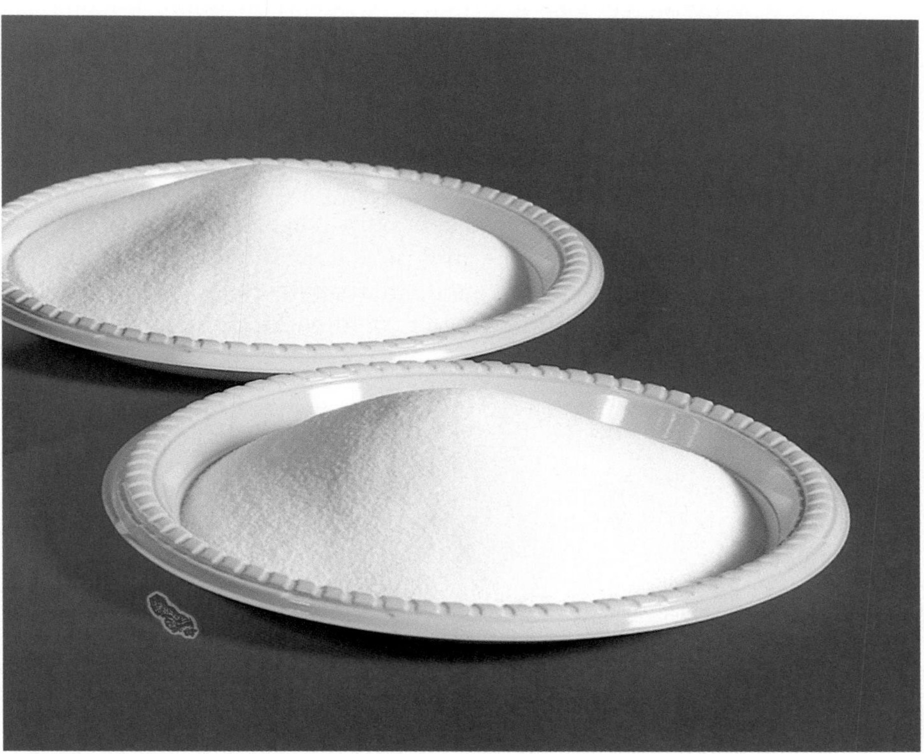

Figure 6.5 One of these compounds is vanillin, $C_8H_8O_3$, and one is glucose, $C_6H_{12}O_6$. How could a chemist use percentage composition to find out which is which?

Section Review

1. (C) Acetylene, C_2H_2, is the fuel in a welder's torch. It contains an equal number of carbon and hydrogen atoms. Explain why acetylene is not 50% carbon by mass.

2. (K/U) When determining percentage composition, why is it acceptable to work with either molar quantities, expressed in grams, or average molecular (or atomic or formula unit) quantities, expressed in atomic mass units?

3. (I) Indigo, $C_{16}H_{10}N_2O_2$, is the common name of the dye that gives blue jeans their characteristic colour. Calculate the mass of oxygen in 25.0 g of indigo.

Unit Investigation Prep

Before you design your experiment to find the composition of a mixture, think about using mass percents in analysis. If you wanted to determine the percent by mass of each component in a mixture, what would you need to do first? Compare this situation to finding percentage composition of a pure substance.

4 I Potassium perchlorate, $KClO_4$, is used extensively in explosives. Calculate the mass of oxygen in a 24.5 g sample of potassium perchlorate.

5 I 18.4 g of silver oxide, Ag_2O, is decomposed into silver and oxygen by heating. What mass of silver will be produced?

6 MC The label on a box of baking soda (sodium hydrogen carbonate, $NaHCO_3$) claims that there are 137 mg of sodium per 0.500 g of baking soda. Comment on the validity of this claim.

7 I A typical soap molecule consists of a polyatomic anion associated with a cation. The polyatomic anion contains hydrogen, carbon, and oxygen. One particular soap molecule has 18 carbon atoms. It contains 70.5% carbon, 11.5% hydrogen, and 10.4% oxygen by mass. It also contains one alkali metal cation. Identify the cation.

8 I Examine the photographs below. When concentrated sulfuric acid is added to sucrose, $C_{12}H_{22}O_{11}$, a column of pure carbon is formed, as well as some water vapour and other gases. How would you find the mass percent of carbon in sucrose using this reaction? You may assume that all the carbon in the sucrose is converted to carbon. Design an experiment to determine the mass percent of carbon in sucrose, based on this reaction. Do not try to perform this experiment. What difficulties might you encounter?

A

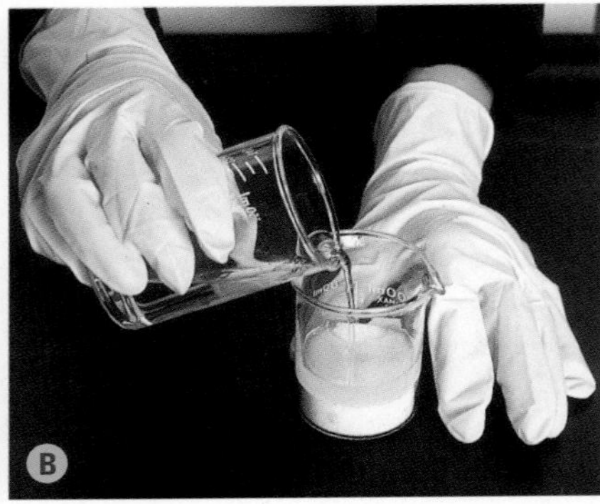
B

C

The Empirical Formula of a Compound

6.2

As part of his atomic theory, John Dalton stated that atoms combine with one another in simple whole number ratios to form compounds. For example, the molecular formula of benzene, C_6H_6, indicates that one molecule of benzene contains 6 carbon atoms and 6 hydrogen atoms. The **empirical formula** (also known as the simplest formula) of a compound shows the lowest whole number ratio of the elements in the compound. The **molecular formula** (also known as the actual formula) describes the number of atoms of each element that make up a molecule or formula unit. Benzene, with a molecular formula of C_6H_6, has an empirical formula of CH. Table 6.1 shows the molecular formulas of several compounds, along with their empirical formulas.

Table 6.1 Comparing Molecular Formulas and Empirical Formulas

Name of compound	Molecular (actual) formula	Empirical (simplest) formula	Lowest ratio of elements
hydrogen peroxide	H_2O_2	HO	1:1
glucose	$C_6H_{12}O_6$	CH_2O	1:2:1
benzene	C_6H_6	CH	1:1
acetylene (ethyne)	C_2H_2	CH	1:1
aniline	C_6H_7N	C_6H_7N	6:7:1
water	H_2O	H_2O	2:1

It is possible for different compounds to have the same empirical formula, as you can see in Figure 6.6. For example, benzene and acetylene both have the empirical formula CH. Each, however, is a unique compound. Benzene, C_6H_6, is a clear liquid with a molar mass of 78 g/mol and a boiling point of 80°C. Acetylene, C_2H_2, has a molar mass of 26 g/mol. It is a highly flammable gas, commonly used in a welder's torch. There is, in fact, no existing compound with the molecular formula CH. The empirical formula of a compound shows the lowest whole number ratio of the atoms in the compound. It does not express the composition of a molecule.

Many compounds have molecular formulas that are the same as their empirical formulas. One example is ammonia, NH_3. Try to think of three other examples.

Figure 6.6 The same empirical formula can represent more than one compound. These two compounds are different—at room temperature, one is a gas and one is a liquid. Yet they have the same empirical formula.

Section Preview/ Specific Expectations

In this section, you will

- **perform** an experiment to determine the percentage composition and the empirical formula of a compound
- **calculate** the empirical formula of a compound using percentage composition
- **communicate** your understanding of the following terms: *empirical formula, molecular formula*

Language LINK

The word "empirical" comes from the Greek word *empeirikos*, meaning, roughly, "by experiment." Why do you think the simplest formula of a compound is called its empirical formula?

Math LINK

In mathematics, you frequently need to reduce an expression to lowest terms. For example, $\frac{4x^2}{x}$ is equivalent to $4x$. A ratio of 5:10 is equivalent to 1:2. In chemistry, however, the "lowest terms" version of a chemical formula is not equivalent to its "real" molecular formula. Why not?

The relationship between the molecular formula of a compound and its empirical formula can be expressed as

Molecular formula subscripts = $n \times$ Empirical formula subscripts, where $n = 1, 2, 3...$

This relationship shows that the molecular formula of a compound is the same as its empirical formula when $n = 1$. What information do you need in order to determine whether the molecular formula of a compound is the same as its empirical formula?

Determining a Compound's Empirical Formula

In the previous section, you learned how to calculate the percentage composition of a compound from its chemical formula. Now you will do the reverse. You will use the percentage composition of a compound, along with the concept of the mole, to calculate the empirical formula of the compound. Since the percentage composition can often be determined by experiment, chemists use this calculation when they want to identify a compound.

The following Sample Problem illustrates how to use percentage composition to obtain the empirical formula of a compound.

mind STRETCH

How do the molar masses of C_6H_6 and C_2H_2 compare with the molar mass of their empirical formula? How does the molar mass of water compare with the molar mass of its empirical formula? Describe the relationship between the molar mass of a compound and the molar mass of the empirical formula of the compound.

Sample Problem

Finding a Compound's Empirical Formula from Percentage Composition: Part A

Problem

Calculate the empirical formula of a compound that is 85.6% carbon and 14.4% hydrogen.

What Is Required?

You need to find the empirical formula of the compound.

What Is Given?

You know the percentage composition of the compound. You have access to a periodic table.

Plan Your Strategy

Since you know the percentage composition, it is convenient to assume that you have 100 g of the compound. This means that you have 85.6 g of carbon and 14.4 g of hydrogen. Convert each mass to moles. The number of moles can then be converted into a lowest terms ratio of the elements to get the empirical formula.

Act on Your Strategy

$$\text{Number of moles of C in 100 g sample} = \frac{85.6 \cancel{\text{g}}}{12.01 \cancel{\text{g}}/\text{mol}} = 7.13 \text{ mol}$$

$$\text{Number of moles of H in 100 g sample} = \frac{14.4 \cancel{\text{g}}}{1.01 \cancel{\text{g}}/\text{mol}} = 14.3 \text{ mol}$$

Continued ...

Continued ...
FROM PAGE 208

Now determine the lowest whole number ratio. Divide both molar amounts by the lowest molar amount.

$$C_{\frac{7.13}{7.13}}H_{\frac{14.3}{7.13}} \rightarrow C_{1.00}H_{2.01} \rightarrow CH_2$$

Alternatively, you can set up your solution as a table.

Element	Mass percent (%)	Grams per 100 g sample (g)	Molar mass (g/mol)	Number of moles (mol)	Molar amount ÷ lowest molar amount
C	85.6	85.6	12.01	7.13	$\frac{7.13}{7.13} = 1$
H	14.4	14.4	1.01	14.3	$\frac{14.3}{7.13} = 2.01$

The empirical formula of the compound is CH_2.

PROBLEM TIP

The fact that 2.01 was rounded to 2 in CH_2 is fine. The percentage composition is often determined by experiment, so it is unlikely to be exact.

Check Your Solution

Work backward. Calculate the percentage composition of CH_2.

$$\text{Mass percent of C} = \frac{12.01\ \cancel{g/mol}}{14.03\ \cancel{g/mol}} \times 100\%$$
$$= 85.6\%$$

$$\text{Mass percent of H} = \frac{2 \times 1.01\ \cancel{g/mol}}{14.03\ \cancel{g/mol}} \times 100\%$$
$$= 14.0\%$$

The percentage composition calculated from the empirical formula closely matches the given data. The formula is reasonable.

Practice Problems

9. A compound consists of 17.6% hydrogen and 82.4% nitrogen. Determine the empirical formula of the compound.

10. Find the empirical formula of a compound that is 46.3% lithium and 53.7% oxygen.

11. What is the empirical formula of a compound that is 15.9% boron and 84.1% fluorine?

12. Determine the empirical formula of a compound made up of 52.51% chlorine and 47.48% sulfur.

Tips for Solving Empirical Formula Problems

In the Sample Problem above, the numbers were rounded at each step to simplify the calculation. To calculate an empirical formula successfully, however, you should not round the numbers until you have completed the calculation. Use the maximum number of significant digits that your calculator will allow, throughout the calculation. Rounding too soon when calculating an empirical formula may result in getting the wrong answer.

Table 6.2 Converting Subscripts in Empirical Formulas

When you see this decimal...	Try multiplying all subscripts by...
$x.80$ ($\frac{4}{5}$)	5
$x.75$ ($\frac{3}{4}$)	4
$x.67$ ($\frac{2}{3}$)	3
$x.60$ ($\frac{3}{5}$)	5
$x.40$ ($\frac{2}{5}$)	5
$x.50$ ($\frac{1}{2}$)	2
$x.33$ ($\frac{1}{3}$)	3
$x.25$ ($\frac{1}{4}$)	4
$x.20$ ($\frac{1}{5}$)	5
$x.17$ ($\frac{1}{6}$)	6

Often only one step is needed to determine the number of moles in an empirical formula. This is not always the case, however. Since you must divide by the lowest number of moles, initially one of your ratio terms will always be 1. If your other terms are quite close to whole numbers, as in the last Sample Problem, you can round them to the closest whole numbers. If your other terms are not close to whole numbers, you will need to do some additional steps. This is because empirical formulas do not always contain the subscript 1. For example, Fe_2O_3 contains the subscripts 2 and 3.

Decimals such as 0.95 to 0.99 can be rounded up to the nearest whole number. Decimals such as 0.01 to 0.05 can be rounded down to the nearest whole number. Other decimals require additional manipulation. What if you have the empirical formula $C_{1.5}H_3O_1$? To convert all subscripts to whole numbers, multiply each subscript by 2. This gives you the empirical formula $C_3H_6O_2$. Thus, a ratio that involves a decimal ending in 0.5 must be doubled. What if a decimal ends in 0.45 to 0.55? Round the decimal so that it ends in .5, and then double the ratio.

Table 6.2 gives you some strategies for converting subscripts to whole numbers. The variable x stands for any whole number. Examine the following Sample Problem to learn how to convert the empirical formula subscripts to the lowest possible whole numbers.

Sample Problem

Finding a Compound's Empirical Formula from Percentage Composition: Part B

Problem

The percentage composition of a fuel is 81.7% carbon and 18.3% hydrogen. Find the empirical formula of the fuel.

What Is Required?

You need to determine the empirical formula of the fuel.

What Is Given?

You know the percentage composition of the fuel. You have access to a periodic table.

Plan Your Strategy

Convert mass percent to mass, then to number of moles. Then find the lowest whole number ratio.

Act on Your Strategy

Element	Mass percent (%)	Grams per 100 g sample (g)	Molar mass (g/mol)	Number of moles (mol)	Molar amount ÷ lowest molar amount
C	81.7	81.7	12.0	6.81	$\frac{6.81}{6.81} = 1$
H	18.3	18.3	1.01	18.1	$\frac{8.1}{6.81} = 2.66$

Continued ...

Continued ...
FROM PAGE 210

You now have the empirical formula $C_1H_{2.66}$. Convert the subscript 2.66 ($\frac{8}{3}$) to a whole number. $C_{1\times3}H_{2.66\times3} = C_3H_8$.

Check Your Solution

Work backward. Calculate the percentage composition of C_3H_8.

$$\text{Mass percent of C} = \frac{3 \times 12.01 \cancel{\text{g/mol}}}{44.09 \cancel{\text{g/mol}}} \times 100\%$$
$$= 81.7\%$$

$$\text{Mass percent of H} = \frac{8 \times 1.008 \cancel{\text{g/mol}}}{44.09 \cancel{\text{g/mol}}} \times 100\%$$
$$= 18.3\%$$

The percentage composition calculated from the empirical formula matches the percentage composition given in the problem.

PROBLEM TIP

Notice that Table 6.2 suggests multiplying by 3 when you obtain a subscript ending in .67, which is very close to .66.

Practice Problems

13. An oxide of chromium is made up of 68.4% chromium and 31.6% oxygen. What is the empirical formula of this oxide?
14. Phosphorus reacts with oxygen to give a compound that is 43.7% phosphorus and 56.4% oxygen. What is the empirical formula of the compound?
15. An inorganic salt is composed of 17.6% sodium, 39.7% chromium, and 42.8% oxygen. What is the empirical formula of this salt?
16. Compound X contains 69.9% carbon, 6.86% hydrogen, and 23.3% oxygen. Determine the empirical formula of compound X.

Determining the Empirical Formula by Experiment

In practice, you can determine a compound's empirical formula by analyzing its percentage composition. There are several different ways to do this. One way is to use a synthesis reaction in which a sample of an element with a known mass reacts with another element to form a compound. Since you know the mass of one of the elements and you can measure the mass of the compound produced, you can calculate the percentage composition.

For example, copper reacts with the oxygen in air to form the green compound copper oxide. Many buildings in Canada, such as the Parliament buildings in Ottawa, have green roofs due to this reaction. (See Figure 6.7.) Imagine you have a 5.0 g sample of copper shavings. You allow the copper shavings to react completely with oxygen. (This would take a long time!) If the resulting compound has a mass of 6.3 g, you know that the compound contains 5.0 g copper and 1.3 g oxygen. Although you can use the periodic table to predict that the formula for copper oxide is CuO, the masses help you confirm your prediction. Try converting the masses given above into an empirical formula.

In Investigation 6-A, you will use a synthesis reaction to determine the empirical formula of magnesium oxide by experiment.

Figure 6.7 The copper in the roof of the Parliament buildings reacts with the oxygen in the air to form copper carbonate, $Cu_2CO_3(OH)_2$.

Investigation 6-A

SKILL FOCUS

- Predicting
- Performing and recording
- Analyzing and interpreting
- Communicating results

Determining the Empirical Formula of Magnesium Oxide

When magnesium metal is heated over a flame, it reacts with oxygen in the air to form magnesium oxide, Mg_xO_y:

$$Mg_{(s)} + O_{2(g)} \rightarrow Mg_xO_{y(s)}$$

In this investigation, you will react a strip of pure magnesium metal with oxygen, O_2, in the air to form magnesium oxide. Then you will measure the mass of the magnesium oxide produced to determine the percentage composition of magnesium oxide. You will use this percentage composition to calculate the empirical formula of magnesium oxide. CAUTION **Do not perform this investigation unless welder's goggles are available.**

Question

What is the percentage composition and empirical formula of magnesium oxide?

Predictions

Using what you have learned about writing formulas, predict the molecular formula and percentage composition of magnesium oxide.

Materials

electronic balance
small square of sandpaper or emery paper
8 cm strip of magnesium ribbon
laboratory burner
sparker
retort stand
ring clamp
clay triangle
clean crucible with lid
crucible tongs
ceramic pad
distilled water
wash bottle
disposal beaker
welder's goggles

Note: Make sure that the mass of the magnesium ribbon is at least 0.10 g.

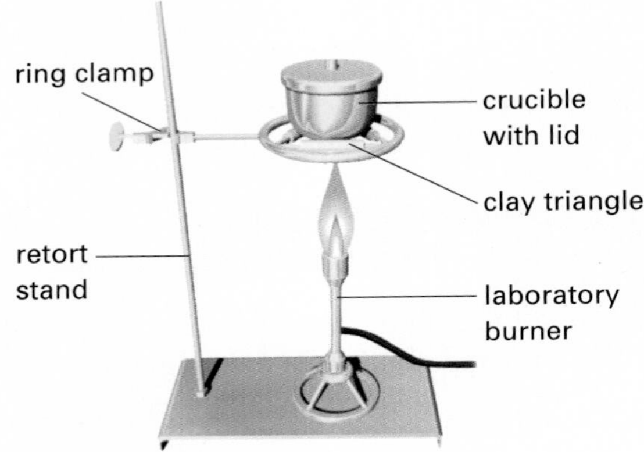

Safety Precautions

- Do not look directly at the burning magnesium.
- Do not put a hot crucible on the bench or the balance.

Procedure

1. Make a table like the one below.

Observations

Mass of clean, empty crucible and lid	
Mass of crucible, lid, and magnesium	
Mass of crucible and magnesium oxide	

2. Assemble the apparatus as shown in the diagram.
3. Obtain a strip of magnesium, about 8 cm long, from your teacher. Clean the magnesium strip with sandpaper or emery paper to remove any oxide coating.
4. Measure and record the mass of the empty crucible and lid. Add the strip of cleaned magnesium to the crucible. Record the mass of the crucible, lid, and magnesium.

5. With the lid off, place the crucible containing the magnesium on the clay triangle. Heat the crucible with a strong flame. Using the crucible tongs, hold the lid of the crucible nearby. CAUTION When the magnesium ignites, quickly cover the crucible with the lid. Continue heating for about 1 min.

6. Carefully remove the lid. CAUTION Heat the crucible until the magnesium ignites once more. Again, quickly cover the crucible. Repeat this heating and covering of the crucible until the magnesium no longer ignites. Heat for a further 4 to 5 min with the lid off.

7. Using the crucible tongs, put the crucible on the ceramic pad to cool.

8. When the crucible is cool enough to touch, put it on the bench. Carefully grind the product into small particles using the glass rod. Rinse any particles on the glass rod into the crucible with distilled water from the wash bottle.

9. Add enough distilled water to the crucible to thoroughly wet the contents. The white product is magnesium oxide. The yellowish-orange product is magnesium nitride.

10. Return the crucible to the clay triangle. Place the lid slightly ajar. Heat the crucible gently until the water begins to boil. Continue heating until all the water has evaporated, and the product is completely dry. Allow the crucible to cool on the ceramic pad.

11. Using the crucible tongs, carry the crucible and lid to the balance. Measure and record the mass of the crucible and lid.

12. Do not put the magnesium oxide in the garbage or in the sink. Put it in the disposal beaker designated by your teacher.

Analysis

1. **(a)** What mass of magnesium did you use in the reaction?
   (b) What mass of magnesium oxide was produced?
   (c) Calculate the mass of oxygen that reacted with the magnesium.
   (d) Use your data to calculate the percentage composition of magnesium oxide.
   (e) Determine the empirical formula of magnesium oxide. Remember to round your empirical formula to the nearest whole number ratio, such as 1:1, 1:2, 2:1, or 3:3.

2. **(a)** Verify your empirical formula with your teacher. Use the empirical formula of magnesium oxide to determine the mass percent of magnesium in magnesium oxide.
   (b) Calculate your percent error (PE) by finding the difference between the experimental mass percent (EP) of magnesium and the actual mass percent (AP) of magnesium. Then you divide the difference by the actual mass percent of magnesium and multiply by 100%.

$$PE = \frac{EP - AP}{AP} \times 100\%$$

3. Why did you need to round the empirical formula you obtained to a whole number ratio?

Conclusion

4. Compare the empirical formula you obtained with the empirical formula you predicted.

Applications

5. Write a balanced chemical equation for the reaction of magnesium with oxygen gas, O_2.

6. **(a)** Suppose that you had allowed some magnesium oxide smoke to escape during the investigation. How would the Mg:O ratio have been affected? Would the ratio have increased, decreased, or remained unchanged? Explain using sample calculations.
   (b) How would your calculated value for the empirical formula of magnesium oxide have been affected if all the magnesium in the crucible had not burned? Support your answer with sample calculations.
   (c) Could either of the situations mentioned in parts (a) and (b) have affected your results? Explain.

Section Wrap-up

In section 6.2, you learned how to calculate the empirical formula of a compound based on percentage composition data obtained by experiment. In section 6.3, you will learn how chemists use the empirical formula of a compound and its molar mass to determine the molecular formula of a compound.

Section Review

1 (a) K/U Why is the empirical formula of a compound also referred to as its simplest formula?

(b) K/U Explain how the empirical formula of a compound is related to its molecular formula.

2 I Methyl salicylate, or oil of wintergreen, is produced by the wintergreen plant. It can also be prepared easily in a laboratory. Methyl salicylate is 63.1% carbon, 5.31% hydrogen, and 31.6% oxygen. Calculate the empirical formula of methyl salicylate.

3 I Determine the empirical formula of the compound that is formed by each of the following reactions.

(a) 0.315 mol chlorine atoms react completely with 1.1 mol oxygen atoms

(b) 4.90 g silicon react completely with 24.8 g chlorine

4 I Muscle soreness from physical activity is caused by a buildup of lactic acid in muscle tissue. Analysis of lactic acid reveals it to be 40.0% carbon, 6.71% hydrogen, and 53.3% oxygen by mass. Calculate the empirical formula of lactic acid.

5 MC Imagine that you are a lawyer. You are representing a client charged with possession of a controlled substance. The prosecutor introduces, as forensic evidence, the empirical formula of the substance that was found in your client's possession. How would you deal with this evidence as a lawyer for the defence?

6 I Olive oil is used widely in cooking. Oleic acid, a component of olive oil, contains 76.54% carbon, 12.13% hydrogen and 11.33% oxygen by mass. What is the empirical formula of oleic acid?

7 I Phenyl valerate is a colourless liquid that is used as a flavour and odorant. It contains 74.13% carbon, 7.92% hydrogen and 17.95% oxygen by mass. Determine the empirical formula of phenyl valerate.

8 I Ferrocene is the common name given to a unique compound that consists of one iron atom sandwiched between two rings containing hydrogen and carbon. This orange, crystalline solid is added to fuel oil to improve combustion efficiency and eliminate smoke. As well, it is used as an industrial catalyst and a high-temperature lubricant.

(a) Elemental analysis reveals ferrocene to be 64.56% carbon, 5.42% hydrogen and 30.02% iron by mass. Determine the empirical formula of ferrocene.

(b) Read the description of ferrocene carefully. Does this description provide enough information for you to determine the molecular formula of ferrocene? Explain your answer.

The Molecular Formula of a Compound

6.3

Determining the identity of an unknown compound is important in all kinds of research. It can even be used to solve crimes. **Forensic scientists** specialize in analyzing evidence for criminal and legal cases. To understand why forensic scientists might need to find out the molecular formula of a compound, consider the following example.

Suppose that a suspect in a theft investigation is a researcher in a biology laboratory. The suspect frequently works with formaldehyde, CH_2O. Police officers find traces of a substance at the crime scene, and send samples to the Centre for Forensic Science. The forensic analysts find that the substance contains a compound that has an empirical formula of CH_2O. Will this evidence help to convict the suspect? Not necessarily.

As you can see from Table 6.3, there are many compounds that have the empirical formula CH_2O. The substance might be formaldehyde, but it could also be lactic acid (found in milk) or acetic acid (found in vinegar). Neither lactic acid nor acetic acid connect the theft to the suspect. Further information is required to prove that the substance is formaldehyde. Analyzing the physical properties of the substance would help to discover whether it is formaldehyde. Another important piece of information is the molar mass of the substance. Continue reading to find out why.

Section Preview/ Specific Expectations

In this section, you will

- **determine** the molecular formula of a compound, given the empirical formula of the compound and some additional information
- **identify** real-life situations in which the analysis of unknown substances is important
- **communicate** your understanding of the following terms: *forensic scientists*

Table 6.3 Six Compounds with the Empirical Formula CH_2O

Name	Molecular formula	Whole-number multiple	M (g/mol)	Use or function
formaldehyde	CH_2O	1	30.03	disinfectant; biological preservative
acetic acid	$C_2H_4O_2$	2	60.05	acetate polymers; vinegar (5% solution)
lactic acid	$C_3H_6O_3$	3	90.08	causes milk to sour; forms in muscles during exercise
erythrose	$C_4H_8O_4$	4	120.10	forms during sugar metabolism
ribose	$C_5H_{10}O_5$	5	150.13	component of many nucleic acids and vitamin B_2
glucose	$C_6H_{12}O_6$	6	180.16	major nutrient for energy in cells

CH_2O $C_2H_4O_2$ $C_3H_6O_3$ $C_4H_8O_4$ $C_5H_{10}O_5$ $C_6H_{12}O_6$

Determining a Molecular Formula

Recall the equation

Molecular formula subscripts = $n \times$ Empirical formula subscripts, where $n = 1, 2, 3...$

Additional information is required to obtain the molecular formula of a compound, given its empirical formula. We can use the molar mass and build on the above equation, as follows:

Molar mass of compound = $n \times$ Molar mass of empirical formula, where $n = 1, 2, 3...$

Careers in Chemistry

Analytical Chemistry

Ben Johnson, Steve Vezina, Eric Lamaze—all of these athletes tested positive for performance-enhancing substances that are banned by the International Olympic Committee (IOC). Who conducts the tests for these substances? Meet Dr. Christiane Ayotte, head of Canada's Doping Control Laboratory since 1991.

The Doping Control Lab

What happens to a urine sample after it arrives at the doping control lab? Technicians and scientists must be careful to ensure careful handling of the sample. Portions of the sample are taken for six different analytical procedures. More than 150 substances are banned by the IOC. These substances are grouped according to their physical and chemical properties. There are two main steps for analyzing a sample:

1. purification, which involves steps such as filtration and extraction using solvents, and
2. analysis by either gas chromatography, mass spectrometry, or high-performance liquid chromatography. Chromatography refers to certain methods by which chemists separate mixtures into pure substances.

For most substances, just their presence in a urine sample means a positive result. Other substances must be present in an amount higher than a certain threshold. According to Dr. Ayotte, a male athlete would have to consume "10 very strong French coffees within 15 min" to go over the 12 mg/L limit for caffeine. Ephedrines and pseudoephedrines, two decongestants that are found in cough remedies and that act as stimulants, have a cut-off level. This allows athletes to take them up to one or two days before a competition.

Challenges

Dr. Ayotte and her team face many challenges. They look for reliable tests for natural substances, develop new analytical techniques, and determine the normal levels of banned substances for male and female athletes. Dr. Ayotte must defend her tests in hearings and with the press, especially when high-profile athletes get positive results. Her dreams include an independent international doping control agency and better drug-risk education for athletes.

For Dr. Ayotte, integrity and a logical mind are essential aspects of being a good scientist.

Make Career Connections

1. For information about careers in analytical chemistry, contact university and college chemistry departments.
2. For information about doping control and the movement for drug-free sport, contact the Canadian Centre for Ethics in Sport (CCES), the World Anti-Doping Agency (WADA), and the Centre for Sport and Law.

Thus, the molar mass of a compound is a whole number multiple of the "molar mass" of the empirical formula.

Chemists can use a mass spectrometer to determine the molar mass of a compound. They can use the molar mass, along with the "molar mass" of a known empirical formula, to determine the compound's molecular formula. For example, the empirical formula CH has a "molar mass" of 13 g/mol. We know, however, that acetylene, C_2H_2, and benzene, C_6H_6, both have the empirical formula CH. Suppose it is determined, through mass spectrometry, that a sample has a molar mass of 78 g/mol. We know that the compound is C_6H_6, since 6×13 g = 78 g, as shown in Table 6.4.

Examine the Sample Problem and Practice Problems that follow to learn how to find the molecular formula of a compound using the empirical formula and the molar mass of the compound.

Table 6.4 Relating Molecular and Empirical Formulas

Formula	Molar Mass (g)	Ratio
C_6H_6 molecular	78	$\frac{78}{13} = 6$
CH empirical	13	

Sample Problem

Determining a Molecular Formula

Problem

The empirical formula of ribose (a sugar) is CH_2O. In a separate experiment, using a mass spectrometer, the molar mass of ribose was determined to be 150 g/mol. What is the molecular formula of ribose?

What Is Required?

You need to find the molecular formula of ribose.

What Is Given?

You know the empirical formula and the molar mass of ribose.

Plan Your Strategy

Divide the molar mass of ribose by the "molar mass" of the empirical formula. The answer you get is the factor by which you multiply the empirical formula.

Act on Your Strategy

The "molar mass" of the empirical formula CH_2O, determined using the periodic table, is

$$12 \text{ g/mol} + 2(1) \text{ g/mol} + 16 \text{ g/mol} = 30 \text{ g/mol}$$

The molar mass of ribose is 150 g/mol.

$$\frac{150 \cancel{\text{g/mol}}}{30 \cancel{\text{g/mol}}} = 5$$

$$\begin{aligned}\text{Molecular formula subscripts} &= 5 \times \text{ Empirical formula subscripts} \\ &= C_{1\times5}H_{2\times5}O_{1\times5} \\ &= C_5H_{10}O_5\end{aligned}$$

Therefore, the molecular formula of ribose is $C_5H_{10}O_5$.

Check Your Solution

Work backward by calculating the molar mass of $C_5H_{10}O_5$.

$(5 \times 12.01 \text{ g/mol}) + (10 \times 1.01 \text{ g/mol}) + (5 \times 16.00 \text{ g/mol}) = 150 \text{ g/mol}$

The calculated molar mass matches the molar mass that is given in the problem. The answer is reasonable.

Continued ...

CHEM FACT

Three classifications of food are proteins, fats, and carbohydrates. Many carbohydrates have the empirical formula CH_2O. This empirical formula looks like a hydrate of carbon, hence the name "carbohydrate." Glucose, fructose, galactose, mannose, and sorbose all have the empirical formula CH_2O since they all have the same molecular formula, $C_6H_{12}O_6$. What makes these sugars different is the way in which their atoms are bonded to one another. In Chapter 13, you will learn more about different compounds with the same formulas.

Continued ...

FROM PAGE 217

Practice Problems

17. The empirical formula of butane, the fuel used in disposable lighters, is C_2H_5. In an experiment, the molar mass of butane was determined to be 58 g/mol. What is the molecular formula of butane?
18. Oxalic acid has the empirical formula CHO_2. Its molar mass is 90 g/mol. What is the molecular formula of oxalic acid?
19. The empirical formula of codeine is $C_{18}H_{21}NO_3$. If the molar mass of codeine is 299 g/mol, what is its molecular formula?
20. A compound's molar mass is 240.28 g/mol. Its percentage composition is 75.0% carbon, 5.05% hydrogen, and 20.0% oxygen. What is the compound's molecular formula?

CHEM FACT

Codeine is a potent pain reliever. It acts on the pain centre in the brain, rather than interrupting pain messages from, for example, a headache or a sore arm. It is potentially habit-forming and classified as a narcotic.

Section Wrap-up

How do chemists obtain the data they use to identify compounds? In Investigation 6-A, you explored one technique for finding the percentage composition, and hence the empirical formula, of a compound containing magnesium and oxygen. In section 6.4, you will learn about another technique that chemists use to determine the empirical formula of compounds containing carbon and hydrogen. You will learn how chemists combine this technique with mass spectrometry to determine the compound's molecular formula. You will also learn about a new type of compound and perform an experiment to determine the molecular formula of one of these compounds.

Section Review

Unit Investigation Prep

Before you design your experiment to find the composition of a mixture, think about the relationships between different compounds. Suppose you have 5.0 g of copper. This copper reacts to form copper(II) chloride, also containing 5.0 g of copper. How can you determine the mass of copper(II) chloride formed?

1. K/U Explain the role that a mass spectrometer plays in determining the molecular formula of an unknown compound.
2. I Tartaric acid, also known as cream of tartar, is used in baking. Its empirical formula is $C_2H_3O_3$. If 1.00 mol of tartaric acid contains 3.61×10^{24} oxygen atoms, what is the molecular formula of tartaric acid?
3. MC Why is the molecular formula of a compound much more useful to a forensic scientist than the empirical formula of the compound?
4. K/U Vinyl acetate, $C_4H_6O_2$, is an important industrial chemical. It is used to make some of the polymers in products such as adhesives, paints, computer discs, and plastic films.
   (a) What is the empirical formula of vinyl acetate?
   (b) How does the molar mass of vinyl acetate compare with the molar mass of its empirical formula?
5. I A compound has the formula $C_{6x}H_{5x}O_x$, where x is a whole number. Its molar mass is 186 g/mol; what is its molecular formula?

Finding Empirical and Molecular Formulas by Experiment

6.4

Section Preview/ Specific Expectations

In this section, you will

- **identify** real-life situations in which the analysis of unknown substances is important
- **determine** the empirical formula of a hydrate through experimentation
- **explain** how a carbon-hydrogen analyzer can be used to determine the empirical formula of a compound
- **communicate** your understanding of the following terms: *carbon-hydrogen combustion analyzer, hydrate, anhydrous*

You have learned how to calculate the percentage composition of a compound using its formula. Often, however, the formula of a compound is not known. Chemists must determine the percentage composition and molar mass of an unknown compound through experimentation. Then they use this information to determine the molecular formula of the compound. Determining the molecular formula is an important step in understanding the properties of the compound and developing a way to synthesize it in a laboratory.

In Investigation 6-A, you reacted a known mass of magnesium with oxygen and found the mass of the product. Then you determined the percentage composition and empirical formula of magnesium oxide. This is just one method for determining percentage composition. It is suitable for analyzing simple compounds that react in predictable ways. Chemists have developed other methods for analyzing different types of compounds, as you will learn in this section.

The Carbon-Hydrogen Combustion Analyzer

A large number of important chemicals are composed of hydrogen, carbon, and oxygen. The **carbon-hydrogen combustion analyzer** is a useful instrument for analyzing these chemicals. It allows chemists to determine the percentage composition of compounds that are made up of carbon, hydrogen, and oxygen. The applications of this instrument include forensic science, food chemistry, pharmaceuticals and academic research—anywhere that an unknown compound needs to be analyzed.

The carbon-hydrogen combustion analyzer works because we know that compounds containing carbon and hydrogen will burn in a stream of pure oxygen, O_2, to yield only carbon dioxide and water. If we can find the mass of the carbon dioxide and water separately, we can determine the mass percent of carbon and hydrogen in the compound.

Examine Figure 6.8 to see how a carbon-hydrogen combustion analyzer works. A sample, made up of only carbon and hydrogen, is placed in a furnace. The sample is heated and simultaneously reacted with a stream

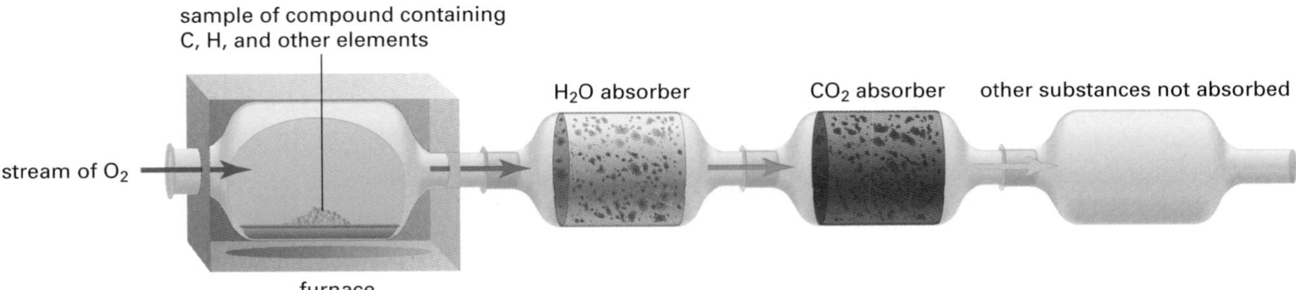

Figure 6.8 A schematic diagram of a carbon-hydrogen combustion analyzer. After the combustion, all the carbon in the sample is contained in the carbon dioxide. All the hydrogen in the sample is contained in the water.

CHECKPOINT

Carbon dioxide reacts with sodium hydroxide to form sodium carbonate and water. Write a balanced chemical equation for this reaction.

of oxygen. Eventually the sample is completely combusted to yield only water vapour and carbon dioxide.

The water vapour is collected by passing it through a tube that contains magnesium perchlorate, $Mg(ClO_4)_2$. The magnesium perchlorate absorbs all of the water. The mass of the tube is determined before and after the reaction. The difference is the mass of the water that is produced in the reaction. We know that all the hydrogen in the sample is converted to water. Therefore, we can use the percentage composition of hydrogen in water to determine the mass of the hydrogen in the sample.

The carbon dioxide is captured in a second tube, which contains sodium hydroxide, NaOH. The mass of this tube is also measured before and after the reaction. The increase in the mass of the tube corresponds to the mass of the carbon dioxide that is produced. We know that all the carbon in the sample reacts to form carbon dioxide. Therefore, we can use the percentage composition of carbon in carbon dioxide to determine the mass of the carbon in the sample.

PROBEWARE

If you have access to probeware, do the Chemistry 11 lab, Determining Molecular Mass, or a similar lab available from a probeware company.

The carbon-hydrogen combustion analyzer can also be used to find the empirical formula of a compound that contains carbon, hydrogen, and one other element, such as oxygen. The difference between the mass of the sample and the mass of the hydrogen and carbon produced is the mass of the third element.

Examine the following Sample Problem to learn how to determine the empirical formula of a compound based on carbon-hydrogen combustion data.

CHECKPOINT

If you have a compound containing carbon, hydrogen, and oxygen, what two instruments would you need to determine its molecular formula?

Sample Problem

Carbon-Hydrogen Combustion Analyzer Calculations

Problem

A 1.000 g sample of a pure compound, containing only carbon and hydrogen, was combusted in a carbon-hydrogen combustion analyzer. The combustion produced 0.6919 g of water and 3.338 g of carbon dioxide.

(a) Calculate the masses of the carbon and the hydrogen in the sample.

(b) Find the empirical formula of the compound.

What Is Required?

You need to find

(a) the mass of the carbon and the hydrogen in the sample

(b) the empirical formula of the compound

What Is Given?

You know the mass of the sample. You also know the masses of the water and the carbon dioxide produced in the combustion of the sample.

Continued ...

Continued ...
FROM PAGE 220

Plan Your Strategy

All the hydrogen in the sample was converted to water. Multiply the mass percent (as a decimal) of hydrogen in water by the mass of the water to get the mass of the hydrogen in the sample.

Similarly, all the carbon in the sample has been incorporated into the carbon dioxide. Multiply the mass percent (as a decimal) of carbon in carbon dioxide by the mass of the carbon dioxide to get the mass of carbon in the sample. Convert to moles and determine the empirical formula.

Act on Your Strategy

(a) Mass of H in sample

$$= \frac{2.02 \text{ g } H_2}{18.02 \cancel{\text{ g } H_2O}} \times 0.6919 \cancel{\text{ g } H_2O} = 0.077\ 56 \text{ g } H_2$$

$$\text{Mass of C in sample} = \frac{12.01 \text{ g C}}{44.01 \cancel{\text{ g } CO_2}} \times 3.338 \cancel{\text{ g } CO_2} = 0.9109 \text{ g C}$$

The sample contained 0.077 56 g of hydrogen and 0.9109 g of carbon.

(b) $\text{Moles of H in sample} = \frac{0.07756 \cancel{\text{g}}}{1.008 \cancel{\text{g}}/\text{mol}} = 0.07694 \text{ mol}$

$$\text{Moles of C in sample} = \frac{0.9109 \cancel{\text{g}}}{12.01 \cancel{\text{g}}/\text{mol}} = 0.07584 \text{ mol}$$

$$\begin{aligned}\text{Empirical formula} &= C_{\frac{0.07584}{0.07584}} H_{\frac{0.07584}{0.07694}} \\ &= C_{1.0}H_{1.0} \\ &= CH\end{aligned}$$

Check Your Solution

The sum of the masses of carbon and hydrogen is 0.077 56 g + 0.9109 g = 0.988 46 g. This is close to the mass of the sample. Therefore your answers are reasonable.

Practice Problems

21. A 0.539 g sample of a compound that contained only carbon and hydrogen was subjected to combustion analysis. The combustion produced 1.64 g of carbon dioxide and 0.807 g of water. Calculate the percentage composition and the empirical formula of the sample.

22. An 874 mg sample of cortisol was subjected to carbon-hydrogen combustion analysis. 2.23 g of carbon dioxide and 0.652 g of water were produced. The molar mass of cortisol was found to be 362 g/mol using a mass spectrometer. If cortisol contains carbon, hydrogen, and oxygen, determine its molecular formula.

Cortisol is an important steroid hormone. It helps your body synthesize protein. Cortisol can also reduce inflammation, and is used to treat allergies and rheumatoid arthritis.

Chemistry Bulletin

Science | Technology | Society | Environment

Accident or Arson?

All chemists who try to identify unknown compounds are like detectives. Forensic chemists, however, actually work with investigators. They use their chemical knowledge to help explain evidence. Forensic chemists are especially helpful in an arson investigation.

Investigating Arson

One of the main jobs of the investigator in a possible arson case is to locate and sample residual traces of accelerants. Accelerants are flammable substances that are used to quickly ignite and spread a fire. They include compounds called hydrocarbons, which contain hydrogen and carbon. Examples of hydrocarbons include petrol, kerosene and diesel.

Portable instruments called *sniffers* can be used to determine the best places to collect samples. These sniffers, however, are not able to determine the type of hydrocarbon present. As well, they can be set off by vapours from burnt plastics. Deciding whether or not a substance is an accelerant is best done by a chemist in a laboratory, using a technique called gas chromatography (GC).

In the Forensic Laboratory

In the laboratory, the sample residue must be concentrated on charcoal or another material. Then the sample is ready for GC analysis. GC is used to separate and detect trace amounts of volatile hydrocarbons and separate them from a mixture. Most accelerants are complex mixtures. They have many components, in different but specific ratios.

GC involves taking the concentrated residue and passing it through a gas column. As the sample residue moves through the column, the different components separate based on their boiling points. The compound with the lowest boiling point emerges from the column and onto a detector first. The other components follow as they reach their boiling points. It is possible to identify each component of a mixture based on the time that it emerges from the column. A detector records this information on a chromatogram. Each component is represented by a peak on a graph. The overall pattern of peaks is always the same for a specific type of accelerant. Therefore, accelerants are identified by their components and the relative proportions of their components.

Only trace amounts of an accelerant need to be collected because current analytical tools are extremely sensitive. If an accelerant is used to start a fire, it is highly likely that there will be trace amounts left over after the fire. The presence of an accelerant at a fire scene strongly suggests that the fire was started intentionally.

Making Connections

1. What other types of crime could be solved by a forensic chemist? Brainstorm a list.
2. What other instruments might a forensic chemist use to identify compounds? Using the Internet or reference books, do some research to find out.

Hydrated Ionic Compounds

You have learned how to find the molecular formula of a compound that contains only hydrogen, carbon, and oxygen. When chemists use this method, they usually have no mass percent data for the compound when they begin. In some cases, however, chemists know most of the molecular formula of a compound, but one significant piece of information is missing.

For example, many ionic compounds crystallize from a water solution with water molecules incorporated into their crystal structure, forming a **hydrate**. Hydrates have a specific number of water molecules chemically bonded to each formula unit. A chemist may know the formula of the ionic part of the hydrate but not how many water molecules are present for each formula unit.

Epsom salts, for example, consist of crystals of magnesium sulfate heptahydrate, $MgSO_4{\cdot}7H_2O$. Every formula unit of magnesium sulfate has seven molecules of water weakly bonded to it. A raised dot in a chemical formula, in front of one or more water molecules, denotes a hydrated compound. Note that the dot does not include multiplication, but rather a weak bond between an ionic compound and one or more water molecules. Some other examples of hydrates are shown in Table 6.4.

Compounds that have no water molecules incorporated into them are called **anhydrous** to distinguish them from their hydrated forms. For example, a chemist might refer to $CaSO_4$ as anhydrous calcium sulfate. This is because it is often found in hydrated form as calcium sulfate dihydrate, shown in Figure 6.9.

mind STRETCH

Suppose there is $MgSO_4{\cdot}7H_2O$ in the chemistry prep room. The experiment you want to do, however, calls for $MgSO_4$. How do you think you might remove the water from $MgSO_4{\cdot}7H_2O$?

Table 6.4 Selected Hydrates

Formula	Chemical name
$CaSO_4{\cdot}2H_2O$	calcium sulfate dihydrate (gypsum)
$CaCl_2{\cdot}2H_2O$	calcium chloride dihydrate
$LiCl_2{\cdot}4H_2O$	lithium chloride tetrahydrate
$MgSO_4{\cdot}7H_2O$	magnesium sulfate heptahydrate (Epsom salts)
$Ba(OH)_2{\cdot}8H_2O$	barium hydroxide octahydrate
$Na_2CO_3{\cdot}10H_2O$	sodium carbonate decahydrate
$KAl(SO_4)_2{\cdot}12H_2O$	potassium aluminum sulfate dodecahydrate (alum)

Figure 6.9 **Alabaster is a compact form of gypsum often used in sculpture. Gypsum is the common name for calcium sulfate dihydrate, $CaSO_4{\cdot}2H_2O$.**

The molar mass of a hydrated compound must include the mass of any water molecules that are in the compound. For example, the molar mass of magnesium sulfate heptahydrate includes the mass of 7 mol of water. It is very important to know whether a compound exists as a hydrate. For example, if a chemical reaction calls for 0.25 mol of copper(II) chloride, you need to know whether you are dealing with anhydrous copper(II) chloride or with copper(II) chloride dihydrate, shown in Figure 6.10. The mass of 0.25 mol of $CuCl_2$ is 33.61 g. The mass of 0.25 mol of $CuCl_2{\cdot}2H_2O$ is 38.11 g.

Calculations involving hydrates involve using the same techniques you have already practised for determining percent by mass, empirical formulas, and molecular formulas.

The following Sample Problem shows how to find the percent by mass of water in a hydrate. It also shows how to determine the formula of a hydrate based on an incomplete chemical formula.

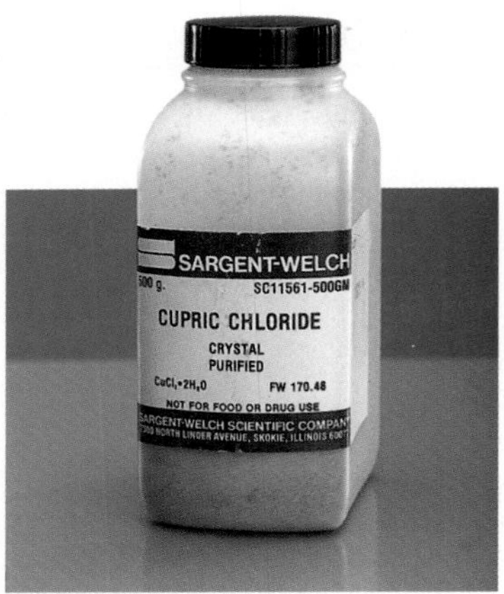

Figure 6.10 If you need 5 mol of $CuCl_2$, how much of the compound above would you use?

Sample Problem

Determining the Formula of a Hydrate

Problem

A hydrate of barium hydroxide, $Ba(OH)_2 \cdot xH_2O$, is used to make barium salts and to prepare certain organic compounds. Since it reacts with CO_2 from the air to yield barium carbonate, $BaCO_3$, it must be stored in tightly stoppered bottles.

(a) A 50.0 g sample of the hydrate contains 27.2 g of $Ba(OH)_2$. Calculate the percent, by mass, of water in $Ba(OH)_2 \cdot xH_2O$.

(b) Find the value of x in $Ba(OH)_2 \cdot xH_2O$.

What Is Required?

(a) You need to calculate the percent, by mass, of water in the hydrate of barium hydroxide.

(b) You need to find how many water molecules are bonded to each formula unit of $Ba(OH)_2$.

What Is Given?

The formula of the sample is $Ba(OH)_2 \cdot xH_2O$.
The mass of the sample is 50.0 g.
The sample contains 27.2 g of $Ba(OH)_2$.

Plan Your Strategy

(a) To find the mass of water in the hydrate, find the difference between the mass of barium hydroxide and the total mass of the sample. Divide by the total mass of the sample and multiply by 100%.

(b) Find the number of moles of barium hydroxide in the sample. Then find the number of moles of water in the sample. To find out how many water molecules bond to each formula unit of barium hydroxide, divide each answer by the number of moles of barium hydroxide.

Act on Your Strategy

(a) Mass percent of water in $Ba(OH)_2 \cdot xH_2O$

$$= \frac{\text{(Total mass of sample)} - \text{(Mass of Ba(OH)}_2\text{ in sample)}}{\text{(Total mass of sample)}} \times 100\%$$

$$= \frac{50.0 \cancel{g} - 27.2 \cancel{g}}{50.0 \cancel{g}} \times 100\%$$

$$= 45.6\%$$

(b) $$\text{Moles of Ba(OH)}_2 = \frac{\text{Mass of Ba(OH)}_2}{\text{Molar mass of Ba(OH)}_2}$$

$$= \frac{27.2 \cancel{g}}{171.3 \cancel{g}/\text{mol}}$$

$$= 0.159 \text{ mol Ba(OH)}_2$$

Continued ...

Continued ...

FROM PAGE 224

$$\text{Moles of } H_2O = \frac{\text{Mass of } H_2O}{\text{Molar mass of } H_2O}$$
$$= \frac{50.0\ \cancel{g} - 27.2\ \cancel{g}}{18.02\ \cancel{g}/\text{mol}}$$
$$= 1.27 \text{ mol } H_2O$$

$$\frac{0.159}{0.159} \text{ mol } Ba(OH)_2 : \frac{1.27}{0.159} \text{ mol } H_2O = 1.0 \text{ mol } Ba(OH)_2 : 8.0 \text{ mol } H_2O$$

The value of x in $Ba(OH)_2{\cdot}xH_2O$ is 8.
Therefore, the molecular formula of the hydrate is $Ba(OH)_2{\cdot}8H_2O$.

PROBLEM TIP

This step is similar to finding an empirical formula based on percentage composition.

Check Your Solution

Work backward.
According to the formula, the percent by mass of water in $Ba(OH)_2{\cdot}8H_2O$ is:

$$\frac{144.16\ \cancel{g/mol}}{315.51\ \cancel{g/mol}} \times 100\% = 45.7\%$$

According to the question, the percent by mass of water in the hydrate of $Ba(OH)_2$ is:

$$\frac{(50.0\ \cancel{g} - 27.2\ \cancel{g})}{50.0\ \cancel{g}} \times 100\% = 45.6\%$$

Therefore, your answer is reasonable.

Practice Problems

23. What is the percent by mass of water in magnesium sulfite hexahydrate, $MgSO_3{\cdot}6H_2O$?

24. A 3.34 g sample of a hydrate has the formula $SrS_2O_3{\cdot}xH_2O$, and contains 2.30 g of SrS_2O_3. Find the value of x.

25. A hydrate of zinc chlorate, $Zn(ClO_3)_2{\cdot}xH_2O$, contains 21.5% zinc by mass. Find the value of x.

CHECKPOINT

Write an equation that shows what happens when you heat magnesium sulfate hexahydrate enough to convert it to its anhydrous form.

Determining the Molecular Formula of a Hydrate

As you have just discovered, calculations involving hydrates usually involve comparing the anhydrous form of the ionic compound to the hydrated form. Many chemicals are available in hydrated form. Usually chemists are only interested in how much of the ionic part of the hydrate they are working with. This is because, in most reactions involving hydrates, the water portion of the compound does not take part in the reaction. Only the ionic portion does.

How do chemists determine how many water molecules are bonded to each ionic formula unit in a hydrate? One method is to heat the compound in order to convert it to its anhydrous form. The bonds that join the water molecules to the ionic compound are very weak compared with the strong ionic bonds within the ionic compound. Heating a hydrate usually removes the water molecules, leaving the anhydrous compound behind. In Investigation 6-B, you will heat a hydrate to determine its formula.

Electronic Learning Partner

A video clip describing hydrated ionic compounds can be found on the Chemistry 11 Electronic Learning Partner.

SKILL FOCUS

Predicting

Performing and recording

Analyzing and interpreting

Communicating results

Investigation 6-B

Determining the Chemical Formula of a Hydrate

Many ionic compounds exist as hydrates. Often you can convert hydrates to anhydrous ionic compounds by heating them. Thus, hydrates are well suited to determining percentage composition experimentally.

In this investigation, you will find the mass percent of water in a hydrate of copper(II) sulfate hydrate, $CuSO_4{\cdot}xH_2O$. You will use copper(II) sulfate hydrate for an important reason: The crystals of the hydrate are blue, while anhydrous copper(II) sulfate is white.

Question

What is the molecular formula of the hydrate of copper(II) sulfate, $CuSO_4{\cdot}xH_2O$?

Prediction

Predict what reaction will occur when you heat the hydrate of copper(II) sulfate.

Materials

400 mL beaker (if hot plate is used)
tongs
scoopula
electronic balance
glass rod
hot pad
3 g to 5 g hydrated copper(II) sulfate

Safety Precautions

Heat the hydrate at a low to medium temperature only.

Procedure

Note: If you are using a hot plate as your heat source, use the 400 mL beaker. If you are using a laboratory burner, use the porcelain evaporating dish.

1. Make a table like the one below, for recording your observations.

Observations

Mass of empty beaker or evaporating dish	
Mass of beaker or evaporating dish + hydrated copper(II) sulfate	
Mass of beaker or evaporating dish + anhydrous copper(II) sulfate	

A hydrate of copper(II) sulfate (far left) is light blue. It loses its colour on heating.

2. Measure the mass of the beaker and stirring rod. Record the mass in your table.
3. Add 3 g to 5 g hydrated copper(II) sulfate to the beaker.
4. Measure the mass of the beaker with the hydrated copper(II) sulfate. Record the mass in your table.
5. If you are using a hot plate, heat the beaker with the hydrated copper(II) sulfate until the crystals lose their blue colour. You may need to stir occasionally with the glass rod. Be sure to keep the heat at a medium setting. Otherwise, the beaker may break.
6. When you see the colour change, stop heating the beaker. Turn off or unplug the hot plate. Remove the beaker with the beaker tongs. Allow the beaker and crystals to cool on a hot pad.
7. Find the mass of the beaker with the white crystals. Record the mass in your table.
8. Return the anhydrous copper(II) sulfate to your teacher when you are finished. Do not put it in the sink or in the garbage.

Analysis

1. (a) Determine the percent by mass of water in your sample of hydrated copper(II) sulfate. Show your calculations clearly.
   (b) Do you expect the mass percent of water that you determined to be similar to the mass percents that other groups determined? Explain.
2. (a) On the chalkboard, write the mass of your sample of hydrated copper(II) sulfate, the mass of the anhydrous copper(II) sulfate, and the mass percent of water that you calculated.
   (b) How do your results compare with other groups' results?

Conclusion

3. Based on your observations, determine the molecular formula of $CuSO_4{\cdot}xH_2O$.

Applications

4. Suppose that you heated a sample of a hydrated ionic compound in a test tube. What might you expect to see inside the test tube, near the mouth of the test tube? Explain.
5. You obtained the mass percent of water in the copper sulfate hydrate.
   (a) Using your observations, calculate the percentage composition of the copper sulfate hydrate.
   (b) In the case of a hydrate, and assuming you know the formula of the associated anhydrous ionic compound, do you think it is more useful to have the mass percent of water in the hydrate or the percentage composition? Explain your answer.
6. Compare the formula that you obtained for the copper sulfate hydrate with the formulas that other groups obtained. Are there any differences? How might these differences have occurred?
7. Suppose that you did not completely convert the hydrate to the anhydrous compound. Explain how this would affect
   (a) the calculated percent by mass of water in the compound
   (b) the molecular formula you determined
8. Suppose the hydrate was heated too quickly and some of it was lost as it spattered out of the container. Explain how this would affect
   (a) the calculated percent by mass of water in the compound
   (b) the molecular formula you determined
9. Suggest a source of error (not already mentioned) that would result in a value of x that is
   (a) higher than the actual value
   (b) lower than the actual value

Section Wrap-up

In section 6.4, you learned several practical methods for determining empirical and molecular formulas of compounds. You may have noticed that these methods work because compounds react in predictable ways. For example, you learned that a compound containing carbon and hydrogen reacts with oxygen to produce water and carbon dioxide. From the mass of the products, you can determine the amount of carbon and hydrogen in the reactant. You also learned that a hydrate decomposes when it is heated to form water and an anhydrous compound. Again, the mass of one of the products of this reaction helps you identify the reactant. In Chapter 7, you will learn more about how to use the information from chemical reactions in order to do quantitative calculations.

Section Review

1 K/U Many compounds that contain carbon and hydrogen also contain nitrogen. Can you find the nitrogen content by carbon-hydrogen analysis, if the nitrogen does not interfere with the combustion reaction? If so, explain how. If not, explain why not.

2 I What would be the mass of a bag of anhydrous magnesium sulfate, $MgSO_4$, if it contained the same amount of magnesium as a 1.00 kg bag of Epsom salts, $MgSO_4 \cdot 7H_2O$? Give your answer in grams.

3 K/U A compound that contains carbon, hydrogen, chlorine, and oxygen is subjected to carbon-hydrogen analysis. Can the mass percent of oxygen in the compound be determined using this method? Explain your answer.

4 C Imagine that you are an analytical chemist. You are presented with an unknown compound, in the form of a white powder, for analysis. Your job is to determine the molecular formula of the compound. Create a flow chart that outlines the questions that you would ask and the analyses you would carry out. Briefly explain why each question or analysis is needed.

5 MC A carbon-hydrogen analyzer uses a water absorber (which contains magnesium perchlorate, $Mg(ClO_4)_2$) and a carbon dioxide absorber (which contains sodium hydroxide, NaOH). The water absorber is always located in front of the carbon dioxide absorber. What does this suggest about the sodium hydroxide that is contained in the CO_2 absorber?

6 I A hydrate of zinc nitrate has the formula $Zn(NO_3)_2 \cdot xH_2O$. If the mass of 1 mol of anhydrous zinc nitrate is 63.67% of the mass of 1 mol of the hydrate, what is the value of x?

7 K/U A 2.524 g sample of a compound contains carbon, hydrogen, and oxygen. The sample is subjected to carbon-hydrogen analysis. 3.703 g of carbon dioxide and 1.514 g of water are collected.

(a) Determine the empirical formula of the compound.

(b) If one molecule of the compound contains 12 atoms of hydrogen, what is the molecular formula of the compound?

CHAPTER 6 Review

Reflecting on Chapter 6

Summarize this chapter in the format of your choice. Here are a few ideas to use as guidelines:

- Determine the mass percent of each element in a compound.
- Predict the empirical formula of a compound using the periodic table, and test your prediction through experimentation.
- Use experimental data to determine the empirical (simplest) formula of a compound.
- Use the molar mass and empirical formula of a compound to determine the molecular (actual) formula of the compound.
- Determine experimentally the percent by mass of water in a hydrate. Use this information to determine its molecular formula.
- Explain how a carbon-hydrogen combustion analyzer can be used to determine the mass percent of carbon, hydrogen, and oxygen in a compound.

Reviewing Key Terms

For each of the following terms, write a sentence that shows your understanding of its meaning.

anhydrous

carbon-hydrogen combustion analyzer

empirical formula

forensic scientists

hydrate

law of definite proportions

mass percent

molecular formula

percentage composition

Knowledge/Understanding

1. When determining the percentage composition of a compound from its formula, why do you base your calculations on a one mole sample?
2. The main engines of the space shuttle burn hydrogen and oxygen, with water as the product. Is this synthetic (human-made) water the same as water found in nature? Explain.
3. (a) What measurements need to be taken during a carbon-hydrogen combustion analysis?
   (b) Acetylene, C_2H_2, and benzene, C_6H_6, both have the same empirical formula. How would their results compare in a carbon-hydrogen combustion analysis? Explain your answer.
4. If you know the molar mass of a substance, and the elements that make up the substance, can you determine its molecular formula? Explain your answer.

Inquiry

5. A 5.00 g sample of borax (sodium tetraborate decahydrate, $Na_2B_4O_7{\cdot}10H_2O$) was thoroughly heated to remove all the water of hydration. What mass of anhydrous sodium tetraborate remained?
6. Determine the percentage composition of each compound.
   (a) freon-12, CCl_2F_2
   (b) white lead, $Pb_3(OH)_2(CO_3)_2$
7. (a) What mass of water is present in 25.0 g of $MgCl_2{\cdot}2H_2O$?
   (b) What mass of manganese is present in 5.00 g of potassium permanganate, $KMnO_4$?
8. Silver nitrate, $AgNO_3$, can be used to test for the presence of halide ions in solution. It combines with the halide ions to form a silver halide precipitate. In medicine, it is used as an antiseptic and an antibacterial agent. Silver nitrate drops are placed in the eyes of newborn babies to protect them against an eye disease.
   (a) Calculate the mass percent of silver in silver nitrate.
   (b) What mass of pure silver is contained in 2.00×10^2 kg of silver nitrate?
9. Barium sulfate, $BaSO_4$, is opaque to X-rays. For this reason, it is sometimes given to patients before X-rays of their intestines are taken. What mass of barium is contained in 45.8 g of barium sulfate?
10. Bismuth nitrate, $Bi(NO_3)_2$, is used in the production of some luminous paints. How many grams of pure bismuth are in a 268 g sample of bismuth nitrate?
11. The molar mass of a compound is approximately 121 g. The empirical formula of the

compound is CH_2O. What is the molecular formula of the compound?

12. A complex organic compound, with the name 2,3,7,8-tetrachlorodibenza-para-dioxin, belongs to a family of toxic compounds called *dioxins*. The empirical formula of a certain dioxin is $C_6H_2OCl_2$. If the molar mass of this dioxin is 322 g/mol, what is its molecular formula?

13. A student obtains an empirical formula of $C_1H_{2.67}$ for a gaseous compound.
    (a) Why is this not a valid empirical formula?
    (b) Use the student's empirical formula to determine the correct empirical formula.

14. Progesterone, a hormone, is made up of 80.2% carbon, 10.18% oxygen, and 9.62% hydrogen. Determine the empirical formula of progesterone.

15. An inorganic salt is composed of 17.6% sodium, 39.7% chromium, and 42.8% oxygen. What is the empirical formula of this salt?

16. What is the empirical formula of a compound that contains 67.6% mercury, 10.8% sulfur, and 21.6% oxygen?

17. (a) An inorganic salt is made up of 38.8% calcium, 20.0% phosphorus, and 41.2% oxygen. What is the empirical formula of this salt?
    (b) On further analysis, each formula unit of this salt is found to contain two phosphate ions. Predict the molecular formula of this salt.

18. Capsaicin is the compound that is responsible for the "hotness" of chili peppers. Chemical analysis reveals capsaicin to contain 71.0% carbon, 8.60% hydrogen, 15.8% oxygen, and 4.60% nitrogen.
    (a) Determine the empirical formula of capsaicin.
    (b) Each molecule of capsaicin contains one atom of nitrogen. What is the molecular formula of capsaicin?

19. A compound has the formula X_2O_5, where X is an unknown element. The compound is 44.0% oxygen by mass. What is the identity of element X?

20. A 1.254 g sample of an organic compound that contains only carbon, hydrogen, and oxygen reacts with a stream of chlorine gas, $Cl_{2(g)}$. After the reaction, 4.730 g of HCl and 9.977 g of CCl_4 are obtained. Determine the empirical formula of the organic compound.

21. A 2.78 g sample of hydrated iron(II) sulfate, $FeSO_4 \cdot xH_2O$, was heated to remove all the water of hydration. The mass of the anhydrous iron(II) sulfate was 1.52 g. Calculate the number of water molecules associated with each formula unit of $FeSO_4$.

22. Citric acid is present in citrus fruits. It is composed of carbon, hydrogen, and oxygen. When a 0.5000 g sample of citric acid was subjected to carbon-hydrogen combustion analysis, 0.6871 g of carbon dioxide and 0.1874 g of water were produced. Using a mass spectrometer, the molar mass of citric acid was determined to be 192 g/mol.
    (a) What are the percentages of carbon, hydrogen, and oxygen in citric acid?
    (b) What is the empirical formula of citric acid?
    (c) What is the molecular formula of citric acid?

23. Methanol, CH_3OH (also known as methyl alcohol), is a common laboratory reagent. It can be purchased at a hardware store under the name "methyl hydrate" or "wood alcohol." If 1.00 g of methanol is subjected to carbon-hydrogen combustion analysis, what masses of carbon dioxide and water are produced?

24. Copper can form two different oxides: copper(II) oxide, CuO, and copper(I) oxide, Cu_2O. Suppose that you find a bottle labelled "copper oxide" in the chemistry prep room. You call this mystery oxide Cu_xO. Design an experiment to determine the empirical formula of Cu_xO. Assume that you have a fully equipped chemistry lab at your disposal. Keep in mind the following information:
    - Both CuO and Cu_2O react with carbon to produce solid copper and carbon dioxide gas:
    $$Cu_xO_{(s)} + C_{(s)} \rightarrow Cu_{(s)} + CO_{2(g)}$$
    This reaction proceeds with strong heating.
    - Carbon reacts with oxygen to produce carbon dioxide gas:
    $$C_{(s)} + O_{2(g)} \rightarrow CO_{2(g)}$$
    This reaction also proceeds with strong heating.
    - Carbon is available in the form of activated charcoal.

(a) State at least one safety precaution that you would take.
(b) State the materials required, and sketch your apparatus.
(c) Outline your procedure.
(d) What data do you need to collect?
(e) State any assumptions that you would make.

25. Magnesium sulfate, $MgSO_4$, is available as anhydrous crystals or as a heptahydrate. Assume that you are given a bottle of $MgSO_4$, but you are not sure whether or not it is the hydrate.
(a) What method could you use, in a laboratory, to determine whether this is the hydrate?
(b) If it is the hydrate, what results would you expect to see?
(c) If it is the anhydrous crystals, what results would you expect to see?

Communication

26. Draw a concept map to relate the following terms: molar mass of an element, molar mass of a compound, percentage composition, empirical formula, and molecular formula. Use an example for each term.

27. Draw a schematic diagram of a carbon-hydrogen combustion analyzer. Write a few sentences to describe each stage of the analysis as dimethyl ether, C_2H_6O, passes through the apparatus.

Making Connections

28. For many years, tetraethyl lead, $Pb(C_2H_5)_4$, a colourless liquid, was added to gasoline to improve engine performance. Over the last 20 years it has been replaced with non-lead-containing additives due to health risks associated with exposure to lead. Tetraethyl lead was added to gasoline up to 2.0 mL per 3.8 L (1.0 US gallon) of gasoline.
(a) Calculate the mass of tetraethyl lead in 1.0 L of gasoline. The density of $Pb(C_2H_5)_4$ is 1.653 g/mL.
(b) Calculate the mass of elemental lead in 1.0 L of gasoline.

29. Natron is the name of the mixture of salts that was used by the ancient Egyptians to dehydrate corpses before mummification. Natron is composed of Na_2CO_3, $NaHCO_3$, NaCl, and $CaCl_2$. The Na_2CO_3 absorbs water from tissues to form $Na_2CO_3{\cdot}7H_2O$.
(a) Name the compound $Na_2CO_3{\cdot}7H_2O$.
(b) Calculate the mass percent of water in $Na_2CO_3{\cdot}7H_2O$.
(c) What mass of anhydrous $NaCO_3$ is required to dessicate (remove all the water) from an 80 kg body that is 78% water by mass?

30. Imagine that you are an analytical chemist at a pharmaceutical company. One of your jobs is to determine the purity of the acetylsalicylic acid (ASA), $C_9H_8O_4$. ASA is prepared by reacting salicylic acid (SA), $C_7H_6O_3$, with acetic anhydride, $C_4H_6O_3$. Acetic acid, $C_2H_3O_2H$, is also produced

$$\underset{\text{SA}}{C_7H_6O_3} + \underset{\text{acetic anhydride}}{C_4H_6O_3} \rightarrow \underset{\text{ASA}}{C_9H_8O_4} + \underset{\text{acetic acid}}{C_2H_3O_2H}$$

ASA often contains unreacted SA. Since it is not acceptable to sell ASA contaminated with SA, one of your jobs is to analyze the ASA to check purity. Both ASA and SA are white powders.
(a) You analyze a sample that you believe to be pure ASA, but which is actually contaminated with some SA. How will this affect the empirical formula that you determine for the sample?
(b) Another sample contains ASA contaminated with 0.35 g SA. The mass of the sample is 5.73 g. What empirical formula will you obtain?

Answers to Practice Problems and Short Answers to Section Review Questions:
Practice Problems: 1. 36% Ca; 64% Cl 2. 48.1% Ni; 16.9% P; 35.0% O 3. 39.5% C; 7.8% H; 52.7% O 4. 26.6% K; 35.4% Cr; 38.0% O 5.(a) 63.65% N (b) 13.24% N (c) 35.00% N (d) 22.23% N 6. 2.06% H; 32.69% S; 62.25% O 7. 47.47% O 8.(a) 63.19% Mn; 36.81% O (b) 158 g 9. NH_3 10. Li_2O 11. BF_3 12. SCl 13. Cr_2O_3 14. P_2O_5 15. $Na_2Cr_2O_7$ 16. $C_{12}H_{14}O_3$ 17. C_4H_{10} 18. $C_2H_2O_4$ 19. $C_{18}H_{21}NO_3$ 20. $C_{15}H_{12}O_3$ 21. C_5H_{12}; 83.3% C; 16.7% H. 22. $C_{21}H_{30}O_5$ 23. 50.9% 24. 5 25. 4
Section Review: 6.1: 3. 3.05 g 4. 11.3 g 5. 17.1 g. 7. Na^+ 6.2: 2. $C_8H_8O_3$ 3.(a) Cl_2O_7 (b) $SiCl_4$ 4. CH_2O 6. $C_9H_{17}O$ 7. $C_{11}H_{14}O_2$ 8.(a) $FeC_{10}H_{10}$ (b) Yes. 6.3: 2. $C_4H_6O_6$ 4.(a) C_2H_3O (b) Double 5. $C_{12}H_{10}O_2$ 6.4: 1. Yes 2. 488 g 4. No 6. 6 7.(a) CH_2O (b) $C_6H_{12}O_6$ 8.(a) CH_2 (b) $CH_{12}H_2$

CHAPTER 7

Quantities in Chemical Reactions

A spacecraft, such as the space shuttle on the left, requires a huge amount of fuel to supply the thrust needed to launch it into orbit. Engineers work very hard to minimize the launch mass of a spacecraft because each kilogram requires additional fuel. As well, each kilogram costs thousands of dollars to launch.

In 1969, the *Apollo 11* space mission was the first to land astronauts on the Moon. The engineers on the project faced a challenge when deciding on a fuel for the lunar module. The lunar module took the astronauts from the Moon, back to the command module that was orbiting the Moon. The engineers chose a fuel consisting of hydrazine, N_2H_4, and dinitrogen tetroxide, N_2O_4. These compounds, when mixed, reacted instantaneously and produced the energy needed to launch the lunar module from the Moon.

How do engineers know how much of each reactant they need for a chemical reaction? In this chapter, you will use the concept of the mole to calculate the amounts of reactants that are needed to produce given amounts of products. You will learn how to predict the amounts of products that will be produced in a chemical reaction. You will also learn how to apply this knowledge to any chemical reaction for which you know the balanced chemical equation. Finally, you will learn how calculated amounts deviate from the amounts in real-life situations.

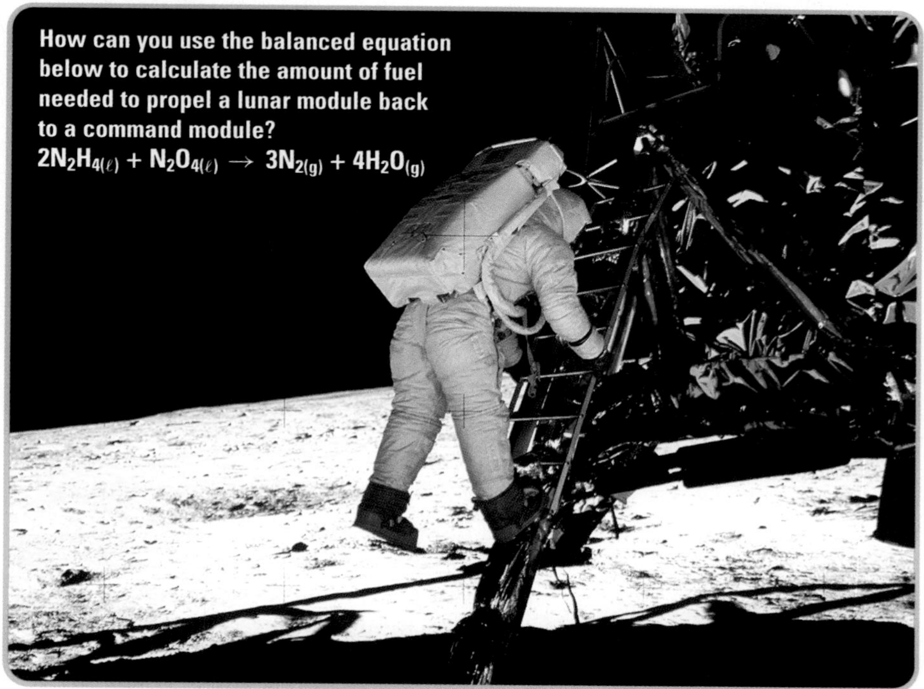
How can you use the balanced equation below to calculate the amount of fuel needed to propel a lunar module back to a command module?
$2N_2H_{4(\ell)} + N_2O_{4(\ell)} \rightarrow 3N_{2(g)} + 4H_2O_{(g)}$

Chapter Preview

Concepts and Skills You Will Need

Before you begin this chapter, review the following concepts and skills:

- balancing chemical equations (Chapter 4, section 4.1)
- understanding the Avogadro constant and the mole (Chapter 5, section 5.2)
- explaining the relationship between the mole and molar mass (Chapter 5, section 5.3)
- solving problems involving number of moles, number of particles, and mass (Chapter 5, section 5.3)

7.1 Stoichiometry

Section Preview/ Specific Expectations

In this section, you will

- **explain** quantitative relationships in a chemical equation, in moles, grams, atoms, or molecules
- **perform** laboratory experiments to determine the meaning of the coefficients in a balanced chemical equation
- **calculate**, for any given reactant or product in a chemical equation, the corresponding mass or quantity of any other reactant or product
- **demonstrate** an awareness of the importance of quantitative chemical relationships in the home or in industry
- **communicate** your understanding of the following terms: *mole ratios, stoichiometry*

Balanced chemical equations are essential for making calculations related to chemical reactions. To understand why, consider the following analogy.

Imagine that you are making salads. You need one head of lettuce, two cucumbers, and five radishes for each salad. Figure 7.1 shows how you can express this as an equation.

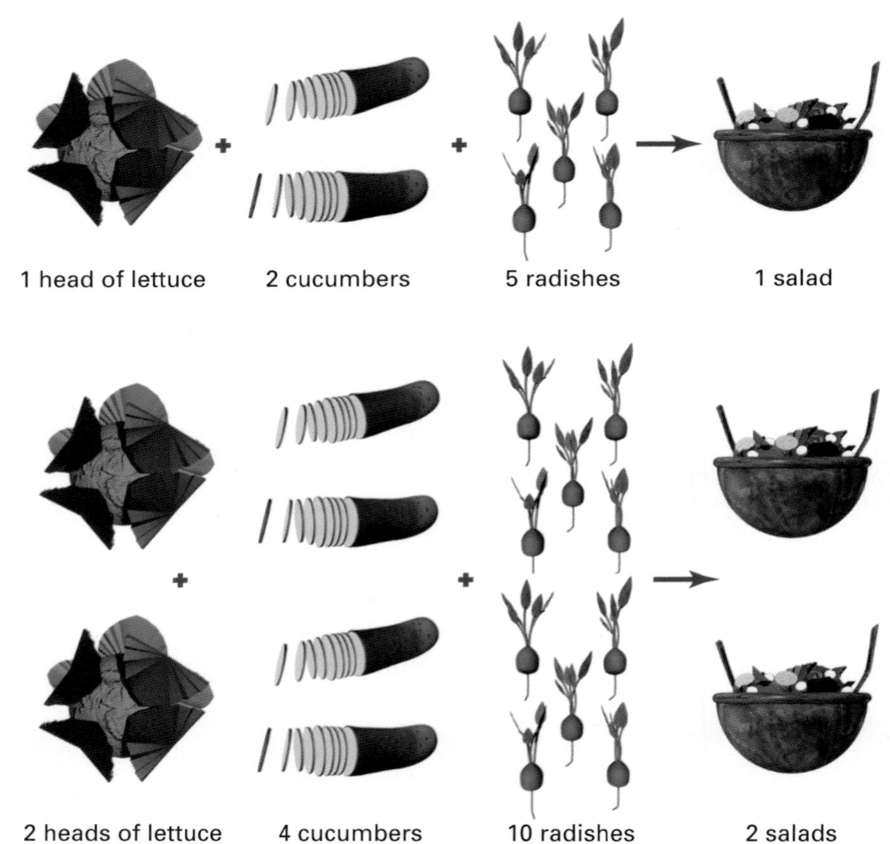

Figure 7.1 **A salad analogy showing how equations can be multiplied**

Now imagine that you are making two salads. How much of each ingredient do you need? You need twice the amount that you used to make one salad, as shown in Figure 7.1.

How many salads can you make if you have three heads of lettuce, six cucumbers, and 15 radishes? According to the salad equation, you can make three salads.

You can get the same kind of information from a balanced chemical equation. In Chapter 4, you learned how to classify chemical reactions and balance the chemical equations that describe them. In Chapters 5 and 6, you learned how chemists relate the number of particles in a substance to the amount of the substance in moles and grams. In this section, you will use your knowledge to interpret the information in a chemical equation, in terms of particles, moles, and mass. Try the following Express Lab to explore the molar relationships between products and reactants.

ExpressLab  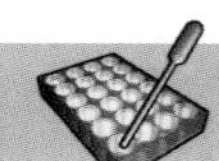 Mole Relationships in a Chemical Reaction

The following balanced equation shows the reaction between sodium hydrogen carbonate, $NaHCO_3$, and hydrochloric acid, HCl.

$$NaHCO_{3(s)} + HCl_{(aq)} \rightarrow CO_{2(g)} + H_2O_{(\ell)} + NaCl_{(aq)}$$

In this Express Lab, you will determine the mole relationships between the products and reactants in the reaction. Then you will compare the mole relationships with the balanced chemical equation.

Safety Precautions

Be careful when using concentrated hydrochloric acid. It burns skin and clothing. Do not inhale its vapour.

Procedure

1. Obtain a sample of sodium hydrogen carbonate that is approximately 1.0 g.
2. Place a 24-well microplate on a balance. Measure and record its mass.
3. Place all the sodium hydrogen carbonate in well A4 of the microplate. Measure and record the mass of the microplate and sample.
4. Fill a thin-stem pipette with 8 mol/L hydrochloric acid solution.
5. Wipe the outside of the pipette. Stand it, stem up, in well A3.
6. Measure and record the total mass of the microplate and sample.
7. Add the hydrochloric acid from the pipette to the sodium hydrogen carbonate in well A4. Allow the gas to escape after each drop.
8. Continue to add the hydrochloric acid until all the sodium hydrogen carbonate has dissolved and the solution produces no more bubbles.
9. Return the pipette, stem up, to well A3. Again find the total mass of the microplate and samples.
10. Dispose of the reacted chemicals as directed by your teacher.

Analysis

1. Calculate the number of moles of sodium hydrogen carbonate used.
2. Find the difference between the total mass of the microplate and samples before and after the reaction. This difference represents the mass of carbon dioxide gas produced.
3. Calculate the number of moles of carbon dioxide produced.
4. Express your answers to questions 1 and 3 as a mole ratio of mol $NaHCO_3$:mol CO_2.
5. According to the balanced equation, how many molecules of sodium hydrogen carbonate react to form one molecule of carbon dioxide?
   (a) Express your answer as a ratio.
   (b) Compare this ratio to your mole ratio in question 4.
6. How many moles of carbon dioxide do you think would be formed from 4.0 mol of sodium hydrogen carbonate?

You can use your understanding of the relationship between moles and number of particles to see how chemical equations communicate information about how many moles of products and reactants are involved in a reaction.

Particle Relationships in a Balanced Chemical Equation

As you learned in Chapter 4, the coefficients in front of compounds and elements in chemical equations tell you how many atoms and molecules participate in a reaction. A chemical equation can tell you much more, however. Consider, for example, the equation that describes the production of ammonia. Ammonia is an important industrial chemical. Several of its uses are shown in Figure 7.2 on the following page.

Figure 7.2 **Ammonia can be applied directly to the soil as a fertilizer. An aqueous (water) solution of ammonia can be used as a household cleaner.**

Ammonia can be prepared industrially from its elements, using a process called the Haber Process. The Haber Process is based on the balanced chemical equation below.

$$N_{2(g)} + 3H_{2(g)} \rightarrow 2NH_{3(g)}$$

This equation tells you that one molecule of nitrogen gas reacts with three molecules of hydrogen gas to form two molecules of ammonia gas.

As you can see in Figure 7.3, there is the same number of each type of atom on both sides of the equation.

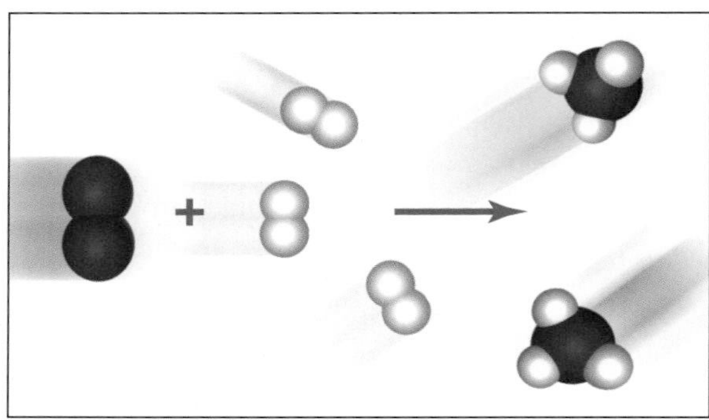

Figure 7.3 **The reaction of nitrogen gas with hydrogen gas.**

You can use a ratio to express the numbers of atoms in the equation, as follows:

$$1 \text{ molecule } N_2 : 3 \text{ molecules } H_2 : 2 \text{ molecules } NH_3$$

What happens if you multiply the ratio by 2? You get

$$2 \text{ molecules } N_2 : 6 \text{ molecules } H_2 : 4 \text{ molecules } NH_3$$

This means that two molecules of nitrogen gas react with six molecules of hydrogen gas to produce four molecules of ammonia gas. Multiplying the original ratio by one dozen gives the following relationship:

$$1 \text{ dozen molecules } N_2 : 3 \text{ dozen molecules } H_2 : 2 \text{ dozen molecules } NH_3$$

Suppose that you want to produce 20 molecules of ammonia. How many molecules of nitrogen do you need? You know that you need one molecule of nitrogen for every two molecules of ammonia produced. In other words, the number of molecules of nitrogen that you need is one half the number of molecules of ammonia that you want to produce.

$$20\ \cancel{\text{molecules NH}_3} \times \frac{1 \text{ molecule N}_2}{2\ \cancel{\text{molecules NH}_3}} = 10 \text{ molecules N}_2$$

Try the following problems to practise working with ratios in balanced chemical equations.

Practice Problems

1. Consider the following reaction.

$$2H_{2(g)} + O_{2(g)} \rightarrow 2H_2O_{(\ell)}$$

(a) Write the ratio of H_2 molecules: O_2 molecules: H_2O molecules.

(b) How many molecules of O_2 are required to react with 100 molecules of H_2, according to your ratio in part (a)?

(c) How many molecules of water are formed when 2478 molecules of O_2 react with H_2?

(d) How many molecules of H_2 are required to react completely with 6.02×10^{23} molecules of O_2?

2. Iron reacts with chlorine gas to form iron(III) chloride, $FeCl_3$.

$$2Fe_{(s)} + 3Cl_{2(g)} \rightarrow 2FeCl_{3(s)}$$

(a) How many atoms of Fe are needed to react with three molecules of Cl_2?

(b) How many molecules of $FeCl_3$ are formed when 150 atoms of Fe react with sufficient Cl_2?

(c) How many Cl_2 molecules are needed to react with 1.204×10^{24} atoms of Fe?

(d) How many molecules of $FeCl_3$ are formed when 1.806×10^{24} molecules of Cl_2 react with sufficient Fe?

3. Consider the following reaction.

$$Ca(OH)_{2(aq)} + 2HCl_{(aq)} \rightarrow CaCl_2 + 2H_2O_{(\ell)}$$

(a) How many formula units of calcium chloride, $CaCl_2$, would be produced by 6.7×10^{25} molecules of hydrochloric acid, HCl?

(b) How many molecules of water would be produced in the reaction in part (a)?

Mole Relationships in Chemical Equations

Until now, you have assumed that the coefficients in a chemical equation represent particles. They can, however, also represent moles. Consider the following ratio to find out why.

$$1 \text{ molecule } N_2 : 3 \text{ molecules } H_2 : 2 \text{ molecules } NH_3$$

You can multiply the above ratio by the Avogadro constant to obtain

$$1 \times N_A \text{ molecules } N_2 : 3 \times N_A \text{ molecules } H_2 : 2 \times N_A \text{ molecules } NH_3$$

This is the same as

$$1 \text{ mol } N_2 : 3 \text{ mol } H_2 : 2 \text{ mol } NH_3$$

So the chemical equation $N_{2(g)} + 3H_{2(g)} \rightarrow 2NH_{3(g)}$ also means that 1 mol of nitrogen molecules reacts with 3 mol of hydrogen molecules to form 2 mol of ammonia molecules. The relationships between moles in a balanced chemical equation are called **mole ratios**. For example, the mole ratio of nitrogen to hydrogen in the equation above is 1 mol N_2 :3 mol H_2. The mole ratio of hydrogen to ammonia is 3 mol H_2 :2 mol NH_3.

You can manipulate mole ratios in the same way that you can manipulate ratios involving molecules. For example, suppose that you want to know how many moles of ammonia are produced by 2.8 mol of hydrogen. You know that you can obtain 2 mol of ammonia for every 3 mol of hydrogen. Therefore, you multiply the number of moles of hydrogen by the mole ratio of ammonia to hydrogen. Another way to think about this is to equate the known mole ratio of hydrogen to ammonia to the unknown mole ratio of hydrogen to ammonia and solve for the unknown.

unknown ratio **known ratio**

$$\frac{n \text{ mol NH}_3}{2.8 \text{ mol H}_2} = \frac{2 \text{ mol NH}_3}{3 \text{ mol H}_2}$$

$$(2.8 \cancel{\text{mol H}_2}) \frac{n \text{ mol NH}_3}{\cancel{2.8 \text{ mol H}_2}} = (2.8 \cancel{\text{mol H}_2}) \frac{2 \text{ mol NH}_3}{3 \cancel{\text{mol H}_2}}$$

$$n \text{ mol NH}_3 = 1.9 \text{ mol NH}_3$$

Try the following Practice Problems to work with mole ratios.

CHEM FACT

Because the coefficients of a balanced chemical equation can represent moles, it is acceptable to use fractions in an equation. For example, you can write the equation

$2H_{2(g)} + O_{2(g)} \rightarrow 2H_2O_{(\ell)}$

as

$H_{2(g)} + \frac{1}{2}O_{2(g)} \rightarrow H_2O_{(\ell)}$

Half an oxygen molecule is an oxygen atom, which does not accurately reflect the reaction. Half a mole of oxygen molecules, however, makes sense.

Practice Problems

4. Aluminum bromide can be prepared by reacting small pieces of aluminum foil with liquid bromine at room temperature. The reaction is accompanied by flashes of red light.

$$2Al_{(s)} + 3Br_{2(\ell)} \rightarrow 2AlBr_{3(s)}$$

How many moles of Br_2 are needed to produce 5 mol of $AlBr_3$, if sufficient Al is present?

5. Hydrogen cyanide gas, $HCN_{(g)}$, is used to prepare clear, hard plastics, such as Plexiglas™. Hydrogen cyanide is formed by reacting ammonia, NH_3, with oxygen and methane, CH_4.

$$2NH_{3(g)} + 3O_{2(g)} + 2CH_{4(g)} \rightarrow 2HCN_{(g)} + 6H_2O_{(g)}$$

(a) How many moles of O_2 are needed to react with 1.2 mol of NH_3?

(b) How many moles of H_2O can be expected from the reaction of 12.5 mol of CH_4? Assume that sufficient NH_3 and O_2 are present.

6. Ethane gas, C_2H_6, is present in small amounts in natural gas. It undergoes complete combustion to produce carbon dioxide and water.

$$2C_2H_{6(g)} + 7O_{2(g)} \rightarrow 4CO_{2(g)} + 6H_2O_{(g)}$$

(a) How many moles of O_2 are required to react with 13.9 mol of C_2H_6?

(b) How many moles of H_2O would be produced by 1.40 mol of O_2 and sufficient ethane?

7. Magnesium nitride reacts with water to produce magnesium hydroxide and ammonia gas, NH_3 according to the balanced chemical equation

$$Mg_3N_{2(s)} + 6H_2O_{(\ell)} \rightarrow 3Mg(OH)_{2(s)} + 2NH_{3(g)}$$

a) How many molecules of water are required to react with 2.3 mol Mg_3N_2?

b) How many molecules of $Mg(OH)_2$ will be expected in part (a)?

Different Ratios of Reactants

The relative amounts of reactants are important. Different mole ratios of the same reactants can produce different products. For example, carbon can combine with oxygen in two different ratios, forming either carbon monoxide or carbon dioxide. In the following reaction, the mole ratio of carbon to oxygen is 2 mol C:1 mol O_2.

$$2C_{(s)} + O_{2(g)} \rightarrow 2CO_{(g)}$$

In the next reaction, the mole ratio of carbon to oxygen is 1 mol C:1 mol O_2.

$$C_{(s)} + O_{2(g)} \rightarrow CO_{2(g)}$$

Thus, carbon dioxide forms if carbon and oxygen are present in a mole ratio of about 1 mol C:1 mol O_2. Carbon dioxide is a product of cellular respiration in animals and humans, and it is a starting material for photosynthesis. It is also one of the products of the complete combustion of a hydrocarbon fuel.

If there is a relative shortage of oxygen, however, and the mole ratio of carbon to oxygen is closer to 2 mol C:1 mol O, carbon monoxide forms. Carbon monoxide is colourless, tasteless, and odourless. It is a highly poisonous gas, that is responsible for the deaths of hundreds of people in Canada and the United States every year. Carbon monoxide can escape from any fuel-burning appliance: furnace, water heater, fireplace, wood stove, or space heater. If you have one of these appliances in your home, make sure that it has a good supply of oxygen to avoid the formation of carbon monoxide.

There are many reactions in which different mole ratios of the reactants result in different products. The following Sample Problem will help you understand how to work with these reactions.

Technology LINK

In many areas, it is mandatory for every home to have a carbon monoxide detector, like the one shown below. If you do not have a carbon monoxide detector in your home, you can buy one at a hardware store for a modest price. It could end up saving your life.

A carbon monoxide detector emits a sound when the level of carbon monoxide exceeds a certain limit. Find out how a carbon monoxide detector works, and where it should be placed. Present your findings as a public service announcement.

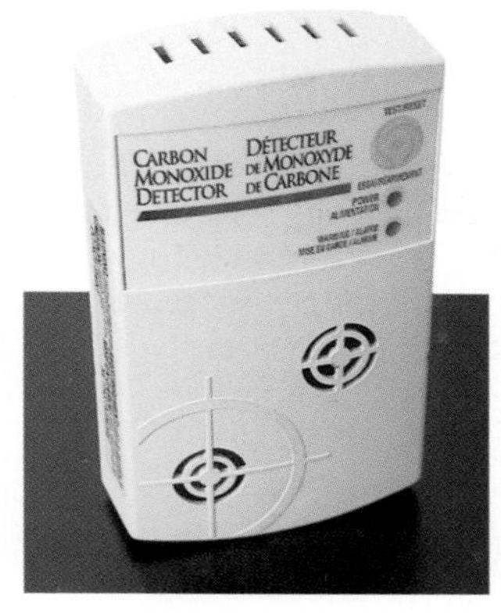

Sample Problem

Mole Ratios of Reactants

Problem

Vanadium can form several different compounds with oxygen, including V_2O_5, VO_2, and V_2O_3. Determine the number of moles of oxygen that are needed to react with 0.56 mol of vanadium to form divanadium pentoxide, V_2O_5.

What Is Required?

You need to find the number of moles of oxygen that are needed to react with 0.56 mol of vanadium to form divanadium pentoxide.

What Is Given?

Reactant: vanadium, V $\rightarrow$ 0.56 mol
Reactant: oxygen, O_2
Product: divanadium pentoxide, V_2O_5

Continued ...

Continued ...

FROM PAGE 239

Plan Your Strategy

Write a balanced chemical equation for the formation of vanadium(V) oxide. Use the known mole ratio of vanadium to oxygen to calculate the unknown amount of oxygen.

Act on Your Strategy

The balanced equation is $4V_{(s)} + 5O_{2(g)} \rightarrow 2V_2O_{5(s)}$

To determine the number of moles of oxygen required, equate the known ratio of oxygen to vanadium from the balanced equation to the unknown ratio from the question.

$$\underset{\textbf{unknown ratio}}{\frac{n \text{ mol } O_2}{0.56 \text{ mol V}}} = \underset{\textbf{known ratio}}{\frac{5 \text{ mol } O_2}{4 \text{ mol V}}}$$

Multiply both sides of the equation by 0.56 mol V.

$$(\cancel{0.56 \text{ mol V}}) \frac{n \text{ mol } O_2}{\cancel{0.56 \text{ mol V}}} = (0.56 \text{ mol V}) \frac{5 \text{ mol } O_2}{4 \text{ mol V}}$$

$$n \text{ mol } O_2 = (0.56 \cancel{\text{ mol V}}) \frac{5 \text{ mol } O_2}{4 \cancel{\text{ mol V}}}$$

$$= 0.70 \text{ mol } O_2$$

Check Your Solution

The units are correct. The mole ratio of vanadium to oxygen is 4 mol V:5 mol O_2. Multiply 0.70 mol by 4/5, and you get 0.56 mol. The answer is therefore reasonable.

Practice Problems

8. Refer to the Sample Problem above.

(a) How many moles of V are needed to produce 7.47 mol of VO_2? Assume that sufficient O_2 is present.

(b) How many moles of V are needed to react with 5.39 mol of O_2 to produce V_2O_3?

9. Nitrogen, N_2, can combine with oxygen, O_2, to form several different oxides of nitrogen. These oxides include NO_2, NO, and N_2O.

(a) How many moles of O_2 are required to react with 9.35×10^{-2} moles of N_2 to form N_2O?

(b) How many moles of O_2 are required to react with 9.35×10^{-2} moles of N_2 to form NO_2?

10. When heated in a nickel vessel to 400°C, xenon can be made to react with fluorine to produce colourless crystals of xenon tetrafluoride.

a) How many moles of fluorine gas, F_2, would be required to react with 3.54×10^{-1} mol of xenon?

b) Under somewhat similar reaction conditions, xenon hexafluoride can also be obtained. How many moles of fluorine would be required to react with the amount of xenon given in part (a) to produce xenon hexafluoride?

CHECKPOINT

Do you think that xenon could be made to react with bromine or iodine under the same conditions outlined in Practice Problem 10? Explain why or why not, using your understanding of periodic trends.

Mass Relationships in Chemical Equations

As you have learned, the coefficients in a balanced chemical equation represent moles as well as particles. Therefore, you can use the molar masses of reactants and products to determine the mass ratios for a reaction. For example, consider the equation for the formation of ammonia:

$$N_{2(g)} + 3H_{2(g)} \rightarrow 2NH_{3(g)}$$

You can find the mass of each substance using the equation $m = M \times n$ as follows:

$$1 \text{ mol } N_2 \times 28.0 \text{ g/mol } N_2 = 28.0 \text{ g } N_2$$
$$3 \text{ mol } H_2 \times 2.02 \text{ g/mol } H_2 = 6.1 \text{ g } H_2$$
$$2 \text{ mol } NH_3 \times 17.0 \text{ g/mol } NH_3 = 34.1 \text{ g } NH_3$$

CHECKPOINT

Refer back to Chapter 5. Calculate the molar mass of N_2, H_2, and NH_3.

In Table 7.1, you can see how particles, moles, and mass are related in a chemical equation. Notice that the mass of the product is equal to the total mass of the reactants. This confirms the law of conservation of mass.

Table 7.1 What a Balanced Chemical Equation Tells You

Balanced equation	$N_{2(g)} + 3H_{2(g)}$	→	$2NH_{3(g)}$
Number of particles (molecules)	1 molecule N_2 + 3 molecules H_2	→	2 molecules NH_3
Amount (mol)	1 mol N_2 + 3 mol H_2	→	2 mol NH_3
Mass (g)	28.0 g N_2 + 6.1 g H_2	→	34.1 g NH_3
Total mass (g)	34.1 g reactants	→	34.1 g product

Stoichiometric Mass Calculations

You now know what a balanced chemical equation tells you in terms of number of particles, number of moles, and mass of products and reactants. How do you use this information? Because reactants and products are related by a fixed ratio, if you know the number of moles of one substance, the balanced equation tells you the number of moles of all the other substances. In Chapters 5 and 6, you learned how to convert between particles, moles, and mass. Therefore, *if you know the amount of one substance in a chemical reaction (in particles, moles, or mass), you can calculate the amount of any other substance in the reaction (in particles, moles, or mass), using the information in the balanced chemical equation.*

You can see that a balanced chemical equation is a powerful tool. It allows chemists to predict the amount of products that will result from a reaction involving a known amount of reactants. As well, chemists can use a balanced equation to calculate the amount of reactants they will need to produce a desired amount of products. They can also use it to predict the amount of one reactant they will need to react completely with another reactant.

Language LINK

The word "stoichiometry" is derived from two Greek words: *stoikheion*, meaning "element," and *metron*, meaning "to measure." What other words might be derived from the Greek word *metron*?

Stoichiometry is the study of the relative quantities of reactants and products in chemical reactions. Stoichiometric calculations are used for many purposes. One purpose is determining how much of a reactant is needed to carry out a reaction. This kind of knowledge is useful for any chemical reaction, and it can even be a matter of life or death.

In a spacecraft, for example, carbon dioxide is produced as the astronauts breathe. To maintain a low level of carbon dioxide, air in the cabin is passed continuously through canisters of lithium hydroxide granules. The carbon dioxide reacts with the lithium hydroxide in the following way:

$$CO_{2(g)} + 2LiOH_{(s)} \rightarrow Li_2CO_{3(s)} + H_2O_{(g)}$$

The canisters are changed periodically as the lithium hydroxide reacts. Engineers must calculate the amount of lithium hydroxide needed to ensure that the carbon dioxide level is safe. As you learned earlier, every kilogram counts in space travel. Therefore, a spacecraft cannot carry much more than the minimum amount.

History LINK

The concept of stoichiometry was first described in 1792 by the German scientist Jeremias Benjamin Richter (1762–1807). He stated that "stoichiometry is the science of measuring the quantitative proportions or mass ratios in which chemical elements stand to one another." Can you think of another reason why Richter was famous?

Figure 7.4 **A spacecraft is a closed system. All chemical reactions must be taken into account when engineers design systems to keep the air breathable.**

To determine how much lithium hydroxide is needed, engineers need to ask and answer two important questions:

- How much carbon dioxide is produced per astronaut each day?
- How much lithium hydroxide is needed per kilogram of carbon dioxide?

Engineers can answer the first question by experimenting. To answer the second question, they need to do stoichiometric calculations. Examine the following Sample Problems to see how these calculations would be done.

Sample Problem

Mass to Mass Calculations for Reactants

Problem

Carbon dioxide that is produced by astronauts can be removed with lithium hydroxide. The reaction produces lithium carbonate and water. An astronaut produces an average of 1.00×10^3 g of carbon dioxide each day. What mass of lithium hydroxide should engineers put on board a spacecraft, per astronaut, for each day?

What Is Required?

You need to find the mass of lithium that is needed to react with 1.00×10^3 g of carbon dioxide.

What Is Given?

Reactant: carbon dioxide, $CO_2 \rightarrow 1.00 \times 10^3$ g
Reactant: lithium hydroxide, LiOH
Product: lithium carbonate, Li_2CO_3
Product: water, H_2O

Plan Your Strategy

Step 1 Write a balanced chemical equation.

Step 2 Convert the given mass of carbon dioxide to the number of moles of carbon dioxide.

Step 3 Calculate the number of moles of lithium hydroxide based on the mole ratio of lithium hydroxide to carbon dioxide.

Step 4 Convert the number of moles of lithium hydroxide to grams.

Act on Your Strategy

The balanced chemical equation is

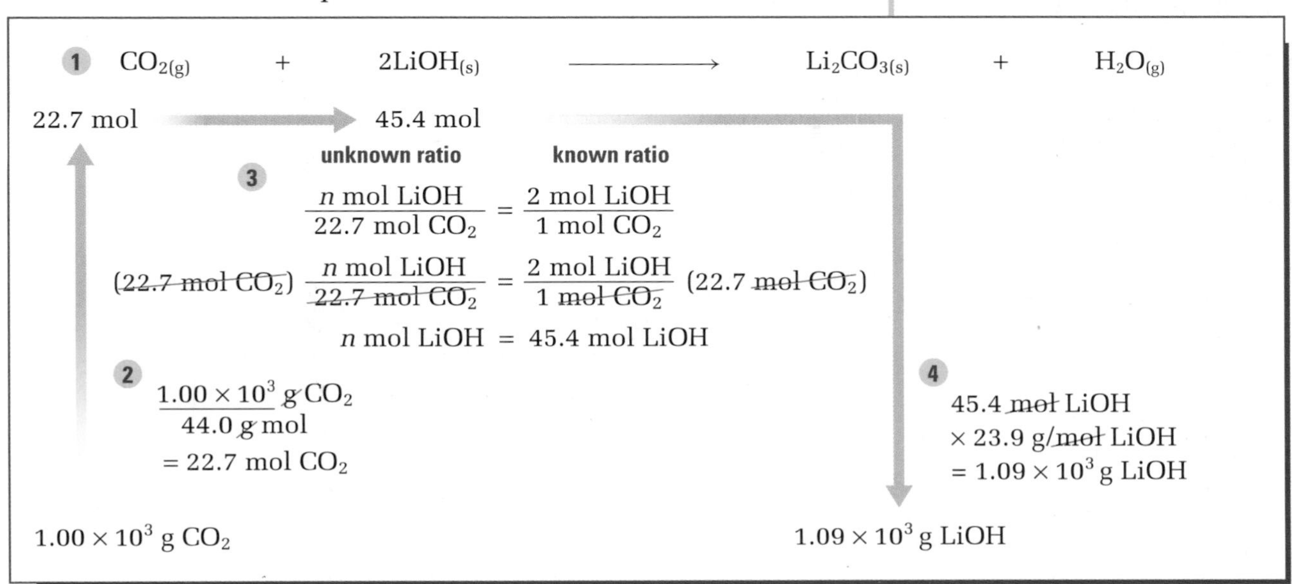

Continued ...

The Group 18 elements in the periodic table are currently called the noble gases. In the past, however, they were referred to as the inert gases. They were believed to be totally unreactive. Scientists have found that this is not true. Some of them can be made to react with reactive elements, such as fluorine, under the proper conditions. In 1962, the synthesis of the first compound that contained a noble gas was reported. Since then, a number of noble gas compounds have been prepared, mostly from xenon. A few compounds of krypton, radon, and argon have also been prepared.

In the early 1960s, Neil Bartlett, of the University of British Columbia, synthesized the first compound that contained a noble gas.

Continued ...
FROM PAGE 243

Therefore, 1.09×10^3 g LiOH are required.

Check Your Solution

The units are correct. Lithium hydroxide has a molar mass that is about half of carbon dioxide's molar mass, but there are twice as many moles of lithium hydroxide. Therefore it makes sense that the mass of lithium hydroxide required is about the same as the mass of carbon dioxide produced.

Practice Problems

11. Ammonium sulfate, $(NH_4)_2SO_4$, is used as a source of nitrogen in some fertilizers. It reacts with sodium hydroxide to produce sodium sulfate, water and ammonia.

$$(NH_4)_2SO_{4(s)} + 2NaOH_{(aq)} \rightarrow Na_2SO_{4(aq)} + 2NH_{3(g)} + 2H_2O_{(\ell)}$$

What mass of sodium hydroxide is required to react completely with 15.4 g of $(NH_4)_2SO_4$?

12. Iron(III) oxide, also known as rust, can be removed from iron by reacting it with hydrochloric acid to produce iron(III) chloride and water.

$$Fe_2O_{3(s)} + 6HCl_{(aq)} \rightarrow 2FeCl_{3(aq)} + 3H_2O_{(\ell)}$$

What mass of hydrogen chloride is required to react with 1.00×10^2 g of rust?

13. Iron reacts slowly with hydrochloric acid to produce iron(II) chloride and hydrogen gas.

$$Fe_{(s)} + 2HCl_{(aq)} \rightarrow FeCl_{2(aq)} + H_{2(g)}$$

What mass of HCl is required to react with 3.56 g of iron?

14. Dinitrogen pentoxide is a white solid. When heated it decomposes to produce nitrogen dioxide and oxygen.

$$N_2O_{5(s)} \rightarrow 2NO_{2(g)} + O_{2(g)}$$

How many grams of oxygen gas will be produced in this reaction when 2.34 g of NO_2 are made?

Sample Problem

Mass to Mass Calculations for Products and Reactants

Problem

In the Chapter 7 opener, you learned that a fuel mixture consisting of hydrazine, N_2H_4, and dinitrogen tetroxide, N_2O_4, was used to launch a lunar module. These two compounds react to form nitrogen gas and water vapour. If 50.0 g of hydrazine reacts with sufficient dinitrogen tetroxide, what mass of nitrogen gas is formed?

Continued ...

Continued ...
FROM PAGE 244

What Is Required?

You need to find the mass of nitrogen gas that is formed from 50.0 g of hydrazine.

What Is Given?

Reactant: hydrazine, $N_2H_4 \rightarrow$ 150.0 g
Reactant: dinitrogen tetroxide, N_2O_4
Product: nitrogen, N_2
Product: water, H_2O

Plan Your Strategy

Step 1 Write a balanced chemical equation.

Step 2 Convert the mass of hydrazine to the number of moles of hydrazine.

Step 3 Calculate the number of moles of nitrogen, using the mole ratio of hydrazine to nitrogen.

Step 4 Convert the number of moles of nitrogen to grams.

Act on Your Strategy

The balanced chemical equation is

1 $2N_2H_{4(\ell)} + N_2O_{4(\ell)} \longrightarrow 3N_{2(g)} + 4H_2O_{(g)}$

4.679 mol → 7.019 mol

3 unknown ratio = known ratio

$$\frac{n \text{ mol } N_2}{4.680 \text{ mol } N_2H_4} = \frac{3 \text{ mol } N_2}{2 \text{ mol } N_2H_4}$$

$$(4.679 \cancel{\text{mol } N_2H_4}) \frac{n \text{ mol } N_2}{4.679 \cancel{\text{mol } N_2H_4}} = \frac{3 \text{ mol } N_2}{2 \cancel{\text{mol } N_2H_4}} (4.679 \cancel{\text{mol } N_2H_4})$$

$$= 7.019 \text{ mol } N_2$$

2

$$\frac{150.0 \cancel{\text{g}} N_2H_4}{32.06 \cancel{\text{g}}/\text{mol}} = 4.6 \text{ mol}$$

4

$7.019 \cancel{\text{mol } N_2}$
$\times 28.01 \text{ g}/\cancel{\text{mol } N_2}$
$= 196.6$ g

150.0 g N_2H_4 → 196.6 g N_2

Therefore, 196.6 g of nitrogen are formed.

Check Your Solution

The units are correct. Nitrogen has a molar mass that is close to hydrazine's molar mass. Therefore, to estimate the amount of nitrogen from the mass of hydrazine, multiply the mole ratio of nitrogen to hydrazine (3:2) by hydrazine's mass (150 g) to get 225 g, which is close to the calculated answer, 196.6 g. The answer is reasonable.

Electronic Learning Partner

Go to your Chemistry 11 Electronic Learning Partner for a video clip showing an experiment that uses stoichiometry.

Continued ...
FROM PAGE 245

Practice Problems

15. Powdered zinc reacts rapidly with powdered sulfur in a highly exothermic reaction.

$$8Zn_{(s)} + S_{8(s)} \rightarrow 8ZnS_{(s)}$$

What mass of zinc sulfide is expected when 32.0 g of S_8 reacts with sufficient zinc?

16. The addition of concentrated hydrochloric acid to manganese(IV) oxide leads to the production of chlorine gas.

$$4HCl_{(aq)} + MnO_{2(g)} \rightarrow MnCl_{2(aq)} + Cl_{2(g)} + 2H_2O_{(\ell)}$$

What mass of chlorine can be obtained when 4.76×10^{-2} g of HCl react with sufficient MnO_2?

17. Aluminum carbide, Al_4C_3, is a yellow powder that reacts with water to produce aluminum hydroxide and methane.

$$Al_4C_{3(s)} + 12H_2O_{(\ell)} \rightarrow 4Al(OH)_{3(s)} + 3CH_{4(g)}$$

What mass of water is required to react completely with 25.0 g of aluminum carbide?

18. Magnesium oxide reacts with phosphoric acid, H_3PO_4, to produce magnesium phosphate and water.

$$3MgO_{(s)} + 2H_3PO_{4(aq)} \rightarrow Mg_3(PO_4)_{2(s)} + 3H_2O_{(\ell)}$$

How many grams of magnesium oxide are required to react completely with 33.5 g of phosphoric acid?

Canadians  in Chemistry

As a chemist with Environment Canada's Atmospheric Science Division in Dartmouth, Nova Scotia, Dr. Stephen Beauchamp studies toxic chemicals, such as mercury. Loons in Nova Scotia's Kejimkujik National Park are among the living creatures that he studies. Kejimkujik loons have higher blood mercury levels (5 µg Hg/1 g blood) than any other North American loons (2 µg Hg/1 g blood). Mercury is also found in high levels in the fish the loons eat. Mercury causes behavioural problems in the loons. As well, it may affect the loons' reproductive success and immune function.

Bacteria convert environmental mercury into methyl mercury, CH_3Hg. This is the form that is most easily absorbed into living organisms. Beauchamp examines forms and concentrations of mercury in the air, soil, and water.

Mercury emission sources include electrical power generation, manufacturing, and municipal waste incineration. Sources such as these, however, do not totally account for the high mercury levels found in Kejimkujik loons and other area wildlife. Beauchamp is working to discover what other factors are operating so that he will be able to recommend ways to improve the situation.

Dr. Stephen Beauchamp in Halifax Harbour. The flux chamber beside him helps him measure the changing concentrations of mercury in the air and water.

A General Process for Solving Stoichiometric Problems

You have just solved several stoichiometric problems. In these problems, masses of products and reactants were given, and masses were also required for the answers. Chemists usually need to know what mass of reactants they require and what mass of products they can expect. Sometimes, however, a question requires you to work with the number of moles or particles. Use the same process for solving stoichiometric problems, whether you are working with mass, moles, or particles:

Step 1 Write a balanced chemical equation.

Step 2 If you are given the mass or number of particles of a substance, convert it to the number of moles.

Step 3 Calculate the number of moles of the required substance based on the number of moles of the given substance, using the appropriate mole ratio.

Step 4 Convert the number of moles of the required substance to mass or number of particles, as directed by the question.

Examine the following Sample Problem to see how to work with mass and particles.

Sample Problem

Mass and Particle Stoichiometry

Problem

Passing chlorine gas through molten sulfur produces liquid disulfur dichloride. How many molecules of chlorine react to produce 50.0 g of disulfur dichloride?

What Is Required?

You need to determine the number of molecules of chlorine gas that produce 50.0 g of disulfur dichloride.

What Is Given?

Reactant: chlorine, Cl_2
Reactant: sulfur, S
Product: disulfur dichloride, $S_2Cl_2 \rightarrow$ 50.0 g

Plan Your Strategy

Step 1 Write a balanced chemical equation.

Step 2 Convert the given mass of disulfur dichloride to the number of moles.

Step 3 Calculate the number of moles of chlorine gas using the mole ratio of chlorine to disulfur dichloride.

Step 4 Convert the number of moles of chlorine gas to the number of particles of chlorine gas.

Continued ...

Continued ...
FROM PAGE 247

Act on Your Strategy

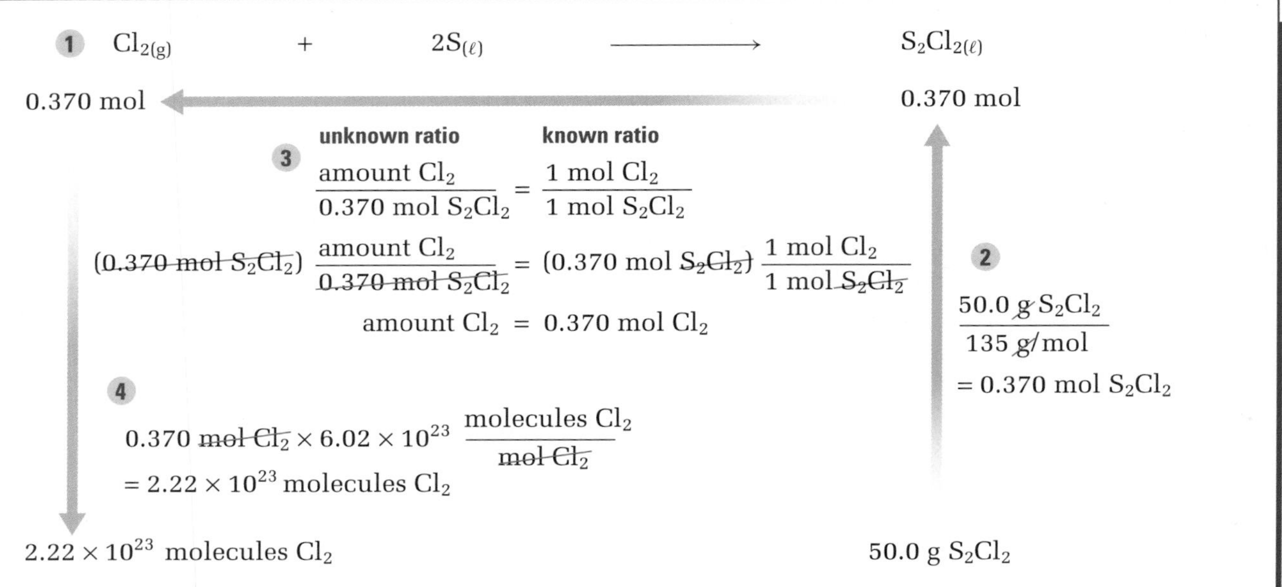

1 $Cl_{2(g)}$ + $2S_{(\ell)}$ → $S_2Cl_{2(\ell)}$
0.370 mol
0.370 mol
unknown ratio
known ratio
3 $\frac{\text{amount } Cl_2}{0.370 \text{ mol } S_2Cl_2} = \frac{1 \text{ mol } Cl_2}{1 \text{ mol } S_2Cl_2}$
$(0.370 \text{ mol } S_2Cl_2)\frac{\text{amount } Cl_2}{0.370 \text{ mol } S_2Cl_2} = (0.370 \text{ mol } S_2Cl_2)\frac{1 \text{ mol } Cl_2}{1 \text{ mol } S_2Cl_2}$
amount Cl_2 = 0.370 mol Cl_2
2 $\frac{50.0 \text{ g } S_2Cl_2}{135 \text{ g/mol}}$
$= 0.370 \text{ mol } S_2Cl_2$
4 $0.370 \text{ mol } Cl_2 \times 6.02 \times 10^{23} \frac{\text{molecules } Cl_2}{\text{mol } Cl_2}$
$= 2.22 \times 10^{23}$ molecules Cl_2
2.22×10^{23} molecules Cl_2
50.0 g S_2Cl_2

Therefore, 2.22×10^{23} molecules of chlorine gas are required.

Check Your Solution

The units are correct. 2.0×10^{23} is about 1/3 of a mole, or 0.33 mol. One-third of a mole of disulfur dichloride has a mass of 45 g, which is close to 50 g. The answer is reasonable.

The thermite reaction generates enough heat to melt the elemental iron that is produced.

Practice Problems

19. Nitrogen gas is produced in an automobile air bag. It is generated by the decomposition of sodium azide, NaN_3.

$$2NaN_{3(s)} \rightarrow 3N_{2(g)} + 2Na_{(s)}$$

(a) To inflate the air bag on the driver's side of a certain car, 80.0 g of N_2 is required. What mass of NaN_3 is needed to produce 80.0 g of N_2?

(b) How many atoms of Na are produced when 80.0 g of N_2 are generated in this reaction?

20. The reaction of iron(III) oxide with powdered aluminum is known as the thermite reaction.

$$2Al_{(s)} + Fe_2O_{3(s)} \rightarrow Al_2O_{3(s)} + 2Fe_{(\ell)}$$

(a) Calculate the mass of aluminum oxide, Al_2O_3, that is produced when 1.42×10^{24} atoms of Al react with Fe_2O_3.

(b) How many molecules of Fe_2O_3 are needed to react with 0.134 g of Al?

Continued ...

Continued ...

FROM PAGE 248

21. The thermal decomposition of ammonium dichromate is an impressive reaction. When heated with a Bunsen burner or propane torch, the orange crystals of ammonium dichromate slowly decompose to green chromium(III) oxide in a volcano-like display. Colourless nitrogen gas and water vapour are also given off.

$$(NH_4)_2Cr_2O_{7(s)} \rightarrow Cr_2O_{3(s)} + N_{2(g)} + 4H_2O_{(g)}$$

(a) Calculate the number of molecules of Cr_2O_3 that is produced from the decomposition of 10.0 g of $(NH_4)_2Cr_2O_7$.

(b) In a different reaction, 16.9 g of N_2 is produced when a sample of $(NH_4)_2Cr_2O_7$ is decomposed. How many water molecules are also produced in this reaction?

(c) How many molecules of $(NH_4)_2Cr_2O_7$ are needed to produce 1.45 g of H_2O?

22. Ammonia gas reacts with oxygen to produce water and nitrogen oxide. This reaction can be catalyzed, or sped up, by Cr_2O_3, produced in the reaction in problem 21.

$$4NH_{3(g)} + 5O_{2(g)} \rightarrow 4NO_{(g)} + 6H_2O_{(\ell)}$$

(a) How many molecules of oxygen are required to react with 34.0 g of ammonia?

(b) What mass of nitrogen oxide is expected from the reaction of 8.95×10^{24} molecules of oxygen with sufficient ammonia?

Section Wrap-up

You have learned how to do stoichiometric calculations, using balanced chemical equations to find amounts of reactants and products. In these calculations, you assumed that the reactants and products occurred in the exact molar ratios shown by the chemical equation. In real life, however, reactants are often not present in these exact ratios. Similarly, the amount of product that is predicted by stoichiometry is not always produced. In the next two sections, you will learn how chemists deal with these challenges.

Section Review

1 K/U Why is a balanced chemical equation needed to solve stoichiometric calculations?

2 K/U The balanced chemical equation for the formation of water from its elements is sometimes written as

$$H_{2(g)} + \frac{1}{2}O_{2(g)} \rightarrow H_2O_{(\ell)}$$

Explain why it is acceptable to use fractional coefficients in a balanced chemical equation.

Unit Investigation Prep

Before you design your quantitative analysis investigation at the end of Unit 2, decide how you will make use of the concepts you learned in this section. Assume that you know the identity of reactants and you know what products will be formed in the reaction. If you can measure how much product is formed in the reaction, can you determine how much reactant was initially present? Explain how, using an example.

3 C In the following reaction, does 1.0 g of sodium react completely with 0.50 g of chlorine? Explain your answer.

$$Na_{(s)} + \frac{1}{2}Cl_{2(g)} \rightarrow NaCl$$

4 K/U Sulfur and oxygen can combine to form sulfur dioxide, SO_2, and sulfur trioxide, SO_3.

(a) Write a balanced chemical equation for the formation of SO_2 from S and O_2.

(b) Write a balanced chemical equation for the formation of SO_3.

(c) How many moles of O_2 must react with 1 mol of S to form 1 mol of SO_3?

(d) What mass of O_2 is needed to react with 32.1 g of S to form SO_3?

5 K/U The balanced chemical equation for the combustion of propane is

$$C_3H_{8(g)} + 5O_{2(g)} \rightarrow 3CO_{2(g)} + 4H_2O_{(g)}$$

(a) Write the mole ratios for the reactants and products in the combustion of propane.

(b) How many moles of O_2 are needed to react with 0.500 mol of C_3H_8?

(c) How many molecules of O_2 are needed to react with 2.00 mol of C_3H_8?

(d) If 3.00 mol of C_3H_8 burn completely in O_2, how many moles of CO_2 are produced?

6 I Phosphorus pentachloride, PCl_5, reacts with water to form phosphoric acid, H_3PO_4, and hydrochloric acid, HCl.

$$PCl_{5(s)} + 4H_2O_{(\ell)} \rightarrow H_3PO_{4(aq)} + 5HCl_{(aq)}$$

(a) What mass of PCl_5 is needed to react with an excess quantity of H_2O to produce 23.5 g of H_3PO_4?

(b) How many molecules of H_2O are needed to react with 3.87 g of PCl_5?

7 I A chemist has a beaker containing lead nitrate, $Pb(NO_3)_2$, dissolved in water. The chemist adds a solution containing sodium iodide, NaI, and a bright yellow precipitate is formed. The chemist continues to add NaI until no further yellow precipitate is formed. The chemist filters the precipitate, dries it in an oven, and finds it has a mass of 1.43 g.

(a) Write a balanced chemical equation to describe what happened in this experiment. Hint: compounds with sodium ions are always soluble.

(b) Use the balanced chemical equation to determine what mass of lead nitrate, $Pb(NO_3)_2$, was dissolved in the water in the beaker.

8 MC The Apollo-13 mission overcame an astonishing number of difficulties on its return to Earth. One problem the astronauts encountered was removing carbon dioxide from the air they were breathing. Do some research to find out:

(a) What happened to lead to an unexpected accumulation of carbon dioxide?

(b) What did the astronauts do to overcome this difficulty?

The Limiting Reactant

7.2

A balanced chemical equation shows the mole ratios of the reactants and products. To emphasize this, the coefficients of equations are sometimes called **stoichiometric coefficients**. Reactants are said to be present in **stoichiometric amounts** when they are present in a mole ratio that corresponds exactly to the mole ratio predicted by the balanced chemical equation. This means that when a reaction is complete, there are no reactants left. In practice, however, there often *are* reactants left.

In the previous section, you looked at an "equation" for making a salad. You looked at situations in which you had the right amounts of ingredients to make one or more salads, with no leftover ingredients.

1 head of lettuce + 2 cucumbers + 5 radishes → 1 salad

What if you have two heads of lettuce, 12 cucumbers, and 25 radishes, as in Figure 7.5? How many salads can you make? Because each salad requires two heads of lettuce, you can make only two salads. Here the amount of lettuce limits the number of salads you can make. Some of the other two ingredients are left over.

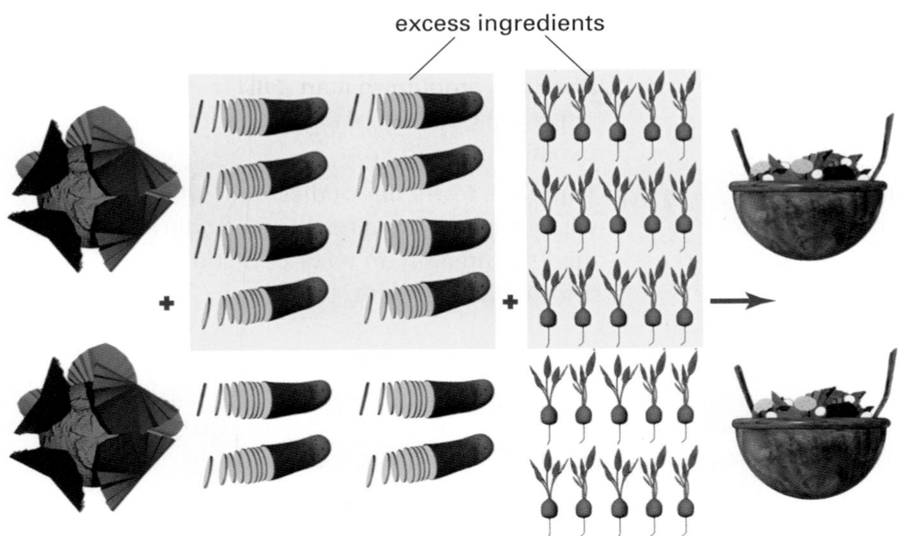

Figure 7.5 **Which ingredient limits how many salads can be made?**

Chemical reactions often work in the same way. For example, consider the first step in extracting zinc from zinc oxide:

$$ZnO_{(s)} + C_{(s)} \rightarrow Zn_{(s)} + CO_{(g)}$$

If you were carrying out this reaction in a laboratory, you could obtain samples of zinc oxide and carbon in a 1:1 mole ratio. In an industrial setting, however, it is impractical to spend time and money ensuring that zinc oxide and carbon are present in stoichiometric amounts. It is also unnecessary. In an industrial setting, engineers add more carbon, in the form of charcoal, than is necessary for the reaction. All the zinc oxide reacts, but there is carbon left over.

Section Preview/ Specific Expectations

In this section, you will

- **calculate,** for any given reactant or product in a chemical equation, the corresponding mass or quantity (in moles or molecules) of any other reactant or product
- **perform** an investigation to determine the limiting reactant in a chemical reaction
- **assess** the importance of determining the limiting reactant
- **solve** problems involving percentage yield and limiting reactants
- **communicate** your understanding of the following terms: *stoichiometric coefficients, stoichiometric amounts, limiting reactant, excess reactant*

Figure 7.6 All the gasoline in this car's tank has reacted. Thus, even though there is still oxygen available in the air, the combustion reaction cannot proceed.

Having one or more reactants in excess is very common. Another example is seen in gasoline-powered vehicles. Their operation depends on the reaction between fuel and oxygen. Normally, the fuel-injection system regulates how much air enters the combustion chamber, and oxygen is the limiting reactant. When the fuel is very low, however, fuel becomes the limiting reactant and the reaction cannot proceed, as in Figure 7.6.

In nature, reactions almost never have reactants in stoichiometric amounts. Think about respiration, represented by the following chemical equation:

$$C_6H_{12}O_{6(s)} + 6O_{2(g)} \rightarrow 6CO_{2(g)} + 6H_2O_{(\ell)}$$

When an animal carries out respiration, there is an unlimited amount of oxygen in the air. The amount of glucose, however, depends on how much food the animal has eaten.

ThoughtLab  The Limiting Item

Imagine that you are in the business of producing cars. A simplified "equation" for making a car is
1 car body + 4 wheels + 2 wiper blades → 1 car

Procedure

1. Assume that you have 35 car bodies, 120 wheels, and 150 wiper blades in your factory. How many complete cars can you make?
2. (a) Which item "limits" the number of complete cars that you can make? Stated another way, which item will "run out" first?
   (b) Which items are present in excess amounts?
   (c) How much of each "excess" item remains after the "reaction"?

Analysis

1. Does the *amount* that an item is in excess affect the quantity of the product that is made? Explain.
2. There are fewer car bodies than wheels and wiper blades. Explain why car bodies are not the limiting item, in spite of being present in the smallest amount.

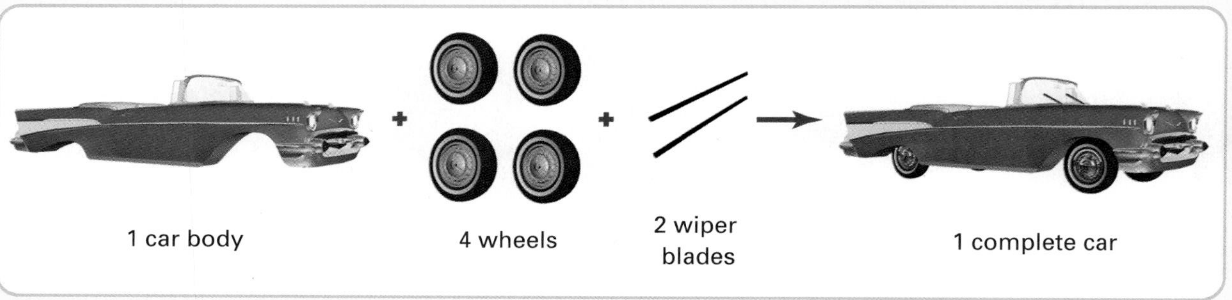

Determining the Limiting Reactant

The reactant that is completely used up in a chemical reaction is called the **limiting reactant**. In other words, the limiting reactant determines how much product is produced. When the limiting reactant is used up, the reaction stops. In real-life situations, there is almost always a limiting reactant.

A reactant that remains after a reaction is over is called the **excess reactant**. Once the limiting reactant is used, no more product can be made, regardless of how much of the excess reactants may be present.

When you are given amounts of two or more reactants to solve a stoichiometric problem, you first need to identify the limiting reactant. One way to do this is to find out how much product would be produced by each reactant if the other reactant were present in excess. The reactant that produces the least amount of product is the limiting reactant. Examine the following Sample Problem to see how to use this approach to identify the limiting reactant.

Sample Problem

Identifying the Limiting Reactant

Problem

Lithium nitride reacts with water to form ammonia and lithium hydroxide, according to the following balanced chemical equation:

$$Li_3N_{(s)} + 3H_2O_{(\ell)} \rightarrow NH_{3(g)} + 3LiOH_{(aq)}$$

If 4.87 g of lithium nitride reacts with 5.80 g of water, find the limiting reactant.

What Is Required?

You need to determine whether lithium nitride or water is the limiting reactant.

What Is Given?

Reactant: lithium nitride, $Li_3N \rightarrow 4.87$ g
Reactant: water, $H_2O \rightarrow 5.80$ g
Product: ammonia, NH_3
Product: lithium hydroxide, LiOH

Plan Your Strategy

Convert the given masses into moles. Use the mole ratios of reactants and products to determine how much ammonia is produced by each amount of reactant. The limiting reactant is the reactant that produces the smaller amount of product.

Act on Your Strategy

$$n \text{ mol } Li_3N = \frac{4.87 \cancel{\text{g}}\ Li_3N}{34.8 \cancel{\text{g}}/\text{mol}}$$
$$= 0.140 \text{ mol } Li_3N$$

$$n \text{ mol } H_2O = \frac{5.80 \cancel{\text{g}}\ H_2O}{18.0 \cancel{\text{g}}/\text{mol}}$$
$$= 0.322 \text{ mol } H_2O$$

Calculate the amount of NH_3 produced, based on the amount of Li_3N.

$$n \text{ mol of } NH_3 = \frac{1 \text{ mol } NH_3}{1 \cancel{\text{mol } Li_3N}} (0.140 \cancel{\text{mol } Li_3N})$$
$$= 0.140 \text{ mol } NH_3$$

Calculate the amount of NH_3 produced, based on the amount of H_2O.

Continued ...

PROBLEM TIP

To determine the limiting reactant, you can calculate how much of either ammonia or lithium hydroxide would be produced by the reactants. In this problem, ammonia was chosen because only one mole is produced, simplifying the calculation.

FROM PAGE 253

$$n \text{ mol } NH_3 = \frac{1 \text{ mol } NH_3}{3 \text{ mol } H_2O} \times (0.322 \text{ mol } H_2O)$$
$$= 0.107 \text{ mol } NH_3$$

The water would produce less ammonia than the lithium nitride. Therefore, the limiting reactant is water. Notice that there is more water than lithium nitride, in terms of mass and moles. Water is the limiting reactant, however, because 3 mol of water are needed to react with 1 mol of lithium nitride.

Check Your Solution

According to the balanced chemical equation, the ratio of lithium nitride to water is 1/3. The ratio of lithium nitride to water, based on the mole amounts calculated, is 0.14:0.32. Divide this ratio by 0.14 to get 1.0:2.3. For each mole of lithium nitride, there are only 2.3 mol water. However, 3 mol are required by stoichiometry. Therefore, water is the limiting reactant.

Practice Problems

23. The following balanced chemical equation shows the reaction of aluminum with copper(II) chloride. If 0.25 g of aluminum reacts with 0.51 g of copper(II) chloride, determine the limiting reactant.

$$2Al_{(s)} + 3CuCl_{2(aq)} \rightarrow 3Cu_{(s)} + 2AlCl_{3(aq)}$$

24. Hydrogen fluoride, HF, is a highly toxic gas. It is produced by the double displacement reaction of calcium fluoride, CaF_2, with concentrated sulfuric acid, H_2SO_4.

$$CaF_{2(s)} + H_2SO_{4(\ell)} \rightarrow 2HF_{(g)} + CaSO_{4(s)}$$

Determine the limiting reactant when 10.0 g of CaF_2 reacts with 15.5 g of H_2SO_4.

25. Acrylic, a common synthetic fibre, is formed from acrylonitrile, C_3H_3N. Acrylonitrile can be prepared by the reaction of propylene, C_3H_6, with nitric oxide, NO.

$$4C_3H_{6(g)} + 6NO_{(g)} \rightarrow 4C_3H_3N_{(g)} + 6H_2O_{(g)} + N_{2(g)}$$

What is the limiting reactant when 126 g of C_3H_6 reacts with 175 g of NO?

26. 3.76 g of zinc reacts with 8.93×10^{23} molecules of hydrogen chloride. Which reactant is present in excess?

You now know how to use a balanced chemical equation to find the limiting reactant. Can you find the limiting reactant by experimenting? You know that the limiting reactant is completely consumed in a reaction, while any reactants in excess remain after the reaction is finished. In Investigation 7-A, you will observe a reaction and identify the limiting reactant, based on your observations.

SKILL FOCUS

Predicting

Performing and recording

Analyzing and interpreting

Communicating results

Investigation 7-A

Limiting and Excess Reactants

In this investigation, you will predict and observe a limiting reactant. You will use the single replacement reaction of aluminum with aqueous copper(II) chloride:

$$2Al_{(s)} + 3CuCl_{2(aq)} \rightarrow 3Cu_{(s)} + 2AlCl_{3(aq)}$$

Note that copper(II) chloride, $CuCl_2$, is light blue in aqueous solution. This is due to the $Cu^{2+}{}_{(aq)}$ ion. Aluminum chloride, $AlCl_{3(aq)}$, is colourless in aqueous solution.

Question

How can observations tell you which is the limiting reactant in the reaction of aluminum with aqueous copper(II) chloride?

Prediction

Your teacher will give you a beaker that contains a 0.25 g piece of aluminum foil and 0.51 g of copper(II) chloride. Predict which one of these reactants is the limiting reactant.

Materials

100 mL beaker or 125 mL Erlenmeyer flask
stirring rod
0.51 g $CuCl_2$
0.25 g Al foil

Safety Precautions

The reaction mixture may get hot. Do not hold the beaker as the reaction proceeds.

Procedure

1. To begin the reaction, add about 50 mL of water to the beaker that contains the aluminum foil and copper(II) chloride.
2. Record the colour of the solution and any metal that is present at the beginning of the reaction.
3. Record any colour changes as the reaction proceeds. Stir occasionally with the stirring rod.
4. When the reaction is complete, return the beaker, with its contents, to your teacher for proper disposal. Do not pour anything down the drain.

Analysis

1. According to your observations, which reactant was present in excess? Which reactant was the limiting reactant?
2. How does your prediction compare with your observations?
3. Do stoichiometric calculations to support your observations of the limiting reactant. Refer to the previous ThoughtLab if you need help.
4. If your prediction of the limiting reactant was incorrect, explain why.

Conclusions

5. Write a conclusion to explain how your experimental observation supported your theoretical calculations.

Applications

6. Magnesium ($Mg_{(s)}$) and hydrogen chloride ($HCl_{(aq)}$) react according to the following skeleton equation:

$$Mg_{(s)} + HCl_{(aq)} \rightarrow MgCl_{2(aq)} + H_{2(g)}$$

(a) Balance the skeleton equation.

(b) Examine the equation carefully. What evidence would you have that a reaction was taking place between the hydrochloric acid and the magnesium?

(c) You have a piece of magnesium of unknown mass, and a beaker of water in which is dissolved an unknown amount of hydrogen chloride. Design an experiment to determine which reactant is the limiting reactant.

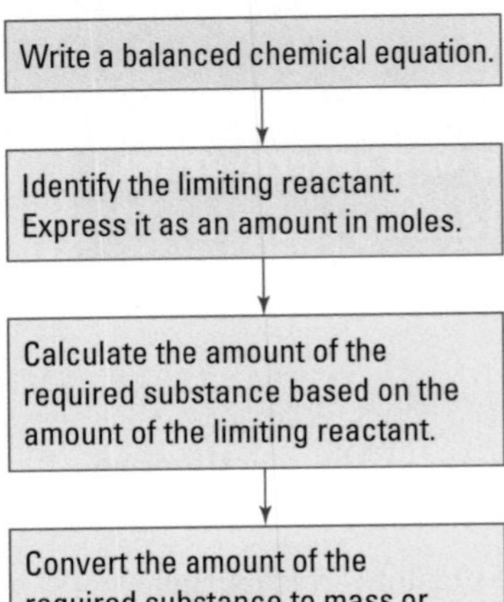
Write a balanced chemical equation.
Identify the limiting reactant. Express it as an amount in moles.
Calculate the amount of the required substance based on the amount of the limiting reactant.
Convert the amount of the required substance to mass or number of particles, as directed by the question.

Figure 7.7 **Be sure to determine the limiting reactant in any stoichiometric problem before you solve it.**

PROBEWARE

If you have access to probeware, do the Chemistry 11 lab, Stoichiometry, now.

The Limiting Reactant in Stoichiometric Problems

You are now ready to use what you know about finding the limiting reactant to predict the amount of product that is expected in a reaction. This type of prediction is a routine part of a chemist's job, both in academic research and industry. To produce a compound, for example, chemists need to know how much product they can expect from a given reaction. In analytical chemistry, chemists often analyze an impure substance by allowing it to react in a known reaction. They predict the expected mass of the product(s) and compare it with the actual mass of the product(s) obtained. Then they can determine the purity of the compound.

Since chemical reactions usually occur with one or more of the reactants in excess, you often need to determine the limiting reactant before you carry out stoichiometric calculations. You can incorporate this step into the process you have been using to solve stoichiometric problems, as shown in Figure 7.7.

Sample Problem

The Limiting Reactant in a Stoichiometric Problem

Problem

White phosphorus consists of a molecule made up of four phosphorus atoms. It burns in pure oxygen to produce tetraphosphorus decaoxide.

$$P_{4(s)} + 5O_{2(g)} \rightarrow P_4O_{10(s)}$$

A 1.00 g piece of phosphorus is burned in a flask filled with 2.60×10^{23} molecules of oxygen gas. What mass of tetraphosphorus decaoxide is produced?

What Is Required?

You need to find the mass of tetraphosphorus decaoxide that is produced.

What Is Given?

You know the balanced chemical equation. You also know the mass of phosphorus and the number of oxygen molecules that reacted.

Plan Your Strategy

First convert each reactant to moles and find the limiting reactant. Using the mole to mole ratio of the limiting reactant to the product, determine the number of moles of tetraphosphorus decaoxide that is expected. Convert this number of moles to grams.

Act on Your Strategy

$$n \text{ mol } P_4 = \frac{1.00 \text{ g } P_4}{123.9 \text{ g/mol } P_4}$$
$$= 8.07 \times 10^{-3} \text{ mol } P_4$$

Continued ...

Continued ...
FROM PAGE 256

$$n \text{ mol } O_2 = \frac{2.60 \times 10^{23} \text{ molecules}}{6.02 \times 10^{23} \text{ molecules/mol}} = 0.432 \text{ mol } O_2$$

Calculate the amount of P_4O_{10} that would be produced by the P_4.

$$\frac{n \text{ mol } P_4O_{10}}{8.07 \times 10^{-3} \text{ mol } P_4} = \frac{1 \text{ mol } P_4O_{10}}{1 \text{ mol } P_4}$$

$$(8.07 \times 10^{-3} \text{ mol } P_4)\frac{n \text{ mol } P_4O_{10}}{8.07 \times 10^{-3} \text{ mol } P_4} = \frac{1 \text{ mol } P_4O_{10}}{1 \text{ mol } P_4}(8.07 \times 10^{-3} \text{ mol } P_4) = 8.07 \times 10^{-3} \text{ mol } P_4O_{10}$$

Calculate the amount of P_4O_{10} that would be produced by the O_2.

$$\frac{n \text{ mol } P_4O_{10}}{0.432 \text{ mol } O_2} = \frac{1 \text{ mol } P_4O_{10}}{5 \text{ mol } O_2}$$

$$(0.432 \text{ mol } O_2)\frac{n \text{ mol } P_4O_{10}}{0.432 \text{ mol } O_2} = \frac{1 \text{ mol } P_4O_{10}}{5 \text{ mol } O_2}(0.432 \text{ mol } O_2) = 8.64 \times 10^{-2} \text{ mol } P_4O_{10}$$

Since P_4 would produce less P_4O_{10} than O_2 would, P_4 is the limiting reactant.

$$P_{4(s)} + 5O_{2(g)} \longrightarrow P_4O_{10(s)}$$

0.00807 mol → 0.00807 mol

unknown ratio = known ratio

$$\frac{n \text{ mol } P_4O_{10}}{0.00807 \text{ mol } P_4} = \frac{1 \text{ mol } P_4O_{10}}{1 \text{ mol } P_4}$$

$$(0.00807 \text{ mol } P_4)\frac{n \text{ mol } P_4O_{10}}{0.00807 \text{ mol } P_4} = (0.00807 \text{ mol } P_4)\frac{1 \text{ mol } P_4O_{10}}{1 \text{ mol } P_4} = 0.00807 \text{ mol } P_4O_{10}$$

$0.00807 \text{ mol } P_4O_{10} \times 284 \text{ g/mol } P_4O_{10}$

2.29 g P_4O_{10}

Check Your Solution

There were more than 5 times as many moles of O_2 as moles of P_4, so it makes sense that P_4 was the limiting reactant. An expected mass of 2.29 g of tetraphosphorus decaoxide is reasonable. It is formed in a 1:1 ratio from phosphorus. It has a molar mass that is just over twice the molar mass of phosphorus.

Practice Problems

27. Chloride dioxide, ClO_2, is a reactive oxidizing agent. It is used to purify water.

$$6ClO_{2(g)} + 3H_2O_{(\ell)} \rightarrow 5HClO_{3(aq)} + HCl_{(aq)}$$

(a) If 71.00 g of ClO_2 is mixed with 19.00 g of water, what is the limiting reactant?

(b) What mass of $HClO_3$ is expected in part (a)?

(c) How many molecules of HCl are expected in part (a)?

Continued ...

Continued ...
FROM PAGE 257

28. Hydrazine, N_2H_4, reacts exothermically with hydrogen peroxide, H_2O_2

$$N_2H_{4(\ell)} + 7H_2O_{2(aq)} \rightarrow 2HNO_{3(g)} + 8H_2O_{(g)}$$

(a) 120 g of N_2H_4 reacts with an equal mass of H_2O_2. Which is the limiting reactant?

(b) What mass of HNO_3 is expected?

(c) What mass, in grams, of the excess reactant remains at the end of the reaction?

29. In the textile industry, chlorine is used to bleach fabrics. Any of the toxic chlorine that remains after the bleaching process is destroyed by reacting it with a sodium thiosulfate solution, $Na_2S_2O_{3(aq)}$.

$$Na_2S_2O_{3(aq)} + 4Cl_{2(g)} + 5H_2O_{(\ell)} \rightarrow 2NaHSO_{4(aq)} + 8HCl_{(aq)}$$

135 kg of $Na_2S_2O_3$ reacts with 50.0 kg of Cl_2 and 238 kg of water. How many grams of $NaHSO_4$ are expected?

30. Manganese(III) fluoride can be formed by the reaction of manganese(II) iodide with fluorine.

$$2MnI_{2(s)} + 13F_{2(g)} \rightarrow 2MnF_{3(s)} + 4IF_{5(\ell)}$$

(a) 1.23 g of MnI_2 reacts with 25.0 g of F_2. What mass of MnF_3 is expected?

(b) How many molecules of IF_5 are produced in part (a)?

(c) What reactant is in excess? How much of it remains at the end of the reaction?

You will use the concepts of stoichiometry and limiting reactants in the Chemistry Course Challenge. If you have two reactants and you want to use up all of one reactant, which is the limiting reactant?

Carbon disulfide, CS_2, is an extremely volatile and flammable substance. It is so flammable that it can ignite when exposed to boiling water! Because carbon disulfide vapour is more than twice as dense as air, it can "blanket" the floor of a laboratory. There have been cases where the spark from an electrical motor has ignited carbon disulfide vapour in a laboratory, causing considerable damage. For this reason, specially insulated electrical motors are required in laboratory refrigerators and equipment.

Section Wrap-up

You now know how to identify a limiting reactant. This allows you to predict the amount of product that will be formed in a reaction. Often, however, your prediction will not accurately reflect reality. When a chemical reactions occurs—whether in a laboratory, in nature, or in industry—the amount of product that is formed is often different from the amount that was predicted by stoichiometric calculations. You will learn why this happens, and how chemists deal with it, in section 7.3.

Section Review

1 (C) Why do you not need to consider reactants that are present in excess amounts when carrying out stoichiometric calculations? Use an everyday analogy to explain the idea of excess quantity.

2 (a) (C) Magnesium reacts with oxygen gas, O_2, from the air. Which reactant do you think will be present in excess?

(b) C Gold is an extremely unreactive metal. Gold does react, however, with *aqua regia* (a mixture of concentrated nitric acid, $HNO_{3(aq)}$, and hydrochloric acid, $HCl_{(aq)}$). The complex ion $AuCl_4^-$, as well as NO_2 and H_2O, are formed. This reaction is always carried out with *aqua regia* in excess. Why would a chemist not have the gold in excess?

(c) C In general, what characteristics or properties of a chemical compound or atom make it suitable to be used as an *excess* reactant?

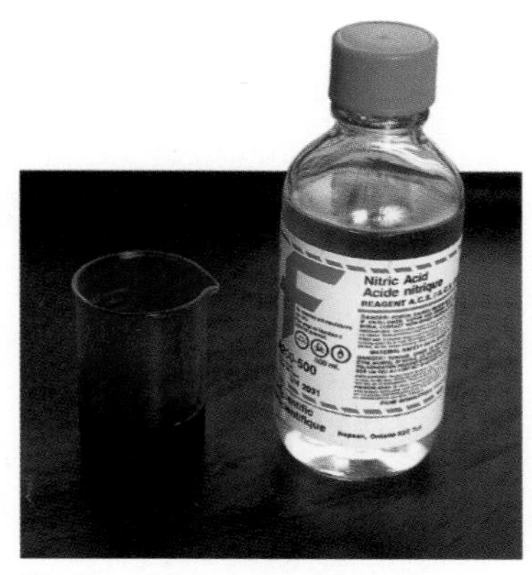

Nitric acid reacts with copper metal to produce nitrogen monoxide, NO. Upon reaction with oxygen in the air, the nitrogen monoxide, NO, is converted to poisonous, brown nitrogen dioxide, NO_2, gas.

3 I Copper is a relatively inert metal. It is unreactive with most acids. It does, however, react with nitric acid.

$$3Cu_{(s)} + 8HNO_{3(aq)} \rightarrow 3Cu(NO_3)_{2(aq)} + 2NO_{(g)} + 4H_2O_{(\ell)}$$

What mass of NO is produced when 57.4 g of Cu reacts with 165 g of HNO_3?

4 I Iron can be produced when iron(III) oxide reacts with carbon monoxide gas.

$$Fe_2O_{3(s)} + 3CO_{(g)} \rightarrow 2Fe_{(s)} + 3CO_{2(g)}$$

11.5 g of Fe_2O_3 reacts with 2.63×10^{24} molecules of CO. What mass of Fe is expected?

5 I The reaction of an aqueous solution of iron(III) sulfate with aqueous sodium hydroxide produces aqueous sodium sulfate and a solid precipitate, iron(III) hydroxide.

$$Fe_2(SO_4)_{3(aq)} + 6NaOH_{(aq)} \rightarrow 3Na_2SO_{4(aq)} + 2Fe(OH)_{3(s)}$$

What mass of $Fe(OH)_3$ is produced when 10.0 g of $Fe_2(SO_4)_3$ reacts with an equal mass of NaOH?

6 I Carbon disulfide is used as a solvent for water-insoluble compounds, such as fats, oils, and waxes. Calculate the mass of carbon disulfide that is produced when 17.5 g of carbon reacts with 225 g of sulfur dioxide according to the following equation:

$$5C_{(s)} + 2SO_{2(g)} \rightarrow CS_{2(\ell)} + 4CO_{(g)}$$

7 I A chemist adds some zinc shavings to a beaker containing a blue solution of copper chloride. The contents of the beaker are stirred. After about an hour, the chemist observes that the blue colour has not completely disappeared.

(a) Write a balanced chemical equation to describe this reaction.

(b) What other observations would you expect the chemist to make?

(c) According to the chemist's observations, which reactant was the limiting reactant?

(d) The beaker contained 3.12 g of copper chloride dissolved in water. What does this tell you, quantitatively, about the amount of zinc that was added?

Unit Investigation Prep

Consider what you have learned about limiting reactants when you design your quantitative analysis experiment at the end of Unit 2. Imagine you add one reactant (A) to an unknown amount of a second reactant (B). You intend to analyze the products (C) in order to calculate the amount of B. In this case, which reactant should be the limiting reactant, A or B? How do you know which reactant is the limiting reactant when you do not know the amount of reactant B?

7.3 Percentage Yield

Section Preview/ Specific Expectations

In this section, you will

- **solve** problems involving percentage yield and limiting reactants
- **compare**, using laboratory results, the theoretical yield of a reaction with the actual yield
- **calculate** the percentage yield of a reaction, and suggest sources of experimental error
- **solve** stoichiometric problems involving the percentage purity of the reactants
- **communicate** your understanding of the following terms: *theoretical yield, actual yield, competing reaction, percentage yield, percentage purity*

When you write an examination, the highest grade that you can earn is usually 100%. Most people, however, do not regularly earn a grade of 100%. A percentage on an examination is calculated using the following equation:

$$\text{Percentage grade} = \frac{\text{Marks earned}}{\text{Maximum possible marks}} \times 100\%$$

Similarly, a batter does not succeed at every swing. A batter's success rate is expressed as a decimal fraction. The decimal can be converted to a percent by multiplying by 100%, as shown in Figure 7.8. In this section, you will learn about a percentage that chemists use to predict and express the "success" of reactions.

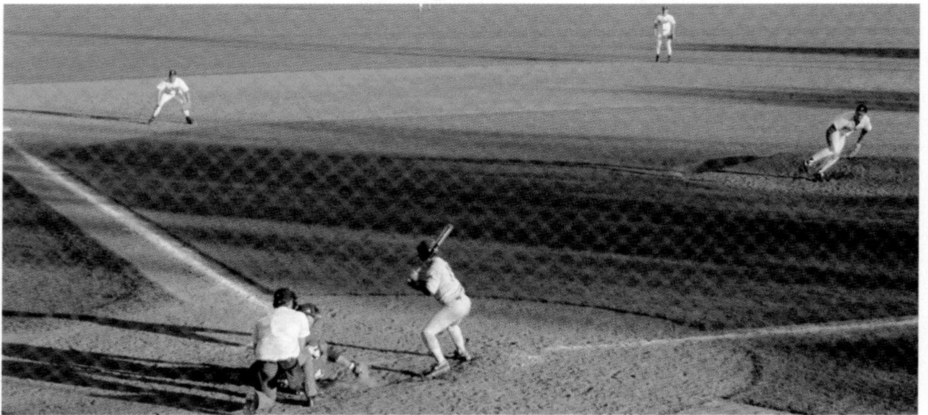

Figure 7.8 A baseball player's batting average is calculated as hits/attempts. For example, a player with 6 hits for 21 times at bat has a batting average of 6/21 = 0.286. This represents a success rate of 28.6%.

Theoretical Yield and Actual Yield

Chemists use stoichiometry to predict the amount of product that can be expected from a chemical reaction. The amount of product that is predicted by stoichiometry is called the **theoretical yield**. This predicted yield, however, is not always the same as the amount of product that is actually obtained from a chemical reaction. The amount of product that is obtained in an experiment is called the **actual yield**.

CHEM FACT

Actual yield is a *measured* quantity. Theoretical yield is a *calculated* quantity.

Why Actual Yield and Theoretical Yield Are Often Different

The actual yield of chemical reactions is usually less than the theoretical yield. This is caused by a variety of factors. For example, sometimes less than perfect collection techniques contribute to a lower than expected yield.

A reduced yield may also be caused by a **competing reaction**: a reaction that occurs at the same time as the principal reaction and involves its reactants and/or products. For example, phosphorus reacts with chlorine to form phosphorus trichloride. Some of the phosphorus trichloride, however, can then react with chlorine to form phosphorus pentachloride.

Here are the chemical equations for these competing reactions:

$$2P_{(s)} + 3Cl_{2(g)} \rightarrow 2PCl_{3(\ell)}$$
$$PCl_{3(\ell)} + Cl_{2(g)} \rightarrow PCl_{5(s)}$$

Therefore, not all the phosphorus is converted to phosphorus trichloride. So the actual yield of phosphorus trichloride is less than the theoretical yield.

Experimental design and technique may affect the actual yield, as well. For example, suppose that you need to obtain a product by filtration. Some of the product may remain in solution and therefore not be caught on the filter paper.

Another common cause of reduced yield is impure reactants. The theoretical yield is calculated based on the assumption that reactants are pure. You will learn about the effects of impure reactants on page 265.

Calculating Percentage Yield

The **percentage yield** of a chemical reaction compares the mass of product obtained by experiment (the actual yield) with the mass of product determined by stoichiometric calculations (the theoretical yield). It is calculated as follows:

$$\text{Percentage yield} = \left(\frac{\text{Actual yield}}{\text{Theoretical yield}}\right) \times 100\%$$

In section 7.1, you looked at the reaction of hydrogen and nitrogen to produce ammonia. You assumed that all the nitrogen and hydrogen reacted. Under certain conditions of temperature and pressure, this is a reasonable assumption. When ammonia is produced industrially, however, temperature and pressure are manipulated to maximize the speed of production. Under these conditions, the actual yield is much less than the theoretical yield. Examine the next Sample Problem to learn how to calculate percentage yield.

Sample Problem

Calculating Percentage Yield

Problem

Ammonia can be prepared by reacting nitrogen gas, taken from the atmosphere, with hydrogen gas.

$$N_{2(g)} + 3H_{2(g)} \rightarrow 2NH_{3(g)}$$

When 7.5×10^1 g of nitrogen reacts with sufficient hydrogen, the theoretical yield of ammonia is 9.10 g. (You can verify this by doing the stoichiometric calculations.) If 1.72 g of ammonia is obtained by experiment, what is the percentage yield of the reaction?

What Is Required?

You need to find the percentage yield of the reaction.

What Is Given?

actual yield = 1.72 g
theoretical yield = 9.10 g

Continued ...

Continued ...
FROM PAGE 261

Plan Your Strategy

Divide the actual yield by the theoretical yield, and multiply by 100%.

Act on Your Strategy

$$\text{Percentage yield} = \frac{\text{Actual yield}}{\text{Theoretical yield}} \times 100\%$$
$$= \frac{1.72\ \cancel{g}}{9.10\ \cancel{g}} \times 100\%$$
$$= 18.9\%$$

The percentage yield of the reaction is 18.9%.

Check Your Solution

By inspection, you can see that 1.72 g is roughly 20% of 9.10 g.

Practice Problems

31. 20.0 g of bromic acid, $HBrO_3$, is reacted with excess HBr.

$$HBrO_{3(aq)} + 5HBr_{(aq)} \rightarrow 3H_2O_{(\ell)} + 3Br_{2(aq)}$$

(a) What is the theoretical yield of Br_2 for this reaction?

(b) If 47.3 g of Br_2 is produced, what is the percentage yield of Br_2?

32. Barium sulfate forms as a precipitate in the following reaction:

$$Ba(NO_3)_{2(aq)} + Na_2SO_{4(aq)} \rightarrow BaSO_{4(s)} + 2NaNO_{3(aq)}$$

When 35.0 g of $Ba(NO_3)_2$ is reacted with excess Na_2SO_4, 29.8 g of $BaSO_4$ is recovered by the chemist.

(a) Calculate the theoretical yield of $BaSO_4$.

(b) Calculate the percentage yield of $BaSO_4$.

33. Yeasts can act on a sugar, such as glucose, $C_6H_{12}O_6$, to produce ethyl alcohol, C_2H_5OH, and carbon dioxide.

$$C_6H_{12}O_6 \rightarrow 2C_2H_5OH + 2CO_2$$

If 223 g of ethyl alcohol are recovered after 1.63 kg of glucose react, what is the percentage yield of the reaction?

Sometimes chemists know what percentage yield to expect from a chemical reaction. This is especially true of an industrial reaction, where a lot of experimental data are available. As well, the reaction has usually been carried out many times, with large amounts of reactants. Examine the next Sample Problem to learn how to predict the actual yield of a reaction from a known percentage yield.

Sample Problem

Predicting Actual Yield Based on Percentage Yield

Problem

Calcium carbonate can be thermally decomposed to calcium oxide and carbon dioxide.

$$CaCO_{3(s)} \rightarrow CaO_{(s)} + CO_{2(g)}$$

Under certain conditions, this reaction proceeds with a 92.4% yield of calcium oxide. How many grams of calcium oxide can the chemist expect to obtain if 12.4 g of calcium carbonate is heated?

What Is Required?

You need to calculate the amount of calcium oxide, in grams, that will be formed in the reaction.

What Is Given?

Percentage yield CaO = 92.4%
m $CaCO_3$ = 12.4 g

Plan Your Strategy

Calculate the theoretical yield of calcium oxide using stoichiometry. Then multiply the theoretical yield by the percentage yield to predict the actual yield.

Act on Your Strategy

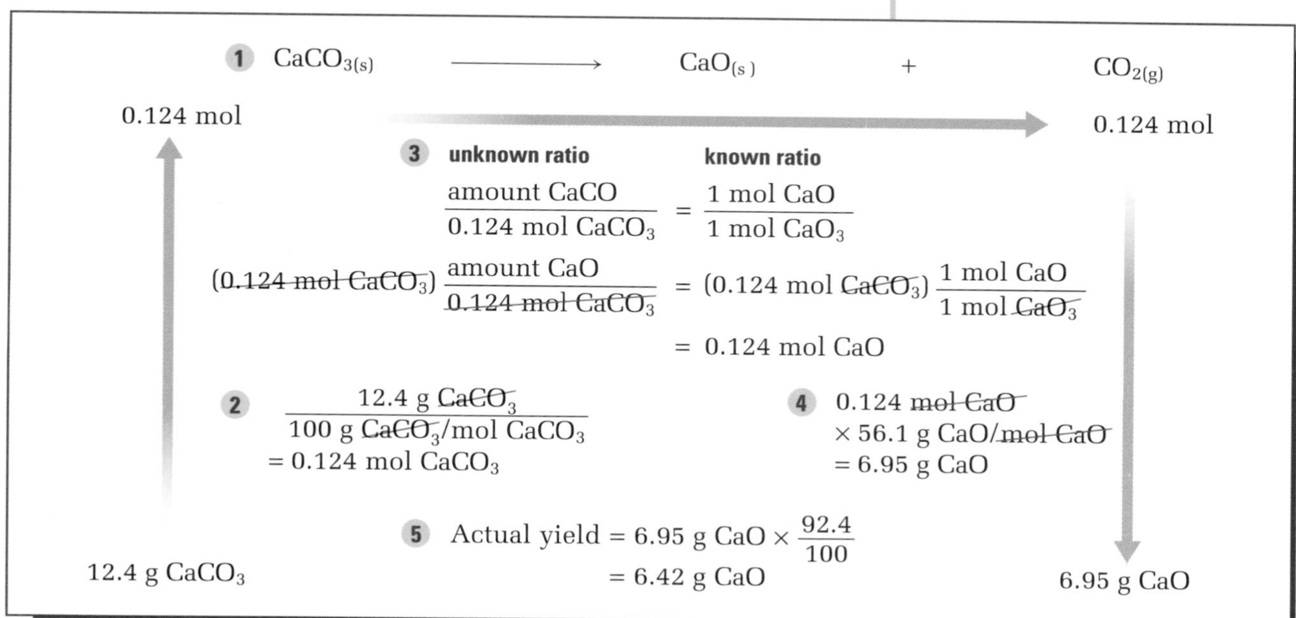

1 $CaCO_{3(s)} \longrightarrow CaO_{(s)} + CO_{2(g)}$
0.124 mol
0.124 mol
3 unknown ratio
known ratio
$\frac{\text{amount CaCO}}{0.124 \text{ mol } CaCO_3} = \frac{1 \text{ mol CaO}}{1 \text{ mol } CaO_3}$
$(\cancel{0.124 \text{ mol } CaCO_3}) \frac{\text{amount CaO}}{\cancel{0.124 \text{ mol } CaCO_3}} = (0.124 \text{ mol } \cancel{CaCO_3}) \frac{1 \text{ mol CaO}}{1 \text{ mol } \cancel{CaO_3}}$
$= 0.124 \text{ mol CaO}$
2 $\frac{12.4 \text{ g } \cancel{CaCO_3}}{100 \text{ g } \cancel{CaCO_3}/\text{mol } CaCO_3}$
$= 0.124 \text{ mol } CaCO_3$
4 $0.124 \cancel{\text{ mol CaO}}$
$\times 56.1 \text{ g CaO}/\cancel{\text{mol CaO}}$
$= 6.95 \text{ g CaO}$
5 Actual yield $= 6.95 \text{ g CaO} \times \frac{92.4}{100}$
$= 6.42 \text{ g CaO}$
12.4 g $CaCO_3$
6.95 g CaO

Check Your Solution

92.5% of 6.95 g is about 6.4 g. The answer is reasonable.

Continued ...

Continued ...
FROM PAGE 263

Practice Problems

34. The following reaction proceeds with a 70% yield.

$$C_6H_{6(\ell)} + HNO_{3(aq)} \rightarrow C_6H_5NO_{2(\ell)} + H_2O_{(\ell)}$$

Calculate the mass of $C_6H_5NO_2$ expected if 12.8 g of C_6H_6 reacts with excess HNO_3.

35. The reaction of toluene, C_7H_8, with potassium permanganate, $KMnO_4$, gives less than a 100% yield.

$$C_7H_{8(\ell)} + 2KMnO_{4(aq)} \rightarrow KC_7H_5O_{2(aq)} + 2MnO_{2(s)} + KOH_{(aq)} + H_2O_{(\ell)}$$

(a) 8.60 g of C_7H_8 is reacted with excess $KMnO_4$. What is the theoretical yield, in grams, of $KC_7H_5O_2$?

(b) If the percentage yield is 70.0%, what mass of $KC_7H_5O_2$ can be expected?

(c) What mass of C_7H_8 is needed to produce 13.4 g of $KC_7H_5O_2$, assuming a yield of 60%?

36. Marble is made primarily of calcium carbonate. When calcium carbonate reacts with hydrogen chloride, it reacts to form calcium chloride, carbon dioxide and water. If this reaction occurs with 81.5% yield, what mass of carbon dioxide will be collected if 15.7 g of $CaCO_3$ is added to sufficient hydrogen chloride?

37. Mercury, in its elemental form or in a chemical compound is highly toxic. Water-soluble mercury compounds, such as mercury(II) nitrate, can be removed from industrial wastewater by adding sodium sulfide to the water, which forms a precipitate of mercury(II) sulfide, which can then be filtered out.

$$Hg(NO_3)_{2(aq)} + Na_2S_{(aq)} \rightarrow HgS_{(s)} + 2NaNO_{3(aq)}$$

If 3.45×10^{23} formula units of $Hg(NO_3)_2$ are reacted with excess Na_2S, what mass of HgS can be expected if this process occurs with 97.0% yield?

mind STRETCH

You hear a great deal about the fuel consumption of automobiles. What about air consumption? Your challenge is to determine the information you need to answer the following question: What mass of air does an automobile require to travel from Thunder Bay, Ontario, to Smooth Rock Falls, Ontario? This is a distance of 670 km.

When you are finished, go to question 23 on page 273. You can check your answer and solve the problem, too.

COURSE CHALLENGE

How would you determine the percentage yield of a double displacement reaction that produces a precipitate? Consider this question to prepare for your Chemistry Course Challenge.

Applications of Percentage Yield

The percentage yield of chemical reactions is extremely important in industrial chemistry and the pharmaceutical industry. For example, the synthesis of certain drugs involves many sequential chemical reactions. Often each reaction has a low percentage yield. This results in a tiny overall yield. Research chemists, who generally work with small quantities of reactants, may be satisfied with a poor yield. Chemical engineers, on the other hand, work with very large quantities. They may use hundreds or even thousands of kilograms of reactants! A difference of 1% in the yield of a reaction can translate into thousands of dollars.

The work of a chemist in a laboratory can be likened to making spaghetti for a family. The work of a chemical engineer, by contrast, is like making spaghetti for 10 000 people! Learn more about chemical engineers in Careers in Chemistry on the next page. Then perform an investigation to determine the percentage yield of a reaction on page 266.

Careers 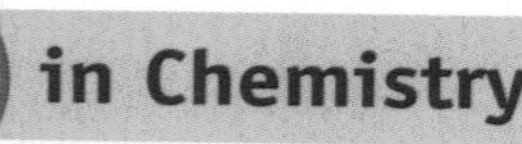 in Chemistry

Chemical Engineer

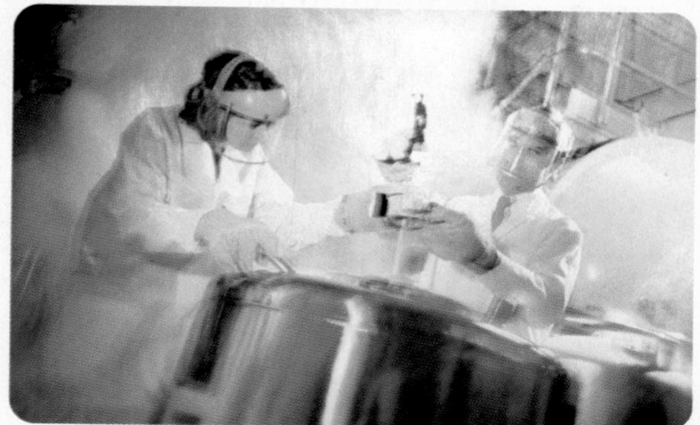

Chemical engineers are sometimes described as "universal engineers" because of their unique knowledge of math, physics, engineering, and chemistry. This broad knowledge allows them to work in a variety of areas, from designing paint factories to developing better tasting, more nutritious foods. Canadian chemical engineers are helping to lead the world in making cheap, long-lasting, and high-quality CDs and DVDs. In addition to designing and operating commercial plants, chemical engineers can be found in university labs, government agencies, and consulting firms.

Producing More for Less

Once chemists have developed a product in a laboratory, it is up to chemical engineers to design a process to make the product in commercial quantities as efficiently as possible. "Scaling up" production is not just a matter of using larger beakers. Chemical engineers break down the chemical process into a series of smaller "unit operations" or processes and techniques. They use physics, chemistry, and complex mathematical models. For example, making liquid pharmaceutical products (such as syrups, solutions, and suspensions) on a large scale involves adding specific amounts of raw materials to large mixing tanks. Then the raw materials are heated to a set temperature and mixed at a set speed for a given amount of time. The final product is filtered and stored in holding tanks. Chemical engineers ensure that each process produces the maximum amount of product.

Becoming a Chemical Engineer

To become a chemical engineer, you need a bachelor's degree in chemical engineering. Most provinces also require a Professional Engineer (P. Eng.) designation. Professional engineers must have at least four years of experience and must pass an examination. As well, they must commit to continuing their education to keep up with current developments. Chemical engineers must be able to work well with people and to communicate well.

Make Career Connections

1. Discuss engineering studies and careers with working engineers, professors, and engineering students. Look for summer internship programs and job shadowing opportunities. Browse the Internet. Contact your provincial engineering association, engineering societies, and universities for more information.
2. Participate in National Engineering Week in Canada in March of each year. This is when postsecondary institutions, companies, science centres, and other organizations hold special events, including engineering contests and workshops.

Percentage Purity

Often impure reactants are the cause of a percentage yield of less than 100%. Impurities cause the mass data to be incorrect. For example, suppose that you have 1.00 g of sodium chloride and you want to carry out a reaction with it. You think that the sodium chloride may have absorbed some water, so you do not know exactly how much pure sodium chloride you have. If you calculate a theoretical yield for your reaction based on 1.00 g of sodium chloride, your actual yield will be less. There is not 1.00 g of sodium chloride in the sample.

Investigation 7-B

SKILL FOCUS

- Predicting
- Performing and recording
- Analyzing and interpreting
- Communicating results

Determining the Percentage Yield of a Chemical Reaction

The percentage yield of a reaction is determined by numerous factors: The nature of the reaction itself, the conditions under which the reaction was carried out, and the nature of the reactants used.

In this investigation, you will determine the percentage yield of the following chemical reaction:

$$Fe_{(s)} + CuCl_{2(aq)} \rightarrow FeCl_{2(aq)} + Cu_{(s)}$$

You will use steel wool, since it is virtually pure iron.

Question

What is the percentage yield of the reaction of iron and copper chloride when steel wool and copper chloride dihydrate are used as reactants?

Predictions

Predict the mass of copper that will be produced if 1.00 g of iron (steel wool) reacts *completely* with a solution containing excess $CuCl_2$. Also predict the maximum possible yield.

Materials

2 beakers (250 mL)
stirring rod
electronic balance, accurate to two decimal places
distilled water
wash bottle with distilled water
drying oven or heat lamp
about 1.00 g rust-free, degreased steel wool
5.00 g copper chloride dihydrate, $CuCl_2 \cdot 2H_2O$
15 mL 1 mol/L hydrochloric acid, HCl

Safety Precautions

If you get either $CuCl_2$ or the HCl solution on your skin, flush with plenty of cold water.

Procedure

1. Label a clean, dry 250 mL beaker with your initials. Use a glass marker, or write with pencil on the frosted area of the beaker. Do not use tape, since the beaker will be dried in an oven later.

2. Copy the table below into your notebook. Record the mass of the labelled beaker in your table.

Observations

Mass of empty beaker	
Mass of steel wool	
Mass of beaker containing clean, dry copper	

3. Put about 50 mL of distilled water in the beaker. Add 5.00 g of $CuCl_2 \cdot 2H_2O$ to the water. Stir to dissolve.

4. Record the mass of the steel wool in your table.

5. Add the steel wool to the $CuCl_2$ solution in the beaker. Allow it to sit until all the steel wool has reacted. This could take up to 20 min.

6. When the reaction is complete, decant the solution into a 250 mL beaker, as shown in the diagram.

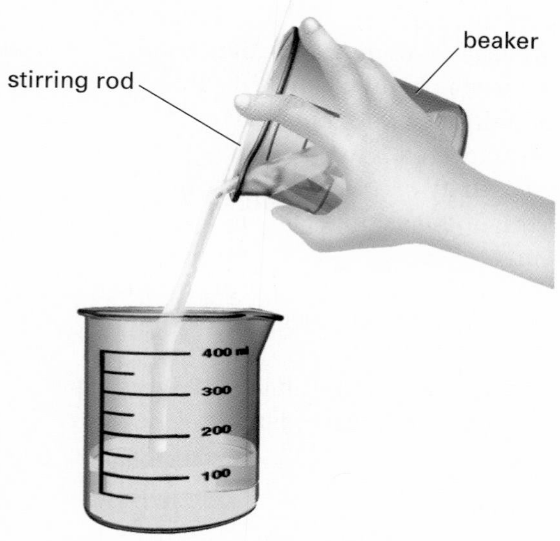

Pouring *down* a stirring rod ensures that no liquid dribbles down the outside of the beaker. The glove in this illustration is omitted so you can clearly see where to place your fingers. Always wear gloves when handling chemicals in the laboratory.

7. Using a wash bottle, rinse the copper several times with distilled water. Decant the water as shown in the diagram.
8. Add 10 to 15 mL of 1 mol/L HCl to further wash the copper. Decant the HCl, and wash the copper again with distilled water. (If the copper is still not clean, wash it again with the HCl. Remember to do a final wash with distilled water.)
9. Place your labelled reaction beaker, containing the cleaned copper, in a drying oven overnight.
10. Find the mass of the beaker containing the dry copper.
11. Return the beaker, containing the copper, to your teacher for proper disposal.

Analysis

1. **(a)** Using the mass of the iron (steel wool) you used, calculate the theoretical yield of the copper, in grams.
   (b) How does the mass of the copper you collected compare with the expected theoretical yield?
2. Based on the amount of iron that you used, prove that the 5.00 g of $CuCl_2 \cdot 2H_2O$ was the excess reactant.

Conclusion

3. Calculate the percentage yield for this reaction.

Applications

4. If your percentage yield was not 100%, suggest sources of error.
5. How would you attain an improved percentage yield if you performed this reaction again? Consider your technique and materials.
6. Do some research to find out the percent by mass of iron in steel wool. Predict what your percentage yield would be if you had used pure iron in this reaction. Would it make a difference?

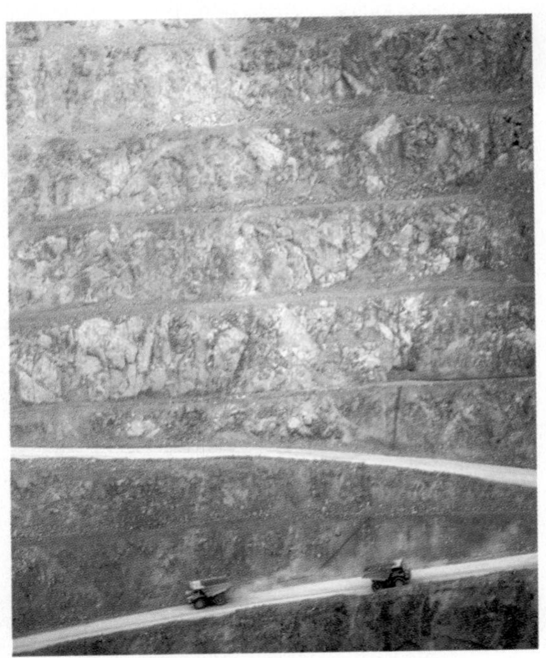

Figure 7.9 Copper is removed from mines like this one in the form of an ore. There must be sufficient copper in the ore to make the mine economically viable.

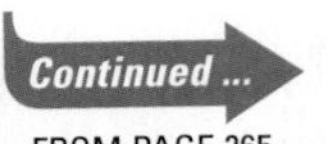

Continued ... FROM PAGE 265

In the mining industry, metals are usually recovered in the form of an ore. An ore is a naturally occurring rock that contains a high concentration of one or more metals. Whether an ore can be profitably mined depends on several factors: the cost of mining and refining the ore, the price of the extracted metal, and the cost of any legal and environmental issues related to land use. The inaccurate chemical analysis of an ore sample can cost investors millions of dollars if the ore deposit does not yield what was expected.

The **percentage purity** of a sample describes what proportion, by mass, of the sample is composed of a specific compound or element. For example, suppose that a sample of gold has a percentage purity of 98%. This means that every 100 g of the sample contains 98 g of gold and 2 g of impurities.

You can apply your knowledge of stoichiometry and percentage yield to solve problems related to percentage purity.

Sample Problem

Finding Percentage Purity

Problem

Iron pyrite, FeS_2, is known as "fool's gold" because it looks similar to gold. Suppose that you have a 13.9 g sample of *impure* iron pyrite. (The sample contains a non-reactive impurity.) You heat the sample in air to produce iron(III) oxide, Fe_2O_3, and sulfur dioxide, SO_2.

$$4FeS_{2(s)} + 11O_{2(g)} \rightarrow 2Fe_2O_{3(s)} + 8SO_{2(g)}$$

If you obtain 8.02 g of iron(III) oxide, what was the percentage of iron pyrite in the original sample? Assume that the reaction proceeds to completion. That is, all the available iron pyrite reacts completely.

What Is Required?

You need to determine the percentage purity of the iron pyrite sample.

What Is Given?

The mass of Fe_2O_3 is 8.02 g. The reaction proceeds to completion. You can assume that sufficient oxygen is present.

Plan Your Strategy

Steps 1–4 Use your stoichiometry problem-solving skills to find the mass of Fe_2S expected to have produced 8.02 g Fe_2O_3.

Step 5 Determine percentage purity of the Fe_2S using the following formula:

$$\frac{\text{theoretical mass (g)}}{\text{sample size (g)}} \times 100\%$$

Continued ...

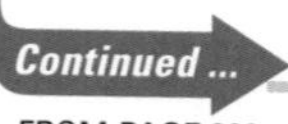
FROM PAGE 268

Act on Your Strategy

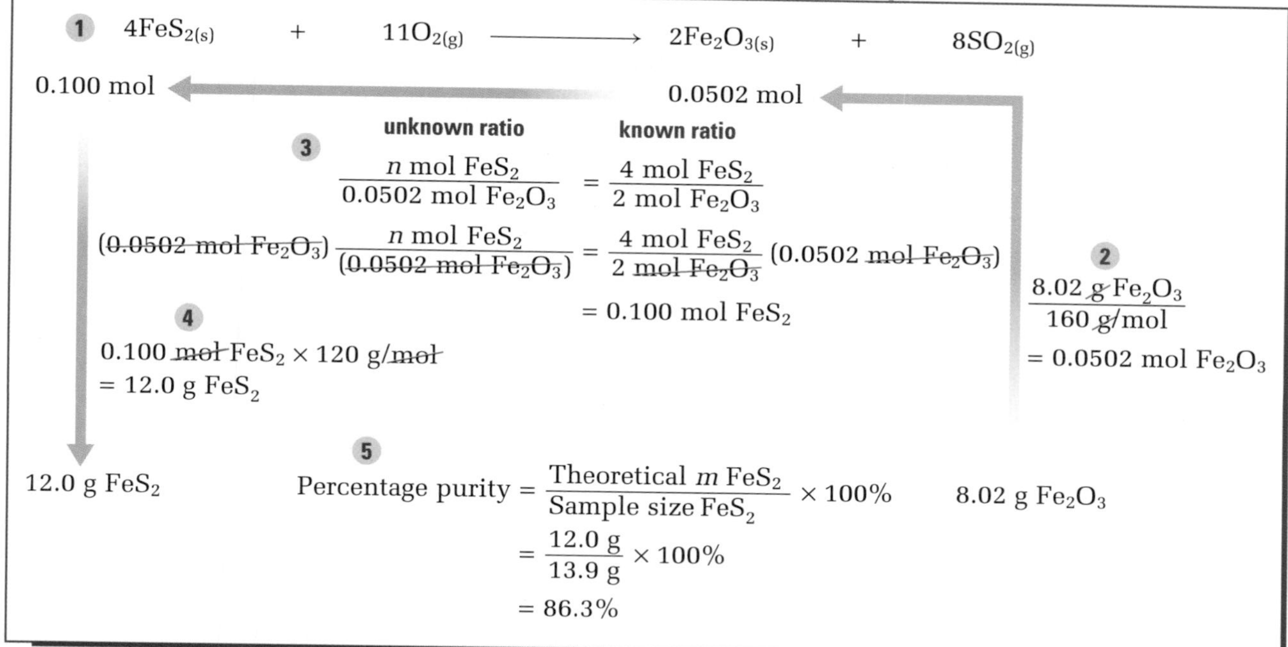

Therefore, the percentage purity of the iron pyrite is 86.3%.

Check Your Solution

The units are correct. The molar mass of iron pyrite is 3/4 the molar mass of iron(III) oxide. Mutiplying this ratio by the mole ratio of iron pyrite to iron(III) oxide (4/2) and 8 g gives 12 g. The answer is reasonable.

Practice Problems

38. An impure sample of silver nitrate, $AgNO_3$, has a mass 0.340 g. It is dissolved in water and then treated with excess hydrogen chloride, $HCl_{(aq)}$. This results in the formation of a precipitate of silver chloride, AgCl.

$$AgNO_{3(aq)} + HCl_{(aq)} \rightarrow AgCl_{(s)} + HNO_{3(aq)}$$

The silver chloride is filtered, and any remaining hydrogen chloride is washed away. Then the silver chloride is dried. If the mass of the dry silver chloride is measured to be 0.213 g, what mass of silver nitrate was contained in the original (impure) sample?

39. Copper metal is mined as one of several copper-containing ores. One of these ores contains copper in the form of malachite. Malachite exists as a double salt, $Cu(OH)_2{\cdot}CuCO_3$. It can be thermally decomposed at 200°C to yield copper(II) oxide, carbon dioxide gas, and water vapour.

$$Cu(OH)_2{\cdot}CuCO_{3(s)} \rightarrow 2CuO_{(s)} + CO_{2(g)} + H_2O_{(g)}$$

Continued ...

Continued ...
FROM PAGE 269

(a) 5.000 kg of malachite ore, containing 5.20% malachite, $Cu(OH)_2 \cdot CuCO_3$, is thermally decomposed. Calculate the mass of copper(II) oxide that is formed. Assume 100% reaction.

(b) Suppose that the reaction had a 78.0% yield, due to incomplete decomposition. How many grams of CuO would be produced?

40. Ethylene oxide, C_2H_4O, is a multi-purpose industrial chemical used, among other things, as a rocket propellant. It can be prepared by reacting ethylene bromohydrin, C_2H_5OBr, with sodium hydroxide.

$$C_2H_5OBr + NaOH \rightarrow C_2H_4O + NaBr + H_2O$$

If this reaction proceeds with an 89% yield, what mass of C_2H_4O can be obtained when 3.61×10^{23} molecules of C_2H_5OBr react with excess sodium hydroxide?

Section Wrap-up

In this section, you have learned how the amount of products formed by experiment relates to the theoretical yield predicted by stoichiometry. You have learned about many factors that affect actual yield, including the nature of the reaction, experimental design and execution, and the purity of the reactants. Usually, when you are performing an experiment in a laboratory, you want to maximize your percentage yield. To do this, you need to be careful not to contaminate your reactants or lose any products. Either might affect your actual yield.

Section Review

1 K/U When calculating the percentage yield of a reaction, what units should you use: grams, moles, or number of particles? Explain.

2 I Methyl salicylate, otherwise known as oil of wintergreen, is produced by the wintergreen plant. It can also be synthesized by heating salicylic acid, $C_7H_6O_3$, with methanol, CH_3OH.

$$C_7H_6O_{3(s)} + CH_3OH_{(\ell)} \rightarrow C_8H_8O_{3(\ell)} + H_2O_{(\ell)}$$

A chemist reacts 3.50 g of salicylic acid with excess methanol. She calculates the theoretical yield of methyl salicylate to be 3.86 g. If 2.84 g of methyl salicylate are recovered, what is the percentage yield of the reaction?

3 C Unbeknownst to a chemist, the limiting reactant in a certain chemical reaction is impure. How will this affect the percentage yield of the reaction? Explain.

4 I You have a sample of copper that is impure, and you wish to determine its purity. You have some silver nitrate, $AgNO_3$, at your disposal. You also have some copper that you know is 100% pure.

(a) Design an experiment to determine the purity of the copper sample.

(b) Even with pure copper, the reaction may not proceed with 100% yield. How will you address this issue?

CHAPTER 7 Review

Reflecting on Chapter 7

Summarize this chapter in the format of your choice. Here are a few ideas to use as guidelines:

- Use the coefficients of a balanced chemical equation to determine the mole ratios between reactants and products.
- Predict quantities required or produced in a chemical reaction.
- Calculate the limiting reactant in cases where the amount of various reactants was given.
- Calculate the percentage yield of a chemical reaction based on the amount of product(s) obtained relative to what was predicted by stoichiometry.
- Use the percentage yield of a reaction to predict the amount of product(s) formed.
- Determine the percentage purity of a reactant based on the actual yield of a reaction.

Reviewing Key Terms

For each of the following terms, write a sentence that shows your understanding of its meaning.

actual yield	competing reaction
excess reactant	limiting reactant
mole ratios	percentage purity
percentage yield	stoichiometric amounts
stoichiometric coefficients	stoichiometry
	theoretical yield

Knowledge/Understanding

1. Explain the different interpretations of the coefficients in a balanced chemical equation.
2. Why is a *balanced* chemical equation needed for stoichiometric calculations?
3. In what cases would it not be necessary to determine the limiting reactant before beginning any stoichiometric calculations?
4. Why was the concept of percentage yield introduced?
5. A student is trying to determine the mass of aluminum oxide that is produced when aluminum reacts with excess oxygen.
$$4Al_{(s)} + 3O_{2(g)} \rightarrow 2Al_2O_{3(s)}$$
The student states that 4 g of aluminum reacts with 3 g of oxygen to produce 2 g of aluminum oxide. Is the student's reasoning correct? Explain your answer.

Inquiry

6. A freshly exposed aluminum surface reacts with oxygen to form a tough coating of aluminum oxide. The aluminum oxide protects the metal from further corrosion.
$$4Al_{(s)} + 3O_{2(g)} \rightarrow 2Al_2O_{3(s)}$$
How many grams of oxygen are needed to react with 0.400 mol of aluminum?
7. Calcium metal reacts with chlorine gas to produce calcium chloride.
$$Ca_{(s)} + Cl_{2(g)} \rightarrow CaCl_{2(s)}$$
How many formula units of $CaCl_2$ are expected from 5.3 g of calcium and excess chlorine?
8. Propane is a gas at room temperature, but it exists as a liquid under pressure in a propane tank. It reacts with oxygen in the air to form carbon dioxide and water vapour.
$$C_3H_{8(\ell)} + 5O_{2(g)} \rightarrow 3CO_{2(g)} + 4H_2O_{(g)}$$
What mass of carbon dioxide gas is expected when 97.5 g of propane reacts with sufficient oxygen?
9. Powdered zinc and sulfur react in an extremely rapid, exothermic reaction. The zinc sulfide that is formed can be used in the phosphor coating on the inside of a television tube.
$$Zn_{(s)} + S_{(s)} \rightarrow ZnS_{(s)}$$
A 6.00 g sample of Zn is allowed to react with 3.35 g of S.
   (a) Determine the limiting reactant.
   (b) Calculate the mass of ZnS expected.
   (c) How many grams of the excess reactant will remain after the reaction?
10. Titanium(IV) chloride reacts violently with water vapour to produce titanium(IV) oxide and hydrogen chloride gas. Titanium(IV) oxide, when finely powdered, is extensively used in paint as a white pigment.
$$TiCl_{4(s)} + H_2O_{(\ell)} \rightarrow TiO_{2(s)} + 4HCl_{(g)}$$
The reaction has been used to create smoke screens. In moist air, the $TiCl_4$ reacts to produce a thick smoke of suspended TiO_2 particles. What mass of TiO_2 can be expected when 85.6 g of $TiCl_4$ is reacted with excess water vapour?

11. Silver reacts with hydrogen sulfide gas, which is present in the air. (Hydrogen sulfide has the odour of rotten eggs.) The silver sulfide, Ag_2S, that is produced forms a black tarnish on the silver.

$$4Ag_{(s)} + 2H_2S_{(g)} + O_{2(g)} \rightarrow 2Ag_2S_{(s)} + 2H_2O_{(g)}$$

How many grams of silver sulfide are formed when 1.90 g of silver reacts with 0.280 g of hydrogen sulfide and 0.160 g of oxygen?

12. 20.8 g of calcium phosphate, $Ca_3(PO_4)_2$, 13.3 g of silicon dioxide, SiO_2, and 3.90 g of carbon react according to the following equation:

$$2Ca_3(PO_4)_{2(s)} + 6SiO_{2(s)} + 10C_{(s)} \rightarrow P_{4(s)} + 6CaSiO_{3(s)} + 10CO_{(g)}$$

Determine the mass of calcium silicate, $CaSiO_3$, that is produced.

13. 1.56 g of As_2S_3, 0.140 g of H_2O, 1.23 g of HNO_3, and 3.50 g of $NaNO_3$ are reacted according to the equation below:

$$3As_2S_{3(s)} + 4H_2O_{(\ell)} + 10HNO_{3(aq)} + 18NaNO_{3(aq)} \rightarrow 9Na_2SO_{4(aq)} + 6H_3AsO_{4(aq)} + 28NO_{(g)}$$

What mass of H_3AsO_4 is produced?

14. 2.85×10^2 g of pentane, C_5H_{12}, reacts with 3.00 g of oxygen gas, according to the following equation:

$$C_5H_{12(\ell)} + 8O_{2(g)} \rightarrow 5CO_{2(g)} + 6H_2O_{(\ell)}$$

What mass of carbon dioxide gas, is produced?

15. Silica (also called silicon dioxide), along with other silicates, makes up about 95% of Earth's crust—the outermost layer of rocks and soil. Silicon dioxide is also used to manufacture transistors. Silica reacts with hydrofluoric acid to produce silicon tetrafluoride and water vapour.

$$SiO_{2(s)} + 4HF_{(aq)} \rightarrow SiF_{4(g)} + 2H_2O_{(g)}$$

(a) 12.2 g of SiO_2 is reacted with a small excess of HF. What is the theoretical yield, in grams, of H_2O?
(b) If the actual yield of water is 2.50 g, what is the percentage yield of the reaction?
(c) Assuming the yield obtained in part (b), what mass of SiF_4 is formed?

16. An impure sample of barium chloride, $BaCl_2$, is added to an aqueous solution that contains 4.36 g of sodium sulfate, Na_2SO_4.

$$BaCl_{2(s)} + Na_2SO_{4(aq)} \rightarrow BaSO_{4(s)} + 2NaCl_{(aq)}$$

After the reaction is complete, the solid barium sulfate, $BaSO_4$, is filtered and dried. Its mass is found to be 2.62 g. What is the percentage purity of the original barium chloride?

17. Benzene reacts with bromine to form bromobenzene, C_6H_5Br.

$$C_6H_{6(\ell)} + Br_{2(\ell)} \rightarrow C_6H_5Br_{(\ell)} + HBr_{(g)}$$

(a) What is the maximum amount of C_6H_5Br that can be formed from the reaction of 7.50 g of C_6H_6 with excess Br_2?
(b) A competing reaction is the formation of dibromobenzene, $C_6H_4Br_2$.

$$C_6H_{6(\ell)} + 2Br_{2(\ell)} \rightarrow C_6H_4Br_{2(\ell)} + 2HBr_{(g)}$$

If 1.25 g of $C_6H_4Br_2$ was formed by the competing reaction, how much C_6H_6 was *not* converted to C_6H_5Br?
(c) Based on your answer to part (b), what was the actual yield of C_6H_5Br? Assume that all the C_6H_5Br that formed was collected.
(d) Calculate the percentage yield of C_6H_5Br.

18. Refer to Practice Problem 39. Design an experiment to determine the mole to mole ratio of pure malachite to copper(II) oxide. Include an outline of the procedure and any safety precautions. Clearly indicate which data need to be recorded.

19. A chemist wishes to prepare a compound called compound E. The molar mass of compound E is 100 g/mol. The synthesis requires four consecutive reactions, each with a yield of 60%.

$A \rightarrow B$ $\quad C \rightarrow D$
$B \rightarrow C$ $\quad D \rightarrow E$

(a) The chemist begins the synthesis with 50 g of starting material, called compound A. If the molar mass of compound A is 200 g/mol, how many grams of compound E will be produced?
(b) How many grams of compound A are needed to produce 70 g of compound E?

Communication

20. Develop a new analogy for the concept of limiting and excess reactant.

21. Examine the balanced chemical "equation"

$$2A + B \rightarrow 3C + D$$

Using a concept map, explain how to calculate the number of grams of C that can be obtained when a given mass of A reacts with a certain number of molecules of B. Assume that you know the molar mass of A and C. Include proper units. For simplicity, assume that A is limiting, but don't forget to show how to determine the limiting reactant.

22. Assume that your friend has missed several chemistry classes and that she has asked you to help her prepare for a stoichiometry test. Unfortunately, because of other commitments, you do not have time to meet face to face. You agree to email your friend a set of point-form instructions on how to solve stoichiometry problems, including those that involve a limiting reactant. She also needs to understand the concept of percentage yield. Write the text of this email. Assume that your friend has a good understanding of the mole concept.

Making Connections

23. How many grams of air are required for an automobile to travel from Thunder Bay, Ontario, to Smooth Rock Falls, Ontario? This is a distance of 670 km. Assume the following:
    - Gasoline is pure octane, C_8H_{18}. (Gasoline is actually a mixture of hydrocarbons.)
    - The average fuel consumption is 10 L per 100 km.
    - Air has a density of 1.21 g/L.
    - The density of the gasoline is 0.703 g/mL.

    The balanced chemical equation for the complete combustion of octane is

    $$2C_8H_{18(\ell)} + 25O_{2(g)} \rightarrow 16CO_{2(g)} + 18H_2O_{(g)}$$

24. You must remove mercury ions present as mercury(II) nitrate in the waste water of an industrial facility. You have decided to use sodium sulfide in the reaction below. Write a short essay that addresses the following points. Include a well-organized set of calculations where appropriate.

    $$Hg(NO_3)_{2(aq)} + Na_2S_{(aq)} \rightarrow HgS_{(s)} + 2NaNO_{3(aq)}$$

    (a) Explain why the chemical reaction above can be used to remove mercury ions from the waste water. What laboratory technique must be used in order that this reaction is as effective as possible for removing mercury from the waste stream?

    (b) Why is mercury(II) sulfide less of an environmental concern than mercury(II) nitrate?

    (c) What assumptions are being made regarding the toxicity of sodium sulfide and sodium nitrate relative to either mercury nitrate or mercury sulfide?

    (d) Every litre of waste water contains approximately 0.03 g of $Hg(NO_3)_2$. How many kg of Na_2S will be required to remove the soluble mercury ions from 10 000 L of waste water?

    (e) What factors would a company need to consider in adopting any method of cleaning its wastewater?

Answers to Practice Problems and Short Answers to Section Review Questions

Practice Problems: **1.(a)** 2:1:2 **(b)** 50 **(c)** 4956 **(d)** 1.20×10^{24} **2.(a)** 2 **(b)** 150 **(c)** 1.806×10^{24} **(d)** 1.204×10^{24} **3.(a)** 3.4×10^{25} **(b)** 6.7×10^{25} **4.** 7.5 mol **5.(a)** 1.8 mol **(b)** 37.5 mol **6.(a)** 48.7 mol **(b)** 1.20 mol **7.(a)** 8.3×10^{24} **(b)** 4.2×10^{24} **8.(a)** 7.47 mol **(b)** 7.19 mol **9.(a)** 4.68×10^{-2} mol **(b)** 0.187 mol **10.(a)** 0.708 mol **(b)** 1.06 mol **11.** 9.32 g **12.** 137 g **13.** 4.65 g **14.** 0.814 g **15.** 97.2 g **16.** 2.31×10^{-2} g **17.** 37.6 g **18.** 20.7 g **19.(a)** 124 g **(b)** 1.14×10^{24} **20.(a)** 120 g **(b)** 1.49×10^{21} **21.(a)** 2.39×10^{22} **(b)** 1.45×10^{24} **(c)** 1.21×10^{22} **22.(a)** 1.50×10^{24} **(b)** 357 g **23.** $CuCl_2$ **24.** CaF_2 **25.** C_3H_6 **26.** Zn **27.(a)** ClO_2 **(b)** 74.1 g **(c)** 1.06×10^{23} **28.(a)** H_2O_2 **(b)** 63.5 g **(c)** 104 g **29.** 4.23×10^4 g **30.(a)** 0.446 g **(b)** 4.80×10^{21} **(c)** F_2, 24.0 g **31.(a)** 74.4 g **(b)** 63.6% **32.(a)** 31.3 g **(b)** 95.2% **33.** 26.7% **34.** 14.1 g **35.(a)** 15.0 g **(b)** 10.5 g **(c)** 12.8 g **36.** 5.63 g **37.** 129 g **38.** 0.252 g **39.(a)** 187 g **(b)** 146 g **40.** 23.5 g

Section Review: **7.1:** **4.(a)** $S + O_2 \rightarrow SO_2$ **(b)** $2S + 3O_2 \rightarrow 2SO_3$ **(c)** 1.5 mol **(d)** 48.0 g **5.(a)** 1:5:3:4 **(b)** 2.50 mol **(c)** 6.02×10^{24} **(d)** 9.00 mol **6.(a)** 49.9 g **(b)** 4.48×10^{22} **7.(a)** $Pb(NO_3)_{2(aq)} + 2NaI_{(aq)} \rightarrow PbI_{2(s)} + 2NaNO_{3(aq)}$ **(b)** 1.03 g **7.2:** **2.(a)** oxygen **3.** 18.1 g **4.** 8.04 g **5.** 5.34 g **6.** 22.2 g **7.(a)** $Zn_{(s)} + CuCl_{2(aq)} \rightarrow ZnCl_{2(aq)} + Cu_{(s)}$ **(b)** zinc gone **(c)** zinc **(d)** less than 1.52 g Zn **7.3:** **2.** 73.6%

Design Your Own Investigation

SKILL FOCUS

- Initiating and planning
- Performing and recording
- Analyzing and interpreting
- Communicating results

Analyzing a Mixture Using Stoichiometry

Background

Analytical chemists are employed in industrial research, academic research, and forensic science. Their job usually involves two different types of work: *qualitative analysis* and *quantitative analysis*. A qualitative analysis determines which substances are present. A quantitative analysis determines how much of a specific substance is present.

Many other chemists also need to use the principles of analytical chemistry. An environmental chemist, for example, tests water or soil for the presence and amounts of impurities. In this investigation, you will design a procedure to analyze a sample of contaminated sand.

This analyst is collecting run-off water for testing.

Pre-Lab Focus

A shipment of sand has become contaminated with barium chloride, $BaCl_2$, and sodium chloride, NaCl. As an analytical chemist, your job is to design a procedure to determine the percent by mass of each substance in the sand.

You will need to use the following double displacement reaction. This reaction produces a precipitate of barium sulfate.

$$BaCl_{2(aq)} + Na_2SO_{4(aq)} \rightarrow BaSO_{4(s)} + 2NaCl_{(aq)}$$

To design your procedure, carefully consider the physical and chemical properties of the two substances in the mixture.

Question

How can you determine the percent by mass of barium chloride and the percent by mass of sodium chloride in the sand?

Hypothesis

Formulate a hypothesis in response to the question above.

Materials

electronic balance
retort stand
ring clamp
filter funnel
filter paper (to fit filter funnel)
glass rod
3 beakers (250 mL)
drying oven (optional)
wash bottle with distilled water
sample of sand containing $BaCl_2$ and NaCl
solid sodium sulfate, Na_2SO_4

Safety Precautions

- Be careful not to inhale grains of the sodium sulfate, Na_2SO_4.
- Wear gloves at all times. Wash your hands thoroughly after you have finished the investigation.

Procedure

1. Design a rubric, an assessment checklist, or some other means of assessing your experimental design and procedure.
2. Using the materials, design a procedure that will help you solve the problem. Remember that you will need to perform several trials.
3. Before beginning your procedure, show it to your teacher for approval.
4. Obtain a sample of the mixture from your teacher. Record the number of the sample. Carry out your procedure.
5. When you have finished, dispose of the materials as directed by your teacher.
6. Use the following questions to write a complete laboratory report. Include all the equations and calculations that you used.

Analysis

Here are some questions that you will need to answer in your laboratory report.

1. What mass of sand was in the sample? What property of sand did you use to determine this mass?
2. Explain why you used sodium sulfate, Na_2SO_4, in your procedure.
3. What physical and chemical properties of the reactants and products did you use in your procedure?
4. What assumption did you make about the reactivity of sodium chloride with sodium sulfate?
5. What possible sources of error did your procedure introduce? How might they have affected your results?
6. Would your procedure have worked if you had used contaminated sugar instead of contaminated sand? Explain. How could you have changed your procedure?

Assessment

After you complete this investigation:

- Assess your procedure by having a classmate try to duplicate your results.
- Use the rubric you developed to assess the success of your experimental design.
- List ways you would improve your procedure if you were to perform this investigation again.
- Assess your presentation based on the clarity of chemical concepts to be conveyed.

Conclusion

7. Write a statement that gives the percent by mass of barium chloride, $BaCl_2$, and the percent by mass of sodium chloride, NaCl, in the sample of contaminated sand.

UNIT 2 Review

Knowledge/Understanding

True/False

In your notebook, indicate whether each statement is true or false. If a statement is false, rewrite it to make it true.

1. The molecular formula of a compound is the same as its empirical formula.
2. A 2.02 g sample of hydrogen, H_2, contains the same number of molecules as 32.0 g of oxygen, O_2.
3. The average atomic mass of an element is equal to the mass of its most abundant isotope.
4. The numerical value of the molar mass of a compound (expressed in atomic mass units) is the same as its molar mass (expressed in grams).
5. The fundamental unit for chemical quantity is the gram.
6. The mass of 1.00 mol of any chemical compound is always the same.
7. 1.00 mol of any chemical compound or element contains 6.02×10^{23} particles.
8. The value of the Avogadro constant depends on temperature.
9. The empirical formula of an unknown compound must be determined by experiment.
10. The actual yield of most chemical reactions is less than 100%.
11. The theoretical yield of a chemical reaction must be determined by experiment.
12. Stoichiometric calculations are used to determine the products of a chemical reaction.

Multiple Choice

In your notebook, write the letter for the best answer to each question.

13. The number of molecules in 2.0 mol of nitrogen gas, $N_{2(g)}$, is
    (a) 1.8×10^{24}
    (b) 2.4×10^{23}
    (c) 1.2×10^{24}
    (d) 1.2×10^{23}
    (e) 4.0×10^{23}
14. The molar mass of a compound with the empirical formula CH_2O has a mass of approximately 121 g. What is the molecular formula of the compound?
    (a) $C_4H_8O_4$
    (b) $C_2H_6O_2$
    (c) $C_3H_3O_6$
    (d) $C_3H_6O_3$
    (e) CH_6O
15. Read the following statements about balancing chemical equations. Which of these statements is true?
    (a) To be balanced, an equation must have the same number of moles on the left side and the right side.
    (b) A chemical formula may be altered in order to balance a chemical equation.
    (c) To be balanced, a chemical equation must have the same number of each type of atom on both sides.
    (d) It is unacceptable to use fractional coefficients when balancing a chemical equation.
    (e) A skeleton equation contains all the spectator ions.
16. What is the molar mass of ammonium dichromate, $(NH_4)_2Cr_2O_7$?
    (a) 248 g/mol
    (b) 234 g/mol
    (c) 200 g/mol
    (d) 252 g/mol
    (e) 200 g/mol
17. A sample of benzene, C_6H_6, contains 3.0×10^{23} molecules of benzene. How many atoms are in the sample?
    (a) 36×10^{24}
    (b) 1.8×10^{23}
    (c) 3.6×10^{24}
    (d) 2.5×10^{22}
    (e) 3.0×10^{23}
18. What is the molar mass of zinc sulfate heptahydrate, $ZnSO_4 \cdot 7H_2O$?
    (a) 161 g/mol
    (b) 288 g/mol
    (c) 182 g/mol
    (d) 240 g/mol
    (e) 312 g/mol

19. The molecular formula of citric acid monohydrate is $C_6H_8O_7 \cdot H_2O$. Its molecular mass is as follows:
(a) 192 g/mol
(b) 210 g/mol
(c) 188 g/mol
(d) 206 g/mol
(e) 120 g/mol

20. The relative mass of one isotope of sulfur is 31.9721 u. Its abundance is 95.02%. Naturally occurring elemental sulfur has a relative atomic mass of 32.066. The mass number of the one other isotope of sulfur is
(a) 31
(b) 32
(c) 33
(d) 34
(e) 35

21. A sample of ethane, C_2H_6, has a mass of 9.3 g. It contains the same number of atoms as
(a) 23.0 g of sodium, Na
(b) 32.0 g of oxygen, O_2
(c) 48.0 g of ozone, O_3
(d) 30.0 g of formaldehyde, CH_2O
(e) 14.0 g of nitrogen gas, N_2

22. A sample of ozone, O_3, has a mass of 48.0 g. It contains the same number of atoms as
(a) 58.7 g of nickel
(b) 27.0 g of aluminum
(c) 38.0 g of fluorine
(d) 3.02 g of hydrogen
(e) 32.0 g of oxygen

23. Which substance contains 9.03×10^{23} atoms?
(a) 16.0 g of oxygen, O_2
(b) 4.00 g of helium, He
(c) 28.0 g of nitrogen, N_2
(d) 22.0 g of carbon dioxide, CO_2
(e) 8.0 g of methane, CH_4

24. Examine the following formulas. Which formula is an empirical formula?
(a) C_2H_4
(b) C_6H_6
(c) C_2H_2
(d) H_2O_2
(e) $Na_2Cr_2O_7$

25. A sample of sulfur trioxide, SO_3, has a mass of 20 g. How many moles are in the sample?
(a) 0.20
(b) 0.25
(c) 0.50
(d) 0.75
(e) 0.80

26. How many molecules are in 1.00 mg of glucose, $C_6H_{12}O_6$?
(a) 2.18×10^{18}
(b) 3.34×10^{18}
(c) 2.18×10^{21}
(d) 3.34×10^{21}
(e) 3.34×10^{20}

27. A sample that contains carbon, hydrogen, and oxygen is analyzed in a carbon-hydrogen combustion analyzer. All the oxygen in the sample is
(a) converted to the oxygen in carbon dioxide
(b) converted to oxygen in water
(c) mixed with the excess oxygen used to combust the sample
(d) converted to oxygen in carbon dioxide and/or water
(e) both (c) and (d)

28. A compound that contains carbon, hydrogen, and oxygen is going to be analyzed in a carbon-hydrogen combustion analyzer. Before beginning the analysis, which of the following steps must be carried out?
I. Find the mass of the unknown sample.
II. Add the precise amount of oxygen that is needed for combustion.
III. Find the mass of the carbon dioxide and water absorbers.
(a) I only
(b) I and II only
(c) I, II, and III
(d) I and III only
(e) none of the above

Short Answer

29. Answer the following questions, related to the concept of the mole.
    (a) How many N_2 molecules are in a 1.00 mol sample of N_2? How many N atoms are in this sample?
    (b) How many PO_4^{3-} ions are in 2.5 mol of $Ca_3(PO_4)_2$?
    (c) How many O atoms are in 0.47 mol of $Ca_3(PO_4)_2$?
30. Explain how a balanced chemical equation follows the law of conservation of mass. Use an example to illustrate your explanation.
31. List all the information that can be obtained from a balanced chemical equation.
32. Answer the following questions, related to the limiting reactant.
    (a) Explain the concept of the limiting reactant. Use a real-life analogy that is not used in this textbook.
    (b) What is the opposite of a limiting reactant?
    (c) Explain why, in many chemical reactions, the reactants are not present in stoichiometric amounts.
33. Consider a 7.35 g sample of propane, C_3H_8.
    (a) How many moles of propane are in this sample?
    (b) How many molecules of propane are in this sample?
    (c) How many atoms of carbon are in this sample?
34. How many atoms are in 10.0 g of white phosphorus, P_4?
35. A 2.00 g sample of the mineral troegerite, $(UO_2)_3(AsO_4)_2 \cdot 12H_2O$, has 1.38×10^{21} uranium atoms. How many oxygen atoms are present in 2.00 g of troegerite?
36. Fuels that contain hydrogen can be classified according to their mass percent of hydrogen. Which of the following compounds has the greatest mass percent of hydrogen: ethanol, C_2H_5OH, or cetyl palmitate, $C_{32}H_{64}O_2$? Explain your answer.
37. Methyl tertiary butyl ether, or MTBE, is currently used as an octane booster in gasoline. It has replaced the environmentally unsound tetraethyl lead. MTBE has the formula $C_5H_{12}O$. What is the percentage composition of each element in MTBE?
38. Ammonia can be produced in the laboratory by heating ammonium chloride with calcium hydroxide.

    $$2NH_4Cl_{(s)} + Ca(OH)_{2(s)} \rightarrow CaCl_{2(s)} + 2NH_{3(g)} + 2H_2O_{(g)}$$

    8.93 g of ammonium chloride is heated with 7.48 g of calcium hydroxide. What mass of ammonia, NH_3, can be expected? Assume that the reaction has 100% yield.

Inquiry

39. Design an experiment to determine the value of x in sodium thiosulfate, $Na_2S_2O_3 \cdot xH_2O$. Include an outline of your procedure. Describe the data that you need to collect. What assumptions do you need to make?
40. Design an experiment to determine the mole-to-mole ratio of lead(II) nitrate, $Pb(NO_3)_2$, to potassium iodide, KI, in the reaction:
    $Pb(NO_3)_{2(aq)} + KI_{(aq)} \rightarrow PbI_{2(s)} + KNO_{3(aq)}$
    Assume that you have solutions of lead(II) nitrate and potassium iodide. Both of these solutions contain 0.0010 mol of solute per 10 mL of solution.
41. The following reaction can be used to obtain lead(II) chloride, $PbCl_2$. Lead(II) chloride is moderately soluble in warm water.
    $Pb(NO_3)_{2(aq)} + 2NaCl_{(aq)} \rightarrow PbCl_{2(s)} + 2NaNO_{3(aq)}$
    Explain why carrying out this reaction in a warm aqueous solution is unlikely to produce a 100% yield of lead(II) chloride.
42. Imagine that you are given a sheet of aluminum foil that measures 10.0 cm × 10.0 cm. It has a mass of 0.40 g.
    (a) The density of aluminum is 2.70 g/cm^3. Determine the thickness of the aluminum foil, in millimeters.
    (b) Using any of the above information, determine the radius of an aluminum atom, in nanometers. Assume that each aluminum atom is cube-shaped.
    (c) How will your answer to part (b) change if you assume that each aluminum atom is spherical?

(d) What question(s) do your answers to parts (b) and (c) raise?

43. Consider the double displacement reaction below.

$$CaCl_{2(aq)} + Na_2SO_{4(aq)} \rightarrow CaCO_{3(s)} + 2NaCl_{(aq)}$$

(a) Design an experiment to determine the percentage yield of this reaction. Clearly indicate the measurements that need to be taken, along with suggested amounts.

(b) How could the skills of a chemist influence the outcome of this experiment?

Communication

44. It is impossible for a single atom of neon, with a mass of exactly 20.18 u, to exist. Explain why.

45. The molecular mass of a compound is measured in atomic mass units but its molar mass is measured in grams. Explain why this is true.

46. Explain the relationship between an empirical formula and a molecular formula. Use sodium tartrate, $Na_2C_4H_4O_6$, and cyanocolabamin, $C_{63}H_{88}C{\cdot}N_{14}O_{14}P$ (vitamin B_{12}), to illustrate your answer.

47. Explain why an empirical formula can represent many different molecules.

48. Chemists need to know the percentage yield of a reaction. Why is this true, particularly for industrial reactions?

49. Examine the following reaction. List the steps needed to calculate the number of grams of C that can be expected when a given mass of A reacts with a given mass of B. Include proper units for each step. Express the answer in terms of A, B, C, and/or D as necessary.

$$2A + 3B \rightarrow 4C + D$$

Making Connections

50. Reread the Unit 2 opener.

(a) Suppose that you ate a dessert containing poppy seeds. As a result, you tested positive for opiates when you applied for a summer job. What can you do?

(b) Now suppose that you are a policy-writer for a manufacturing company that uses large, dangerous machines. What do you need to consider when you write a policy that deals with employee drug testing? What factors influence whether drug testing is warranted, how often it is warranted, and what substances should be tested for? How will you decide on levels that are acceptable? Do the federal and provincial Human Rights Commissions have anything to say about these issues?

51. The combustion of gasoline in an automobile engine can be represented by the equation

$$2C_8H_{18(g)} + 25O_{2(g)} \rightarrow 16CO_{2(g)} + 18H_2O_{(g)}$$

(a) In a properly tuned engine with a full tank of gas, what reactant do you think is limiting? Explain your reasoning.

(b) A car that is set to run properly at sea level will run poorly at higher altitudes, where the air is less dense. Explain why.

(c) The reaction of atmospheric oxygen with atmospheric nitrogen to form nitrogen monoxide, NO, occurs along with the combustion of fuel.

$$N_{2(g)} + O_{2(g)} \rightarrow 2NO_{(g)}$$

What adjustments need to be made to a vehicle's carburetor or fuel injectors (which control the amount of fuel and air that are mixed) to compensate for this reaction? Explain your answer.

COURSE CHALLENGE

Planet Unknown

Consider the following as you continue to plan for your Chemistry Course Challenge:

- In a reaction that produces a precipitate that you want to recover, what techniques would you use to maximize your percentage yield?
- When you have a given amount of one reactant and you decide the amount of a second reactant, which reactant should be the limiting reactant if you want to maximize your percentage yield?
- Write a balanced chemical equation for the reaction between sodium phosphate and calcium nitrate. Recall that ionic compounds containing sodium are soluble.

UNIT 3

Solutions and Solubility

UNIT 3 CONTENTS

UNIT 3 OVERALL EXPECTATIONS

- What are the properties of solutions, and what methods are used to describe their concentrations?
- What skills are involved in working experimentally with solutions, and in solving quantitative solution problems?
- How can a scientific understanding of solutions help us interpret and make decisions about environmental issues involving solutes and solvents?

Unit Issue Prep

Look ahead to the issue at the end of Unit 3. Start preparing for this issue now by listing the kinds of questions you will need to address, and the ways in which you may like to present your responses.

Keeping fish as pets seems like a simple and straightforward task. You start off with a suitable fish tank, some water, and some fish. Once you have added a few water-plants for decoration, all you have to do is feed the fish daily. It sounds simple.

In fact, providing a safe, life-sustaining environment for fish involves a variety of complex, interdependent factors. Many of these are directly related to chemistry. For example, the presence of ammonia, NH_3, is one of the most common causes of death in poorly maintained fish tanks. Ammonia is produced when uneaten fish food decays. This compound dissolves readily in water. It is toxic to fish in even minute concentrations.

Chlorine is another chemical that can harm fish when it is dissolved in the tank water. Unfortunately, almost all tap water is treated with chlorine. Other factors that affect the quality of water for fish include its acidity (pH), its hardness, and its temperature. For example, warm water contains too little dissolved oxygen. This will cause the fish to suffocate.

Water quality is essential to every living thing. In this unit, you will find out more about water quality. You will learn to identify important solutions in your life, and how to calculate their concentrations. You will also investigate what can happen when acids and bases interact with water and with each other.

CHAPTER 8

Solutions and Their Concentrations

Your environment is made up of many important solutions, or homogeneous mixtures. The air you breathe and the liquids you drink are solutions. So are many of the metallic objects that you use every day. The quality of a solution, such as tap water, depends on the substances that are dissolved in it. "Clean" water may contain small amounts of dissolved substances, such as iron and chlorine. "Dirty" water may have dangerous chemicals dissolved in it.

The difference between clean water and undrinkable water often depends on concentration: the amount of a dissolved substance in a particular quantity of a solution. For example, tap water contains a low concentration of fluoride to help keep your teeth healthy. Water with a high concentration of fluoride, however, could be harmful to your health.

Water is a good solvent for many substances. You may have noticed, however, that grease-stained clothing cannot be cleaned by water alone. Grease is one substance that does not dissolve in water. Why doesn't it dissolve? In this chapter, you will find out why. You will learn how solutions form. You will explore factors that affect a substance's ability to dissolve. You will find out more about the concentration of solutions, and you will have a chance to prepare your own solutions as well.

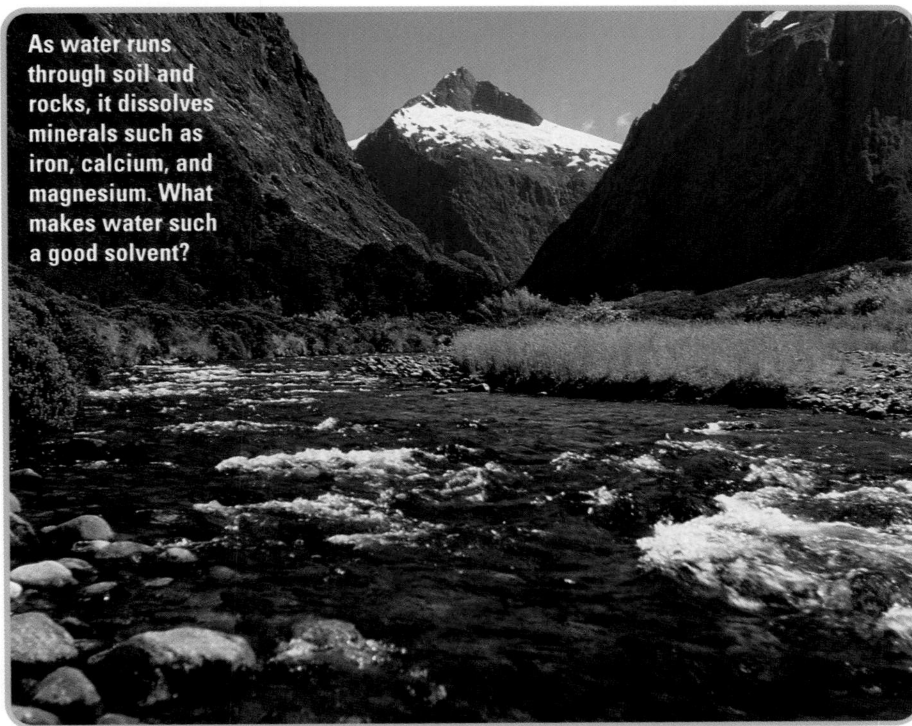

As water runs through soil and rocks, it dissolves minerals such as iron, calcium, and magnesium. What makes water such a good solvent?

Chapter Preview

Concepts and Skills You Will Need

Before you begin this chapter, review the following concepts and skills:

- classifying mixtures (Chapter 1, section 1.3)
- predicting molecular polarity (Chapter 3, section 3.3)
- distinguishing between intermolecular and intramolecular forces (Chapter 3, section 3.2)
- describing the shape and bonding of the water molecule (Chapter 3, section 3.3)
- calculating molar mass (Chapter 5, section 5.3)
- calculating molar amounts (Chapter 5, section 5.3)

8.1 Types of Solutions

Section Preview/ Specific Expectations

In this section, you will

- **explain** solution formation, referring to polar and non-polar solvents
- **identify examples** of solid, liquid, and gas solutions from everyday life
- **communicate** your understanding of the following terms: *solution, solvent, solutes, variable composition, aqueous solution, miscible, immiscible, alloys, solubility, saturated solution, unsaturated solution*

As stated in the chapter opener, a **solution** is a homogeneous mixture. It is uniform throughout. If you analyze any two samples of a solution, you will find that they contain the same substances in the same relative amounts. The simplest solutions contain two substances. Most common solutions contain many substances.

A **solvent** is any substance that has other substances dissolved in it. In a solution, the substance that is present in the largest amount (whether by volume, mass, or number of moles) is usually referred to as the solvent. The other substances that are present in the solution are called the **solutes**.

Pure substances (such as pure water, H_2O) have fixed composition. You cannot change the ratio of hydrogen, H, to oxygen, O, in water without producing an entirely new substance. Solutions, on the other hand, have **variable composition**. This means that different ratios of solvent to solute are possible. For example, you can make a weak or a strong solution of sugar and water, depending on how much sugar you add. Figure 8.1 shows a strong solution of tea and water on the left, and a weak solution of tea and water on the right. The ratio of solvent to solute in the strong solution is different from the ratio of solvent to solute in the weak solution.

Figure 8.1 How can a solution have variable composition yet be uniform throughout?

COURSE CHALLENGE

How can you find out what solutes are dissolved in a sample of water? What physical properties might be useful? You will need answers to these questions when you do your Chemistry Course Challenge.

When a solute dissolves in a solvent, no chemical reaction occurs. Therefore, the solute and solvent can be separated using physical properties, such as boiling point or melting point. For example, water and ethanol have different boiling points. Using this property, a solution of water and ethanol can be separated by the process of distillation. Refer back to Chapter 1, section 1.2. What physical properties, besides boiling point, can be used to separate the components of solutions and other mixtures?

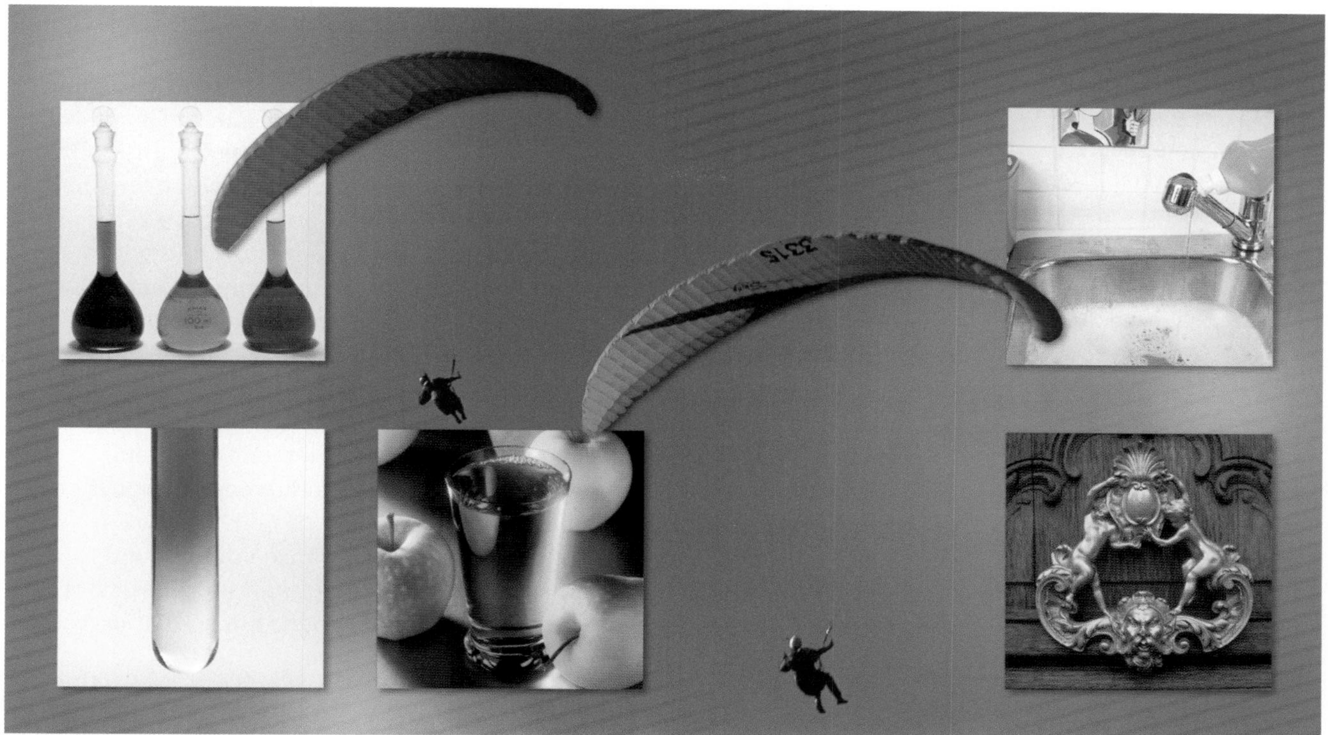

Figure 8.2 **Can you identify the components of some of these solutions?**

A solution can be a gas, a liquid, or a solid. Figure 8.2 shows some examples of solutions. Various combinations of solute and solvent states are possible. For example, a gas can be dissolved in a liquid, or a solid can be dissolved in another solid. Solid, liquid, and gaseous solutions are all around you. Steel is a solid solution of carbon in iron. Juice is a liquid solution of sugar and flavouring dissolved in water. Air is an example of a gaseous solution. The four main components of dry air are nitrogen (78%), oxygen (21%), argon (0.9%), and carbon dioxide (0.03%). Table 8.1 lists some other common solutions.

Table 8.1 Types of Solutions

Original state of solute	Solvent	Examples
gas	gas	air; natural gas; oxygen-acetylene mixture used in welding
gas	liquid	carbonated drinks; water in rivers and lakes containing oxygen
gas	solid	hydrogen in platinum
liquid	gas	water vapour in air; gasoline-air mixture
liquid	liquid	alcohol in water; antifreeze in water
liquid	solid	amalgams, such as mercury in silver
solid	gas	mothballs in air
solid	liquid	sugar in water; table salt in water; amalgams
solid	solid	alloys, such as the copper-nickel alloy used to make coins

mind STRETCH

Take another look at the four components of dry air. Which component would you call the solvent? Which components are the solutes?

Electronic Learning Partner

Your Chemistry 11 Electronic Learning Partner can help you learn more about the properties of water.

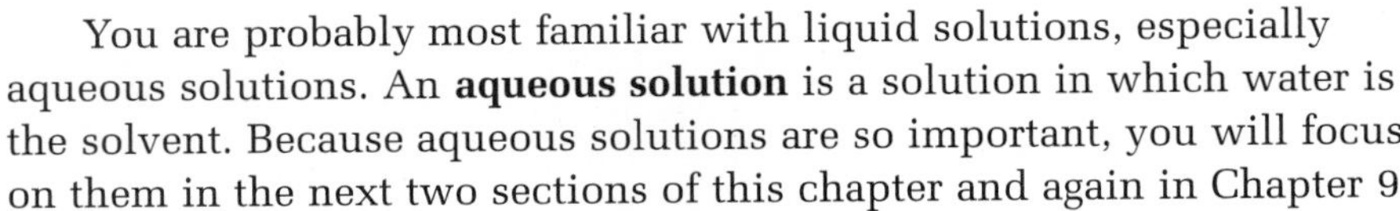

You are probably most familiar with liquid solutions, especially aqueous solutions. An **aqueous solution** is a solution in which water is the solvent. Because aqueous solutions are so important, you will focus on them in the next two sections of this chapter and again in Chapter 9.

Some liquids, such as water and ethanol, dissolve readily in each other in any proportion. That is, any amount of water dissolves in any amount of ethanol. Similarly, any amount of ethanol dissolves in any amount of water. Liquids such as these are said to be **miscible** with each other. Miscible liquids can be combined in any proportions. Thus, either ethanol or water can be considered to be the solvent. Liquids that do not readily dissolve in each other, such as oil and water, are said to be **immiscible**.

CHEM FACT

An alloy that is made of a metal dissolved in mercury is called an *amalgam*. A traditional dental amalgam, used to fill cavities in teeth, contains 50% mercury. Due to concern over the use of mercury, which is toxic, dentists now use other materials, such as ceramic materials, to fill dental cavities.

As you know from Chapter 4, solid solutions of metals are called **alloys**. Adding even small quantities of another element to a metal changes the properties of the metal. Technological advances throughout history have been linked closely to the discovery of new alloys. For example, bronze is an alloy of copper and tin. Bronze contains only about 10% tin, but it is much stronger than copper and more resistant to corrosion. Also, bronze can be melted in an ordinary fire so that castings can be made, as shown in Figure 8.3.

Solubility and Saturation

The ability of a solvent to dissolve a solute depends on the forces of attraction between the particles. There is always some attraction between solvent and solute particles, so some solute always dissolves. The **solubility** of a solute is the amount of solute that dissolves in a given quantity of solvent, at a certain temperature. For example, the solubility of sodium chloride in water at 20°C is 36 g per 100 mL of water.

A **saturated solution** is formed when no more solute will dissolve in a solution, and excess solute is present. For example, 100 mL of a saturated solution of table salt (sodium chloride, NaCl) in water at 20°C contains 36 g of sodium chloride. The solution is saturated with respect to sodium chloride. If more sodium chloride is added to the solution, it will not dissolve. The solution may still be able to dissolve other solutes, however.

An **unsaturated solution** is a solution that is not yet saturated. Therefore, it can dissolve more solute. For example, a solution that contains 20 g of sodium chloride dissolved in 100 mL of water at 20°C is unsaturated. This solution has the potential to dissolve another 16 g of salt, as Figure 8.4 demonstrates.

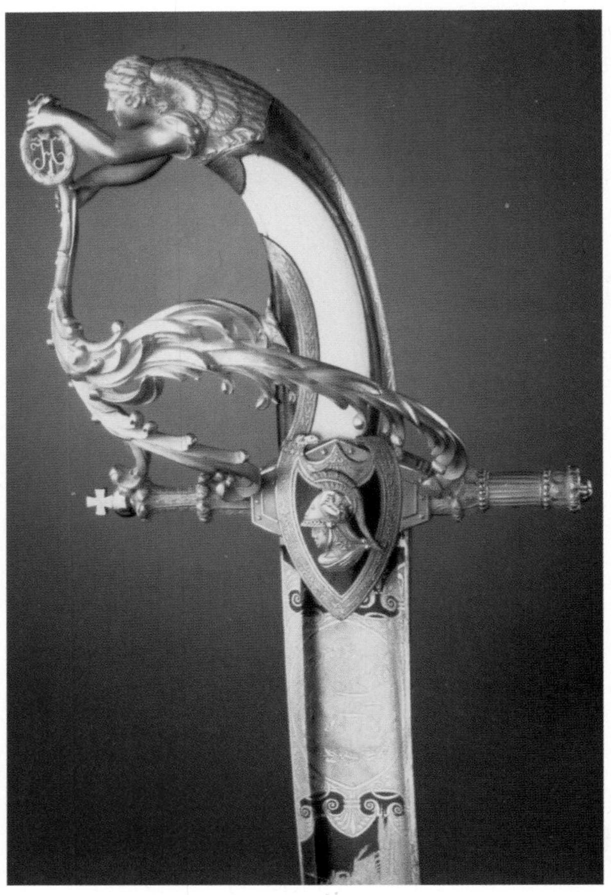

Figure 8.3 **The introduction of the alloy bronze around 3000 BCE led to the production of better-quality tools and weapons.**

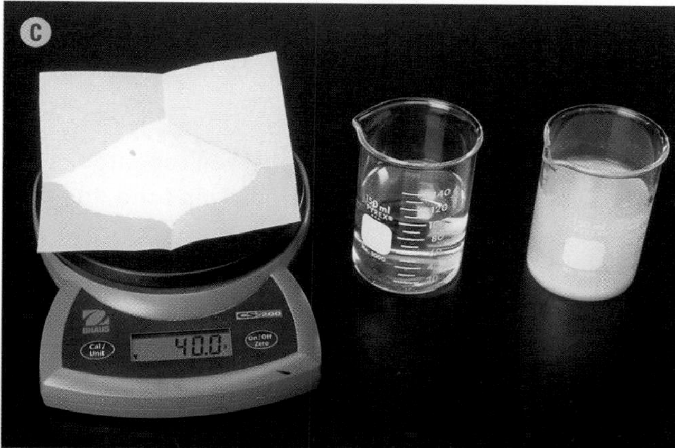

Figure 8.4 **At 20°C, the solubility of table salt in water is 36 g/100 mL.**

A 20 g of NaCl dissolve to form an unsaturated solution.

B 36 g of NaCl dissolve to form a saturated solution.

C 40 g of NaCl are added to 100 mL of water. 36 g dissolve to form a saturated solution. 4 g of undissolved solute are left.

Suppose that a solute is described as *soluble* in a particular solvent. This generally means that its solubility is greater than 1 g per 100 mL of solvent. If a solute is described as *insoluble*, its solubility is less than 0.1 g per 100 mL of solvent. Substances with solubility between these limits are called *sparingly soluble*, or *slightly soluble*. Solubility is a relative term, however. Even substances such as oil and water dissolve in each other to some extent, although in very tiny amounts.

The general terms that are used to describe solubility for solids and liquids do not apply to gases in the same way. For example, oxygen is described as soluble in water. Oxygen from the air dissolves in the water of lakes and rivers. The solubility of oxygen in fresh water at 20°C is only 9 mg/L, or 0.0009 g/100 mL. This small amount of oxygen is enough to ensure the survival of aquatic plants and animals. A solid solute with the same solubility, however, would be described as insoluble in water.

mind STRETCH

Imagine that you are given a filtered solution of sodium chloride. How can you decide whether the solution is saturated or unsaturated?

Identifying Suitable Solvents

Water is a good solvent for many compounds, but it is a poor solvent for others. If you have grease on your hands after adjusting a bicycle chain, you cannot use water to dissolve the grease and clean your hands. You need to use a detergent, such as soap, to help dissolve the grease in the water. You can also use another solvent to dissolve the grease. How can you find a suitable solvent? How can you predict whether a solvent will dissolve a particular solute? Try the Thought Lab on the next page to find out for yourself.

Electronic Learning Partner

Go to your Chemistry 11 Electronic Learning Partner to find out more about the properties of two solvents: water and benzene.

ThoughtLab  Matching Solutes and Solvents

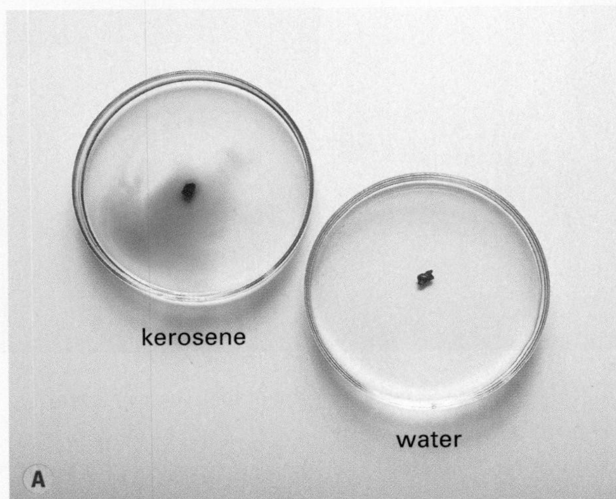

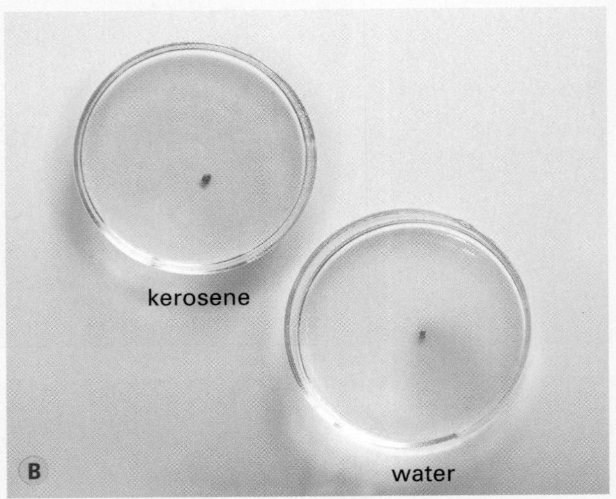

Although there is a solvent for every solute, not all mixtures produce a solution. Table salt dissolves in water but not in kerosene. Oil dissolves in kerosene but not in water. What properties must a solvent and a solute share in order to produce a solution?

In an investigation, the bottom of a Petri dish was covered with water, as shown in photo A. An equal amount of kerosene was added to a second Petri dish. When a crystal of iodine was added to the water, it did not dissolve. When a second crystal of iodine was added to the kerosene, however, it did dissolve.

Procedure

Classify each compound as ionic (containing ions), polar (containing polar molecules), or non-polar (containing non-polar molecules).

(a) iodine, I_2

(b) cobalt(II) chloride, $CoCl_2$

(c) sucrose, $C_{12}H_{22}O_{11}$ (**Hint:** Sucrose contains 8 O–H bonds.)

In photo B, the same experiment was repeated with crystals of cobalt(II) chloride. This time, the crystal dissolved in the water but not in the kerosene.

Analysis

1. Water is a polar molecule. Therefore, it acts as a polar solvent.
   (a) Think about the compounds you classified in the Procedure. Which compounds are soluble in water?
   (b) Assume that the interaction between solutes and solvents that you examine here applies to a wide variety of substances. Make a general statement about the type of solute that dissolves in polar solvents.
2. Kerosene is non-polar. It acts as a non-polar solvent.
   (a) Which of the compounds you classified in the Procedure is soluble in kerosene?
   (b) Make a general statement about the type of solute that dissolves in non-polar solvents.

Section Wrap-up

In this section, you learned the meanings of several important terms, such as *solvent*, *solute*, *saturated solution*, *unsaturated solution*, *aqueous solution*, and *solubility*. You need to know these terms in order to understand the material in the rest of the chapter. In section 8.2, you will examine the factors that affect the rate at which a solute dissolves in a solvent. You will also learn about factors that affect solubility.

Section Review

1 K/U Name the two basic components of a solution.

2 K/U Give examples of each type of solution.

(a) solid solution

(b) liquid solution

(c) gaseous solution (at room temperature)

3 K/U Explain the term "homogeneous mixture."

4 C How do the properties of a homogeneous mixture differ from the properties of a heterogeneous mixture, or mechanical mixture? Use diagrams to explain.

5 K/U Give examples of each type of mixture.

(a) homogeneous mixture

(b) mechanical mixture (heterogeneous mixture)

6 K/U Distinguish between the following terms: soluble, miscible, and immiscible.

7 K/U Distinguish between an alloy and an amalgam. Give one example of each.

8 K/U What type of solute dissolves in a polar solvent, such as water? Give an example.

9 I Potassium bromate, $KBrO_3$, is sometimes added to bread dough to make it easier to work with. Suppose that you are given an aqueous solution of potassium bromate. How can you determine if the solution is saturated or unsaturated?

10 K/U Two different clear, colourless liquids were gently heated in an evaporating dish. Liquid A left no residue, while liquid B left a white residue. Which liquid was a solution, and which was a pure substance? Explain your answer.

11 I You are given three liquids. One is a pure substance, and the second is a solution of two miscible liquids. The third is a solution composed of a solid solute dissolved in a liquid solvent. Describe the procedure you would follow to distinguish between the three solutions.

12 MC In 1989, the oil tanker Exxon Valdez struck a reef in Prince William Sound, Alaska. The accident released 40 million litres of crude oil. The oil eventually covered 26 000 km^2 of water.

(a) Explain why very little of the spilt oil dissolved in the water.

(b) The density of crude oil varies. Assuming a value of 0.86 g/mL, estimate the average thickness of the oil slick that resulted from the Exxon Valdez disaster.

(c) How do you think most of the oil from a tanker accident is dispersed over time? Why would this have been a slow process in Prince William Sound?

13 MC Food colouring is often added to foods such as candies, ice cream, and icing. Are food colouring dyes more likely to be polar or non-polar molecules? Explain your answer.

8.2 Factors That Affect Rate of Dissolving and Solubility

Section Preview/ Specific Expectations

In this section, you will

- **explain** some important properties of water
- **explain** solution formation in terms of intermolecular forces between polar, ionic, and non-polar substances
- **describe** the relationship between solubility and temperature for solids, liquids, and gases
- **communicate** your understanding of the following terms: *rate of dissolving, dipole, dipole-dipole attraction, hydrogen bonding, ion-dipole attractions, hydrated, electrolyte, non-electrolytes*

As you learned in section 8.1, the solubility of a solute is the *amount* of solute that dissolves in a given volume of solvent at a certain temperature. Solubility is determined by the intermolecular attractions between solvent and solute particles. You will learn more about solubility and the factors that affect it later in this section. First, however, you will look at an important property of a solution: the **rate of dissolving**, or how quickly a solute dissolves in a solvent.

The rate of dissolving depends on several factors, including temperature, agitation, and particle size. You have probably used these factors yourself when making solutions like the fruit juice shown in Figure 8.5.

Figure 8.5 Fruit juice is soluble in water. The concentrated juice in this photograph, however, will take a long time to dissolve. Why?

Factors That Affect the Rate of Dissolving

You may have observed that a solute, such as sugar, dissolves faster in hot water than in cold water. In fact, *for most solid solutes, the rate of dissolving is greater at higher temperatures.* At higher temperatures, the solvent molecules have greater kinetic energy. Thus, they collide with the undissolved solid molecules more frequently. This increases their rate of dissolving.

Suppose that you are dissolving a spoonful of sugar in a cup of hot coffee. How can you make the sugar dissolve even faster? You can stir the coffee. *Agitating a mixture by stirring or by shaking the container increases the rate of dissolving.* Agitation brings fresh solvent into contact with undissolved solid.

Finally, you may have noticed that a large lump of solid sugar dissolves more slowly than an equal mass of powdered sugar. *Decreasing the size of the particles increases the rate of dissolving.* When you break up a large mass into many smaller masses, you increase the surface area that is in contact with the solvent. This allows the solid to dissolve faster. Figure 8.6 shows one way to increase the rate of dissolving.

Figure 8.6 Chemists often grind solids into powders using a mortar and pestle. This increases the rate of dissolving.

Solubility and Particle Attractions

By now, you are probably very familiar with the process of dissolving. You already know what it looks like when a solid dissolves in a liquid. Why, however, does something dissolve? What is happening at the molecular level?

The reasons why a solute may or may not dissolve in a solvent are related to the forces of attraction between the solute and solvent particles. These forces include the attractions between two solute particles, the attractions between two solvent particles, and the attractions between a solute particle and a solvent particle. When the forces of attraction between *different* particles in a mixture are stronger than the forces of attraction between *like* particles in the mixture, a solution forms. The strength of each attraction influences the solubility, or the amount of a solute that dissolves in a solvent.

To make this easier to understand, consider the following three steps in the process of dissolving a solid in a liquid.

The Process of Dissolving at the Molecular Level

Step 1 The forces between the particles in the solid must be broken. This step always requires energy. In an ionic solid, the forces that are holding the ions together must be broken. In a molecular solid, the forces between the molecules must be broken.

Step 2 Some of the intermolecular forces between the particles in the liquid must be broken. This step also requires energy.

Step 3 There is an attraction between the particles of the solid and the particles of the liquid. This step always gives off energy.

The solid is more likely to dissolve in the liquid if the energy change in step 3 is greater than the sum of the energy changes in steps 1 and 2. (You will learn more about energy and dissolving in Unit 5.)

Math LINK

Calculate the surface area of a cube with the dimensions 5.0 cm × 5.0 cm × 5.0 cm. Now imagine cutting this cube to form smaller cubes with the dimensions 1.0 cm × 1.0 cm × 1.0 cm. How many smaller cubes could you make? Calculate their total surface area.

Polar and Non-Polar Substances

In the Thought Lab in section 8.1, you observed that solid iodine is insoluble in water. Only a weak attraction exists between the non-polar iodine molecules and the polar water molecules. On the other hand, the intermolecular forces between the water molecules are very strong. As a result, the water molecules remain attracted to each other rather than attracting the iodine molecules.

You also observed that iodine is soluble in kerosene. Both iodine and kerosene are non-polar substances. The attraction that iodine and kerosene molecules have for each other is greater than the attraction between the iodine molecules in the solid and the attraction between the kerosene molecules in the liquid.

The Concept Organizer shown on the next page summarizes the behaviour of polar and non-polar substances in solutions. You will learn more about polar and non-polar substances later in this section.

CHECKPOINT

Remember that the *rate* at which a solute dissolves is different from the *solubility* of the solute. In your notebook, explain briefly and clearly the difference between rate of dissolving and solubility.

Concept Organizer Polar and Non-Polar Compounds

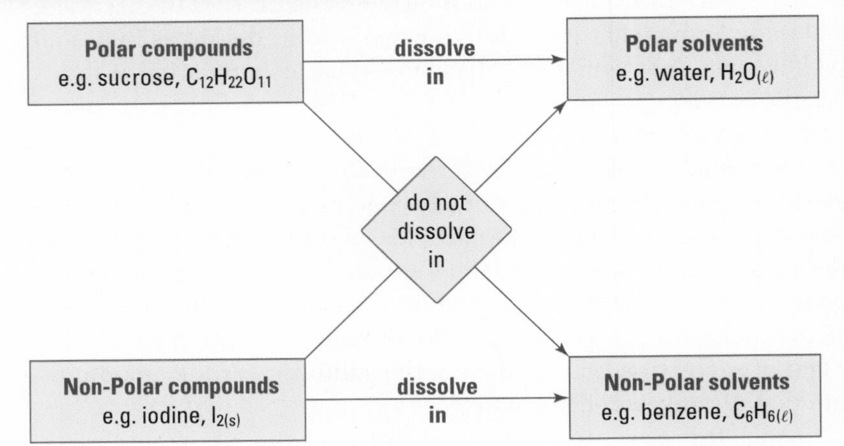

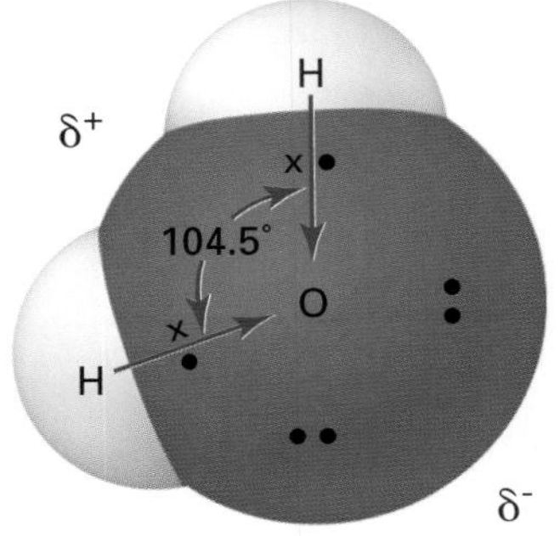

Figure 8.7 **The bent shape and polar bonds of a water molecule give it a permanent dipole.**

Solubility and Intermolecular Forces

You have learned that solubility depends on the forces between particles. Thus, polar substances dissolve in polar solvents, and non-polar substances dissolve in non-polar solvents. What are these forces that act between particles?

In Chapter 3, section 3.3, you learned that a water molecule is polar. It has a relatively large negative charge on the oxygen atom, and positive charges on both hydrogen atoms. Molecules such as water, which have charges separated into positive and negative regions, are said to have a permanent dipole. A **dipole** consists of two opposite charges that are separated by a short distance. Figure 8.7 shows the dipole of a water molecule.

Dipole-Dipole Attractions

The attraction between the opposite charges on two different polar molecules is called a **dipole-dipole attraction**. Dipole-dipole attractions are intermolecular. This means that they act *between* molecules. Usually they are only about 1% as strong as an ionic or covalent bond. In water, there is a special dipole-dipole attraction called **hydrogen bonding**. It occurs between the oxygen atom on one molecule and the hydrogen atoms on a nearby molecule. Hydrogen bonding is much stronger than an ordinary dipole-dipole attraction. It is much weaker, however, than the covalent bond between the oxygen and hydrogen atoms in a water molecule. Figure 8.8 illustrates hydrogen bonding between water molecules.

Electronic Learning Partner

Go to the Chemistry 11 Electronic Learning Partner to find out how hydrogen bonding leads to water's amazing surface tension.

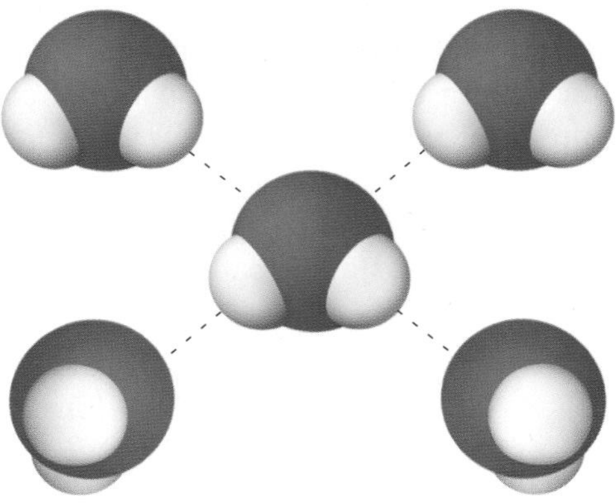

Figure 8.8 **Hydrogen bonding between water molecules is shown as dotted lines. The H atoms on each molecule are attracted to O atoms on other water molecules.**

Ion-Dipole Attractions

Ionic crystals consist of repeating patterns of oppositely charged ions, as shown in Figure 8.9. What happens when an ionic compound comes in contact with water? The negative end of the dipole on some water molecules attracts the cations on the surface of the ionic crystal. At the same time, the positive end of the water dipole attracts the anions. These attractions are known as **ion-dipole attractions**: attractive forces between an ion and a polar molecule. If ion-dipole attractions can replace the ionic bonds between the cations and anions in an ionic compound, the compound will dissolve. *Generally an ionic compound will dissolve in a polar solvent.* For example, table salt (sodium chloride, NaCl) is an ionic compound. It dissolves well in water, which is a polar solvent.

When ions are present in an aqueous solution, each ion is **hydrated**. This means that it is surrounded by water molecules. Hydrated ions can move through a solution and conduct electricity. A solute that forms an aqueous solution with the ability to conduct electricity is called an **electrolyte**. Figure 8.10 shows hydrated sodium chloride ions, which are electrolytes.

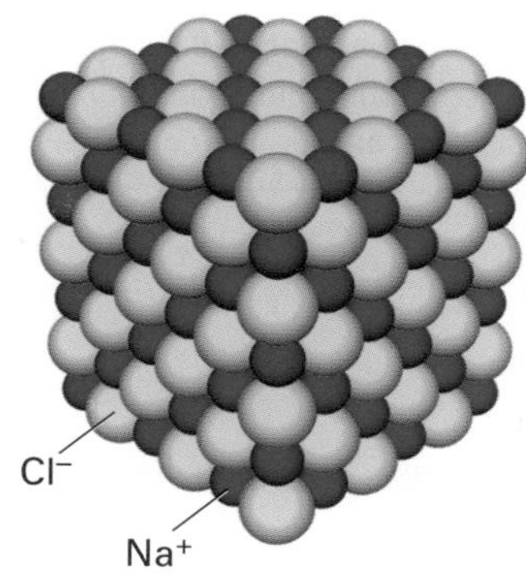

Figure 8.9 Ionic crystals have very ordered structures.

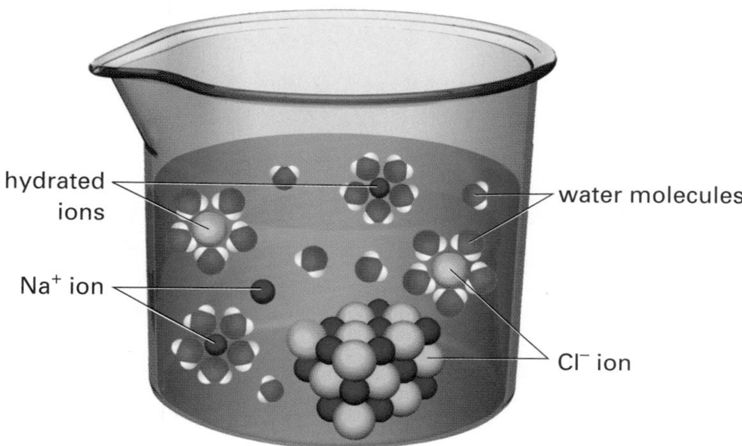

Figure 8.10 Ion-dipole attractions help to explain why sodium chloride dissolves in water.

An Exception: Insoluble Ionic Compounds

Although most ionic compounds are soluble in water, some are not very soluble at all. The attraction between ions is difficult to break. As a result, compounds with very strong ionic bonds, such as silver chloride, tend to be less soluble in water than compounds with weak ionic bonds, such as sodium chloride.

Predicting Solubility

You can predict the solubility of a binary compound, such as mercury(II) sulfide, HgS, by comparing the electronegativity of each element in the compound. If there is a large difference in the two electronegativities, the bond between the elements is polar or even ionic. This type of compound probably dissolves in water. If there is only a small difference in the two electronegativities, the bond is not polar or ionic. This type of compound probably does not dissolve in water. For example, the electronegativity of mercury is 1.9. The electronegativity of sulfur is 2.5. The difference in these two electronegativities is small, only 0.6. Therefore, you can predict that mercury(II) sulfide is insoluble in water. In Chapter 9, you will learn another way to predict the solubility of ionic compounds in water.

CHECKPOINT

Look back at the Concept Organizer on page 292. Where do ionic compounds belong in this diagram?

The Solubility of Covalent Compounds

Many covalent compounds do not have negative and positive charges to attract water molecules. Thus they are not soluble in water. There are some exceptions, however. Methanol (a component of windshield washer fluid), ethanol (the "alcohol" in alcoholic beverages), and sugars (such as sucrose) are examples of covalent compounds that are extremely soluble in water. These compounds dissolve because their molecules contain polar bonds, which are able to form hydrogen bonds with water.

For example, sucrose molecules have a number of sites that can form a hydrogen bond with water to replace the attraction between the sucrose molecules. (See Figure 8.11.) The sucrose molecules separate and become hydrated, just like dissolved ions. The molecules remain neutral, however. As a result, sucrose and other soluble covalent compounds do not conduct electricity when dissolved in water. They are **non-electrolytes**.

Electronic Learning Partner

The Chemistry 11 Electronic Learning Partner contains a video clip describing how water dissolves ionic and some covalent compounds. This will be useful if you are having difficulty visualizing particle attractions.

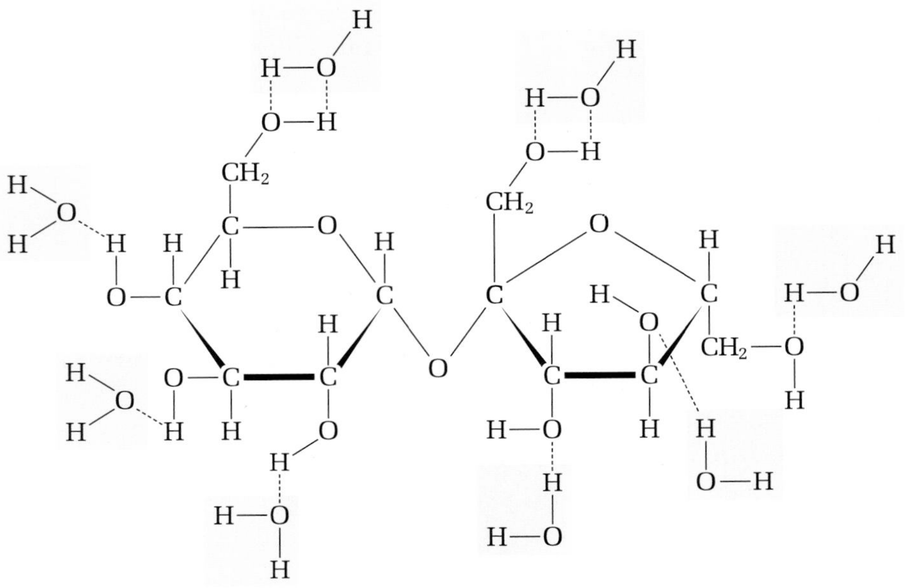

Figure 8.11 **A sucrose molecule contains several O–H atom connections. The O–H bond is highly polar, with the H atom having the positive charge. The negative charges on water molecules form hydrogen bonds with a sucrose molecule, as shown by the dotted lines.**

Insoluble Covalent Compounds

The covalent compounds that are found in oil and grease are insoluble in water. They have no ions or highly polar bonds, so they cannot form hydrogen bonds with water molecules. Non-polar compounds tend to be soluble in non-polar solvents, such as benzene or kerosene. The forces between the solute molecules are replaced by the forces between the solute and solvent molecules.

In general, *ionic solutes and polar covalent solutes both dissolve in polar solvents. Non-polar solutes dissolve in non-polar solvents.* The phrase *like dissolves like* summarizes these observations. It means that solutes and solvents that have similar properties form solutions.

If a compound has both polar and non-polar components, it may dissolve in both polar and non-polar solvents. For example, acetic acid, CH_3COOH, is a liquid that forms hydrogen bonds with water. It is fully miscible with water. Acetic acid also dissolves in non-polar solvents, such as benzene and carbon tetrachloride, because the CH_3 component is non-polar.

Factors That Affect Solubility

You have taken a close look at the attractive forces between solute and solvent particles. Now that you understand why solutes dissolve, it is time to examine the three factors that affect solubility: molecule size, temperature, and pressure. Notice that these three factors are similar to the factors that affect the rate of dissolving. Be careful not to confuse them.

Molecule Size and Solubility

Small molecules are often more soluble than larger molecules. Methanol, CH_3OH, and ethanol, CH_3CH_2OH, are both completely miscible with water. These compounds have OH groups that form hydrogen bonds with water. Larger molecules with the same OH group but more carbon atoms, such as pentanol, $CH_3CH_2CH_2CH_2CH_2OH$, are far less soluble. All three compounds form hydrogen bonds with water, but the larger pentanol is less polar overall, making it less soluble. Table 8.2 compares five molecules by size and solubility.

Table 8.2 Solubility and Molecule Size

Name of compound	methanol	ethanol	propanol	butanol	pentanol
Chemical formula	CH_3OH	CH_3CH_2OH	$CH_3CH_2CH_2OH$	$CH_3(CH_2)_3OH$	$CH_3(CH_2)_4OH$
Solubility	infinitely soluble	infinitely soluble	very soluble	9 g/100 mL (at 25°C)	3 g/100 mL (at 25°C)

Temperature and Solubility

At the beginning of this section, you learned that temperature affects the rate of dissolving. Temperature also affects solubility. You may have noticed that solubility data always include temperature. The solubility of a solute in water, for example, is usually given as the number of grams of solute that dissolve in 100 mL of water at a specific temperature. (See Table 8.2 for two examples.) Specifying temperature is essential, since the solubility of a substance is very different at different temperatures.

When a solid dissolves in a liquid, energy is needed to break the strong bonds between particles in the solid. At higher temperatures, more energy is present. Thus, *the solubility of most solids increases with temperature.* For example, caffeine's solubility in water is only 2.2 g/100 mL at 25°C. At 100°C, however, caffeine's solubility increases to 40 g/100 mL.

The bonds between particles in a liquid are not as strong as the bonds between particles in a solid. When a liquid dissolves in a liquid, additional energy is not needed. Thus, *the solubility of most liquids is not greatly affected by temperature.*

Gas particles move quickly and have a great deal of kinetic energy. When a gas dissolves in a liquid, it loses some of this energy. At higher temperatures, the dissolved gas gains energy again. As a result, the gas comes out of solution and is less soluble. Thus, *the solubility of gases decreases with higher temperatures.*

In the next investigation, you will observe and graph the effect of temperature on the solubility of a solid dissolved in a liquid solvent, water. As you have learned, most solid solutes become more soluble at higher temperatures. By determining the solubility of a solute at various temperatures, you can make a graph of solubility against temperature. The curve of best fit, drawn through the points, is called the *solubility curve.* You can use a solubility curve to determine the solubility of a solute at any temperature in the range shown on the graph.

The link between cigarette smoking and lung cancer is well known. Other cancers are also related to smoking. It is possible that a smoker who consumes alcohol may be at greater risk of developing stomach cancer. When a person smokes, a thin film of tar forms inside the mouth and throat. The tar from cigarette smoke contains many carcinogenic (cancer-causing) compounds. These compounds are non-polar and do not dissolve in saliva. They are more soluble in alcohol, however. As a result, if a smoker drinks alcohol, carcinogenic compounds can be washed into the stomach.

Investigation 8-A

SKILL FOCUS

Predicting

Performing and recording

Analyzing and interpreting

Plotting Solubility Curves

In this investigation, you will determine the temperature at which a certain amount of potassium nitrate is soluble in water. You will then dilute the solution and determine the solubility again. By combining your data with other students' data, you will be able to plot a solubility curve.

Question

What is the solubility curve of KNO_3?

Prediction

Draw a sketch to show the shape of the curve you expect for the solubility of a typical solid dissolving in water at different temperatures. Plot solubility on the *y*-axis and temperature on the *x*-axis.

Safety Precautions

- Before lighting the Bunsen burner, check that there are no flammable solvents nearby. If you are using a Bunsen burner, tie back long hair and loose clothing. Be careful of the open flame.
- After turning it on, be careful not to touch the hot plate.

Materials

large test tube
balance
stirring wire
two-hole stopper to fit the test tube, with a thermometer inserted in one hole
400 mL beaker
graduated cylinder or pipette or burette
hot plate or Bunsen burner with ring clamps and wire gauze
retort stand and thermometer clamp
potassium nitrate, KNO_3
distilled water

Procedure

1. Read through the steps in this Procedure. Prepare a data table to record the mass of the solute, the initial volume of water, the total volume of water after step 9, and the temperatures at which the solutions begin to crystallize.
2. Put the test tube inside a beaker for support. Place the beaker on a balance pan. Set the reading on the balance to zero. Then measure 14.0 g of potassium nitrate into the test tube.
3. Add one of the following volumes of distilled water to the test tube, as assigned by your teacher: 10.0 mL, 15.0 mL, 20.0 mL, 25.0 mL, 30.0 mL. (If you use a graduated cylinder, remember to read the volume from the bottom of the water meniscus. You can make a more accurate volume measurement using either a burette or a pipette.)
4. Pour about 300 mL of tap water into the beaker. Set up a hot-water bath using a hot plate, retort stand, and thermometer clamp. Alternatively, use a Bunsen burner, retort stand, ring clamp, thermometer clamp, and wire gauze.
5. Put the stirring wire through the second hole of the stopper. Insert the stopper, thermometer, and wire into the test tube. Make sure that the thermometer bulb is below the surface of the solution. (Check the diagram on the next page to make sure that you have set up the apparatus properly thus far.)
6. Place the test tube in the beaker. Secure the test tube and thermometer to the retort stand, using clamps. Begin heating the water bath gently.
7. Using the stirring wire, stir the mixture until the solute completely dissolves. Turn the heat source off, and allow the solution to cool.

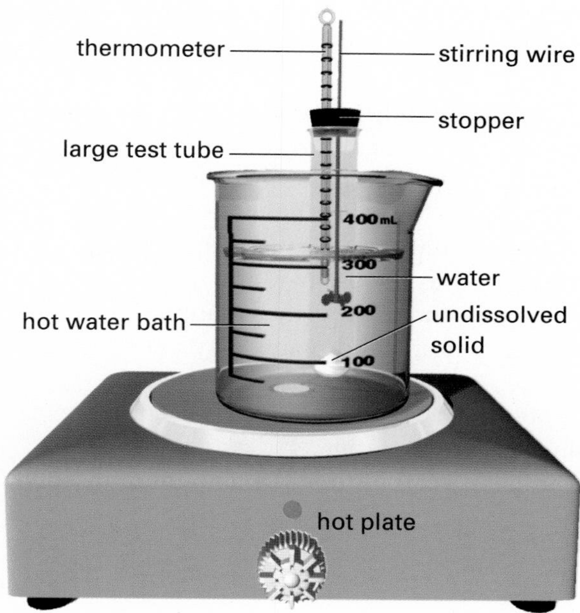

8. Continue stirring. Record the temperature at which crystals begin to appear in the solution.
9. Remove the stopper from the test tube. Carefully add 5.0 mL of distilled water. The solution is now more dilute and therefore more soluble. Crystals will appear at a lower temperature.
10. Put the stopper, with the thermometer and stirring wire, back in the test tube. If crystals have already started to appear in the solution, begin warming the water bath again. Repeat steps 7 and 8.
11. If no crystals are present, stir the solution while the water bath cools. Record the temperature at which crystals first begin to appear.
12. Dispose of the aqueous solutions of potassium nitrate into the labelled waste container.

Analysis

1. Use the volume of water assigned by your teacher to calculate how much solute dissolved in 100 mL of water. Use the following equation to help you:

$$\frac{x\text{ g}}{100\text{ mL}} = \frac{14.0\text{ g}}{\text{your volume}}$$

This equation represents the solubility of KNO_3 at the temperature at which you recorded the first appearance of crystals. Repeat your calculation to determine the solubility after the solution was diluted. Your teacher will collect and display all the class data for this investigation.

2. Some of your classmates were assigned the same volume of water that you were assigned. Compare the temperatures they recorded for their solutions with the temperatures you recorded. Comment on the precision of the data. Should any data be removed before averaging?
3. Average the temperatures at which crystal formation occurs for solutions that contain the same volume of water. Plot these data on graph paper. Set up your graph sideways on the graph paper (landscape orientation). Plot solubility on the vertical axis. (The units are grams of solute per 100 mL of water.) Plot temperature on the horizontal axis.
4. Draw the best smooth curve through the points. (Do not simply join the points.) Label each axis. Give the graph a suitable title.

Conclusions

5. Go back to the sketch you drew to predict the solubility of a typical solid dissolving in water at different temperatures. Compare the shape of your sketch with the shape of your graph.
6. Use your graph to *interpolate* the solubility of potassium nitrate at
   (a) 60°C **(b)** 40°C
7. Use your graph to *extrapolate* the solubility of potassium nitrate at
   (a) 80°C **(b)** 20°C

Application

8. At what temperature can 40 mL of water dissolve the following quantities of potassium nitrate?
   (a) 35.0 g **(b)** 20.0 g

Figure 8.12 This image shows the result of heat pollution. Warmer water contains less dissolved oxygen.

Heat Pollution: A Solubility Problem

For most solids, and almost all ionic substances, solubility increases as the temperature of the solution increases. Gases, on the other hand, always become *less* soluble as the temperature increases. This is why a refrigerated soft drink tastes fizzier than the same drink at room temperature. The warmer drink contains less dissolved carbon dioxide than the cooler drink.

This property of gases makes heat pollution a serious problem. Many industries and power plants use water to cool down overheated machinery. The resulting hot water is then returned to local rivers or lakes. Figure 8.12 shows steam rising from a "heat-polluted" river. Adding warm water into a river or lake does not seem like actual pollution. The heat from the water, however, increases the temperature of the body of water. As the temperature increases, the dissolved oxygen in the water decreases. Fish and other aquatic wildlife and plants may not have enough oxygen to breathe.

The natural heating of water in rivers and lakes can pose problems, too. Fish in warmer lakes and rivers are particularly vulnerable in the summer. When the water warms up even further, the amount of dissolved oxygen decreases.

ExpressLab  The Effect of Temperature on Soda Water

In this Express Lab, you will have a chance to see how a change in temperature affects the dissolved gas in a solution. You will be looking at the pH of soda water. A low pH (1–6) indicates that the solution is acidic. You will learn more about pH in Chapter 10.

Safety Precautions

- If using a hot plate, avoid touching it when it is hot.
- If using a Bunsen burner, check that there are no flammable solvents nearby.

Procedure

1. Open a can of cool soda water. (Listen for the sound of excess carbon dioxide escaping.) Pour about 50 mL into each of two 100 mL beakers. Note the rate at which bubbles form. Record your observations.
2. Add a few drops of universal indicator to both beakers. Record the colour of the solutions. Then estimate the pH.
3. Measure and record the mass of each beaker. Measure and record the temperature of the soda water.
4. Place one beaker on a heat source. Heat it to about 50°C. Compare the rate of formation of the bubbles with the rate of formation in the beaker of cool soda water. Record any change in colour in the heated solution. Estimate its pH.
5. Allow the heated solution to cool. Again record any change in colour in the solution. Estimate its pH.
6. Measure and record the masses of both beakers. Determine any change in mass by comparing the final and initial masses.

Analysis

1. Which sample of soda water lost the most mass? Explain your observation.
2. Did the heated soda water become more or less acidic when it was heated? Explain why you think this change happened.

Pressure and Solubility

The final factor that affects solubility is pressure. Changes in pressure have hardly any effect on solid and liquid solutions. Such changes do affect the solubility of a gas in a liquid solvent, however. The solubility of the gas is directly proportional to the pressure of the gas above the liquid. For example, the solubility of oxygen in lake water depends on the air pressure above the lake.

When you open a carbonated drink, you can observe the effect of pressure on solubility. Figure 8.13 shows this effect. Inside a soft drink bottle, the pressure of the carbon dioxide gas is very high: about 400 kPa. When you open the bottle, you hear the sound of escaping gas as the pressure is reduced. Carbon dioxide gas escapes quickly from the bottle, since the pressure of the carbon dioxide in the atmosphere is much lower: only about 0.03 kPa. The solubility of the carbon dioxide in the liquid soft drink decreases greatly. Bubbles begin to rise in the liquid as gas comes out of solution and escapes. It takes a while for all the gas to leave the solution, so you have time to enjoy the taste of the soft drink before it goes "flat."

Figure 8.14 illustrates another example of dissolved gases and pressure. As a scuba diver goes deeper underwater, the water pressure increases. The solubility of nitrogen gas, which is present in the lungs, also increases. Nitrogen gas dissolves in the diver's blood. As the diver returns to the surface, the pressure acting on the diver decreases. The nitrogen gas in the blood comes out of solution. If the diver surfaces too quickly, the effect is similar to opening a soft drink bottle. Bubbles of nitrogen gas form in the blood. This leads to a painful and sometimes fatal condition known as "the bends." You will learn more about gases and deep-sea diving in Chapter 11.

Figure 8.13 What happens when the pressure of the carbon dioxide gas in a soft drink bottle is released? The solubility of the gas in the soft drink solution decreases.

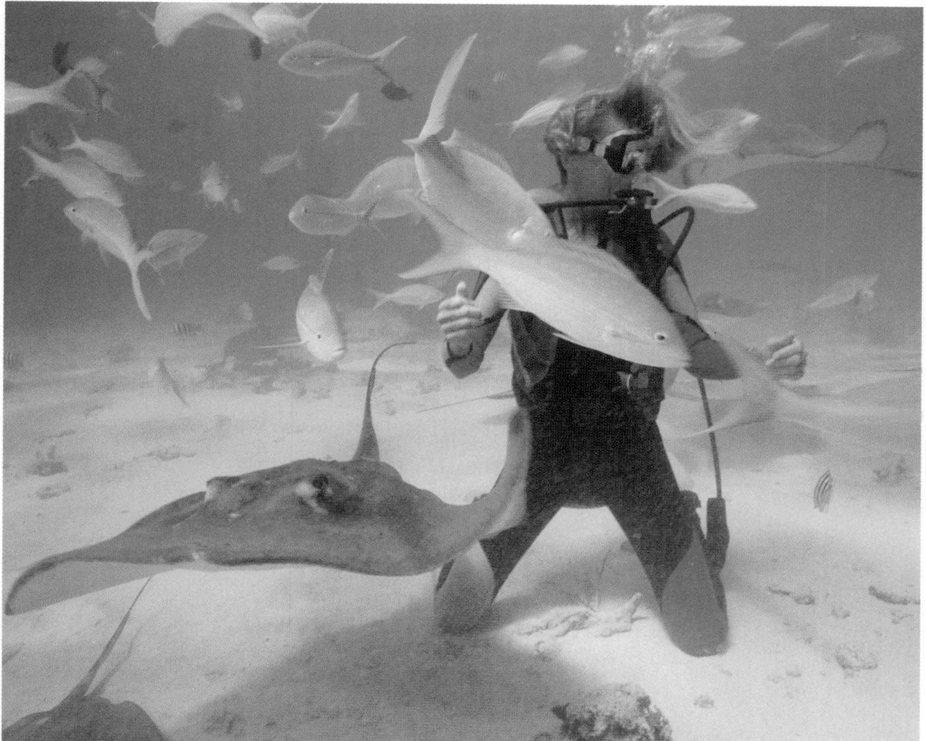

Figure 8.14 Scuba divers must heed the effects of decreasing water pressure on dissolved nitrogen gas in their blood. They must surface slowly to avoid "the bends."

Do you crack your knuckles? The sound you hear is another example of the effect of pressure on solubility. Joints contain fluid. When a joint is suddenly pulled or stretched, the cavity that holds the fluid gets larger. This causes the pressure to decrease. A bubble of gas forms, making the sound you hear. You cannot repeatedly crack your knuckles because it takes some time for the gas to re-dissolve.

Chemistry Bulletin

Science | Technology | Society | Environment

Solvents and Coffee: What's the Connection?

The story of coffee starts with the coffee berry. First the pulp of the berry is removed. This leaves two beans, each containing 1% to 2% caffeine. The beans are soaked in water and natural enzymes to remove the outer parchment husk and to start a slight fermentation process. Once the beans have been fermented, they are dried and roasted. Then the coffee is ready for grinding. Grinding increases the surface area of the coffee. Thus, finer grinds make it easier to dissolve the coffee in hot water.

Decaffeinated coffee satisfies people who like the smell and taste of coffee but cannot tolerate the caffeine. How is caffeine removed from coffee?

All the methods of extracting caffeine take place before the beans are roasted. Caffeine and the other organic compounds that give coffee its taste are mainly non-polar. (Caffeine does contain some polar bonds, however, which allows it to dissolve in hot water.) Non-polar solvents, such as benzene and trichloroethene, were once used to dissolve and remove caffeine from the beans. These chemicals are now considered to be too hazardous. Today most coffee manufacturers use water or carbon dioxide as solvents.

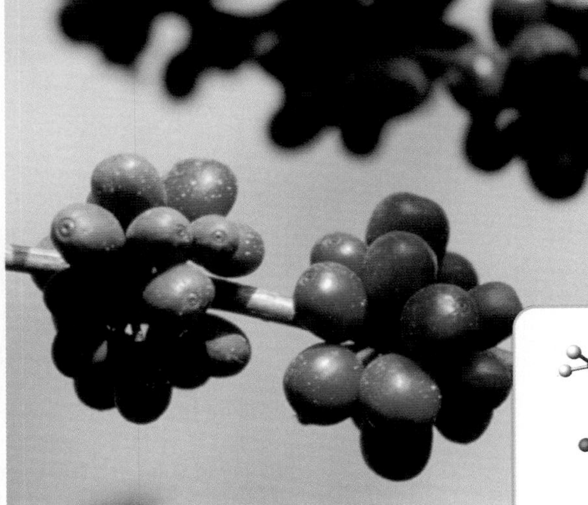

In the common Swiss Water Process, coffee beans are soaked in hot water. This dissolves the caffeine and the flavouring compounds from the beans. The liquid is passed through activated carbon filters. The filters retain the caffeine, but let the flavouring compounds pass through. The filtered liquid, now caffeine-free, is sprayed back onto the beans. The beans reabsorb the flavouring compounds. Now they are ready for roasting.

Carbon dioxide gas is a normal component of air. In the carbon dioxide decaffeination process, the gas is raised to a temperature of at least 32°C. Then it is compressed to a pressure of about 7400 kPa. At this pressure, it resembles a liquid but can flow like a gas. The carbon dioxide penetrates the coffee beans and dissolves the caffeine. When the pressure returns to normal, the carbon dioxide reverts to a gaseous state. The caffeine is left behind.

What happens to the caffeine that is removed by decaffeination? Caffeine is so valuable that it is worth more than the cost of taking it out of the beans. It is extensively used in the pharmaceutical industry, and for colas and other soft drinks.

Making Connections

1. As you have read, water is a polar liquid and the soluble fractions of the coffee grounds are non-polar. Explain, in chemical terms, how caffeine and the coffee flavour and aroma are transferred to hot coffee.
2. Why does hot water work better in the brewing process than cold water?
3. In chemical terms, explain why fine grinds of coffee make better coffee.

How can caffeine form hydrogen bonds with water?

Section Wrap-up

In this section, you examined the factors that affect the rate of dissolving: temperature, agitation, and particle size. Next you looked at the forces between solute and solvent particles. Finally, you considered three main factors that affect solubility: molecule size, temperature, and pressure. In section 8.3, you will learn about the effects of differing amounts of solute dissolved in a certain amount of solvent.

Section Review

1 K/U Describe the particle attractions that occur as sodium chloride dissolves in water.

2 K/U When water vaporizes, which type of attraction, intramolecular or intermolecular, is broken? Explain.

3 K/U Describe the effect of increasing temperature on the solubility of

(a) a typical solid in water

(b) a gas in water

4 K/U Sugar is more soluble in water than salt. Why does a salt solution (brine) conduct electricity, while a sugar solution does not?

5 K/U Dissolving a certain solute in water releases heat. Dissolving a different solute in water absorbs heat. Explain why.

6 I The graph below shows the solubility of various substances plotted against the temperature of the solution.

(a) Which substance decreases in solubility as the temperature increases?

(b) Which substance is least soluble at room temperature? Which substance is most soluble at room temperature?

(c) The solubility of which substance is least affected by a change in temperature?

(d) At what temperature is the solubility of potassium chlorate equal to 40 g/100 g of water?

(e) 20 mL of a saturated solution of potassium nitrate at 50°C is cooled to 20°C. Approximately what mass of solid will precipitate from the solution? Why is it not possible to use the graph to interpolate an accurate value?

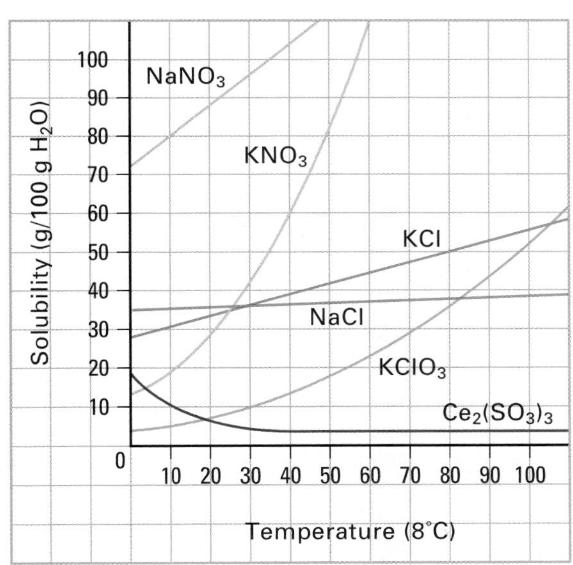

7 I A saturated solution of potassium nitrate was prepared at 70°C and then cooled to 55°C. Use your graph from Investigation 8-A to predict the fraction of the dissolved solute that crystallized out of the solution.

8 MC Would you expect to find more mineral deposits near a thermal spring or near a cool mountain spring? Explain.

Unit Issue Prep

Think about how the properties of water affect its behaviour in the environment. Look ahead to the Unit 3 Issue. How could water's excellent ability as a solvent become a problem?

8.3 The Concentration of Solutions

Section Preview/ Specific Expectations

In this section, you will

- **solve problems** involving the concentration of solutions
- **express** concentration as grams per 100 mL, mass and volume percents, parts per million and billion, and moles per litre
- **communicate** your understanding of the following terms: *concentration, mass/volume percent, mass/mass percent, volume/volume percent, parts per million, parts per billion, molar concentration*

Material Safety Data Sheet

Component Name	CAS Number
PHENOL, 100% Pure	108952

SECTION III: Hazards Identification

- Very hazardous in case of ingestion, inhalation, skin contact, or eye contact.
- Product is corrosive to internal membranes when ingested.
- Inhalation of vapours may damage central nervous system. Symptoms: nausea, headache, dizziness.
- Skin contact may cause itching and blistering.
- Eye-contact may lead to corneal damage or blindness.
- Severe over-exposure may lead to lung-damage, choking, or coma.

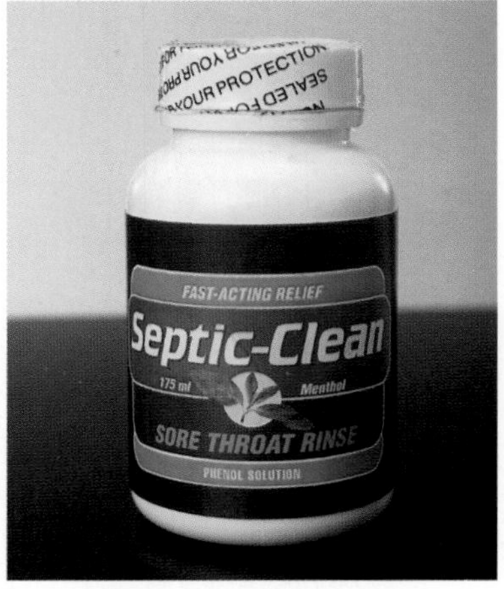

Figure 8.15 Should phenol be banned from drugstores?

Phenol is a hazardous liquid, especially when it is at room temperature. It is a volatile chemical. Inhaling phenol adversely affects the central nervous system, and can lead to a coma. Inhalation is not the only danger. Coma and death have been known to occur within 10 min after phenol has contacted the skin. Also, as little as 1 g of phenol can be fatal if swallowed.

Would you expect to find such a hazardous chemical in over-the-counter medications? Check your medicine cabinet at home. You may find phenol listed as an ingredient in throat sprays and in lotions to relieve itching. You may also find it used as an antiseptic or disinfectant. Is phenol a hazard or a beneficial ingredient in many medicines? This depends entirely on **concentration**: the amount of solute per quantity of solvent. At high concentrations, phenol can kill. At low concentrations, it is a safe component of certain medicines.

Modern analytical tests allow chemists to detect and measure almost any chemical at extremely low concentrations. In this section, you will learn about various ways that chemists use to express the concentration of a solution. As well, you will find the concentration of a solution by experiment.

Concentration as a Mass/Volume Percent

Recall that the solubility of a compound at a certain temperature is often expressed as the mass of the solute per 100 mL of solvent. For example, you know that the solubility of sodium chloride is 36 g/100 mL of water at room temperature. The final volume of the sodium chloride solution may or may not be 100 mL. It is the volume of the solvent that is important.

Chemists often express the concentration of an *unsaturated* solution as the mass of solute dissolved per volume of the *solution*. This is different from solubility. It is usually expressed as a percent relationship. A **mass/volume percent** gives the mass of solute dissolved in a volume of solution, expressed as a percent. The mass/volume percent is also referred to as the *percent (m/v)*.

$$\text{Mass/volume percent} = \frac{\text{Mass of solute (in g)}}{\text{Volume of solution (in mL)}} \times 100\%$$

Suppose that a hospital patient requires an intravenous drip to replace lost body fluids. The intravenous fluid may be a saline solution that contains 0.9 g of sodium chloride dissolved in 100 mL of solution, or 0.9% (m/v). *Notice that the number of grams of solute per 100 mL of solution is numerically equal to the mass/volume percent.* Explore this idea further in the following problems.

PROBEWARE

If you have access to probeware, do the Chemistry 11 lab, "Concentration of Solutions" now.

Sample Problem

Solving for a Mass/Volume Percent

Problem

A pharmacist adds 2.00 mL of distilled water to 4.00 g of a powdered drug. The final volume of the solution is 3.00 mL. What is the concentration of the drug in g/100 mL of solution? What is the percent (m/v) of the solution?

What Is Required?

You need to calculate the concentration of the solution, in grams of solute dissolved in 100 mL of solution. Then you need to express this concentration as a mass/volume percent.

What Is Given?

The mass of the dissolved solute is 4.00 g. The volume of the solution is 3.00 mL.

Plan Your Strategy

There are two possible methods for solving this problem.

Method 1

Use the formula

$$\text{Mass/volume percent} = \frac{\text{Mass of solute (in g)}}{\text{Volume of solution (in mL)}} \times 100\%$$

Method 2

Let x represent the mass of solute dissolved in 100 mL of solution. The ratio of the dissolved solute, x, in 100 mL of solution must be the same as the ratio of 4.00 g of solute dissolved in 3.00 mL of solution. The concentration, expressed in g/100 mL, is numerically equal to the percent (m/v) of the solution.

Act on Your Strategy

Method 1

$$\begin{aligned}\text{Percent (m/v)} &= \frac{4.00\text{ g}}{3.00\text{ mL}} \times 100\% \\ &= 133\%\end{aligned}$$

Continued ...

Continued ...

FROM PAGE 303

Method 2

$$\frac{x}{100\text{ mL}} = \frac{4.00\text{ g}}{3.00\text{ mL}}$$

$$\frac{x}{100\text{ mL}} = 1.33\text{ g/mL}$$

$$x = 100\ \cancel{\text{mL}} \times 1.33\text{ g/}\cancel{\text{mL}}$$

$$= 133\text{ g}$$

The concentration of the drug is 133 g/100 mL of solution, or 133% (m/v).

Check Your Solution

The units are correct. The numerical answer is large, but this is reasonable for an extremely soluble solute.

Sample Problem

Finding Mass for an (m/v) Concentration

Problem

Many people use a solution of trisodium phosphate, Na_3PO_4 (commonly called TSP), to clean walls before putting up wallpaper. The recommended concentration is 1.7% (m/v). What mass of TSP is needed to make 2.0 L of solution?

What Is Required?

You need to find the mass of TSP needed to make 2.0 L of solution.

What Is Given?

The concentration of the solution should be 1.7% (m/v). The volume of solution that is needed is 2.0 L.

Plan Your Strategy

There are two different methods you can use.

Method 1

Use the formula for (m/v) percent. Rearrange the formula to solve for mass. Then substitute in the known values.

Method 2

The percent (m/v) of the solution is numerically equal to the concentration in g/100 mL. Let x represent the mass of TSP dissolved in 2.0 L of solution. The ratio of dissolved solute in 100 mL of solution must be the same as the ratio of the mass of solute, x, dissolved in 2.0 L (2000 mL) of solution.

Continued ...

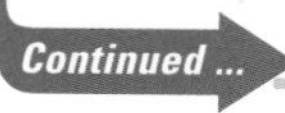
Continued ...
FROM PAGE 304

Act on Your Strategy

Method 1

$$\text{(m/v) percent} = \frac{\text{Mass of solute (in g)}}{\text{Volume of solution (in mL)}} \times 100\%$$

$$\therefore \text{Mass of solute} = \frac{\text{(m/v) percent} \times \text{Volume of solution}}{100\%}$$

$$= \frac{17\% \times 2000 \text{ mL}}{100\%}$$

$$= 34 \text{ g}$$

Method 2

A TSP solution that is 1.7% (m/v) contains 1.7 g of solute dissolved in 100 mL of solution.

$$\frac{1.7 \text{ g}}{100 \text{ mL}} = \frac{x}{2000 \text{ mL}}$$

$$0.017 \text{ g/mL} = \frac{x}{2000 \text{ mL}}$$

$$x = 0.017 \text{ g/mL} \times 2000 \text{ mL}$$

$$= 34 \text{ g}$$

Therefore, 34 g of TSP are needed to make 2.0 L of cleaning solution.

Check Your Solution

The units are appropriate for the problem. The answer appears to be reasonable.

Practice Problems

1. What is the concentration in percent (m/v) of each solution?
   - **(a)** 14.2 g of potassium chloride, KCl (used as a salt substitute), dissolved in 450 mL of solution
   - **(b)** 31.5 g of calcium nitrate, $Ca(NO_3)_2$ (used to make explosives), dissolved in 1.80 L of solution
   - **(c)** 1.72 g of potassium permanganate, $KMnO_4$ (used to bleach stone-washed blue jeans), dissolved in 60 mL of solution
2. A solution of hydrochloric acid was formed by dissolving 1.52 g of hydrogen chloride gas in enough water to make 24.1 mL of solution. What is the concentration in percent (m/v) of the solution?
3. At 25°C, a saturated solution of carbon dioxide gas in water has a concentration of 0.145% (m/v). What mass of carbon dioxide is present in 250 mL of the solution?
4. Ringer's solution contains three dissolved salts in the same proportions as they are found in blood. The salts and their concentrations (m/v) are as follows: 0.86% NaCl, 0.03% KCl, and 0.033% $CaCl_2$. Suppose that a patient needs to receive 350 mL of Ringer's solution by an intravenous drip. What mass of each salt does the pharmacist need to make the solution?

Concentration as a Mass/Mass Percent

The concentration of a solution that contains a solid solute dissolved in a liquid solvent can also be expressed as a mass of solute dissolved in a mass of solution. This is usually expressed as a percent relationship. A **mass/mass percent** gives the mass of a solute divided by the mass of solution, expressed as a percent. The mass/mass percent is also referred to as the *percent (m/m)*, or the *mass percent*. It is often inaccurately referred to as a weight (w/w) percent, as well. Look at your tube of toothpaste, at home. The percent of sodium flouride in the toothpaste is usually given as a w/w percent. This can be confusing, since weight (*w*) is not the same as mass (*m*). In fact, this concentration should be expressed as a mass/mass percent.

$$\text{Mass/mass percent} = \frac{\text{Mass of solute (in g)}}{\text{Mass of solution (in g)}} \times 100\%$$

For example, 100 g of seawater contains 0.129 g of magnesium ion (along with many other substances). The concentration of Mg^{2+} in seawater is 0.129 (m/m). *Notice that the number of grams of solute per 100 g of solution is numerically equal to the mass/mass percent.*

The concentration of a solid solution, such as an alloy, is usually expressed as a mass/mass percent. Often the concentration of a particular alloy may vary. Table 8.3 gives typical compositions of some common alloys.

Table 8.3 The Composition of Some Common Alloys

Alloy	Uses	Typical percent (m/m) composition
brass	ornaments, musical instruments	Cu (85%) Zn (15%)
bronze	statues, castings	Cu (80%) Zn (10%) Sn (10%)
cupronickel	"silver" coins	Cu (75%) Ni (25%)
dental amalgam	dental fillings	Hg (50%) Ag (35%) Sn (15%)
duralumin	aircraft parts	Al (93%) Cu (5%) other (2%)
pewter	ornaments	Sn (85%) Cu (7%) Bi (6%) Sb ((2%)
stainless steel	cutlery, knives	Fe (78%) Cr (15%) Ni (7%)
sterling silver	jewellery	Ag (92.5%) Cu (7.5%)

Figure 8.16, on the following page, shows two objects made from brass that have distinctly different colours. The difference in colours reflects the varying concentrations of the copper and zinc that make up the objects.

Figure 8.16 Brass can be made using any percent from 50% to 85% copper, and from 15% to 50% zinc. As a result, two objects made of brass can look very different.

Sample Problem

Solving for a Mass/Mass Percent

Problem

Calcium chloride, $CaCl_2$, can be used instead of road salt to melt the ice on roads during the winter. To determine how much calcium chloride had been used on a nearby road, a student took a sample of slush to analyze. The sample had a mass of 23.47 g. When the solution was evaporated, the residue had a mass of 4.58 g. (Assume that no other solutes were present.) What was the mass/mass percent of calcium chloride in the slush? How many grams of calcium chloride were present in 100 g of solution?

What Is Required?

You need to calculate the mass/mass percent of calcium chloride in the solution (slush). Then you need to use your answer to find the mass of calcium chloride in 100 g of solution.

What Is Given?

The mass of the solution is 23.47 g. The mass of calcium chloride that was dissolved in the solution is 4.58 g.

Plan Your Strategy

There are two methods that you can use to solve this problem.

Method 1

Use the formula for mass/mass percent.

$$\text{Mass/mass percent} = \frac{\text{Mass of solute (in g)}}{\text{Mass of solution (in g)}} \times 100\%$$

Continued ...

Continued ...

FROM PAGE 307

The mass of calcium chloride in 100 g of solution will be numerically equal to the mass/mass percent.

Method 2

Use ratios, as in the previous Sample Problems.

Act on Your Strategy

Method 1

$$\text{Mass/mass percent} = \frac{4.58\ \cancel{g}}{23.47\ \cancel{g}} \times 100\%$$
$$= 19.5\%$$

Method 2

$$\frac{x\ \text{g}}{100\ \text{g}} = \frac{4.58\ \text{g}}{23.47\ \text{g}}$$
$$\frac{x\ \cancel{g}}{100\ \cancel{g}} = 0.195$$
$$x = 19.5\%$$

The mass/mass percent was 19.5% (m/m). 19.5 g of calcium chloride was dissolved in 100 g of solution.

Check Your Solution

The mass units divide out properly. The final answer has the correct number of significant digits. It appears to be reasonable.

Practice Problems

5. Calculate the mass/mass percent of solute for each solution.

(a) 17 g of sulfuric acid in 65 g of solution

(b) 18.37 g of sodium chloride dissolved in 92.2 g of water
Hint: Remember that a solution consists of both solute and solvent.

(c) 12.9 g of carbon tetrachloride dissolved in 72.5 g of benzene

6. If 55 g of potassium hydroxide is dissolved in 100 g of water, what is the concentration of the solution expressed as mass/mass percent?

7. Steel is an alloy of iron and about 1.7% carbon. It also contains small amounts of other materials, such as manganese and phosphorus. What mass of carbon is needed to make a 5.0 kg sample of steel?

8. Stainless steel is a variety of steel that resists corrosion. Your cutlery at home may be made of this material. Stainless steel must contain at least 10.5% chromium. What mass of chromium is needed to make a stainless steel fork with a mass of 60.5 g?

9. 18-carat white gold is an alloy. It contains 75% gold, 12.5% silver, and 12.5% copper. A piece of jewellery, made of 18-carat white gold, has a mass of 20 g. How much pure gold does it contain?

Concentration as a Volume/Volume Percent

When mixing two liquids to form a solution, it is easier to measure their volumes than their masses. A **volume/volume percent** gives the volume of solute divided by the volume of solution, expressed as a percent. The volume/volume percent is also referred to as the *volume percent concentration*, *volume percent*, *percent (v/v)*, or the *percent by volume*. You can see this type of concentration on a bottle of rubbing alcohol from a drugstore. (See Figure 8.17.)

$$\text{Volume/volume percent} = \frac{\text{Volume of solute (in mL)}}{\text{Volume of solution (in mL)}} \times 100\%$$

Read through the Sample Problem below, and complete the Practice Problems that follow. You will then have a better understanding of how to calculate the volume/volume percent of a solution.

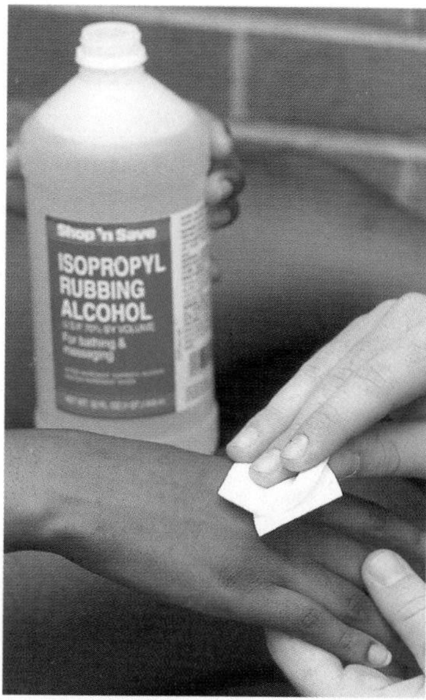

Figure 8.17 The concentration of this solution of isopropyl alcohol in water is expressed as a volume/volume percent.

Sample Problem

Solving for a Volume/Volume Percent

Problem

Rubbing alcohol is commonly used as an antiseptic for small cuts. It is sold as a 70% (v/v) solution of isopropyl alcohol in water. What volume of isopropyl alcohol is used to make 500 mL of rubbing alcohol?

What Is Required?

You need to calculate the volume of isopropyl alcohol (the solute) used to make 500 mL of solution.

What Is Given?

The volume/volume percent is 70% (v/v). The final volume of the solution is 500 mL.

Plan Your Strategy

Method 1

Rearrange the following formula to solve for the volume of the solute. Then substitute the values that you know into the rearranged formula.

$$\text{Volume/volume percent} = \frac{\text{Volume of solute}}{\text{Volume of solution}} \times 100\%$$

Method 2

Use ratios to solve for the unknown volume.

CHEM FACT

Archaeologists can learn a lot about ancient civilizations by chemically analyzing the concentration of ions in the soil where the people lived. When crops are grown, the crops remove elements such as nitrogen, magnesium, calcium, and phosphorus from the soil. Thus, soil with a lower-than-average concentration of these elements may have held ancient crops. Chlorophyll, which is present in all plants, contains magnesium ions. In areas where ancient crops were processed, the soil has a higher concentration of Mg^{2+}.

Continued ...
FROM PAGE 309

Act on Your Strategy

Method 1

$$\text{Volume/volume percent} = \frac{\text{Volume of solute}}{\text{Volume of solution}} \times 100\%$$

$$\begin{aligned}\text{Volume of solute} &= \frac{\text{Volume/volume percent} \times \text{Volume of solution}}{100\%} \\ &= \frac{70\% \times 500 \text{ mL}}{100\%} \\ &= 350 \text{ mL}\end{aligned}$$

Method 2

$$\frac{x \text{ mL}}{500 \text{ mL}} = \frac{70 \text{ mL}}{100 \text{ mL}}$$

$$\begin{aligned}x &= 0.7 \times 500 \text{ mL} \\ &= 350 \text{ mL}\end{aligned}$$

Therefore, 350 mL of isopropyl alcohol is used to make 500 mL of 70% (v/v) rubbing alcohol.

Check Your Solution

The answer seems reasonable. It is expressed in appropriate units.

Practice Problems

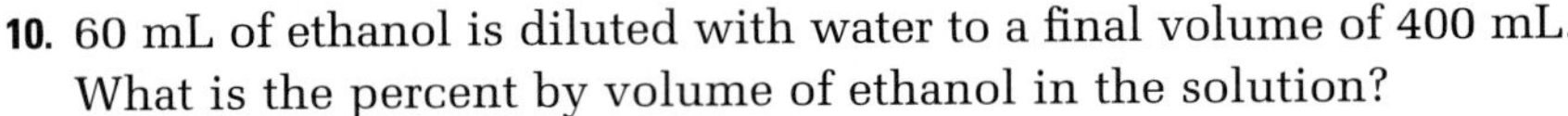

10. 60 mL of ethanol is diluted with water to a final volume of 400 mL. What is the percent by volume of ethanol in the solution?

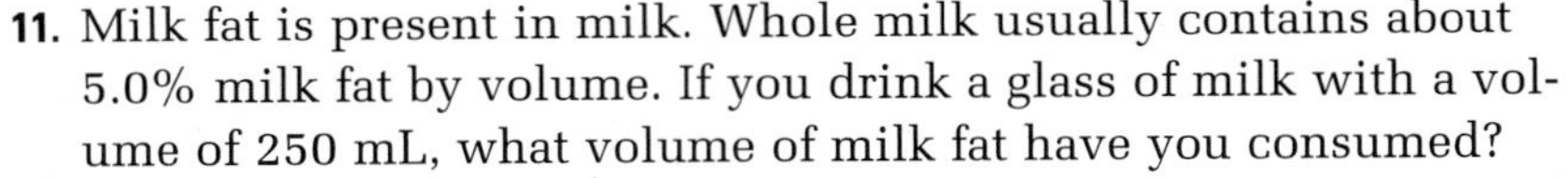

11. Milk fat is present in milk. Whole milk usually contains about 5.0% milk fat by volume. If you drink a glass of milk with a volume of 250 mL, what volume of milk fat have you consumed?

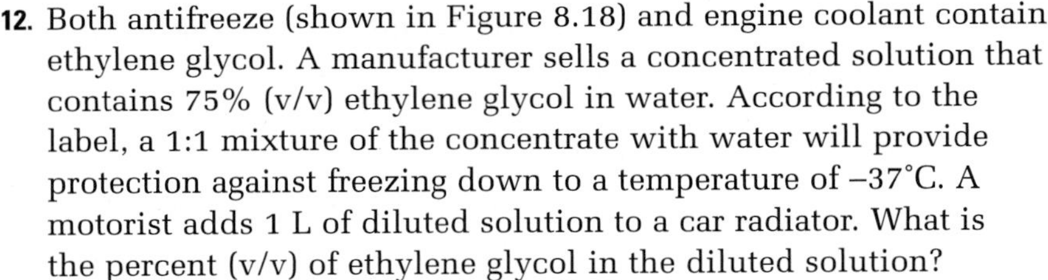

12. Both antifreeze (shown in Figure 8.18) and engine coolant contain ethylene glycol. A manufacturer sells a concentrated solution that contains 75% (v/v) ethylene glycol in water. According to the label, a 1:1 mixture of the concentrate with water will provide protection against freezing down to a temperature of –37°C. A motorist adds 1 L of diluted solution to a car radiator. What is the percent (v/v) of ethylene glycol in the diluted solution?

13. The average adult human body contains about 5 L of blood. Of this volume, only about 0.72% consists of leukocytes (white blood cells). These essential blood cells fight infection in the body. What volume of pure leukocyte cells is present in the body of a small child, with only 2.5 L of blood?

14. Vinegar is sold as a 5% (v/v) solution of acetic acid in water. How much water should be added to 15 mL of pure acetic acid (a liquid at room temperature) to make a 5% (v/v) solution of acetic acid? **Note:** Assume that when water and acetic acid are mixed, the total volume of the solution is the sum of the volumes of each.

Figure 8.18 Antifreeze is a solution of ethylene glycol and water.

Concentration in Parts per Million and Parts per Billion

The concentration of a very small quantity of a substance in the human body, or in the environment, can be expressed in **parts per million (ppm)** and **parts per billion (ppb)**. Both parts per million and parts per billion are usually mass/mass relationships. They describe the amount of solute that is present in a solution. Notice that parts per million does not refer to the number of particles, but to the *mass* of the solute compared with the *mass* of the solution.

$$\text{ppm} = \frac{\text{Mass of solute}}{\text{Mass of solution}} \times 10^6$$

or

$$\frac{\text{Mass of solute}}{\text{Mass of solution}} = \frac{x\text{ g}}{10^6\text{ g of solution}}$$

$$\text{ppb} = \frac{\text{Mass of solute}}{\text{Mass of solution}} \times 10^9$$

or

$$\frac{\text{Mass of solute}}{\text{Mass of solution}} = \frac{x\text{ g}}{10^9\text{ g of solution}}$$

One part per million is equal to 1¢ in $10 000. One part per billion is equal to 1 s in almost 32 years.

What distance (in km) would you travel if 1 cm represented 1 ppm of your journey?

A swimming pool has the dimensions 10 m × 5 m × 2 m. If the pool is full of water, what volume of water (in cm^3) would represent 1 ppb of the water in the pool?

Sample Problem

Parts per Billion in Peanut Butter

Problem

A fungus that grows on peanuts produces a deadly toxin. When ingested in large amounts, this toxin destroys the liver and can cause cancer. Any shipment of peanuts that contains more than 25 ppb of this dangerous fungus is rejected. A company receives 20 t of peanuts to make peanut butter. What is the maximum mass (in g) of fungus that is allowed?

What Is Required?

You need to find the allowed mass (in g) of fungus in 20 t of peanuts.

What Is Given?

The allowable concentration of the fungus is 25 ppb. The mass of the peanut shipment is 20 t.

Plan Your Strategy

Method 1

Convert 20 t to grams. Rearrange the formula below to solve for the allowable mass of the fungus.

$$\text{ppb} = \frac{\text{Mass of fungus}}{\text{Mass of peanuts}} \times 10^9$$

Method 2

Use ratios to solve for the unknown mass.

Continued ...

Continued ...
FROM PAGE 311

Act on Your Strategy

Method 1

First convert the mass in tonnes into grams.

$20\ t \times 1000\ kg/t \times 1000\ g/kg = 20 \times 10^6\ g$

Next rearrange the formula and find the mass of the fungus.

$$ppb = \frac{\text{Mass of fungus}}{\text{Mass of peanuts}} \times 10^9$$

$$\therefore \text{Mass of fungus} = \frac{ppb \times \text{Mass of peanuts}}{10^9}$$

$$= \frac{25\ ppb \times (20 \times 10^6\ g)}{10^9}$$

$$= 0.5\ g$$

Method 2

$$\frac{x\ g\ \text{solute}}{20 \times 10^6\ g\ \text{solution}} = \frac{25\ g\ \text{solute}}{1 \times 10^9\ g\ \text{solution}}$$

$$x\ g = (20 \times 10^6\ g\ \text{solution}) \times \frac{25\ \cancel{g}\ \text{solute}}{1 \times 10^9\ \cancel{g}\ \text{solution}}$$

$$= 0.5\ g$$

The maximum mass of fungus that is allowed is 0.5 g.

Check Your Solution

The answer appears to be reasonable. The units divided correctly to give grams. **Note:** Parts per million and parts per billion have no units. The original units, g/g, cancel out.

Practice Problems

15. Symptoms of mercury poisoning become apparent after a person has accumulated more than 20 mg of mercury in the body.

(a) Express this amount as parts per million for a 60 kg person.

(b) Express this amount as parts per billion.

(c) Express this amount as a (m/m) percent.

16. The use of the pesticide DDT has been banned in Canada since 1969 because of its damaging effect on wildlife. In 1967, the concentration of DDT in an average lake trout, taken from Lake Simcoe in Ontario, was 16 ppm. Today it is less than 1 ppm. What mass of DDT would have been present in a 2.5 kg trout with DDT present at 16 ppm?

17. The concentration of chlorine in a swimming pool is generally kept in the range of 1.4 to 4.0 mg/L. The water in a certain pool has 3.0 mg/L of chlorine. Express this value as parts per million. (**Hint:** 1 L of water has a mass of 1000 g.)

18. Water supplies with dissolved calcium carbonate greater than 500 mg/L are considered unacceptable for most domestic purposes. Express this concentration in parts per million.

Careers in Chemistry

Product Development Chemist

A solvent keeps paint liquefied so that it can be applied to a surface easily. After the paint has been exposed to the air, the solvent evaporates and the paint dries. Product development chemists develop and improve products such as paints. To work in product development, they require at least one university chemistry degree.

Chemists who work with paints must examine the properties of many different solvents. They must choose solvents that dissolve paint pigments well, but evaporate quickly and pose a low safety hazard.

Product development chemists must consider human health and environmental impact when choosing between solvents. Many solvents that have been used in the past, such as benzene and carbon tetrachloride, are now known to be harmful to the health and/or the environment. A powerful new solvent called *d-limonene* has been developed from the peel of oranges and lemons. This solvent is less harmful than many older solvents. It has been used successfully as a cleaner for airport runways and automotive parts, and as a pesticide. Chemists are now studying new applications for d-limonene.

Make Career Connections

1. Use reference books or the Internet to find the chemical structure of d-limonene. What else can you discover about d-limonene?
2. To learn more about careers involving work with solvents, contact the Canadian Chemical Producers Association (CCPA).

Molar Concentration

The most useful unit of concentration in chemistry is molar concentration. **Molar concentration** is the number of moles of solute that can dissolve in 1 L of solution. Notice that the volume of the *solution* in *litres* is used, rather than the volume of the *solvent* in *millilitres*. Molar concentration is also known as *molarity*.

$$\text{Molar concentration (in mol/L)} = \frac{\text{Amount of solute (in mol)}}{\text{Volume of solution (in L)}}$$

This formula can be shortened to give

$$C = \frac{n}{V}$$

Molar concentration is particularly useful to chemists because it is related to the number of particles in a solution. None of the other measures of concentration are related to the number of particles. If you are given the molar concentration and the volume of a solution, you can calculate the amount of dissolved solute in moles. This allows you to solve problems involving quantities in chemical reactions, such as the ones on the following pages.

Sample Problem

Calculating Molar Concentration

Problem

A saline solution contains 0.90 g of sodium chloride, NaCl, dissolved in 100 mL of solution. What is the molar concentration of the solution?

What Is Required?

You need to find the molar concentration of the solution in mol/L.

What Is Given?

You know that 0.90 g of sodium chloride is dissolved in 100 mL of solution.

Plan Your Strategy

Step 1 To find the amount (in mol) of sodium chloride, first determine its molar mass. Then divide the amount of sodium chloride (in g) by its molar mass (in g/mol).

Step 2 Convert the volume of solution from mL to L using this formula:

$$\text{Volume (in L)} = \text{Volume (in mL)} \times \frac{1.000\ \text{L}}{1000\ \text{mL}}$$

Step 3 Use the following formula to calculate the molar concentration:

$$\text{Molar concentration (in mol/L)} = \frac{\text{Amount of solute (in mol)}}{\text{Volume of solution (in L)}}$$

Act on Your Strategy

Step 1 Molar mass of NaCl = 22.99 + 35.45
= 58.44 g/mol

$$\text{Amount of NaCl} = \frac{0.90\ \text{g}}{58.44\ \text{g/mol}} = 1.54 \times 10^{-2}\ \text{mol}$$

Step 2 Convert the volume from mL to L.

$$\text{Volume} = 100\ \cancel{\text{mL}} \times \frac{1.000\ \text{L}}{1000\ \cancel{\text{mL}}} = 0.100\ \text{L}$$

Step 3 Calculate the molar concentration.

$$\text{Molar concentration} = \frac{1.54 \times 10^{-2}\ \text{mol}}{0.100\ \text{L}} = 1.54 \times 10^{-1}\ \text{mol/L}$$

The molar concentration of the saline solution is 0.15 mol/L.

Check Your Solution

The answer has the correct units for molar concentration.

Sample Problem

Using Molar Concentration to Find Mass

Problem

At 20°C, a saturated solution of calcium sulfate, $CaSO_4$, has a concentration of 0.0153 mol/L. A student takes 65 mL of this solution and evaporates it. What mass (in g) is left in the evaporating dish?

What Is Required?

You need to find the mass (in g) of the solute, calcium sulfate.

What Is Given?

The molar concentration is 0.0153 mol/L. The volume of the solution is 65 mL.

Plan Your Strategy

Step 1 Convert the volume from mL to L using the formula

$$\text{Volume (in L)} = \text{Volume (in mL)} \times \frac{1.000 \text{ L}}{1000 \text{ mL}}$$

Step 2 Rearrange the following formula to solve for the amount of solute (in mol).

$$\text{Molar concentration (in mol/L)} = \frac{\text{Amount of solute (in mol)}}{\text{Volume of solution (in L)}}$$

Step 3 Determine the molar mass of calcium sulfate. Use the molar mass to find the mass in grams, using the formula below:

$$\text{Mass (in g) of } CaSO_4$$
$$= \text{Amount (in mol)} \times \text{Molar mass of } CaSO_4 \text{ (in g/mol)}$$

Act on Your Strategy

Step 1 Convert the volume from mL to L.

$$\text{Volume} = 65 \cancel{\text{mL}} \times \frac{1.000 \text{ L}}{1000 \cancel{\text{mL}}}$$
$$= 0.065 \text{ L}$$

Step 2 Rearrange the formula to solve for the amount of solute.

$$\text{Molar concentration} = \frac{\text{Amount of solute}}{\text{Volume of solution}}$$

$$\therefore \text{Amount of solute} = \text{Molar concentration} \times \text{Volume of solution}$$
$$= 0.0153 \text{ mol/L} \times 0.065 \text{ L}$$
$$= 9.94 \times 10^{-4} \text{ mol}$$

Step 3 Determine the molar mass. Then find the mass in grams.

$$\text{Molar mass of } CaSO_4 = 40.08 + 32.07 + (4 \times 16.00)$$
$$= 136.15 \text{ g/mol}$$

$$\text{Mass (in g) of } CaSO_4 = 9.94 \times 10^{-4} \text{ mol} \times 136 \text{ g/mol}$$
$$= 0.135 \text{ g}$$

Continued ...

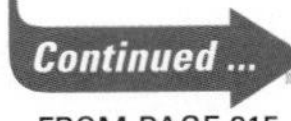

FROM PAGE 315

Therefore, 0.14 g of calcium sulfate are left in the evaporating dish.

Check Your Solution

The answer has the correct units and the correct number of significant figures.

Practice Problems

19. What is the molar concentration of each solution?

(a) 0.50 mol of NaCl dissolved in 0.30 L of solution

(b) 0.289 mol of iron(III) chloride, $FeCl_3$, dissolved in 120 mL of solution

(c) 0.0877 mol of copper(II) sulfate, $CuSO_4$, dissolved in 70 mL of solution

(d) 4.63 g of sugar, $C_{12}H_{22}O_{11}$, dissolved in 16.8 mL of solution

(e) 1.2 g of $NaNO_3$ dissolved in 80 mL of solution

20. What mass of solute is present in each aqueous solution?

(a) 1.00 L of 0.045 mol/L calcium hydroxide, $Ca(OH)_2$, solution

(b) 500 mL of 0.100 mol/L silver nitrate, $AgNO_3$, solution

(c) 2.5 L of 1.00 mol/L potassium chromate, K_2CrO_4, solution

(d) 40 mL of 6.0 mol/L sulfuric acid, H_2SO_4, solution

(e) 4.24 L of 0.775 mol/L ammonium nitrate, NH_4NO_3, solution

21. A student dissolves 30.46 g of silver nitrate, $AgNO_3$, in water to make 500 mL of solution. What is the molar concentration of the solution?

22. What volume of 0.25 mol/L solution can be made using 14 g of sodium hydroxide, NaOH?

23. A 100 mL bottle of skin lotion contains a number of solutes. One of these solutes is zinc oxide, ZnO. The concentration of zinc oxide in the skin lotion is 0.915 mol/L. What mass of zinc oxide is present in the bottle?

24. Formalin is an aqueous solution of formaldehyde, HCHO, used to preserve biological specimens. What mass of formaldehyde is needed to prepare 1.5 L of formalin with a concentration of 10 mol/L?

You have done many calculations for the concentration of various solutions. Now you are in a position to do some hands-on work with solution concentration. In the following investigation, you will use what you have learned to design your own experiment to determine the concentration of a solution.

Investigation 8-B

SKILL FOCUS

Initiating and planning

Performing and recording

Communicating results

Predicting

Determining the Concentration of a Solution

Your teacher will give you a sample of a solution. Design and perform an experiment to determine the concentration of the solution. Express the concentration as

(a) mass of solute dissolved in 100 mL of *solution*

(b) mass of solute dissolved in 100 g of *solvent*

(c) amount of solute (in mol) dissolved in 1 L of *solution*

Safety Precautions

When you have designed your investigation, think about the safety precautions you will need to take.

Materials

any apparatus in the laboratory

solution containing a solid dissolved in water

Note: Your teacher will tell you the name of the solute.

Procedure

1. Think about what you need to know in order to determine the concentration of a solution. Then design your experiment so that you can measure each quantity you need. Assume that the density of pure water is 1.00 g/mL.
2. Write the steps that will allow you to measure the quantities you need. Design a data table for your results. Include a space for the name of the solute in your solution.
3. When your teacher approves your procedure, complete your experiment.
4. Dispose of your solution as directed by your teacher.

Analysis

1. Express the concentration of the solution you analyzed as

   (a) mass of solute dissolved in 100 mL of *solution*

   (b) mass of solute dissolved in 100 g of *solvent*

   (c) molar concentration

   Show your calculations.

Conclusions

2. List at least two important sources of error in your measurements.
3. List at least two important ways that you could improve your procedure.
4. Did the solute partially decompose on heating, producing a gas and another solid? If so, how do you think this affected the results of your experiment?

Concept Organizer

Summary of Sections 8.1 to 8.3

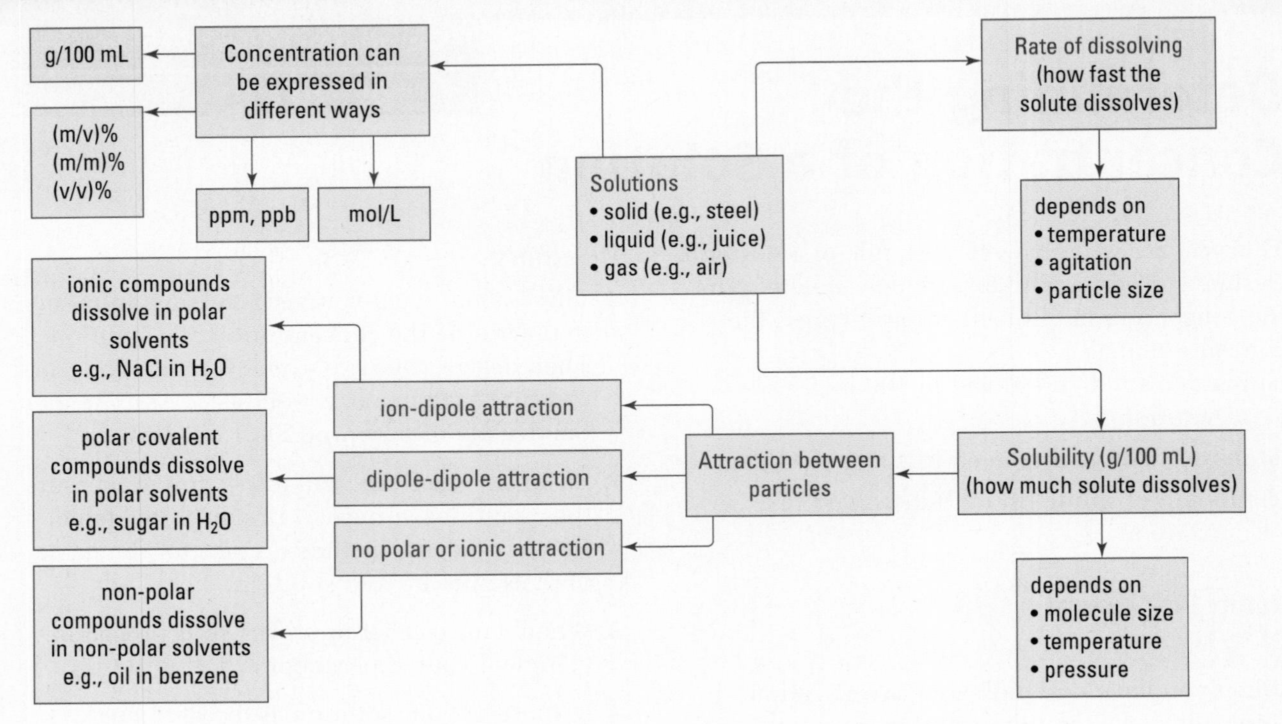

Section Wrap-up

You have learned about several different ways in which chemists express concentration: mass/volume, mass/mass, and volume/volume percent; parts per million and parts per billion; and molar concentration. The Concept Organizer above summarizes what you have learned in this chapter so far.

In section 8.4, you will learn how standard solutions of known concentration are prepared. You will also learn how to dilute a standard solution.

Section Review

Unit Issue Prep

The concentration of a pesticide or other contaminant dissolved in water determines the level of pollution. If you wish to work on your Unit 3 Issue now, do some research on pesticides such as DDT, diazinon, and MCPA. What concentrations of these pesticides are unacceptable in water systems?

1. (K/U) Ammonium chloride, NH_4Cl, is a very soluble salt. 300 g of ammonium chloride are dissolved in 600 mL of water. What is the percent (m/m) of the solution?

2. (K/U) A researcher measures 85.1 mL of a solution of liquid hydrocarbons. The researcher then distills the sample to separate the pure liquids. If 20.3 mL of the hydrocarbon hexane are recovered, what is its percent (v/v) in the sample?

3. (K/U) A stock solution of phosphoric acid is 85.0% (m/v) H_3PO_4 in water. What is its molar concentration?

4. (MC) Cytosol is an intracellular solution containing many important solutes. Research this solution. Write a paragraph describing the function of cytosol, and the solutes it contains.

Preparing Solutions

What do the effectiveness of a medicine, the safety of a chemical reaction, and the cost of an industrial process have in common? They all depend on solutions that are made carefully with known concentrations. A solution with a known concentration is called a **standard solution**. There are two ways to prepare an aqueous solution with a known concentration. You can make a solution by dissolving a measured mass of pure solute in a certain volume of solution. Alternatively, you can dilute a solution of known concentration.

Section Preview/ Specific Expectations

In this section, you will

- **prepare** solutions by dissolving a solid solute and then diluting a concentrated solution
- **communicate** your understanding of the following terms: *standard solution, volumetric flask*

Using a Volumetric Flask

A **volumetric flask** is a pear-shaped glass container with a flat bottom and a long neck. Volumetric flasks like the ones shown in Figure 8.19 are used to make up standard solutions. They are available in a variety of sizes. Each size can measure a fixed volume of solution to ±0.1 mL at a particular temperature, usually 20°C. When using a volumetric flask, you must first measure the mass of the pure solute. Then you transfer the solute to the flask using a funnel, as shown in Figure 8.20. At this point, you add the solvent (usually water) to dissolve the solute, as in Figure 8.21. You continue adding the solvent until the bottom of the meniscus appears to touch the line that is etched around the neck of the flask. See Figure 8.22. This is the volume of the solution, within ±0.1 mL. If you were performing an experiment in which significant digits and errors were important, you would record the volume of a solution in a 500 mL volumetric flask as 500.0 mL ±0.1 mL. Before using a volumetric flask, you need to rinse it several times with a small quantity of distilled water and discard the washings. *Standard solutions are never stored in volumetric flasks.* Instead, they are transferred to another bottle that has a secure stopper or cap.

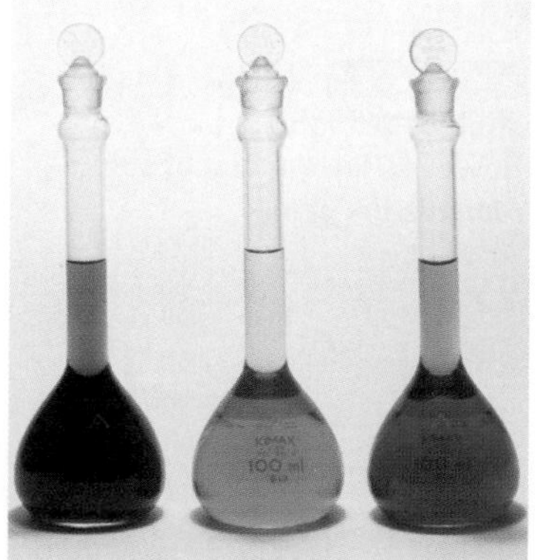

Figure 8.19 These volumetric flasks, from left to right, contain solutions of chromium(III) salts, iron(III) salts, and cobalt(II) salts.

Figure 8.20 Transfer a known mass of solid solute into the volumetric flask. Alternatively, dissolve the solid in a small volume of solvent. Then add the liquid to the flask.

Figure 8.21 Add distilled water until the flask is about half full. Swirl the mixture around in order to dissolve the solute completely. Rinse the beaker that contained the solute with solvent. Add the rinsing to the flask.

Figure 8.22 Add the rest of the water slowly. When the flask is almost full, add the water drop by drop until the bottom of the meniscus rests at the etched line.

Diluting a Solution

You can make a less concentrated solution of a known solution by adding a measured amount of additional solvent to the standard solution. The number of molecules, or moles, of solute that is present remains the same before and after the dilution. (See Figure 8.23.)

To reinforce these ideas, read through the Sample Problem below. Then try the Practice Problems that follow.

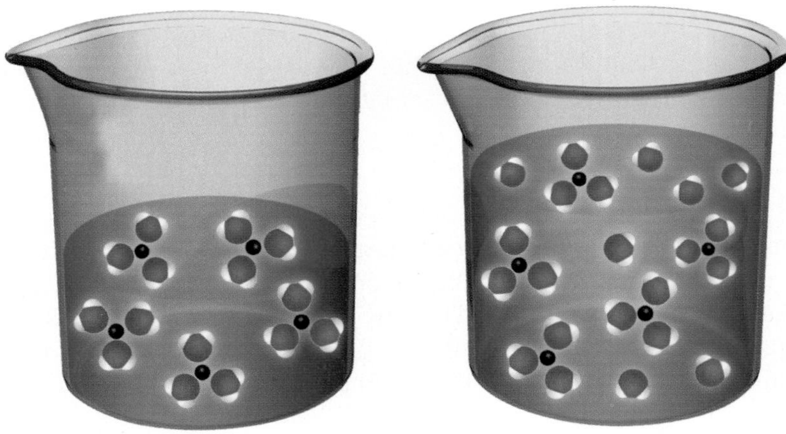

Figure 8.23 When a solution is diluted, the volume increases. However, the amount of solute remains the same.

Sample Problem

Diluting a Standard Solution

Problem

For a class experiment, your teacher must make 2.0 L of 0.10 mol/L sulfuric acid. This acid is usually sold as an 18 mol/L concentrated solution. How much of the concentrated solution should be used to make a new solution with the correct concentration?

What Is Required?

You need to find the volume of concentrated solution to be diluted.

What Is Given?

Initial concentration = 18 mol/L
Concentration of diluted solution = 0.10 mol/L
Volume of diluted solution = 2.0 L

Plan Your Strategy

Note: Amount of solute (mol) after dilution = Amount of solute (mol) before dilution

Step 1 Calculate the amount of solute (in mol) that is needed for the final dilute solution.

Step 2 Calculate the volume of the concentrated solution that will provide the necessary amount of solute.

Continued ...

FROM PAGE 320

Act on Your Strategy

Step 1 Calculate the amount of solute that is needed for the final dilute solution.

$$\text{Molar concentration (in mol/L)} = \frac{\text{Amount of solute (in mol)}}{\text{Volume of solution (in L)}}$$

$\therefore$ Amount of solute = Molar concentration × Volume of solution

For the final dilute solution,

$$\begin{aligned}\text{Amount of solute} &= 0.10\ \text{mol/L} \times 2.0\ \text{L}\\ &= 0.20\ \text{mol}\end{aligned}$$

Step 2 Calculate the volume of the original concentrated solution that is needed.

Rearrange and use the molar concentration equation. Substitute in the amount of solute you calculated in step 1.

$$\begin{aligned}\text{Volume of solution (in L)} &= \frac{\text{Amount of solute (in mol)}}{\text{Molar concentration (in mol/L)}}\\ &= \frac{0.20\ \cancel{\text{mol}}}{18\ \cancel{\text{mol}}/\text{L}}\\ &= 0.011\ \text{L}\end{aligned}$$

Therefore, 0.011 L, or 11 mL, of the concentrated 18 mol/L solution should be used to make 2.0 L of 0.10 mol/L sulfuric acid.

Check Your Solution

The units are correct. The final solution must be much less concentrated. Thus, it is reasonable that only a small volume of concentrated solution is needed.

Practice Problems

25. Suppose that you are given a solution of 1.25 mol/L sodium chloride in water, $NaCl_{(aq)}$. What volume must you dilute to prepare the following solutions?

(a) 50 mL of 1.00 mol/L $NaCl_{(aq)}$

(b) 200 mL of 0.800 mol/L $NaCl_{(aq)}$

(c) 250 mL of 0.300 mol/L $NaCl_{(aq)}$

26. What concentration of solution is obtained by diluting 50.0 mL of 0.720 mol/L aqueous sodium nitrate, $NaNO_{3(aq)}$, to each volume?

(a) 120 mL **(b)** 400 mL **(c)** 5.00 L

27. A solution is prepared by adding 600 mL of distilled water to 100 mL of 0.15 mol/L ammonium nitrate. Calculate the molar concentration of the solution. Assume that the volume quantities can be added together.

Now that you understand how to calculate standard solutions and dilution, it is time for you to try it out for yourself. In the following investigation, you will prepare and dilute standard solutions.

SKILL FOCUS

Initiating and planning

Performing and recording

Analyzing and interpreting

Investigation 8-C

Estimating Concentration of an Unknown Solution

Copper(II) sulfate, $CuSO_4$, is a soluble salt. It is sometimes added to pools and ponds to control the growth of fungi. Solutions of this salt are blue in colour. The intensity of the colour increases with increased concentration. In this investigation, you will prepare copper(II) sulfate solutions with known concentrations. Then you will estimate the concentration of an unknown solution by comparing its colour intensity with the colour intensities of the known solutions.

Copper(II) sulfate pentahydrate is a *hydrate*. Hydrates are ionic compounds that have a specific amount of water molecules associated with each ion pair.

Question

How can you estimate the concentration of an unknown solution?

Part 1 Making Solutions with Known Concentrations

Safety Precautions

- Copper(II) sulfate is poisonous. Wash your hands at the end of this investigation.
- If you spill any solution on your skin, wash it off immediately with copious amounts of cool water.

Materials

graduated cylinder
6 beakers
chemical balance
stirring rod
copper(II) sulfate pentahydrate, $CuSO_4{\cdot}5H_2O$
distilled water
labels or grease marker

Procedure

1. With your partner, develop a method to prepare 100 mL of 0.500 mol/L aqueous $CuSO_4{\cdot}5H_2O$ solution. Include the water molecules that are hydrated to the crystals, as given in the molecular formula, in your calculation of the molar mass. Show all your calculations. Prepare the solution.
2. Save some of the solution you prepared in step 1, to be tested in Part 2. Use the rest of the solution to make the dilutions in steps 3 to 5. Remember to label the solutions.
3. Develop a method to dilute part of the 0.500 mol/L $CuSO_4$ solution, to make 100 mL of 0.200 mol/L solution. Show your calculations. Prepare the solution.
4. Show your calculations to prepare 100 mL of 0.100 mol/L solution, using the solution you prepared in step 3. (You do not need to describe the method because it will be similar to the method you developed in step 3. Only the volume diluted will be different.) Prepare the solution.
5. Repeat step 4 to make 100 mL of 0.050 mol/L $CuSO_4$, by diluting part of the 0.100 mol/L solution you made. Then make 50 mL of 0.025 mol/L solution by diluting part of the 0.050 mol/L solution.

Part 2 Estimating the Concentration of an Unknown Solution

Materials

paper towels
6 clean, dry, identical test tubes
medicine droppers
5 prepared solutions from Part 1
10 mL of copper(II) sulfate, $CuSO_4$, solution with an unknown concentration

Procedure

1. You should have five labelled beakers containing $CuSO_4$ solutions with the following concentrations: 0.50 mol/L, 0.20 mol/L, 0.10 mol/L, 0.05 mol/L, and 0.025 mol/L. Your teacher will give you a sixth solution of unknown concentration. Record the letter or number that identifies this solution.
2. Label each test tube, one for each solution. Pour a sample of each solution into a test tube. The height of all the test tubes should be the same. Use a medicine dropper to add or take away solution as needed. (Be careful not to add water, or a solution of different concentration, to a test tube.)
3. The best way to compare colour intensity is by looking down through the test tube. Wrap each test tube with a paper towel to stop light from entering the side. Arrange the solutions of known concentration in order.
4. Place the solutions over a diffuse light source such as a lightbox. Compare the colour of the unknown solution with the colours of the other solutions.
5. Use your observations to estimate the concentration of the unknown solution.
6. Pour the solutions of $CuSO_4$ into a beaker supplied by your teacher. Wash your hands.

Analysis

1. Describe any possible sources of error for Part 1 of this investigation.
2. What is your estimate of the concentration of the unknown solution?

Conclusion

3. Obtain the concentration of the unknown solution from your teacher. Calculate the percentage error in your estimate.

Applications

4. Use your estimated concentration of the unknown solution to calculate the mass of $CuSO_4 \cdot 5H_2O$ that your teacher would need to prepare 500 mL of this solution.
5. If your school has a spectrometer or colorimeter, you can measure the absorption of light passing through the solutions. By measuring the absorption of solutions of copper(II) sulfate with different concentrations, you can draw a graph of absorption against concentration.

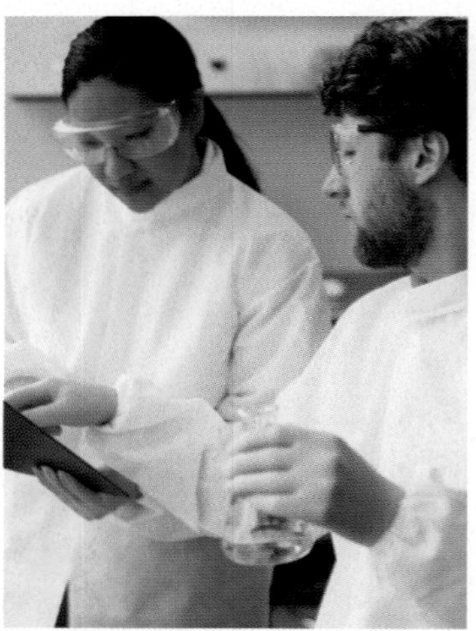

Figure 8.24 **When diluting acid, always add the acid to the water—never the reverse. Rubber gloves, a lab coat, and safety goggles or a face shield protect against acid splashes.**

Diluting Concentrated Acids

The acids that you use in your investigations are bought as concentrated standard solutions. Sulfuric acid is usually bought as an 18 mol/L solution. Hydrochloric acid is usually bought as a 12 mol/L solution. These acids are far too dangerous for you to use at these concentrations. Your teacher dilutes concentrated acids, following a procedure that minimizes the hazards involved.

Concentrated acids should be diluted in a fume hood because breathing in the fumes causes acid to form in air passages and lungs. Rubber gloves must be used to protect the hands. A lab coat is needed to protect clothing. Even small splashes of a concentrated acid will form holes in fabric. Safety goggles, or even a full-face shield, are essential.

Mixing a strong, concentrated acid with water is a very exothermic process. A concentrated acid is denser than water. Therefore, when it is poured into water, it sinks into the solution and mixes with the solution. The heat that is generated is spread throughout the solution. This is the only safe way to mix an acid and water. If you added water to a concentrated acid, the water would float on top of the solution. The heat generated at the acid-water layer could easily boil the solution and splatter highly corrosive liquid. The sudden heat generated at the acid-water boundary could crack the glassware and lead to a very dangerous spill. Figure 8.24 illustrates safety precautions needed to dilute a strong acid.

Section Wrap-up

In this section, you learned how to prepare solutions by dissolving a solid solute and then diluting a concentrated solution. In the next chapter, you will see how water is used as a solvent in chemistry laboratories. Many important reactions take place in water. You will also learn more about water pollution and water purification.

Section Review

1. What mass of potassium chloride, KCl, is used to make 25.0 mL of a solution with a concentration of 2.00 mol/L?

2. A solution is prepared by dissolving 42.5 g of silver nitrate, $AgNO_3$ in a 1 L volumetric flask. What is the molar concentration of the solution?

3. The solution of aqueous ammonia that is supplied to schools has a concentration of 14 mol/L. Your class needs 3.0 L of a solution with a concentration of 0.10 mol/L.
   - **(a)** What procedure should your teacher follow to make up this solution?
   - **(b)** Prepare an instruction sheet or a help file for your teacher to carry out this dilution.

4. 47.9 g of potassium chlorate, $KClO_3$, is used to make a solution with a concentration of 0.650 mol/L. What is the volume of the solution?

5. Water and 8.00 mol/L potassium nitrate solution are mixed to produce 700 mL of a solution with a concentration of 6.00 mol/L. What volumes of water and potassium nitrate solution are used?

CHAPTER 8 Review

Reflecting on Chapter 8

Summarize this chapter in the format of your choice. Here are a few ideas to use as guidelines:

- Describe the difference between a saturated and an unsaturated solution.
- Explain how you can predict whether a solute will dissolve in a solvent.
- What factors affect the rate of dissolving?
- What factors affect solubility?
- How does temperature affect the solubility of a solid, a liquid, and a gas?
- Describe how particle attractions affect solubility.
- Explain how to plot a solubility curve.
- Write the formulas for (m/v) percent, (m/m) percent, (v/v) percent, ppm, ppb, and molar concentration.
- Explain how you would prepare a standard solution using a volumetric flask.

Reviewing Key Terms

For each of the following terms, write a sentence that shows your understanding of its meaning.

solution	hydrogen bonding
solvent	ion-dipole attractions
solutes	hydrated
variable composition	electrolyte
aqueous solution	non-electrolytes
miscible	concentration
immiscible	mass/volume percent
alloys	mass/mass percent
solubility	volume/volume percent
saturated solution	parts per million
unsaturated solution	parts per billion
rate of dissolving	molar concentration
dipole	standard solution
dipole-dipole attraction	volumetric flask

Knowledge/Understanding

1. Identify at least two solutions in your home that are
   (a) beverages
   (b) found in the bathroom or medicine cabinet
   (c) solids
2. How is a solution different from a pure compound? Give specific examples.
3. Mixing 2 mL of linseed oil and 4 mL of turpentine makes a binder for oil paint. What term is used to describe liquids that dissolve in each other? Which liquid is the solvent?
4. How does the bonding in water molecules account for the fact that water is an exellent solvent?
5. Why does an aqueous solution of an electrolyte conduct electricity, but an aqueous solution of a non-electrolyte does not?
6. Use the concept of forces between particles to explain why oil and water are immiscible.
7. Explain the expression "like dissolves like" in terms of intermolecular forces.
8. What factors affect the rate of dissolving of a solid in a liquid?
9. Which of the following substances would you expect to be soluble in water? Briefly explain each answer.
   (a) potassium chloride, KCl
   (b) carbon tetrachloride, CCl_4
   (c) sodium sulfate, Na_2SO_4
   (d) butane, C_4H_{10}
10. Benzene, C_6H_6, is a liquid at room temperature. It is sometimes used as a solvent. Which of the following compounds is more soluble in benzene: naphthalene, $C_{10}H_8$, or sodium fluoride, NaF? Would you expect ethanol, CH_3CH_2OH, to be soluble in benzene? Explain your answers.

Inquiry

11. Boric acid solution is used as an eyewash. What mass of boric acid is present in 250 g of solution that is 2.25% (m/m) acid in water?
12. 10% (m/m) sodium hydroxide solution, $NaOH_{(aq)}$, is used to break down wood fibre to make paper.
    (a) What mass of solute is needed to make 250 mL of 10% (m/m) solution?
    (b) What mass of solvent is needed?
    (c) What is the molar concentration of the solution?
13. What volume of pure ethanol is needed to make 800 mL of a solution of ethanol in water that is 12% (v/v)?

14. Some municipalities add sodium fluoride, NaF, to drinking water to help protect the teeth of children. The concentration of sodium fluoride is maintained at 2.9×10^{-5} mol/L. What mass (in mg) of sodium fluoride is dissolved in 1 L of water? Express this concentration in ppm.

15. A saturated solution of sodium acetate, $NaCH_3COO$, can be prepared by dissolving 4.65 g in 10.0 mL of water at 20°C. What is the molar concentration of the solution?

16. What is the molar concentration of each of the following solutions?
  (a) 7.25 g of silver nitrate, $AgNO_3$, dissolved in 100 mL of solution
  (b) 80 g of glucose, $C_6H_{12}O_6$, dissolved in 70 mL of solution

17. Calculate the mass of solute that is needed to prepare each solution below.
  (a) 250 mL of 0.250 mol/L calcium acetate, $Ca(CH_3COO)_2$
  (b) 1.8 L of 0.35 mol/L ammonium sulfate, $(NH_4)_2SO_4$

18. Calculate the molar concentration of each solution formed after dilution.
  (a) 20 mL of 6.0 mol/L hydrochloric acid, $HCl_{(aq)}$, diluted to 70 mL
  (b) 300 mL of 12.0 mol/L ammonia, $NH_{3(aq)}$, diluted to 2.50 L

19. Calculate the molar concentration of each solution. Assume that the volumes can be added.
  (a) 85.0 mL of 1.50 mol/L ammonium chloride, $NH_4Cl_{(aq)}$, added to 250 mL of water
  (b) a 1:3 dilution of 1.0 mol/L calcium phosphate (that is, one part stock solution mixed with three parts water)

20. A standard solution of 0.250 mol/L calcium ion is prepared by dissolving solid calcium carbonate in an acid. What mass of calcium carbonate is needed to prepare 1.00 L of the solution?

21. Suppose that your teacher gives you three test tubes. Each test tube contains a clear, colourless liquid. One liquid is an aqueous solution of an electrolyte. Another liquid is an aqueous solution of a non-electrolyte. The third liquid is distilled water. Outline the procedure for an experiment to identify which liquid is which.

22. Fertilizers for home gardeners may be sold as aqueous solutions. Suppose that you want to begin a company that sells an aqueous solution of potassium nitrate, KNO_3, fertilizer. You need a solubility curve (a graph of solubility versus temperature) to help you decide what concentration to use for your solution. Describe an experiment that you might perform to develop a solubility curve for potassium nitrate. State which variables are controlled, which are varied, and which must be measured.

23. Potassium alum, $KAl(SO_4)_2{\cdot}12H_2O$, is used to stop bleeding from small cuts. The solubility of potassium alum, at various temperatures, is given in the following table.

Solubility of Potassium Alum

Solubility (g/100 g water)	Temperature (°C)
4	0
10	10
15	20
23	30
31	40
49	50
67	60
101	70
135	80

  (a) Plot a graph of solubility against temperature.
  (b) From your graph, interpolate the solubility of potassium alum at 67°C.
  (c) By extrapolation, estimate the solubility of potassium alum at 82°C.
  (d) Look at your graph. At what temperature will 120 g of potassium alum form a saturated solution in 100 g of water?

24. Use the graph on the next page to answer questions 24 and 25. At 80°C, what mass of sodium chloride dissolves in 1.0 L of water?

25. What minimum temperature is required to dissolve 24 g of potassium nitrate in 40 g of water?

26. A teacher wants to dilute 200 mL of 12 mol/L hydrochloric acid to make a 1 mol/L solution. What safety precautions should the teacher take?

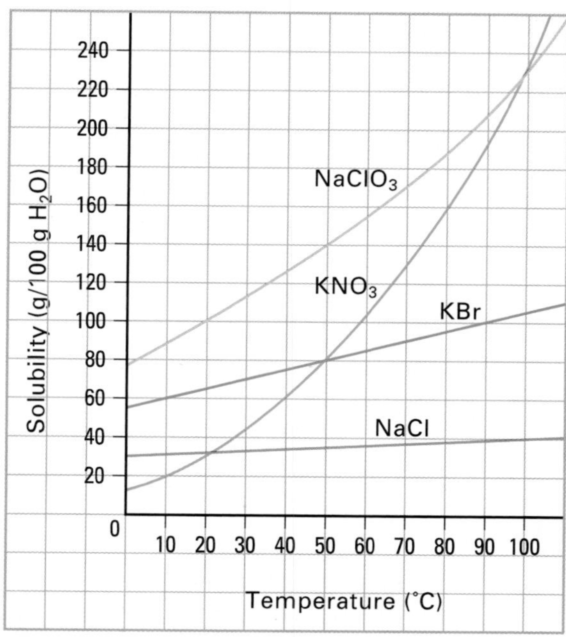

This graph shows the solubility of four salts at various temperatures. Use it to answer questions 24 and 25.

Communication

27. Suppose that you make a pot of hot tea. Later, you put a glass of the tea in the refrigerator to save it for a cool drink. When you take it out of the refrigerator some hours later, you notice that it is cloudy. How could you explain this to a younger brother or sister?
28. Define each concentration term.
    (a) percent (m/v)
    (b) percent (m/m)
    (c) percent (v/v)
    (d) parts per million, ppm
    (e) parts per billion, ppb
29. The concentration of iron in the water that is supplied to a town is 0.25 mg/L. Express this in ppm and ppb.
30. Ammonia is a gas at room temperature and pressure, but it can be liquefied easily. Liquid ammonia is probably present on some planets. Scientists speculate that it might be a good solvent. Explain why, based on the structure of the ammonia molecule shown above.

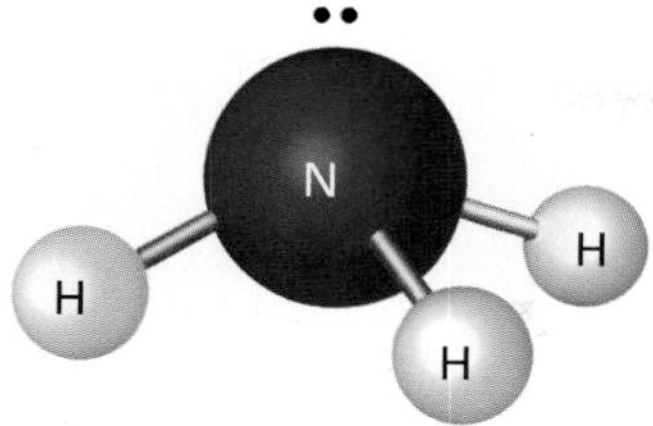

31. At 20°C, the solubility of oxygen in water is more than twice that of nitrogen. A student analyzed the concentration of dissolved gases in an unpolluted pond. She found that the concentration of nitrogen gas was greater than the concentration of oxygen. Prepare an explanation for the student to give to her class.
32. What is the concentration of pure water?

Making Connections

33. A bright red mineral called cinnabar has the chemical formula HgS. It can be used to make an artist's pigment, but it is a very insoluble compound. A saturated solution at 25°C has a concentration of 2×10^{-27} mol/L. In the past, why was heavy metal poisoning common in painters? Why did painters invariably waste more cinnabar than they used?
34. Vitamin A is a compound that is soluble in fats but not in water. It is found in certain foods, including yellow fruit and green vegetables. In parts of central Africa, children frequently show signs of vitamin A deficiency, although their diet contains a good supply of the necessary fruits and vegetables. Why?

Answers to Practice Problems and Short Answers to Section Review Questions:

Practice Problems: 1.(a) 3.16% (b) 1.75% (c) 2.9% 2. 6.31% 3. 0.362 g 4. 3.0 g, 0.1 g, 0.12 g 5.(a) 26% (b) 16.6% (c) 15.1% 6. 35% 7. 85 g 8. 6.35 g 9. 15 g 10. 15% 11. 12 mL 12. 38% 13. 18 mL 14. 285 mL 15.(a) 0.33 ppm (b) 3.3×10^2 ppb (c) 0.000033% 16. 0.040 g 17. 3.0 ppm 18. 500 ppm 19.(a) 1.7 mol/L (b) 2.41 mol/L (c) 1.2 mol/L (d) 0.805 mol/L (e) 0.18 mol/L 20.(a) 3.3 g (b) 8.49 g (c) 4.9×10^2 g (d) 24 g (e) 263 g 21. 0.359 mol/L 22. 1.4 L 23. 7.45 g 24. 4.5×10^2 g 25.(a) 40 mL (b) 128 mL (c) 60.0 mL 26.(a) 0.300 mol/L (b) 9.00×10^{-2} mol/L (c) 7.20×10^{-3} mol/L 27. 2.1×10^{-2} mol/L

Section Review: 8.1: 1. solute, solvent 8. polar, ionic 13. non-polar **8.2:** 3.(a) increases (b) decreases 6.(a) $Ce_2(SO_4)_3$ (b) $Ce_2(SO_4)_3$, $NaNO_3$ (c) NaCl (d) 84°C (e) 10 g **8.3:** 1. 33.3 2. 23.8% 3. 8.67 mol/L **8.4:** 1. 3.73 g 2. 0.25 mol/L 4. 601 mL 5. 175 mL, 525 mL

500 mL
± 5%
400
300

CHAPTER 9 Aqueous Solutions

In 1966, a Russian scientist who was lecturing in England stunned the scientific community with the report of a new discovery. He said that a colleague had isolated a new form of liquid water. Dubbed "polywater," this substance was prepared in a laboratory by heating and then condensing $H_2O_{(g)}$ in very narrow glass capillary tubes. Over the next seven years, scientists around the world conducted studies and published over 500 papers about polywater's properties. For example, its boiling point, density, and viscosity were greater than those of ordinary water. Its freezing point was lower.

Then the popular media took an interest. They turned a simmering scientific curiosity into a boiling concern. Polywater apparently had a powerful capacity for hydrogen bonding. What if it escaped from a laboratory? There were dire predictions that polywater would "take over" Earth's water resources. The imagined consequences for life on Earth were grim.

Eventually further studies revealed that polywater was simply a concentrated solution of silicon (silicic acid) and several ionic compounds in ordinary water. The glass tubing was the source of these solutes. They had leached into the water.

The polywater event occurred because scientists overlooked, for awhile, water's remarkable power as a solvent. In this chapter, you will learn how to predict which compounds are soluble in water. As well, you will consider how chemical reactions in aqueous (water) solutions are useful in industry and in protecting the quality of our water supplies.

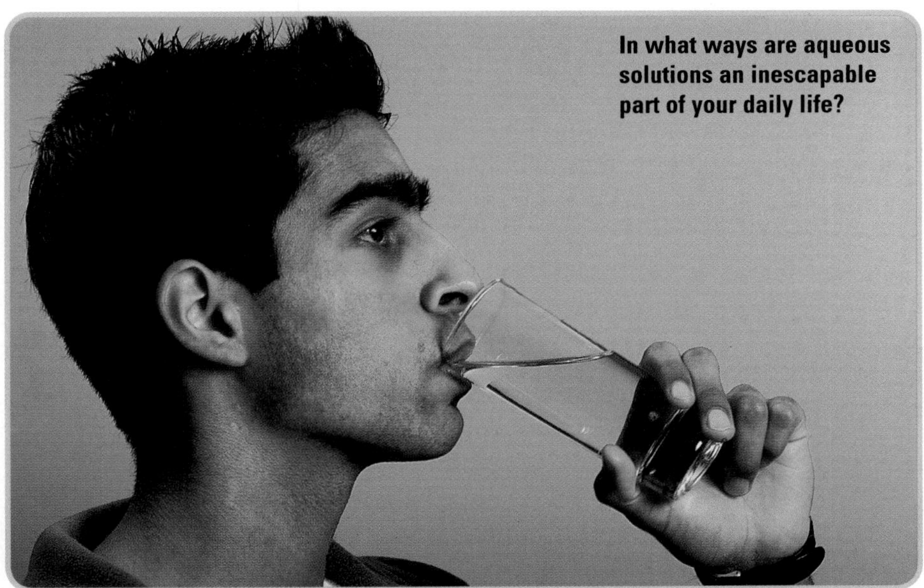

In what ways are aqueous solutions an inescapable part of your daily life?

Chapter Preview

Concepts and Skills You Will Need

Before you begin this chapter, review the following concepts and skills:

- identifying factors that affect solubility (Chapter 8, section 8.2)
- performing stoichiometry calculations (Chapter 7, section 7.1)

9.1 Making Predictions About Solubility

Section Preview/ Specific Expectations

In this section, you will

- **describe** and **identify** combinations of aqueous solutions that result in the formation of precipitates
- **communicate**, using appropriate scientific vocabulary, ideas related to solubility and aqueous solutions
- **communicate** your understanding of the following terms: *precipitate, general solubility guidelines*

In Chapter 8, you examined factors that affect the solubility of a compound. As well, you learned that the terms "soluble" and "insoluble" are relative, because no substance is completely insoluble in water. "Soluble" generally means that more than about 1 g of solute will dissolve in 100 mL of water at room temperature. "Insoluble" means that the solubility is less than 0.1 g per 100 mL. While many ionic compounds are soluble in water, many others are not. Cooks, chemists, farmers, pharmacists, and gardeners need to know which compounds are soluble and which are insoluble. (See Figure 9.1.)

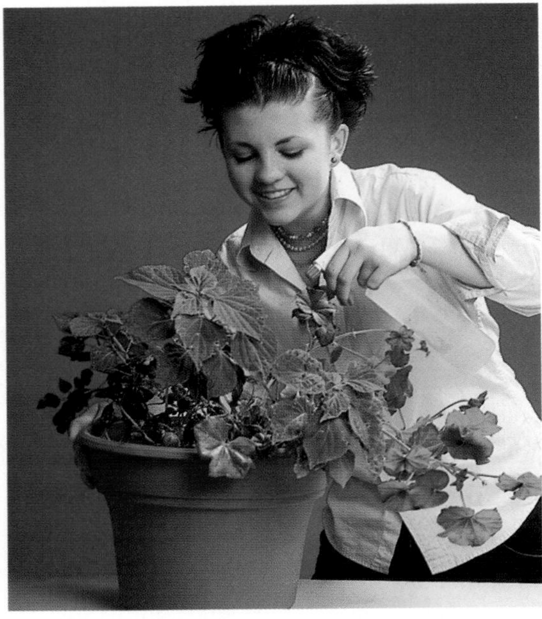

Figure 9.1 Plant food may come in the form of a liquid or a powder. Gardeners dissolve the plant food in water, and then either spray or water the plant with the resulting solution.

Factors That Affect the Solubility of Ionic Substances

Nearly all alkali metal compounds are soluble in water. Sulfide and phosphate compounds are usually insoluble. How, then, do you account for the fact that sodium sulfide and potassium phosphate are soluble, while iron sulfide and calcium phosphate are insoluble? Why do some ions form soluble compounds, while other ions form insoluble compounds?

The Effect of Ion Charge on Solubility

Compounds of ions with small charges tend to be soluble. Compounds of ions with large charges tend to be insoluble. Why? Increasing the charge increases the force that holds the ions together. For example, phosphates (compounds of PO_4^{3-}) tend to be insoluble. On the other hand, the salts of alkali metals are soluble. Alkali metal cations have a single positive charge, so the force that holds the ions together is less.

CHEM FACT

Because the terms "soluble" and "insoluble" are relative, some textbooks give different definitions and units of concentration to describe them. Here are a few examples:

Soluble	Partly or slightly soluble	Insoluble
more than 1 g in 100 mL *or* greater than 0.1 mol/L	between 1 g and 0.1 g in 100 mL *or* between 0.1 mol/L and 0.01 mol/L	less than 0.1 g in 100 mL *or* less than 0.01 mol/L

The Effect of Ion Size on Solubility

When an atom gives up or gains an electron, the size of the ion that results is different from the size of the original atom. In Figure 9.2A, for example, you can see that the sodium ion is smaller than the sodium atom. In general, the ions of metals tend to be smaller than their corresponding neutral atoms. For the ions of non-metals, the reverse is true. The ions of non-metals tend to be larger than their corresponding neutral atoms.

Small ions bond more closely together than large ions. Thus, the bond between small ions is stronger than the bond between large ions with the same charge. As a result, compounds with small ions tend to be less soluble than compounds with large ions. Consider the ions of elements from Group 17 (VIIA), for example. Recall, from Chapters 2 and 3, that the size of the ions increases as you go down a family in the periodic table. (See Figure 9.2B.) Therefore, you would expect that fluoride compounds are less soluble than chloride, bromide, and iodide compounds. This tends to be the case.

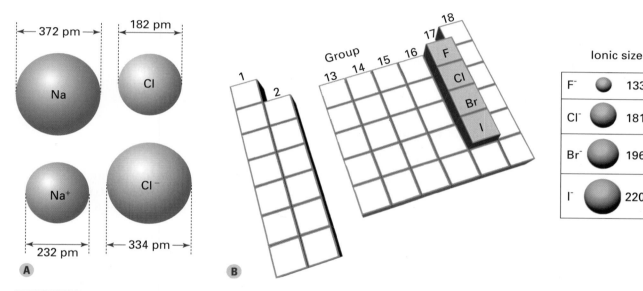

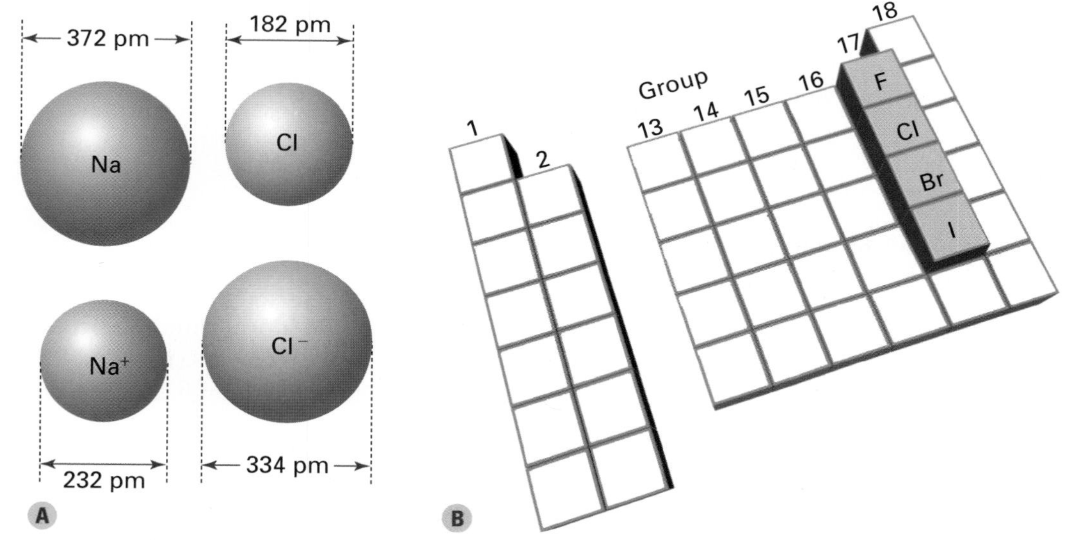

Ionic size

F^-		133 pm
Cl^-		181 pm
Br^-		196 pm
I^-		220 pm

Figure 9.2A The radius of the cation of a metal, such as sodium, tends to be smaller than the radius of the atom from which it was formed. The radius of the anion of a non-metal, such as phosphorus, tends to be larger than the radius of its atom.

Figure 9.2B In Group 17 (VIIA), fluoride ions are smaller than chloride, bromide, and iodide ions. How does this periodic trend affect the solubility of compounds that are formed from these elements?

Making Predictions About Solubility

Sulfides (compounds of S^{2-}) and oxides (compounds of O^{2-}) are influenced by both ion size and ion charge. These compounds tend to be insoluble because their ions have a double charge *and* are relatively small. Even so, a few sulfides and oxides are soluble, as you will discover in Investigation 1-A.

Many interrelated factors affect the solubility of substances in water. This makes it challenging to predict which ionic substances will dissolve in water. By performing experiments, chemists have developed guidelines to help them make predictions about solubility. In Investigation 9-A, you will perform your own experiments to develop quidelines about the solubility of ionic compounds in water.

mind STRETCH

Sketch an outline of the periodic table. Add labels and arrows to indicate what you think are the trends for ionic size (radius) across a period and down a group. Then suggest reasons for these trends. Specifically, identify the factors that you think are responsible for the differences in the sizes of ions and their "parent" atoms.

SKILL FOCUS

Predicting

Performing and recording

Analyzing and interpreting

Communicating results

Investigation 9-A

The Solubility of Ionic Compounds

In this investigation, you will work with a set of solutions. You will chemically combine small quantities, two at a time. This will help you determine which combinations react to produce a **precipitate**. A precipitate is an insoluble solid that may result when two aqueous solutions chemically react. *The appearance of a precipitate indicates that an insoluble compound is present.* Then you will compile your data with the data from other groups to develop some guidelines about the solubility of several ionic compounds.

Problem

How can you develop guidelines to help you predict the solubility of ionic compounds in water?

Prediction

Read the entire Procedure. Predict which combination of anions and cations will likely be soluble and which combination will likely be insoluble. Justify your prediction by briefly explaining your reasoning.

Materials

12-well or 24-well plate, or spot plate
toothpicks
cotton swabs
wash bottle with distilled water
piece of black paper
piece of white paper
labelled dropper bottles of aqueous solutions that contain the following cations:
Al^{3+}, NH_4^+, Ba^{2+}, Ca^{2+}, Cu^{2+}, Fe^{2+}, Mg^{2+}, Ag^+, Na^+, Zn^{2+}
labelled dropper bottles of aqueous solutions that contain the following anions:
CH_3COO^-, Br^-, CO_3^{2-}, Cl^-, OH^-, PO_4^{3-}, SO_4^{2-}, S^{2-}

Safety Precautions

- Do not contaminate the dropper bottles. The tip of a dropper should not make contact with either the plate or another solution. Put the cap back on the bottle immediately after use.
- Dispose of solutions as directed by your teacher.
- Make sure that you are working in a well-ventilated area.
- If you accidentally spill any of the solutions on your skin, wash the area immediately with plenty of cool water.

Procedure

1. Your teacher will give you a set of nine solutions to test. Each solution includes one of five cations or four anions. Design a table to record the results of all the possible combinations of cations with anions in your set of solutions.
2. Decide how to use the well plate or spot plate to test systematically all the combinations of cations with anions in your set. If your plate does not have enough wells, you will need to clean the plate before you can test all the possible combinations. To clean the plate, first discard solutions into the container provided by your teacher. Then rinse the plate with distilled water, and clean the wells using a cotton swab.
3. To test each combination of anion and cation, add one or two drops into your well plate or spot plate. Then stir the mixture using a toothpick. Rinse the toothpick with running water before each stirring. Make sure that you keep track of the combinations of ions in each well or spot.

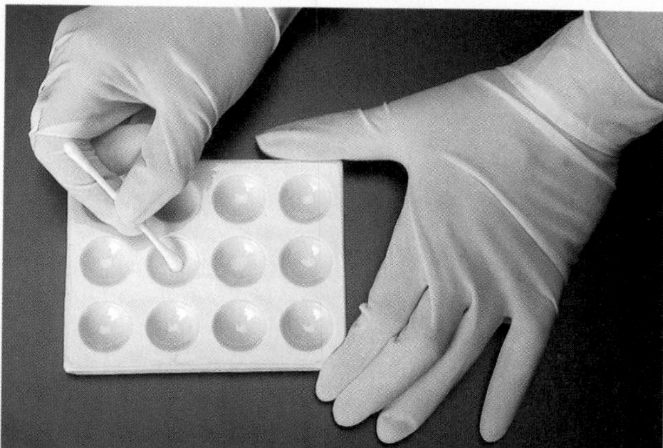

Why is it necessary to clean the well or spot plate as described in step 2?

4. Examine each mixture for evidence of a precipitate. Place the plate on a sheet of white or black paper. (Use whichever colour of paper helps you see a precipitate best.) Any cloudy appearance in the mixture is evidence of a precipitate. Many precipitates are white.
   - If you can see that a precipitate has formed, enter "I" in your table. This indicates that the combination of ions produces an insoluble substance.
   - If you cannot see a precipitate, enter "S" to indicate that the ion you are testing is soluble.
5. Repeat steps 2 to 4 for each cation solution.
6. Discard the solutions and precipitates into the container provided by your teacher. Rinse the plate with water, and clean the wells using a cotton swab.
7. If time permits, your teacher may give you a second set of solutions to test.
8. Add your observations to the class data table. Use your completed copy of the class data table to answer the questions below.

Analysis

1. Identify any cations that
   (a) always appear to form soluble compounds
   (b) always appear to form insoluble compounds
2. Identify any anions that
   (a) always appear to form soluble compounds
   (b) always appear to form insoluble compounds
3. Based on your observations, which sulfates are insoluble?
4. Based on your observations, which phosphates are soluble?
5. Explain why each reagent solution you tested must contain both cations and anions.
6. Your teacher prepared the cation solutions using compounds that contain the nitrate ion. For example, the solution marked Ca^{2+} was prepared by dissolving $Ca(NO_3)_2$ in water. Why were nitrates used to make these solutions?

Conclusions

7. Which group in the periodic table most likely forms cations with salts that are usually soluble?
8. Which group in the periodic table most likely forms anions with salts that are usually soluble?
9. Your answers to questions 7 and 8 represent a preliminary set of guidelines for predicting the solubility of the compounds you tested. Many reference books refer to guidelines like these as "solubility rules." Why might "solubility guidelines" be a better term to use for describing solubility patterns?

Application

10. Predict another combination of an anion and a cation (not used in this investigation) that you would expect to be soluble. Predict another combination that you would expect to be insoluble. Share your predictions, and your reasons, with the class. Account for any agreement or disagreement.

Soluble or Insoluble: General Solubility Guidelines

As you have seen, nearly all salts that contain the ammonium ion or an alkali metal are soluble. This observed pattern does not tell you how soluble these salts are, however. As well, it does not tell you whether ammonium chloride is more or less soluble than sodium chloride. Chemists rely on published data for this information. (See Figure 9.3.)

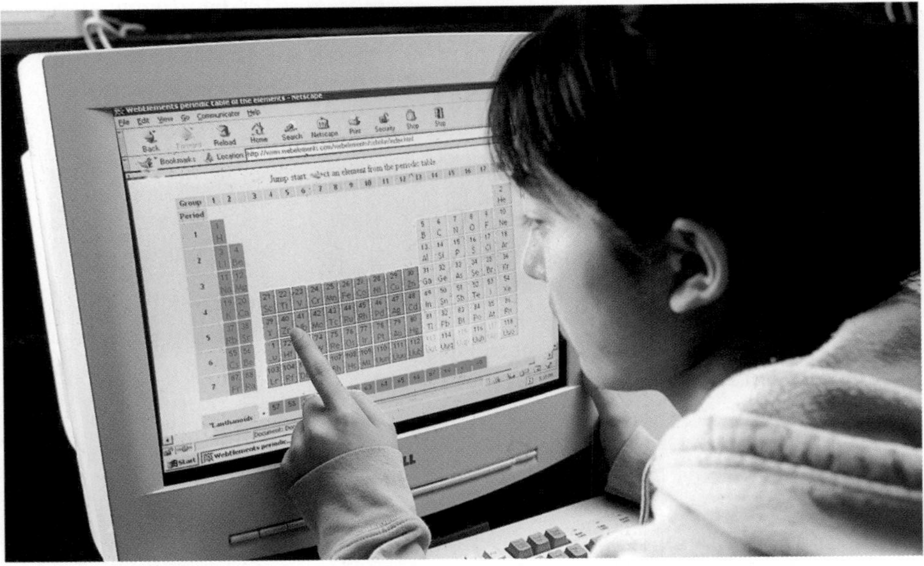

Figure 9.3 Many web sites on the Internet provide chemical and physical data for tens of thousands of compounds. Print resources, such as ***The CRC Handbook of Chemistry and Physics***, provide these data as well.

Many factors affect solubility. Thus, predicting solubility is neither straightforward nor simple. Nevertheless, the **general solubility guidelines** in Table 9.1 are a useful summary of ionic-compound interactions with water. To use Table 9.1, remember that a higher guideline number always takes precedence over a lower guideline number. For example, barium chloride, $BaCl_2$, is a white crystalline powder. The barium ion, Ba^{2+}, is listed in guideline 4 as insoluble. The chloride ion, Cl^-, is listed in guideline 3 as soluble. The higher guideline number takes precedence. Thus, you would predict that barium chloride is soluble.

You will be referring to the general solubility guidelines often in this chapter and in Chapter 10. They will help you identify salts that are soluble and insoluble in aqueous solutions. Always keep in mind, however, that water is a powerful solvent. Even an "insoluble" salt may dissolve enough to present a serious hazard if it is highly poisonous.

Table 9.1 General Solubility Guidelines

Guideline	Cations	Anions	Result	Exceptions
1	Li^+, Na^+, K^+, Rb^+, Cs^+, NH_4^+	NO_3^-, CH_3COO^-, ClO_3^-	soluble	$Ca(ClO_3)_2$ is insoluble
2	Ag^+, Pb^{2+}, Hg^+	CO_3^{2-}, PO_4^{3-}, O^{2-}, S^{2-}, OH^-	insoluble	BaO and $Ba(OH)_2$ are soluble. Group 2 sulfides tend to decompose.
3		Cl^-, Br^-, I^-	soluble	
4	Ba^{2+}, Ca^{2+}, Sr^{2+}		insoluble	
5	Mg^{2+}, Cu^{2+}, Zn^{2+}, Fe^{2+}, Fe^{3+}, Al^{3+}	SO_4^{2-}	soluble	

Practice Problems

1. Decide whether each of the following salts is soluble or insoluble in distilled water. Give reasons for your answer.
   - (a) lead(II) chloride, $PbCl_2$ (a white crystalline powder used in paints)
   - (b) zinc oxide, ZnO (a white pigment used in paints, cosmetics, and calamine lotion)
   - (c) silver acetate, $AgCH_3COO$ (a whitish powder that is used to help people quit smoking because of the bitter taste it produces)
2. Which of the following compounds are soluble in water? Explain your reasoning for each compound.
   - (a) potassium nitrate, KNO_3 (used to manufacture gunpowder)
   - (b) lithium carbonate, Li_2CO_3 (used to treat people who suffer from depression)
   - (c) lead(II) oxide, PbO (used to make crystal glass)
3. Which of the following compounds are insoluble in water?
   - (a) calcium carbonate, $CaCO_3$ (present in marble and limestone)
   - (b) magnesium sulfate, $MgSO_4$ (found in the hydrated salt, $MgSO_4 \cdot 7H_2O$, also known as Epsom salts; used for the relief of aching muscles and as a laxative)
   - (c) aluminum phosphate, $AlPO_4$ (found in dental cements)

Web LINK

www.school.mcgrawhill.ca/resources/

Different references present solubility guidelines in different ways. During a 1 h "surf session" on the Internet, a student collected ten different versions. See how many versions you can find. Compare their similarities and differences. Which version(s) do you prefer, and why? To start your search, go to the web site above. Go to **Science Resources**, then to **Chemistry 11** to find out where to go next.

Section Wrap-up

In Chapter 8, you focussed mainly on physical changes that involve solutions. In the first section of this chapter, you observed that mixing aqueous solutions of ionic compounds may result in either a physical change (dissolving) or a chemical change (a reaction that forms a precipitate). Chemical changes that involve aqueous solutions, especially ionic reactions, are common. They occur in the environment, in your body, and in the bodies of other organisms. In the next section, you will look more closely at reactions that involve aqueous solutions. As well, you will learn how to represent these reactions using a special kind of chemical equation, called an ionic equation.

Section Review

1 **(a)** K/U Name the two factors that affect the solubility of an ionic compound in water.

(b) Briefly explain how each factor affects solubility.

2 K/U Which would you expect to be less soluble: sodium fluoride, NaF (used in toothpaste), or sodium iodide, NaI (added to table salt to prevent iodine deficiency in the diet)? Explain your answer.

3 K/U Which of the following compounds are soluble in water?

(a) calcium sulfide, CaS (used in skin products)

(b) iron(II) sulfate, $FeSO_4$ (used as a dietary supplement)

(c) magnesium chloride, $MgCl_2$ (used as a disinfectant and a food tenderizer)

4 MC Which of the following compounds are insoluble in water? For each compound, relate its solubility to the use described.

(a) barium sulfate, $BaSO_4$ (can be used to obtain images of the stomach and intestines because it is opaque to X-rays)

(b) aluminum hydroxide, $Al(OH)_3$ (found in some antacid tablets)

(c) zinc carbonate, $ZnCO_3$ (used in suntan lotions)

5 C Calcium nitrate is used in fireworks. Silver nitrate turns dark when exposed to sunlight. When freshly made, both solutions are clear and colourless. Imagine that someone has prepared both solutions but has not labelled them. You do not want to wait for the silver nitrate solution to turn dark in order to identify the solutions. Name a chemical that can be used to precipitate a silver compound with the silver nitrate solution, but will produce no precipitate with the calcium nitrate solution. State the reason for your choice.

6 C Suppose that you discover four dropper bottles containing clear, colourless liquids in your school laboratory. The following four labels lie nearby:

- barium, Ba^{2+}
- chloride, Cl^-
- silver, Ag^+
- sulfate, SO_4^{2-}.

Unfortunately the labels have not been attached to the bottles. You decide to number the bottles 1, 2, 3, and 4. Then you mix the solutions in pairs. Three combinations give white precipitates: bottles 1 and 2, 1 and 4, and 2 and 3. Which ion does each bottle contain?

Language LINK

This poem is by an unknown author. It is not great literature. For decades, however, it has helped many students remember the solubility guidelines. Maybe it will do the same for you. If not, you can try writing your own poem!

Potassium, sodium, and
ammonium salts
Whatever they may be,
Can always be relied upon
For solubility.

Every single sulfate
Is soluble it's said,
Except barium and calcium
And strontium and lead.

Most every chloride's soluble,
That's what we've always read,
Save silver, mercurous mercury,
And (slightly) chloride of lead.

When asked about the nitrates,
The answer's always clear;
They each and all are soluble,
That's all we want to hear.

Metallic bases won't dissolve,
That is, all but three;
Potassium, sodium, and
ammonium
Dissolve quite readily.

But then you must remember,
You must surely not forget,
Calcium and barium
Dissolve a little bit.

Carbonates are insoluble,
It's lucky that it's so,
Or all our marble buildings
Would melt away like snow.

Reactions in Aqueous Solutions 9.2

When you mix two aqueous ionic compounds together, there are two possible outcomes. Either the compounds will remain in solution without reacting, or one aqueous ionic compound will chemically react with the other. How can you predict which outcome will occur? Figure 9.4 shows what happens when an aqueous solution of lead(II) nitrate is added to an aqueous solution of potassium iodide. As you can see, a yellow solid—a precipitate—is forming. This is a double displacement reaction. Recall, from Chapter 4, that a double displacement reaction is a chemical reaction that involves the exchange of ions to form two new compounds. It has the general equation

$$WX + YZ \rightarrow WZ + YX$$

In a double displacement reaction, the cations exchange anions. In the reaction shown in Figure 9.4, for example, the lead cation is exchanged with the iodide anion.

You can usually recognize a double displacement reaction by observing one of these possible results:

- the formation of a precipitate (so that ions are removed from solution as an insoluble solid)
- the formation of a gas (so that ions are removed from solution in the form of a gaseous product)
- the formation of water (so that H^+ and OH^- ions are removed from solution as water)

In this section, you will examine each of these results. At the same time, you will learn how to represent a double displacement reaction using a special kind of chemical equation: an ionic equation.

Section Preview/ Specific Expectations

In this section, you will

- **describe** combinations of aqueous solutions that result in the formation of precipitates
- **perform** a qualitative analysis of ions in solutions
- **represent** double displacement reactions by their net ionic equations
- **write** balanced chemical equations and net ionic equations for double displacement reactions
- **communicate** your understanding of the following terms: *spectator ions, total ionic equation, net ionic equation, qualitative analysis*

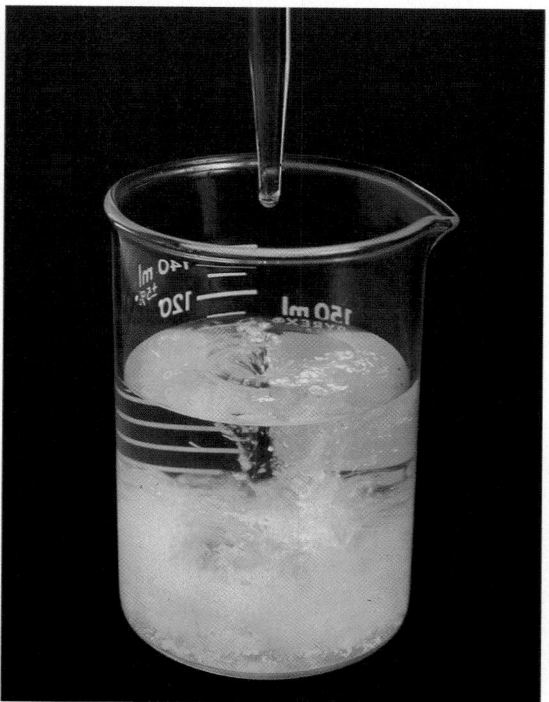

Figure 9.4 **Lead(II) nitrate and potassium iodide are clear, colourless aqueous solutions. Mixing them causes a double displacement reaction. An insoluble yellow precipitate (lead(II) iodide) and a soluble salt (potassium nitrate) are produced.**

Language LINK

Double displacement reactions are also called *metathesis reactions*. The word "metathesis" (pronounced with the stress on the second syllable: meh-TATH-e-sis) means "interchange." In chemistry, a metathesis reaction occurs when ions or atoms are exchanged between different compounds. Chemists are not the only people who use this word. Use a dictionary or encyclopedia to find non-chemistry examples of metathesis.

Double Displacement Reactions That Produce a Precipitate

A double displacement reaction that results in the formation of an insoluble substance is often called a *precipitation reaction*. Figure 9.4 is a clear example of a precipitation reaction. What if you did not have this photograph, however, and you were unable to do an experiment? Could you have predicted that mixing $Pb(NO_3)_{(aq)}$ and $2KI_{(aq)}$ would result in an insoluble compound? Yes. When you are given (on paper) a pair of solutions to be mixed together, start by thinking about the exchange of ions that may occur. Then use the general solubility guidelines (Table 9.1) to predict which compounds, if any, are insoluble.

For example, consider lead(II) nitrate, $Pb(NO_3)_2$, and potassium iodide, KI. Lead(II) nitrate contains Pb^{2+} cations and NO_3^- anions. Potassium iodide contains K^+ cations and I^- anions. Exchanging positive ions results in lead(II) iodide, PbI_2, and potassium nitrate, KNO_3. From the solubility guidelines, you know that all potassium salts and nitrates are soluble. Thus, potassium nitrate is soluble. The Pb^{2+} ion is listed in guideline 2 as an insoluble cation. The I^- ion is listed in guideline 3 as a soluble anion. Remember that a higher guideline number takes precedence over a lower guideline number. Thus, you can predict that lead(II) iodide is insoluble. It will form a precipitate when the solutions are mixed. The balanced chemical equation for this reaction is

$$Pb(NO_3)_{2(aq)} + 2KI_{(aq)} \rightarrow 2KNO_{3(aq)} + PbI_{2(s)}$$

Sample Problem

Predicting the Formation of a Precipitate

Problem

Which of the following pairs of aqueous solutions produce a precipitate when mixed together? Write the balanced chemical equation if you predict a precipitate. Write "NR" if you predict that no reaction takes place.

(a) potassium carbonate and copper(II) sulfate

(b) ammonium chloride and zinc sulfate

What Is Required?

You need to predict whether or not each pair of aqueous solutions forms an insoluble product (a precipitate). If it does, you need to write a balanced chemical equation.

What Is Given?

You know the names of the compounds in each solution.

Plan Your Strategy

Start by identifying the ions in each pair of compounds. Then exchange the positive ions in the two compounds. Compare the resulting compounds against the solubility guidelines, and make your prediction.

Continued ...

Continued ...
FROM PAGE 338

Act on Your Strategy

(a) Potassium carbonate contains K^+ and CO_3^{2-} ions. Copper(II) sulfate contains Cu^{2+} and SO_4^{2-} ions. Exchanging positive ions results in potassium sulfate, K_2SO_4, and copper(II) carbonate, $CuCO_3$.

All potassium salts are soluble, so these ions remain dissolved in solution.

The copper(II) ion is listed in guideline 5 as a soluble cation. The carbonate anion is listed in guideline 2 as insoluble. Because guideline 2 is higher, copper(II) carbonate should be insoluble. So you can predict that a precipitate forms. The balanced chemical equation for this reaction is

$$K_2CO_{3(aq)} + CuSO_{4(aq)} \rightarrow K_2SO_{4(aq)} + CuCO_{3(s)}$$

(b) Ammonium chloride contains NH_4^+ and Cl^- ions. Zinc sulfate consists of Zn^{2+} and SO_4^{2-} ions. Exchanging positive ions results in ammonium sulfate, $(NH_4)_2SO_4$, and zinc chloride, $ZnCl_2$.

Since all ammonium salts are soluble, the ammonium sulfate stays dissolved in solution.

Zinc chloride consists of a guideline 5 soluble cation and a guideline 3 soluble anion. Because guideline 3 is higher, zinc chloride should be soluble. Thus, you can predict that no precipitate forms.

$$NH_4Cl_{(aq)} + ZnSO_{4(aq)} \rightarrow NR$$

Check Your Solution

An experiment is always the best way to check a prediction. If possible, obtain samples of these solutions from your teacher, and mix them together.

Practice Problems

4. Predict the result of mixing each pair of aqueous solutions. Write a balanced chemical equation if you predict that a precipitate forms. Write "NR" if you predict that no reaction takes place.

(a) sodium sulfide and iron(II) sulfate

(b) sodium hydroxide and barium nitrate

(c) cesium phosphate and calcium bromide

(d) sodium carbonate and sulfuric acid

(e) sodium nitrate and copper(II) sulfate

(f) ammonium iodide and silver nitrate

(g) potassium carbonate and iron(II) nitrate

(h) aluminum nitrate and sodium phosphate

(i) potassium chloride and iron(II) nitrate

(j) ammonium sulfate and barium chloride

(k) sodium sulfide and nickel(II) sulfate

(l) lead(II) nitrate and potassium bromide

Figure 9.5 You can easily identify limestone and marble by their reaction with hydrochloric acid. The gas that is produced by this double displacement reaction is carbon dioxide.

Double Displacement Reactions That Produce a Gas

Double displacement reactions are responsible for producing a number of gases. (See Figure 9.5.) These gases include

- hydrogen
- hydrogen sulfide (a poisonous gas that smells like rotten eggs)
- sulfur dioxide (a reactant in forming acid rain)
- carbon dioxide
- ammonia

A Reaction that Produces Hydrogen Gas

The alkali metals form bonds with hydrogen to produce compounds called hydrides. Hydrides react readily with water to produce hydrogen gas. Examine the following equation for the reaction of lithium hydride, LiH, with water. If you have difficulty visualizing the ion exchange that takes place, rewrite the equation for yourself using HOH instead of H_2O.

$$LiH_{(s)} + H_2O_{(\ell)} \rightarrow LiOH_{(aq)} + H_{2(g)}$$

A Reaction that Produces Hydrogen Sulfide Gas

Sulfides react with certain acids, such as hydrochloric acid, to produce hydrogen sulfide gas.

$$K_2S_{(aq)} + 2HCl_{(aq)} \rightarrow 2KCl_{(aq)} + H_2S_{(g)}$$

A Reaction that Produces Sulfur Dioxide Gas

Some reactions produce a compound that, afterward, decomposes into a gas and water. Sodium sulfite is used in photography as a preservative. It reacts with hydrochloric acid to form sulfurous acid. The sulfurous acid then breaks down into sulfur dioxide gas and water. The net reaction is the sum of both changes. If the same compound appears on both sides of an equation (as sulfurous acid, H_2SO_3, does here), it can be eliminated. This is just like eliminating terms from an equation in mathematics.

$$Na_2SO_{3(aq)} + 2HCl_{(aq)} \rightarrow 2NaCl_{(aq)} + \cancel{H_2SO_{3(aq)}}$$

$$\cancel{H_2SO_{3(aq)}} \rightarrow SO_{2(g)} + H_2O_{(\ell)}$$

Therefore, the net reaction is

$$Na_2SO_{3(aq)} + 2HCl_{(aq)} \rightarrow 2NaCl_{(aq)} + SO_{2(g)} + H_2O_{(\ell)}$$

A Reaction that Produces Carbon Dioxide Gas

The reaction of a carbonate with an acid produces carbonic acid. Carbonic acid decomposes rapidly into carbon dioxide and water.

$$Na_2CO_{3(aq)} + 2HCl_{(aq)} \rightarrow 2NaCl_{(aq)} + \cancel{H_2CO_{3(aq)}}$$

$$\cancel{H_2CO_{3(aq)}} \rightarrow CO_{2(g)} + H_2O_{(\ell)}$$

The net reaction is

$$Na_2CO_{3(aq)} + 2HCl_{(aq)} \rightarrow 2NaCl_{(aq)} + CO_{2(g)} + H_2O_{(\ell)}$$

Ammonia gas can be prepared by the reaction of an ammonium salt with a base.

$$NH_4Cl_{(aq)} + NaOH_{(aq)} \rightarrow NaCl_{(aq)} + NH_{3(aq)} + H_2O_{(\ell)}$$

A Reaction that Produces Ammonia Gas

Ammonia gas is very soluble in water. You can detect it easily, however, by its sharp, pungent smell.

Double Displacement Reactions That Produce Water

The neutralization reaction between an acid and a base is a very important double displacement reaction. In a neutralization reaction, water results when an H^+ ion from the acid bonds with an OH^- ion from the base.

$$H_2SO_{4(aq)} + 2NaOH_{(aq)} \rightarrow Na_2SO_{4(aq)} + 2H_2O_{(\ell)}$$

Most metal oxides are bases. Therefore, a metal oxide will react with an acid in a neutralization reaction to form a salt and water.

$$2HNO_{3(aq)} + MgO_{(s)} \rightarrow Mg(NO_3)_{2(aq)} + H_2O_{(\ell)}$$

Non-metal oxides are acidic. Therefore, a non-metal oxide will react with a base. This type of reaction is used in the space shuttle. Cabin air is circulated through canisters of lithium hydroxide (a base) to remove the carbon dioxide before it can reach dangerous levels.

$$2LiOH_{(s)} + CO_{2(g)} \rightarrow Li_2CO_{3(aq)} + H_2O_{(\ell)}$$

Representing Aqueous Ionic Reactions with Net Ionic Equations

Mixing a solution that contains silver ions with a solution that contains chloride ions produces a white precipitate of silver chloride. There must have been other ions present in each solution, as well. You know this because it is impossible to have a solution of just a cation or just an anion. Perhaps the solution that contained silver ions was prepared using silver nitrate or silver acetate. Similarly, the solution that contained chloride ions might have been prepared by dissolving NaCl in water, or perhaps NH_4Cl or another soluble chloride. Any solution that contains $Ag^+_{(aq)}$ will react with any other solution that contains $Cl^-_{(aq)}$ to form a precipitate of $AgCl_{(s)}$. The other ions in the solutions are not important to the net result. These ions are like passive onlookers. They are called **spectator ions**.

The reaction between silver nitrate and sodium chloride can be represented by the following chemical equation:

$$AgNO_{3(aq)} + NaCl_{(aq)} \rightarrow NaNO_{3(aq)} + AgCl_{(s)}$$

This equation does not show the change that occurs, however. It shows the reactants and products as intact compounds. In reality, soluble ionic compounds dissociate into their respective ions in solution. So chemists often use a **total ionic equation** to show the dissociated ions of the soluble ionic compounds.

$$Ag^+_{(aq)} + NO_3{}^-_{(aq)} + Na^+_{(aq)} + Cl^-_{(aq)} \rightarrow Na^+_{(aq)} + NO_3{}^-_{(aq)} + AgCl_{(s)}$$

Notice that the precipitate, AgCl, is still written as an ionic formula. This makes sense because precipitates are insoluble, so they do not dissociate into ions. Also notice that the spectator ions appear on both sides of the equation. Here is the total ionic equation again, with slashes through the spectator ions.

$$Ag^+_{(aq)} + \cancel{NO_3{}^-_{(aq)}} + \cancel{Na^+_{(aq)}} + Cl^-_{(aq)} \rightarrow \cancel{Na^+_{(aq)}} + \cancel{NO_3{}^-_{(aq)}} + AgCl_{(s)}$$

If you eliminate the spectator ions, the equation becomes

$$Ag^+_{(aq)} + Cl^-_{(aq)} \rightarrow AgCl_{(s)}$$

An ionic equation that is written this way, without the spectator ions, is called a **net ionic equation**. Before you try writing your own net ionic equations, examine the guidelines in Table 9.2 below.

Table 9.2 Guidelines for Writing a Net Ionic Equation

1. Include only ions and compounds that have reacted. Do not include spectator ions.
2. Write the soluble ionic compounds as ions. For example, write $NH_4^+{}_{(aq)}$ and $Cl^-{}_{(aq)}$, instead of $NH_4Cl_{(aq)}$.
3. Write insoluble ionic compounds as formulas, not ions. For example, zinc sulfide is insoluble, so you write it as $ZnS_{(s)}$, not Zn^{2+} and S^{2-}.
4. Since covalent compounds do not produce ions in aqueous solution, write their molecular formulas. Water is a common example, because it dissociates only very slightly into ions. When a reaction involves a gas, always include the gas in the net ionic equation.
5. Write strong acids (discussed in the next chapter) in their ionic form. There are six strong acids:
   - hydrochloric acid (write as $H^+{}_{(aq)}$ and $Cl^-{}_{(aq)}$, not $HCl_{(aq)}$)
   - hydrobromic acid (write as $H^+{}_{(aq)}$ and $Br^-{}_{(aq)}$)
   - hydroiodic acid (write as $H^+{}_{(aq)}$ and $I^-{}_{(aq)}$)
   - sulfuric acid (write as $H^+{}_{(aq)}$ and $SO_{4(aq)}$)
   - nitric acid (write as $H^+{}_{(aq)}$ and $NO_3^-{}_{(aq)}$)
   - perchloric acid (write as $H^+{}_{(aq)}$ and $ClO_4^-{}_{(aq)}$)

   All other acids are weak and form few ions. Therefore, write them in their molecular form.
6. Finally, check that the net ionic equation is balanced for charges as well as for atoms.

Sample Problem

Writing Net Ionic Equations

Problem

A chemical reaction occurs when the following aqueous solutions are mixed: sodium sulfide and iron(II) sulfate. Identify the spectator ions. Then write the balanced net ionic equation.

What Is Required?

You need to identify the spectator ions and write a balanced net ionic equation for the reaction between sodium sulfide and iron(II) sulfate.

What Is Given?

You know the chemical names of the compounds.

Continued ...

Continued ...
FROM PAGE 342

Plan Your Strategy

Step 1 Start by writing the chemical formulas of the given compounds.

Step 2 Then write the complete chemical equation for the reaction, using your experience in predicting the formation of a precipitate.

Step 3 Once you have the chemical equation, you can replace the chemical formulas of the soluble ionic compounds with their dissociated ions.

Step 4 This will give you the total ionic equation. Next you can identify the spectator ions (the ions that appear on both sides of the equation).

Step 5 Finally, by rewriting the total ionic equation *without* the spectator ions, you will have the net ionic equation.

Act on Your Strategy

Steps 1 and 2 The chemical equation for the reaction is

$Na_2S_{(aq)} + FeSO_{4(aq)} \rightarrow Na_2SO_{4(aq)} + FeS_{(s)}$

Step 3 The total ionic equation is

$\cancel{2Na^{+}}_{(aq)} + S^{2-}{}_{(aq)} + Fe^{2+}{}_{(aq)} + \cancel{SO_4{}^{2-}}_{(aq)} \rightarrow \cancel{2Na^{+}}_{(aq)} + \cancel{SO_4{}^{2-}}_{(aq)} + FeS_{(s)}$

Step 4 Therefore, the spectator ions are $Na^{+}{}_{(aq)}$ and $SO_4{}^{2-}{}_{(aq)}$.

Step 5 The net ionic equation is

$Fe^{2+}{}_{(aq)} + S^{2-}{}_{(aq)} \rightarrow FeS_{(s)}$

Check Your Solution

Take a final look at your net ionic equation to make sure that no ions are on both sides of the equation.

Practice Problems

5. Mixing each pair of aqueous solutions results in a chemical reaction. Identify the spectator ions. Then write the balanced net ionic equation.

(a) sodium carbonate and hydrochloric acid

(b) sulfuric acid and sodium hydroxide

6. Identify the spectator ions for the reaction that takes place when each pair of aqueous solutions is mixed. Then write the balanced net ionic equation.

(a) ammonium phosphate and zinc sulfate

(b) lithium carbonate and nitric acid

(c) sulfuric acid and barium hydroxide

Identifying Ions in Aqueous Solution

Suppose that you have a sample of water. You want to know what, if any, ions are dissolved in it. Today technological devices, such as the mass spectrometer, make this investigative work fairly simple. Before such devices, however, chemists relied on *wet chemical techniques*: experimental tests, such as submitting a sample to a series of double displacement reactions. Chemists still use wet chemical techniques. With each reaction, insoluble compounds precipitate out of the solution. (See Figure 9.6.) This enables the chemist to determine, eventually, the identity of one or several ions in the solution. This ion-identification process is an example of **qualitative analysis**.

Chemists use a range of techniques for qualitative analysis. For example, the colour of an aqueous solution can help to identify one of the ions that it contains. Examine Table 9.3. However, the intensity of ion colour varies with its concentration in the solution. Also keep in mind that many ions are colourless in aqueous solution. For example, the cations of elements from Groups 1 (IA) and 2 (IIA), as well as aluminum, zinc, and most anions, are colourless. So there are limits to the inferences you can make if you rely on solution colour alone.

Another qualitative analysis technique is a flame test. A dissolved ionic compound is placed in a flame. Table 9.4 lists the flame colours associated with several ions. Notice that all the ions are metallic. The flame test is only useful for identifying metallic ions in aqueous solution.

Qualitative analysis challenges a chemist's creative imagination and chemical understanding. Discover this for yourself in Investigation 9-B.

Table 9.3 The Colour of Some Common Ions in Aqueous Solution

	Ions	Symbol	Colour
Cations	chromium (II) copper(II)	Cr^{2+} Cu^{2+}	blue
	chromium(III) copper(I) iron(II) nickel(II)	Cr^{3+} Cu^{+} Fe^{2+} Ni^{2+}	green
	iron(III)	Fe^{3+}	pale yellow
	cobalt(II) manganese(II)	Co^{2+} Mn^{2+}	pink
Anions	chromate	CrO_4^{2-}	yellow
	dichromate	$Cr_2O_7^{2-}$	orange
	permanganate	MnO_4^-	purple

Table 9.4 The Flame Colour of Selected Metallic Ions

Ion	Symbol	Colour
lithium	Li^+	red
sodium	Na^+	yellow
potassium	K^+	violet
cesium	Cs^+	violet
calcium	Ca^{2+}	red
strontium	Sr^{2+}	red
barium	Ba^{2+}	yellowish-green
copper	Cu^{2+}	bluish-green
boron	B^{2+}	green
lead	Pb^{2+}	bluish-white

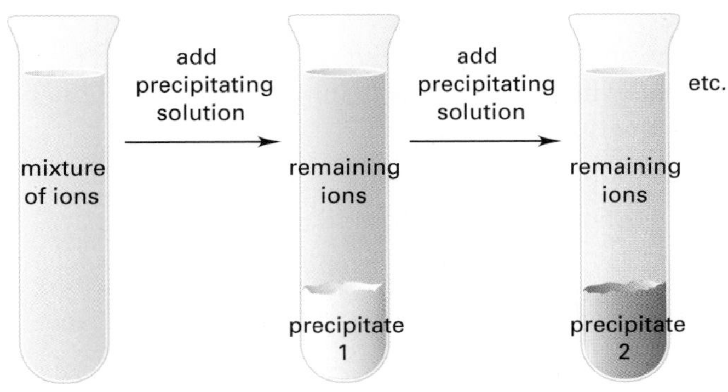

Figure 9.6 This illustration shows the basic idea behind a qualitative analysis for identifying ions in an aqueous solution. At each stage, the resulting precipitate is removed.

Investigation 9-B

SKILL FOCUS
Predicting
Performing and recording
Analyzing and interpreting

Qualitative Analysis

In this investigation, you will apply your knowledge of chemical reactions and the general solubility guidelines to identify unknown ions.

Question

How can you identify ions in solution?

Predictions

Read the entire Procedure. Can you predict the results of any steps? Write your predictions in your notebook. Justify each prediction.

Materials

Part 1

12-well or 24-well plate, or spot plate
toothpicks
cotton swabs
unknowns: 4 dropper bottles (labelled A, B, C, and D) of solutions that include $Na^{+}{}_{(aq)}$, $Ag^{+}{}_{(aq)}$, $Ca^{2+}{}_{(aq)}$, and $Cu^{2+}{}_{(aq)}$
reactants: 2 labelled dropper bottles, containing dilute $HCl_{(aq)}$ and dilute $H_2SO_{4(aq)}$

Part 2

cotton swabs
Bunsen burner
heat-resistant pad
unknowns: 4 dropper bottles, containing the same unknowns that were used in Part 1
reactants: 4 labelled dropper bottles, containing $Na^{+}{}_{(aq)}$, $Ag^{+}{}_{(aq)}$, $Ca^{2+}{}_{(aq)}$, and $Cu^{2+}{}_{(aq)}$

Part 3

12-well or 24-well plate, or spot plate
toothpicks
cotton swabs
unknowns: 3 dropper bottles (labelled X, Y, and Z), containing solutions of $SO_4{}^{2-}{}_{(aq)}$, $CO_3{}^{2-}{}_{(aq)}$, and $I^{-}{}_{(aq)}$
reactants: 3 labelled dropper bottles, containing $Ba^{2+}{}_{(aq)}$, $Ag^{+}{}_{(aq)}$, and $HCl_{(aq)}$

Safety Precautions

- Be careful not to contaminate the dropper bottles. The tip of a dropper should not make contact with either the plate or another solution. Put the cap back on the bottle immediately after use.
- Hydrochloric acid and sulfuric acid are corrosive. Wash any spills on your skin with plenty of cool water. Inform your teacher immediately.
- Part 2 of this investigation requires an open flame. Tie back long hair, and confine any loose clothing.

Procedure

Part 1 Using Acids to Identify Cations

1. Read steps 2 and 3 below. Design a suitable table for recording your observations.
2. Place one or two drops of each unknown solution into four different wells or spots. Add one or two drops of hydrochloric acid to each unknown. Record your observations.
3. Repeat step 2. This time, test each unknown solution with one or two drops of sulfuric acid. Record your observations.
4. Answer Analysis questions 1 to 5.

Part 2 Using Flame Tests to Identify Cations

Note: Your teacher may demonstrate this part or provide you with an alternative version.

1. Design tables to record your observations.
2. Observe the appearance of each known solution. Record your observations. Repeat for each unknown solution. Some cations have a characteristic colour. (Refer to Table 9.4.) If you think that you can identify one of the unknowns, record your identification.

3. Flame tests can identify some cations. Set up the Bunsen burner and heat-resistant pad. Light the burner. Adjust the air supply to produce a hot flame with a blue cone.

4. Place a few drops of solution containing $Na^{+}_{(aq)}$ on one end of a cotton swab.

   CAUTION Carefully hold the saturated tip so it is just in the Bunsen burner flame, near the blue cone. You may need to hold it in this position for as long as 30 s to allow the solution to vaporize and mix with the flame. Record the colour of the flame.

5. Not all cations give colour to a flame. The sodium ion *does* give a distinctive colour to a flame, however. It is often present in solutions as a contaminant. For a control, repeat step 4 with water and record your observations. You can use the other end of the swab for a second test. Dispose of used swabs in the container your teacher provides.

6. Repeat the flame test for each of the other known solutions. Then test each of the unknown solutions.

7. Answer Analysis question 6.

Part 3 Identifying Anions

1. Place one or two drops of each unknown solution into three different wells or spots. Add one or two drops of $Ba^{2+}_{(aq)}$ to each unknown solution. Stir with a toothpick. Record your observations.

2. Add a drop of hydrochloric acid to any well or spot where you observed a precipitate in step 1. Stir and record your observations.

3. Repeat step 1, adding one or two drops of $Ag^{+}_{(aq)}$ to each unknown solution. Record the colour of any precipitate that forms.

4. Answer Analysis questions 7 to 9.

Analysis

1. (a) Which of the cations you tested should form a precipitate with hydrochloric acid? Write the net ionic equation.
   (b) Did your results support your predictions? Explain.

2. (a) Which cation(s) should form a precipitate when tested with sulfuric acid? Write the net ionic equation.
   (b) Did your results support your predictions? Explain.

3. Which cation(s) should form a soluble chloride and a soluble sulfate?

4. Which cation has a solution that is not colourless?

5. Based on your analysis so far, tentatively identify each unknown solution.

6. Use your observations of the flame tests to confirm or refute the identifications you made in question 5. If you are not sure, check your observations and analysis with other students. If necessary, repeat some of your tests.

7. Which anion(s) should form a precipitate with Ba^{2+}? Write the net ionic equation.

8. Which precipitate should react when hydrochloric acid is added? Give reasons for your prediction.

9. Tentatively identify each anion. Check your observations against the results you obtained when you added hydrochloric acid. Were they what you expected? If not, check your observations and analysis with other students. If necessary, repeat some of your tests.

Conclusion

10. Identify the unknown cations and anions in this investigation. Explain why you do, or do not, have confidence in your decisions. What could you do to be more confident?

Section Wrap-up

Qualitative analysis helps you identify ions that may be present in a solution. It does not, however, tell you how much of these ions are present. In other words, it does not provide any quantitative information about the quantity or concentration of ions in solution. In the next section, you will find out how to calculate this quantitative information, using techniques you learned in Unit 2.

Section Review

1. C Briefly compare the relationships among a chemical formula, a total ionic equation, and a net ionic equation. Use sentences or a graphic organizer.
2. K/U Write a net ionic equation for each double displacement reaction in aqueous solution.
   - **(a)** tin(II) chloride with potassium phosphate
   - **(b)** nickel(II) chloride with sodium carbonate
   - **(c)** chromium(III) sulfate with ammonium sulfide
3. K/U For each reaction in question 2, identify the spectator ions.
4. K/U Would you expect a qualitative analysis of a solution to give you the amount of each ion present? Explain why or why not.
5. I A solution of limewater, $Ca(OH)_{2(aq)}$, is basic. It is used to test for the presence of carbon dioxide. Carbon dioxide is weakly acidic and turns limewater milky. Use a chemical equation to explain what happens during the test. What type of reaction occurs?
6. K/U State the name and formula of the precipitate that forms when aqueous solutions of copper(II) sulfate and sodium carbonate are mixed. Write the net ionic equation for the reaction. Identify the spectator ions.
7. I All the solutions in this photograph have the same concentration: 0.1 mol/L. Use Table 9.3 to infer which ion causes the colour in each solution. How much confidence do you have in your inferences? What could you do to increase your confidence?

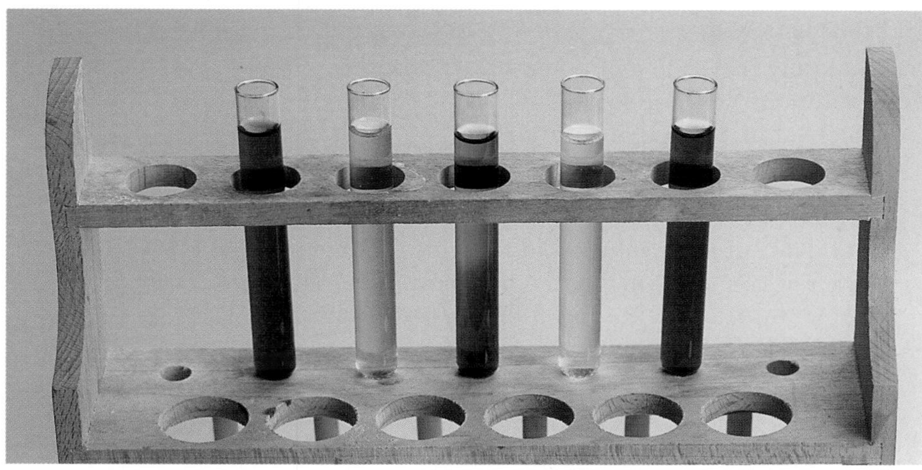

9.3 Stoichiometry in Solution Chemistry

Section Preview/ Specific Expectations

In this section, you will

- **represent** double displacement reactions by their net ionic equations
- **write** balanced chemical equations for double displacement reactions
- **solve** stoichiometry problems that involve solutions
- **describe** and **work with** solutions that have known concentrations

Recall that stoichiometry involves calculating the amounts of reactants and products in chemical reactions. If you know the atoms or ions in a formula or a reaction, you can use stoichiometry to determine the amounts of these atoms or ions that react. Solving stoichiometry problems in solution chemistry involves the same strategies you learned in Unit 2. Calculations involving solutions sometimes require a few additional steps, however. For example, if a precipitate forms, the net ionic equation may be easier to use than the chemical equation. Also, some problems may require you to calculate the amount of a reactant, given the volume and concentration of the solution.

Take your time working through the next three Sample Problems. Make sure that you understand how to arrive at the solutions. Then try the Practice Problems on page 352.

Sample Problem

The Concentration of Ions

Problem

Calculate the concentration (in mol/L) of chloride ions in each solution.

(a) 19.8 g of potassium chloride dissolved in 100 mL of solution

(b) 26.5 g of calcium chloride dissolved in 150 mL of solution

(c) a mixture of the two solutions in parts (a) and (b), assuming that the volumes are additive

What Is Required?

(a) and **(b)** You need to find the concentration (in mol/L) of chloride ions in two different solutions.

(c) You need to find the concentration of chloride ions when the two solutions are mixed.

What Is Given?

You know that 19.8 g of potassium chloride is dissolved in 100 mL of solution. You also know that 26.5 g of calcium chloride is dissolved in 150 mL of solution.

Plan Your Strategy

(a) and **(b)** For each solution, determine the molar mass. Find the amount (in mol) using the mass and the molar mass. Write equations for the dissociation of the substance. (That is, write the total ionic equation.)

Continued ...

FROM PAGE 348

Use the coefficients in the dissociation equation to determine the amount (in mol) of chloride ions present. Calculate the concentration (in mol/L) of chloride ions from the amount and volume of the solution.

(c) Add the amounts of chloride ions in the two solutions to find the total. Add the volumes of the solutions to find the total volume. Calculate the concentration of chloride ions (in mol/L) using the total amount (in mol) divided by the total volume (in L).

Act on Your Strategy

(a) and **(b)**

Solution	KCl	$CaCl_2$
Molar mass	$39.10 + 35.45 = 74.55$ g	$40.08 + (2 \times 35.45) = 110.98$ g
Amount (mol)	$19.8 \text{ g} \times \frac{1 \text{ mol}}{74.55 \text{ g}} = 0.266 \text{ mol}$	$26.5 \text{ g} \times \frac{1 \text{ mol}}{110.98 \text{ g}} = 0.239 \text{ mol}$
Dissociation equation	$KCl_s \rightarrow K^+_{(aq)} + Cl^-_{(aq)}$	$CaCl_{2(s)} \rightarrow Ca^{2+}_{(aq)} + 2Cl^-_{(aq)}$
Amount of Cl^-	$0.266 \text{ mol KCl} \times \frac{1 \text{ mol Cl}^-}{1 \text{ mol KCl}} = 0.266 \text{ mol}$	$0.239 \text{ mol CaCl}_2 \times \frac{2 \text{ mol Cl}^-}{1 \text{ mol CaCl}_2} = 0.478 \text{ mol}$
Concentration of Cl^-	$\frac{0.266 \text{ mol}}{0.100 \text{ L}} = 2.66 \text{ mol/L}$	$\frac{0.478 \text{ mol}}{0.150 \text{ L}} = 3.19 \text{ mol/L}$

The concentration of chloride ions when 19.8 g of potassium chloride is dissolved in 100 mL of solution is 2.66 mol/L. The concentration of chloride ions when 26.5 g of calcium chloride is dissolved in 150 mL of solution is 3.19 mol/L.

(c) $$\text{Total amount of } Cl^-_{(aq)} = 0.266 + 0.478 \text{ mol} = 0.744 \text{ mol}$$

$$\text{Total volume of solution} = 0.100 + 0.150 \text{ L} = 0.250 \text{ L}$$

$$\text{Total concentration of } Cl^-_{(aq)} = \frac{0.744 \text{ mol}}{0.250 \text{ L}} = 2.98 \text{ mol/L}$$

The concentration of chloride ions when the solutions are mixed is 2.98 mol/L

Check Your Solution

The units for amount and concentration are correct. The answers appear to be reasonable. When the solutions are mixed, the concentration of the chloride ions is not a simple average of the concentrations of the two solutions. Why? The volumes of the two solutions were different.

Sample Problem

The Mass Percent of Ions

Problem

The leaves of a rhubarb plant contain a relatively high concentration of oxalate ions, $C_2O_4^{2-}$. Oxalate ions are poisonous, causing respiratory failure. To determine the percent of oxalate ions, a student measured the mass of some leaves. Then the student ground up the leaves and added excess calcium chloride solution to precipitate calcium oxalate. The student tested 238.6 g of leaves. The dried mass of calcium oxalate was 0.556 g. What was the mass percent of oxalate ions in the leaves?

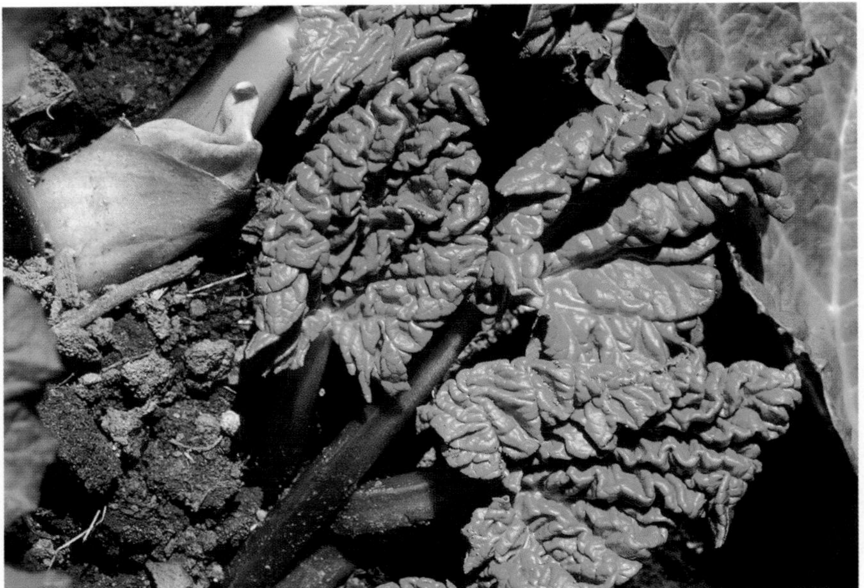

Although rhubarb stalks are safe to eat, the leaves are poisonous.

What Is Required?

You need to find the mass percent of oxalate ions in the leaves.

What Is Given?

You know the mass of the leaves is 238.6 g. You also know the mass of dried calcium oxalate is 0.556 g.

Plan Your Strategy

Determine the molar mass of calcium oxalate. Use the mass of calcium oxalate and its molar mass to find the amount (in mol) of calcium oxalate. Write the net ionic equation for the formation of calcium oxalate. From the coefficients in the net ionic equation, find the amount of oxalate ions (in mol). Calculate the mass of oxalate ions from the amount of oxalate ions (in mol) and the molar mass. Calculate the mass percent of oxalate ions in rhubarb leaves from the mass of the leaves and the mass of the oxalate ions present.

Continued ...

Continued ...
FROM PAGE 350

Act on Your Strategy

Molar mass of CaC_2O_4 = 128.1 g

Amount of $CaC_2O_4 = 0.556 \text{ g} \times \dfrac{1 \text{ mol}}{128.1 \text{ g}}$

$= 0.004\ 34$ mol

The net ionic equation is

$Ca^{2+}{}_{(aq)} + C_2O_4{}^{2-}{}_{(aq)} \rightarrow CaC_2O_{4(s)}$

The mole ratio of $CrO_4{}^{2-}$ to CaC_2O_4 is 1:1. Therefore, the leaves must have contained 0.004 34 mol of $C_2O_4{}^{2-}$.

Molar mass of $C_2O_4{}^{2-}$ = 88.02 g

Mass of $C_2O_4{}^{2-} = 0.004\ 34 \cancel{\text{mol}} \times \dfrac{88.02 \text{ g}}{1 \cancel{\text{mol}}}$

$= 0.382$ g

Mass percent of $C_2O_4{}^{2-}$ in the leaves $= \dfrac{0.382 \cancel{\text{g}}}{238.6 \cancel{\text{g}}} \times 100\%$

$= 0.160\%$

The mass percent of oxalate ions in the leaves is 0.160%.

Check Your Solution

Since the units divide out properly, you can be fairly confident that the answer is correct. The final value appears to be reasonable.

PROBLEM TIP

Review the method for solving stoichiometry problems you learned in Chapter 7, Section 7.1.

Sample Problem

Finding the Minimum Volume to Precipitate

Problem

Aqueous solutions that contain silver ions are usually treated with chloride ions to recover silver chloride. What is the minimum volume of 0.25 mol/L magnesium chloride, $MgCl_{2(aq)}$, needed to precipitate all the silver ions in 60 mL of 0.30 mol/L silver nitrate, $AgNO_{3(aq)}$? Assume that silver chloride is completely insoluble in water.

What Is Required?

You need to find the minimum volume of magnesium chloride that will precipitate all the silver ions.

What Is Given?

You know the volumes and concentrations of the silver nitrate (volume = 60 mL; concentration = 0.30 mol/L). The concentration of the magnesium chloride solution is 0.25 mol/L.

Plan Your Strategy

Find the amount (in mol) of silver nitrate from the volume and concentration of solution. Write a balanced chemical

Continued ...

Continued ...
FROM PAGE 351

equation for the reaction. Use mole ratios from the coefficients in the equation to determine the amount (in mol) of magnesium chloride that is needed. Use the amount (in mol) of magnesium chloride and the concentration of solution to find the volume that is needed.

Act on Your Strategy

$$\text{Amount AgNO}_3 = 0.060\,\cancel{L} \times 0.30\ \text{mol}/\cancel{L} = 0.018\ \text{mol}$$

$$MgCl_{2(aq)} + 2AgNO_{3(aq)} \rightarrow 2AgCl_{(s)} + Mg(NO_3)_{2(aq)}$$

The mole ratio of $MgCl_2$ to $AgNO_3$ is 1:2.

$$n\ \text{mol MgCl}_2 = 0.018\ \text{mol}\ \cancel{AgNO_3} \times \frac{1\ \text{mol MgCl}_2}{2\ \text{mol}\ \cancel{AgNO_3}} = 0.0090\ \text{mol}$$

$$\text{Volume of 0.25 mol/L MgCl}_2 \text{ needed} = \frac{0.0090\ \cancel{\text{mol}}}{0.25\ \cancel{\text{mol}}/\text{L}} = 0.036\ \text{L}$$

The minimum volume of 0.25 mol/L magnesium chloride that is needed is 36 mL.

Check Your Solution

The answer is in millilitres, an appropriate unit of volume. The amount appears to be reasonable.

Practice Problems

7. Food manufacturers sometimes add calcium acetate to puddings and sweet sauces as a thickening agent. What volume of 0.500 mol/L calcium acetate, $Ca(CH_3COO)_{2(aq)}$, contains 0.300 mol of acetate ions?

8. Ammonium phosphate can be used as a fertilizer. 6.0 g of ammonium phosphate is dissolved in sufficient water to produce 300 mL of solution. What are the concentrations (in mol/L) of the ammonium ions and phosphate ions present?

9. An aqueous solution of a certain salt contains chloride ions. A sample of this solution was made by dissolving 17.59 g of the salt in a 1 L volumetric flask. Then 25.00 mL of the solution was treated with excess silver nitrate. The precipitate, $AgCl_{(s)}$, was filtered and dried. If the mass of the dry precipitate was 47.35 g, what was the mass percent of chloride ions in the solution?

10. The active ingredient in some rat poisons is thallium(I) sulfate, Tl_2SO_4. A chemist takes a 500 mg sample of thallium(I) sulfate and adds potassium iodide, to precipitate yellow thallium(I) iodide. When the precipitate is dried, its mass is 200 mg. What is the mass percent of Tl_2SO_4 in the rat poison?

Limiting Reactant Problems in Aqueous Solutions

In Chapter 7, you learned how to solve limiting reactant problems. You can always recognize a limiting reactant problem because you are always given the amounts of both reactants. A key step in a limiting reactant problem is determining which one of the two reactants is limiting. In aqueous solutions, this usually means finding the amount of a reactant, given the volume and concentration of the solution.

Sample Problem

Finding the Mass of a Precipitated Compound

Problem

Mercury salts have a number of important uses in industry and in chemical analysis. Because mercury compounds are poisonous, however, the mercury ions must be removed from the waste water. Suppose that 25.00 mL of 0.085 mol/L aqueous sodium sulfide is added to 56.5 mL of 0.10 mol/L mercury(II) nitrate. What mass of mercury(II) sulfide, $HgS_{(s)}$, precipitates?

What Is Required?

You need to find the mass of mercury(II) sulfide that precipitates.

What Is Given?

You know the volumes and concentrations of the sodium sulfide and mercury(II) nitrate solutions.

Plan Your Strategy

Write a balanced chemical equation for the reaction. Find the amount (in mol) of each reactant, using its volume and concentration. Identify the limiting reactant. Determine the amount (in mol) of mercury(II) sulfide that forms. Calculate the mass of mercury(II) sulfide that precipitates.

Act on Your Strategy

The chemical equation is

$$Hg(NO_3)_{2(aq)} + Na_2S_{(aq)} \rightarrow 2NaNO_{3(aq)} + HgS_{(s)}$$

Calculate the amount (in mol) of each reactant.

$$\begin{aligned}\text{Amount of } Hg(NO_3)_2 &= 0.0565 \cancel{L} \times 0.10 \text{ mol}/\cancel{L} \\ &= 0.005\ 65 \text{ mol}\end{aligned}$$

$$\begin{aligned}\text{Amount of } Na_2S &= 0.0250 \cancel{L} \times 0.085 \text{ mol}/\cancel{L} \\ &= 0.002\ 12 \text{ mol}\end{aligned}$$

The reactants are in a 1:1 ratio. Because Na_2S is present in the smallest amount, it is the limiting reactant.
The equation indicates that each mol of Na_2S reacts to produce the same amount of $HgS_{(s)}$ precipitate. This amount is 0.002 12 mol.

Continued ...

PROBLEM TIP

Chemists solve limiting reactant problems in different ways. The method used here is different from the one you used in Chapter 7, Section 7.2

Continued ...
FROM PAGE 353

Molar mass of HgS = 200.6 + 32.1
= 232.7

Mass of $HgS_{(s)}$ = Amount × Molar mass

$$= 0.002\ 12\ \cancel{\text{mol}} \times \frac{232.7\ \text{g}}{1\ \cancel{\text{mol}}}$$

= 0.493 g

The mass of mercury(II) sulfide that precipitates is 0.49 g

Check Your Solution

The answer has appropriate units of mass. This answer appears to be reasonable, given the values in the problem.

Sample Problem

Finding the Mass of Another Precipitated Compound

Problem

Silver chromate, Ag_2CrO_4, is insoluble. It forms a brick-red precipitate. Calculate the mass of silver chromate that forms when 50.0 mL of 0.100 mol/L silver nitrate reacts with 25.0 mL of 0.150 mol/L sodium chromate.

What Is Required?

You need to find the mass of silver chromate that precipitates.

What Is Given?

You know the volumes and concentrations of the silver nitrate and sodium chromate solutions.

Plan Your Strategy

Write a balanced chemical equation for the reaction. Find the amount (in mol) of each reactant, using its volume and concentration. Identify the limiting reactant. Determine the amount (in mol) of silver chromate that forms. Calculate the mass of silver chromate that precipitates.

Act on Your Strategy

The chemical equation is

$$2AgNO_{3(aq)} + Na_2CrO_{4(aq)} \rightarrow Ag_2CrO_{4(s)} + 2NaNO_{3(aq)}$$

Calculate the amount (in mol) of each reactant.

Amount of $AgNO_3$ = 0.0500 L × 0.100 mol/L
= 5.00×10^{-3} mol

Amount of Na_2CrO_4 = 0.0250 L × 0.150 mol/L
= 3.75×10^{-3} mol

Continued ...

FROM PAGE 354

To identify the limiting reactant, divide by the coefficient in the equation and find the smallest result.

$AgNO_3$: $5.00 \times \frac{10^{-3}\text{ mol}}{2} = 2.50 \times 10^{-3}$ mol

Na_2CrO_4: $3.75 \times \frac{10^{-3}\text{ mol}}{1} = 3.75 \times 10^{-3}$ mol

Since the smallest result is given for $AgNO_3$, this reactant is the limiting reactant.

Using the coefficients in the balanced equation, 2 mol of $AgNO_3$ react for each mole of Ag_2CrO_4 formed.

$$\text{Amount of } Ag_2CrO_4 = 5.00 \times 10^{-3}\text{ mol } AgNO_3 \times \frac{1\text{ mol } Ag_2CrO_4}{2\text{ mol } AgNO_3}$$
$$= 2.50 \times 10^{-3}\text{ mol } Ag_2CrO_4$$

The molar mass of Ag_2CrO_4 is 331.7 g/mol.

$$\text{Mass of precipitate} = 2.50 \times 10^{-3}\cancel{\text{mol}} \times \frac{331.7\text{ g}}{1\cancel{\text{mol}}}$$
$$= 0.829\text{ g}$$

The mass of silver chromate that precipitates is 0.829 g.

Check Your Solution

The answer has appropriate units of mass. The answer appears to be reasonable, given the values in the problem.

Practice Problems

11. 8.76 g of sodium sulfide is added to 350 mL of 0.250 mol/L lead(II) nitrate solution. Calculate the maximum mass of precipitate that can form.

12. 25.0 mL of 0.400 mol/L $Pb(NO_3)_{2(aq)}$ is mixed with 300 mL of 0.220 mol/L $KI_{(aq)}$. What is the maximum mass of precipitate that can form?

13. A student mixes 15.0 mL of 0.250 mol/L aqueous sodium hydroxide with 20.0 mL of 0.400 mol/L aqueous aluminum nitrate.

(a) Write the chemical equation for the reaction.

(b) Calculate the maximum mass of precipitate that forms.

Section Wrap-up

In the last two sections, you have used qualitative and quantitative techniques to investigate ions in aqueous solution. Every drop of water that comes into your home contains a variety of such ions. It also contains other substances, in various concentrations. You have investigated water quality in previous grades. In the next section, you will consider the chemistry of water quality.

Section Review

1. **I** Equal volumes of 0.120 mol/L potassium nitrate and 0.160 mol/L iron(III) nitrate are mixed together. What is the concentration of nitrate ions in the mixture?

2. **I** Suppose that you want to remove the barium ions from 120 mL of 0.0500 mol/L aqueous barium nitrate solution. What is the minimum mass of sodium carbonate that you should add?

3. **I** An excess of aluminum foil is added to a certain volume of 0.675 mol/L aqueous copper(II) sulfate solution. The mass of solid copper that precipitates is measured and found to be 4.88 g. What was the volume of the copper(II) sulfate solution?

4. **I** To generate hydrogen gas, a student adds 5.77 g of mossy zinc to 80.1 mL of 4.00 mol/L hydrochloric acid in an Erlenmeyer flask. When the reaction is over, what is the concentration of aqueous zinc chloride in the flask?

5. **I** Copper can be recovered from scrap metal by adding sulfuric acid. Soluble copper sulfate is formed. The copper sulfate then reacts with metallic iron in a single displacement reaction. To simulate this reaction, a student places 1.942 g of iron wool in a beaker that contains 136.3 mL of 0.0750 mol/L aqueous copper(II) sulfate. What mass of copper is formed?

6. **I** Your stomach secretes hydrochloric acid to help you digest the food you have eaten. If too much HCl is secreted, however, you may need to take an antacid to neutralize the excess. One antacid product contains the compound magnesium hydroxide, $Mg(OH)_2$.

   (a) Predict the reaction that takes place when magnesium hydroxide reacts with hydrochloric acid. (**Hint:** This is a double-displacement reaction.)

   (b) Imagine that you are a chemical analyst testing the effectiveness of antacids. If 0.10 mol/L HCl serves as your model for stomach acid, how many litres will react with an antacid that contains 0.10 g of magnesium hydroxide?

7. Even though lead is toxic, many lead compounds are still used as paint pigments (colourings). What volume of 1.50 mol/L lead(II) acetate contains 0.400 Pb^{2+} ions.

Aqueous Solutions and Water Quality

9.4

Water, as you know, is the most abundant substance on Earth's surface. Unfortunately there is more than enough water in some places, but not enough in others. As well, because water is used and re-used, and because it is such a powerful solvent, it is easily polluted.

Canada, with less than 1% of the world's population, has 22% of its fresh water. So, in most regions of our country, the *quality* of water is usually of greater concern than the quantity available. In this section, you will examine factors that affect water quality. You will also study some of the reactions that make polluted water safe and fit to drink.

Acceptable Concentrations of Substances in Drinking Water

The federal government assesses the health risks of specific substances in drinking water. Guidelines for acceptable concentrations of each substance are then established in partnership with provincial and territorial governments. See Table 9.5. These governments require water suppliers to meet the guidelines. Corrective action may be taken when guidelines are violated.

Table 9.5 Acceptable Concentrations of Selected Ions and Compounds in Drinking Water

Ion or compound	Maximum Acceptable Concentration (MAC) (mg/L)	Interim Maximum Acceptable Concentration (mg/L)	Aesthetic Objectives (AO) (mg/L)
aldrin and dieldrin (organic insecticides*)	0.0007		
aluminum		<0.1 or <0.2**	
arsenic		0.025	
benzene (organic component of gasoline)	0.005		
cadmium (component of batteries)	0.005		
chloride			≤250
fluoride	1.5		
iron			≤0.3
lead	0.010		
malathion (organic insecticide)	0.19		
mercury	0.001		
selenium	0.01		
sulfide (as H_2S)			≤0.05
toluene (organic solvent)			≤0.024
uranium	0.1		

*As you will learn in Unit 5, the term "organic" refers to most compounds structured around the element carbon. Toluene belongs to a large class of petroleum-related compounds called hydrocarbons.

** Health-based guidelines have not yet been established. The concentrations that are listed depend on the method of treatment. They are noted as a precautionary measure.

Section Preview/ Specific Expectations

In this section, you will

- **explain** the origins of pollutants in natural waters
- **identify** the allowed concentrations of pollutants in drinking water
- **explain** the origins and consequences of water hardness, and **outline** methods for softening hard water
- **describe** the technology involved in purifying drinking water and treating waste water
- **communicate** your understanding of the following terms: *water treatment, hard water, soft water, ion exchange, waste-water treatment*

CHECKPOINT

You are probably familiar with the water cycle: the natural process that ensures a continual supply of Earth's water. Sketch, from memory, a schematic drawing of the water cycle. Add labels to indicate the states and energy changes involved. Then infer the main solutes and suspended materials in water at each of the following locations:

- the atmosphere
- Earth's surface (at an urban and a rural location)
- below Earth's surface (that is, in ground water)
- a fresh-water lake
- an ocean

For each location, suggest sources of the solutes and materials you inferred.

Sources That Compromise Water Quality

Pure water does not exist in nature. All water naturally contains dissolved substances or ions. For example, rainwater is naturally acidic. This is because water droplets dissolve atmospheric gases, such as carbon dioxide, to form carbonic acid, H_2CO_3. As water filters through soil and rock, it tends to dissolve (leach) certain ions and compounds, such as $Ca^{2+}_{(aq)}$, $Mg^{2+}_{(aq)}$, $Fe^{2+}_{(aq)}$, $Fe^{3+}_{(aq)}$, and $SO_4^{2-}_{(aq)}$. In general, these substances pose little or no threat to plants, animals, or you. In contrast, many of the substances listed in Table 9.5 may pose a threat. They result mainly from human activities, such as manufacturing, food and materials processing, farming, and garbage disposal. The sources of these substances can be classified in the following three broad categories.

- *Point Sources*: A manufacturing or processing plant that discharges untreated or insufficiently treated waste water into a river or lake is a point source of pollution. Point sources include wrecked tankers that leak oil and factories that discharge metallic ions, organic compounds, acids, and bases. Water itself can be a point source of thermal pollution, when thermal power plants discharge warm water into a lake from their cooling towers. Point sources can spread pollution over huge areas.
- *Diffuse Sources*: Pollution that comes from a wide range of sources, not from a single source, is said to come from a diffuse source. (See Figure 9.7.) A heavy downpour can cause run-off from farm fields to enter rivers and lakes. This run-off often carries undesirable pollutants, including fecal matter, pesticides, and fertilizer compounds, such as nitrates and phosphates.
- *Indirect Sources*: Air, water, and soil can become polluted from a variety of indirect sources. Motor vehicles and factory smokestacks release gases that can indirectly cause many different types of pollution. The acidic gases that are produced (sulfur dioxide and nitrogen oxides) dissolve and contribute to the formation of acid rain. The concentrations of ions in ground water are affected by how acidic the water is. For example, aluminum compounds are commonly found in soil. Acidic water increases the leaching of aluminum ions into rivers and lakes.

CHEM FACT

Drinking water guidelines exist for microscopic organisms, such as bacteria and protozoans, as well as for chemicals. These guidelines were compromised in the town of Walkerton, Ontario, in 2000. Seven people died and about 2000 people became ill as a result of exposure to *E. coli* bacteria that were present in the town's water supply. The maximum acceptable concentration for *E. coli* bacteria is zero organisms detectable in 100 mL of drinking water. To achieve this target, chlorine is commonly used to kill *E. coli* and other organisms in drinking water. Despite receiving two warnings from the Ministry of the Environment, the town's water manager chose not to increase the concentration of chlorine in the water supply. As a result of this incident, the Ontario government drafted legislation to enforce strict adherence to water quality guidelines.

Figure 9.7 ***Landfill leachate*** **is rainwater that has percolated through landfill wastes. It dissolves numerous compounds, many of them toxic. Modern landfills have containment and treatment procedures to prevent this diffuse source of pollution from contaminating ground water.**

Treating Water for Your Home

If you live on a farm or in a remote area, you probably obtain your water directly from a well on your property. If you live in an urban community, you probably obtain your water through a municipal or regional water authority. Before the water is made available to you, it is processed at a **water treatment** plant to remove pollutants. Lake, river, or reservoir water enters the treatment plant, where a number of physical and chemical processes take place. Figure 9.8 summarizes these processes.

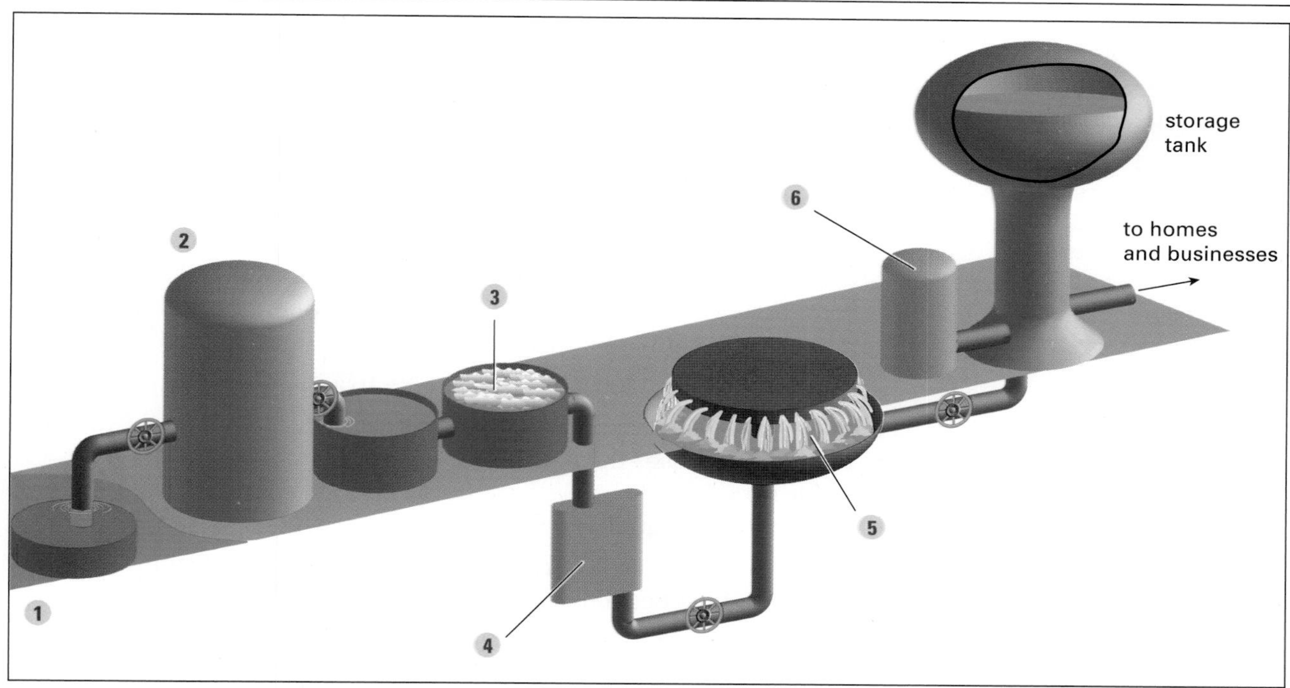

1. A coarse screen made of metal bars filters large particles and trash such as bottles and cans.
2. Chlorine is added to kill bacteria and viruses. It also helps to remove dissolved hydrogen sulfide:
$H_2S_{(aq)} + 4Cl_{2(aq)} + 4H_2O_{(l)} \rightarrow 8Cl^-_{(aq)} + SO_4{}^{2-}{}_{(aq)} + 10H^+_{(aq)}$
3. A process called *flocculation* removes suspended particles such as clay and microorganisms. Lime (CaO) and aluminum sulfate ($Al_2(SO_4)_3$) are added. They react together to form aluminum hydroxide:
$3CaO_{(s)} + 3H_2O_{(l)} + Al_2(SO_4)_{3(s)} \rightarrow 2Al(OH)_{3(s)} + 3Ca^{2+}_{(aq)} + 3SO_4{}^{2-}{}_{(aq)}$
Aluminum hydroxide is a sticky gel that traps the finely suspended particles. Lime is a basic oxide, so it decreases the acidity of the water, and it precipitates some calcium carbonate. The aluminum hydroxide with its trapped particles settles to the bottom of the tank.
4. The water is passed through a bed of graded gravel and sand in a filtering tank.
5. Water is often saturated with oxygen by spraying it into the air, which helps to remove volatile organic compounds and improves the taste and odour.
6. The water receives a second treatment with chlorine to kill bacteria. Ammonia is added to make the chlorine last longer in the piping. Some municipalities add compounds such as sodium fluoride.

Figure 9.8 The following processes take place at a water treatment plant. These processes must ensure that the water meets all the allowable concentrations.

CHEM FACT

Why is hard water so hard to lather? Ordinary soap is often sodium stearate, $NaC_{17}H_{35}O_2$. Sodium stearate consists of the sodium salt of a compound that contains many carbon atoms. The calcium ions in hard water displace the sodium ions, and the new salts are insoluble:

$$Ca^{2+}_{(aq)} + 2NaC_{17}H_{35}O_{2(aq)} \rightarrow Ca(C_{17}H_{35}O_2)_{2(s)} + 2Na^+_{(aq)}$$

As a result, the effectiveness of the soap is reduced. The insoluble salts form a scum on clothing and a grey slimy deposit in sinks and bathtubs: "bathtub ring."

Hard Water and Soft Water

The water that flows through your faucet has been treated to remove, or limit, a large number of pollutants. It is far from pure, however. For example, it still contains dissolved ions, such as $Ca^{2+}_{(aq)}$, $Mg^{2+}_{(aq)}$, $Fe^{2+}_{(aq)}$, $Fe^{3+}_{(aq)}$, and $SO_4{}^{2-}_{(aq)}$. These ions, especially calcium and magnesium, make it difficult to form lather with soap. Water with high concentrations of these ions is called **hard water**, partly because it is "hard" to lather. Water with relatively low concentrations of these ions lathers well. It is called **soft water**.

Ground water is usually harder than surface water in the same region. The extent of the hardness depends on the types of rocks through which the water flows. It also depends on the length of time that the water is in contact with the rocks. The Rocky Mountains, the Canadian Shield, and most of the Maritimes have very insoluble bedrock. Thus, the water in these regions is usually soft. The sedimentary rocks of the Niagara Peninsula and the Prairies are more soluble, resulting in water that ranges from moderately hard to very hard.

The most common type of rock to cause hard water is limestone (calcium carbonate). Limestone is usually considered insoluble. The small amount that does dissolve forms low concentrations of important ions.

$$CaCO_{3(s)} + H_2O_{(\ell)} \rightarrow Ca^{2+}_{(aq)} + HCO^-_{3(aq)} + OH^-_{(aq)}$$

When the water contains dissolved acids (often H_2CO_3 from rainwater), the $H^+_{(aq)}$ ion increases the concentrations of calcium and hydrogen carbonate ions.

$$CaCO_{3(s)} + H^+_{(aq)} \rightarrow Ca^{2+}_{(aq)} + HCO^{3-}_{(aq)}$$

Hydrogen carbonate ions can be economically costly. Solutions that contain these ions decompose when heated to form carbonates.

$$2HCO_3{}^-_{(aq)} \rightarrow H_2O_{(\ell)} + CO_{2(g)} + CO_3{}^{2-}_{(aq)}$$

The carbonate ions recombine with calcium ions to form calcium carbonate deposits. These deposits form a coating on heating elements in kettles and boilers, and build up inside hot water pipes. The coating is commonly called boiler scale. (See Figure 9.9.) It not only reduces the flow of water in pipes, but it also increases the cost of heating the water.

A simple way to remove boiler scale from the inside of a kettle or a coffee maker is to add vinegar. Acetic acid in vinegar reacts with calcium (and magnesium) carbonates to form soluble salts. (Remember that all acetates are soluble.)

$$CaCO_{3(s)} + 2CH_3COOH_{(aq)} \rightarrow Ca^{2+}_{(aq)} + 2CH_3COO^-_{(aq)} + H_2O_{(\ell)} + CO_{2(g)}$$

Figure 9.9 This photomicrograph shows the crystalline structure of precipitated boiler scale.

Treating Water at Home

The ions that cause hard water are not a health hazard but they can be a nuisance. They are not always removed at municipal treatment plants. If you wish, you can remove some of these ions (mainly $Ca^{2+}_{(aq)}$, $Mg^{2+}_{(aq)}$, and $Fe^{2+}_{(aq)}$) yourself. For relatively small volumes of water (such as enough to fill a bathtub or a washing machine), you can add sodium carbonate decahydrate, $Na_2CO_3{\cdot}10H_2O$. This compound is commonly called washing soda. It is an inexpensive way to add carbonate ion to the water. The carbonate ions precipitate the unwanted ions.

$$Ca^{2+}{}_{(aq)} + CO_3{}^{2-}{}_{(aq)} \rightarrow CaCO_{3(s)}$$

The sodium ions in the washing soda behave as spectator ions, leaving the water soft.

For the large volume of water needed on a daily basis, people often install an **ion exchange** water softener. This apparatus exchanges one kind of ion for another. (See Figure 9.10.) The hard water passes through a column that is packed with beads. The beads are made from an insoluble plastic material and are coated with sodium salts, often NaCl. (The salt-coated beads are referred to as an ion exchange resin.) As the hard water passes through the column, the ions in the water displace the sodium ions on the resin. After most of the sodium ions have been exchanged for calcium ions (and other hardness-causing ions), the resin is regenerated. This is done by passing a very concentrated solution of sodium chloride (brine) through the column. The calcium ions are flushed out of the system, along with excess sodium chloride solution.

Biology LINK

A vivid red dye is produced from a small insect, called the cochineal, that lives in the Peruvian Andes. Two chemists from Simon Fraser University in British Columbia, Dr. Cam Oehlschlager and Dr. Eva Czyzewska, have developed an improved process for extracting the natural red colouring agent from the insects. The ions that cause hard water change the colour of the dye, however. So the first step in the process uses demineralized hot water. After precipitating the dye, it is dried and used for a variety of products. Some of your classmates may be wearing examples of these products right now. Use the library or the Internet to find out what they are.

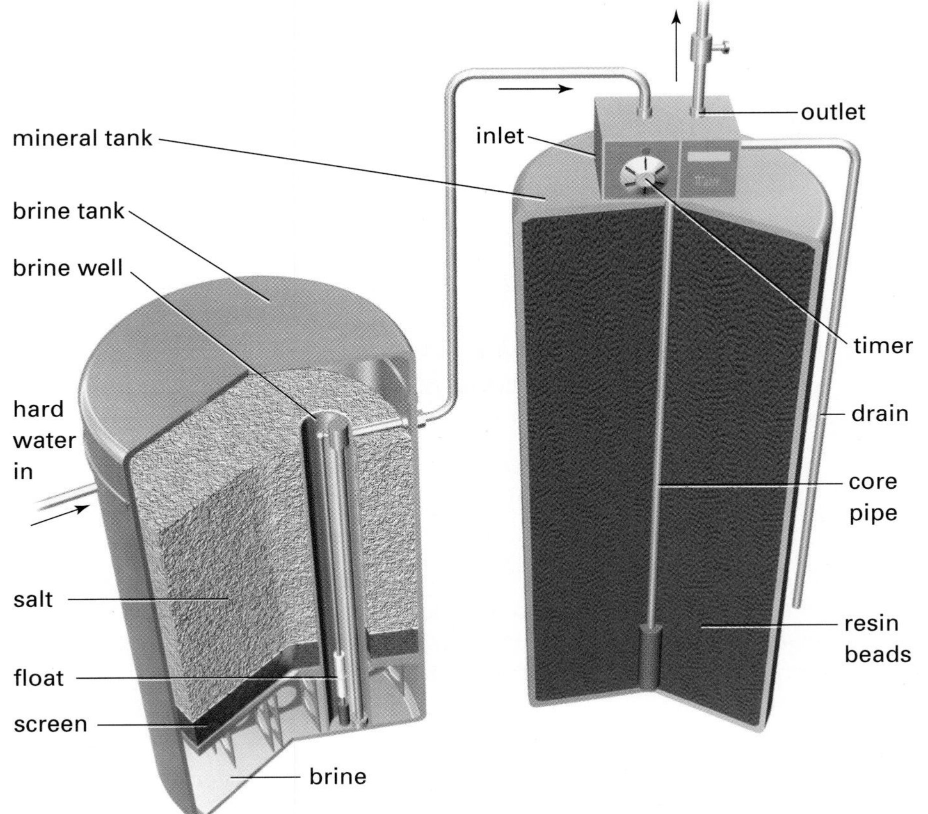

Electronic Learning Partner

Your Chemistry 11 Electronic Learning Partner has a movie detailing several methods for purifying water.

Figure 9.10 An ion exchange water softener adds significant amounts of sodium to the water. Thus, it may not be the best choice for drinking water. This is especially true for people on sodium-restricted diets.

Waste-Water Treatment

The treatment of waste water (sewage) is often divided into three types: primary, secondary, and tertiary treatment. (See Figure 9.11.)

- Primary treatment mainly involves removing solids from waste water physically, using filters and settling tanks.
- Secondary treatment involves using bacteria to chemically decompose dissolved and suspended organic compounds.
- Tertiary treatment involves chemical treatments to remove the majority of remaining ions and disease-causing micro-organisms.

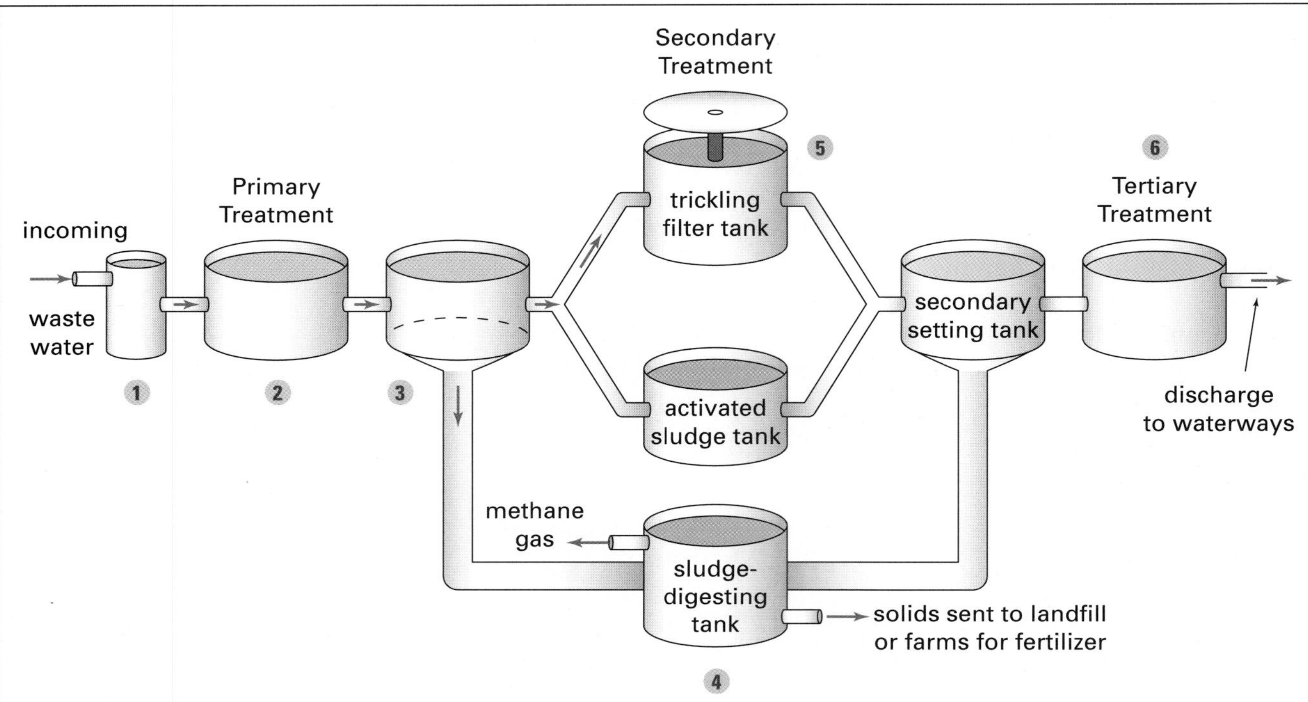

Primary Treatment

1. Screens remove larger solids.
2. As waste water slowly passes through a grit tank, smaller particles (such as gravel, sand, and food wastes) settle out.
3. In a sedimentation tank, finer-sized particles settle out slowly to form sludge.
4. Sludge, oils, and grease from the sedimentation tank move to a sludge-digesting tank. There they are chemically decomposed ("digested") by anaerobic bacteria.

Secondary Treatment

5. Bacterial decomposition of dissolved organic matter takes place in trickling filter tanks and activated sludge tanks.

Tertiary Treatment

6. The resulting waste water may be returned to waterways or used by industries or for irrigation. Processes such as precipitation and filtration remove most remaining pollutants. Treatment with chlorine or ozone kills most of the disease-causing micro-organisms. The resulting water is safe for most purposes, but not for drinking.

Figure 9.11 The main steps involved in the treatment of waste water: Municipalities may use one, two, or all three treatment methods, depending on their needs and finances.

ThoughtLab  Testing Hard Water and Soft Water

Water is hard if it contains high concentrations of calcium and magnesium ions, Ca^{2+} and Mg^{2+}. If these ions are present in lower concentrations, the water is considered to be soft. Distilled water has few ions of any kind, and no Ca^{2+} or Mg^{2+}. Sodium oxalate, $Na_2C_2O_4$, is a compound that causes Ca^{2+} to precipitate as calcium oxalate, CaC_2O_4, and Mg^{2+} to precipitate as magnesium oxalate, MgC_2O_4.

Procedure

1. A student added 1 mL of hard water to one test tube, 1 mL of soft water to another, and 1 mL of distilled water to a third. The hard water sample contained Ca^{2+} and Mg^{2+}. The soft water contained lower concentrations of these ions.
2. Into each test tube, the student put two drops of 0.1 mol/L sodium oxalate solution. Then the student mixed the contents.

Analysis

1. Infer what the student observed in each test tube. Write a net ionic equation if you predict that a precipitate formed. Write "NR" if you think that no reaction occurred.
2. Imagine that you have the three water samples the student used, but no sodium oxalate solution. How else could you test the validity of your predictions?
3. With your teacher's permission, test your predictions. CAUTION Sodium oxalate is poisonous. Wear an apron and safety goggles. Handle all the solutions carefully.

Canadians 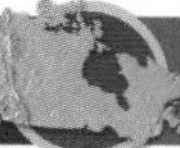 in Chemistry

Dr. Jiangning Wu: Cleaning the World's Water

Around the world, researchers are working to reduce the pollutants in our lakes, rivers, and ground water. Both industry and agriculture too often produce waste water that contains everything from animal droppings and excess fertilizer to artificial colours and other chemicals. Yet, without industry and agriculture, we would not have many of the modern comforts we expect and enjoy.

Jiangning Wu was born and raised in Nanjing, China. She came to Canada to complete her studies in chemical engineering at the University of Windsor. After earning her doctorate in 1992, she turned her attention to the use of enzymes for purifying waste water.

Enzymes are naturally occurring substances that speed up chemical reactions without being used up or changed by the reactions. They are basic to living organisms. Thus it is no surprise that they are well-suited to removing organic pollutants from waste water. For instance, an enzyme that is found in horseradish helps to accelerate the oxidation of soluble organic pollutants known as phenols. The resulting compounds are less soluble and more easily removed from the waste water.

In 1993, Dr. Wu found that, with a single additive, the same enzyme was able to remove phenols from waste water. Her strategy worked even for the low concentrations of phenols typical of industrial waste water. Moreover, significantly lower concentrations of the enzyme were needed! Dr. Wu's discovery brought researchers one step closer to a commercial process, which would mean cleaner waters and less stress on the environment.

Dr. Wu is a professor in the School of Chemical Engineering at Ryerson Polytechnical University in Toronto. She is currently investigating the use of ozone for waste-water treatment and food preservation.

Section Wrap-up

In this section, you applied your understanding of aqueous solutions to explore the chemistry of water quality. Some of the chemicals you considered belong to a group of compounds that are called acids and bases. In the next chapter, you will investigate the properties and chemical behaviour of these important compounds.

CHEM FACT

Chlorine was first used to disinfect water in Britain in 1904, after a typhoid epidemic. (Typhoid is a water-borne, contagious illness that is caused by a species of *Salmonella* bacteria.) Strict limits are necessary because chlorine is ineffective when its concentration is less than 0.1 mg/L. It gives water an unpleasant taste at concentrations above 1.0 mg/L. Chlorine has a disadvantage, however. It can react with other chemicals in the water to form poisonous compounds, such as chloroform, $CHCl_3$. These chemicals may remain in solution even after the entire treatment process.

Section Review

1. K/U In what ways can water in the environment become polluted? Give at least two examples for each.
2. K/U Which chemical in Table 9.5 has the lowest acceptable concentration? Which chemical has the highest acceptable concentration? Rearrange the chemicals in the table so they are organized by concentration, from lowest to highest, rather than alphabetically by name.
3. MC What is the source of the water you use at home? Based on your experience with this water, which ions do you think it contains? Explain your answer.
4. C Use a graphic organizer to outline the main steps involved in treating drinking water and waste water.
5. MC An alternative to ion exchange water softening is a process called *reverse osmosis*. Use your knowledge of osmosis to infer, in general terms, how this method might work. Consult print or electronic resources to modify or expand on your ideas. Then use suitable software to communicate your findings as a graphic organizer or a virtual slide presentation.
6. MC One component of the waste water that enters a waste water plant for treatment is urea, $(NH_2)_2CO$.
   - **(a)** Use reference materials to find out how urea is chemically changed to nitrate ions. (There are several steps.)
   - **(b)** Write chemical equations to represent the steps involved in this process.
   - **(c)** What happens to aqueous nitrates when they are discharged to waterways in the environment?

CHAPTER 9 Review

Reflecting on Chapter 9

Summarize this chapter in the format of your choice. Here are a few ideas to use as guidelines:

- Predict combinations of aqueous solutions that result in the formation of precipitates.
- Describe your experiences with a qualitative analysis of ions in solution.
- Represent a double-displacement reaction using its net ionic equation.
- Write balanced chemical equations and net ionic equations for double-displacement reactions.
- Apply your understanding of stoichiometry to solve quantitative problems involving solutions.
- Identify the origins of pollutants in drinking water, and the allowed concentrations of some of these pollutants.
- Examine the causes and effects of water hardness. Considered several methods for softening hard water.
- Compare the chemistry and the technology of water treatment and waste-water treatment.

Reviewing Key Terms

For each of the following terms, write a sentence that shows your understanding of its meaning.

general solubility guidelines
hard water
net ionic equation
qualitative analysis
spectator ions
waste-water treatment
ion exchange
precipitate
soft water
total ionic equation
water treatment

Knowledge/Understanding

1. In your own words, define the terms "spectator ion" and "net ionic equation."
2. Identify the spectator ions in the following skeleton equation. Then write the balanced ionic equation for the reaction.
$Al(NO_3)_{3(aq)} + NH_4OH_{(aq)} \rightarrow Al(OH)_{3(s)} + NH_4NO_{3(aq)}$
3. Hydrogen sulfide gas can be prepared by the reaction of sulfuric acid with sodium sulfide.
$H_2SO_{4(aq)} + Na_2S_{(aq)} \rightarrow H_2S_{(g)} + Na_2SO_{4(aq)}$
Write the net ionic equation for this reaction.
4. Each of the following combinations of reagents results in a double displacement reaction. In your notebook, complete the chemical equation. Then identify the spectator ions, and write the net ionic equation.
(a) copper(II) chloride$_{(aq)}$ + ammonium phosphate$_{(aq)}$ $\rightarrow$
(b) aluminum nitrate$_{(aq)}$ + barium hydroxide$_{(aq)}$ $\rightarrow$
(c) sodium hydroxide$_{(aq)}$ + magnesium chloride$_{(aq)}$ $\rightarrow$
5. Use the general solubility guidelines to name three reagents that will combine with each ion below to form a precipitate. Assume that the reactions take place in aqueous solution. For each reaction, write the net ionic equation.
(a) bromide ion
(b) carbonate ion
(c) lead(II) ion
(d) iron(III) ion
6. The transition metals form insoluble sulfides, often with a characteristic colour. Write the net ionic equation for the precipitation of each ion by the addition of an aqueous solution of sodium sulfide.
(a) $Cr^{3+}{}_{(aq)}$ (**Note:** $Cr_2S_{3(s)}$ is brown-black.)
(b) $Ni^{2+}{}_{(aq)}$ (**Note:** $NiS_{(s)}$ is black.)
(c) $Mn^{2+}{}_{(aq)}$ (**Note:** $MnS_{2(s)}$ is green or red, depending on the arrangement of ions in the solid.)
7. Identify three cations and three anions that are commonly found in ground water. Suggest at least one likely source for each.
8. Briefly describe two steps in the primary treatment of waste water, one involving a physical change and the other involving a chemical change.
9. **(a)** Many liquid antacids contain magnesium hydroxide, $Mg(OH)_2$. Why must the bottle be shaken before a dose is poured?
(b) Stomach acid contains hydrochloric acid. Excess acid that backs up into the esophagus is the cause of "heartburn." Write the chemical equation and the net ionic equation for the reaction that takes place when someone with heartburn swallows a dose of liquid antacid.

10. Aqueous solutions of iron(III) chloride and ammonium sulfide react in a double displacement reaction.
  (a) Write the name and formula of the substance that precipitates.
  (b) Write the chemical equation for the reaction.
  (c) Write the net ionic equation.

Inquiry

11. A reference book states that the solubility of silver sulfate is 0.57 g in 100 mL of cold water. You decide to check this by measuring the mass of a silver salt precipitated from a known volume of saturated silver sulfate solution. Solubility data show that silver chloride is much less soluble than silver nitrate. Explain why you should not use barium chloride to precipitate the silver ions. Suggest a different reagent, and write the net ionic equation for the reaction.
12. The presence of copper(II) ions in solution can be tested by adding an aqueous solution of sodium sulfide. The appearance of a black precipitate indicates that the test is positive. A solution of copper(II) bromide is tested this way. What precipitate is formed? Write the net ionic equation for the reaction.
13. An old home-gardening "recipe" for fertilizer suggests adding 15 g of Epsom salts (magnesium sulfate heptahydrate, $MgSO_4{\cdot}7H_2O$) to 4 L of water. What will be the concentration of magnesium ions?
14. Calculate the concentration (in mol/L) of each aqueous solution.
  (a) 7.37 g of table sugar, $C_{12}H_{22}O_{11}$, dissolved in 125 mL of solution
  (b) 15.5 g of ammonium phosphate, $(NH_4)_3PO_4$, dissolved in 180 mL of solution
  (c) 76.7 g of glycerol, $C_3H_8O_3$, dissolved in 1.20 L of solution
15. 50.0 mL of 0.200 mol/L $Ca(NO_3)_{2(aq)}$ is mixed with 200 mL of 0.180 mol/L $K_2SO_{4(aq)}$. What is the concentration of sulfate ions in the final solution?
16. Suppose that 1.00 L of 0.200 mol/L $KNO_{3(aq)}$ is mixed with 2.00 L of 0.100 mol/L $Ca(NO_3)_{2(aq)}$. Determine the concentrations of the major ions in the solution.
17. Equal masses of each of the following salts are dissolved in equal volumes of water: sodium chloride, calcium chloride, and iron(III) chloride. Which salt produces the largest concentration of chloride ions?
18. Imagine that you are the chemist at a cement factory. You are responsible for analyzing the factory's waste water. If a 50.0 mL sample of waste water contains 0.090 g of $Ca^{2+}{}_{(aq)}$ and 0.029 g of $Mg^{2+}{}_{(aq)}$, calculate
  (a) the concentration of each ion in mol/L
  (b) the concentration of each ion in ppm
19. The concentration of calcium ions, Ca^{2+}, in blood plasma is about 2.5×10^{-3} mol/L. Calcium ions are important in muscle contraction and in regulating heartbeat. If the concentration of calcium ions falls too low, death is inevitable. In a television drama, a patient is brought to hospital after being accidentally splashed with hydrofluoric acid. The acid readily penetrates the skin, and the fluoride ions combine with the calcium ions in the blood. If the patient's volume of blood plasma is 2.8 L, what amount (in mol) of fluoride ions would completely combine with all the calcium ions in the patient's blood?
20. A double displacement reaction occurs in aqueous solution when magnesium phosphate reacts with lead(II) nitrate. If 20.0 mL of 0.750 mol/L magnesium phosphate reacts, what is the maximum mass of precipitate that can be formed?

Communication

21. Phosphate ions act as a fertilizer. They promote the growth of algae in rivers and lakes. They can enter rivers and lakes from fields that are improperly fertilized or from untreated waste water that contains phosphate detergents. How can the water be treated to remove the phosphate ions?
22. A chemist analyzes the sulfate salt of an unknown alkaline earth metal. The chemist adds 1.273 g of the salt to excess barium chloride solution. After filtering and drying, the mass of precipitate is found to be 2.468 g.
  (a) Use the formula MSO_4 to represent the unknown salt. Write the molecular and net ionic equations for the reaction.

(b) Calculate the amount (in mol) of MSO_4 used in the reaction.

(c) Determine the molar mass of the unknown salt.

(d) What is the likely identity of the unknown metal cation? What test might the chemist perform to help confirm this conclusion?

23. The same volume of solution is made using the same masses of two salts: rubidium carbonate and calcium carbonate. Which salt gives the larger concentration of aqueous carbonate ions?

24. A prospector asks you to analyze a bag of silver ore. You measure the mass of the ore and add excess nitric acid to it. Then you add excess sodium chloride solution. You filter and dry the precipitate. The mass of the ore is 856.1 g, and 1.092 g of silver chloride is collected.

(a) Why did you first treat the ore with excess nitric acid?

(b) Calculate the mass percent of silver in the ore. The ore that is extracted at a silver mine typically contains about 0.085% silver by mass. Should the prospector keep looking or begin celebrating?

Making Connections

25. Think about your activities yesterday. Which activity required the most use of water? Estimate the volume of water you used. Would it make sense to have two supplies of water to your home, one for drinking and a second of lower purity for every other activity that uses water? Give reasons for your answer.

26. List three different household wastes that are commonly discarded and have the potential to contaminate ground water if rain leaches through your local landfill site. What chemical(s) does each contain? For each waste, identify an alternative to dumping it in a landfill site.

27. Water is essential for crops. Improper irrigation over a number of years, however, can result in farmed land becoming laden with toxic chemical compounds. Research how this happens. Find out whether this is a concern to farmers near where you live.

Answers to Practice Problems and Short Answers to Section Review Questions:

Practice Problems: 1.(a) insoluble (b) insoluble (c) soluble 2.(a) soluble (b) soluble (c) insoluble 3.(a) insoluble (b) soluble (c) insoluble 4.(a) $Na_2S_{(aq)} + FeSO_{4(aq)} \rightarrow Na_2SO_{4(aq)} + FeS_{(s)}$ (b) $NaOH_{(aq)} + Ba(NO_3)_{2(aq)} \rightarrow NR$ (c) $2Cs_3PO_{4(aq)} + 3CaBr_{2(aq)} \rightarrow 6CsBr_{(aq)} + Ca_3(PO_4)_{2(s)}$ (d) $Na_2CO_{3(aq)} + H_2SO_{4(aq)} \rightarrow NR$ (e) $NaNO_{3(aq)} + CuSO_{4(aq)} \rightarrow NR$ (f) $NH_4I_{(aq)} + AgNO_{3(aq)} \rightarrow AgI_{(s)} + NH_4NO_{3(aq)}$ (g) $K_2CO_{3(aq)} + Fe(NO_3)_{2(aq)} \rightarrow FeCO_{3(s)} + 2KNO_{3(aq)}$ (h) $Al(NO_3)_{2(aq)} + Na_3PO_{4(aq)} \rightarrow AlPO_{4(s)} + 3NaNO_{3(aq)}$ (i) $KCl_{(aq)} + Fe(NO_3)_{2(aq)} \rightarrow NR$ (j) $(NH_4)_2SO_{4(aq)} + BaCl_{2(aq)} \rightarrow BaSO_{4(s)} + 2NH_4Cl_{(aq)}$ (k) $Na_2S_{(aq)} + NiSO_{4(aq)} \rightarrow NiS_{(s)} + Na_2SO_{4(aq)}$ (l) $Pb(NO_3)_2 + 2KBr_{(aq)} \rightarrow PbBr_{2(s)} + 2KNO_{3(aq)}$ 5.(a) spectator ions: $Na^+_{(aq)}$ and $Cl^-_{(aq)}$; net ionic equation: $CO_3{}^{2-}{}_{(aq)} + 2H^+_{(aq)} \rightarrow CO_{2(g)} + H_2O_{(l)}$ (b) spectator ions: $Na^+_{(aq)}$ and $SO_4{}^{2-}{}_{(aq)}$; net ionic equation: $H^+_{(aq)} + OH^-_{(aq)} \rightarrow H_2O_{(l)}$ 6.(a) spectator ions: $NH_4{}^+{}_{(aq)}$ and $SO_4{}^{2-}{}_{(aq)}$; net ionic equation: $Zn^{2+}{}_{(aq)} + PO_4{}^{3-}{}_{(aq)} \rightarrow Zn_3(PO_4)_{2(s)}$ (b) spectator ions: $Li^+_{(aq)}$ and $NO^{3-}{}_{(aq)}$; net ionic equation: $CO_3{}^{2-}{}_{(aq)} + 2H^+_{(aq)} \rightarrow CO_{2(g)} + H_2O_{(l)}$ (c) spectator ions: none; net ionic equation: $2H^+_{(aq)} + SO_4^{2-}{}_{(aq)} + Ba^{2-}{}_{(aq)} + 2OH^-_{(aq)} \rightarrow BaSO_{4(s)} + H_2O_{(l)}$ 7. 300 mL 8. concentration of $NH_4{}^+{}_{(aq)}$ is 0.40 mol/L; concentration of $PO_4{}^{3-}{}_{(aq)}$ is 0.13 mol/L 9. 66.57% 10. 30.47% 11. 20.9 g of PbS 12. 4.61 g PbI_2 13.(a) $3NaOH_{(aq)} + Al(NO_3)_{3(aq)} \rightarrow Al(OH)_{3(s)} + 3NaNO_{3(aq)}$ (b) 0.0975 g

Section Review: 9.1: 2. NaF less soluble, because F^- is smaller than I^-. 3.(a) insoluble (b) soluble (c) soluble. 4. all insoluble. 5. Any reagent containing Cl^-, Br^-, or I^- will precipitate silver ion but leave calcium ion in solution. 6. 1 = Ag^+, 2 = $SO_4{}^{2+}$, 3 = Ba^{2+}, 4 = Cl^-. **9.2:** 2.(a) $3Sn^{2+}{}_{(aq)} + 2PO_4{}^{3-}{}_{(aq)} \rightarrow Sn_3(PO_4)_{2(s)}$ (b) $Ni^{2+}{}_{(aq)} + CO_3{}^{2-(aq)} \rightarrow NiCO_{3(s)}$ (c) $2Cr^{3+}{}_{(aq)} + 3S^{2-}{}_{(aq)} \rightarrow Cr_2S_{3(s)}$ 3.(a) $Cl^-_{(aq)}$ and $K^+_{(aq)}$ (b) $Cl^-_{(aq)}$ and $Na^+_{(aq)}$ (c) $NH_4{}^+{}_{(aq)}$ and $SO_4{}^{2-}{}_{(aq)}$. 6. copper(II) carbonate, $CuCO_3$; $Cu^{2+}{}_{(aq)} + CO_3{}^{2-}{}_{(aq)} \rightarrow CuCO_{3(s)}$; spectator ions: $SO_4{}^{2-}{}_{(aq)}$ and $Na^+_{(aq)}$. **9.3:** 1. 0.300 mol/L nitrate ion. 2. 0.636 g $Na_2CO_{3(s)}$. 3. 114 mL. 4. 1.09 mol/L $ZnCl_2$. 5. 0.650 g Cu 6. 3.4×10^{-2} L 7. 0.267 L

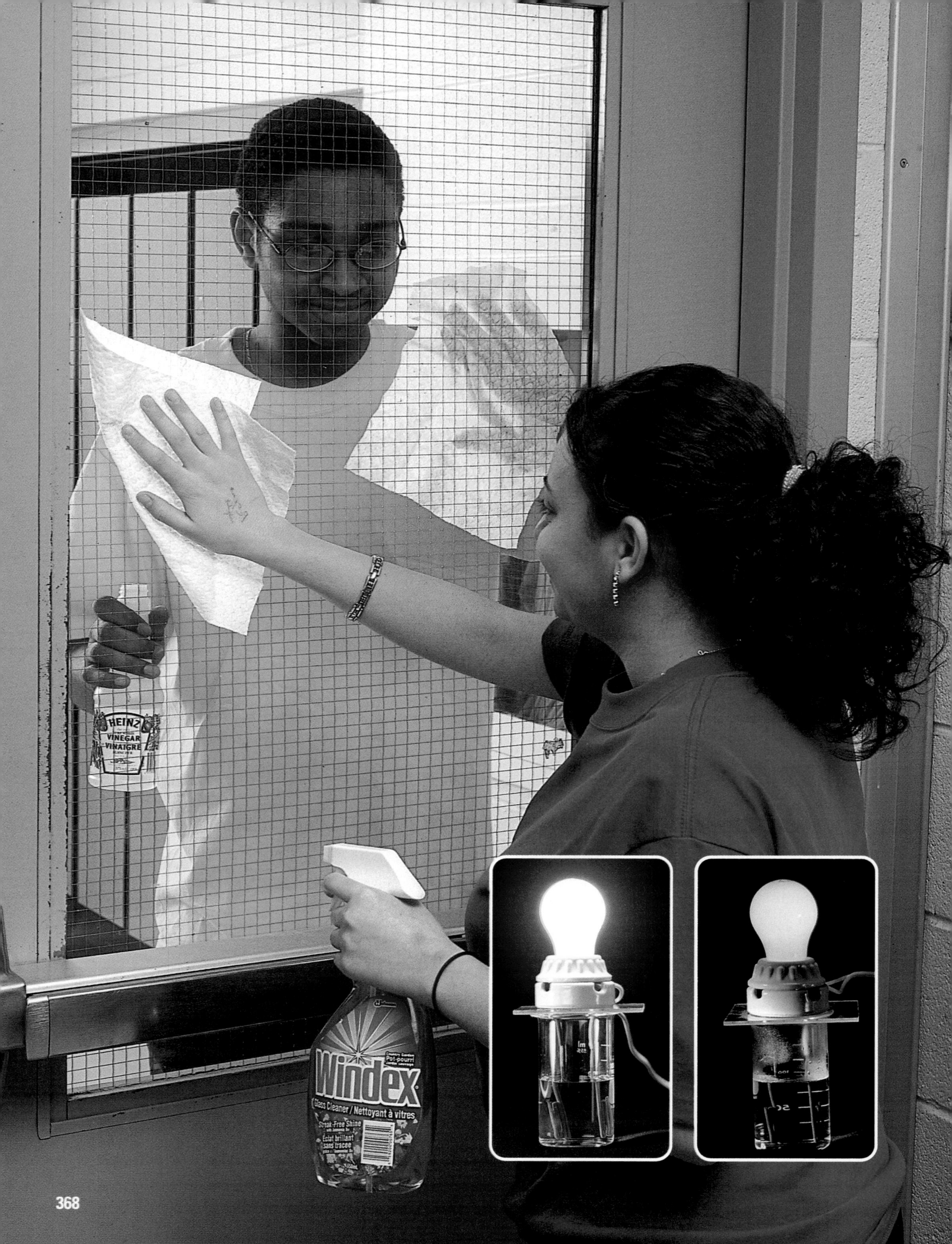
HEINZ
VINEGAR
VINAIGRE
Windex
Glass Cleaner / Nettoyant à vitres
Streak-Free Shine
Éclat brillant sans traces

CHAPTER 10

Acids and Bases

What do cheese, stomach juices, baking soda, oven cleaner, and underarm odour have in common? They are all acidic or basic. Do you know which are acidic and which are basic?

Acids and bases are very important chemicals. They have been used for thousands of years. Vinegar is an acidic solution that is common in many food and cleaning products. It was discovered long ago—before people invented the skill of writing to record its use. Today, acids are also used to manufacture fertilizers, explosives, plastics, motor vehicles, and computer circuit boards.

Like acids, bases have numerous uses in the home and in chemical industries. Nearly 5000 years ago, in the Middle East, the Babylonians made soap using the bases in wood ash. Today, one of Canada's most important industries, the pulp and paper industry, uses huge quantities of a base called sodium hydroxide. Sodium hydroxide is also used to manufacture soaps, detergents, dyes, and many other compounds.

In this chapter, you will learn about the properties of acids and bases. You will learn how these properties change when acids and bases react together. As well, you will have a chance to estimate and measure the acidity of aqueous solutions.

Chapter Preview

Concepts and Skills You Will Need

Before you begin this chapter, review the following concepts and skills:

- **describing** and **calculating** the concentration of solutions (Chapter 8, section 8.2)
- **performing** stoichiometry calculations (Chapter 7, section 7.2)
- **naming** and **identifying** polyatomic ions and their formulas (Chapter 3, section 3.4)

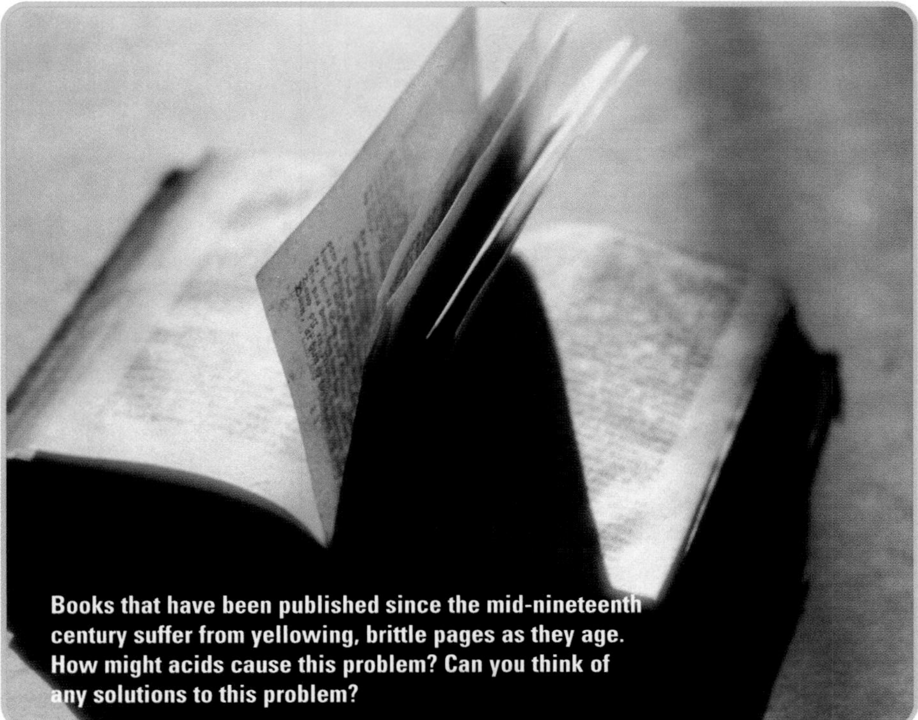

Books that have been published since the mid-nineteenth century suffer from yellowing, brittle pages as they age. How might acids cause this problem? Can you think of any solutions to this problem?

10.1 Acid-Base Theories

Section Preview/ Specific Expectations

In this section, you will

- **describe** and **compare** the Arrhenius and Brønsted-Lowry theories of acids and bases
- **identify** conjugate acid-base pairs
- **communicate** your understanding of the following terms: *Arrhenius theory of acids and bases, hydronium ion, Brønsted-Lowry theory of acids and bases, conjugate acid-base pair, conjugate base, conjugate acid*

As you can see in Table 10.1, acids and bases are common products in the home. It is easy to identify some products as acids. Often the word "acid" appears in the list of ingredients. Identifying bases is more difficult. Acids and bases have different properties, however, that enable you to distinguish between them.

Table 10.1 Common Acids and Bases in the Home

Acids	
Product	**Acid(s) contained in the product**
citrus fruits (such as lemons, limes, oranges and tomatoes)	citric acid and ascorbic acid
dairy products (such as cheese, milk, and yogurt)	lactic acid
vinegar	acetic acid
soft drinks	carbonic acid; may also contain phosphoric acid and citric acid
underarm odour	3-methyl-2-hexenoic acid
Bases	
Product	**Base contained in the product**
oven cleaner	sodium hydroxide
baking soda	sodium hydrogen carbonate
washing soda	sodium carbonate
glass cleaner (some brands)	ammonia

Language LINK

The word "acid" comes from the Latin *acidus*, meaning "sour tasting." As you will learn in this chapter, bases are the "base" (the foundation) from which many other compounds form. A base that is soluble in water is called an alkali. The word "alkali" comes from an Arabic word meaning "ashes of a plant." In the ancient Middle East, people rinsed plant ashes with hot water to obtain a basic solution. The basic solution was then reacted with animal fats to make soap.

Properties of Acids and Bases

One way to distinguish acids from bases is to describe their observable properties. For example, acids taste sour, and they change colour when mixed with coloured dyes called indicators. Bases taste bitter and feel slippery. They also change colour when mixed with indicators.

CAUTION You should never taste or touch acids, bases, or any other chemicals. Early chemists used their senses of taste and touch to observe the properties of many chemicals. This dangerous practice often led to serious injury, and sometimes death.

Another property that can be used to distinguish acids from bases is their conductivity in solution. As you can see in Figure 10.1, aqueous solutions of acids and bases conduct electricity. This is evidence that ions are present in acidic and basic solutions. Some of these solutions, such as hydrochloric acid and sodium hydroxide (a base), cause the bulb to glow brightly. Most acidic and basic solutions, however, cause the bulb to glow dimly.

pure water

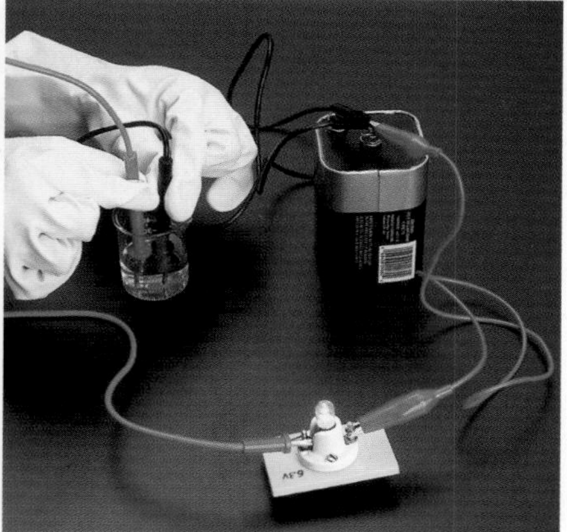

hydrochloric acid, $HCl_{(aq)}$ (1 mol/L)

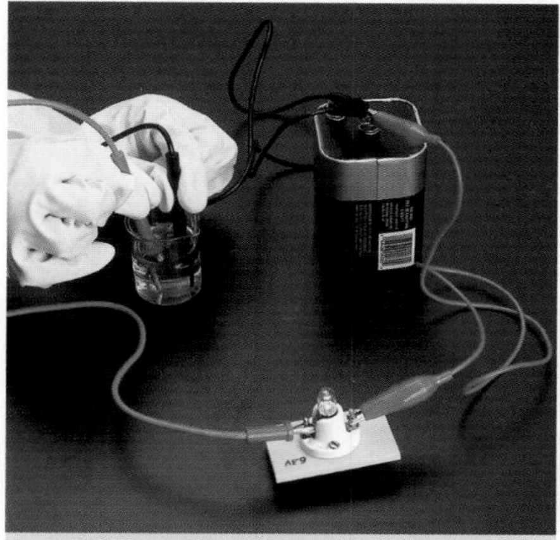

acetic acid, $CH_3COOH_{(aq)}$ (1 mol/L)

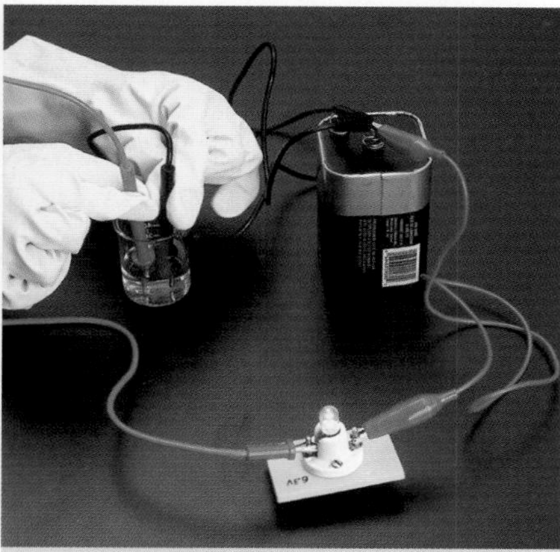

sodium hydroxide, $NaOH_{(aq)}$ (1 mol/L)

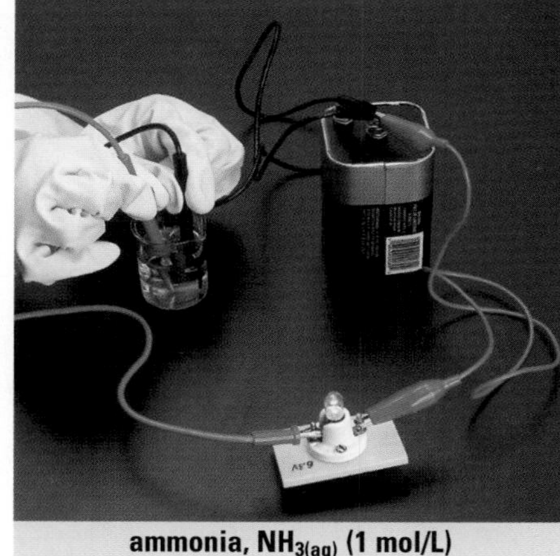

ammonia, $NH_{3(aq)}$ (1 mol/L)

Figure 10.1 **Aqueous solutions of acids and bases can be tested using a conductivity tester. The brightness of the bulb is a clue to the concentration of ions in the solution. Which of these solutions have higher concentrations of ions? Which have lower concentrations?**

mind STRETCH

An oxide is a compound of oxygen with a metal or non-metal. Most metal oxides react with water to form basic solutions. For example, calcium oxide is a metal oxide that is important in the construction industry as an ingredient of cement. Calcium oxide reacts with water to form a basic solution of calcium hydroxide.

$$CaO_{(s)} + H_2O_{(\ell)} \rightarrow Ca(OH)_{2(aq)}$$

Many municipal water treatment plants use calcium hydroxide to soften very hard water before releasing it for public use. Most non-metal oxides react with water to form acidic solutions. For example, sulfur dioxide gas dissolves in water to form sulfurous acid.

$$SO_{2(g)} + H_2O_{(\ell)} \rightarrow H_2SO_{3(aq)}$$

The metallic character of the elements in the periodic table, and the acid-base properties of their oxides, show a distinct trend across periods and down groups. Infer what this trend is. In other words, state what you think happens to the acid-base properties of oxides as you go across a period and down a group. Make a quick sketch of the periodic table to illustrate this trend. How would you describe the acid-base properties of the metalloids? (Use your knowledge of the physical properties of the metalloids to help you make your inference.)

Table 10.2 on the next page summarizes the observable properties of acids and bases. These observable properties include their physical characteristics and their chemical behaviour. The Express Lab on page 373 provides you with an opportunity to compare some of these properties. What are acids and bases, however? How does chemical composition determine whether a substance is acidic or basic? You will consider one possible answer to this question starting on page 373.

Table 10.2 Some Observable Properties of Acids and Bases

	Property		
	Taste	**Electrical conductivity in solution**	**Feel of solution**
ACIDS	taste sour	conduct electricity	have no characteristic feel
BASES	taste bitter 	conduct electricity 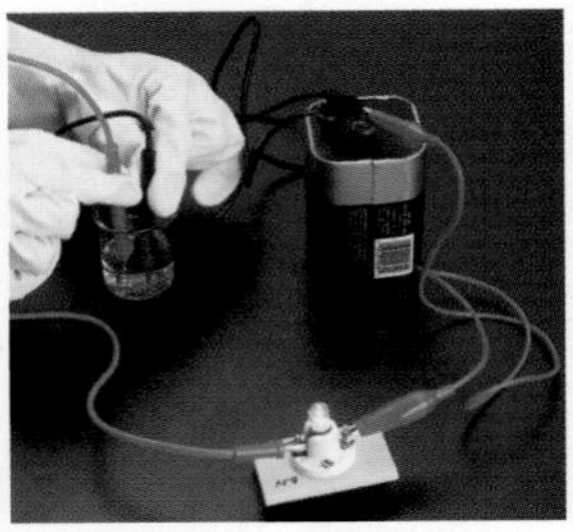	feel slippery 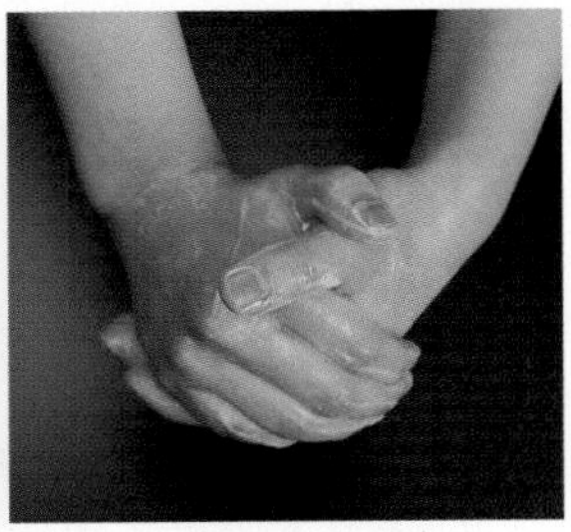

	Property		
	Reaction with litmus paper	**Reaction with active metals**	**Reaction with carbonate compounds**
ACIDS	Acids turns blue litmus red	produce hydrogen gas	produce carbon dioxide gas
BASES	Bases turn red litmus blue 	do not react 	do not react 

ExpressLab  Clean a Penny

Many cleaning products contain an acid or a base. For example, some window cleaners contain vinegar (acetic acid). Other window cleaners contain ammonia (a base). Oven cleaners, however, contain only bases. This activity will help you infer why.

Safety Precautions

Materials

water
vinegar
100 mL graduated cylinder
spoon or scoopula
baking soda
3 small beakers (about 200 mL)
3 tarnished pennies

Procedure

1. Predict which solution(s) will clean the penny best. Give reasons for your prediction.
2. In one beaker, mix 50 mL of vinegar with about 150 mL of water. In a second beaker, mix about 20 mL to 30 mL spoonfuls of baking soda with 150 mL of water. In the third beaker, put only 150 mL of water.
3. Place a tarnished penny in each beaker. Observe what happens for about 15 min.

Analysis

1. Which solution was the best cleaner? How did your observations compare with your prediction?
2. What results would you expect if you tried cleaning a penny in a solution of lemon juice? What if you used a dilute solution of ammonia? **Note:** If you want to test your predictions, ask your teacher for the concentrations of the solutions you should use.
3. The base that is often used in oven cleaners is sodium hydroxide. This base is very corrosive, and it can burn skin easily. A corrosive acid, such as hydrochloric acid, could also remove baked-on grease and grime from ovens. Why are bases a better choice for oven cleaners?

The Arrhenius Theory of Acids and Bases

In Figure 10.1, you saw evidence that ions are present in solutions of acids and bases. When hydrogen chloride dissolves in water, for example, it dissociates (breaks apart) into hydrogen ions and chloride ions.

$$HCl_{(aq)} \rightarrow H^{+}{}_{(aq)} + Cl^{-}{}_{(aq)}$$

When sodium hydroxide dissolves in water, it dissociates to form sodium ions and hydroxide ions.

$$NaOH_{(aq)} \rightarrow Na^{+}{}_{(aq)} + OH^{-}{}_{(aq)}$$

The dissociations of other acids and bases in water reveal a pattern. This pattern was first noticed in the late nineteenth century by a Swedish chemist named Svanté Arrhenius. (See Figure 10.2.)

$$HBr_{(aq)} \rightarrow \mathbf{H^{+}}{}_{(aq)} + Br^{-}{}_{(aq)}$$

$$H_2SO_{4(aq)} \rightarrow \mathbf{H^{+}}{}_{(aq)} + HSO^{-}{}_{4(aq)}$$

$$HClO_{4(aq)} \rightarrow \mathbf{H^{+}}{}_{(aq)} + ClO^{-}{}_{4(aq)}$$

acids dissociating in water, and their resulting ions

$$LiOH_{(aq)} \rightarrow Li^{+}{}_{(aq)} + \mathbf{OH^{-}}{}_{(aq)}$$

$$KOH_{(aq)} \rightarrow K^{+}{}_{(aq)} + \mathbf{OH^{-}}{}_{(aq)}$$

$$Ba(OH)_{2(aq)} \rightarrow Ba^{+}{}_{(aq)} + \mathbf{2OH^{-}}{}_{(aq)}$$

bases dissociating in water, and their resulting ions

Figure 10.2 Svanté Arrhenius (1859–1927).

Web LINK

www.school.mcgrawhill.ca/resources/

Scientists did not embrace the Arrhenius theory when they first heard about it during the 1880s. Why were they unimpressed with this theory? What was necessary to convince them? Arrhenius is featured on several web sites on the Internet. To link with these web sites, go to the web site above. Go to **Science Resources,** then to **Chemistry 11** to find out where to go next.

In 1887, Arrhenius published a theory to explain the nature of acids and bases. It is called the **Arrhenius theory of acids and bases**.

The Arrhenius Theory of Acids and Bases

- An acid is a substance that dissociates in water to produce one or more hydrogen ions, H^+.
- A base is a substance that dissociates in water to form one or more hydroxide ions, OH^-.

According to the Arrhenius theory, acids increase the concentration of H^+ in aqueous solutions. Thus, an Arrhenius acid must contain hydrogen as the source of H^+. You can see this in the dissociation reactions for acids on the previous page.

Bases, on the other hand, increase the concentration of OH^- in aqueous solutions. An Arrhenius base must contain the hydroxyl group, —OH, as the source of OH^-. You can see this in the dissociation reactions for bases on the previous page.

Limitations of the Arrhenius Theory

The Arrhenius theory is useful if you are interested in the ions that result when an acid or a base dissociates in water. It also helps explain what happens when an acid and a base undergo a neutralization reaction. In such a reaction, an acid combines with a base to form an ionic compound and water. Examine the following reactions:

$$HCl_{(aq)} + NaOH_{(aq)} \rightarrow NaCl_{(aq)} + H_2O_{(\ell)}$$

The net ionic equation for this reaction shows the principal ions in the Arrhenius theory.

$$H^+{}_{(aq)} + OH^-{}_{(aq)} \rightarrow H_2O_{(\ell)}$$

Since acids and bases produce hydrogen ions and hydroxide ions, water is an inevitable result of acid-base reactions.

Problems arise with the Arrhenius theory, however. One problem involves the ion that is responsible for acidity: H^+. Look again at the equation for the dissociation of hydrochloric acid.

$$HCl_{(aq)} \rightarrow H^+{}_{(aq)} + Cl^-{}_{(aq)}$$

This dissociation occurs in aqueous solution, but chemists often leave out H_2O as a component of the reaction. They simply assume that it is there. What happens if you put H_2O into the equation?

$$HCl_{(aq)} + H_2O_{(\ell)} \rightarrow H^+{}_{(aq)} + Cl^-{}_{(aq)} + H_2O_{(\ell)}$$

Notice that the water is unchanged when the reaction is represented this way. However, you learned earlier that water is a polar molecule. The O atom has a partial negative charge, and the H atoms have partial positive charges. Thus, H_2O must interact in some way with the ions H^+ and Cl^-. In fact, chemists made a discovery in the early twentieth century. They realized that protons do not exist in isolation in aqueous solution. (The hydrogen ion is simply a proton. It is a positively charged nuclear particle.) Instead, protons are always *hydrated*: they are attached to water molecules. A hydrated proton is called a **hydronium ion,** $H_3O^+{}_{(aq)}$. (See Figure 10.3.)

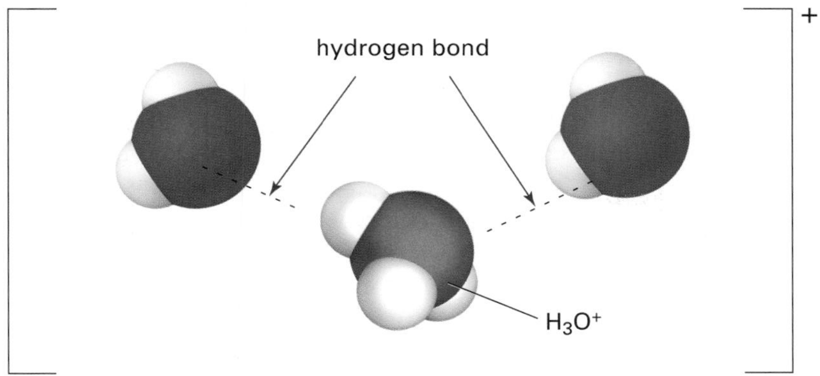

Figure 10.3 For convenience, chemists often use $H^+_{(aq)}$ as a shorthand notation for the hydronium ion, $H_3O^+_{(aq)}$. Hydronium ions do not exist independently. Instead, they form hydrogen bonds with other water molecules. Thus, a more correct formula is $H^+(H_2O)_n$, where *n* is usually 4 or 5.

There is another problem with the Arrhenius theory. Consider the reaction of ammonia, NH_3, with water.

$$NH_{3(g)} + H_2O_{(\ell)} \rightarrow NH^+_{4(aq)} + OH^-_{(aq)}$$

Ammonia is one of several substances that produce basic solutions in water. As you can see, ammonia does not contain hydroxide ions. However, it does produce these ions when it reacts with water. Ammonia also undergoes a neutralization reaction with acids. The Arrhenius theory cannot explain the basic properties of ammonia. Nor can it explain the fact that certain other substances, such as salts that contain carbonate ions, also have basic properties.

There is yet another problem with the Arrhenius theory. It is limited to acid-base reactions in a single solvent, water. Many acid-base reactions take place in other solvents, however.

CHECKPOINT

Use the idea of the hydronium ion to complete the following equation:

$$HCl_{(aq)} + H_2O_{(\ell)} \rightarrow$$

The Brønsted-Lowry Theory of Acids and Bases

In 1923, two chemists working independently of each other, proposed a new theory of acids and bases. (See Figure 10.4.) Johannes Brønsted in Copenhagen, Denmark, and Thomas Lowry in London, England, proposed what is called the **Brønsted-Lowry theory of acids and bases.** This theory overcame the problems related to the Arrhenius theory.

The Brønsted-Lowry Theory of Acids and Bases

- An acid is a substance from which a proton (H^+ ion) can be removed.
- A base is a substance that can remove a proton (H^+ ion) from an acid.

Figure 10.4 Johannes Brønsted (1879–1947), left, and Thomas Lowry (1874–1936), right. Brønsted published many more articles about ions in solution than Lowry did. Thus, some chemistry resources refer to the "Brønsted theory of acids and bases."

In many chemistry references, Brønsted-Lowry acids are called "proton donors." Brønsted-Lowry bases are called "proton acceptors." Although these terms are common, they create a false impression about the energy that is involved in acid-base reactions. Breaking bonds always requires energy. For example, removing a proton from a hydrochloric acid molecule requires 1.4×10^3 kJ/mol. This is far more energy than the word "donor" implies.

Like an Arrhenius acid, a Brønsted-Lowry acid must contain H in its formula. This means that all Arrhenius acids are also Brønsted-Lowry acids. However, any negative ion (not just OH^-) can be a Brønsted-Lowry base. In addition, water is not the only solvent that can be used.

According to the Brønsted-Lowry theory, there is only one requirement for an acid-base reaction. One substance must provide a proton, and another substance must receive the same proton. In other words, *an acid-base reaction involves the transfer of a proton.*

The idea of proton transfer has major implications for understanding the nature of acids and bases. According to the Brønsted-Lowry theory, any substance can behave as an acid, but only if another substance behaves as a base at the same time. Similarly, any substance can behave as a base, but only if another substance behaves as an acid at the same time.

For example, consider the reaction between hydrochloric acid and water shown in Figure 10.5. In this reaction, hydrochloric acid is an acid because it provides a proton (H^+) to the water. The water molecule receives the proton. Therefore, according to the Brønsted-Lowry theory, water is a base in this reaction. When the water receives the proton, it becomes a hydronium ion (H_3O^+). Notice the hydronium ion on the right side of the equation.

proton transfer — conjugate acid

$$HCl_{(aq)} + H_2O_{(\ell)} \rightarrow H_3O^+{}_{(aq)} + Cl^-{}_{(aq)}$$

acid — base — conjugate base

Figure 10.5 The reaction between hydrochloric acid and water, according to the Brønsted-Lowry theory

Two molecules or ions that are related by the transfer of a proton are called a **conjugate acid-base pair.** (Conjugate means "linked together.") The **conjugate base** of an acid is the particle that remains when a proton is removed from the acid. The **conjugate acid** of a base is the particle that results when the base receives the proton from the acid. In the reaction between hydrochloric acid and water, the hydronium ion is the conjugate acid of the base, water. The chloride ion is the conjugate base of the acid, hydrochloric acid.

According to the Brønsted-Lowry theory, every acid has a conjugate base, and every base has a conjugate acid. The conjugate base and conjugate acid of an acid-base pair are linked by the transfer of a proton. The conjugate base of the acid-base pair has one less hydrogen than the acid. It also has one more negative charge than the acid. The conjugate acid of the acid-base pair has one more hydrogen than the base. It also has one less negative charge than the base.

These ideas about acid-base reactions and conjugate acid-base pairs will become clearer as you study the following Sample Problems and Practice Problems.

Sample Problem

Conjugate Acid-Base Pairs

Problem

Hydrogen bromide is a gas at room temperature. It is soluble in water, forming hydrobromic acid. Identify the conjugate acid-base pairs.

What Is Required?

You need to identify two sets of conjugate acid-base pairs.

What Is Given?

You know that hydrogen bromide forms hydrobromic acid in aqueous solution.

Plan Your Strategy

- Write a balanced chemical equation.
- On the left side of the equation, identify the acid as the molecule that provides the proton. Identify the base as the molecule that accepts the proton.
- On the right side of the equation, identify the particle that has one proton less than the acid on the left side as the conjugate base of the acid. Identify the particle on the right side that has one proton more than the base on the left side as the conjugate acid of the base.

Act on Your Strategy

Hydrogen bromide provides the proton, so it is the Brønsted-Lowry acid in the reaction. Water receives the proton, so it is the Brønsted-Lowry base. The conjugate acid-base pairs are HBr/Br^- and H_2O/H_3O^+.

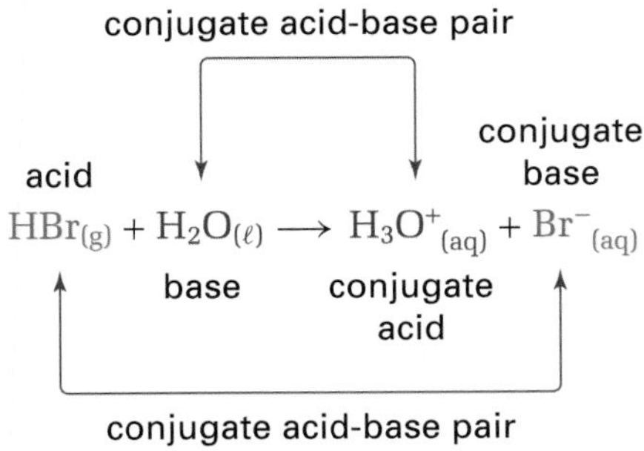

Check Your Solution

The formulas of the conjugate pairs differ by one proton, H^+, as expected.

Sample Problem

More Conjugate Acid-Base Pairs

Problem

Ammonia is a pungent gas at room temperature. Its main use is in the production of fertilizers and explosives. It is very soluble in water. It forms a basic solution that is used in common products, such as glass cleaners. Identify the conjugate acid-base pairs in the reaction between aqueous ammonia and water.

$$NH_{3(g)} + H_2O_{(\ell)} \rightarrow NH_4{}^+{}_{(aq)} + OH^-{}_{(aq)}$$

What Is Required?

You need to identify the conjugate acid-base pairs.

What Is Given?

The chemical equation is given.

Plan Your Strategy

- Identify the proton-provider on the left side of the equation as the acid. Identify the proton-remover (or proton-receiver) as the base.
- Identify the conjugate acid and base on the right side of the equation by the difference of a single proton from the acid and base on the left side.

Act on Your Strategy

The conjugate acid base pairs are $NH_4{}^+/NH_3$ and H_2O/OH^-.

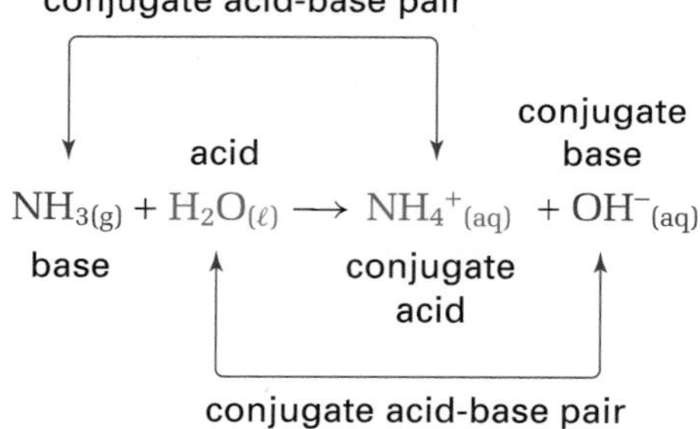

Check Your Solution

The formulas of the conjugate pairs differ by a proton, as expected.

PROBLEM TIP

In the previous Sample Problem, water acted as a base. In this Sample Problem, water acts as an acid. **Remember:** Water can act as a proton-provider (an acid) in some reactions and as a proton-receiver (a base) in others.

Practice Problems

1. Hydrogen cyanide is a poisonous gas at room temperature. When this gas dissolved in water, the following reaction occurs:

$$HCN_{(aq)} + H_2O_{(\ell)} \rightarrow H_3O^+{}_{(aq)} + CN^-{}_{(aq)}$$

Identify the conjugate acid-base pairs.

Continued ...

Continued ...
FROM PAGE 378

2. Sodium acetate is a good electrolyte. In water, the acetate ion reacts as follows:

$$CH_3COO^-_{(aq)} + H_2O_{(\ell)} \rightarrow CH_3COOH_{(aq)} + OH^-_{(aq)}$$

Identify the conjugate acid-base pairs.

3. Write equations to show how the hydrogen sulfide ion, HS^-, can react with water. First show the ion acting as an acid. Then show the ion acting as a base.

Section Wrap-up

The two theories that you have considered in this section attempt to explain the chemical nature of acids and bases. Table 10.3 summarizes the key points of these theories.

In both the Arrhenius theory and the Brønsted-Lowry theory, acids and bases form ions in solution. Many characteristics of acid-base behaviour are linked to the number of ions that form from a particular acid or base. One of these characteristics is strength.

In section 10.2, you will learn why a dilute solution of vinegar is safe to ingest, while the same molar concentration of hydrochloric acid would be extremely poisonous.

Table 10.3 Comparing the Arrhenius Theory and the Brønsted-Lowry Theory

Theory	Arrhenius	Brønstead-Lowry
Acid	any substance that dissociatesto form H^+ in aqueous solution	any substance that provides a proton to another substance (or any substance from which a proton may be removed)
Base	any substance that dissociates to form OH^- in aqueous solution	any substance that receives a proton from an acid (or any substance that removes a proton from an acid)
Example	$HCl_{(aq)} \rightarrow H^+_{(aq)} + Cl^-_{(aq)}$	$HCl_{(aq)} + H_2O_{(\ell)} \rightarrow H_3O^+_{(aq)} + Cl^-_{(aq)}$

Section Review

1 (I) Suppose that you have four unknown solutions, labelled A, B, C, and D. You use a conductivity apparatus to test their conductivity, and obtain the results shown below. Use these results to answer the questions that follow.

Solution	Results of Conductivity Test
A	the bulb glows dimly
B	the bulb glows strongly
C	the bulb does not glow
D	the bulb glows strongly

(a) Which of these solutions has a high concentration of dissolved ions? What is your evidence?

(b) Which of these solutions has a low concentration of dissolved ions? What is your evidence?

(c) Which of the four unknowns are probably aqueous solutions of acids or bases?
(d) Based on these tests alone, can you distinguish the acidic solution(s) from the basic solutions(s)? Why or why not?
(e) Suggest one way that you could distinguish the acidic solution(s) from the basic solution(s).

2 (a) K/U Define an acid and a base according to the Arrhenius theory.
(b) Give two examples of observations that the Arrhenius theory can explain.
(c) Give two examples of observations that the Arrhenius theory can *not* explain.

3 (a) K/U Define an acid and a base according to the Brønsted-Lowry theory.
(b) What does the Brønsted-Lowry theory have in common with the Arrhenius theory? In what ways is it different?
(c) Which of the two acid-base theories is more comprehensive? (In other words, which explains a broader body of observations?)

4 (a) K/U What is the conjugate acid of a base?
(b) What is the conjugate base of an acid?
(c) Use an example to illustrate your answers to parts (a) and (b) above.

5 K/U Write the formula for the conjugate acid of the following:
(a) the hydroxide ion, OH^-
(b) the carbonate ion, CO_3^{2-}

6 K/U Write the formula for the conjugate base of the following:
(a) nitric acid, HNO_3
(b) the hydrogen sulfate ion, HSO_4^-

7 K/U Which of the following compounds is an acid according to the Arrhenius theory?
(a) H_2O
(b) $Ca(OH)_2$
(c) H_3PO_3
(d) HF

8 K/U Which of the following compounds is a base according to the Arrhenius theory?
(a) KOH
(b) $Ba(OH)_2$
(c) HClO
(d) H_3PO_4

9 C Hydrofluoric acid dissociates in water to form fluoride ions.
(a) Write a balanced chemical equation for this reaction.
(b) Identify the conjugate acid-base pairs.
(c) Explain how you know whether or not you have correctly identified the conjugate acid-base pairs.

10 I Identify the conjugate acid-base pairs in the following reactions:
(a) $H_2PO_4^-{}_{(aq)} + CO^{2-}{}_{3(aq)} \rightarrow HPO_4^{2-}{}_{(aq)} + HCO_3^-{}_{(aq)}$
(b) $HCOOH_{(aq)} + CN^-{}_{(aq)} \rightarrow HCOO^-{}_{(aq)} + HCN_{(aq)}$
(c) $H_2PO_4^-{}_{(aq)} + OH^-{}_{(aq)} \rightarrow HPO_4^{2-}{}_{(aq)} + H_2O_{(\ell)}$

Strong and Weak Acids and Bases

10.2

Re-examine Figure 10.1 on page 371. Look at the photographs of hydrochloric acid and acetic acid. The conductivity tester is testing the same concentrations of both acids. As you can see, the bulb glows brightly in the hydrochloric acid. The bulb glows dimly in the acetic acid. How can these different results be explained?

Section Preview/ Specific Expectations

In this section, you will

- **explain**, in terms of the degree to which they dissociate, the difference between strong and weak acids and bases
- **distinguish** between binary acids and oxoacids
- **define** pH, and experimentally determine the effect on pH of diluting an acidic solution
- **communicate** your understanding of the following terms: *strong acid, weak acid, strong base, weak base, binary acid, oxoacid, pH*

Strong Acids and Weak Acids

You know that ions are present in an aqueous solution of an acid. These ions result from the dissociation of the acid. An acid that dissociates completely into ions in water is called a **strong acid.** For example, hydrochloric acid is a strong acid. *All* the molecules of hydrochloric acid in an aqueous solution dissociate into H^+ and Cl^- ions. The H^+ ions, as you know, bond with surrounding water molecules to form hydronium ions, H_3O^+. (See Figure 10.6.) *The concentration of hydronium ions in a dilute solution of a strong acid is equal to the concentration of the acid.* Thus, a 1.0 mol/L solution of hydrochloric acid contains 1.0 mol/L of hydronium ions. Table 10.4 lists the strong acids.

Table 10.4 Strong Acids

Strong Acids
hydrochloric acid, HCl
hydrobromic acid, HBr
hydroiodic acid, HI
nitric acid, HNO_3
sulfuric acid, H_2SO_4
perchloric acid, $HClO_4$

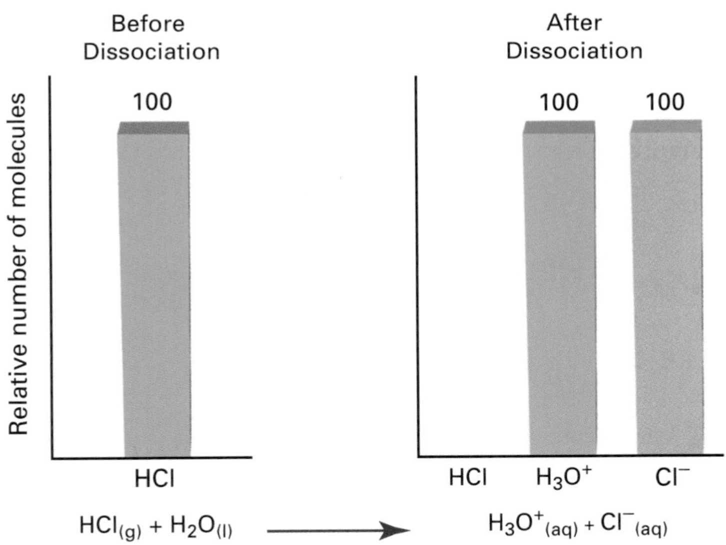

Figure 10.6 **When hydrogen chloride molecules enter an aqueous solution, 100% of the hydrogen chloride molecules dissociate. As a result, the solution contains the same percent of H^+ ions (in the form of H_3O^+) and Cl^- ions: 100%.**

A **weak acid** is an acid that dissociates very slightly in a water solution. Thus, only a small percentage of the acid molecules break apart into ions. Most of the acid molecules remain intact. For example, acetic acid is a weak acid. On average, only about 1% (one in a hundred) of the acetic acid molecules dissociate at any given moment in a 0.1 mol/L solution. (The number of acid molecules that dissociate depends on the concentration and temperature of the solution.) In fact, *the concentration of hydronium ions in a solution of a weak acid is always less than the concentration of the dissolved acid.* (See Figure 10.7.)

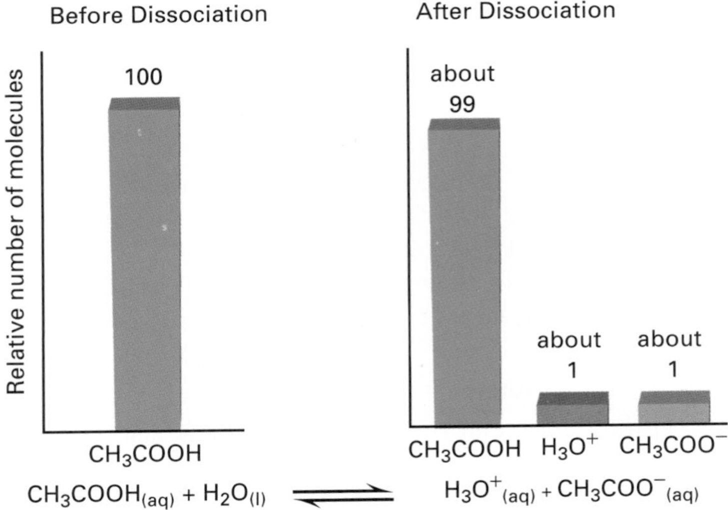

Figure 10.7 When acetic acid molecules enter an aqueous solution, only about 1% of them dissociate. Thus, the number of acetic acid molecules in solution is far greater than the number of hydronium ions and acetate ions.

Notice the arrow that is used in the equation in Figure 10.7. It points in both directions, indicating that the reaction is *reversible*. In other words, the products of the reaction also react to produce the original reactants. In this reaction, molecules of acetic acid dissociate just as quickly and as often as the dissociated ions re-associate to produce acetic acid molecules. (Figure 10.8 will help you visualize what happens.)

Most acids are weak acids. Whenever you see a reversible chemical equation involving an acid, you can safely assume that the acid is weak.

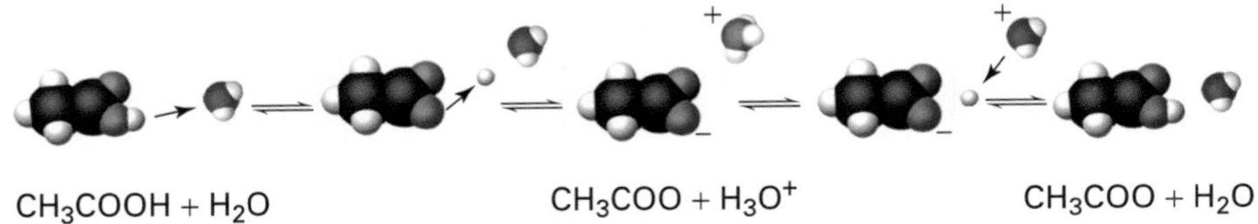

Figure 10.8 When acetic acid dissolves in water, acetic acid molecules dissociate and re-associate at the same time and at the same rate.

A few acids contain only a single hydrogen ion that can dissociate. These acids are called *monoprotic acids*. (The prefix mono- means "one." The root -protic refers to "proton.") Hydrochloric acid, hydrobromic acid, and hydroiodic acid are strong monoprotic acids. Hydrofluoric acid, HF, is weak monoprotic acid.

Many acids contain two or more hydrogen ions that can dissociate. For example, sulfuric acid, $H_2SO_{4(aq)}$, has two hydrogen ions that can dissociate. As you know from Table 10.4, sulfuric acid is a strong acid. This is true only for its first dissociation, however.

$$H_2SO_{4(aq)} \rightarrow H^+_{(aq)} + HSO_4{}^-_{(aq)}$$

The resulting aqueous hydrogen sulfate ion, HSO^-_4, is a weak acid. It dissociates to form the sulfate ion in the following reversible reaction:

$$HSO_4{}^-_{(aq)} \rightleftharpoons H^+_{(aq)} + SO_4{}^{2-}_{(aq)}$$

Thus, acids that contain two hydrogen ions dissociate to form two anions. These acids are sometimes called *diprotic acids.* (The prefix di-, as you know, means "two.") The acid that is formed by the first dissociation is stronger than the acid that is formed by the second dissociation.

Acids that contain three hydrogen ions are called *triprotic acids.* Phosphoric acid, $H_3PO_{4(aq)}$, is a triprotic acid. It gives rise to three anions, as follows:

$$H_3PO_{4(aq)} + H_2O_{(\ell)} \rightleftharpoons H^+_{(aq)} + H_2PO_4{}^-{}_{(aq)}$$
$$H_2PO_4{}^-{}_{(aq)} + H_2O_{(\ell)} \rightleftharpoons H^+_{(aq)} + HPO_4{}^{2-}{}_{(aq)}$$
$$HPO_4{}^{2-}{}_{(aq)} + H_2O_{(\ell)} \rightleftharpoons H^+_{(aq)} + PO_4{}^{3-}{}_{(aq)}$$

Here again, the acid that is formed by the first dissociation is stronger than the acid that is formed by the second dissociation. This acid is stronger than the acid that is formed by the third dissociation. Keep in mind, however, that all three of these acids are weak because only a small proportion of them dissociates.

Strong Bases and Weak Bases

Like a strong acid, a **strong base** dissociates completely into ions in water. All oxides and hydroxides of the alkali metals—Group 1 (IA)—are strong bases. The oxides and hydroxides of the alkaline earth metals—Group 2 (IIA)—below beryllium are also strong bases.

Recall that the concentration of hydronium ions in a dilute solution of a strong acid is equal to the concentration of the acid. Similarly, the concentration of hydroxide ions in a dilute solution of a strong base is equal to the concentration of the base. For example, a 1.0 mol/L solution of sodium hydroxide (a strong base) contains 1.0 mol/L of hydroxide ions.

Table 10.5 lists some common strong bases. Barium hydroxide, $Ba(OH)_2$, and strontium hydroxide, $Sr(OH)_2$, are strong bases that are soluble in water. Magnesium oxide, MgO, and magnesium hydroxide, $Mg(OH)_2$, are also strong bases, but they are considered to be insoluble. A small amount of these compounds does dissolve in water, however. Virtually all of this small amount dissociates completely.

Most bases are weak. A **weak base** dissociates very slightly in a water solution. The most common weak base is aqueous ammonia. In a 0.1 mol/L solution, only about 1% of the ammonia molecules react with water to form hydroxide ions. This reversible reaction is represented in Figure 10.9.

Table 10.5 Common Strong Bases

Common Strong Bases
sodium hydroxide, NaOH
potassium hydroxide, KOH
calcium hydroxide, $Ca(OH)_2$
strontium hydroxide, $Sr(OH)_2$
barium hydroxide, $Ba(OH)_2$

$$NH_3 + H_2O \rightleftharpoons NH_4{}^+ + OH^-$$

Figure 10.9 Ammonia does not contain hydroxide ions, so it is not an Arrhenius base. As you can see, however, an ammonia molecule can remove a proton from water, leaving a hydroxide ion behind. Thus, ammonia is a Brønsted-Lowry weak base.

Naming Acids and Their Anions

There are two main kinds of acids: binary acids and oxoacids. A **binary acid** is composed of two elements: hydrogen and a non-metal. Two examples of binary acids are hydrofluoric acid and hydrochloric acid. All binary acids have the general formula $HX_{(aq)}$. The H represents one or more hydrogen atoms. The X represents the non-metal. As you can see in Table 10.6, the names of binary acids are made up of the following parts:

- the prefix hydro-
- a root that is formed from the name of the non-metal
- the suffix -ic
- the word "acid" at the end

CHECKPOINT

The terms "strong" and "concentrated" have very different meanings when describing solutions of acids, bases, or salts. Similarly, the terms "weak" and "dilute" have very different meanings. In your own words, summarize the difference between "strong" and "concentrated." Give examples that illustrate this difference. Then summarize the difference between "weak" and "dilute," again giving examples.

Table 10.6 Examples of Naming Binary Acids

Binary acid	Prefix	Non-metal root	Suffix
hydrofluoric acid, $HF_{(aq)}$	hydro-	-fluor-	-ic
hydrochloric acid, $HCl_{(aq)}$	hydro-	-chlor-	-ic
hydrosulfuric acid, H_2S	hydro-	-sulfur-	-ic

As you know, anions are formed when binary acids dissociate. The names of these anions end in the suffix -ide. For example, hydroflouric acid forms the anion fluoride, F^-. Hydrochloric acid forms the anion chloride, Cl^-.

An **oxoacid** is an acid formed from a polyatomic ion that contains oxygen, hydrogen, and another element. (Oxoacids are called oxyacids in some chemistry textbooks). In Chapter 3, you learned the names of common polyatomic ions and their valences (oxidation numbers). The names of oxoacids are similar to the names of their polyatomic oxoanions. Only the suffix is different. Study the three rules and examples for naming oxoacids below. Then try the Practice Problems that follow.

1. For anions that end in -ate, the suffix of the acid is -ic. For example, the acid of the chlorate anion ClO_3^-, is chloric acid, $HClO_3$.
2. For anions that end in -ite, the suffix of the acid is -ous. For example, the acid of the chlorite anion, ClO_2^-, is chlorous acid, $HClO_2$.
3. The prefixes hypo- and per- remain as part of the acid name. For example, the acid of the perchlorate anion, ClO_4^-, is perchloric acid, $HClO_4$. The acid of the hypochlorite anion, ClO^-, is hypochlorous acid, HClO.

mind STRETCH

The conjugate base of a strong acid is always a weak base. Conversely, the conjugate base of a weak acid is always a strong base. Explain this inverse relationship.

Practice Problems

4. (a) Write the chemical formula for hydrobromic acid. Then write the name and formula for the anion that it forms.
   (b) Hydrosulfuric acid, H_2S, forms two anions. Name them and write their formulas.
5. Write the chemical formulas for the following acids. Then name and write the formulas for the oxoanions that form from each acid. Refer to Chapter 3, Table 3.5, Names and Valences of Some Common Polyatomic Ions, as necessary.
   (a) nitric acid
   (b) nitrous acid
   (c) hyponitrous acid
   (d) phosphoric acid
   (e) phosphorous acid
   (f) periodic acid

Describing Acid and Base Strength Quantitatively: pH

You are probably familiar with the term "pH" from a variety of sources. Advertisers talk about the "pH balance" of products such as soaps, shampoos, and skin creams. People who own aquariums and swimming pools must monitor the pH of the water. (See Figure 10.10.) Gardeners and farmers use simple tests to determine the pH of the soil. They know that plants and food crops grow best within a narrow range of pH. Similarly, the pH of your blood must remain within narrow limits for you to stay healthy.

Figure 10.10 **Maintaining a safe environment for an aquarium or a swimming pool requires measuring the pH of the water and knowing how to adjust it.**

pH is clearly related to health, and to the proper functioning of products and systems. (Notice that the "p" is always lower case, even at the start of a sentence.) What exactly is pH? How is it measured? To answer these questions, consider a familiar substance: water.

The Power of Hydrogen in Water

As you know, all aqueous solutions contain ions. Even pure water contains a few ions that are produced by the dissociation of water molecules. **Remember:** The double arrow in the equation shows that the reaction is reversible. The ions recombine to form water molecules.

$$H_2O_{(\ell)} + H_2O_{(\ell)} \rightleftharpoons H_3O^+_{(aq)} + OH^-_{(aq)}$$

On average, at 25°C, only about two water molecules in a billion are dissociated at any given moment. As you know, it is the ions in solution that conduct electricity. If there are virtually no ions, no electricity is conducted. This is why pure water is such a poor conductor. Chemists have determined that the concentration of hydronium ions in neutral water, at 25°C, is only 1.0×10^{-7} mol/L. The dissociation of water also produces the same, very small number of hydroxide ions. Therefore, the concentration of hydroxide ions is also 1.0×10^{-7} mol/L.

Chemists sometimes use square brackets around a chemical formula. This shorthand notation means "the concentration of" the chemical inside the brackets. For example, $[H_3O^+]$ is read as "the concentration of hydronium ions." Thus, the concentration of hydronium ions and hydroxide ions in neutral water can be written as

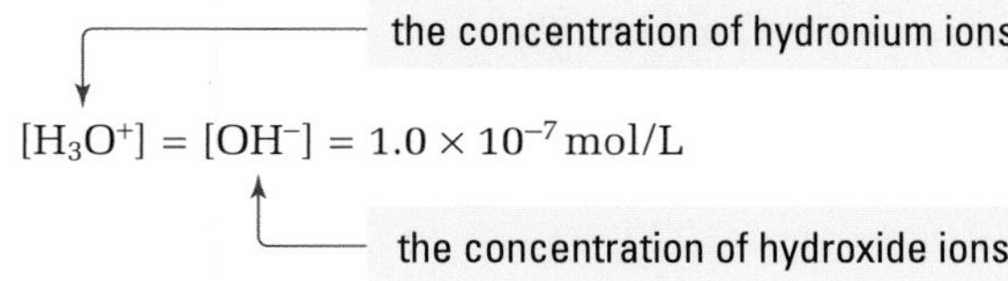

$$[H_3O^+] = [OH^-] = 1.0 \times 10^{-7} \text{ mol/L}$$

Compared with neutral water, acidic solutions contain a higher concentration of hydronium ions. Basic solutions contain a lower concentration of hydronium ions. Therefore, the dissociation of water provides another way of thinking about acids and bases. *An acid is any compound that increases [H_3O^+] when it is dissolved in water. A base is any compound that increases [OH^-] when it is dissolved in water.* (See Figure 10.11.)

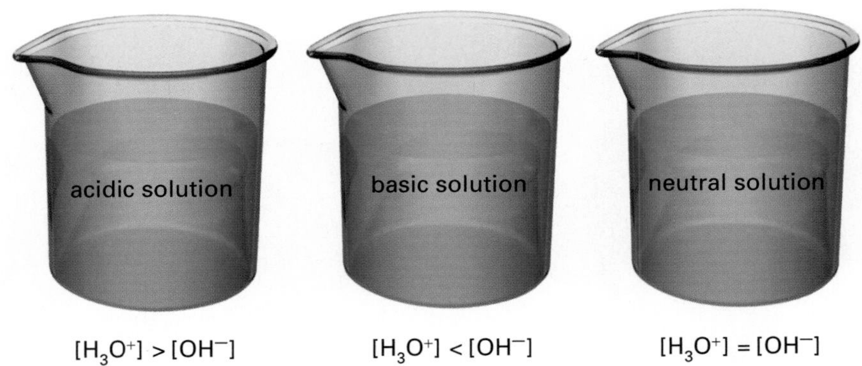

Figure 10.11 **The relationship between the concentrations of hydronium ions and hydroxide ions in a solution determines whether the solution is acidic, basic, or neutral.**

The pH Scale: Measuring by Powers of Ten

The concentration of hydronium ions ranges from about 10 mol/L for a concentrated strong acid to about 10^{-15} mol/L for a concentrated strong base. This wide range of concentrations, and the negative powers of 10, are not very convenient to work with. In 1909, a Danish biochemist, Søren Sørensen, suggested a method for converting concentrations to positive numbers. His method involved using the numerical system of logarithms.

The logarithm of a number is the power to which you must raise 10 to equal that number. For example, the logarithm of 10 is 1, because $10^1 = 10$. The logarithm of 100 is 2, because $10^2 = 100$. (See Appendix E for more information about exponents and logarithms.)

Sørensen defined **pH** as $-\log [H^+]$. Since Sørensen did not know about hydronium ions, his definition of pH is based on Arrhenius' hydrogen ion. Many chemistry references reinterpret the H so that it refers to the Brønsted-Lowry hydronium ion, H_3O^+, instead. This textbook adopts the hydronium ion usage. Thus, the definition for pH becomes $pH = -\log [H_3O^+]$. Recall, though, that chemists use [H^+] as a shorthand notation for [H_3O^+]. As a result, both equations give the same product.

As you can see in Figure 10.12, the "p" in pH stands for the word "power." The power referred to is exponential power: the power of 10. The "H" stands for the concentration of hydrogen ions (or H_3O^+ ions), measured in mol/L.

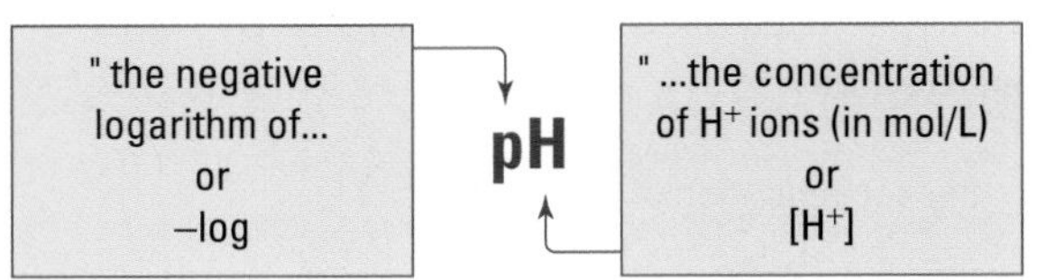

Figure 10.12 **The concept of pH makes working with very small values, such as 0.000 000 000 000 01, much easier.**

The concept of pH allows hydronium (or hydrogen) ion concentrations to be expressed as positive numbers, rather than negative exponents. For example, recall that $[H_3O^+]$ of neutral water at 25°C is 1.0×10^{-7}.

$$\begin{aligned}\therefore pH &= -\log [H_3O^+] \\ &= -\log (1.0 \times 10^{-7}) \\ &= -(-7.00) \\ &= 7.00\end{aligned}$$

$[H_3O^+]$ in acidic solutions is greater than $[H_3O^+]$ in neutral water. For example, if $[H_3O^+]$ in an acid is 1.0×10^{-4} mol/L, this is 1000 times greater than $[H_3O^+]$ in neutral water. Use Table 10.7 to make sure that you understand why. The pH of the acid is 4.00. All acidic solutions have a pH that is less than 7.

Table 10.7 Understanding pH

Range of acidity and basicity	$[H_3O^+]$ (mol/L)	Exponential notation (mol/L)	log	pH (-log $[H_3O^+]$)
strong acid	1	1×10^{0}	0	0
	0.1	1×10^{-1}	−1	1
	0.01	1×10^{-2}	−2	2
	0.001	1×10^{-3}	−3	3
	0.000 1	1×10^{-4}	−4	4
	0.000 01	1×10^{-5}	−5	5
	0.000 001	1×10^{-6}	−6	6
neutral $[H^+] = [OH^-] = 1.0 \times 10^{-7}$	0.000 000 1	1×10^{-7}	−7	7
	0.000 000 01	1×10^{-8}	−8	8
	0.000 000 001	1×10^{-9}	−9	9
	0.000 000 000 1	1×10^{-10}	−10	10
	0.000 000 000 01	1×10^{-11}	−11	11
	0.000 000 000 001	1×10^{-12}	−12	12
	0.000 000 000 000 1	1×10^{-13}	−13	13
strong base	0.000 000 000 000 01	1×10^{-14}	−14	14

$[H_3O^+]$ in basic solutions is less than $[H_3O^+]$ in pure water. For example, if $[H_3O^+]$ in a base is 1.0×10^{-11} mol/L, this is 10 000 times less than $[H_3O^+]$ in neutral water. The pH of the base is 11.00. All basic solutions have a pH that is greater than 7.

The relationship among pH, $[H_3O^+]$, and the strength of acids and bases is summarized in the Concept Connection on the next page. Use the following Sample Problem and Practice Problems to assess your understanding of this relationship. Then, in Investigation 10-A, you will look for a pattern involving the pH of a strong acid, a weak acid, and dilutions of both.

Science LINK

Using logarithms is a convenient way to count a wide range of values by powers of 10. Chemists are not the only scientists who use such logarithms, however. Audiologists (scientists who study human hearing) use logarithms, too. Research the decibel scale to find out how it works. Present your findings in the medium of your choice.

Math LINK

How do you determine the number of significant digits in a pH? You count only the digits to the right of the decimal point. For example, suppose that the concentration of hydronium ions in a sample of orange juice is 2.5×10^{-4} mol/L. This number has two significant digits: the 2 and the 5. The power of 10 only tells us where to place the decimal: 0.000 25. The pH of the sample is $-\log (2.5 \times 10^{-4}) = 3.602\ 059$. The digit to the left of the decimal (the 3) is derived from the power of 10. Therefore, it is not considered to be a significant digit. Only the two digits to the right of the decimal are significant. Thus, the pH value is rounded off to 3.60.

Identify the significant digits in each pH value below.

1. The pH of drain cleaner is 13.1.
2. The pH of milk is 6.4.
3. The pH of vinegar is 2.85.
4. The pH of lemon juice is 2.310.

Concept Organizer pH, $[H_3O^+]$, and the Strength of Acids and Bases

Type of solution	$[H_3O+]$ (mol/L)	Concentration of hydronium and hydroxide ions	pH at 25°C
acidic solution	greater than 1×10^{-7}	$[H_3O^+] > [OH^-]$	< 7.00
neutral solution	1×10^{-7}	$[H_3O^+] = [OH^-]$	7.00
basic solution	less than 1×10^{-7}	$[H_3O^+] < [OH^-]$	> 7.00

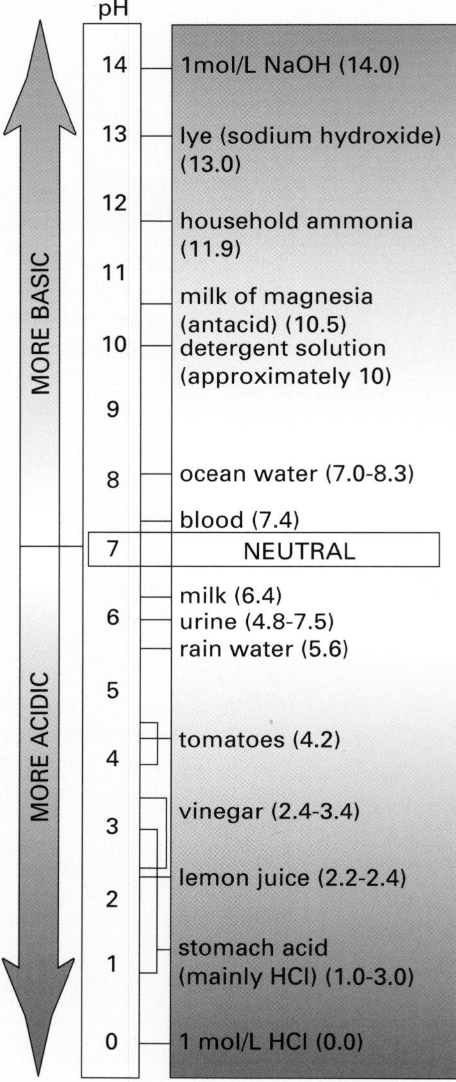

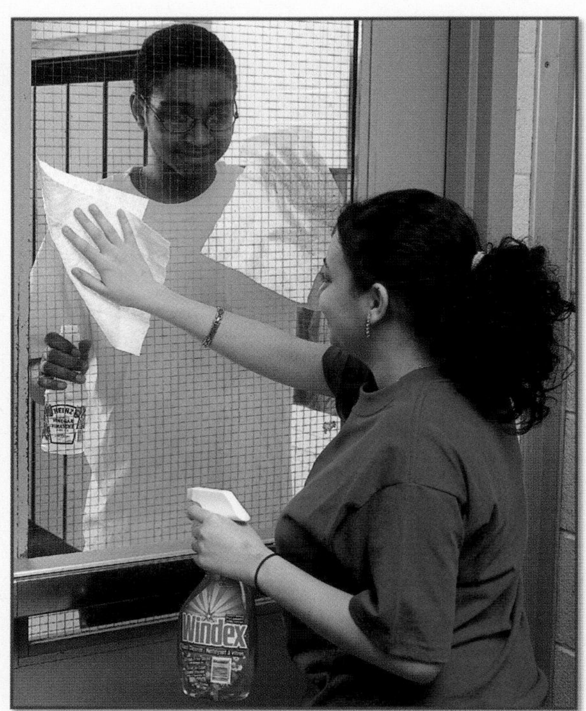

The pH values of many common solutions fall within a range from 0 to 14, as shown on this pH scale. The table above the pH scale relates the positive pH values to their hydronium ion concentrations and their logarithms.

Sample Problem

Calculating the pH of a Solution

Problem

Calculate the pH of a solution with $[H_3O^+] = 3.8 \times 10^{-3}$ mol/L.

What Is Required?

You need to calculate the pH, given $[H_3O^+]$.

What Is Given?

You know that $[H_3O^+]$ is 3.8×10^{-3} mol/L.

Plan Your Strategy

Use the equation $pH = -\log [H_3O^+]$ to solve for the unknown.

Act on Your Strategy

$$\begin{aligned} pH &= -\log (3.8 \times 10^{-3}) \\ &= 2.42 \end{aligned}$$

Check Your Solution

$[H_3O^+]$ is greater than 1.0×10^{-7} mol/L. Therefore, the pH should be less than 7.00. The solution is acidic, as you would expect.

PROBLEM TIP

Appendix E, "Math and Chemistry", explains how you can do these calculations with a calculator.

Practice Problems

6. Calculate the pH of each solution, given the hydronium ion concentration.
- **(a)** $[H_3O^+] = 0.0027$ mol/L
- **(b)** $[H_3O^+] = 7.28 \times 10^{-8}$ mol/L
- **(c)** $[H_3O^+] = 9.7 \times 10^{-5}$ mol/L
- **(d)** $[H_3O^+] = 8.27 \times 10^{-12}$

7. $[H_3O^+]$ in a cola drink is about 5.0×10^{-3} mol/L. Calculate the pH of the drink. State whether the drink is acidic or basic.

8. A glass of orange juice has $[H_3O^+]$ of 2.9×10^{-4} mol/L. Calculate the pH of the juice. State whether the result is acidic or basic.

9. (a) $[H_3O^+]$ in a dilute solution of nitric acid, HNO_3, is 6.3×10^{-3} mol/L. Calculate the pH of the solution.

(b) $[H_3O^+]$ of a solution of sodium hydroxide is 6.59×10^{10} mol/L. Calculate the pH of the solution.

SKILL FOCUS

Predicting

Performing and recording

Analyzing and interpreting

Investigation 10-A

The Effect of Dilution on the pH of an Acid

In this investigation, you will compare the effects of diluting a strong acid and a weak acid.

In Part 1, you will measure the pH of a strong acid. Then you will perform a series of ten-fold dilutions. That is, each solution will be one-tenth as dilute as the previous solution. You will measure and compare the pH after each dilution.

In Part 2, you will measure the pH of a weak acid with the same initial concentration as the strong acid. Then you will perform a series of ten-fold dilutions with the weak acid. Again, you will measure and compare the pH after each dilution.

Problem

How does the pH of dilutions of a strong acid compare with the pH of dilutions of a weak acid?

Prediction

Predict each pH, and explain your reasoning.

(a) the pH of 0.10 mol/L hydrochloric acid

(b) the pH of the hydrochloric acid after one ten-fold dilution

(c) the pH of the hydrochloric acid after each of six more ten-fold dilutions

(d) the pH of 0.10 mol/L acetic acid, compared with the pH of 0.10 mol/L hydrochloric acid

(e) the pH of the acetic acid after one ten-fold dilution

Safety Precautions

Hydrochloric acid is corrosive. Wash any spills on skin or clothing with plenty of cool water. Inform your teacher immediately.

Materials

100 mL graduated cylinder
100 mL beaker
2 beakers (250 mL)
universal indicator paper and glass rod
pH meter
0.10 mol/L hydrochloric acid (for Part 1)
0.10 mol/L acetic acid (for Part 2)
distilled water

Procedure

Part 1 The pH of Solutions of a Strong Acid

1. Copy the table below into your notebook. Record the pH you predicted for each dilution.
2. Pour about 40 mL of 0.10 mol/L hydrochloric acid into a clean, dry 100 mL beaker. Use the end of a glass rod to transfer a drop of solution to a piece of universal pH paper into the acid. Compare the colour against the colour chart to determine the pH. Record the pH. Then measure and record the pH of the acid using a pH meter. Rinse the electrode with distilled water afterward.

Data Table for Part 1

$[HCl_{(aq)}]$ mol/L	Predicted pH	pH measured with universal indicator	pH measured with pH meter
1×10^{-1}			
1×10^{-2}			
1×10^{-3}			
1×10^{-4}			
1×10^{-5}			
1×10^{-6}			
1×10^{-7}			
1×10^{-8}			

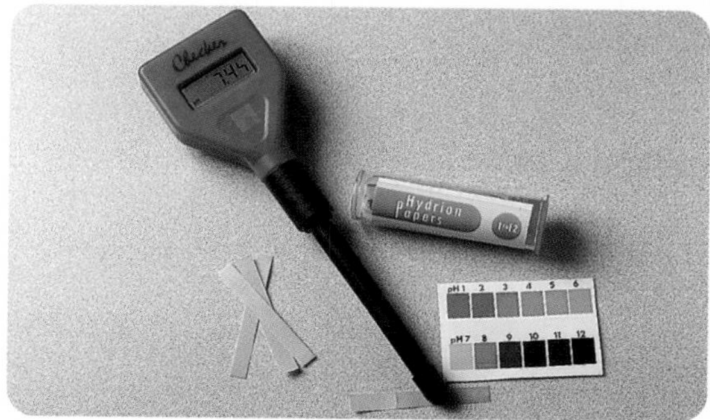

3. Measure 90 mL of distilled water in a 100 mL graduated cylinder. Add 10 mL of the acid from step 2. The resulting 100 mL of solution is one-tenth as concentrated as the acid from step 2. Pour the dilute solution into a clean, dry 250 mL beaker. Use universal pH paper and a pH meter to measure the pH. Record your results.

4. Repeat step 3. Pour the new dilute solution into a second clean, dry beaker. Dispose of the more concentrated acid solution as directed by your teacher. Rinse and dry the beaker so you can use it for the next dilution.

5. Make further dilutions and pH measurements until the hydrochloric acid solution is 1.0×10^{-8} mol/L

Part 2 The pH of Solutions of a Weak Acid

1. Design a table to record your predictions and measurements for 0.10 mol/L and 0.010 mol/L concentrations of acetic acid.

2. Use the same procedure that you used in Part 1 to measure and record the pH of a 0.10 mol/L sample of acetic acid. Then dilute the solution to 0.010 mol/L. Measure the pH again.

Analysis

1. Which do you think gave the more accurate pH: the universal indicator paper or the pH meter? Explain.

2. For the strong acid, compare the pH values you predicted with the measurements you made. How can you explain any differences for the first few dilutions?

3. What was the pH of the solution that had a concentration of 1.0×10^{-8} mol/L? Explain the pH you obtained.

4. Compare the pH of 0.10 mol/L acetic acid with the pH of 0.10 mol/L hydrochloric acid. Why do you think the pH values are different, even though the concentrations of the acids were the same?

5. What effect does a ten-fold dilution of a strong acid (hydrochloric acid) have on the pH of the acid? What effect does the same dilution of a weak acid (acetic acid) have on its pH? Compare the effects for a strong acid and a weak acid. Account for any differences.

Conclusion

6. Use evidence from your investigation to support the conclusion that a weak acid dissociates less than a strong acid of identical concentration.

7. Why is the method for calculating the pH of a strong acid (if it is not too dilute) not appropriate for a weak acid?

Applications

8. Nicotinic acid is a B vitamin. The pH of a 0.050 mol/L solution of this acid is measured to be 3.08. Is it a strong acid or a weak acid? Explain. What would be the pH of a solution of nitric acid having the same concentration?

9. Would you expect to be able to predict the pH of a weak base, given its concentration? Design an experiment you could perform to check your answer.

Chemistry Bulletin

Science | Technology | Society | Environment

The Chemistry of Oven Cleaning

Oven cleaning is not a job that most people enjoy. Removing baked-on grease from inside an oven requires serious scrubbing. Any chemical oven cleaners that help to make the job easier are usually welcome. Like all chemicals, however, the most effective oven cleaners require attention to safety.

Cleaners that contain strong bases are the most effective for dissolving grease and grime. Bases are effective because they produce soaps when they react with the fatty acids in grease. When a strong base (such as sodium hydroxide, NaOH, or potassium hydroxide, KOH) is used on a dirty oven, the fat molecules that make up the grease are split into smaller molecules. Anions from the base then bond with some of these molecules to form soap.

One end of a soap molecule is non-polar (uncharged), so it is soluble in dirt and grease, which are also non-polar. The other end of a soap molecule is polar (charged), so it is soluble in water. Because of its two different properties, soap acts like a "bridge" between the grease and the water. Soap enables grease to dissolve in water and be washed away, thus allowing the cleaner to remove the grease from the oven surface.

Cleaners that contain sodium hydroxide and potassium hydroxide are very effective. They are also caustic and potentially very dangerous. For example, sodium hydroxide, in the concentrations that are used in oven cleaners, can irritate the skin and cause blindness if it gets in the eyes. As well, it is damaging to paints and fabrics.

There are alternatives to sodium hydroxide and other strong base cleaners. One alternative involves using ammonia, NH_3, which is a weak base. If a bowl of dilute ammonia solution is placed in an oven and left for several hours, most of the grease and grime can be wiped off.

Ammonia does not completely ionize in water. Only a small portion dissociates. Although an ammonia solution is less caustic than sodium hydroxide, it can be toxic if inhaled directly. As well, ammonia vapours can cause eye, lung, and skin irritations. At higher concentrations, ammonia can be extremely toxic.

Baking soda is a non-toxic alternative, but it is much less effective. Therefore, it requires even more scrubbing. An abrasive paste can be made by mixing baking soda and water. The basic properties of baking soda also have a small effect on grease and grime if it is applied to the oven and left for several hours.

Making Connections

1. Survey the cleaners in your home or school. Which cleaners contain bases and which contain acids? What cleaning jobs can an acid cleaner perform well? How do most acid cleaners work?
2. Some companies claim to make environmentally sensitive cleaners. Investigate these cleaners. What chemicals do they contain? See if you can infer how they work. You might like to design a controlled experiment to test the effectiveness of several oven cleaners. **CAUTION** Obtain permission from your teacher before performing such an experiment.

Section Wrap-up

In this section, you considered the relationship among the strength of acids and bases, the concentration of hydronium and hydroxide ions, and pH. Much of the time, you examined acids and bases acting independently of each other. However, acids and bases often interact. In fact, acid-base reactions have many important applications in the home, as well as in the laboratory. In section 10.3, you will investigate acid-base reactions.

Section Review

1 K/U Distinguish, in terms of degree of dissociation, between a strong acid and a weak acid, and a strong base and a weak base.

2 K/U Give one example of the following:

(a) a weak acid

(b) a strong acid

(c) a strong base

(d) a weak base

3 K/U Formic acid, HCOOH, is responsible for the painful bites of fire ants. Is formic acid strong or weak? Explain.

4 K/U $KMnO_4$ is an intense purple-coloured solid that can be made into a solution to kill bacteria. What is the name of this compound? Give the name and the formula of the acid that forms when $KMnO_4$ combines with water.

5 K/U State the name or the formula for the acid that forms from each of the following anions:

(a) hydrogen sulfate

(b) F^-

(c) HS^-

(d) bromite

6 K/U Explain the meaning of pH, both in terms of hydrogen ions and hydronium ions.

7 K/U Arrange the following foods in order of increasing acidity: beets, pH = 5.0; camembert cheese, pH = 7.4; egg white, pH = 8.0; sauerkraut, pH = 3.5; yogurt, pH = 4.5.

8 I Calculate the pH of each body fluid, given the concentration of hydronium ions.

(a) tears, $[H_3O^+] = 4.0 \times 10^{-8}$ mol/L

(b) stomach acid, $[H_3O^+] = 4.0 \times 10^{-2}$ mol/L

9 I Calculate the pH of the solution that is formed by diluting 50 mL of 0.025 mol/L hydrochloric acid to a final volume of 1.0 L

10 C What is $[H_3O^+]$ in a solution with pH = 0? Why do chemists not usually use pH to describe $[H_3O^+]$ when the pH value would be a negative number?

Unit Issue Prep

As you investigate the contamination of Prince Edward Island's soils with sodium arsenite from pesticides, investigate the link between the pH and solubility. For example, water polluted with sodium arsenite may be treated with lime (calcium oxide), CaO. What is the purpose of this treatment?

10.3 Acid-Base Reactions

Section Preview/ Specific Expectations

In this section, you will

- **perform** calculations involving neutralization reactions
- **determine** the concentration of an acid in solution by conducting a titration
- **communicate** your understanding of the following terms: *neutralization reaction, salt, acid-base indicator, titration, equivalence point, end-point*

Is there a box of baking soda in your refrigerator at home? Baking soda is sodium hydrogen carbonate. (It is also commonly called sodium bicarbonate.) Baking soda removes the odours caused by spoiling foods. The smelly breakdown products of many foods are acids. Baking soda, a base, eliminates the odours by neutralizing the characteristic properties of the acids.

Adding a base to an acid neutralizes the acid's acidic properties. This type of reaction is called a **neutralization reaction.**

There are many different acids and bases. Being able to predict the results of reactions between them is important. Bakers, for example, depend on neutralization reactions to create light, fluffy baked goods. Gardeners and farmers depend on these reactions to modify the characteristics of the soil. Industrial chemists rely on these reactions to produce the raw materials that are used to make a wide variety of chemicals and chemical products. (See Figure 10.13.)

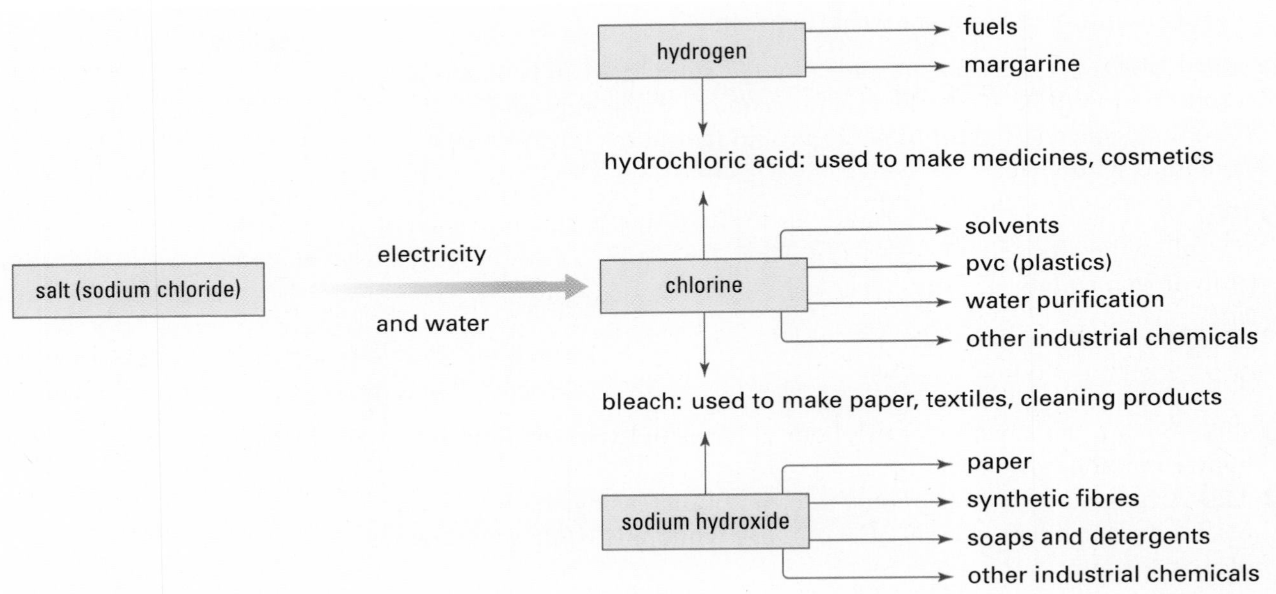

Figure 10.13 You know sodium chloride as common table salt. As you can see here, however, sodium chloride is anything but common. Sodium chloride is a product of an acid-base reaction between hydrochloric acid and sodium hydroxide.

Neutralization Reactions

The reaction between an acid and a base produces an ionic compound (a salt) and water.

$$\text{acid} + \text{base} \rightarrow \text{a salt} + \text{water}$$

A **salt** is an ionic compound that is composed of the anion from an acid and a cation from a base. For example, sodium nitrate is a salt that is found in many kitchens. It is often added to processed meat to preserve the colour and to slow the rate of spoiling by inhibiting bacterial growth. Sodium nitrate can be prepared in a laboratory by reacting nitric acid with sodium hydroxide, as shown on the next page.

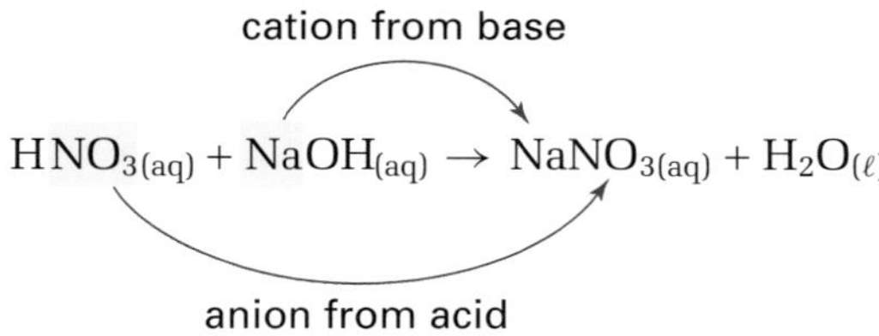

The balanced chemical equation for this reaction shows that 1 mol of nitric acid reacts with 1 mol of sodium hydroxide. If equal molar quantities of nitric acid and sodium hydroxide are used, the result is a neutral (pH 7) aqueous solution of sodium nitrate. In fact, *when any strong acid reacts with any strong base in the mole ratio from the balanced chemical equation, a neutral aqueous solution of a salt is formed.* Reactions between acids and bases of different strengths usually do not result in neutral solutions.

For most neutralization reactions, there are no visible signs that a reaction is occurring. How can you determine that a neutralization reaction is taking place? One way is to use an **acid-base indicator.** This is a substance that changes colour in acidic and basic solutions. Most acid-base indicators are weak, monoprotic acids. The undissociated weak acid is one colour. Its conjugate base is a different colour.

colour 1

$$H(indicator)_{(aq)} \rightleftharpoons H^+ + (indicator)^-_{(aq)}$$

colour 2

In an acidic solution, the indicator does not dissociate very much. It appears as colour 1. In a basic solution, the indicator dissociates much more. It appears as colour 2. Often a single drop of indicator causes a dramatic change in colour. For example, phenolphthalein is an indicator that chemists often use for reactions between a strong acid and a strong base. It is colourless between pH 0 and pH 8. It turns pink between pH 8 and pH 10. (See Figure 10.14.)

Figure 10.14 A good indicator, such as the phenolphthalein shown here, must give a vivid colour change.

If a small quantity of an acid or a base is spilled in a laboratory, you can use a neutralization reaction to minimize the hazard. To neutralize a basic solution spill, you can add solid sodium hydrogen sulfate or citric acid. For an acidic solution spill, you can use sodium hydrogen carbonate (baking soda). Note that you cannot use a strong acid or base to clean up a spill. This would result in another hazardous spill. As well, the neutralization reaction would generate a lot of heat, and thus produce a very hot solution.

CHECKPOINT

Show that the net ionic equation for the reaction between HNO_3 (a strong acid) and NaOH (a strong base) results in the formation of water.

CHEM FACT

An old remedy to relieve the prickly sting of a nettle plant is to rub the area with the leaf of a dock plant. The sting contains an acid. This acid is neutralized by a base that is present in the dock leaf. Bees and ants also have an acidic sting. You can wash the sting with soap, because soap is basic. You can also apply baking soda (a base) to the skin for more effective relief. If you are stung by a wasp, however, you should apply vinegar. The sting of a wasp contains a base.

Calculations Involving Neutralization Reactions

Suppose that a solution of an acid reacts with a solution of a base. You can determine the concentration of one solution if you know the concentration of the other. (This assumes that the volumes of both are accurately measured.) Use the concentration and volume of one solution to determine the amount (in moles) of reactant that it contains. The balanced chemical equation for the reaction describes the mole ratio in which the compounds combine. In the following Sample Problems and Practice Problems, you will see how to do these calculations.

Sample Problem

Finding Concentration

Problem

13.84 mL of hydrochloric acid, $HCl_{(aq)}$, just neutralizes 25.00 mL of a 0.1000 mol/L solution of sodium hydroxide, $NaOH_{(aq)}$. What is the concentration of the hydrochloric acid?

What Is Required?

You need to find the concentration of the hydrochloric acid.

What Is Given?

Volume of hydrochloric acid, HCl = 13.84 mL
Volume of sodium hydroxide, NaOH = 25.00 mL
Concentration of sodium hydroxide, NaOH = 0.1000 mol/L

Plan Your Strategy

Step 1 Write the balanced chemical equation for the reaction.
Step 2 Calculate the amount (in mol) of sodium hydroxide added, based on the volume and concentration of the sodium hydroxide solution.
Step 3 Determine the amount (in mol) of hydrochloric acid needed to neutralize the sodium hydroxide.
Step 4 Find $[HCl_{(aq)}]$, based on the amount and volume of hydrochloric acid solution needed.

Act on Your Strategy

Step 1 The balanced chemical equation is

$$HCl_{(aq)} + NaOH_{(aq)} \rightarrow NaCl_{(aq)} + H_2O_{(\ell)}$$

Step 2

$$\text{Amount (in mol)} = \text{Concentration (in mol/L)} \times \text{Volume (in L)}$$

$$\text{Amount NaOH (in mol) added} = 0.1000 \text{ mol/}\cancel{\text{L}} \times 0.02500 \cancel{\text{L}} = 2.500 \times 10^{-3} \text{ mol}$$

Continued ...

PROBLEM TIPS

1. Make sure that the values you use in your calculations refer to the same reactant. For example, you can use the concentration and volume of sodium hydroxide to find the amount of sodium hydroxide in this problem. You cannot use the concentration of sodium hydroxide and the volume of hydrochloric acid.
2. In the solution, the volumes are converted to litres. If all the volumes are expressed in the same unit, the conversion step is not necessary.
3. Do not drop significant digits, even zeros, during your calculations.

FROM PAGE 396

Step 3 HCl reacts with NaOH in a 1:1 ratio, so there must be 2.500×10^{-3} mol HCl.

Step 4 $\text{Concentration (in mol/L)} = \dfrac{\text{Amount (in mol)}}{\text{Volume (in L)}}$

$$[HCl_{(aq)}] = \frac{2.500 \times 10^{3}\text{ mol}}{0.01384\text{ L}}$$
$$= 0.1806\text{ mol/L}$$

Therefore, the concentration of hydrochloric acid is 0.1806 mol/L.

Check Your Solution

$[HCl_{(aq)}]$ is greater than $[NaOH_{(aq)}]$. This is reasonable because a smaller volume of hydrochloric acid was required. As well, the balanced equation shows a 1:1 mole ratio between these reactants.

Sample Problem

Finding Volume

Problem

What volume of 0.250 mol/L sulfuric acid, $H_2SO_{4(aq)}$, is needed to react completely with 37.2 mL of 0.650 mol/L potassium hydroxide, $KOH_{(aq)}$?

What Is Required?

You need to find the volume of sulfuric acid.

What Is Given?

Concentration of sulfuric acid, H_2SO_4 = 0.250 mol/L
Concentration of potassium hydroxide, KOH = 0.650 mol/L
Volume of potassium hydroxide, KOH = 37.2 mL.

Plan Your Strategy

Step 1 Write the balanced chemical equation for the reaction.

Step 2 Calculate the amount (in mol) of potassium hydroxide, based on the volume and concentration of the potassium hydroxide solution.

Step 3 Determine the amount (in mol) of sulfuric acid that is needed to neutralize the potassium hydroxide.

Step 4 Find the volume of the sulfuric acid, based on the amount and concentration of sulfuric acid needed.

Act on Your Strategy

Step 1 The balanced chemical equation is

$$H_2SO_{4(aq)} + 2KOH \rightarrow K_2SO_{4(aq)} + 2H_2O_{(\ell)}$$

Step 2 Amount (in mol) of KOH $= 0.650\text{ mol/}\cancel{\text{L}} \times 0.0372\ \cancel{\text{L}}$
$= 0.02418$ mol

Continued ...

Continued ...
FROM PAGE 397

Step 3 H_2SO_4 reacts with KOH in a 1:2 mole ratio. The amount of H_2SO_4 needed is

$$\frac{1 \text{ mol } H_2SO_4}{2 \text{ mol KOH}} = \frac{0.01209 \text{ mol } H_2SO_4}{0.2418 \text{ mol KOH}}$$

$$\frac{0.024\ 18 \text{ mol KOH} \times 1 \text{ mol } H_2SO_4}{2 \text{ mol KOH}} = 0.01209 \text{ mol } H_2SO_4$$

Step 4 Amount (in mol) H_2SO_4

$$= 0.01209 \text{ mol}$$

$$= 0.250 \text{ mol/L} \times \text{Volume } H_2SO_{4(aq)} \text{ (in L)}$$

$$\text{Volume } H_2SO_{4(aq)} = \frac{0.01209 \text{ mol}}{0.250 \text{ mol/L}}$$

$$= 0.04836 \text{ L}$$

Therefore, the volume of sulfuric acid that is needed is 48.4 mL.

Check Your Solution

The balanced chemical equation shows that half the amount of sulfuric acid will neutralize a given amount of potassium hydroxide. The concentration of sulfuric acid, however, is less than half the concentration of potassium hydroxide. Therefore, the volume of sulfuric acid should be greater than the volume of potassium hydroxide.

Practice Problems

10. 17.85 mL of nitric acid neutralizes 25.00 mL of 0.150 mol/L $NaOH_{(aq)}$. What is the concentration of the nitric acid?

11. What volume of 1.015 mol/L magnesium hydroxide is needed to neutralize 40.0 mL of 1.60 mol/L hydrochloric acid?

12. What volume of 0.150 mol/L hydrochloric acid is needed to neutralize each solution below?

(a) 25.0 mL of 0.135 mol/L sodium hydroxide

(b) 20.0 mL of 0.185 mol/L ammonia solution

(c) 80 mL of 0.0045 mol/L calcium hydroxide

13. What concentration of sodium hydroxide solution is needed for each neutralization reaction?

(a) 37.82 mL of sodium hydroxide neutralizes 15.00 mL of 0.250 mol/L hydrofluoric acid.

(b) 21.56 mL of sodium hydroxide neutralizes 20.00 mL of 0.145 mol/L sulfuric acid.

(c) 14.27 mL of sodium hydroxide neutralizes 25.00 mL of 0.105 mol/L phosphoric acid.

Acid-Base Titration

In the previous Sample Problems and Practice Problems, you were given the concentrations and volumes you needed to solve the problems. What if you did not have some of this information? Chemists often need to know the concentration of an acidic or basic solution. To acquire this information, they use an experimental procedure called a titration. In a **titration**, the concentration of one solution is determined by quantitatively observing its reaction with a solution of known concentration. The solution of known concentration is called a *standard solution*. The aim of a titration is to find the point at which the number of moles of the standard solution is stoichiometrically equal to the original number of moles of the unknown solution. This point is referred to as the **equivalence point.** At the equivalence point, all the moles of hydrogen ions that were present in the original volume of one solution have reacted with an equal number of moles of hydroxide ions from the other solution.

Precise volume measurements are needed when you perform a titration. Chemists use special glass apparatus to collect these measurements. (See Figure 10.15.) As well, an acid-base indicator is needed to monitor changes in pH during the titration.

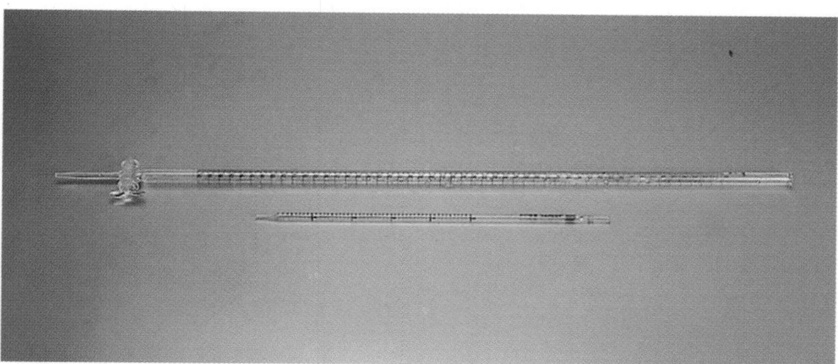

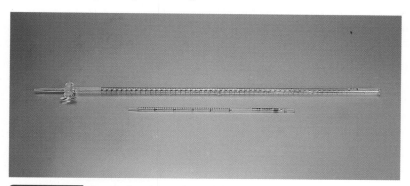

Figure 10.15 **A transfer pipette (bottom) measures a fixed volume of liquid, such as 10.00 mL, 25.00 mL, or 50.0 mL. A burette (top) measures a variable volume of liquid.**

In a titration, a pipette is used to measure a precise volume of standard solution into a flask. The flask sits under a burette that contains the solution of unknown concentration. After adding a few drops of indicator, you take an initial burette reading. Then you start adding the known solution, slowly, to the flask. The **end-point** of the titration occurs when the indicator changes colour. The indicator is chosen so that it matches its equivalance point.

Titration Step by Step

The following pages outline the steps that you need to follow to prepare for a titration. Review these steps carefully. Then observe as your teacher demonstrates them for you. At the end of this section, in Investigation 10-B, you will perform your own titration of a common substance: vinegar.

www.school.mcgrawhill.ca/resources

Sometimes a moving picture is worth a thousand words. To enhance your understanding, your teacher will demonstrate the titration procedure described in this textbook. In addition, some web sites provide downloadable or real-time titration movies to help students visualize the procedure and its techniques. Go to the web site above, then to Science Resources and to Chemistry 11 to see where to go next. Compare the different demonstrations you can find and observe, including your teacher's. Prepare your own set of "Titration Tips" to help you recall important details.

PROBEWARE

If you have access to probeware, try the Chemistry 11 lab, Titrating an Unknown, or a similar lab from a probeware company

TITRATION TIP

Never use your mouth in place of a suction bulb to draw a liquid into a pipette. The liquid could be corrosive or poisonous. As well, you will contaminate the glass stem.

TITRATION TIP

Practice removing the bulb and replacing it with your index finger (or thumb). You need to be able to perform this action quickly and smoothly.

Rinsing the Pipette

A pipette is used to measure and transfer a precise volume of liquid. You rinse a pipette with the solution whose volume you are measuring. This ensures that any drops that remain inside the pipette will form part of the measured volume.

1. Pour a sample of standard solution into a clean, dry beaker.
2. Place the pipette tip in a beaker of distilled water. Squeeze the suction bulb. Maintain your grip while placing it over the stem of the pipette. Do not insert the stem into the bulb. (If your suction bulbs have valves, your teacher will show you how to use them.)
3. Relax your grip on the bulb to draw up a small volume of distilled water.
4. Remove the bulb and discard the water by letting it drain out.
5. Rinse the pipette by drawing several millilitres of solution from the beaker into it. Rotate and rock the pipette to coat the inner surface with solution. Discard the rinse. Rinse the pipette twice in this way. It is now ready to fill with standard solution.

Filling the Pipette

6. Place the tip of the pipette below the surface of the solution.
7. Hold the suction bulb loosely on the end of the glass stem. Use the suction bulb to draw liquid up just past the etched volume mark. (See Figure 10.16.)
8. As quickly and smoothly as you can, slide the bulb off and place your index finger over the end of the glass stem.
9. Gently roll your finger slightly away from end of the stem to let solution drain slowly out.
10. When the bottom of the meniscus aligns with the etched mark, as in Figure 10.17, press your finger back over the end of the stem. This will prevent more solution from draining out.
11. Touch the tip of the pipette to the side of the beaker to remove any clinging drop. See Figure 10.18. The measured volume inside the pipette is now ready to transfer to an Erlenmeyer flask or a volumetric flask.

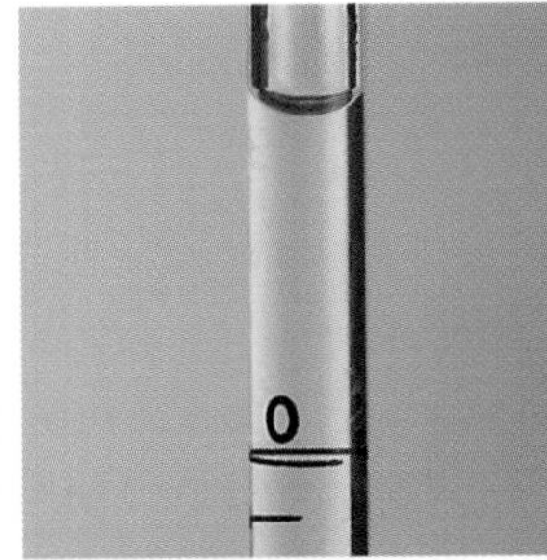

Figure 10.16 Draw a bit more liquid than you need into the pipette. It is easier to reduce this volume than it is to add more solution to the pipettes.

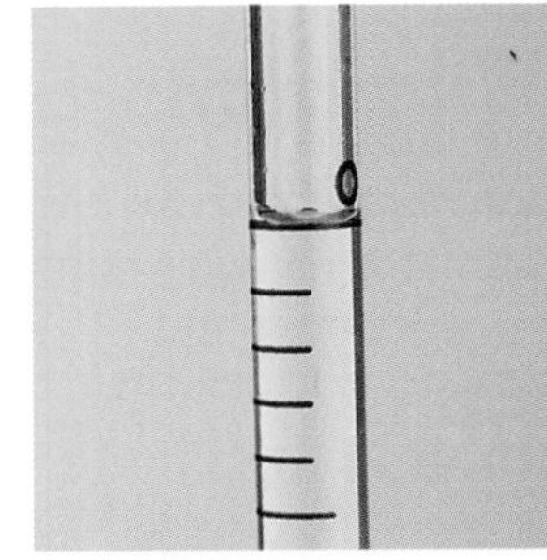

Figure 10.17 The bottom of the meniscus must align exactly with the etched mark.

Figure 10.18 You can prevent a "stubborn" drop from clinging to the pipette tip by touching the tip to the inside of the glass surface.

Transferring the Solution

12. Place the tip of the pipette against the inside glass wall of the flask. Let the solution drain slowly, by removing your finger from the stem.
13. After the solution drains, wait several seconds, then touch the tip to the inside wall of the flask to remove any drop on the end. Note: You may notice a small amount of liquid remaining in the tip. The pipette was calibrated to retain this amount. Do not try to remove it.

Adding the Indicator

14. Add two or three drops of indicator to the flask and its contents. Do not add too much indicator. Using more does not make the colour change easier to see. Also, indicators are usually weak acids. Too much can change the amount of base needed for neutralization. You are now ready to prepare the apparatus for the titration.

TITRATION TIP

If you are right-handed, the tap should be on your right as you face the burette. Use your left hand to operate the tap. Use your right hand to swirl the liquid in the Erlenmeyer flask. If you are left-handed, reverse this arrangement.

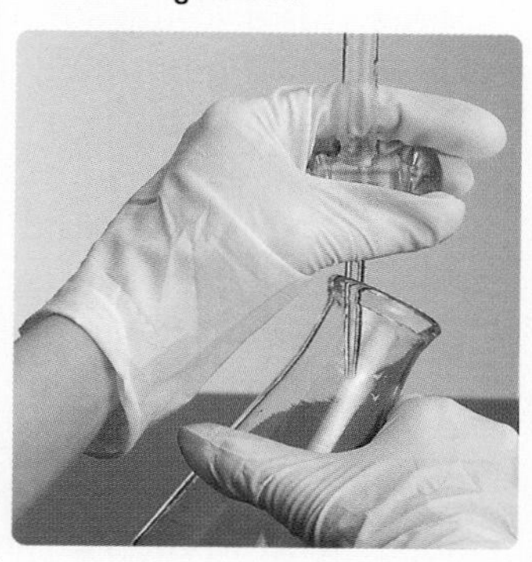

Rinsing the Burette

A burette is used to accurately measure the volume of liquid added during a titration experiment. It is a graduated glass tube with a tap at one end.

15. To rinse the burette, close the tap and add about 10 mL of distilled water from a wash bottle.
16. Tip the burette to one side and roll it gently back and forth so that the water comes in contact with all inner surfaces.
17. Hold the burette over a sink. Open the tap, and let the water drain out. While you do this, check that the tap does not leak. Make sure that it turns smoothly and easily.
18. Rinse the burette with 5 mL to 10 mL of the solution that will be measured. Remember to open the tap to rinse the lower portion of the burette. Rinse the burette twice, discarding the liquid each time.

Filling the Burette

19. Assemble a retort stand and burette clamp to hold the burette. Place a funnel in the top of the burette.
20. With the tap closed, add solution until the liquid is above the zero mark. Remove the funnel. Carefully open the tap. Drain the liquid into a beaker until the bottom of the meniscus is at or below the zero mark.
21. Touch the tip of the burette against the beaker to remove any clinging drop. Check that the portion of the burette that is below the tap is filled with liquid and contains no air bubbles.
22. Record the initial burette reading in your notebook.
23. Replace the beaker with the Erlenmeyer flask that you prepared earlier. Place a sheet of white paper under the Erlenmeyer to help you see the indicator colour change that will occur near the end-point.

TITRATION TIP

Near the end-point, when you see the indicator change colour as liquid enters the flask from the burette, slow the addition of liquid. The end-point can occur very quickly.

Reading the Burette

24. A meniscus reader is a small white card with a thick black line on it. Hold the card behind the burette, with the black line just under the meniscus, as in Figure 10.20. Record the volume added from the burette to the nearest 0.05 mL.

TITRATION TIP

Observe the level of solution in the burette so that your eye is level with the bottom of the meniscus.

Figure 10.20 A meniscus reader helps you read the volume of liquid in the burette more easily

SKILL FOCUS

Predicting

Performing and recording

Analyzing and interpreting

Investigation 10-B

The Concentration of Acetic Acid in Vinegar

Vinegar is a dilute solution of acetic acid, CH_3COOH. Only the hydrogen atom that is attached to an oxygen atom is acidic. Thus, acetic acid is monoprotic. As a consumer, you can buy vinegar with different concentrations. For example, the concentration of table vinegar is different from the concentration of the vinegar that is used for pickling foods. To maintain consistency and quality, manufacturers of vinegar need to determine the percent concentration of acetic acid in the vinegar. In this investigation, you will determine the concentration of acetic acid in a sample of vinegar.

Prediction

Which do you predict has the greater concentration of acetic acid: table vinegar or pickling vinegar? Give reasons for your prediction.

Materials

pipette
suction bulb
retort stand
burette
burette clamp
3 beakers (250 mL)
3 Erlenmeyer flasks (250 mL)
labels
meniscus reader
sheet of white paper
funnel
table vinegar
pickling vinegar
sodium hydroxide solution
distilled water
dropper bottle containing phenolphthalein

Safety Precautions

Both vinegar and sodium hydroxide solutions are corrosive. Wash any spills on skin or clothing with plenty of water. Inform your teacher immediately.

Procedure

1. Record the following information in your notebook. Your teacher will tell you the concentration of the sodium hydroxide solution.
   - concentration of $NaOH_{(aq)}$ (in mol/L)
   - type of vinegar solution
   - volume of pipette (in mL)
2. Copy the table below into your notebook, to record your observations.

Burette Readings for the Titration of Acetic Acid

Reading (mL)	Trial 1	Trial 2	Trial 3
final reading			
initial reading			
volume added			

3. Label a clean, dry beaker for each liquid: $NaOH_{(aq)}$, vinegar, and distilled water. Obtain each liquid. Record the type of vinegar you will be testing.
4. Obtain a pipette and a suction bulb. Record the volume of the pipette for trial 1. Rinse it with distilled water, and then with vinegar.
5. Pipette some vinegar into the first Erlenmeyer flask. Record this amount. Add approximately 50 mL of water. Also add two or three drops of phenolphthalein indicator.
6. Set up a retort stand, burette clamp, burette, and funnel. Rinse the burette first with distilled water. Then rinse it with sodium hydroxide solution. Make sure that there are no air bubbles in the burette. Also make sure

that the liquid fills the tube below the glass tap. Remove the funnel before beginning the titration.

7. Place a sheet of white paper under the Erlenmeyer flask. Titrate sodium hydroxide into the Erlenmeyer flask while swirling the contents. The end-point of the titration is reached when a permanent pale pink colour appears. If you are not sure whether you have reached the end-point, take the burette reading. Add one drop of sodium hydroxide, or part of a drop. Observe the colour of the solution. If you go past the end-point, the solution will become quite pink.

8. Repeat the titration twice more. Record your results for each of these trials.

9. When you have finished all three trials, dispose of the chemicals as directed by your teacher. Rinse the pipette and burette with distilled water. Leave the burette tap open.

Analysis

1. Average the two closest burette readings. Average all three readings if they agree within about ±0.2 mL.

2. Write the chemical equation for the reaction of acetic acid with sodium hydroxide.

3. Calculate the concentration of acetic acid in your vinegar sample. Use the average volume and concentration of sodium hydroxide, and the volume of vinegar.

4. Find the molar mass of acetic acid. Then calculate the mass of acid in the volume of vinegar you used.

5. The density of vinegar is 1.01 g/mL. (The density of the more concentrated vinegar solution is greater than the density of the less concentrated solution. You can ignore the difference, however.) Calculate the mass of the vinegar sample. Find the percent by mass of acetic acid in the sample.

Conclusions

6. Compare your results with the results of other students who used the same type of vinegar. Then compare the concentration of acetic acid in table vinegar with the concentration in pickling vinegar. How did your results compare with your prediction?

7. List several possible sources of error in this investigation.

Application

9. Most shampoos are basic. Why do some people rinse their hair with vinegar after washing it?

Explain how you decided.

compare? Explain.

Section Wrap-up

In this section, as in much of Unit 3, you combined liquid solutes and

(b) If samples of each acid were used in separate titration experiments with 0.50 mol/L sodium hydroxide solution, how would the volume of acid required for neutralization compare? State your reasoning.

11. Write balanced chemical equations for the following reactions:
(a) calcium oxide with hydrochloric acid
(b) magnesium with sulfuric acid
(c) sodium carbonate with nitric acid

12. Domestic bleach is typically a 5% solution of sodium hypochlorite, $NaOCl_{(aq)}$. It is made by bubbling chlorine gas through a solution of sodium hydroxide.
(a) Write a balanced chemical equation showing the reaction that takes place.
(b) In aqueous solution, the hypochlorite ion combines with $H^{+}_{(aq)}$ present in water to form hypochlorous acid. Write the equation for this reaction. Is the hypochlorite ion acting as an acid or a base?

13. In this chapter, you are told that $[H_3O^+]$ in pure water is 1.0×10^{-7} mol/L at 25°C. Thus, two out of every one billion water molecules have dissociated. Check these data by answering the following questions.
(a) What is the mass (in g) of 1.0 L of water?
(b) Calculate the amount (in mol) of water in 1.0 L. This is the concentration of water in mol/L.
(c) Divide the concentration of hydronium ions by the concentration of water. Your answer should be about 2 ppb.

Inquiry

14. 80.0 mL of 4.00 mol/L, H_2SO_4 are diluted to 400.0 mL by adding water. What is the molar concentration of the sulfuric acid after dilution?

15. In a titration experiment, 25.0 mL of an aqueous solution of sodium hydroxide was required to neutralize 50.0 mL of 0.010 mol/L hydrochloric acid. What is the molar concentration of the sodium hydroxide solution?

16. A burette delivers 20 drops of solution per 1.0 mL. What amount (mol) of $H^{+}_{(aq)}$ is present in one drop of a 0.20 mol/L HCl solution?

17. How is a 1.0 mol/L solution of hydrochloric acid different from a 1.0 mol/L solution of acetic acid? Suppose that you added a strip of magnesium metal to each acid. Would you observe any differences in the reactions? Explain your answer so that grade 9 students could understand it.

Communication

18. Commercial processors of potatoes remove the skin by using a 10-20% by mass solution of sodium hydroxide. The potatoes are soaked in the solution for a few minutes at 60-70°C, after which the peel can be sprayed off using fresh water. You work in the laboratory at a large food processor and must analyse a batch of sodium hydroxide solution. You pipette 25.00 mL of $NaOH_{(aq)}$, and find it has a mass of 25.75 g. Then you titrate the basic solution against 1.986 mol/L HCI, and find it requires 30.21 mL of acid to reach an end point.
(a) Inform your supervisor what the molar concentration of the sodium hydroxide is.
(b) The mass percent of NaOH present must be a minimum of 10% for the solution to be used. Advise your supervisor whether or not the solution can be used to process more potatoes, and explain your reasoning.

19. Ammonia is an important base, used to make fertilizers, nylon, and nitric acid. The manufacture of ammonia depends on a process discovered by Fritz Haber (1868-1934). After gathering information from print or electronic resources, write an obituary for Haber. Describe his accomplishments and the effect on society of plentiful supplies of ammonia.

Making Connections

20. Limestone, chalk, and marble are all forms of calcium carbonate. Limestone rock can be used to build roads, but it is a very important basic compound used in large quantities by chemical industries. For example, limestone is used directly to make concrete and cement. It is also used in the manufacture of glass and in agriculture. Limestone is often processed to make quicklime, CaO, and hydrated lime (calcium hydroxide), $Ca(OH)_2$.

(a) Research the uses of quicklime and hydrated lime. Investigate one of these uses further.
(b) Design a poster illustrating the use you decided to research. Your poster should be both informative and visually interesting. Include a bibliography showing the resources you found useful.

21. On several occasions during the past few years, you have studied the environmental issue of acid rain. Now that you have further developed your understanding of acids and bases in this chapter, reflect on your earlier understandings.
(a) List two facts about acid rain that you now understand in a more comprehensive way. Explain what is different between your previous and your current understanding in each case.
(b) Identify three questions that your teacher could assign as a research project on acid rain. The emphasis of the research must be on how an understanding of chemistry can contribute clarifying the questions and possible solutions involved in this issue. Develop a rubric that would be used to assess any student who is assigned this research project.

22. Research the use of hypochlorous acid in the management of swimming pools and write a report on your findings. Include a discussion on the importance of controlling pool water.

Answers to Practice Problems and Short Answers to Section Review Questions

Practice Problems: 1. HCN/CN^- and H_2O/H_3O^+ **2.** H_2O/OH^- and CH_3COO^-/CH_3COOH **3.** as an acid: $HS^-_{(aq)} + H_2O_{(\ell)} \leftrightarrow S^{2-}_{(aq)} + H_3O^+_{(aq)}$; as a base: $HS^-_{(aq)} + H_2O_{(\ell)} \leftrightarrow H_2S_{(aq)} + OH^-_{(aq)}$ **4.(a)** HBr; bromate; Br^- **(b)** hydrogen sulfide, HS^-; sulfide, S_2^- **5.(a)** nitric acid, HNO_3; nitrate, NO_3^- **(b)** nitrous acid, HNO_2; nitrite, NO_2^- **(c)** hyponitrous acid, HNO; hyponitrite, NO^- **(d)** phosphoric acid, H_3PO_4; dihydrogen phosphate, $H_2PO_4^-$, hydrogen phosphate, HPO_4^{2-}, phosphate, PO_4^{3-} **(e)** phosphorous acid, H_3PO_3; dihydrogen phosphite, $H_2PO_3^-$, hydrogen phosphite, HPO_3^{2-}, phosphite, PO_3^{3-} **(f)** periodic acid, HIO_4; periodate, IO_4^- **6.(a)** 2.57 **(b)** 7.138 **(c)** 4.01 **(d)** 11.082 **7.** pH = 2.30; acidic **8.** 3.54; acidic **9.(a)** 2.20 **(b)** 9.181; basic **10.** 0.210 mol/L **11.** 31.5 mL **12.(a)** 22.5 mL **(b)** 24.7 mL **(c)** 4.8 mL **13.(a)** 0.0992 mol/L **(b)** 0.269 mol/L **(c)** 0.552 mol/L

Section Review: 10.1: 1.(a) B and D **(b)** A (C, although unintended, would also be correct) **(c)** A, B, D **(d)** no **(e)** litmus test, for example **5.(a)** H_2O **(b)** HCO_3^- **6.(a)** NO_3^- **(b)** SO_4^{2-} **7.** c and d **8.** a and b **9.(a)** $HF_{(aq)} + H_2O_{(l)} \rightarrow H_3O^+_{(aq)} + F^-_{(aq)}$ **(b)** HF/F^-; H_2O/H_3O^+ **10.(a)** $H_2PO_4^-/HPO_4^{2-}$ and CO_3^{2-}/HCO_3^- **(b)** $HCOOH/HCOO^-$ and CN^-/HCN **(c)** $H_2PO_4^-/HPO_4^{2-}$ and OH^-/H_2O **10.2: 3.** weak **4.** potassium permanganate; permanganic acid, $HMnO_4$ **5.(a)** sulfuric acid, H_2SO_4 **(b)** hydrofluoric acid, HF **(c)** hydrosulfuric acid, H_2S **(d)** bromous acid, HBr_2 **7.** egg white, camembert cheese, beets, yogurt, sauerkraut **8.(a)** 7.4 **(b)** 1.4 **9.** 2.90 **10.** 1.0 mol/L; concentrations greater than this give negative pH values, but this gives no advantage over the actual concentration of H_3O^+ **10.3: 1.** acid + base = salt + water **4.** 0.304 mol/L **5.** 0.107 mol/L

Island at Risk: A Simulation

Background

Would you rather have fish—or chips? People on Prince Edward Island are discovering that their sources of both of these food items are increasingly at risk.

The fertile red soil of Prince Edward Island is famous for potato growing. Agricultural corporations have been increasing their potato plantings on this island province. In the process, they have been eliminating other crops, which has led to a monoculture. ("Monoculture" refers to the growing of a single crop in an area.) Potato pests are posing a huge threat to this profitable cash crop, however. Part of the problem results from the fact that the potato fields are very close together. To combat the pests—potato bugs, in particular—farmers are applying more and more pesticides, fertilizers, and herbicides to the potato fields.

What are the environmental effects of this heavy use of pesticides (insecticides and herbicides)? After heavy rains on the island, large numbers of dead fish have been found in the ocean, and washed up on the shore near the potato farms. The suspected cause is the leaching of sodium arsenite and other pesticides used in the potato fields. Sodium arsenate is a chemical used to kill both weeds and insects. If the poisoning of the ocean waters continues, Prince Edward Island's unique seafood industry will come to an end.

Suggestions for dealing with this problem have included:

- examples of household chemical products and their uses
- creating a mandatory untilled area spanning 20 m next to waterways;
- growing genetically engineered potatoes that are more blight-resistant. These genetically modified potatoes would contain genes from bacteria that have insecticidal properties;
- employing alternative fertilization and pesticide methods to reduce the amount of chemicals used on the potato crop; and
- rotating crops (that is, alternating types of crops from year to year) so that the spread of potato bugs and the overuse of soil would be less of a problem.

Plan and Present

1. Look up the properties of sodium arsenate in either the CRC Handbook or the Merck Manual. As well, search the Internet for an MSDS sheet about sodium arsenate.
2. Using your knowledge of solutions and dissolving, explain how the chemicals from the potato fields end up in the water with the fish. Make a diagram, a poster, or a PowerPoint™ presentation that clearly illustrates the processes involved in this type of aquatic pollution.
3. Provide a minimum of two advantages and two disadvantages of each of the four possible solutions to the problem.
4. The provincial government in Charlottetown has decided that action must be taken, in some form or another. The following people and organizations have been invited to participate on a board to develop an action plan:

- the P. E. I. Potato Board
- the scientists who helped develop the genetically engineered, blight-resistant potato variety
- the corporations who want to buy the potatoes and process them
- people who fish for their living
- potato farmers

(a) Your class will be divided into groups, and each group will research the point of view of one of these interest groups. Consider questions such as these:
- To what extent do landowners have the right to determine how they use their land?
- What recourse should the fishers on P.E.I. have available to them to respond to the threat posed to their livelihood?
- What possible consumers' rights need to be assessed in analyzing this issue?

(b) Before developing an action plan, each group should decide on and state clearly their motive and their mandate. In other words, what is the reason for their existence as a group and to whom is the group accountable? For example:
- Are the scientists working for biotechnology companies or are they doing pure research?
- What is each group's present policy regarding issues such as environmental protection and sustainability?
- What points might be included in each group's short-term and long-term action plan?

(c) Make a one-page summary of your points to hand out to other members at the meeting. A large diagram, poster, overhead transparency, or PowerPoint™ slide will help your group illustrate how your action plan:
- will affect the environment
- acknowledges the scientific realities of the leaching situation
- satisfies the needs of other interest groups

(d) Hold a meeting of the action plan board, to present all points of view, and to try to reach a consensus on an action plan.

Evaluate the Results

1 **(a)** Take notes on every point of view presented during the meeting.

(b) Decide what you think the best solution is, for everyone involved. Why do you consider it to be the best solution?

Assessment

After you analyze this issue,
- Assess the quality and practicality of your solution.
- Assess your solution based on how clearly the chemistry concepts are expressed and communicated.

2 From what you learned in Unit 3 about solutions and water treatment, do you think there could be other chemical-based solutions to this problem in the future? Give a chemical explanation of how the leachates might be contained, if such processing were possible.

3 Try to suggest some safe, alternative methods of reducing weeds, insects, and harmful microorganisms. For example, could boiling water be used as an effective herbicide? Would this method of reducing weeds cause more problems than it would solve? Why or why not?

Web LINK

www.school.mcgrawhill.ca/resources/
Go to McGraw-Hill Ryerson's Chemistry 11 web site at the above address, and then follow the links to access further information.

Knowledge/Understanding

True/False

In your notebook, indicate whether each statement is true or false. Correct each false statement.

1. The solubility of a gas increases with increased temperature.
2. Molar concentration refers to the amount, in moles, of solute dissolved in one kilogram of solvent.
3. The maximum amount of a solute that will dissolve in a solvent at a certain temperature is called its solubility.
4. The term insoluble has a precise meaning.
5. The component of a solution present in the smaller amount is called the solvent.
6. The molar concentration of a solution containing 4 mol of solute dissolved in 2 L of solvent is 2 mol/L.
7. The rate at which a solid solute dissolves in water can usually be increased by increasing the temperature of the solution.
8. Each ten-fold increase in the concentration of hydronium ions in a solution increases the pH of the solution by one unit.
9. Hydrobromic acid is a strong acid.

Matching

10. Match each description in column B with the correct term in column A.

Column A	Column B
(a) Unsaturated solution	A solution with pH = 10
(b) Saturated solution	NH_3
(c) Dilute solution	SO_3^{2-}
(d) Concentrated solution	H_2SO_3
(e) Arrhenius acid	A solution that contains a relatively large amount of solute
(f) Brønsted–Lowry base	A solution with pH = 8
(g) An example of a weak acid	When solute is added to an aqueous solution, the solute does not dissolve
(h) The conjugate base of HSO_3^-	When solute is added to an aqueous solution, the solute dissolves
(i) A solution more acidic than one with pH = 9	A substance that produces H+ when dissolved in water
(j) A solution more basic than one with pH = 9	A solution that contains a relatively small amount of solute

Multiple Choice

In your notebook, write the letter of the best answer for each of the following questions.

11. Which of the following would best indicate that a sample of water is pure?
    (a) Measure its boiling point.
    (b) Measure its pH.
    (c) Add it to a sample of pure water and see if it is miscible (dissolves infinitely).
    (d) Pass an electric current through it to see if it decomposes into hydrogen gas and oxygen gas.
    (e) See if sodium chloride dissolves in it.
12. If 1.00 g of solid sodium chloride is dissolved in enough water to make 350 mL of solution, what is the molar concentration of the solution?
    (a) 5.98 mol/L
    (b) 1.67×10^{-1} mol/L
    (c) 4.89×10^{-2} mol/L
    (d) 5.98×10^{-3} mol/L
    (e) 4.88×10^{-5} mol/L
13. What volume of 5.00×10^{-2} mol/L $Ca(NO_3)_2$ solution will contain 2.50×10^{-2} mol of nitrate ions?
    (a) 200 mL
    (b) 250 mL
    (c) 500 mL
    (d) 750 mL
    (e) 1.00 L
14. If 40.0 mL of 6.00 mol/L sulfuric acid is diluted to 120 mL by the addition of water, what is the molar concentration of the sulfuric acid after dilution?
    (a) 5.00×10^{-2} mol/L
    (b) 7.50×10^{-2} mol/L
    (c) 1.00 mol/L

(d) 2.0 mol/L
(e) 4.0 mol/L

15. When solutions of sodium chloride, NaCl, and silver nitrate, $AgNO_3$, are mixed, what is the net ionic equation for the reaction that results?
(a) $Na^{+}{}_{(aq)} + NO^{-}{}_{3(aq)} \rightarrow NaNO_{3(aq)}$
(b) $Ag^{+}{}_{(aq)} + Cl^{-}{}_{(aq)} \rightarrow AgCl_{(s)}$
(c) $Na^{+}{}_{(aq)} + NO_3{}^{-}(aq) + Ag^{+}{}_{(aq)} + Cl^{-}{}_{(aq)} \rightarrow AgCl_{(s)} + NaNO_{3(aq)}$
(d) $Na^{+}{}_{(aq)} + NO_3 - {}_{(aq)} + Ag +_{(aq)} + Cl^{-}{}_{(aq)} \rightarrow AgCl_{(s)} + Na^{+}{}_{(aq)} + NO_{3(aq)}^{-}$
(e) $Na^{+}{}_{(aq)} + NO_3{}^{-}{}_{(aq)} + Ag^{+}{}_{(aq)} + Cl^{-}{}_{(aq)} \rightarrow NaNO_{3(aq)} + Ag^{+}{}_{(aq)} + Na^{+}{}_{(aq)}$

16. The acidity in a sample of soil could be neutralized by adding:
(a) sodium chloride
(b) ammonium nitrate
(c) potassium sulfate
(d) calcium oxide
(e) magnesium phosphate

Short Answers

In your notebook, write a sentence or a short paragraph to answer each of the following questions:

17. Is a saturated solution always a concentrated solution? Give an example to explain your answer.

18. How can a homogeneous mixture be distinguished from a heterogeneous mixture? Give one example of each.

19. List three different ways in which the concentration of a solution could be described.

20. What would you observe if a saturated solution of sodium carbonate (commonly called washing soda) at room temperature was cooled to 5°C?

21. Explain why calcium hydroxide (solubility 0.165 g per 100 g water at 20°C) is much more soluble than magnesium hydroxide (solubility 0.0009 g per 100 g water at 20°C).

22. Iron concentrations of 0.2 to 0.3 parts per million in water can cause fabric staining when washing clothes. A typical wash uses 12 L of water. What is the maximum mass of iron that can be present so that the clothes will not be stained?

23. High levels of phosphorus are not toxic, but can cause digestive problems. The allowable drinking water concentration is 0.05 ppm. What is the maximum mass of phosphorus that could be present in a 250 mL glass of tap water?

24. Is a 1% solution of table salt, $NaCl_{(aq)}$, more concentrated, less concentrated, or at the same concentration as a 1% solution of sugar, $C_{12}H_{22}O_{11(aq)}$? Explain.

25. Bones and teeth consist mostly of a compound called hydroxyapatite, $Ca_{10}(PO_4)_6(OH)_2$. This compound contains $PO_4{}^{3-}$ and OH^- ions.
(a) Do you expect hydroxyapatite will be an acid or a base?
(b) Foods that contain sucrose form lactic acid in the mouth and the pH drops. As a result, eating candy promotes a reaction between hydroxyapatite and $H^{+}{}_{(aq)}$. Balance the skeleton reaction:
$Ca_{10}(PO_4)_6(OH)_2 + H^{+}{}_{(aq)} \rightarrow CaHPO_{4(s)} + H_2O_{(\ell)} + Ca^{2+}{}_{(aq)}$
(c) At lower pH values, the $CaHPO_{4(s)}$ also reacts with $H^{+}{}_{(aq)}$:
$CaHPO_{4(s)} + H^{+}{}_{(aq)} \rightarrow Ca^{2+}_{(aq)} + H_2PO_4{}^{-}{}_{(aq)}$
Dentists and toothpaste manufacturers warn that eating candy promotes tooth decay. What chemical evidence have you seen to support this advice?

26. (a) Why does water from different regions vary in its hardness?
(b) Why is filtration not an effective method to remove the hardness from water?
(c) Why should hard water be treated before it is heated in a hot water boiler?

27. Vinegar is added to a kettle with a build-up of scale due to hard water. What would you expect to observe? Explain.

28. Chloroform and diethyl ether were among the first substances used as anaesthetics. Both are non-polar substances.
(a) Would you expect either or both of these substances to be soluble in water? Explain.
(b) Write a sentence or two to describe how you think these substances are able to get from the lungs to the brain.

Inquiry

29. A Chemist has a large beaker containing ice-cold water, and another containing boiling water. The laboratory is well-equipped with other apparatus.
    - (a) Explain how the chemist could maximize the solubility of the following solutes in water (following appropriate safety precautions):
      - (i) magnesium chloride, $MgCl_2$, used to fire-proof wood
      - (ii) benzene, a non-polar liquid used by the industry and found in gasoline
      - (iii) carbon monoxide, CO, a poisonous gas formed by incomplete combustion of hydrocarbons
    - (b) Explain how you could minimize the solubility of the same solutes in water.
30. Design an experiment to collect data on the pH of a stream, over a period of one year. Why might the pH vary at different times of the year?
31. The table below shows the colours of various indicators at different pH values.
    - (a) If a vinegar solution is at pH 5, what colour would you expect the following indicators to show if placed into separate samples of the vinegar?
      - (i) Thymol blue
      - (ii) Bromophenol blue
      - (iii) Phenolphthalein
    - (b) An aqueous solution of sodium acetate used in photographic development makes phenol red indicator red, and phenolphthalein pink. What is the pH of this sodium acetate solution?

Communicating

32. Lead is highly toxic when absorbed into the body, especially for young children. A level of 10 micrograms of lead per decilitre of blood is cause for concern. Do research, then write a report describing the health effects of lead. Include information on the sources of this heavy metal and on how lead might be absorbed by a child.
33. Sulfuric acid is the chemical produced in the largest quantity in the world. Research some of the uses of sulfuric acid. Design a poster for students in a younger grade, illustrating one way in which this acid is used. Your poster should be informative and visually interesting.
34. In the past, scurvy was a disease that killed many sailors. James Lind discovered that eating citrus fruits prevents scurvy. Citrus fruits contain ascorbic acid (vitamin C). Research the life of James Lind, and write a brief biography.

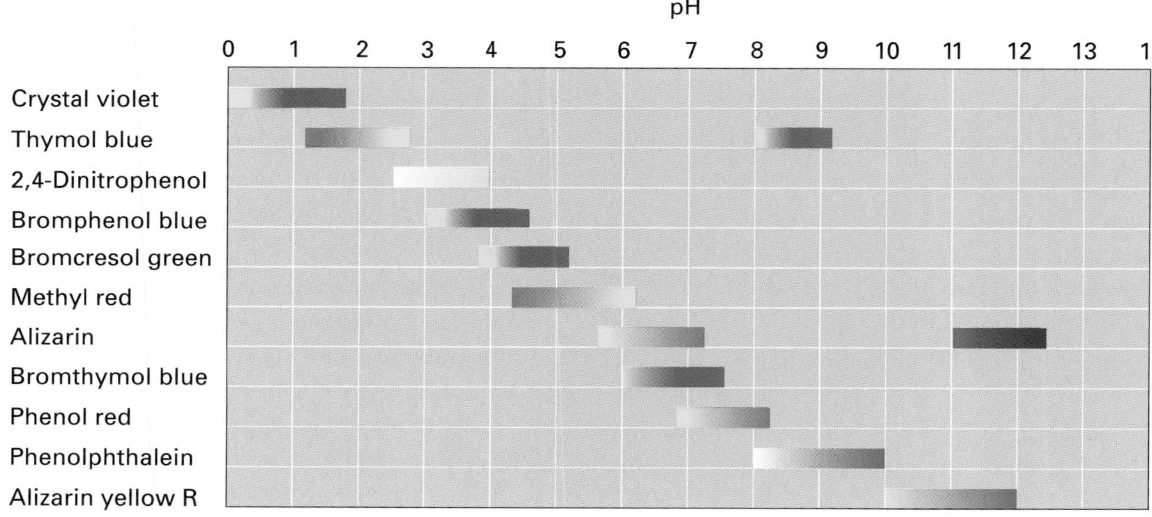

Making Connections

35. The disposal of nuclear waste presents many difficulties. Some proposals suggest burying the waste in glass or ceramic containers deep in the earth.

(a) Why would these containers be superior to containers made of metal or concrete?

(b) What are some of the concerns related to burying radioactive waste?

(c) Make a suggestion on a different way to dispose of radioactive waste. Include a list of the pros and cons of your suggestion.

36. The following is a short list of some weak bases and their uses. Which ones are Brønsted–Lowry bases, but not Arrhenius bases?

(a) ammonia, NH_3, used in the manufacture of fertilizers, plastics, and explosives

(b) zinc oxide, ZnO, a white pigment used in paints and cosmetics

(c) copper(II) hydroxide, $Cu(OH)_2$, used to kill fungi and bacteria

(d) hydrazine, N_2H_4, a colourless liquid that can be used as a rocket fuel

37. In 1963, a treaty was signed by the US, UK, and USSR to ban the atmospheric testing of atomic weapons. Previous testing of atomic weapons had added radioactive isotopes of strontium (Sr-90) and cesium (Cs-137) to the atmosphere. Eventually, these pollutants fell to the ground and may have entered the food chain.

(a) Which would you expect to form more soluble compounds, strontium or cesium? Explain your answer.

(b) State two important factors that might help you to determine the health risks of these isotopes.

COURSE CHALLENGE

Think about these questions as you plan for your Chemistry Course Challenge.

- How might you use your knowledge of solutions to analyze the purity of a sample of water?
- How could you precipitate out contaminants to purify a sample of water?
- What dissolved substances might end up in water as it passes through rock? How might you identify these substances? How could you extract them from the contaminated water?

UNIT 4

Gases and Atmospheric Chemistr

UNIT 4 CONTENTS

UNIT 4 OVERALL EXPECTATIONS

- What laws govern the behaviour of gases?
- What are the relationships between the temperature, pressure, volume, and number of moles of a gas?
- What modern technologies depend on gas chemistry?
- What environmental phenomena and issues are related to gases?

Unit Issue Prep

You probably ride in a car or a bus several times a week, if not every day. What gases do these vehicles produce? How can we prevent environmental harm from vehicle emissions? Think through this issue as you study the material in Unit 4. You will have a chance to debate your point of view with your classmates at the end of the unit.

Did you realize that air turbulence is caused by the behaviour of gases? Turbulence can be caused by changes in air pressure in the atmosphere. Airplanes are designed, however, to adjust to the behaviour of atmospheric gases. The combustion in the engines, the shape of the airplane, and the air mixture that passengers breathe are all designed to take gas behaviour into account.

Gas behaviour and gas reactions are part of everyday life. For example, photosynthesis is a reaction that uses carbon dioxide gas, along with other reactants, to produce oxygen gas. High in the atmosphere, ozone gas interacts with dangerous ultra-violet light to protect us from the Sun's harsh rays.

Scientists study gases to understand how they react. This understanding can help prevent further damage to the environment. It also leads to advances in gas-related technology. Carbonated soft drinks, air bags, medical anaesthetics, scuba equipment, and rocket engines are all based on gas chemistry.

In this unit, you will learn about the relationships between the pressure, temperature, volume, and number of moles of a gas. You will see how all these relationships are combined in the ideal gas law: an equation that predicts the behaviour of a gas in almost any situation.

CHAPTER 11

The Behaviour of Gases

How many gases are shown in the main photograph on the opposite page? The only gas you can actually *see* is the water vapour in the clouds. In fact, however, many gases are present. Our atmosphere is made up of a mixture of different gases. The most important of these gases is oxygen, which we need to breathe. But did you know that most of the air you breathe is composed of nitrogen gas? This fact is well known to deep-sea divers, who encounter problems with nitrogen gas when diving far below the surface of the ocean.

Gases are important in many different areas, from medical technology to the food industry. In this chapter, you will learn how particles behave in the gaseous, liquid, and solid states. You will also learn about laws that predict the behaviour of gases under different conditions of pressure, temperature, and volume. As well, you will discover some of the important ways in which gases are used in everyday life.

Chapter Preview

Concepts and Skills You Will Need

Before you begin this chapter, review the following concepts and skills:

- significant digits (Chapter 1, Section 1.2)
- unit analysis problem solving (Appendix D)
- the behaviour of atoms and molecules (Chapter 3, Sections 3.1, 3.2, and 3.3)

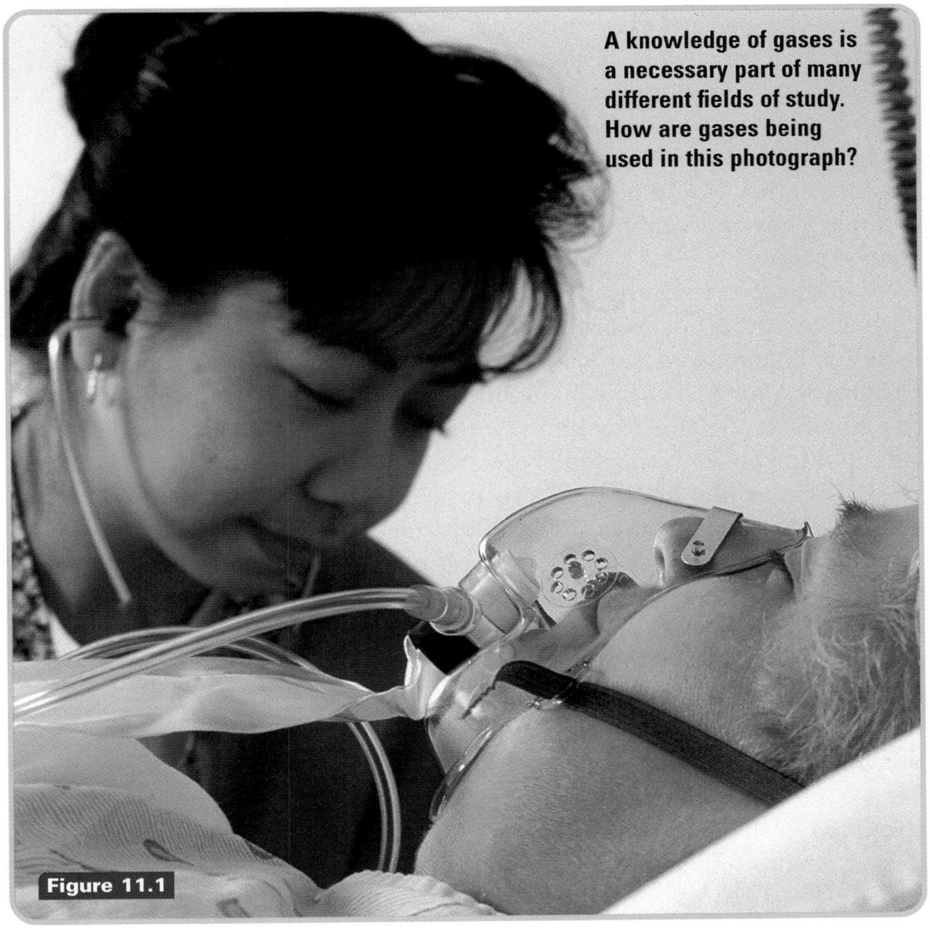

A knowledge of gases is a necessary part of many different fields of study. How are gases being used in this photograph?

Figure 11.1

11.1 States of Matter and the Kinetic Molecular Theory

Section Preview/ Specific Expectations

In this section, you will

- **explain** states of matter in terms of intermolecular forces and the motion of particles
- **perform** an ExpressLab to determine if gases occupy space
- **describe** a gas using the kinetic molecular theory
- **communicate** your understanding of the following terms and concepts: *condensation, kinetic molecular theory of gases, ideal gas*

Most of the universe is composed of plasma, a state of matter that exists at incredibly high temperatures (>5000°C). Under normal conditions, matter on Earth can only exist in the other three physical states, namely, the solid, liquid, or gaseous states. As you learned in an earlier course, the particle theory describes matter in all states as being composed of tiny invisible particles, which can be atoms, ions, or molecules. In this section, you will learn how these particles behave in each state. You will also learn about the forces that cause their behaviour.

Solids and Liquids

In previous courses, you learned about the properties of the different states of matter. You may recall that both solids and liquids are incompressible. That is, the particles cannot squeeze closer together, or compress. The incompressible nature of solids and liquids is not due to the fact that particles are touching. On the contrary, the particle theory states that there is empty space between all particles of matter. The incompressibility of solids and liquids arises instead from the fact that these particles cannot move independently of each other. That is, the movement of one particle affects the movement of other particles, or is restricted by them.

This is especially true for solids. The particles of a solid are held together in a framework, called a crystal lattice. In a crystal lattice, the positions of solid particles are relatively fixed. This explains why solids have definite shapes: the particles are unable to slip past each other and thus change the shape of the solid.

Like solids, liquid particles cannot move independently of one another. They can slip past each other enough, however, to flow and change shape.

Another property of states of matter is their motion. According to the particle theory, all particles that make up matter are in constant motion. In solids, the range of motion of the particles is the most restricted. Each particle of a solid is only able to vibrate around a fixed point in the lattice (see Figure 11.2). This is called *vibrational motion*. Since solid particles are fixed in space, the degree of disorder is very low.

Figure 11.2 The arrangement of particles in a solid. Particles vibrate in a fixed position relative to one another. They are unable to move past each other.

Particles in the liquid state can move more freely than particles in the solid state, although not entirely independently. Liquid particles move with *rotational motion* as well as vibrational motion. This means that the particles can rotate and change position. It explains why liquids are able to flow and change shape, but keep the same volume. Since liquid particles move around more than solid particles, the liquid state has a higher degree of disorder, as shown in Figure 11.3.

To summarise the discussion, particles in solids and liquids are incompressible and thus have definite volumes. The particles in each state cannot move independently of each other. Therefore they are relatively restricted in their motion. How are the properties of gases different from those of solids and liquids?

The Gas State

Unlike solids and liquids, particles in the gaseous state are able to move independently of one another. Gas particles are able to move from one point in space to another. This is called *translational motion*. Thus, gas particles move with all three types of motion: vibrational, rotational, and translational. Gas particles move through space in random fashion. However, they do travel in straight lines until their course is altered by collisions with other particles. Because gas particles move freely, there is a high degree of disorder in a gaseous state.

Gas particles move much faster than liquid particles. Liquids always flow to the lowest point because they are still greatly influenced by gravity. Because gas particles move so quickly, gravity does not affect them as much. Gases flow in all directions, including upward against gravity, until all of the available empty space is occupied. This is why gases expand to fill a container. (See Figure 11.4.)

Gases can be compressed, unlike both solids and liquids. What is different about their particle arrangement that allows for this? The space between gas particles is much larger than the space between liquid or solid particles. Even if gas particles are moved closer together through compression, the distance between each particle is still very large. The particles remain in the gaseous state. When gas molecules are compressed further, eventually the forces between molecules become strong enough to hold the gas molecules together. At this point, the gas changes to the liquid state. This is known as **condensation**.

Figure 11.3 **Particles in liquids are not held in a fixed position relative to other particles. They can slide over and past one another.**

Forces Between Particles

You have examined some properties of solids, liquids, and gases. You have seen how the motion of the particles affects these properties. Now you will examine how the particles affect each other.

The particle theory states that there are attractive forces between particles. The weaker the attractive force is between particles, the freer the particles are to move. Therefore, attractive forces between particles are at their strongest in the solid state. Attractive forces are at their weakest in the gaseous state.

The strength of attractive forces between particles in any physical state depends on two major factors: type of force and temperature. The effect of temperature, or kinetic energy, on the state of a substance will be covered in greater detail later on in this section.

Attractions Between Charged Particles

What types of attractive forces exist between particles? In Chapter 3, you learned that oppositely charged particles attract each other due to *electrostatic attraction*. Ionic bonding is one example of electrostatic attraction. A positive ion (an atom or molecule that has lost electrons) is attracted to a negative ion (an atom or molecule that has gained electrons). Ions form very strong *ionic bonds*. Since these attractive forces are so strong, ionic compounds usually exist in nature as solids. For example, table salt (sodium chloride, NaCl) is a solid crystalline substance. It has a high melting point and a high boiling point.

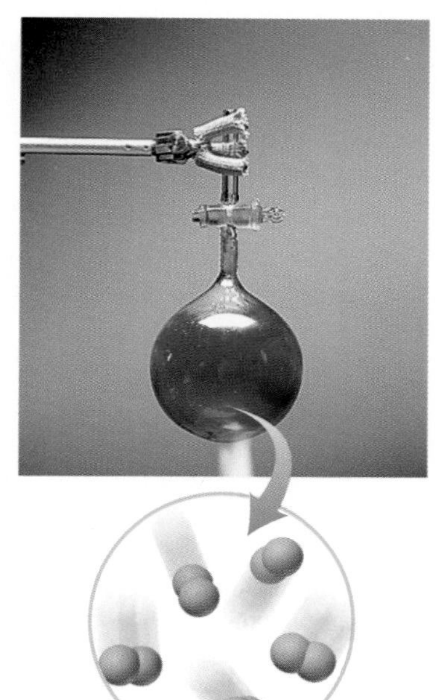

Figure 11.4 **Gas particles move freely in all directions, bouncing off each other as well as off the walls of their container.**

CHECKPOINT

What is a dipole? Go back to Chapter 3 or Chapter 8 to refresh your memory.

Attractions Between Polar Molecules

Not all particles are charged, but attractions can still form between them. You learned about *intermolecular forces* in Chapter 3. Intermolecular forces are forces that exist between neutral molecules, or between molecules and ions.

You know that some molecules are polar due to their asymmetrical shapes. Sulfur dioxide (SO_2) is one example of a polar molecule. These molecules have a permanent dipole effect. This means that one end of the molecule is more positive, and the other end is more negative.

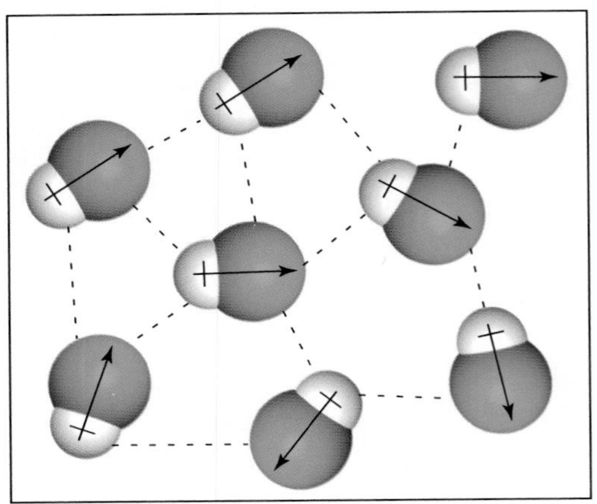

Figure 11.5 **Dipole–dipole interactions between polar HCl molecules**

Polar molecules attract ions and other polar molecules. The partially positive end of one molecule is attracted to the partially negative end of another molecule. This pattern continues throughout the substance. These *dipole–dipole* forces of attraction are not as strong as ionic bonds. Thus substances made up of polar molecules can exist as liquids and gases. For example, ethanol (CH_3CH_2OH) is polar, and exists as a liquid. You know this liquid as rubbing alcohol. Hydrogen chloride (HCl) is also polar. It is a gas under normal conditions as shown in Figure 11.5.

Attractions Between Non-Polar Molecules

What about substances made of non-polar molecules? You learned in Chapter 3 that *weak dispersion forces* form between non-polar molecules. As temporary dipoles form, they cause molecules to move closer together. However, these attractions are temporary and weak. Thus, most small non-polar molecules do not hold together long enough to maintain their solid or liquid forms. As a result, most small non-polar molecules exist as gases at room temperature. For example, carbon dioxide (CO_2) is a gas at normal temperatures.

CHECKPOINT

Hydrogen bonding is a very strong type of dipole-dipole interaction. Go back to Chapter 8 to review how this type of intermolecular force works.

The Relationship Between Size and State

Dispersion forces are also the primary forces of attraction between large non-polar molecules. However, as these molecules increase in size, their melting and boiling points rise. For example, methane (CH_4) is a small non-polar molecule. It has a very low boiling point and exists as a gas at room temperature. Pentane (C_5H_{12}) is a larger non-polar molecule. It has a higher boiling point, so it exists as a liquid at room temperature. Pentane has more sites along its length than methane does where temporary dipoles can form. The dispersion forces add up, so that it takes more energy overall to separate the molecules. This leads to a higher boiling point.

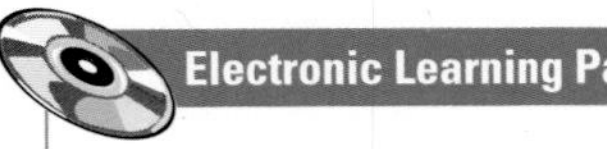

If you are having difficulty visualizing molecules in the different states of matter, go to the Chemistry 11 Electronic Learning Partner.

To summarize, the state of a substance depends on the forces between the particles of that substance. If the forces are very strong, that substance is likely to exist as a solid. If the forces are weaker, that substance will exist as a liquid, or as a gas. The state of a non-polar substance also depends on the size of the molecule. Smaller non-polar molecules are more likely to be gases. Larger non-polar molecules will probably exist as liquids or even solids. Table 11.1 shows the forces discussed in this section ranked in order of strength.

Table 11.1 Attractive Forces

strong forces ← **weak forces**

Force	Ionic	Polar (dipole–dipole)	Dispersion
Type of force	between ions (intramolecular)	between molecules (intermolecular)	between molecules (intermolecular)
State	usually solid	liquid or gas (can also be solid)	liquid or gas
Example	$NaCl_{(s)}$	$CH_3CH_2OH_{(\ell)}$, $HCl_{(g)}$	$C_5H_{12(\ell)}$, $CH_{4(g)}$, $CO_{2(g)}$

The Effect of Kinetic Energy on the State of a Substance

There is one more factor that affects the state of a substance: temperature, which is related to kinetic energy. A hotter substance with high kinetic energy is more likely to overcome attractive forces between molecules, and exist as a gas. A cooler substance with low kinetic energy is more likely to be a solid or a liquid. This explains why heating a substance causes a change in state. When a solid is heated, it gains kinetic energy. Eventually it will melt, and become a liquid. When you add kinetic energy by heating a liquid, it will boil and become a gas. Earlier in this section, you learned that gases move much more quickly than liquids or solids. This is because gases have high kinetic energy.

CHECKPOINT

What kind of molecular forces would you expect for KI, SCl_6, and SiO_2? Use diagrams to explain your answer. Compare their melting and boiling points.

Kinetic Molecular Theory of Gases

The particle theory of matter does not discuss the kinetic energy of particles. Kinetic energy is important, however, when describing the unique properties of gases.

The **kinetic molecular theory of gases** makes the following assumptions:

- The volume of an individual gas molecule is negligible compared to the volume of the container holding the gas. This means that individual gas molecules, with virtually no volume of their own, are extremely far apart and most of the container is "empty" space.
- There are neither attractive nor repulsive forces between gas molecules,
- Gas molecules have high translational energy. They move randomly in all directions, in straight lines. (See Figure 11.6, on page 423.)
- When gas molecules collide with each other or with a container wall, the collisions are perfectly elastic. This means that when gas molecules collide, somewhat like billiard balls, there is no loss of kinetic energy.
- The average kinetic energy of gas molecules is directly related to the temperature. The greater the temperature, the greater the average motion of the molecules and the greater their average kinetic energy.

The kinetic molecular theory describes a hypothetical gas called an **ideal gas**. In an ideal gas, the gas particles take up hardly any space. Also, the particles of an ideal gas do not attract each other.

ExpressLab  Do Gases Take Up Space?

Not everything can be seen with the unaided eye. Looking at a solid or a liquid, it is easy to see that they have mass and volume. Most gases are colourless. How can we "see" the volume of a gas?

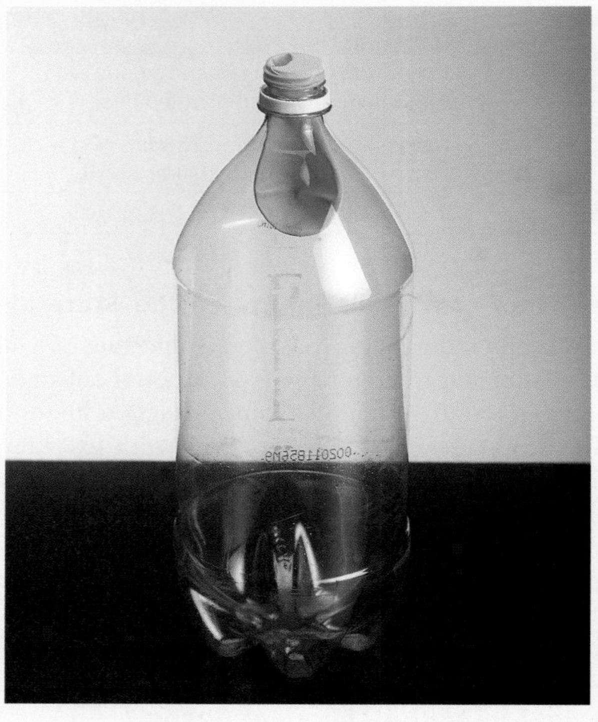

Materials

1 L or 2 L clean plastic soft drink or juice bottle
round balloon
pointed scissors

Safety Precautions

Be careful with the sharp point of the scissors when piercing the plastic bottle.

Procedure

1. Insert the balloon into the bottle, holding the open end. Stretch the open end of the balloon over the lip of the bottle.
2. Step 3 will ask you to inflate the balloon as large as you can. Before you do Step 3, predict how much you will be able to inflate the balloon. Record your prediction in your notebook.
3. Inflate the balloon inside the bottle. How large did it get? Record your observations in your notebook.
4. Using the sharp end of a pair of scissors, puncture a hole in the middle of the bottom of the plastic bottle. Inflate the balloon again. Record your observations in your notebook.

Analysis

1. Was your prediction in Step 2 verified in Step 3? If you had problems inflating the balloon in Step 3, explain why.
2. Was there a difference in how much you were able to inflate the balloon after you punctured a hole in the bottle? If there was, explain why.
3. From your observations in this activity, do gases take up space? Explain your answer.

Why Use the Kinetic Molecular Theory?

How and why did scientists formulate the kinetic molecular theory? Experiments into gas behaviour demonstrate that, under normal temperatures and pressures, nearly all gases behave in similar and predictable ways. The properties and behaviours of real gases can be generalized into a theory of an ideal gas. This generalization makes it possible for us to calculate mathematically, with a high degree of accuracy, how real gases will behave under varying conditions.

Of course, no gas is really "ideal." The ideal gas theory ignores certain facts about real gases. For example, an ideal gas particle does not take up any space. In fact, you know that all particles of matter must take up space. Gas particles are small and far apart, however. Thus the space occupied by the particles is insignificant compared to the total volume of the container. You will learn more about the behaviour of real gases in Chapter 12.

Figure 11.6 This diagram shows the possible path of one gas molecule inside a volleyball. In a sample of gas, there are countless molecules moving in straight lines. They rebound off each other and the inner wall of the volleyball.

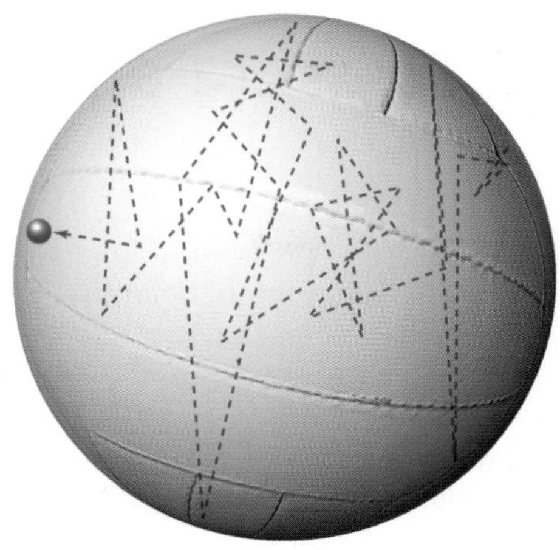

Section Wrap-up

The molecular-level interpretation of gas behaviour given by the kinetic molecular theory helps to explain the macroscopic, or "larger picture," properties of gases in the real world. One of the most important properties of gases is their compressibility—how they react to the application of an external force. In the next section, you will observe how gases behave under pressure. Later in this chapter, you will learn about some interesting applications of pressurized gases.

Oxygen molecules in the atmosphere, at room temperature, travel at an average speed of 443 m/s. This is approximately 1600 km/h!

Section Review

1. K/U Using the kinetic molecular theory of matter, explain each of the following observations.
   - (a) Gases are more compressible than liquids.
   - (b) The density of gases is less than that of solids.
2. K/U In your own words, describe the characteristics of an ideal gas.
3. MC Using your knowledge of intermolecular forces, predict the state of each substance at room temperature. Explain your answer.
   - (a) hexane (C_6H_{14})
   - (b) hydrogen fluoride (HF)
   - (c) potassium chloride (KCl)
4. I Explain each of the following observations.
   - (a) Metals expand when heated, yet contract in cold weather.
   - (b) Gases have no fixed volume.
   - (c) A certain amount of moles of water occupies much more space as a gas than as a liquid.
5. C How does the degree of disorder of a gas compare to that of a liquid or a solid? Explain your answer.
6. K/U Describe the motion of a gas particle.
7. K/U What effect does heating have on the particles of a liquid?
8. (a) C Draw five boxes in your notebook. Inside them, illustrate the motion of gas particles according to the kinetic molecular theory.
   - (b) Draw another five boxes underneath the five boxes in (a). Illustrate how you think the molecules of a real gas might move in comparison.

11.2 Gas Pressure and Volume

Section Preview/ Specific Expectations

In this section, you will

- **perform** experiments to determine the quantitative and graphical relationships between pressure and volume in an ideal gas
- **solve** problems using Boyle's law
- **review** your understanding of the following terms and concepts: *newton, pascal, kilopascal, pressure, volume*
- **interconvert** units of pressure
- **communicate** your understanding of the following terms: *closed system, pressure, pascal, kilopascals, mm Hg, torr, atmospheres, standard atmospheric pressure, Boyle's law*

The earliest use of pressure in English referred to a burden or worry troubling a person's mind. Scientists found this a useful mental model to picture what happens when force is applied to a specific area. They adopted the word pressure to describe any application of force over an area.

Throughout the rest of this chapter, you will discover how gases behave when they are under pressure in a closed system. A **closed system** is one with a constant amount of moles of a substance. It is not open to the atmosphere. Gases in closed systems, from CO_2 in fire extinguishers to O_2 in oxygen tanks, perform important functions in our lives. Understanding the behaviour of gases in closed systems is essential to our safe and effective use of gases.

How is Pressure Calculated?

As you learned in previous studies, **pressure** is defined in physical terms as the force exerted on an object per unit of surface area ($P = F/A$). One commonly used SI unit of pressure is the **pascal (Pa)**, equal to 1 N/m^2. More often, pressure is reported in **kilopascals (kPa)**, equal to 1000 Pa. (You will learn about other units of pressure later in this section.)

Assume a student with a mass of 51.0 kg is sitting on a chair. The force the student applies to the chair is 500.0 N. If the surface area of the chair seat is 0.05 m^2, the pressure the student exerts is

$$P = \frac{F}{A} = \frac{500.0\text{ N}}{0.05\text{ m}^2} = 10\ 000\text{ N/m}^2 = 10\ 000\text{ Pa} = 10.0\text{ kPa}$$

Figure 11.7 shows how a decrease in surface area can dramatically increase pressure.

Figure 11.7 **A woman with a mass of 50.0 kg exerts a pressure of about 21 kPa on the floor as she walks. If another woman with an equal mass is wearing high heels, she will exert a pressure of about 5000 kPa as her heel hits the floor. This pressure is approximately 240 times greater than if she were wearing flat shoes!**

How does a gas exert pressure? In a sense, it cannot exert measurable pressure in the same way that a solid or liquid can. The pressure of a gas is determined by the kinetic motion of its component molecules. Suppose hundreds of billions of gas molecules are in random motion, striking the entire inner surface of their container. Each collision exerts a force on the container's inner surface.

Picture inflating a basketball. As you add more and more air to it, more molecules collide against the inside wall of the basketball. Each collision exerts a force on the basketball's inner surface area. The collective number of collisions as well as the strength of the force form the net or overall gas pressure. Since the molecules move in all directions, the net pressure exerted will be equal throughout. (Figure 11.8 illustrates this.)

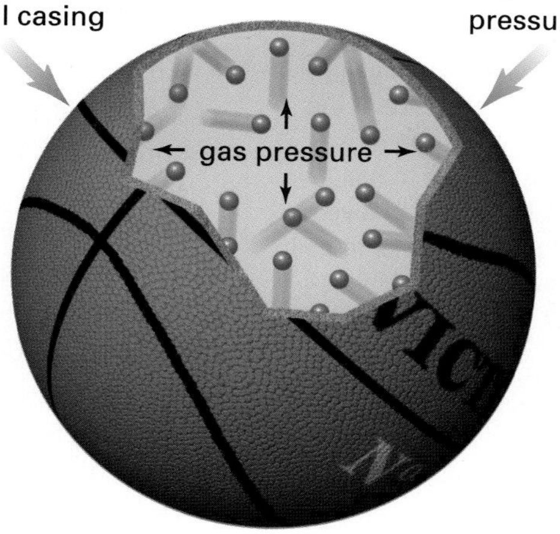

pressure of ball casing

atmospheric pressure

gas pressure

Figure 11.8 This diagram shows gas particles exerting pressure as they bounce off the inner surface of a basketball

Atmospheric Pressure

Despite the popular expression, you can't carry the world on your shoulders! Scientists estimate that the lithosphere, or solid Earth, has a mass of 6.0×10^{24} kg. The hydrosphere, or the portion of Earth covered by water, has an estimated mass of 1.4×10^{21} kg.

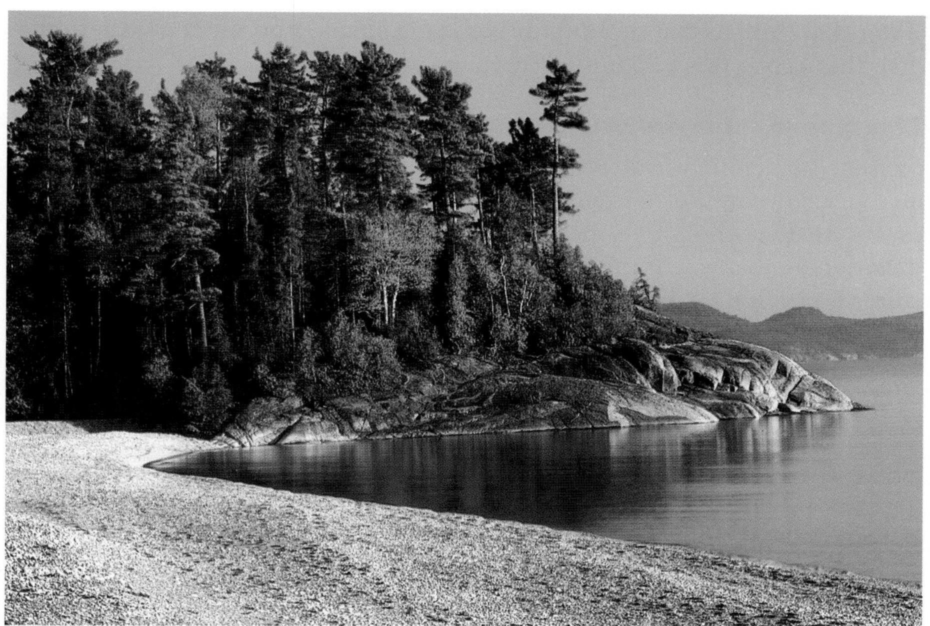

Figure 11.9 The column of air above one square metre (1 m^2) at sea level and 0°C exerts a pressure of 101 325 Pa (1 Pa = 1 N/m^2). This is equivalent to a mass of about 10000 kg over as area of 1 m^2!

You do constantly experience the pressure exerted by Earth's atmosphere. Scientists estimate that the atmosphere has a mass of 5.1×10^{18} kg. Thus air molecules, which have mass, are being pulled down by gravity and are exerting pressure on all objects on Earth. Figure 11.9 shows how much pressure is exerted by the atmosphere over an area of 1 m^2.

COURSE CHALLENGE

What are the main factors determining atmospheric pressure on Earth? In your Chemistry Course Challenge, you will consider the atmosphere on a newly discovered planet. Why might the atmospheric pressure on another planet be different?

mind STRETCH

Using the Internet or another source, find the average atmospheric pressure at the top of Mount Everest. If it takes four minutes to boil a soft-boiled egg at sea level, how long would it take to boil an egg at the top of Mount Everest?

Early Studies of Atmospheric Pressure

In the early seventeenth century, the Italian scientist Galileo Galilei (1564–1642) developed a suction pump. It used air to lift water up to the surface from about 10 m underground. When drawn from greater depths, the column of water collapsed before it reached ground level. Galileo concluded that the water could not be pumped higher because it had reached the "limit of vacuum." Pumping from any depth beyond 10 m required a greater pressure from the suction pump than was provided by the atmosphere. However, Galileo did not know exactly how water was being moved up the tube.

From 1641 to 1642, Evangelista Torricelli (1608–1647) served as Galileo's secretary. He continued Galileo's experiments and concluded that the weight of air was pushing down on the rest of the water. The weight of the air pushed water up the column. This was a logical conclusion since gas molecules, like all matter on Earth, are pulled down by the force of gravity.

Torricelli did further calculations involving the weight of the atmosphere pressing on the water. He then improved upon the experiment by using mercury, which has a density 13.6 times greater than water. He designed the apparatus which we now know as the barometer, shown in Figure 11.10. Torricelli filled a glass tube of 1 cm diameter, closed at one end, with mercury. He then inverted the tube in a dish of mercury. Some of the mercury ran out of the tube. But about 760 mm of mercury remained in the tube. This is about 13.6 times less than the height of water that Galileo could pump. It is the air pushing on the mercury in the dish that keeps this 760 mm of mercury in the tube.

Changes in Atmospheric Pressure

At first, Torricelli considered his experiment a failure. The height of the mercury column did not remain constant at 760 mm, but changed slightly as the weather and air temperature changed. As you learned in Grade 10, these small changes in the height of the mercury column provide us with valuable information. We can predict the weather, in part, by looking at changes in atmospheric pressure. For example, if the atmospheric pressure decreases suddenly, a storm may be on the way.

Atmospheric pressure affects us in other ways, too. People who live at high elevations, such as in the Rocky Mountains, have less mass of air above them. At lower atmospheric pressure, the boiling point of water decreases. Because the water boils at a lower temperature, it takes longer to cook food in boiling water on a mountain than at sea level.

The following ExpressLab demonstrates atmospheric pressure in a dramatic way, showing you firsthand the tremendous pressure that the air around us can exert.

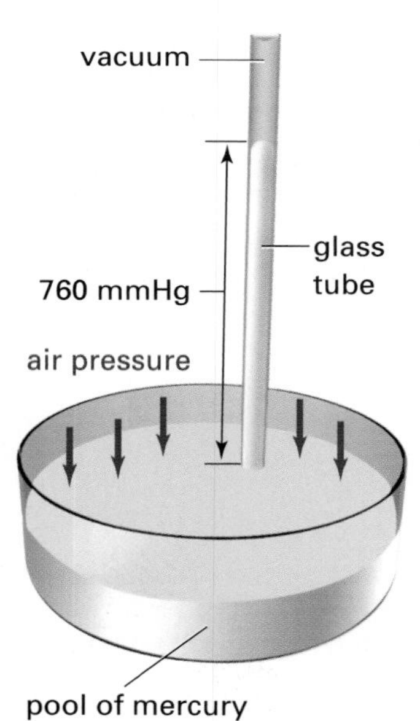

Figure 11.10 Torricelli's barometer

ExpressLab  Observing Atmospheric Pressure in Action

In this ExpressLab, you will see just how powerful atmospheric pressure can be.

Materials

empty, clean soft drink can
hot plate
beaker tongs
large beaker of ice water
10 mL graduated cylinder
5 mL of water

Safety Precautions

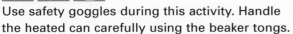

Use safety goggles during this activity. Handle the heated can carefully using the beaker tongs.

Procedure

1. Pour 5 mL of water into the soft drink can.
2. Heat the can on the hot plate until steam begins rising from the opening of the can.
3. Using the beaker tongs, quickly invert the can into the large beaker of ice water so that the opening of the can is just under the surface of the water. Observe carefully.

Analysis

1. What happened to the air and water molecules inside the can when it was heated?
2. Explain what happened to the can when it was placed in the ice water.

Extension

3. Calculate the surface area of the outside of the can exposed to the atmosphere. Assuming an atmospheric pressure of 100.0 kPa, how much force was applied to the can by the atmosphere?

Tools  & Techniques

High Pressure Injectors

When gas molecules under high pressure are allowed to escape and expand, they release kinetic energy. This energy has been harnessed to benefit human health by powering high-pressure injectors, better known as jet injectors.

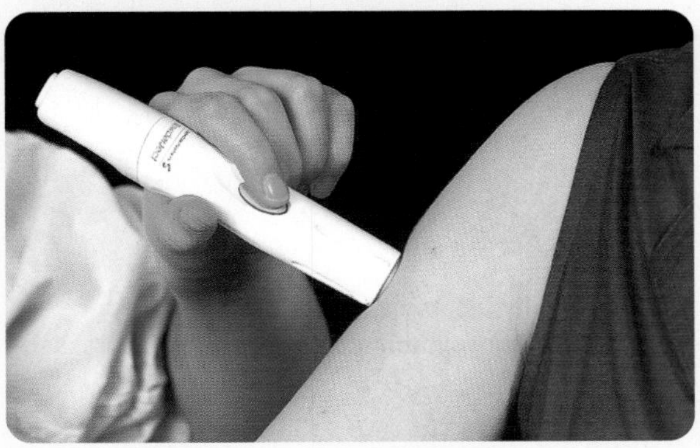

Jet injectors are hypodermic syringes that use high-pressure gas instead of a needle to inject vaccines under a patient's skin. Dr. Robert Higson (1913–1996) developed the "Hypospray" in the 1940s. This device used spring pressure against a plunger to force a vaccine through a tiny nozzle at 1000 km/h. The pressure was enough to drive the vaccine into the tissue of a patient's arm without breaking the skin.

Today jet injectors use a tank of compressed gas and an automatic vaccine dispenser that works through a pistol-like injector. When triggered, the new device releases a measured dose of vaccine into a sterile chamber and a small volume of gas through a hose. As the compressed gas expands, it forces the vaccine at high velocity through the injector's nozzle. The process is fast and simple, ideal for performing mass vaccinations. It also eliminates the problems of disposing of used syringes. Also, it protects medical personnel from infection through accidental needle pricks.

In 1958, Higson led a team that inoculated about 90 000 people in Asia and Africa against polio, typhoid, and cholera using the Hypospray. In 1965, the United Nation's World Health Organization used the jet injector. The organization freeze-dried vaccines for its successful worldwide smallpox eradication program. For its key role in eliminating smallpox from the list of human diseases, the jet injector earned a new name—the "Peace Pistol."

Math LINK

You may encounter tire pressure gauges that are calibrated in pounds per square inch (psi). The conversion factor between kPa and psi is 101.3 kPa = 14.7 psi. To convert a tire pressure of 28.0 psi:

$$\frac{x \text{ KPa}}{28.0 \text{ psi}} = \frac{101.3 \text{ kPa}}{14.7 \text{ psi}}$$

$$x = 28.0 \text{ psi} \times \frac{101.3 \text{ kPa}}{14.7 \text{ psi}}$$

$$= 193 \text{ kPa}$$

What would a tire pressure of 27.3 psi be in kPa? What would the pressure of 198.7 kPa be in psi?

Units of Pressure

For many years, atmospheric pressure was measured in millimetres of mercury (**mm Hg**). In the British Commonwealth and the United States, inches of mercury were used. Standard atmospheric pressure, the pressure of the atmosphere at sea level and 0°C, is 760 mm Hg. More recently, in honour of the work of Torricelli, standard atmospheric pressure has been defined as 760 torr. **1 torr** represents a column of mercury 1 mm in height at 0°C. Another common unit for measuring pressure is **atmospheres (atm)**, where 1 atm is equivalent to 760 torr. While mm Hg, torr, and atm are still used to measure pressure, especially in technological and medical applications, the SI units are pascals (Pa) or kilopascals (kPa).

In other words, **standard atmospheric pressure** at 0°C is equivalent to:

$$760 \text{ mm Hg} = 760 \text{ torr} = 1 \text{ atm} = 101.3 \text{ kPa}$$

Using this relationship, we can convert from one unit to another. For example, a pressure of 100.0 kPa is equivalent to

$$100.0 \text{ kPa} \times \frac{760.0 \text{ torr}}{101.3 \text{ kPa}} = 750.2 \text{ torr}$$

The Relationship Between Pressure and Volume

Figure 11.11 shows a meteorologist preparing to release a weather balloon partially filled with helium gas. As the balloon rises, atmospheric pressure decreases. The volume of the balloon increases.

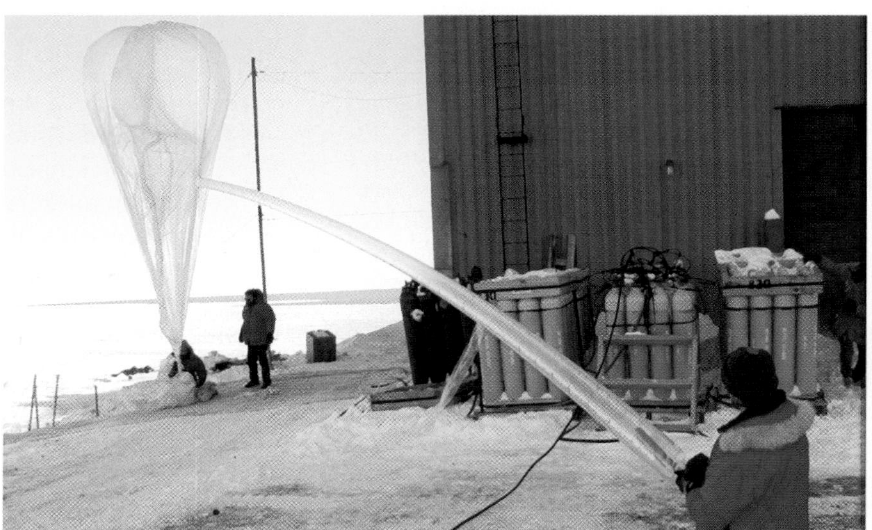

Figure 11.11 **Weather balloons are partially inflated with helium. They carry specialised instruments to measure varying atmospheric conditions such as pressure, temperature, and humidity.**

Since the helium atoms inside the balloon move randomly in all directions, they constantly bombard all the area inside the walls of the balloon, exerting a pressure. With decreasing atmospheric pressure, there are fewer air molecules to collide with the outside of the balloon. As the pressure outside the balloon becomes less than that inside the balloon, the balloon expands. Given an expandable container, such as a balloon, the volume occupied by the gas will increase when external pressure decreases. As external pressure *increases*, gas molecules are forced closer together. The volume of gas then *decreases.*

When we refer to the *volume of a gas*, we are in fact talking about the *volume of the container*. The definition of the volume of a gas is *the space available for gas molecules to move around in*. The kinetic molecular theory of gases assumes that the volume of each gas molecule is essentially zero. Thus, the amount of space for them to move around in is the volume of the container. For all gases, V_{gas} = the volume of the container holding the gas. (Do not confuse this with the *molar volume* of gases. You will learn about molar volume in Chapter 12.)

Think about how the relationship between pressure and area for solids would apply if you were testing a gas in a three-dimensional container. You know that according to the kinetic molecular theory, gas molecules exert pressure over the entire inside surface of their container. If the volume of the container is halved, what would happen to the pressure of the gas inside the container?

Robert Boyle (1627–1691) was an Irish scientist with an interest in chemistry. He investigated the relationship between pressure and volume of gases at constant temperatures. By making careful measurements of the volume of a trapped gas, he was able to describe what happened when the pressure exerted on the gas was increased. Figure 11.12 shows Boyle's experiment. Boyle measured the length of the column of trapped air compared to the length of the column of mercury. Since the length of the mercury column is directly related to its volume, Boyle was able to deduce the relationship between pressure and volume.

Go to the Chemistry 11 Electronic Learning Partner for a demonsration of Boyle's law.

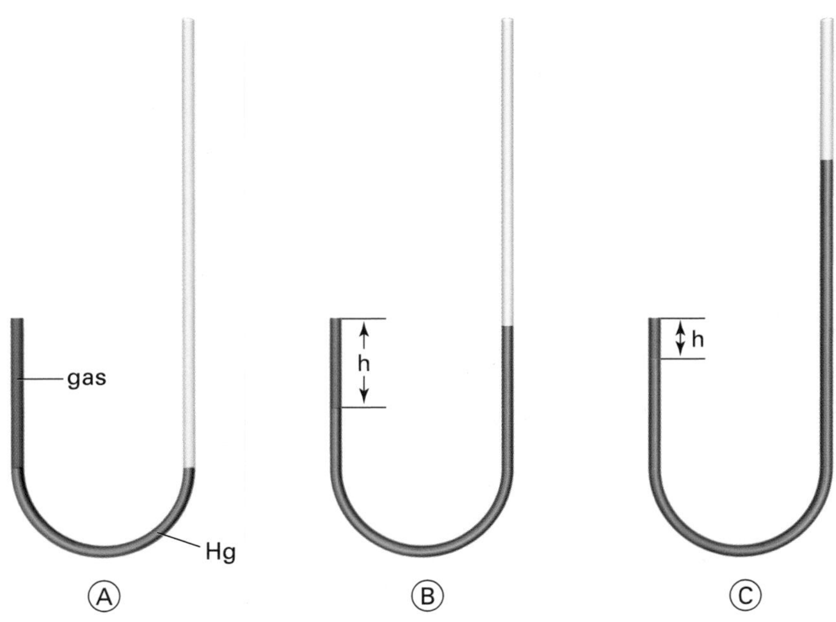

Figure 11.12 **When liquid mercury is added to the open tube, the pressure caused by the weight of mercury on the trapped gas increases. The volume of the trapped gas (b and c) decreases.**

Since mercury is a poisonous element, you will use a different method in Investigation 11-A to examine the relationship between the pressure and the volume of a gas.

Investigation 11-A

SKILL FOCUS

Performing and recording

Analyzing and interpreting

Identifying variables

The Relationship Between the Pressure and the Volume of a Gas

Boyle measured the variance in length of a column of trapped air. Since length is directly proportional to volume in a column with a regular diameter, this gave him an indirect measure of changes in volume of the air with increased pressure. In this activity, air is trapped inside a sealed plastic pipette. You can measure the volume of the trapped gas in terms of the length of the column of air, the way that Boyle did more than three hundred years ago. You will measure the applied pressure in terms of the number of turns of a clamp rather than in kPa.

Question

What is the relationship between the pressure and volume of a fixed amount of gas at a constant temperature?

Materials

thin stem plastic pipette with bulb
small C-clamp
metric ruler (with mm)
match or Bunsen burner
craft stick
coloured water

Safety Precautions

- Be very careful when sealing the end of the plastic pipette with a flame. The plastic will melt and may begin to burn. Hot, molten plastic can burn your skin.
- Do not inhale any of the fumes from the plastic.
- Before lighting the Bunsen burner, check that there are no flammable solvents nearby.

Procedure

1. Squeezing the bulb, draw enough coloured water into a pipette so that the water fills the bulb and extends about 2 to 3 cm down the stem. The rest of the stem should be filled with air.
2. Using a flame from a Bunsen burner, carefully seal the end of the pipette completely. Allow the pipette to cool for at least three minutes before completing the rest of the procedure.
3. Copy out the data table into your notes.

P (no. of turns)	V (mm)	1/V (mm)	P × V (turns · mm)
0			
1			

4. Break a craft stick in two. Place one half of the stick on either side of the pipette bulb. Tighten a small C-clamp around the bulb of the pipette so that the clamp just holds the bulb snugly (see the diagram).
5. Using a ruler, measure the length of trapped gas (the "volume") in millimetres. Record this in your data table.
6. Increase the "pressure" on the bulb by turning the handle of the clamp one half or one complete turn, depending on the size of your clamp. Record the "volume" of the trapped gas in your data table.
7. Repeat step 6 until you have made at least five complete turns.
8. Complete the data table. Calculate an average value for the $P \times V$ column.
9. Plot a graph of P (y-axis) versus V (x-axis).
10. Plot a graph of P (y-axis) versus $1/V$ (x-axis).

Analysis

1. What is the independent variable in this investigation? What is the dependent variable?
2. What relationship exists between volume and pressure, based on the data collected and the graphs produced?
3. Express this relationship mathematically. To help you to do this, look at the mathematical form of Boyle's law, located after this investigation.
4. Calculate the slope of the P vs. $1/V$ graph. How does this value compare with the average $P \times V$ value? Of what significance are these two values?
5. What changes in temperature occurred during the experiment? In the amount of trapped air? Explain how this may have affected your results.

Conclusion

In your own words, state the relationship between pressure and the volume of a gas.

Extension

Using pressure probes and a graphing calculator or computer interface, investigate the relationship between the pressure and the volume of a gas. Produce a data table and graphical interpretation of these results.

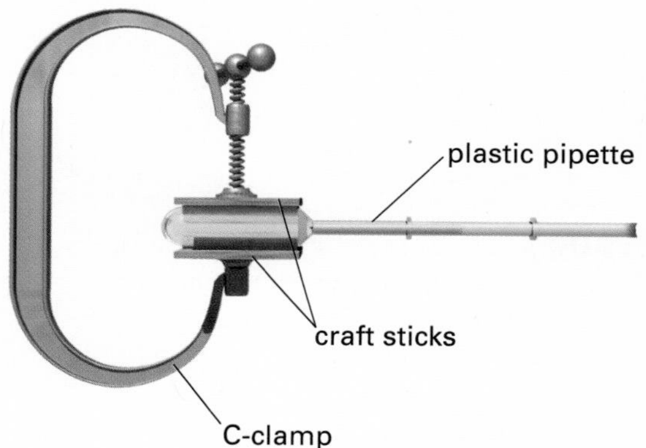

Boyle's Law

In 1662, Robert Boyle stated that *the volume of a given amount of gas, at a constant temperature, varies inversely with the applied pressure.* In other words, as external pressure on a gas increases, the volume of the gas decreases by the same factor. This statement is known as **Boyle's law.** Figure 11.13 illustrates Boyle's law using a bicycle pump.

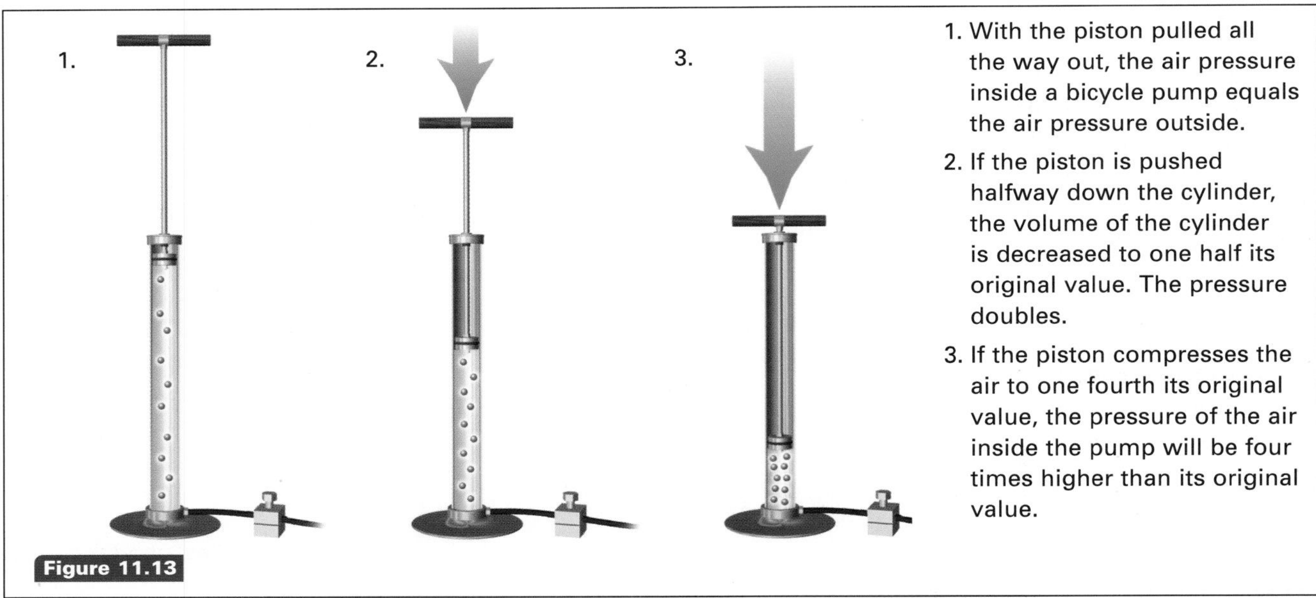

1. With the piston pulled all the way out, the air pressure inside a bicycle pump equals the air pressure outside.
2. If the piston is pushed halfway down the cylinder, the volume of the cylinder is decreased to one half its original value. The pressure doubles.
3. If the piston compresses the air to one fourth its original value, the pressure of the air inside the pump will be four times higher than its original value.

Figure 11.13

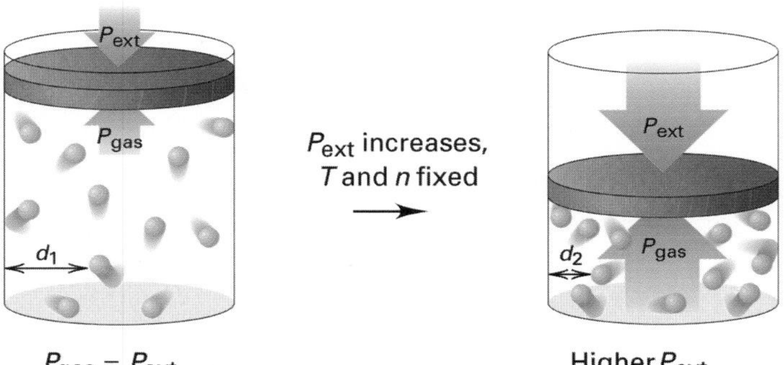

Figure 11.14 At a given temperature, gas molecules travel an average distance (d_1) before they collide with the container wall. When the volume is decreased, the gas molecules travel a shorter distance (d_2) before striking the wall.

Boyle found that this relationship held true for all gases as long as the temperature remained unchanged. As external pressure on a gas increases, the volume of the gas decreases. The gas molecules are forced closer together. However, if the volume of a gas decreases, then the gas molecules have to travel a shorter distance before they strike the container walls, as shown in Figure 11.14. Since they travel a shorter distance, gas molecules will strike the container walls more often per unit time. This increases the internal pressure of the gas. (With an increased volume, there are fewer collisions per unit time and a lower gas pressure is exerted.)

In other words, as the pressure of a closed system increases, its volume decreases. If the pressure is decreased by half, the volume doubles. We can write this relationship mathematically by using the proportionality symbol, α. V α $1/P$ means that volume is inversely proportional to the pressure.

Mathematically, the proportionality sign (α) can be removed by introducing a proportionality constant (k).

$$V \,\alpha\, \frac{1}{P}$$

$$V = \frac{1}{P} \times k \qquad \text{or} \qquad PV = k$$

In Investigation 11-A, when you plotted a graph of P versus $1/V$, you obtained a straight line. The slope of this line gives the value of the proportionality constant, k. If the pressure is tripled, the volume will decrease to one third of its original volume, such that $P \times V = k$. The value of k differs depending on the gas sample and the temperature. Remember, this mathematical relationship only applies if the temperature remains constant. A graph of P versus $1/V$ is shown in Figure 11.15.

For the gas sample at its initial conditions (i)

$$P_iV_i = k$$

If the gas sample is then subjected to a change in pressure, at its final conditions (f)

$$P_fV_f = k$$

Since the slope of the line (k) is constant, and since initial and final conditions are both equal to k, we can write

$$P_iV_i = P_fV_f$$

This mathematical relationship is another way of stating Boyle's law.

CHEM FACT

Scientific discoveries often happen simultaneously. Edmé Mariotte, a French scientist, investigated the pressure–volume relationship of gases independently of Boyle. He did not publish his work until 1676, fourteen years after Boyle had. In many European countries, the mathematical relationship between gas pressure and volume is known as Mariotte's Law.

Sample Problem

Boyle's Law: Calculating Volume

Problem

A sample of helium gas is collected at room temperature in a 4.50 L balloon at standard atmospheric pressure. The balloon is then submerged in a tub of water, also at room temperature, such that the external pressure is increased to 110.2 kPa. What will the final volume of the balloon become?

What Is Required?

You need to find the volume of the balloon after the pressure on the balloon has been increased. ($V_f = ?$)

What Is Given?

- You know the initial pressure and volume, and the final pressure.
  Initial pressure (P_i) = 101.3 kPa
  Initial volume (V_i) = 4.50 L
  Final pressure (P_f) = 110.2 kPa
- You know that the temperature does not change.

Plan Your Strategy

Algebraic method

- Since temperature is constant and pressure and volume have been given, you will need to use the Boyle's law formula.
- You can substitute numbers and units for the variables in the formula to solve for the unknown (V_f).

Continued ...

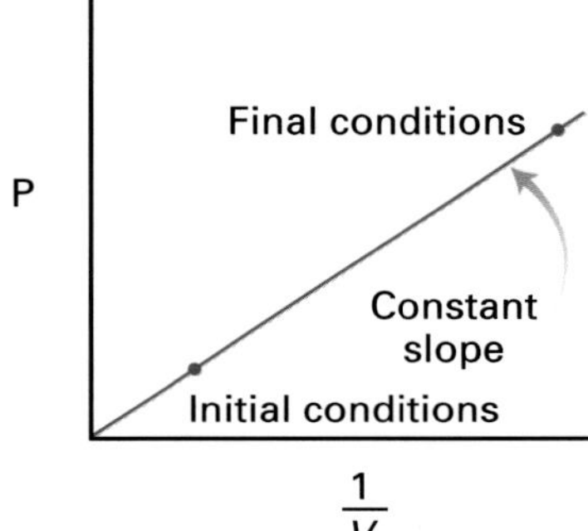

Figure 11.15 If a sample of gas at initial conditions has a change of pressure applied to it, its volume decreases proportionally, such that in its final state, $P \times V = k$.

PROBLEM TIP

In this Sample Problem, you will see two different methods of solving the problem: the algebraic method and the ratio method. Choose the method you prefer to solve this type of problem.

Technology LINK

In 1998, a weather balloon carrying instruments to measure the ozone layer drifted off course. It veered into transatlantic air routes, where it posed a serious danger. By this time, the balloon had expanded in size to about the same volume as Toronto's Sky Dome. Two Canadian Air Force CF-18 jets directed over 1000 rounds of cannon fire at it, but could not bring the balloon down. It finally landed on an island off the coast of Finland. University of Toronto physicists have developed a new mechanism to prevent such an event from recurring. When an experiment has been completed, a parachute will return the scientific instruments to the ground. As the instruments fall, they will pull panels from the side of the balloon, causing it to plummet rapidly.

Can you think of other ways to solve the problem that faced the University of Toronto scientists? Create a set of blueprints for another technological solution, using CAD software if you have access to it.

Continued ...
FROM PAGE 433

Ratio method

- The pressure on the balloon increases. When this happens, if temperature remains the same, you know according to Boyle's law that the volume of the balloon will decrease.
- To find the final volume, you can multiply the initial volume of the balloon by a ratio of the two pressures that is less than one (i.e. $\frac{101.3 \text{ kPa}}{110.2 \text{ kPa}}$).

Act on Your Strategy

Algebraic method

$P_iV_i = P_fV_f$

$(101.3 \text{ kPa})(4.50 \text{ L}) = (110.2 \text{ kPa})(V_f)$

To isolate V_f, you need to divide both sides of the equation by 110.2 kPa.

$$\frac{(101.3 \cancel{\text{kPa}})(4.50 \text{ L})}{(110.2 \cancel{\text{kPa}})} = \frac{\cancel{(110.2 \text{ kPa})}(V_f)}{\cancel{(110.2 \text{ kPa})}}$$

$$\frac{(101.3 \cancel{\text{kPa}})(4.50 \text{ L})}{(110.2 \cancel{\text{kPa}})} = (V_f)$$

$$V_f = 4.137 \text{ L}$$

Ratio method

$V_f = 4.50 \text{ L} \times \text{pressure ratio}$

$= 4.50 \text{ L} \times \frac{101.3 \cancel{\text{kPa}}}{110.2 \cancel{\text{kPa}}}$

$= 4.137 \text{ L}$

Since the least number of significant digits in the question is three, the answer is:

$$V_f = 4.14 \text{ L}$$

Check Your Solution

- The units for the answer are in litres.
- When units cancel out, L remains.
- The volume of the balloon has decreased due to the increase in pressure.

Practice Problems

1. A 50.0 cm^3 sample of nitrogen gas is collected at 101.3 kPa. If the volume is reduced to 5.0 cm^3, and the temperature remains constant, what will the final pressure of the nitrogen be?

Continued ...

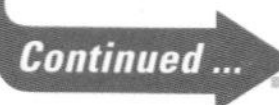

Continued ...
FROM PAGE 434

2. A weather balloon has a volume of 1000 L at a pressure of 740.0 torr. The balloon rises to a height of 1000 m where the atmospheric pressure is measured as 450.0 torr. Assuming there is no change in temperature, what is the final volume of the weather balloon?

3. A 45.0 cm^3 sample of nitrogen gas is collected at 1.0 atm. The nitrogen is compressed to a pressure of 10.0 atm. What is the final volume of the nitrogen if the temperature remains constant?

4. A 45.6 mL sample of gas at 490 torr is compressed to a certain volume at 3 atm. What is the new volume, in litres?

CHEM FACT

Now that you've finished practising Boyle's law problems, take a deep breath and relax. You have just illustrated Boyle's law! When you inhale, muscles in your torso expand your rib cage. The volume of your lungs increases. Since the pressure inside your lungs is decreased with the expansion in volume, outside air under higher pressure rushes in.

Section Wrap-up

In this section, you learned about the relationship between pressure and volume of a gas. This relationship is stated in Boyle's law. With knowledge of gas properties and behaviours, we are able to devise and improve upon technologies used everyday. You will learn about some of these important technologies later in this chapter. In the meantime, the next section examines how gases respond to changes in yet another variable: temperature.

Section Review

1 C Using the relationship 760 mm Hg = 760 torr = 1 atm = 101.3 kPa, convert each of the following units:

(a) 2.03 atm to kPa

(b) 85.2 kPa to atm

(c) 1.50 atm to torr

(d) 600 torr to kPa

2 K/U Use the kinetic molecular theory. Explain why the air pressure inside a capped syringe increases if the volume decreases from 15 cm^3 to 10 cm^3.

3 K/U Explain, using the kinetic molecular theory, why pressure is exerted by gases in all directions.

4 I A 1.00 L helium balloon is floating in the air on a day when the atmospheric pressure is 102.5 kPa and the temperature is 20.0°C. Suddenly, clouds appear and the pressure rapidly drops to 98.6 kPa at a temperature of 20.0°C. What is the new volume of the balloon?

5 I 0.750 L of oxygen gas is trapped at 101.3 kPa in a cylinder with a moveable piston. The piston is moved and the gas is compressed to a volume of 0.500 L. What is the final pressure applied to the oxygen gas if the temperature remains unchanged?

6 MC A student produces 38.3 mL of oxygen gas in a burette. The next day, there are 40.2 mL of gas in the burette at a pressure of 103 kPa. What was the pressure on the previous day, in torr? What might be happening to the weather in the student's neighbourhood?

11.3 Gases and Temperature Changes

Section Preview/ Specific Expectations

In this section, you will

- **perform** laboratory experiments investigating the effects of temperature changes on the volume of gases
- **solve** problems using Charles' law and Gay-Lussac's law
- **convert** units between the Celsius and Kelvin temperature scales
- **communicate** your understanding of the following terms: *Kelvin scale, absolute zero, Charles' law, Gay-Lussac's law, pressure-relief valve, fusible plugs*

CHECKPOINT

As you perform Investigation 11-B, keep in mind the ratio of 1/273 discovered by Charles. Can you recall from your previous studies a special significance for the number 273?

As you learned in Section 11.1, the average kinetic energy of gas molecules is directly related to the temperature. The greater the temperature, the greater the average motion of the molecules and the greater their average kinetic energy. In other words, the temperature of a substance is defined as the measure of the average kinetic energy of the molecules in that substance.

When substances are cooled, they lose kinetic energy. How does this affect their volume? Remember, for an ideal gas, we can think of its particles as having mass but no volume. When you perform Investigation 11-B on page 438, imagine air behaving as an ideal gas. How do you think air, in an expandable container will react to temperature changes?

Temperature and Volume

Sometimes, scientific discoveries are made well before any technological application of them can be envisioned. At other times, the desire to develop new technologies leads to experiments from which discoveries are made.

Jacques Charles (1746–1823), a French scientist, was the first to fill a balloon with hydrogen. He was also interested in hot-air balloons, which were being developed in France at the time. Charles investigated the expansion rates of nitrogen, oxygen, hydrogen, and carbon dioxide. He found that these gases all expanded by the same ratio. For each degree Celsius increase in temperature, all of these gases would expand by a certain fraction. This fraction was 1/273rd of their volume at 0°C. For each degree Celsius decrease in temperature, their volume would decrease by the same fraction. Thus, if a gas at 0°C were to be heated to 273°C, its volume would double. This held true when the pressure and the amount of gas remained constant. Figure 11.16 shows the expansion of the volume of gas in a hot-air balloon as the air inside the balloon is heated.

Figure 11.16 These photographs show the gradual expansion of a hot air balloon. Since hot air is less dense than cooler air, the balloon rises.

Chemistry Bulletin

Science | Technology | Society | Environment

Gas Temperature and Cryogenics

Some people wish to be frozen when they die, trusting that future technologies will be able to revive them. This untested process is called *cryonics*. It was first suggested by Robert C.W. Ettinger in 1962. Ettinger has since set up his own Cryonics Institute in Michigan, where people can have their bodies frozen and stored. We still cannot freeze and re-animate higher animals, such as people. Cryogenic freezing, however, can suspend the life of tissues and organs used for transplants. Cryogenics has also made many other technologies possible. It has proven that science and science fiction are often closer than we think!

The term *cryogenics* can be applied to all temperatures below the normal boiling point of oxygen, or –183°C.

When substances encounter temperatures this low, they often behave strangely. Liquid helium, for example, becomes a "superfluid" at temperatures below –270.97°C. The principle of "superconductivity" is just as interesting Cryogenically cooled rings made out of metals such as lead and aluminum become "superconductors." They can keep currents travelling in circles for hours even after scientists have removed the original source of electricity.

Scientists have made use of the strange things that happen to matter at low temperatures. Superconductors have been used to make huge electromagnets, like the one at the Argonne National Laboratory near Chicago. Argonne's electromagnet can produce a magnetic field 134 000 times as strong as Earth's. It operates on relatively little power because of its superconducting capabilities. Such magnets are used in nuclear power research to find new nuclear particles. And while cryogenics has advanced nuclear science, it also helps scientists to study the effects of nuclear radiation. Scientists study cryogenically frozen atoms suspended in an irradiated state to understand how nuclear radiation can harm human health.

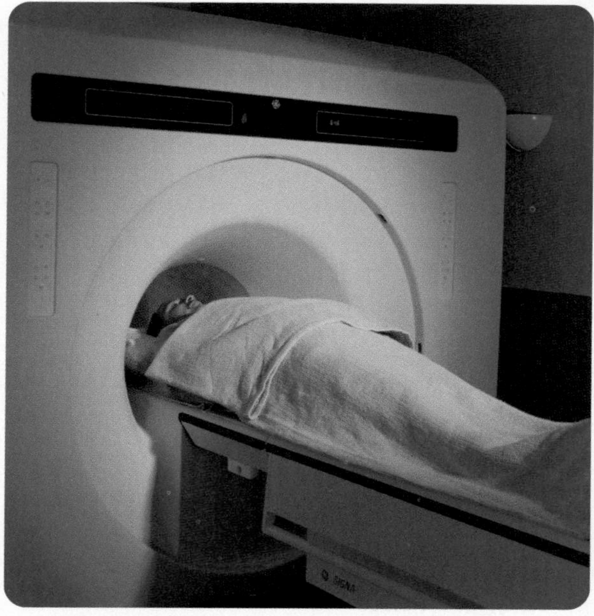

Cryogenically cooled magnets are used in MRI technology.

Cryogenics produces large-scale amounts of nitrogen and oxygen. Scuba divers and astronauts use compressed oxygen tanks that provide a six to eight hour supply of oxygen. Rocket engines use liquid oxygen as fuel. We use liquid nitrogen to make ammonia for fertilizers, to keep frozen foods cold during transport, and to fast-freeze these foods.

The list of applications of cryogenics is long and varied, and research continues. Maybe Ettinger is right and one day all of the occupants of the Cryonics Institute will live again!

Making Connections

1. What do you think might be some possible future applications of cryogenics?
2. Scientists use Dewar Flasks to contain cryogenic fluids. How do these flasks work? Do some research to find out.

SKILL FOCUS

Performing and recording

Analyzing and interpreting

Identifying variables

Investigation 11-B

The Relationship Between Temperature and Volume of a Gas

As you learned in Investigation 11-A, the length of a column of trapped air is directly proportional to its volume. In this investigation, you will see the effect of temperature changes on the volume of a gas, also measured in terms of the length of a column of trapped air.

Question

What is the relationship between the temperature and volume of a fixed amount of gas at a constant external pressure?

Materials

thin stem plastic pipette
metric ruler (with mm)
Celsius thermometer
400 mL beaker
ice
hot plate
match or Bunsen burner
2 elastics
scissors
coloured water
tap water

Safety Precautions

- Be very careful when sealing the end of the plastic pipette with a flame. The plastic will melt and may begin to burn. Hot, molten plastic can burn your skin.
- Do not inhale any of the fumes from the plastic.
- Before lighting the Bunsen burner, check that there are no flammable solvents nearby.

Procedure

1. Squeezing the pipette bulb, draw enough water into the pipette to form a small plug. The rest of the pipette should contain air.
2. Using a flame from a match or Bunsen burner, carefully seal the open end of the pipette completely. Allow the pipette to cool for at least 3 min before carrying on with the rest of the procedure.
3. Using scissors, cut off the tip of the bulb of the pipette.
4. Carefully attach the pipette to the ruler, using a rubber band, so that the bottom of the tube is even with the 1.0 cm mark of the ruler.
5. Fill a 400 mL beaker about two thirds full of tap water and add 3 or 4 ice cubes. Place the thermometer in the water. Then put the ruler with attached pipette into the water. Allow the ruler and pipette to sit for 5 min.

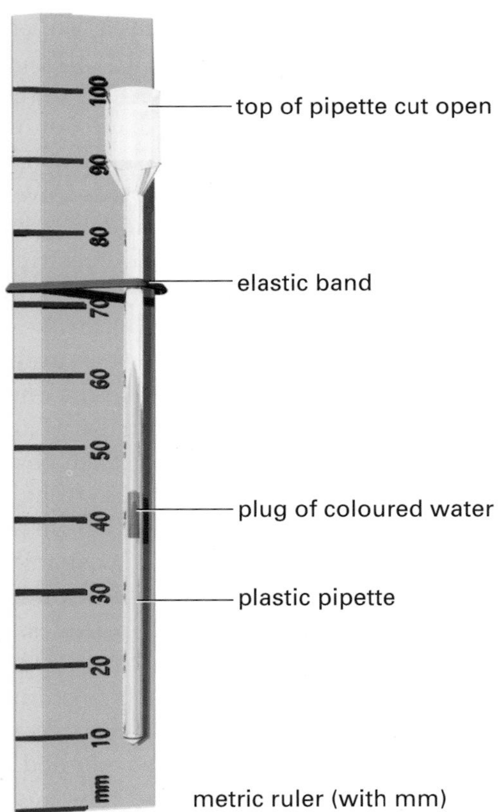

6. Copy the data table into your notebook. Your table should have at least eight rows for data.

Temp. (°C)	V (length of trapped air in mm)

7. After 5 min, measure the length (or "volume") of the trapped gas in mm. Remember that the bottom of the pipette stem is set at the 1.0 cm mark. Record these values in your data table.
8. Place the beaker on the hot plate and **slowly** heat the water in the beaker. Measure the length ("V") and temperature of the trapped gas at every 10°C to 15°C. Measure the length and temperature to a maximum of 60°C.
9. Clean the apparatus and dispose of the pipette as directed.
10. Complete the data table. Find the average of the V/T column.
11. Plot a graph of V (mm) versus T. The horizontal axis (temperature) must extend from −300°C to 100°C.
12. Draw a line of best fit. Extrapolate this line to the x-intercept.

Analysis

1. What is the independent variable in this investigation? What is the dependent variable?
2. What relationship did you notice between temperature and volume?
3. When Jacques Charles did an activity similar to this one, he obtained an x-intercept of −273°C. What is significant about this value?
4. If the value obtained by Charles was correct, what is the percentage error in your x-intercept?

Conclusion

5. In your own words, state the qualitative relationship that exists between the temperature and volume of a fixed amount of gas at constant pressure.

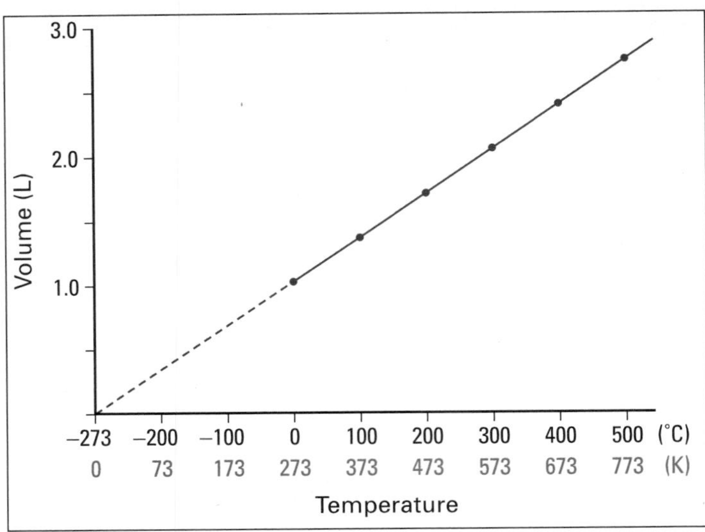

Figure 11.17 Absolute zero for an ideal gas is –273.15°C or 0 K, the point at which all molecular motion theoretically ceases.

The Kelvin Scale and Absolute Zero

Charles found that, regardless of the gas tested, the *x*-intercept on a graph would always be –273°C. In 1848, Lord Kelvin (1824–1907), a Scottish scientist, realised the significance of this finding. He reasoned that at –273°C, molecular motion would cease. At this temperature, kinetic energy would be zero. The volume of a gas would, hypothetically, also be zero.

Of course, real gas molecules do have volume. Also, at low temperatures all gases will condense and change state. Still, Kelvin used this reasoning as the basis for a new temperature scale, the **Kelvin scale**. The starting point for the scale, 0 K, is called **absolute zero**. Figure 11.17 shows how absolute zero can be hypothesized, based on data from experiments. The modern accepted value for 0 K, derived with equipment more sophisticated than that available to Charles, is –273.15°C. Each unit in the Kelvin scale is exactly the same as a unit in the Celsius scale.

There are no degree signs used in the Kelvin scale. More importantly, there are no negative values. What would happen if you tried to calculate a temperature twice as warm as –5°C? Mathematically, the answer would be –10°C, but this is a *colder* temperature. When mathematical manipulations are involved in studying gas behaviour, you need to convert degree Celsius temperatures into kelvins. This is done using the relationship

$$T_K = {}^\circ\text{C} + 273.15 \textbf{ or } {}^\circ\text{C} + 273$$

Most often, you will round off and use 273 as the conversion factor relating K and °C.

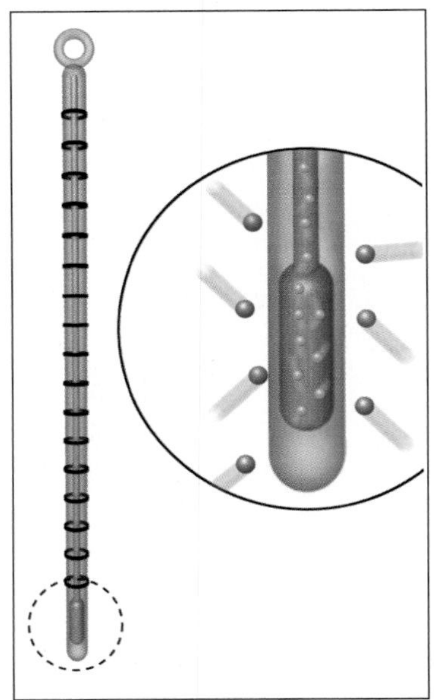

Figure 11.18 As temperature increases, particles move more rapidly, striking the outside of the thermometer with greater force and frequency. The kinetic energy of the particles is transferred to the particles inside the tube of the thermometer. The volume of the liquid inside the tube (usually mercury or coloured alcohol) expands.

Charles' Law

Although Charles discovered that the volume of a fixed amount of gas at constant pressure was proportional to its temperature, he never published this finding. In 1802, Joseph Louis Gay-Lussac (1778–1850), a French scientist, made reference to Charles' work in a published paper. The relationship between temperature and volume has since become known as **Charles' law**. Charles' law states that *the volume of a fixed mass of gas is proportional to its temperature when the pressure is kept constant.*

As you can see from Figure 11.19 on page 442, the volume of a gas increases or decreases by a fixed increment when subjected to a change in temperature. The algebraic statement of Charles' law depends on using absolute, or Kelvin, temperatures. This law is stated as $V \propto T$, where T is measured in kelvins. (Figure 11.18 uses Charles' law to explain how a thermometer works.)

Introducing a proportionality constant (k_1), this relationship can be restated as

$$V = k_1 T \qquad \text{or} \qquad \frac{V}{T} = k_1$$

This relationship only applies if pressure is kept constant and temperature is given in kelvins. If a sample of gas is collected at initial conditions (i), this relationship can be rewritten as

$$\frac{V_i}{T_i} = k_1$$

Suppose the gas sample is subjected to a change in temperature. Under the final conditions (f), there will be a volume change such that

$$\frac{V_f}{T_f} = k_1$$

Since both initial and final conditions are equal to the proportionality constant, Charles' law can be written as

$$\frac{V_i}{T_i} = \frac{V_f}{T_f}$$

Thus, Charles' law can be restated as: *the volume of a fixed mass of gas at constant pressure is directly proportional to its Kelvin temperature.* In the next ThoughtLab, you will convert your temperature findings from Investigation 11-B into kelvins and see why you must use Kelvin temperature when performing calculations with gases.

Electronic Learning Partner

Go to the Chemistry 11 Electronic Learning Partner for a demonstration of Charles' law.

ThoughtLab  Charles' Law and Kelvin Temperature

As you have learned, when making calculations involving temperature in gas samples, Kelvin temperatures must be used. You will see why for yourself in this ThoughtLab.

Materials

data table from Investigation 11-B
graph paper

Procedure

1. Make a new data table like the one below. You should include at least eight rows for data.

Temp (°C)	Temp (K)	V (mm)	$\frac{V \text{ (mm)}}{T \text{(°C)}}$	$\frac{V \text{ (mm)}}{T \text{(K)}}$

2. Fill in columns 1, 3, and 4 with your data from Investigation 11-B.
3. Convert temperature data in column 1 from °C to kelvins. Enter the new values in columns 2 and 5.
4. Plot a graph of V (mm) versus T (K). The horizontal axis (T) must extend from 0 K to 400 K.
5. Draw a line of best fit. Extrapolate this line to the x-intercept.

Analysis

1. What did you notice about the values of V/T (K) in the data table?
2. How do the values of V/T (K) compare to the values of V/T (°C) in the data table? Explain the significance of these two sets of data.
3. What mathematical relationship seems to exist between volume and temperature when temperature is recorded in °C? In kelvins?

Extension

4. Make a new data table like the one below. Include at least eight rows for data.

At 100 kPa		At 163 kPa		At 346 kPa	
T (K)	V (mm)	T (K)	V (mm)	T (K)	V (mm)

In columns 1 and 2, enter your data from columns 2 and 3 of the data table you made above. Carry the data from column 1 into columns 3 and 5.

5. Using the Boyle's law formula ($P_iV_i = P_fV_f$ at constant T and n), calculate the "volume" of the gas sample at 163 kPa and at 346 kPa.
6. Plot a graph of your new data table. Use a different colour for each line. Extrapolate each line to the x-intercept.
7. Of what significance are the results obtained from this graph?

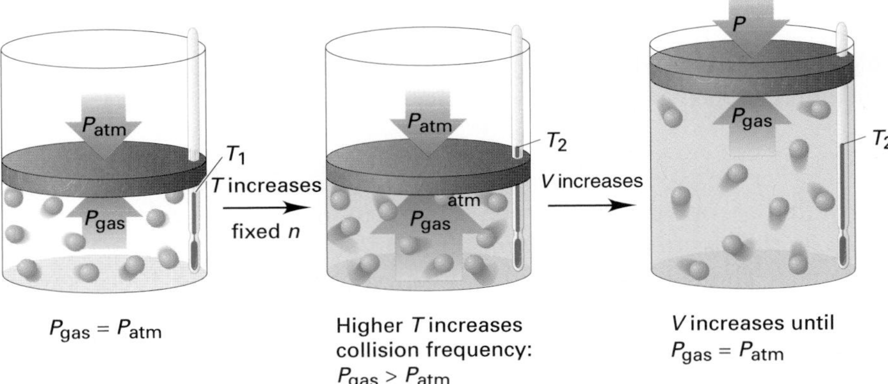

Figure 11.19 **As a gas is heated from T_i to T_f, the molecules move faster and collide with the container walls more frequently, increasing the pressure applied by the gas (P_{gas}). This added pressure increases the volume of the container until the pressure exerted by the gas is equal to the pressure exerted by the atmosphere.**

As helium gas is cooled below –268.95°C, it forms a liquid. At –270.97°C, helium still looks like a liquid, but a liquid with unusual properties. Suddenly, liquid density drops and this "liquid" gains the ability to move through very small holes that helium gas cannot pass through. It flows up the walls of its container defying gravity, and has zero viscosity. Below –270.97°C, helium becomes a superfluid, the only one discovered so far. Helium never changes to a solid.

Sample Problem

Charles' Law: Calculating Volume

Problem

Using a glass syringe, a scientist draws exactly 25.5 cm^3 of dry oxygen at 20.0°C from a metal cylinder. She needs to heat the oxygen for an experiment, so she places the syringe in an oven at 65.0°C and leaves it there for 30 min. Assuming the atmospheric pressure remains the same, what volume will the oxygen occupy?

What Is Required?

You need to find the volume of the oxygen in the syringe after it has been heated for 30 min. ($V_f = ?$)

What Is Given?

- You know the initial volume and temperature.
  Initial volume (V_i) = 25.5 cm^3
  Initial temperature (T_i) = 20.0°C
- You know the final temperature.
  Final temperature (T_f) = 65.0°C
- You know that the pressure does not change.

Continued ...

Continued ...
FROM PAGE 442

Plan Your Strategy

Algebraic method

- Since the pressure is constant and the temperature of the gas increases, you will need to use the Charles' law formula to find the final volume of the gas sample.
- Since T is given in °C, you will need to convert it to kelvins.
- You can substitute numbers and units for the variables in the formula to solve for the unknown (V_f).

Ratio method

- Since the pressure is constant and the gas is subjected to an increase in temperature, you know that, according to Charles' law, the volume will increase.
- Since T is given in °C, you will need to convert it to kelvins.
- To find the final volume, you can multiply the initial volume by a ratio of the kelvin temperatures that is greater than one.

Act on Your Strategy

Algebraic method

$$T_i = (20.0°\text{C} + 273) = 293\ \text{K} \qquad T_f = (65.0°\text{C} + 273) = 338\ \text{K}$$

$$\frac{V_i}{T_i} = \frac{V_f}{T_f}$$

$$\frac{25.5\ \text{cm}^3}{293\ \text{K}} = \frac{V_f}{338\ \text{K}}$$

To isolate V_f, you need to multiply both sides of the equation by 338 K.

$$\frac{25.5\ \text{cm}^3}{293\ \text{K}} \times 338\ \text{K} = \frac{V_f}{\cancel{338\ \text{K}}} \times \cancel{338\ \text{K}}$$

$$\frac{(25.5\ \text{cm}^3)(338\ \cancel{\text{K}})}{(293\ \cancel{\text{K}})} = V_f$$

$$V_f = 29.42\ \text{cm}^3$$

Ratio method

$$T_i = (20.0°\text{C} + 273) = 293\ \text{K} \qquad T_f = (65.0°\text{C} + 273) = 338\ \text{K}$$

$$V_f = V_i \times \text{ temperature ratio}$$
$$= 25.5°\text{C} \times \frac{338\ \cancel{\text{K}}}{293\ \cancel{\text{K}}}$$
$$= 29.42\ \text{cm}^3$$

Continued ...

PROBLEM TIP

In these Sample Problems, you will see two different methods of solving the problem: the algebraic method and the ratio method. Choose the method you prefer to solve this type of problem.

Continued ...
FROM PAGE 443

Since the least number of significant digits in the question is three, the final volume will be reported to three significant digits.

$$V_f = 29.4 \text{ cm}^3$$

Check Your Solution

- The units for the answer are in cubic centimetres.
- When the units cancel out, cm^3 remains.
- The volume of the oxygen gas increased due to an increase in temperature.

Sample Problem

Charles' Law: Calculating Temperature

Problem

A balloon is filled with 2.50 L of dry helium at 23.5°C. The balloon is placed in a freezer overnight. The next morning, the balloon is removed and the volume is found to be 2.15 L. What was the temperature (in °C) inside the freezer if the pressure remained constant?

What Is Required?

You need to find the temperature of the freezer in °C. ($T_f = ?$)

What Is Given?

- You know the initial volume and temperature.
  Initial volume (V_i) = 2.50 L
  Initial temperature (T_i) = 23.5°C
- You know the final volume.
  Final volume (V_f) = 2.15 L
- You know that the pressure does not change.

Plan Your Strategy

Algebraic method

- Since pressure is constant and the volume and temperature change, you will need to use the Charles' law formula to find the final temperature of the gas sample.
- Since T is given in °C, you need to convert it to kelvins.
- You can substitute numbers and units for the variables in the formula to solve for the unknown (T_f).

Continued ...

Continued ...

FROM PAGE 444

Ratio method

- Since pressure remains constant and the volume of the balloon decreases, you know that according to Charles' law, the temperature of the gas must also decrease.
- Since T is given in °C, you need to convert it to kelvins.
- To find the final temperature inside the freezer, you can multiply the initial temperature by a volume ratio that is less than one.

Act on Your Strategy

Algebraic method

$$T_i = (23.5°\text{C} + 273)$$
$$= 297\ \text{K}$$

$$\frac{V_i}{T_i} = \frac{V_f}{T_f}$$

$$\frac{2.50\ \text{L}}{297\ \text{K}} = \frac{2.15\ \text{L}}{T_f}$$

To simplify the equation and make it easier to solve, you can first cross-multiply the above equation.

$$(2.50\ \text{L})(T_f) = (296.5\ \text{K})(2.15\ \text{L})$$

To isolate T_f, you need to divide both sides of the equation by 2.50 L.

$$\frac{(2.50\ \cancel{\text{L}})(T_f)}{(2.50\ \cancel{\text{L}})} = \frac{(297\ \text{K})(2.15\ \text{L})}{(2.50\ \text{L})}$$

$$T_f = \frac{(297\ \text{K})(2.15\ \cancel{\text{L}})}{(2.50\ \cancel{\text{L}})}$$
$$= 255.42\ \text{K}$$

Since the question asks for the temperature in °C, you need to convert kelvins to °C. To do this, subtract 273 from the answer.

$$T_f = (255.42\ \text{K} - 273)$$
$$= -17.6°\text{C}$$

Ratio method

$$T_i = (23.5°\text{C} + 273)$$
$$= 297\ \text{K}$$

$$T_f = 297\ \text{K} \times \text{volume ratio}$$
$$= 297\ \text{K} \times \frac{2.15\ \cancel{\text{L}}}{2.50\ \cancel{\text{L}}}$$
$$= 255.42\ \text{K}$$

Since the question asks for the temperature in °C, you need to convert kelvins to °C. To do this, subtract 273 from the answer.

$$T_f = (255.42\ \text{K} - 273)$$
$$= -17.6°\text{C}$$

Continued ...

Continued ...

FROM PAGE 445

mind STRETCH

Using the formula $y = mx + b$, see if you can derive the Charles' law formula from the graph that you produced in the ThoughtLab on page 441. For help in deriving equations from graphs, see Appendix E.

Since the least number of significant digits in the question is three significant digits, the final temperature will be reported to the same significant digits.

$T_f = -17.6°C$

Check Your Solution

- The unit for the answer is in kelvins.
- When the units cancel out, kelvins remain.
- Kelvins have been converted to °C.
- The temperature of the balloon has decreased, which is reflected in its decrease in volume.

Practice Problems

5. Convert the following temperatures to the Kelvin scale.

(a) 25°C

(b) 37°C

(c) 150°C

6. Convert the following temperatures to degrees Celcius.

(a) 373 K

(b) 98 K

(c) 425 K

7. Give an example of something that might be at each temperature in question 5.

8. A sample of nitrogen gas surrounding a circuit board occupies a volume of 300 mL at 17°C and 100 kPa. What volume will the nitrogen occupy at 100.0°C if the pressure remains constant?

9. A 2.5 L balloon is completely filled with helium indoors at a temperature of 24.2°C. The balloon is taken out on a cold winter day (–17.5°C). What will the volume of the balloon become, assuming a constant pressure?

10. 10.0 L of neon at 20.0°C is expanded to a volume of 30.0 L. If the pressure remains constant, what must the final temperature be (in °C)?

11. A 14.5 cm^3 sample of oxygen gas at 24.3°C is drawn into a syringe with a maximum volume of 60 cm^3. What is the maximum change in temperature that the oxygen can be subjected to before the plunger pops out of the syringe?

12. Methane gas can be condensed by cooling and increasing the pressure. A 600 L sample of methane gas at 25°C and 100 kPa is cooled to –20°C. In a second step, the gas is compressed until the pressure is quadrupled. What will the final volume be? (**Hint:** Use both Boyle's law and Charles' law to answer this question.)

PROBLEM TIP

In question 8, the smallest number of significant digits is two (17°C). However, before doing your calculations, you must convert this value to kelvins (290 K). This value now has *three* significant digits. Round off your final answer to *three* significant digits.

Gay-Lussac's Law

Aside from balloons and syringes, most containers that are used to store gases have a fixed volume. You know that temperature is a measure of the average kinetic energy of the molecules making up a substance. If the temperature of a gas increases, but the volume of its container cannot increase, what happens to the pressure of the gas inside?

Extending the work of Charles, Joseph Louis Gay-Lussac discovered the relationship between temperature and pressure acting on a fixed volume of a gas. (Remember that for gases, V_{gas} = volume of container holding the gas.) As you will learn later in this section, this relationship is very important for the safe handling of gases under pressure in steel tanks or aerosol cans. **Gay-Lussac's law** states that *the pressure of a fixed amount of gas, at constant volume, is directly proportional to its Kelvin temperature.* (See Figure 11.20.)

$P \alpha T$
(if T is given in kelvins and volume and amount of gas is constant)

Introducing a new proportionality constant (k_2), this relationship can be restated as

$$P = k_2T \qquad \text{or} \qquad \frac{P}{T} = k_2$$

If we assign P_i and T_i as the initial conditions, and P_f and T_f for the final conditions, the above relationship can be rewritten as

$$\frac{P_i}{T_i} = \frac{P_f}{T_f}$$

As you will notice, this mathematical relationship is very similar to that of Charles' law.

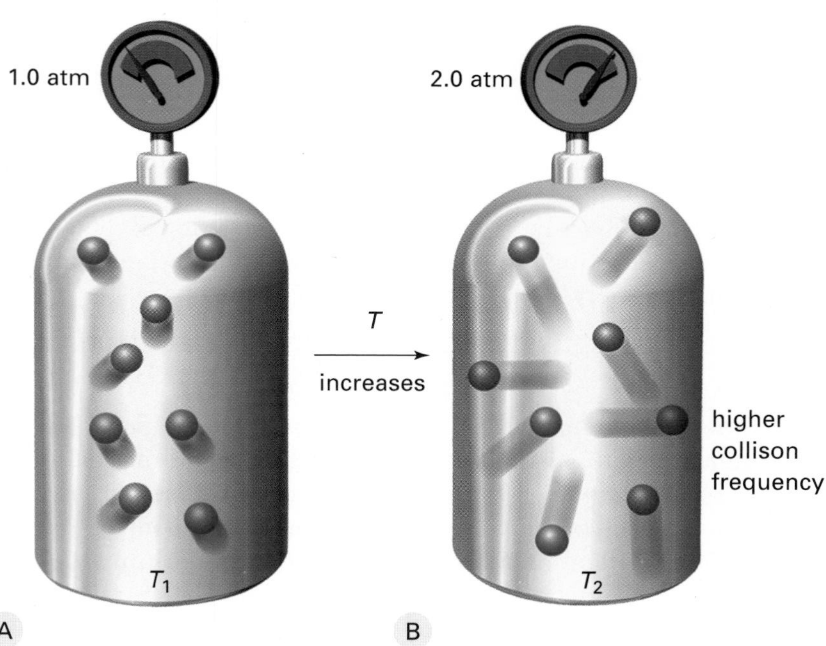

Figure 11.20 **When temperature increases from T_i to T_f in a rigid container with a constant volume, the average speed of the gas molecules increases. Since the molecules move faster, they collide with each other and with the walls of the container more forcefully and more frequently. The gas pressure increases.**

Technology LINK

Underinflated vehicle tires contribute to unsafe road handling and to lower fuel economy. Based on what you know about Gay-Lussac's law, why should you measure the pressure in vehicle tires before driving the vehicle for a long distance?

Web LINK

www.school.mcgrawhill.ca/resources/
Refrigerators and air conditioners function because of the relationship between pressure and temperature. Research the Joule-Thomson Effect. Go to the web site above. Go to **Science Resources**, then to **Chemistry 11** to find out where to go next.

Sample Problem

Gay-Lussac's Law: Calculating Pressure

Problem

A cylinder of chlorine gas (Cl_2) is stored in a concrete-lined room for safety. The cylinder is designed to withstand 50 atm of pressure. The pressure gauge reads 35.0 atm at 23.2°C. An accidental fire in the room next door causes the temperature in the storage room to increase to 87.5°C. What will the pressure gauge read at this temperature?

What Is Required?

You need to find the pressure of the oxygen once the temperature has been increased. (T_f = ?)

What Is Given?

- You know the initial pressure and temperature.
  Initial pressure (P_i) = 35.0 atm
  Initial temperature (T_i) = 23.2°C
- You know the final temperature.
  Final temperature (T_f) = 87.5°C
- You know that the volume of the rigid metal tank will not change appreciably.

PROBLEM TIP

In these Sample Problems, you will see two different methods of solving the problem: the algebraic method and the ratio method. Choose the method you prefer to solve this type of problem.

Plan Your Strategy

Algebraic method

- Since the volume of the cylinder is essentially constant, and the temperature increases, you will need to use the Gay-Lussac's law formula to find the final pressure of the gas sample.
- Since T is given in °C, you need to convert it to kelvins.
- You can substitute numbers and units for the variables in the formula to solve for the unknown. (P_f = ?)

Ratio method

- Since the volume of the cylinder is essentially constant, and the temperature increases, you know that according to Gay-Lussac's law, the pressure exerted by the gas increases as well.
- Since T is given in °C, you need to convert it to kelvins.
- To find the final pressure of the gas, you can multiply the initial pressure by a temperature ratio that is greater than one.

Continued ...

Continued ...
FROM PAGE 448

Act on Your Strategy

Algebraic method

$T_i = (23.2°C + 273) = 296\ K$ $T_f = (87.5°C + 273) = 360\ K$

$$\frac{P_i}{T_i} = \frac{P_f}{T_f}$$

$$\frac{35.0\ atm}{296\ K} = \frac{P_f}{361\ K}$$

To simplify the equation and make it easier to solve, you can first cross-multiply the above equation.

$(35.0\ atm)(361\ K) = (P_f)(296\ K)$

To isolate P_f, you need to divide both sides of the equation by 296 K.

$$\frac{(35.0\ atm)(361\ K)}{(296\ K)} = \frac{(P_f)\cancel{(296\ K)}}{\cancel{(296\ K)}}$$

$$\frac{(35.0\ atm)(361\ \cancel{K})}{(296\ \cancel{K})} = P_f$$

$$P_f = 42.69\ atm$$

Ratio method

$T_i = (23.2°C + 273) = 296\ K$ $T_f = (87.5°C + 273) = 361\ K$

$$P_f = 35.0\ atm \times \text{temperature ratio}$$
$$= 35.0\ atm \times \frac{361\ \cancel{K}}{296\ \cancel{K}}$$
$$= 42.69\ atm$$

Since the least number of significant digits in the question is three, the final pressure will be rounded off to the same number of significant digits.

$$P_f = 42.7\ atm$$

Check Your Solution

- The unit for the answer is in atmospheres.
- When the units cancel out, atm remains.
- Kelvins have been converted to °C.
- The pressure inside the cylinder has increased, as would be expected when the temperature increases.

Practice Problems

13. An unknown gas is collected in a 250.0 mL flask and sealed. Using electronic devices, it is found that the gas inside the flask exerts a pressure of 135.5 kPa at 15°C. What pressure will the gas exert if the temperature (in Kelvins) is doubled?

Continued ...

Continued ...
FROM PAGE 449

14. At 18°C, a sample of helium gas stored in a metal cylinder exerts a pressure of 17.5 atm. What will the pressure become if the tank is placed in a closed room where the temperature increases to 40°C?

15. A gaseous refrigerant, enclosed in copper tubes, surrounds the freezer in a small refrigerator. The gas is found to exert a pressure of 110 kPa at 45°C. The refrigerant is allowed to expand through a nozzle into an expansion chamber such that the exerted pressure decreases to 89 kPa. What is the temperature inside the freezer?

16. Before leaving on a trip to Florida, you measure the pressure inside the tires of your car at a gas station. At −7.5°C the tire pressure is found to be 206.5 kPa. When you arrive in Florida, you stop for dinner. Before leaving, you once again measure the tire pressure at a gas station beside the restaurant. Most pressure gauges in the United States are calibrated in psi. You find the tire pressure to be 34.3 psi. What is the approximate temperature in Florida? (**Hint:** See the MathLink on page 428 to find out how to convert psi to kPa.)

Compressed Gases and Safety Concerns

The Gay-Lussac's law Sample Problem in this section indicates how carefully gases under pressure must be handled. Chlorine gas can cause serious respiratory problems and irritate the skin and mucous membranes. In extreme cases, death from suffocation could result from exposure to this gas. Yet chlorine is an important industrial product. Compounds of chlorine are used in bleaches, oxidizing agents, and solvents, and as intermediates in the manufacture of other substances.

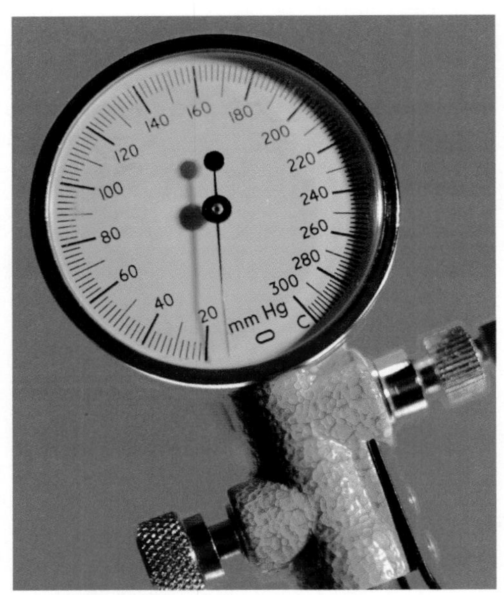

Figure 11.21 Gas cylinders have pressure valves to help the user regulate the amount of gas escaping when the cylinder is opened.

Compressed gases are commonly stored in thick-walled metal cylinders designed especially for this purpose. All cylinders must comply with Canadian Transport Commission (CTC) regulations. Containers must be permanently marked with a serial number and specifications for the volume of the cylinder and the maximum pressure it can withstand. Containers must be tested every five to ten years, with the date of the test stamped on the cylinder.

Figure 11.22 on the next page shows a typical compressed gas cylinder. You can see that it is built to withstand high pressures. There are other safety precautions as well, however.

Most cylinders used to store gases have safety devices regulating the internal gas pressure. The most common of these is a **pressure relief valve**. If the pressure inside the cylinder increases to a dangerous level, a spring allows the valve to open and release excess gas until the internal pressure returns to a safe level. Some pressure relief valves will close once excess gas is released. These valves are relatively expensive compared to non-reclosing valves. Non-reclosing valves are found on common household products such as aerosol hairsprays.

Cylinders used to store gases such as acetylene (C_2H_2), that could cause an explosive chemical reaction at high temperatures, are fitted with **fusible plugs**. These plugs are designed to melt and allow gas to escape at temperatures lower than those at which hazardous reactions can start. Fusible plugs for acetylene cylinders are made of a metal alloy that melts at 100°C.

Not all compressed gas cylinders are fitted with pressure-relief valves or fusible plugs. Cylinders containing toxic gases such as chlorine or phosgene ($COCl_2$) are one example. These gases could cause serious harm to health if released into the air in sufficient quantities. Therefore, these gases, like all compressed gases, must be handled with great care. They must be stored in a well-ventilated, dry area. The surrounding storage area must be fire-resistant, and proper fire-fighting equipment must be immediately available. Gas cylinders should never be stored near electrical circuits that might spark or near other ignition sources such as open flames. Material Data Safety Sheets must be available, and all cylinders must be clearly labelled with WHMIS warning signs.

Figure 11.22 Pressure relief valves prevent a compressed gas cylinder from exploding if the temperature, and thus the pressure increases.

Section Wrap-up

In this section, you learned about the relationship between volume and temperature (Charles' law). You also learned about the relationship between pressure and temperature (Gay-Lussac's law). In the next section, you will see how these relationships can be combined with Boyle's law to produce one equation that works in all three situations.

Section Review

1. K/U What safety concerns and precautions should be taken with compressed gases? Use what you know about the movement of particles to explain these precautions.

2. (a) I A gas at 107 kPa and 300 K is cooled to 146 K at the same volume. What is the new pressure?

   (b) I 17 L of gas at 300 K are cooled to 146 K at the same pressure. What is the new volume?

3. MC Describe the relationship between your answers in parts (a) and (b) of question 2.

4. C Explain in terms of molecular motion why, when the temperature is increased:

   (a) a gas increases in volume

   (b) the pressure of a gas increases

5. K/U A balloon at a party drifts above a hot stove, and explodes. Why did this happen?

11.4 Combined Gas Law Calculations

Section Preview/ Specific Expectations

In this section, you will

- **solve** problems involving the combined gas law and Dalton's law of partial pressures
- **identify** the components of Earth's atmosphere
- **communicate** your understanding of the following terms: *standard temperature, standard temperature and pressure, standard ambient temperature and pressure, combined gas law, Dalton's law of partial pressures*

You may have heard a common joke about Canadian weather: "If you don't like it, wait an hour and it will change." While this is an exaggeration, atmospheric pressure and temperature rarely remain constant for any extended period of time. Since the volume of gases changes when pressure and temperature change, standards have been designed to allow a comparison of different gas volumes.

The average pressure of the atmosphere at sea level is taken as standard pressure (760 mm Hg = 760 torr = 1 atm = 101.3 kPa). The freezing point of water (0°C or 273 K) is defined as **standard temperature**. Together, these conditions are referred to as **standard temperature and pressure (STP)**. (See Figure 11.23.) The normal conditions under which we live are referred to as **standard ambient temperature and pressure**. These conditions are known as **SATP**, defined as 25°C and 100 kPa.

How could we find what the volume of a gas, measured under different conditions, would be when changed to STP or SATP?

Figure 11.23 This photograph illustrates typical STP conditions: 0°C at sea level.

Boyle established that pressure and volume are inversely proportional:

$$P_iV_i = P_fV_f$$

Charles found that volume and temperature are directly proportional:

$$\frac{V_i}{T_i} = \frac{V_f}{T_f}$$

Gay-Lussac discovered that pressure has the same relationship to temperature as volume does:

$$\frac{P_i}{T_i} = \frac{P_f}{T_f}$$

Do you notice a pattern here?

Pressure and volume are directly related to temperature, and inversely related to each other. Write this as one law, and it is possible to calculate situations in which three variables change at the same time. The mathematics work out just as consistently as in the two-variable equations.

CHECKPOINT

Under what conditions might you use standard ambient temperature and pressure as a reference rather than standard temperature and pressure ?

The Combined Gas Law

Boyle's law can be used to solve for changes in volume when pressure changes. The gas must be in a closed system and the temperature must remain constant. You can use Charles' law to solve for changes in volume with temperature changes. This law works only in a closed system in which pressure remains constant. Gay-Lussac's law can solve problems in which the amount and volume of gas remain constant while the temperature and pressure change.

As you learned, Boyle's law ($V \propto 1/P$) is expressed mathematically as $PV = k$. Charles' law ($V \propto T$) is expressed mathematically as $V/T = k_1$ when temperature is recorded in kelvins. Combining these two expressions gives:

$$V \propto \frac{1}{P} \times T \qquad \text{or} \qquad V \propto \frac{T}{P}$$

Introducing a new proportionality constant (k_3), we can write

$$V = \frac{T}{P} \times k_3 \qquad \text{or} \qquad \frac{PV}{T} = k_3$$

This mathematical relationship is the **combined gas law**.

If a sample of a gas is trapped at a measured set of initial conditions, the combined gas law can be rewritten as

$$\frac{P_i V_i}{T_i} = k_3$$

As this gas is then subjected to changes in pressure and temperature, the final condition of the gas can be described mathematically as

$$\frac{P_f V_f}{T_f} = k_3$$

Since both expressions are equal to the same proportionality constant (k_3), the combined gas law can be written as

$$\frac{P_i V_i}{T_i} = \frac{P_f V_f}{T_f}$$

Sample Problem

Finding Volume: The Combined Gas Law

Problem

Sandra is having a birthday party on a mild winter's day. The weather changes and a higher-pressure (103.0 kPa) cold front (–25°C) rushes into town. The original air temperature was –2°C and the pressure was 100.8 kPa. What will happen to the volume of the 4.2 L balloons that were tied to the front of the house?

What Is Required?

You need to find the volume of the balloons under the new conditions of temperature and pressure. (V_f = ?)

Continued ...

Continued ...
FROM PAGE 453

What Is Given?

- You know the initial pressure, volume, and temperature.
  Initial pressure $(P_i) = 100.8$ kPa
  Initial volume $(V_i) = 4.2$ L
  Initial temperature $(T_i) = -2^{\circ}$C
- You know the final pressure and temperature.
  Final pressure $(P_f) = 103.0$ kPa
  Final temperature $(T_f) = -25^{\circ}$C

Plan Your Strategy

Algebraic method

- Since pressure and temperature both change, use the combined gas law to find the final volume of the balloon.
- Since T is given in °C, you need to convert it to kelvins.
- You can rearrange the combined gas law and substitute numbers and units for the variables in the formula to solve for V_f.

Ratio method

- Since pressure and temperature change, you know that the volume of the balloon will also change.
- Since T is given in °C, you need to convert it to kelvins.
- To find the new volume based on the increase in pressure, you need to multiply the initial volume by a pressure ratio that is less than one.
- To find the new volume based on the decrease in temperature, you need to multiply the initial volume by a temperature ratio that is less than one.
- To find the new volume based on both pressure and temperature changes, you can multiply the initial volume by the pressure and temperature ratios.

PROBLEM TIP

In these Sample Problems, you will see two different methods of solving the problem: the algebraic method and the ratio method. Choose the method you prefer to solve this type of problem.

Act on Your Strategy

Algebraic method

$T_i = (-2^{\circ}\text{C} + 273) = 271 \text{ K}$ $\qquad T_f = (-25^{\circ}\text{C} + 273) = 248 \text{ K}$

$$\frac{P_iV_i}{T_f} = \frac{P_fV_f}{T_f}$$

Solve the combined gas law, in this case for V_f.

Continued ...

Continued ...
FROM PAGE 454

To isolate V_f, move T_f and P_f. Multiply T_f up to the numerator on the left side, and divide P_f down to the denominator. This leaves

$$\frac{P_iV_iT_f}{T_iP_f} = V_f$$

$$\frac{(100.8\ \cancel{kPa})(4.2\ L)(248\ \cancel{K})}{(271\ \cancel{K})(103.0\ \cancel{kPa})} = V_f$$

$$V_f = 3.76\ L$$

$$\text{So} \quad V_f = 3.8\ L$$

Ratio method

$T_i = (-2°C + 273) = 271\ K$ $\quad$ $T_f = (-25°C + 273) = 248\ K$

$V_f = V_i \times$ pressure ratio $\times$ temperature ratio

$$= 4.2\ L \times \frac{100.8\ \cancel{kPa}}{103.0\ \cancel{kPa}} \times \frac{248\ \cancel{K}}{271\ \cancel{K}}$$

$$= 3.76\ L \cong 3.8\ L$$

Remember to change the answer to correct significant digits. The least number of digits in the question was two, so the answer must have only two significant digits. $V_f = 3.8$ L.

Check Your Solution

- The unit for the answer is in litres.
- When the units cancel out, L remains.
- The volume of the balloon has decreased, as would be expected when pressure increases and temperature decreases.

Sample Problem

Combined Gas Law

Problem

An automated instrument has been developed to help drug-research chemists determine the amount of nitrogen in a compound. Any compound containing carbon, nitrogen, and hydrogen is reacted with copper(II) oxide to produce CO_2, H_2O, and N_2 gases. The gases are collected separately and analyzed.

In an analysis of 39.8 mg of caffeine using this instrument, 10.1 mL of N_2 gas is produced at 23°C and 746 torr. What must the new temperature of nitrogen be, in °C, if the volume is increased to 12.0 mL, and the pressure is increased to 780 torr?

What Is Required?

You need to find the temperature of the nitrogen under the new conditions of volume and pressure. ($T_f = ?$)

Continued ...

Continued ...

FROM PAGE 455

What Is Given?

- You know the initial pressure, volume, and temperature.
  Initial pressure $(P_i) = 746$ torr
  Initial volume $(V_i) = 10.1$ mL
  Initial temperature $(T_i) = 23$°C
- You know the final pressure and volume.
  Final pressure $(P_f) = 780$ torr
  Final volume $(V_f) = 12.0$ mL

Plan Your Strategy

Algebraic method

- Since pressure and volume both change, you will need to use the combined gas law formula to find the final temperature of the nitrogen.
- Since T is given in °C, you need to convert it to kelvins.
- You can rearrange the combined gas law and substitute numbers and units for the variables in the formula to solve for T_f.

Ratio method

- Since pressure and volume change, you know that the temperature of the nitrogen will also change.
- Since T is given in °C, you need to convert it to kelvins.
- To find the new temperature based on the increase in pressure, you need to multiply the initial volume by a pressure ratio that is greater than one.
- To find the new temperature based on an increase in volume, you need to multiply the initial temperature by a volume ratio that is less than one.
- To find the new temperature based on both pressure and volume changes, you can multiply the initial temperature by the pressure and volume ratios.

Act on Your Strategy

Algebraic method

$T_i = (23°\text{C} + 273) = 296\text{ K}$

$$\frac{P_iV_i}{T_i} = \frac{P_fV_f}{T_f}$$

Divide both sides by P_fV_f to isolate T_f.

$$\frac{P_iV_i}{T_iP_fV_f} = \frac{1}{T_f}$$

Now flip both sides of the equation over:

$$\frac{T_iP_fV_f}{P_iV_i} = \frac{T_f}{1}$$

Continued ...

FROM PAGE 456

Alternatively, you could have divided down P_iV_i and multiplied both T_f and T_i up. This is an equally correct procedure and the result would be the same. Put in the numbers:

$$\frac{(296\text{ K})(780\ \cancel{\text{torr}})(12.0\ \cancel{\text{ml}})}{(746\ \cancel{\text{torr}})(10.1\ \cancel{\text{ml}})} = T_f$$

$$T_f = 368\text{ K}$$

Since the question asks for the temperature in °C, you need to convert kelvins to °C. To do this, subtract 273 from the answer.

$T_f = (368 - 273)$
$= 95°\text{C}$ (Note the correct number of significant digits.)

Ratio method

$T_i = (23°\text{C} + 273) = 296\text{ K}$

$T_f = T_i \times$ pressure ratio $\times$ temperature ratio

$$= 296\text{ K} \times \frac{780\text{ torr}}{746\text{ torr}} \times \frac{12.0\text{ mL}}{10.1\text{ mL}}$$

$= 368\text{ K}$

Since the question asks for the temperature in °C, you need to convert kelvins to °C. To do this, subtract 273 from the answer.

$T_f = (368 - 273)$
$= 95°\text{C}$ (Note the correct number of significant digits.)

Check Your Solution

- The unit for the answer is in degrees Celcius.
- The temperature of the nitrogen has increased, as would be expected when pressure and volume increase.

Practice Problems

17. A sample of gas has a volume of 150 mL at 260 K and 92.3 kPa. What will the new volume be at 376 K and 123 kPa?

18. A cylinder at 48 atm pressure and 290 K releases 35 mL of carbon dioxide gas into a 4.0 L container at 297 K. What is the pressure inside the container?

19. In a large syringe, 48 mL of ammonia gas at STP is compressed to 24 mL and 110 kPa. What must the new temperature of the gas be?

20. A 100 W light bulb has a volume of 180.0 cm^3 at STP. The light bulb is turned on and the heated glass expands slightly, changing the volume of the bulb to 181.5 cm^3 with an internal pressure of 214.5 kPa. What is the temperature of the light bulb (in °C)?

21. Sulfur hexafluoride, $SF_{6(g)}$, is used as a chemical insulator. A 5.0 L sample of this gas is collected at 205.0°C and 350 kPa. What pressure must be applied to this gas sample to reduce its volume to 1.7 L at 25°C?

Gases and Natural Phenomena

Gases under pressure are responsible for natural phenomena such as volcanoes and geysers. On March 20, 1980, residents in the northern part of Washington state heard a rumble from the mountains. They were told it was only a minor earthquake. By March 27, seismologists were sure that something more was involved. Deep inside one of the mountains, Earth's crust was moving. The lower portion of the crust was melting into hot liquid called *magma*. The magma started rising up through cracks in the crust. Trapped water quickly turned into superheated steam. Trapped gases added to the increasing pressure inside the mountain. Then, on May 18, 1980, the top half of Mount St. Helens blew away in a gigantic explosion. It released all of the built-up pressure, along with many tonnes of rock and ash, twenty kilometres into the atmosphere. It also released steam, and gases such as carbon dioxide, nitrogen, and sulfur dioxide. (See Figure 11.24.)

Figure 11.24 Before a volcano erupts, there is a tremendous build-up of fluid and gas pressures inside the volcano due to magma, steam, and gases.

Other gases released in volcanic explosions include the oxides, sulfides, and chlorides of carbon, sulfur, hydrogen, chlorine, and fluorine. These include CO_2, CO, SO_2, SO_3, H_2, H_2S, Cl_2, HCl, and F_2. It is estimated that the eruption of Mt. Pinatubo in the Philippines on June 15, 1991 released about 1.8×10^{10} kg of gases and ash into the atmosphere.

After a volcanic blast, or in an area of volcanic activity, geysers may form. Cool water from Earth's surface trickles down between rocks. It drains deep into Earth's crust, into regions where hot magma is still present. As the water is heated, it forms steam. The steam rises back up though fissures and cracks, meeting more water and heating it up as well. More and more boiling water accumulates, until suddenly, through a narrow opening, the pressurized hot water is violently ejected. As the water cools in the air, it falls back and the process starts again. Yellowstone National Park in the United States has one of the most famous geysers in the world, *Old Faithful*. It erupts approximately once every sixty-five minutes. (Figure 11.25 shows Old Faithful erupting.)

Figure 11.25 Geysers form in areas of early volcanic activity or after a volcanic blast. They should be approached with caution since the hot water that erupts can cause severe burns.

In Chapter 12, you will learn more about the gas chemistry of our atmosphere.

Dalton's Law of Partial Pressures

What if there are two gases in one container, or as in the case of the atmosphere, a mixture of many gases? How does this affect the pressure?

The English scientist John Dalton did a very thorough analysis of the atmosphere. He concluded that it comprised about 79% nitrogen and 21% oxygen. (See Table 11.2 to find out how close he really was.) Dalton noticed that the water vapour, however, seemed quite variable, so he did further experiments. He obtained some very dry air, and measured the pressure in the container. He then introduced some water vapour. The pressure increased! Dalton repeated and adjusted his experiment time after time, always with the same results. He concluded that *the total pressure of a mixture of gases is the sum of the pressures of each of the individual gases.* This is called **Dalton's Law of partial pressures.** (See Figure 11.27 below.)

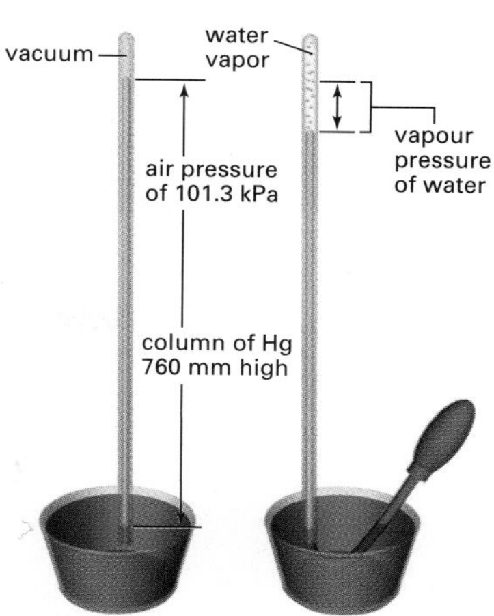

Figure 11.26 A method of determining water vapour pressure. A drop of water vapourizes in the vacuum and the water vapour exerts a pressure on the mercury in the tube.

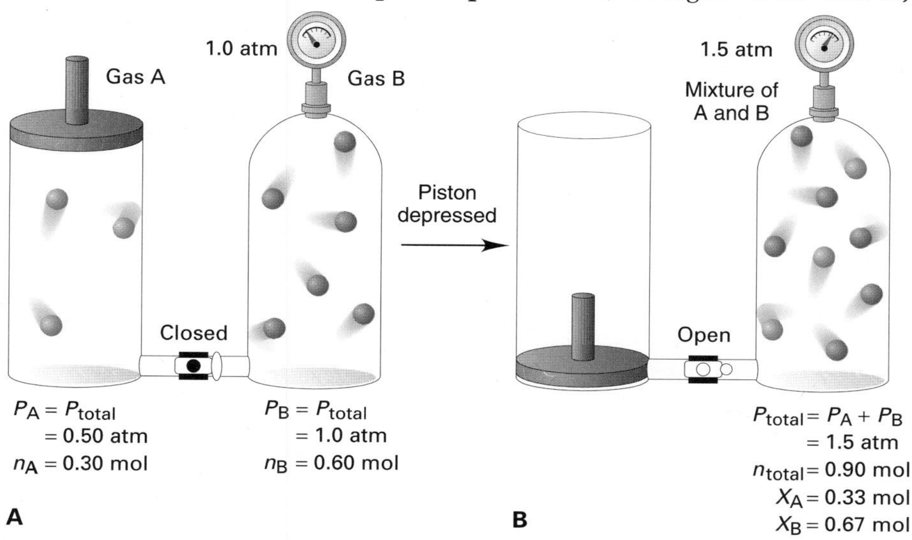

Figure 11.27 Gas A is stored at 0.5 atm and gas B at 1.0 atm. When gas A is added to gas B, the total pressure exerted is now 1.5 atm while the volume of the flask remains the same.

Table 11.2 Components of the Dry Atmosphere

Components	Percentage
Nitrogen (N_2)	78.08
Oxygen (O_2)	20.95
Argon (Ar)	0.93
Carbon dioxide (CO_2)	0.03
Neon (Ne)	0.002
Other gases	0.008

The action of winds mixes the atmosphere so that the composition of the dry atmosphere is fairly constant over the entire Earth. Water vapour, though an important component of the atmosphere, is not listed in the table. The quantity of water vapour in the air, or *humidity*, is variable. In desert climates, the quantity of water vapour will be very small (low humidity). In tropical areas, or near large bodies of water, the quantity of water vapour in the air can be quite substantial (high humidity).

Dalton extended this idea to enhance what he had discovered about the composition of the atmosphere. If the atmosphere is 79% nitrogen, and the atmospheric pressure on a certain day is, for example, 101.3 kPa, then he concluded that the nitrogen itself must be contributing

$$\frac{79}{100} \times 101.3 \text{ kPa} = 80 \text{ kPa}$$

The other gases in the atmosphere contribute pressures corresponding to their percentage of the total composition of air.

The generalized form of Dalton's law of partial pressures is

$$P_{total} = P_1 + P_2 + P_3 + \ldots + P_n$$

This law can be applied to any mixture of gases. Study the following Sample Problem and try the Practice Problems to verify your understanding of how Dalton's law works. Figure 11.26, above, shows one way of determining the vapour pressure of water in the laboratory.

Sample Problem

Applying Dalton's Law of Partial Pressures

Problem

What is the pressure contribution of CO_2 to the atmospheric pressure on a very dry day when the barometer read 0.98 atm? Convert your answer into three different units.

What Is Required?

You must find the contribution of CO_2 to the atmospheric pressure, and then convert the units.

What Is Given?

- According to Table 11.2, CO_2 contributes 0.03% of the atmospheric pressure. The total atmospheric pressure is 0.98 atm. The conversion factors of pressure are
  1 atm = 760 torr = 760 mm Hg = 101.3 kPa

Plan Your Strategy

- You need to multiply the total atmospheric pressure times CO_2's contribution to the total.
- Then you can multiply the answer in atm by the conversion factors to cancel out atm and obtain other units.

Act on Your Strategy

$$\frac{0.03}{100} \times 0.98 \text{ atm} = 2.9 \times 10^{-4} \text{ atm}$$

$$2.9 \times 10^{-4} \cancel{\text{atm}} \times \frac{101.3 \text{ kPa}}{1 \cancel{\text{atm}}} = 0.030 \text{ kPa}$$

$$2.9 \times 10^{-4} \cancel{\text{atm}} \times \frac{760 \text{ torr}}{1 \cancel{\text{atm}}} = 0.22 \text{ torr} = 0.22 \text{ mm Hg}$$

Check Your Solution

- The answers given are all very small. This is expected, as CO_2 only contributes a small percentage of gas molecules to the atmosphere.

Practice Problems

22. To speed up a reaction in a vessel pressurized at 98.0 kPa, a chemist added 202.65 kPa of hydrogen gas. What was the resulting pressure?

23. A gas mixture contains 12% Ne, 23% He, and 65% Rn. If the total pressure is 116 kPa, what is the partial pressure of each gas?

24. The partial pressure of argon gas, making up 40% of a mixture, is 325 torr. What is the total pressure of the mixture in kPa?

25. A mixture of nitrogen and carbon dioxide gas is at a pressure of 1.00 atm and a temperature of 278 K. If 30% of the mixture is nitrogen, what is the partial pressure of the carbon dioxide?

Molar Mass and Gas Behaviour

In the following ThoughtLab, you will examine one more factor affecting gas behaviour:

ThoughtLab Boiling Points of Gases

What effect does molar mass have on the behaviour of a gas? In fact, molar mass helps to determine whether a compound is gaseous or not. You will examine the connection between boiling point and molar mass here.

Table 11.3 The Boiling Points of Various Gases

Gas	Boiling Point (°C)
He	−269
Ne	−246
N_2	−196
Ar	−186
O_2	−183
Kr	−153
Xe	−108

Materials

graph paper; periodic table

Procedure

1. Using a periodic table, record the molar mass of each of the gases listed in the table.
2. Plot a graph of boiling points of gases (*y*-axis) versus their molar masses (*x*-axis).

Analysis

1. From the graph obtained, what relationship exists between the molar mass of a gas and its boiling point?
2. Using the graph, what would be the boiling points of gases such as hydrogen (H_2), fluorine (F_2) and radon (R_n)?
3. Predict whether a substance such as bromine (Br_2) would be a gas or liquid at room temperature. Use a reference book to check your answer.

Section Wrap-up

In this section, you learned how to use the combined gas law for gas calculations. You also learned about natural phenomena that are related to gases. Finally, you learned about Dalton's law of partial pressures. In Chapter 12, you will learn more about gas laws. First, however, you will take a closer look at some technological applications of the gas laws.

Section Review

1. K/U In a large syringe, 48 mL of ammonia gas at STP is compressed to 24 mL and 110 kPa. What must the new temperature of the gas be? Explain the result in terms of kinetic molecular theory.
2. I Design an investigation to test Dalton's law of partial pressure. Write a procedure and list the materials and equipment you need.
3. C Explain Dalton's law of partial pressure.
4. K/U What are the main components of the atmosphere?
5. MC Popcorn pops as the water in the kernel is vapourized. The building pressure becomes too much for the shell of the kernel, and it explodes. Explain, using your knowledge of the behaviour of gases, what happens to the temperature just before and just after the kernel pops. (Use a diagram to help you to visualize the situation.)

11.5 Gas Applications

Section Preview/ Specific Expectations

In this section, you will

- **identify** technological products based on compressed gas
- **describe** how a knowledge of gases is used in other areas of study
- **communicate** your understanding of the following term: *fuel cell*

In the previous section, you learned about the colourless, odorless mixture of gases that make up the atmosphere. It is easy to take the air around us for granted. You know that you need oxygen to breathe, but does it serve any other purposes? Nitrogen is not required for respiration, but it composes about four-fifths of our atmosphere. Is nitrogen a "useless" gas? In this section, you will learn about how gases are used. You will find out more about oxygen and nitrogen, and discover the importance of gases in deep-sea exploration.

Compressed Oxygen

You may never really think about breathing—until it becomes hard to do. In certain situations, normal human respiratory functions are disrupted. Hospitals use compressed oxygen for patients with respiratory disorders such as emphysema, pneumonia, or lung cancer.

You may have heard of hyperbaric oxygen (HBO) chambers being used to treat sports injuries. A high oxygen concentration in the blood causes blood vessels to constrict. This lowers the swelling that can occur in injured tissues. How is this high oxygen content delivered to the blood? In an HBO chamber such as that shown in Figure 11.28, air is compressed to three times normal atmospheric pressure. The patient breathes pure oxygen ($O_{2(g)}$) through a mask. The highly compressed air forces $O_{2(g)}$ in the lungs into the blood. This increases of dissolved $O_{2(g)}$ in the bloodstream. Consequently, oxygen is delivered rapidly to all the blood vessels of the body. In fact, oxygen can reach the injured tissue at 15 times the normal rate!

The use of compressed oxygen can benefit not only athletes, but also vulnerable premature babies. Premature babies can be afflicted with hyaline-membrane disease. This condition prevents the alveoli in their lungs from inflating, which leads to serious breathing difficulties. Placing these babies in an oxygen-rich environment such as an HBO or an incubator (Figure 11.28) helps inflate the alveoli. This increases the infants' chances for survival.

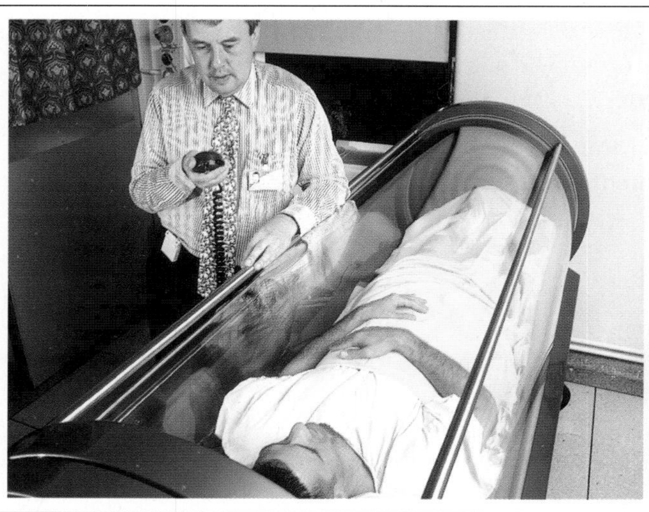

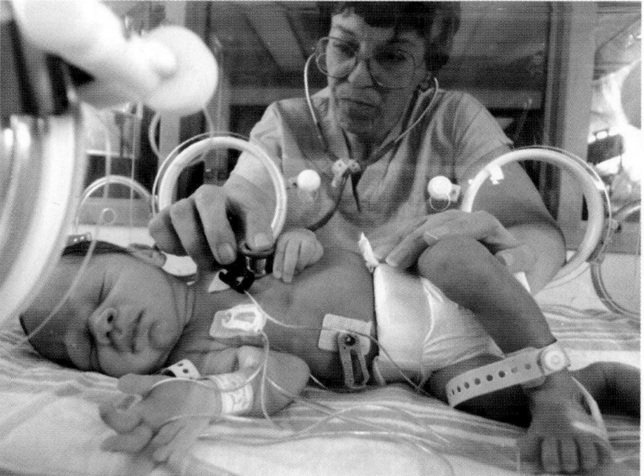

Figure 11.28 Medical applications of pressurized oxygen can help the very strong, the very weak, and everyone in between.

Oxygen at High Altitudes

At other times, respiratory functions may be normal, but the environment poses problems. Commercial jet planes fly at an altitude of around 9.5 km. Airplanes carry pressurized oxygen for their passengers to breathe as they fly at high altitudes. Mountain climbers also need compressed oxygen when climbing to great heights. Those exploring ocean depths need oxygen as well. Scuba divers and submarine crews need compressed oxygen to breathe when they are underwater. You will learn more about undersea exploration later in this section.

Careers  in Chemistry

Cancer Research Specialist

The National Cancer Institute of Canada estimates that in the year 2000 there were more than 132 000 new cases of cancer and 65 000 deaths caused by cancer in Canada. People are desperate for treatment.

Dr. Abdullah Kirumira, President and lead scientist at BioMedica Diagnostics in Windsor, Nova Scotia, has developed a way to speed up drug testing. This method uses compressed carbon dioxide and a dry chemistry diagnostics system including the firefly enzyme, luciferase.

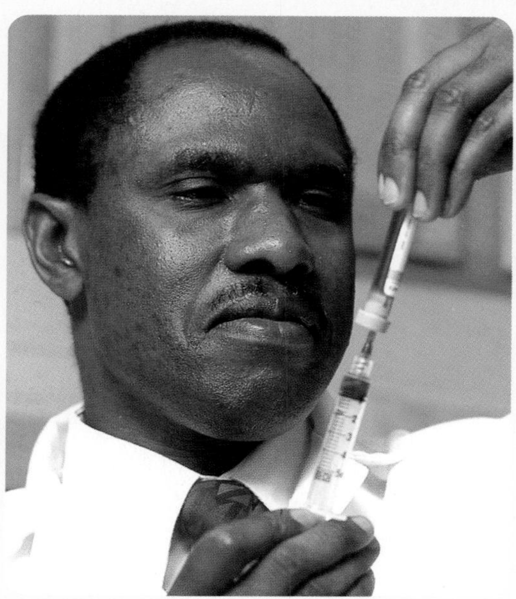

Growing Cancer Cells for Testing

To obtain a consistent supply of cancer cells to test, CO_2 is pumped from a compressed gas cylinder. It is depressurized through a regulator into a specially designed incubator kept at about 37°C, the optimum temperature for cell growth. This CO_2-enriched atmosphere plays two roles:

- it is required for rapid cellular metabolism; and
- it interacts with a bicarbonate buffer in the growth medium to maintain an optimum pH balance.

Global Ties

In 1975, during the reign of dictator Idi Amin, Dr. Kirumira fled his native Uganda to Iraq, where he earned a B.Sc. in food technology. He then travelled to England for a Master's degree in dairy chemistry. From there, he moved to Australia where he earned his Ph.D. in biotechnology. In 1990, he and his wife moved to Nova Scotia in search of a slower-paced lifestyle.

At BioMedica, Dr. Kirumira oversees 12 employees, six of whom have advanced degrees, three technicians and three administrators. In addition, Dr. Kirumira teaches biochemistry and biotechnology at Acadia and Dalhousie Universities.

He frequently visits his home in Uganda. From Uganda, Dr. Kirumira says he can "take back what I have learnt from the West to help the people who probably need it the most."

Make Career Connections

1. To learn more about the biotech applications of using the firefly enzyme, luciferase, you can go to BioMedica's web site.
2. What universities offer degrees in biotechnology? Do research to find two universities near you that offer this program.

Web LINK

www.school.mcgrawhill.ca/resources/

CO_2 is a greenhouse gas, thought to contribute to global warming. Many countries around the world have sought to reduce their emissions of CO_2. International climate control treaties, such as the Montreal Protocol, have set targets to do this.

Since plants cycle CO_2 out of the atmosphere, several countries including Canada have argued that countries with large forest resources should be able to claim them as "carbon sinks." Rather than cutting their CO_2 emissions by the target amount, Canada and other countries maintain that they should be allowed to claim a reduction for the CO_2 absorbed by their large forests.

Research the current debate over "carbon sinks." Do you agree with the idea that CO_2 absorbed by plant life provides a valid strategy for controlling pollution? Go to the web site above. Go to **Science Resources**, then to **Chemistry 11** to find out where to go next.

Oxygen and Combustion

A fire breaks out and smoke begins to fill the room. Breathing becomes difficult. Should you open a window to let the smoke out before you leave through the door? You may have learned you should never open windows before leaving the area of a fire. A fire can only burn as long as it has sufficient oxygen. (See Figure 11.29.)

Oxygen is an extremely important gas because it supports combustion. (You will learn more about combustion reactions in Unit 5.) The more oxygen available, the hotter a fire will burn. Pressurized oxygen is used in the manufacture of steel and specialized alloys. In the manufacture of steel, oxygen is used to remove excess carbon by burning the carbon into carbon monoxide or carbon dioxide. High levels of carbon make steel too brittle for many uses.

A remarkable invention called the **fuel cell** is used by the space industry. Some fuel cells burn hydrogen and oxygen gas, leaving water as a by-product (which astronauts can then drink). Companies such as Ballard Power Systems Inc. of Vancouver, British Columbia are working to make this technology practical for automobiles. Fuel cells are completely non-polluting. However, the current process for making large amounts of hydrogen gas does create a good deal of pollution, in the form of CO_2. Coal or another hydrocarbon is used to heat up and decompose water in a reaction similar to the following:

$$CH_{4(g)} + 2H_2O_{(g)} \xrightarrow{\text{heat}} CO_{2(g)} + 4H_{2(g)}$$

Before astronauts can use the fuel cell in space, they first have to get there! Liquid oxygen is an important component of rocket fuel. Liquid hydrogen and liquid oxygen are mixed and ignited in a combustion chamber. This reaction provides a very rapid expansion of gas. It produces enough energy to lift extremely heavy rockets carrying crewed vessels, such as the Space Shuttle, into orbit (Figure 11.30).

Figure 11.29 Liquid water becomes water vapour when exposed to the heat of a fire. Firefighters must take precautions to avoid steam burns.

Figure 11.30 The large centre tank that is attached to the shuttle orbiter has two compartments, one containing liquid hydrogen and the other containing liquid oxygen. The energy released by the reaction between these two substances provides the thrust needed to propel the orbiter into space.

Nitrogen's Many Uses

Inert gases are not very reactive. The word *inert* is also a synonym for sluggish and slow. Something with the property of inertness hardly sounds useful! Yet this very property makes nitrogen extremely valuable in technological and industrial applications.

Because nitrogen gas reacts with very few substances, it can be used to blanket substances, preventing them from reacting with oxygen. (A reaction with oxygen is called oxidation.) *Blanketing* means covering something with nitrogen, displacing all of the oxygen. This usually happens in a closed container. For example, ground coffee can taste bitter when exposed to oxygen for long periods of time. Packaging coffee in a nitrogen-enhanced container helps coffee keep its flavour. Gaseous nitrogen is also used for the long-term storage of fruits such as apples, allowing us to enjoy a wide variety of produce out of season.

Liquid nitrogen's low boiling point (77 K) means that it can be used in food preservation. Foods frozen quickly in liquid nitrogen retain more nutrients than slowly frozen foods. Less damage is done to cell structure during the quick-freezing process. Freezing foods this way also removes moisture, decreasing their weight and size. Freeze-dried foods are often used by people who require easily portable, lightweight food, such as campers, the military, and space explorers.

In cryosurgery, liquid nitrogen is used to fast-freeze cancerous tissues and warts. This proven new technique kills the cancerous area and allows surgeons to safely remove the dead tissue.

Gases and Undersea Exploration

Throughout history, humans have tried to explore the little-known world of the deep seas. Even with modern diving gear, the deep oceans are a dangerous place for humans. The air that a diver breathes to stay alive underwater can itself be one of the greatest hazards. That's because of the tremendous pressure exerted by the waters above a diver. For roughly every 10 metres of depth, water pressure on the diver's body adds the equivalent of one unit of atmospheric pressure (Figure 11.31).

The air that a diver breathes (mainly nitrogen and oxygen) must equal the pressure of the surrounding water if the diver's lungs are to stay inflated. Of course, this gas pressure increases as a diver goes deeper. As the gas pressure increases, the diver's bloodstream and body tissues absorb higher and higher volumes of gases.

Beyond depths of roughly 40–60 metres, the increased volumes of gas in the diver's body can cause *nitrogen narcosis*. When nitrogen narcosis strikes, normal feelings become dangerously exaggerated. A diver may experience a blissful giddiness, along with disorientation and impaired judgement. This condition is called "rapture of the deep." Divers in this state have been known to remove their breathing apparatus to swim like fish, mistake up for down, or simply lose track of how long they have been underwater.

At the other extreme, a diver's normal sense of caution may degenerate into irrational fear or even panic. A panicked diver may do the very

Figure 11.31 At 50 m underwater, a diver experiences pressure about five times higher than normal.

worst thing—rush to the surface without pausing to decompress on the way up. Why is this dangerous?

The dissolved gases such as nitrogen in the diver's bloodstream are under pressure. This is much like the carbon dioxide that is dissolved in a can of soda. If the diver returns to the surface too quickly, these dissolved gases behave similarly to the gas in soda when a can is opened. The gases escape from the blood as bubbles in the diver's blood vessels. These bubbles can block the flow of blood. They produce a painful and potentially fatal condition called "the bends."

Both nitrogen narcosis and the bends make prolonged deep diving a risky business. They limit both the safe depth and safe duration of human diving. Even skilled professional divers rarely descend beyond about 50 metres. Also, they rarely remain at that depth for much more than 30 minutes.

Electronic Learning Partner

Your Chemistry 11 Electronic Learning Partner has a demonstration of CO_2 fire extinguishers and the chemistry behind their use.

Section Wrap-up

In this section, you learned about some technological products and applications involving gases. You also learned that a knowledge of gases is important in other areas of study, such as medicine and deep-sea exploration.

In Chapter 12, you will find out about the ideal gas law. This law covers the many different gas laws you explored in this chapter. You will also discover a practical application for Dalton's law of partial pressures. You will learn how to do stoichiometric calculations for reactions that consume or produce a gas. In the laboratory, you will have a chance to produce and collect a gas. At the end of the next chapter, you will examine some of the chemistry that takes place in our atmosphere.

Section Review

Unit Issue Prep

What gases are produced as a result of human technologies such as fuel-burning engines?

1. C Why is fuel-cell technology, in its present state, not a pollution-free alternative to the internal combustion gasoline engine?
2. K/U Describe three industrial uses for liquid nitrogen.
3. MC How might covering apples with nitrogen or carbon dioxide gas preserve them longer through the winter?
4. K/U How does a hyperbaric oxygen chamber work to assist human healing?
5. MC Look up the standard mixture of gases for scuba diving tanks. What problem arises in using nitrogen gas in scuba tanks? How is this problem solved?
6. MC What does lightning (or bacteria) do to the nitrogen in the diatomic gas in order to make the atoms useful to plants? Use what you know about compounds and elements to answer this question.
7. MC Interview the science teachers at your school. Are there any compressed gas cylinders in the school? Ask the teachers about the appropriate safety precautions for compressed gases. Then prepare a short safety report on compressed gases in your school.

CHAPTER 11 Review

Reflecting on Chapter 11

Summarize this chapter in the format of your choice. Here are a few ideas to use as guidelines.

- Describe the structure of solids, liquids, and gases in simple terms.
- Compare the intermolecular forces that hold solids together with those that hold liquids and gases together.
- Use the kinetic molecular theory to describe the behaviour of ideal gases.
- Explain Boyle's law.
- Explain Charles' law.
- Explain Gay-Lussac's law.
- Describe how gases under pressure are the cause of some of the natural wonders of our Earth.
- Describe how gases under pressure have many industrial and medical uses.
- Explain Dalton's law of partial pressure.

Reviewing Key Terms

For each of the following terms, write a sentence that shows your understanding of its meaning.

atmospheres
Boyle's law
closed system
Condensation
Dalton's law of partial pressures
fusible plugs
ideal gas
Kelvin scale
kinetic molecular theory of gases
pascal
standard atmospheric pressure
standard ambient temperature and pressure (SATP)
absolute zero
Charles' law
combined gas law
fuel cell
Gay-Lussac's law
kilopascals
mmHg
pressure-relief valve
standard temperature and pressure (STP)
standard temperature
standard atmospheric pressure

Knowledge/Understanding

1. Gases behave differently than solids and liquids. Using what you know about forces between molecules, explain some of these differences.
2. Using the kinetic molecular theory, answer each of the following questions.
   (a) What does it mean when we say that the pressure of a gas has increased?
   (b) What does it mean when we say that the temperature of a gas has decreased?
3. (a) Give a real example that illustrates Boyle's law.
   (b) Give an example that illustrates Charles' law.
4. What is the kinetic molecular theory explaination for 0 K?
5. Explain Gay-Lussac's law in terms of the motion of gas particles.
6. Explain Dalton's law of partial pressures in terms of the motion of gas particles.
7. How is a volcano an example of a natural occurrence in which pressures of gases are of major importance?
8. List four major and four minor components of the atmosphere.
9. How is pure oxygen used medically and industrially?
10. Nitrogen is one of the most important industrial substances produced on Earth. Why is it so important industrially?
11. Would the column of mercury in a barometer be shorter at the top of a mountain or at the base of the mountain? Explain.
12. Using the kinetic molecular theory, explain why the density of a gas is less than that of a liquid. (Density is mass per unit of volume.)
13. A weather balloon containing helium is released into the atmosphere. As it rises, atmospheric pressure and temperature both decrease. Explain why the size of the balloon increases.

Inquiry

14. Below are two diagrams showing a trapped gas. The volume of the gas in container A can be changed to become the same volume as the gas in container B. Describe how you could do this in the laboratory.

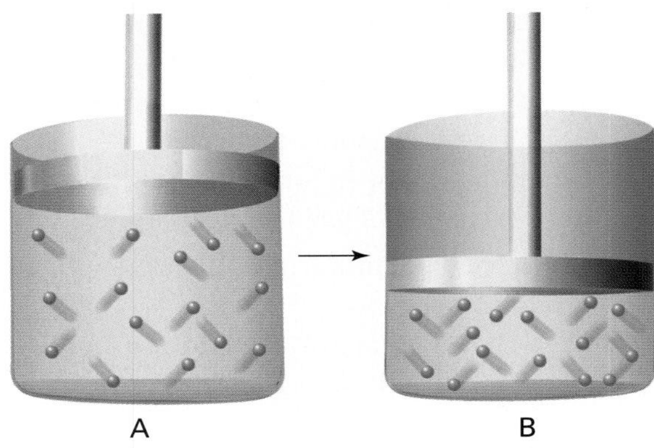

15. You have carried out investigations to determine the relationship between pressure and volume, and between temperature and volume. How could you investigate the relationship between pressure and temperature?
   (a) Write up a procedure to determine the relationship between the pressure and temperature of a gas.
   (b) Identify any problems with your procedure. How could you overcome these problems?
   (c) What materials will you need?
   (d) What variables will you hold constant?
   (e) What variable will you change?
   (f) What variable will you measure?
16. The pressure exerted on 0.25 L of nitrogen is 120 kPa. What volume will this gas occupy at 60 kPa if the temperature and number of moles is constant?
17. Ammonia gas, $NH_{3(g)}$, is used in the production of fertilizer. At 55.0°C, a sample of ammonia gas is found to exert a pressure of 7.5 atm. What pressure will the gas exert if its volume is reduced to one fifth of its original volume at 55.0°C?
18. A 35.0 L sample of dry air at 750 torr is compressed to a volume of 20 L. What is the final pressure exerted on the gas if the temperature remains constant?
19. A 25 g sample of dry air in a large party balloon at 20°C occupies a volume of 20 L. If the temperature is increased to 40°C at constant pressure, how large will the balloon become?
20. A sample of nitrogen gas has a volume of 10 L at 101.3 kPa and 20°C. To preserve biological tissue, the nitrogen gas is cooled to –190°C, almost the temperature of liquid nitrogen, at a pressure of 101.3 kPa. What volume will the nitrogen occupy at this temperature?
21. A 75.3 L sample of oxygen gas at 25.7°C is cooled until its final volume becomes 10 L. If the pressure remains constant, what is the final temperature (in °C)?
22. Calculate the volumes that each of the following gases would occupy at STP.
   (a) 22.4 L of oxygen at 75°C and 700 torr.
   (b) 100 cm^3 of nitrogen at 20°C and 150.0 kPa.
   (c) 45 mL of neon at –50.°C and 200 kPa.
23. A birthday balloon contains 2.0 L of air at STP. What volume will the balloon have at SATP?
24. A mixture of neon and argon gases is collected at 102.7 kPa. If the partial pressure of the neon is 52.5 kPa, what is the partial pressure of argon?
25. A 250 mL glass vessel is filled with krypton gas at a pressure of 700 torr at 25.0°C. If the glass vessel is made to withstand a pressure of 2.0 atm, to what maximum temperature (in °C) can you safely heat the flask?
26. A cylinder with a moveable piston contains hydrogen gas collected at 30°C. The piston is moved until the volume of the hydrogen is halved. The pressure inside the container has increased to 125 kPa at 30°C. What was the initial pressure inside the cylinder?
27. Argon gas is used inside light bulbs because it is a plentiful inert gas. 650 cm^3 of argon gas at STP is heated in order to double its volume at 101.3 kPa. What is the final temperature (in °C)?
28. A truck leaves Yellowknife in early January when the temperature is –30.0°C. The tires of the truck are inflated to 210 kPa. Four days later, the truck arrives in California where the temperature is 30.0°C. What is the air pressure in the tires when the truck arrives at its destination?
29. Methane (CH_4), a natural gas, is stored in a 100 L tank at –10°C and a pressure of 125 atm. The gas is used to provide fuel for a furnace that heats a country home during the winter. The furnace consumes an average of 500 L of methane a day. How long will this supply of methane last if it is burned at 450°C at a pressure of 102 kPa?

30. Neon gas is widely used as the luminous gas in signs. A sample of neon has a volume of 5.5 L at 750 torr at 10.0°C. If the gas is expanded to a volume of 7.5 L at a pressure of 400 torr, what will its final temperature be (in °C)?

31. Halogen lamp bulbs are usually filled with bromine or iodine vapour at 5.0 atm pressure. When turned on, the glass bulb can heat up to more than 1150°C. If room temperature is 20°C, what will the pressure in the bulb be when it reaches its operating temperature?

32. Helium gas is stored in a steel cylinder with a volume of 100 L at 20°C. The pressure gauge on the cylinder indicates a pressure of 25 atm. The cylinder is used to blow up a weather balloon at 25°C. If the final pressure in the cylinder and the balloon is 1.05 atm, how large will the balloon be?

33. A scuba diver is swimming 30.0 m below the surface of Lake Ontario. At this depth, the pressure of the water is 4.0 atm and the temperature is 8.0°C. A bubble of air with a volume of 5.0 mL escapes from the diver's mask. What will the volume of the bubble be when it breaks the surface of the water? The atmospheric pressure is 101.3 kPa and the temperature of the water is 24.0°C.

Communication

34. The weather person on television reports that the barometric pressure is 100.2 kPa. How high a column of mercury will this air pressure support?

35. Convert each of the following pressures:
   (a) 1.5 atm to kPa
   (b) 135.5 kPa to mm Hg
   (c) 750 mm Hg to torr

36. On a visit to your doctor, your blood pressure is taken. The reading is 125.0 mm Hg systolic and 80.0 mm Hg diastolic. What is your blood pressure in kPa?

37. Convert the following temperatures.
   (a) −185.5°C to K
   (b) 125 K to °C

38. Do research to find out how carbonated drinks get their "fizz." Prepare a short PowerPoint™ presentation to the class that describes the manufacture of a typical soft drink.

Making Connections

39. Coal is a fossil fuel that is burned to obtain energy. However, carbon dioxide (CO_2) gas is produced whenever coal is burned. What are the benefits and risks of burning coal to obtain energy? What are the alternatives?
   (a) Do research to answer these questions.
   (b) *Should the government shut down all coal-burning power plants in Canada to help prevent global warming?* Carry out this debate with your class. Divide the class into the executives of a coal-burning power plant, and scientists representing an environmentalist group.

40. Interview a doctor or nurse who works at a hospital. Find out how compressed gases are used at the hospital, and what safety precautions are taken.

41. Suppose you are a member of a consulting firm that is approached by the Ministry of the Environment. You are asked to prepare a report on how the government can help reduce CO_2 and CH_4 gas emissions. Both gases are important greenhouse gases that may be causing global warming.
   (a) Identify the information you will need to prepare this report. This may take the form of a list of questions.
   (b) Research the answers to your questions.
   (c) Prepare a short report giving suggestions on how the Canadian government could help reduce CO_2 and CH_4 gas emissions.

Answers to Practice Problems and Numerical Section Review Questions

Practice Problems: **1.** 1.0×10^3 kPa **2.** 1644 L **3.** 4.5 cm^3 **4.** 0.01 L **5.(a)** 298 K **(b)** 310 K **(c)** 423 K **6.(a)** 100°C **(b)** −175°C **(c)** 152°C **8.** 3.86×10^2 mL **9.** 2.1 L **10.** 606°C **11.** 957°C **12.** 127 L **13.** 2.71×10^2 kPa **14.** 18.8 atm **15.** −16°C **16.** 31°C **17.** 163 mL **18.** 0.43 atm **19.** 148 K **20.** 310°C **21.** 6.4×10^2 kPa **22.** 300.6 kPa **23.** 14 kPa, 27 kPa, 75 kPa **24.** 8.1×10^2 kPa **25.** 0.70 atm
Section Review: **11.1** **3.(a)** liquid **(b)** gas **(c)** solid **11.2** **1.(a)** 206 kPa **(b)** 0.841 atm **(c)** 1.14×10^3 torr **(d)** 80.0 kPa **4.** 1.04 L **5.** 152 kPa **6.** 811 torr **11.3** **2.(a)** 52.1 kPa **(b)** 8.3 L **11.4** **1.** 148 K

CHAPTER 12

Exploring Gas Laws

You are probably used to sharing scientific ideas and observations with your classmates as you work in pairs or in small groups. If you are familiar with Internet chat rooms and e-mail, you may even share your ideas with people around the world. Sharing scientific ideas is an essential part of scientific discovery. Back in the early nineteenth century, the ideas that led to the complete gas laws were shared between colleagues in much the same way that you share scientific ideas with your classmates.

During the nineteenth century, scientists across Europe organized *academies*, or science societies. (This practice still exists today.) Belonging to an academy allowed much more communication among scientists. They read the papers and reports that other scientists had published, wrote letters, and held meetings to discuss ideas. They worked together to develop many important theories and laws, including the gas laws.

In this chapter, you will study more of the gas laws. Although these laws were discovered almost 200 years ago, they are still accepted today. You will learn how the gas laws allow you to solve many problems involving gas behaviour. You will also learn about some of the modern applications and technological advances associated with the gas laws. You will have a chance to test the gas laws through your own experiments and to compare your results with those of early scientists. Finally, you will learn about some of the gas reactions that occur in our atmosphere.

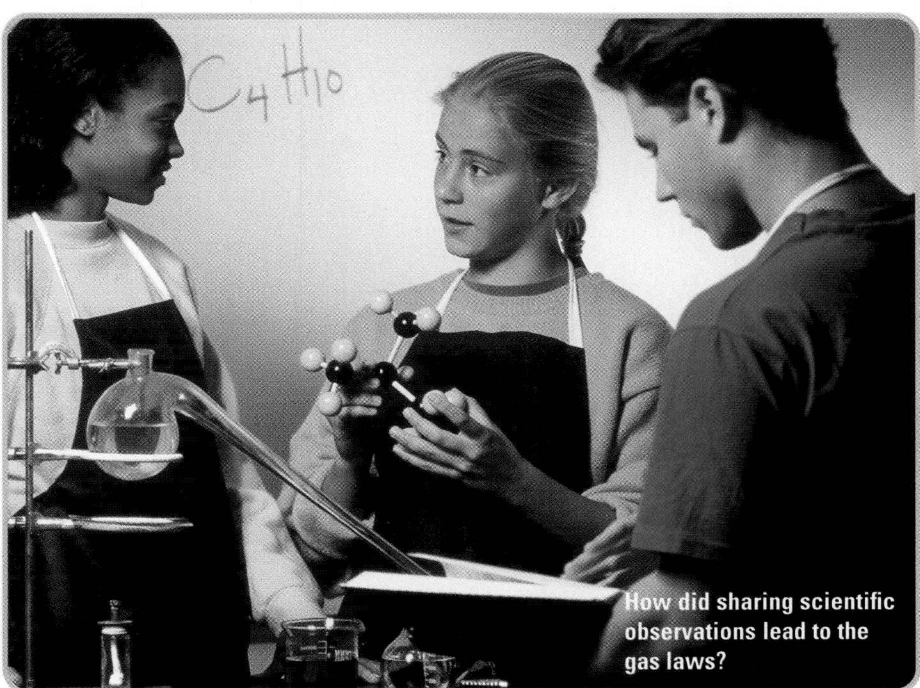

How did sharing scientific observations lead to the gas laws?

Chapter Preview

Concepts and Skills You Will Need

Before you begin this chapter, review the following concepts and skills:

- manipulating equations algebraically, and applying them to solve problems (Chapter 11, section 11.4)
- using correct significant digits in problem solving (Chapter 1, section 1.2)
- converting an amount in grams to an amount in moles, and then to a number of atoms or molecules (Chapter 5, section 5.3)
- solving stoichiometry problems, including writing balanced equations (Chapter 4, section 4.1 and Chapter 7, section 7.1)

12.1 The Ideal Gas Law

Section Preview/ Specific Expectations

In this section, you will

- **state** Avogadro's hypothesis, and **explain** how it contributes to our understanding of the reactions of gases
- **describe** the quantitative relationships that exist among pressure, volume, temperature, and amount of substance, based on the ideal gas law
- **solve** quantitative problems involving the ideal gas law
- **use** and **convert** appropriate units to express pressure and temperature
- **communicate** your understanding of the following terms: *law of combining volumes, law of multiple proportions, Avogadro's hypothesis, molar volume, ideal gas law*

Near the beginning of the nineteenth century, Joseph Gay-Lussac experimented with the volumes of gases. He found that adding two volumes of gas to one volume of gas produced only two volumes of gas. Puzzled, Gay-Lussac tried adding three volumes of gas to one volume. The result was still two! When he tried adding one volume of gas to a second volume of gas, again the result was two. What was going on?

In England, around the same time, John Dalton studied the masses of compounds as they reacted to produce products. After Dalton read about the similar work of other scientists, such as Lavoisier and the British scientist Joseph Priestley, he contacted Gay-Lussac. He described his results and hypotheses to Gay-Lussac. In 1808, both men published their theories. After examining the theories of Dalton and Gay-Lussac, an Italian scientist named Amedeo Avogadro formulated a hypothesis that combined their theories.

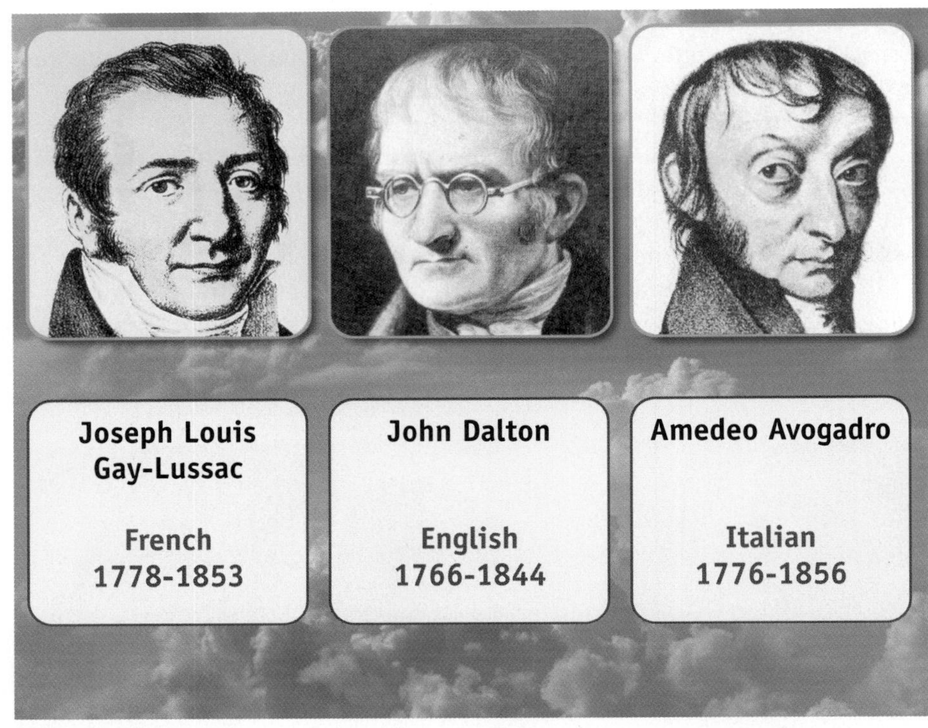

Figure 12.1 **The information that was shared by these three scientists led to the gas laws that we use today.**

The Molar Volume of Gases

Gay-Lussac measured the *volumes* of gases before and after a reaction. His research led him to devise the **law of combining volumes**: When gases react, the volumes of the reactants and the products, measured at equal temperatures and pressures, are always in whole number ratios. For example, 2 volumes of hydrogen gas react with 1 volume of oxygen gas to produce 2 volumes of water vapour.

John Dalton examined the *masses* of compounds before and after a reaction. Dalton's research led him to propose the **law of multiple proportions**: The masses of the elements that combine can be expressed in small whole number ratios.

By combining these ideas, Avogadro related the *volume* of a gas to the *amount* that is present (calculated from the mass). Avogadro divided Dalton's mass ratios by the molar masses of the elements to obtain the mole ratios. He realized that these mole ratios were the same as the volume ratios that Gay-Lussac had obtained. For example, 1 L of hydrogen gas reacts with 1 L of chlorine gas. Avogadro decided that there must be the same number of molecules in each litre of gas. Thus, **Avogadro's hypothesis** was formulated: Equal volumes of all ideal gases at the same temperature and pressure contain the same number of molecules.

Figures 12.2, 12.3, and 12.4 show the three reactions that produced Gay-Lussac's confusing observations. You can see that the mole ratios are the same as the volume ratios. Today our knowledge of atoms and molecules helps us understand Gay-Lussac's results. We know that gases are made of molecules that may contain more than one atom.

Language LINK

The word *symposium* comes from the ancient Greeks. A symposium was a gathering of intellectuals to drink, feast, and talk. All sorts of new ideas arose from these meetings. The participants took away the new ideas to work on them further. Then they brought their findings back to the next symposium.

Many scientific laws are given the name of one person. It is important to remember, however, that most laws are the culmination of many scientists' work over a long time. Science would never move forward without ideas being shared.

Look up the word "symposium" in a dictionary. What meaning does it have today? Do a quick Internet search of the word "symposium." What modern symposiums can you find? What subjects do they cover?

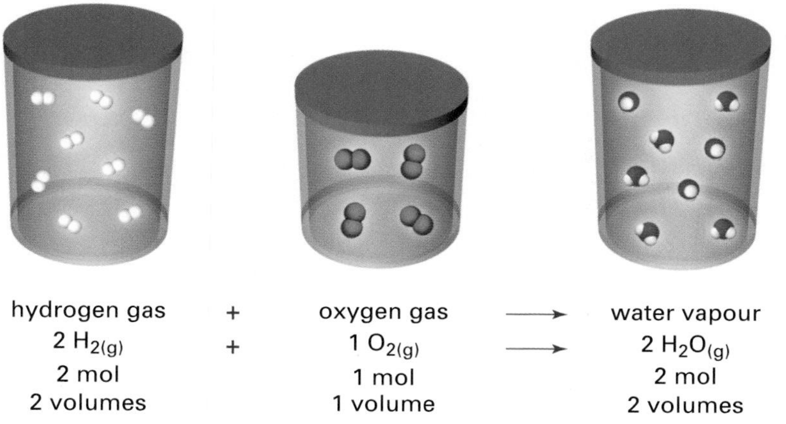

hydrogen gas	+	oxygen gas	→	water vapour
$2\ H_{2(g)}$	+	$1\ O_{2(g)}$	→	$2\ H_2O_{(g)}$
2 mol		1 mol		2 mol
2 volumes		1 volume		2 volumes

Figure 12.2 Hydrogen and oxygen gases combine to form water vapour.

mind STRETCH

As a class, hold a symposium about gas balloons. Research hot air balloons and helium balloons. Also research any other information about balloons, the history of ballons, and balloon travel that interests you. Prepare papers and posters to share your information.

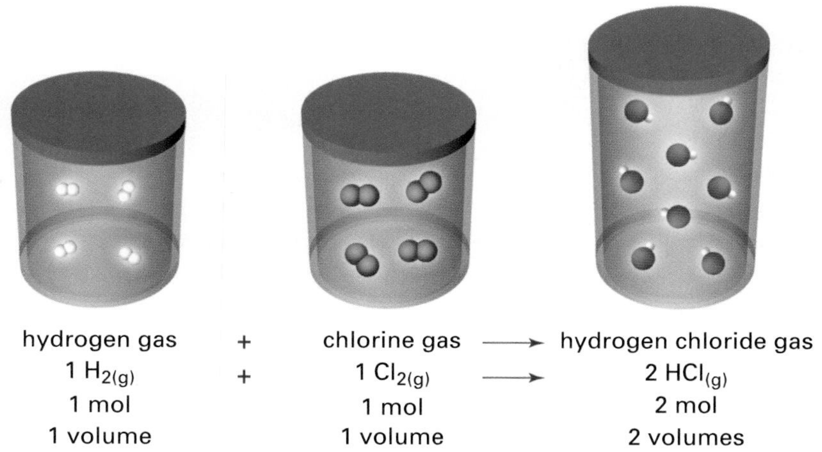

hydrogen gas	+	chlorine gas	→	hydrogen chloride gas
$1\ H_{2(g)}$	+	$1\ Cl_{2(g)}$	→	$2\ HCl_{(g)}$
1 mol		1 mol		2 mol
1 volume		1 volume		2 volumes

Figure 12.3 Hydrogen and chlorine gases combine to form hydrogen chloride gas.

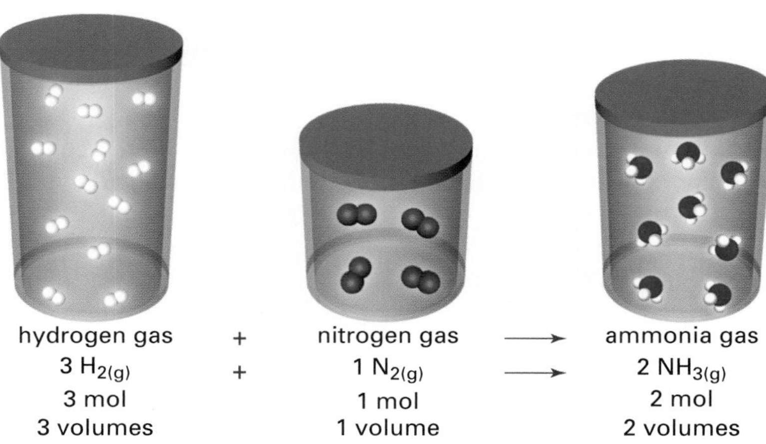

Figure 12.4 **Hydrogen and nitrogen combine to form ammonia gas.**

Avogadro's hypothesis can be written as a mathematical law, shown here.

Avogadro's Law

Avogadro's law gives us a mathematical relationship between the volume of a gas (V) and the number of moles of gas present (n).

$$n \propto V \quad \text{or} \quad n = kV \quad \text{or} \quad \frac{n_1}{V_1} = \frac{n_2}{V_2}$$

where n = number of moles
V = volume
k = a constant

CHECKPOINT

The next Sample Problem involves the combined gas law. Review this law in section 11.4 before continuing.

Based on Avogadro's law, one mole of a gas occupies the same volume as one mole of another gas at the same temperature and pressure. The molar volume of a gas is the space that is occupied by one mole of the gas. **Molar volume** is measured in units of L/mol. You can find the molar volume of a gas by dividing its volume by the number of moles that are present ($\frac{V}{n}$). Look at the Sample Problem below to find out how to calculate molar volume. Then complete the following Thought Lab to find the molar volumes of carbon dioxide gas, oxygen gas, and methane gas at STP.

Sample Problem

The Molar Volume of Nitrogen

Problem

A resealable 1.30 L container has a mass of 4.73 g. Nitrogen gas, $N_{2(g)}$, is added to the container until the pressure is 98.0 kPa at 22.0°C. Together, the container and the gas have a mass of 6.18 g. Calculate the molar volume of nitrogen gas at STP.

What Is Required?

You need to find the volume of one mole of nitrogen (the molar volume) at STP.

Continued ...

FROM PAGE 474

What Is Given?

Set out all the data in a table like the one shown below.

Situation 1: in the container	Situation 2: at STP
$P_i = 98.0$ kPa	$P_f = 101.3$ kPa
$V_i = 1.30$ L	$V_f = ?$
$T_i = 22.0$°C, or 295 K	$T_f = 0$°C, or 273 K
$m_i = 6.18\text{ g} - 4.73\text{ g} = 1.45$ g (Subtract the mass of the container.)	$m_f = 1.45$ g (The mass remains the same.)
$n_i = ?$	$n_f = n_i = ?$

Plan Your Strategy

Algebraic method

Step 1 Calculate the number of moles of nitrogen gas (n_i) by dividing the mass of the nitrogen in the container by the molar mass of nitrogen gas (28.02 g/mol).

$$n = \frac{m}{M}$$

Step 2 Use the combined gas law from Chapter 11 to find the volume of nitrogen at STP (V_f).

Step 3 Use the volume (V_f) and the number of moles ($n_f = n_i$) to find the molar volume ($\frac{V}{n}$). The molar volume is the volume of one mole of gas.

$$\text{Molar volumes} = \frac{V}{n}$$

Ratio method

Step 1 Calculate the number of moles of nitrogen gas (n_i) by dividing the mass of the nitrogen in the container by the molar mass of nitrogen gas (28.02 g/mol).

$$n = \frac{m}{M}$$

Step 2 Since the pressure increases from 98.0 kPa to 101.3 kPa, the volume will decrease. Multiply the initial volume by a pressure ratio that is less than 1. Since the temperature decreases from 295 K to 273 K, the volume will decrease further. Multiply the initial volume by a temperature ratio that is less than 1.

Step 3 Since there is less than 1 mol of nitrogen gas present, the volume of 1 mol of nitrogen gas (the molar volume) will be greater than the volume you calculated in step 2. To find the molar volume, multiply by a mole ratio that is greater than 1.

Continued ...

PROBLEM TIP

In the Sample Problem, you will see two different methods of solving the problem: the algebraic method and the ratio method. Choose the method you prefer to solve this type of problem.

Continued ...
FROM PAGE 475

Act on Your Strategy

Algebraic method

Step 1 Calculate the number of moles.

$$n = \frac{m}{M}$$
$$= \frac{1.45\ \cancel{g}}{28.02\ \cancel{g}/mol}$$
$$= 0.0517\ mol$$

Step 2 Find the volume of nitrogen at STP.

$$\frac{P_iV_i}{T_i} = \frac{P_fV_f}{T_f}$$
$$\therefore V_f = \frac{P_iV_iT_f}{T_iP_f}$$
$$= \frac{98.0\ \cancel{kPa} \times 1.30\ L \times 273\ \cancel{K}}{295\ \cancel{K} \times 101.3\ \cancel{kPa}}$$
$$= 1.16\ L$$

Step 3 Find the molar volume.

$$\text{Molar volume} = V/n$$
$$= \frac{1.16\ L}{0.0517\ mol}$$
$$= 22.4\ L/mol$$

Therefore, the molar volume of nitrogen at STP is 22.4 L/mol.

Ratio method

Step 1 Calculate the number of moles.

$$n = \frac{m}{M}$$
$$= \frac{1.45\ \cancel{g}}{28.02\ \cancel{g}/mol}$$
$$= 0.0517\ mol$$

Step 2 V_f is the final volume of the nitrogen gas.

$$V_f = 1.30\ L \times \frac{98.0\ \cancel{kPa}}{101.3\ \cancel{kPa}} \times \frac{273\ \cancel{K}}{295\ \cancel{K}}$$
$$= 1.16\ L$$

Step 3 V_m is the volume of 1 mol of the nitrogen gas.

$$V_m = 1.16\ L \times \frac{1.00\ \cancel{mol}}{0.0517\ \cancel{mol}}$$
$$= 22.4\ L$$

The molar volume is equal to the volume V_m divided by 1.00 mol. Thus the molar volume is 22.4 L/mol.

Check Your Solution

The answer is expressed in the correct units. It agrees with the accepted value. There are three significant digits in the answer. This is consistent with the least number of significant digits in the question.

Continued ...

Continued ...
FROM PAGE 476

Practice Problems

1. At 19°C and 100 kPa, 0.021 mol of oxygen gas, $O_{2(g)}$, occupy a volume of 0.50 L. What is the molar volume of oxygen gas at this temperature and pressure?
2. What is the molar volume of hydrogen gas, $H_{2(g)}$, at 255°C and 102 kPa, if a 1.09 L volume of the gas has a mass of 0.0513 g?
3. A sample of helium gas, $He_{(g)}$, has a mass of 11.28 g. At STP, the sample has a volume of 63.2 L. What is the molar volume of this gas at 32.2°C and 98.1 kPa?
4. In the Sample Problem, you discovered that the molar volume of nitrogen gas is 22.4 L at STP.
   (a) How many moles of nitrogen are present in 10.0 L at STP?
   (b) What is the mass of this gas sample?

ThoughtLab  Molar Volume of Gases

Two students decided to calculate the molar volumes of carbon dioxide, oxygen, and methane gas. First they measured the mass of an empty 150 mL syringe under vacuum conditions. This ensured that the syringe did not contain any air. Next they filled the syringe with 150 mL of carbon dioxide gas. They measured and recorded the mass of the syringe plus the gas. The students repeated their procedure for oxygen gas and for methane gas.

Finally, the students found the temperature of the room to be 23.0°C (296 K). They found the pressure to be 98.7 kPa. They took these values to be the temperature and pressure of the three gases. The students' results are given in the table.

Procedure

1. Copy the table into your notebook.
2. Calculate the molar volume of carbon dioxide gas at the given temperature and pressure, and at STP. Write your calculations and answers in the table.
3. Do the same calculations for oxygen and methane gas. Write your calculations and answers in the table.

Analysis

1. Compare the three molar volumes at STP. What do you observe?
2. The accepted molar volume of a gas at STP is 22.4 L/mol. Use this value to calculate the percent error in your experimental data for each gas.

Three Gases at 296 K and 98.7 kPa

Gas	carbon dioxide	oxygen	methane
Volume of gas (*V*)	150 mL	150 mL	150 mL
Mass of empty syringe	25.08 g	25.08 g	25.08 g
Mass of gas + syringe	25.34 g	25.27 g	25.18 g
Mass of gas (*m*)			
Molar mass (*M*)			
Number of moles of gas ($n = m/M$)			
Volume of gas at STP (273 K and 101.3 kPa)			
Molar volume at STP ($MV = V/n$)			

In the Thought Lab, you found that the molar volume of one gas is roughly the same as the molar volume of another gas at the same temperature and pressure. In fact, the molar volume of an *ideal* gas at STP is 22.4 L/mol. Figure 12.5 shows a balloon with a volume of 22.4 L compared to some common objects. This is a fairly large volume of gas. For example, a basketball has a volume of only 7.5 L.

Figure 12.5 One mole of any gas at STP occupies 22.4 L (22.4 dm^3). How large is 22.4 L? The other objects are shown for comparison.

CHECKPOINT

What are the temperature and pressure at STP? Go back to section 11.4 to remind yourself.

In Chapter 11, you learned that temperature, pressure, and volume are related. Based on Avogadro's law, the number of moles is related to the temperature, pressure, and volume of a gas. Therefore, Avogadro's law can be applied to solve gas problems involving moles and volume, when the temperature and pressure remain constant. Figure 12.6 explains the relationship among temperature, pressure, volume, and number of moles of a gas.

The following Sample Problems show you how to do gas calculations using Avogadro's law.

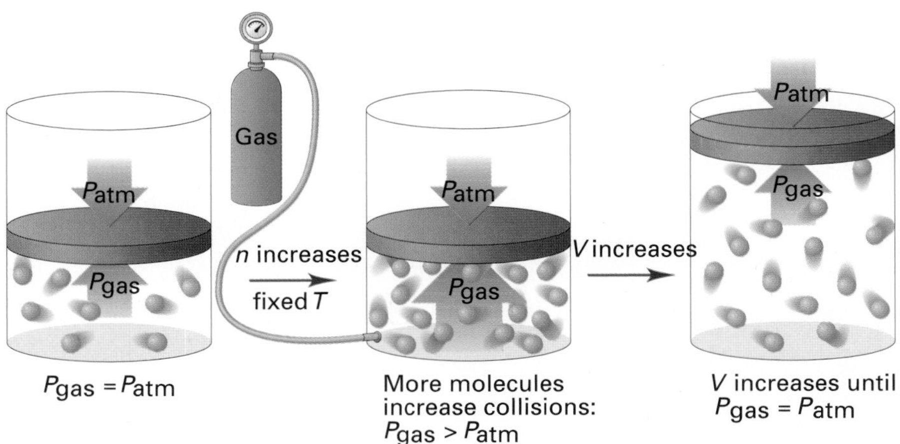

Figure 12.6 At a temperature (T), a given amount of a gas (n) produces a pressure (P). When more gas is added, the number of moles (n) increases. This results in more molecular collisions with the container walls, thus increasing the pressure of the gas. The volume (V) of the gas increases until the pressure is again equal to the pressure of the surroundings.

Sample Problem
Volumes of Gases

Problem
What is the volume of 3.0 mol of nitrous oxide, $NO_{2(g)}$, at STP?

What Is Required?
You need to find the volume of nitrous oxide at STP (V_f).

What Is Given?
You know that 1.00 mol of a gas occupies 22.4 L at STP.

$\therefore n_i = 1.0$ mol

$V_i = 22.4$ L

$n_f = 3.0$ mol of NO_2

Plan Your Strategy

Algebraic method

Use Avogadro's law: $\frac{n_i}{V_i} = \frac{n_f}{V_f}$

Cross multiply to solve for V_f, the unknown volume of NO_2.

Ratio method

There are 3 mol of nitrous oxide. Thus, the volume of nitrous oxide at STP must be larger than the volume of 1 mol of gas at STP. Multiply by a mole ratio that is greater than 1.

Act on Your Strategy

Algebraic method

$$V_f = \frac{n_f V_i}{n_i}$$
$$= \frac{3.0 \cancel{\text{mol}} \times 22.4 \text{ L}}{1.00 \cancel{\text{mol}}}$$
$$= 67 \text{ L}$$

Ratio method

$$V_f = 22.4 \text{ L} \times \frac{3.0 \cancel{\text{mol}}}{1.0 \cancel{\text{mol}}}$$
$$= 67 \text{ L}$$

Therefore, there are 67 L of nitrous oxide.

Check Your Solution
The significant digits and the units are all correct.
The volume of nitrous oxide is three times the volume of 1 mol of gas at STP. This makes sense, since there are 3 mol of nitrous oxide.

PROBLEM TIP

In these Sample Problems, you will see two different methods of solving the problem: the algebraic method and the ratio method. Choose the method you prefer to solve this type of problem.

Sample Problem

Moles of Gas

Problem

Suppose that you have 44.8 L of methane gas at STP.

(a) How many moles are present?

(b) What is the mass (in g) of the gas?

(c) How many molecules of gas are present?

What Is Required?

(a) You need to calculate the number of moles.

(b) You need to calculate the mass of the gas.

(c) You need to calculate the number of molecules.

What Is Given?

The gas is at STP. Thus 1.00 mol of gas has a volume of 22.4 L. You know that one mole contains 6.02×10^{23} molecules. There are 44.8 L of gas.

Plan Your Strategy

Algebraic method

(a) Use Avogadro's law. Solve for the number of moles by cross multiplying.

(b) Multiply the number of moles (n) by the molar mass (M) to find the mass of the gas (m).

$$m = n \times M$$

(c) Multiply the number of moles (n) by the Avogadro constant (6.02×10^{23}) to find the number of molecules.

$$\text{\# of molecules} = \text{n} \times 6.02 \times 10^{23} \text{ molecules/mol}$$

Ratio method

(a) The volume of the unknown gas is 44.8 L. Since the volume is greater than 22.4 L, there is more than 1 mol of gas. To find the unknown number of moles (n), multiply by a volume ratio that is greater than 1.

(b) Multiply the number of moles (n) by the molar mass (M) to find the mass of the gas (m).

$$m = n \times M$$

(c) Multiply the number of moles (n) by the Avogadro constant (6.02×10^{23}) to find the number of molecules.

$$\text{\# of molecules} = \text{n} \times 6.02 \times 10^{23} \text{ molecules/mol}$$

Continued ...

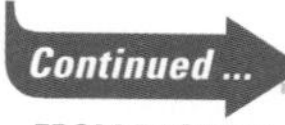

FROM PAGE 480

Act on Your Strategy

Algebraic method

(a) $\dfrac{n_i}{V_i} = \dfrac{n_f}{V_f}$

$$n_f = \frac{n_i V_f}{V_i}$$
$$= \frac{1.00 \text{ mol} \times 44.8 \cancel{\text{L}}}{22.4 \cancel{\text{L}}}$$
$$= 2.00 \text{ mol}$$

(b) Find the molar mass of methane, CH_4.

$1C = 1 \times 12.01$ g/mol

$4H = 4 \times 1.01$ g/mol

$M_{CH_4} = 16.05$ g/mol

$$m = n \times M$$
$$= 2.00 \cancel{\text{mol}} \times 16.05 \text{ g}/\cancel{\text{mol}}$$
$$= 32.1 \text{ g}$$

(c) $\# \text{ molecules} = 2.00 \cancel{\text{mol}} \times 6.02 \times 10^{23} \dfrac{\text{molecules}}{\cancel{\text{mol}}}$

$$= 1.20 \times 10^{24} \text{ molecules}$$

Ratio method

(a) $n = 1.00 \text{ mol} \times \dfrac{44.8 \cancel{\text{L}}}{22.4 \cancel{\text{L}}}$

$$= 2.00 \text{ mol}$$

(b) Find the molar mass of methane, CH_4.

$1C = 1 \times 12.01$ g/mol

$4H = 4 \times 4.04$ g/mol

$M_{CH_4} = 16.05$ g/mol

$$m = n \times M$$
$$= 2.00 \text{ mol} \times 16.05 \text{ g/mol}$$
$$= 32.1 \text{ g}$$

(c) $\# \text{ molecules} = 2.00 \cancel{\text{mol}} \times 6.02 \times 10^{23} \dfrac{\text{molecules}}{\cancel{\text{mol}}}$

$$= 1.20 \times 10^{24} \text{ molecules}$$

Therefore, 2.00 mol of methane are present. The mass of the gas is 32.1 g. 1.20×10^{24} molecules are present.

Check Your Solution

The significant digits are correct.
The volume of methane is double the volume of 1 mol of gas. It makes sense that 2 mol of methane are present. It also makes sense that the number of molecules present is double the Avogadro constant.

PROBLEM TIP

To solve many of these problems, try setting up a proportion and solving by cross multiplication.

Practice Problems

5. A balloon contains 2.0 L of helium gas at STP. How many moles of helium are present?
6. How many moles of gas are present in 11.2 L at STP? How many molecules?
7. What is the volume, at STP, of 3.45 mol of argon gas?
8. A certain set of conditions allows 4.0 mol of gas to be held in a 70 L container. What volume do 6.0 mol of gas need under the same conditions of temperature and pressure?
9. At STP, a container holds 14.01 g of nitrogen gas, 16.00 g of oxygen gas, 66.00 g of carbon dioxide gas, and 17.04 g of ammonia gas. What is the volume of the container?
10. **(a)** What volume do 2.50 mol of oxygen occupy at STP?
    (b) How many molecules are present in this volume of oxygen?
    (c) How many oxygen atoms are present in this volume of oxygen?
11. What volume do 2.00×10^{24} atoms of neon occupy at STP?

Table 12.1 Molar Volume of Several Real Gases at STP

Gas	Molar volume (L/mol)
helium, He	22.398
neon, Ne	22.401
argon, Ar	22.410
hydrogen, H_2	22.430
nitrogen, N_2	22.413
oxygen, O_2	22.414
carbon dioxide, CO_2	22.414
ammonia, NH_3	22.350

Volumes of Real Gases

You know that ideal gases have a volume of 22.4 L at STP. Do *real* gases have the same volume? The volumes of several real gases at STP are given in Table 12.1. All the volumes are very close to 22.4 L/mol, the molar volume of an ideal gas. Scientists have decided that 22.4 L/mol is an acceptable approximation for *any* gas at STP when using gas laws. Although the volumes are the same, one mole of a gas will have a different mass and density than one mole of another gas. (See Figure 12.7.)

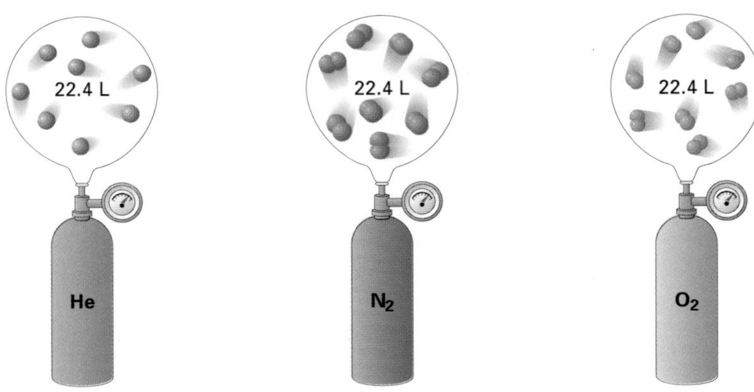

He	N_2	O_2
n = 1 mol	n = 1 mol	n = 1 mol
P = 1 atm (760 torr)	P = 1 atm (760 torr)	P = 1 atm (760 torr)
T = 0°C (273 K)	T = 0°C (273 K)	T = 0°C (273 K)
V = 22.4 L	V = 22.4 L	V = 22.4 L
Number of gas particles = 6.022×10^{23}	Number of gas particles = 6.022×10^{23}	Number of gas particles = 6.022×10^{23}
Mass = 4.003 g	Mass = 28.02 g	Mass = 32.00 g
d = 0.179 g/L	d = 1.25 g/L	d = 1.43 g/L

Figure 12.7 At STP, gases such as helium, nitrogen, and oxygen behave as ideal gases. They have a molar volume of 22.4 L.

A Deeper Look: Real Gas Deviations

Do real gases *always* behave like ideal gases? At STP, you have seen that most real gases behave like ideal gases. At high pressures and/or low temperatures, however, gases no longer behave ideally. To understand why, recall the characteristics of an ideal gas, according to the kinetic molecular theory. (You learned this in Chapter 11.)

- Gas molecules have zero volume of their own.
- Molecules have no attractive forces between each other or with their container.
- Molecules move in perfectly straight lines.
- Collisions are completely elastic and, thus, do not use up energy.

CHECKPOINT

Look back at Table 12.1. It gives more accurate molar volumes for several gases. Using what you have learned about the way real gases behave, explain why these molar volumes are slightly different from the molar volume of an ideal gas.

These characteristics led to the ideal gas law. They are accurate enough for most applications. They are not perfect, however. In fact, most gases fall short of these characteristics in many ways:

- The particles of a real gas have a volume of their own.
- Molecules do attract each other.
- Molecules do not necessarily move in straight lines.
- Collisions are not completely elastic.

At high pressures and/or low temperatures, gases have smaller volumes. This means that the gas molecules are closer together. Since they are closer together, the molecules interact more than when they are far apart. It is no longer true to say that the molecules have no attractive forces between each other or with their container. Part A of Figure 12.8 illustrates how this affects the pressure of a gas. Also, since the total volume is smaller, the amount of space taken up by the gas molecules is more important. You can no longer ignore the volume of the gas molecules. Part B of Figure 12.8 illustrates that gas molecules do occupy part of the volume of the container.

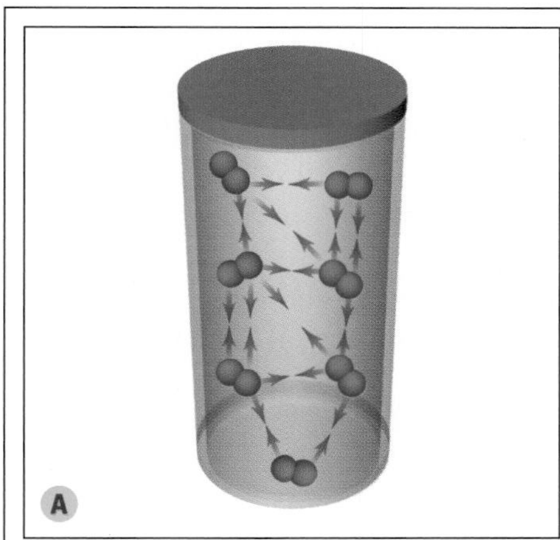

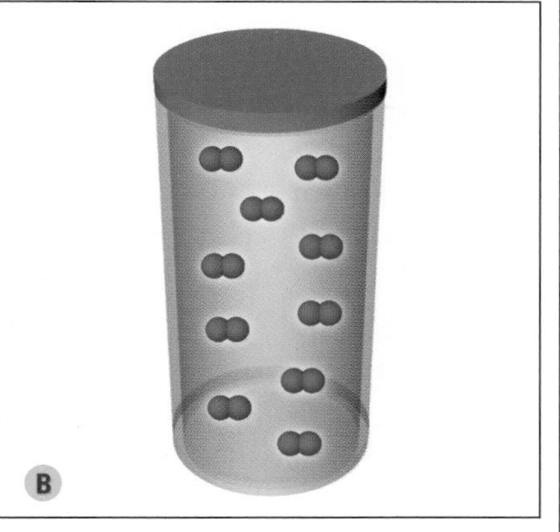

Figure 12.8 **(A) Because particles are attracted to each other, the pressure is reduced. (B) Since particles take up space, the total volume of empty space is smaller than the volume of the container.**

Scientists who want more accuracy in their experiments have adapted the ideal gas law to reflect the behaviour of real gases. In later chemistry courses, you will learn more about this corrected version of the ideal gas law, called the *Van der Waals equation.*

To summarize, at low pressures and high temperatures, most gases behave as ideal gases. Under any conditions that allow attractive forces between molecules to occur, gases no longer behave as an ideal gas. They behave as real gases.

History LINK

The French chemist, Antoine Lavoisier (1743–1794), was the first person to notice the gas-volume relationship. He observed water decomposing to give two volumes of hydrogen and one volume of oxygen. He mentioned this relationship in his 1789 textbook *Eléments de la Chemie.*

Arriving at the Ideal Gas Law

In the Sample Problem earlier, you used three steps to find the molar volume of nitrogen gas. After calculating the number of moles, you calculated the final volume using the combined gas law. Then you used Avogadro's law to find the volume of one mole of nitrogen. There is an easier way to do problems like this—by combining the two gas laws. After Avogadro's work, it did not take scientists long to connect $V \alpha n$ (Avogadro's law) with $V \alpha \frac{T}{P}$ (the combined gas law).

$$V \alpha n$$

$$V \alpha \frac{T}{P}$$

$$\therefore V \alpha \frac{nT}{P}$$

$$\therefore \frac{PV}{nT} = R, \text{ where } R \text{ is a constant}$$

As you know, all gases behave in a similar way. The *universal gas constant* (R), which applies to all gases, was derived for the final equation given above. Examine the calculation below to see how R was derived.

For one mole of gas at STP,

$P = 101.3$ kPa
$T = 273$ K
$V = 22.4$ L
$n = 1.00$ mol

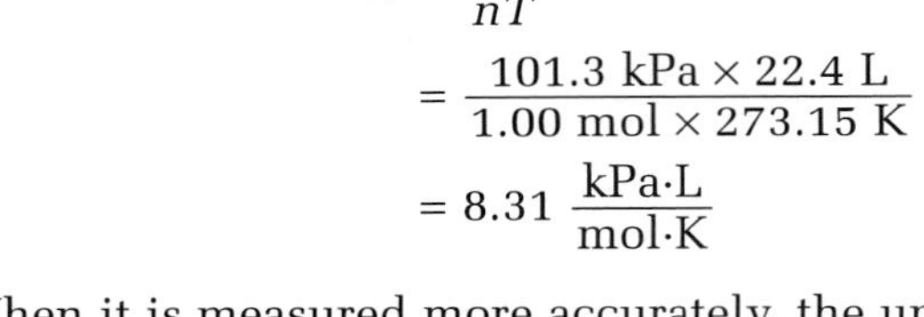

$$R = \frac{PV}{nT} = \frac{101.3 \text{ kPa} \times 22.4 \text{ L}}{1.00 \text{ mol} \times 273.15 \text{ K}} = 8.31 \frac{\text{kPa·L}}{\text{mol·K}}$$

Electronic Learning Partner

Go to the Chemistry 11 Electronic Learning Partner for an interactive simulation on the ideal gas law.

When it is measured more accurately, the universal gas constant (R) has a value of $8.314 \frac{\text{kPa·L}}{\text{mol·K}}$. This was the final piece of the equation. The resulting equation is an efficient tool for solving many problems that involve gases.

The **ideal gas law** states that the pressure multiplied by the volume is equal to the number of moles multiplied by the universal gas constant and the temperature.

$$PV = nRT$$

Guidelines for Using the Ideal Gas Law

- Always convert the temperature to kelvins (K).
- Always convert the masses to moles (mol).
- Always convert the volumes to litres (L).
- Using the ideal gas law will be easier if you always convert the pressures to kilopascals (kPa). Then you can memorize the value of R ($8.314 \frac{\text{kPa·L}}{\text{mol·K}}$) and use it for every calculation. If you happen to forget the value of R, you can calculate it by finding R for 1 mol of gas at STP.

Converting the Units of the Universal Gas Constant

Some people prefer to use a converted value of R, rather than converting all the pressures to kilopascals, as suggested in the guidelines above. What are the units for R if the pressure is given in atmospheres?

$$\frac{8.314\ \cancel{\text{kPa}}\cdot\text{L}}{\text{mol}\cdot\text{K}} \times \frac{1\text{atm}}{101.3\ \cancel{\text{kPa}}} = 0.08206\ \frac{\text{atm}\cdot\text{L}}{\text{mol}\cdot\text{K}}$$

What are the units for R if the pressure is given in torr (mm Hg)?

$$8.314\ \frac{\cancel{\text{kPa}}\cdot\text{L}}{\text{mol}\cdot\text{K}} \times \frac{760\ \text{mmHg}}{101.3\ \cancel{\text{kPa}}} = 62.37\ \frac{\text{mmHg}\cdot\text{L}}{\text{mol}\cdot\text{K}}$$

Since R is an exact constant, always express it with *four* significant digits. Do not worry about changing the significant digits of R in your calculations. Always keep four digits, and then fix the number of significant digits in your answer.

The diagram below illustrates the evolution of the ideal gas law. It is a brief summary of what you have learned so far in this section. The following Sample Problems show you how to do calculations using the ideal gas law.

CHECKPOINT

Go back to section 11.2 to review the units for pressure.

Concept Organizer The Evolution of the Ideal Gas Law

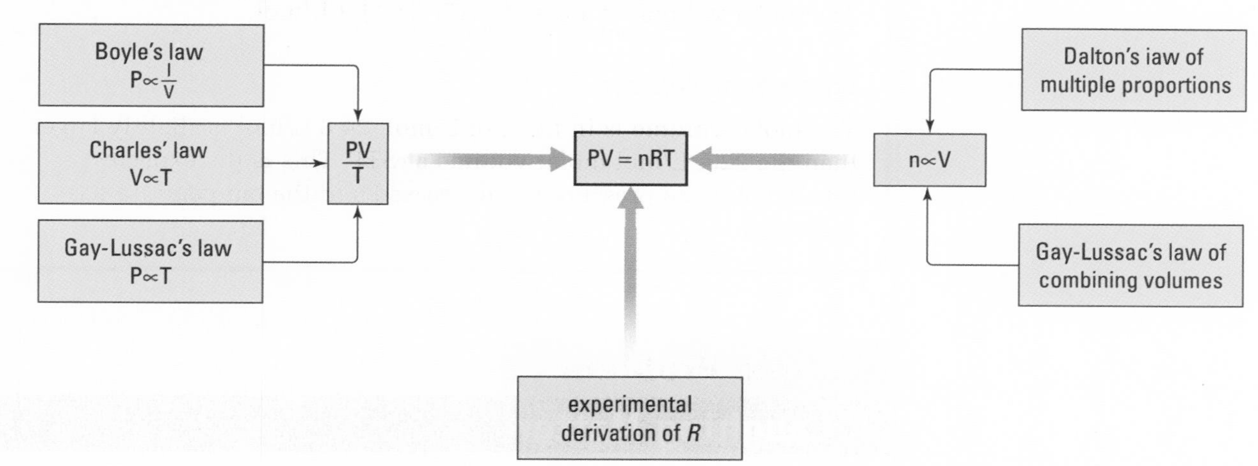

Sample Problem

Calculating Molar Volume Using $PV = nRT$

Problem

Use the ideal gas law to calculate the molar volume of a gas at standard ambient temperature and pressure (SATP). The conditions for SATP are 298 K and 100 kPa.

What Is Required?

You need to find the molar volume, $MV = \frac{V}{n}$, at SATP. The units for your answer will be in mol/L.

Continued ...

Continued ...
FROM PAGE 487

13. How many kilograms of chlorine gas are contained in 87.6 m^3 at 290 K and 2.40 atm? **Hint:** 1 m^3 = 1000 L

14. Calculate the volume of 3.03 g of hydrogen gas at a pressure of 560 torr and a temperature of 139 K.

15. A 6.0×10^2 L reaction tank contains 5.0 mol of oxygen gas and 28 mol of nitrogen gas. If the temperature is 83°C, what is the pressure of the oxygen, in kPa?

Section Wrap-up

In this section, you learned that the ideal gas law was derived from Avogadro's law and the combined gas law. You can use the ideal gas law to calculate the pressure, temperature, volume, or number of moles of a gas. In the next section, you will learn how to apply the ideal gas law to identify unknown gases on the basis of their densities and/or molar masses.

Remember that you can manipulate the ideal gas law to solve for any variable.

$$P = \frac{nRT}{V} \qquad V = \frac{nRT}{P} \qquad n = \frac{PV}{RT} \qquad T = \frac{PV}{nR}$$

The universal gas constant (R) always has the same value: 8.314 kPa·L/mol·K. (You need to use an alternative value of R if pressure is given in atmospheres or in torr.)

Section Review

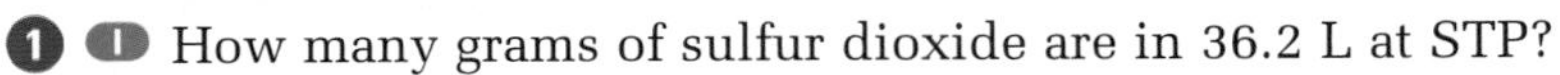

1. I How many grams of sulfur dioxide are in 36.2 L at STP?

PROBEWARE

If you have access to probeware, do the Gas Laws lab, or a similar lab available from a probeware company.

2. I At certain conditions, 20 mol of a gas occupy 498 L. What volume do 35 mol occupy?

3. I What is the pressure of 0.76 mol of gas at 48°C in 8.0 L?

4. I How many grams of bromine gas, Br_2, are in 3.12 L at a pressure of 2.4 atm and a temperature of −20°C? How many molecules of bromine gas are there?

5. I What is the temperature of 6.02 mol of hydrogen sulfide in a 132 L tank at 0.95 atm?

6. K/U What was Avogadro's hypothesis? Explain how Avogadro contributed to our understanding of gases and the relationships among the properties of gases.

7. K/U 1.00 m^3 of regular air at 1.0 atm is compressed to one eighth of the volume at a constant temperature. What is the pressure contribution of the nitrogen? **Hint:** See the Table 11.2 in Chapter 11, section 11.4.

Applications of the Ideal Gas Law

12.2

Studying gases and their properties becomes more interesting when you realize that your own body is a container for gases. Your lungs hold air, an important solution of gases that you need to live. As well, there is the embarrassing type of gas, called methane, that results from the digestive process. How many moles of oxygen do your lungs hold? Using the laws and properties you have learned, you can now calculate this fact about your own body. (See Figure 12.9.)

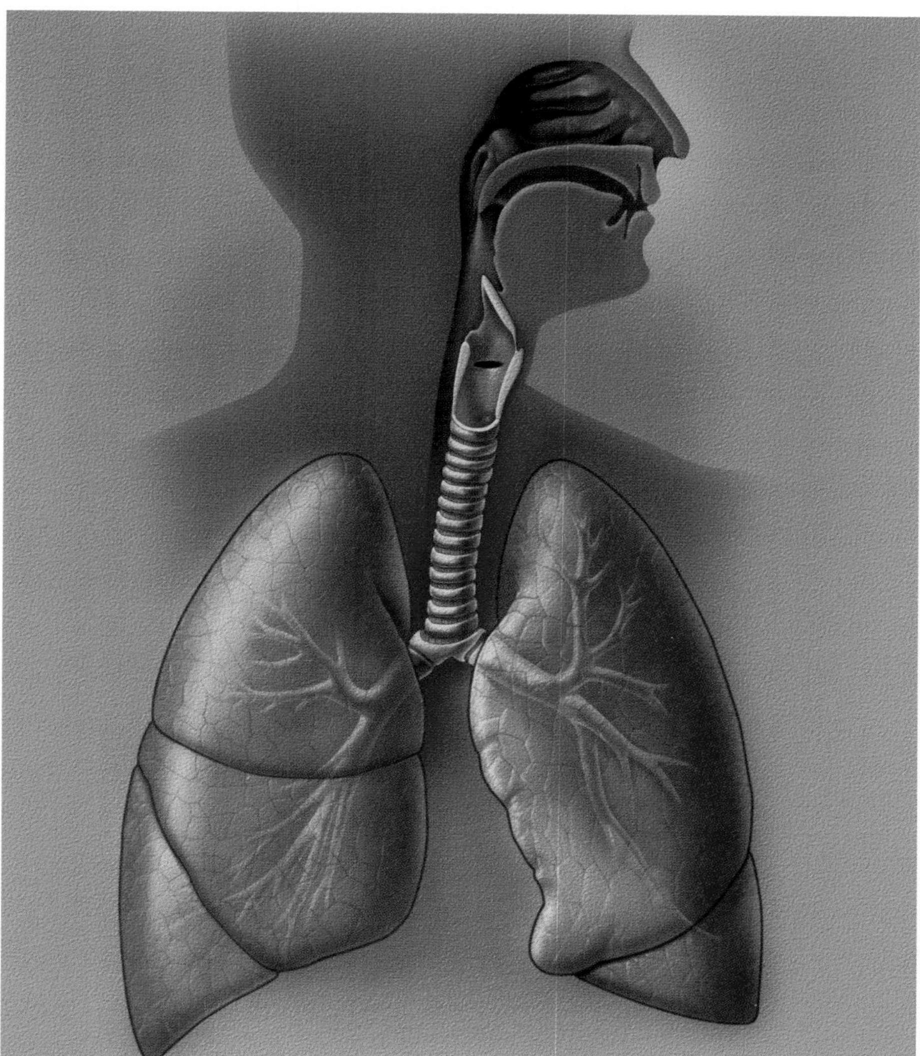

Figure 12.9 **Your lungs hold about 4 L of air. When you breathe out, only 500 mL of air is expelled. The same amount of air is taken in when you inhale one breath. If air contains 20% oxygen gas, how many moles of oxygen gas do your lungs contain at 37°C and 100 kPa?**

In this section, you will learn more about two properties of a gas that are closely related to molar volume: density and molar mass. You have already encountered these properties, but now you will use them to help you with your gas calculations.

Section Preview/ Specific Expectations

In this section, you will

- **solve** more quantitative problems using the ideal gas law
- **determine** the molar mass of an unknown gas through experimentation

Biology LINK

As you breathe in, your diaphragm moves down and your rib cage moves out to increase the volume of your lungs. As the volume increases, the air pressure in your lungs decreases. Air from the outside rushes in to fill the expanded volume. This is an application of Boyle's law. In the same way, your lungs decrease in volume when you exhale. The resulting increase in pressure pushes air out of your lungs.

CHECKPOINT

In this book, we use a capital *M* to symbolize molar mass. When working with solutions, you may see M used to express the molar concentration of a solution (for example, 6 $M_{HCl(aq)}$). Make sure that you do not become confused! Always check to make sure that you understand what *M* is referring to.

Density and Molar Mass

As you learned in the last section, the *molar volume* of a gas is defined as the space that is occupied by one mole of the gas. It is always given in units of L/mol.

The *density* of a gas is similar to the density of a solid or a liquid. Density is found by dividing mass by volume. The density of a gas is usually reported in units of g/L.

The *molar mass* of a gas refers to the mass (in g) of one mole of the gas. You can calculate molar mass by adding the masses of atoms in the periodic table. You can also calculate molar mass by dividing the mass of a sample by the number of moles that are present. Molar mass is always expressed in the units g/mol. Table 12.2 summarizes molar volume, density, and molar mass.

Table 12.2 Molar Volume, Density, and Molar Mass

	Molar volume	Density	Molar mass
Unit	L/mol	g/L	g/mol
Meaning	volume/amount	mass/volume	mass/amount
Calculations	$\text{Molar volume} = \dfrac{\text{Volume}}{\text{Number of moles}}$ $MV = V/n$	$\text{Density} = \dfrac{\text{Mass}}{\text{Volume}}$ $D = m/V$	Molar mass = Sum of molar masses of the atoms in the compound *or* $\text{Molar mass} = \dfrac{\text{Mass}}{\text{Number of moles}}$ $M = m/n$

mind STRETCH

Molar volume (L/mol), density (g/L), and molar mass (g/mol) are closely related. You can calculate each property using the other two. Analyze the units to discover the exact relationship.

Dense Gases Can Be Deadly

You have learned that the volume of a gas at a certain temperature and pressure is the same as the volume of any other gas at the same temperature and pressure. For example, all gases have the approximate volume of 22.4 L at STP. The *molecular masses* of different gases, however, are all different. This means that each gas has a different density, or mass per unit of volume. (Look back at Figure 12.7 to see three examples.)

Understanding the densities of gases can be useful in the everyday world. For example, miners who drill deep into the ground must know which gases are present, and which have the highest densities. (See Figure 12.10.) They must take appropriate safety precautions to avoid explosions, poisoning, or suffocation.

On December 3, 1984, an industrial accident in Bhopal, India, released a large quantity of methylisocyanate, CH_3NCO, a dense gas, into the air. This highly irritating and toxic gas caused the death of more than 3000 people living nearby. The Chemistry Bulletin on the next page gives an example of a natural disaster involving a dense gas. Figure 12.11, on page 492, illustrates a popular use for gases that are less dense than air.

Figure 12.10 Dense gases sit at the bottoms of pits, such as mines and wells. Miners must understand the behaviour of these gases to avoid accidents. For example, dense carbon dioxide gas can cause suffocation.

Chemistry Bulletin

Science | Technology | Society | Environment

The Killing Lakes of Cameroon

On August 15, 1984, a cloud of deadly gas burst from Lake Monoun in Cameroon, a country in western Africa. Thirty-seven people died from suffocation. Two years later, on August 21, 1986, Lake Nyos, a larger and deeper lake, ejected a full cubic kilometre of the same gas. The gas travelled silently into neighbouring villages, killing 1700 people and thousands of livestock. What was this toxic gas?

Lake Nyos and Lake Monoun both sit in volcanic craters. The lakes are hazardous because of their volcanic origin, even though both volcanoes are dormant. Volcanoes are vents through Earth's crust. They carry *magma*, a mixture of molten rock and dissolved gases, to the surface of Earth. When magma rises to Earth's surface in volcanoes, the pressure is decreased. The gases come out of solution and expand.

Magma contains a large quantity of dissolved carbon dioxide and minerals. These are released from the magma into the ground water under a volcano's crater. Lake Nyos and Lake Monoun are both fed at the bottom by volcanic springs of mineral-rich carbonated ground water. In some crater lakes, there is enough water circulation for the carbon dioxide to bubble up from the bottom and be released at the surface. The surface waters of Lake Nyos and Lake Monoun, however, do not mix with deeper waters. The carbon dioxide remains at the bottom. Volcanic springs continue to supply the lakes with carbon dioxide, which remains trapped at the bottom. Fresher surface water sits on this dense lower layer.

Scientists believe that the tragedies at Lake Nyos and Lake Monoun were caused by a disruption of the water layers, perhaps triggered by a landslide, an earthquake, or even a strong wind. Lower carbonated water was suddenly released into the upper water. As it moved up to the surface, the pressure on the gas decreased while the temperature increased. As a result, the carbon dioxide gas rapidly bubbled out of solution. At Lake Nyos, the sudden release of the lower lake water into the upper layers caused a plume of water and gas to rise high into the air.

How did this massive release of carbon dioxide cause so many deaths? Carbon dioxide gas is one-and-a-half times as dense as air. It sinks to the ground and displaces the oxygenated air we need to breathe. The invisible, odourless carbon dioxide that was released from Lake Nyos and Lake Monoun settled on the ground. It travelled rapidly down the slopes into populated regions. People quickly became unconscious and died of suffocation.

Carbon dioxide continues to accumulate at the bottoms of both lakes. An international team of scientists is developing a plan to release this carbon dioxide by controlled degassing. They plan to insert long pipes deep into both lakes and suck up some of the dense bottom water. This will create a pressure difference and cause a fountain of gas-rich water to jet from the pipes.

Making Connections

1. Calculate and compare the density of one mole of $CO_{2(g)}$ with the density of one mole of $N_{2(g)}$ at STP. (Nitrogen gas is the main constituent of air.) Explain the significance of your calculations.
2. How is carbonated water produced artificially, and what is it used for? Do research to find out.

Web LINK

www.school.mcgrawhill.ca/resources/
Many accidents have occurred because of dense gases in mines. Which gases are dangerous? What safety precautions should be taken? To learn more, go to the web site above. Go to **Science Resources**, then to **Chemistry 11** to find out where to go next.

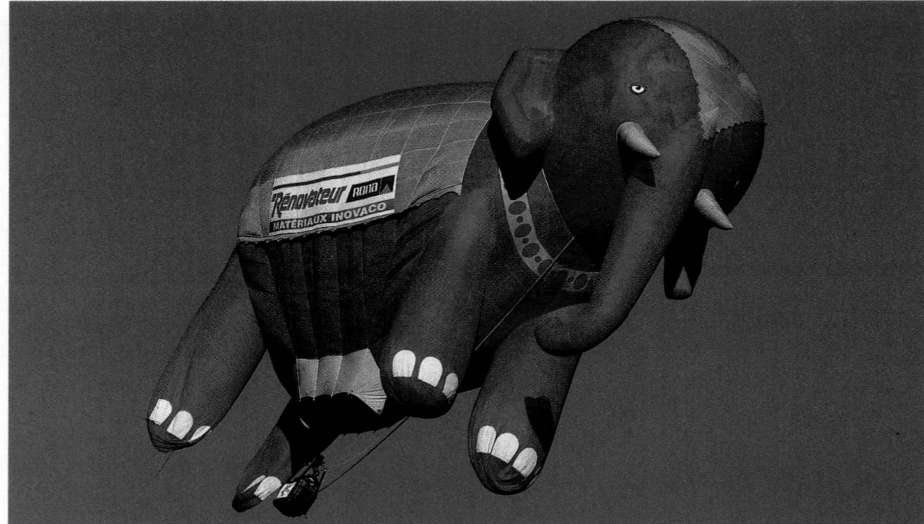

Figure 12.11 **Helium balloons and hot air balloons also depend on the density of gases. These objects are filled with gases that are less dense than air. Therefore, they are able to float.**

In the following Sample Problem, you will use the ideal gas law to find the density of nitrogen gas.

History LINK

In World War I, a highly irritating gas called phosgene, $COCl_2$, was used against the Allied troops. Phosgene is about 3.4 times more dense than air. In concentrations above 50 ppm, it causes the lungs to fill up with fluid. This results in diminished lung capacity, and subsequent collapse of the heart. Phosgene can bring about death within hours.

Sample Problem

Finding the Density of Nitrogen Gas

Problem

Nitrogen gas makes up almost 80% of our atmosphere. What is the density of pure nitrogen gas, in g/L, at 12.50°C and 126.63 kPa?

What Is Required?

Find the density of nitrogen gas, in g/L, at 12.50°C and 126.63 kPa pressure.

What Is Given?

$P = 126.63$ kPa

$T = 12.50$°C

Plan Your Strategy

Step 1 Change the temperature to kelvins. The temperature in this question is given to two decimal places. Therefore, use a conversion factor with two decimal places: + 273.15.

Step 2 Calculate the molar mass (M) of nitrogen gas, N_2, using the molar mass in the periodic table.

Step 3 Since the volume is not given, set it as 1.00 L. Since the pressure is given in kilopascals, use $R = 8.314$ kPa·L/mol·K. Substitute the numbers and units for P, T, R, and V into the ideal gas law equation. Solve for n.

Continued ...

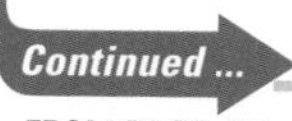

FROM PAGE 492

Step 4 Convert n to the mass of nitrogen (m) by multiplying the number of moles by the molar mass (M) of N_2.

Step 5 Find the density by dividing the mass by the volume (1.00 L).

Act on Your Strategy

Step 1 $T = (12.50°C + 273.15)$
$= 285.65$ K

Step 2 $M_{N_2} = 2 \times 14.01$ g/mol
$= 28.02$ g/mol

Step 3 $PV = nRT$

$$\therefore\ n = \frac{PV}{RT}$$

$$= \frac{(126.63\ \text{kPa})(1.00\ \text{L})}{(8.314\ \text{kPa·L/mol·K})(285.65\ \text{K})}$$

$$= 5.3320 \times 10^{-2}\ \text{mol}$$

Step 4 The mass of N_2 is

$$m = n \times M$$
$$= (5.3348 \times 10^{-2}\ \text{mol})(28.02\ \text{g/mol})$$
$$= 1.4940\ \text{g}$$

Step 5 The mass of 1.00 L of nitrogen gas is 1.494 g. By dividing this by the volume, 1.00 L, we obtain a density of 1.494 g/L.

Check Your Solution

When the units cancel out in the ideal gas equation, mol remains.
When units cancel in the density equation, g/L remain.
The least number of digits in the question is four. Therefore, the answer should have four significant digits, which it does.

Practice Problems

16. Oxygen makes up about 20% of our atmosphere. Find the density of pure oxygen gas, in g/L, for the conditions in the Sample Problem: 12.50°C and 126.63 kPa.

17. Find the density of butane gas, C_4H_{10}, (in g/L) at SATP conditions: 298 K and 100 kPa.

18. The atmosphere of the imaginary planet Xylo is made up entirely of poisonous chlorine gas, Cl_2. The atmospheric pressure of this inhospitable planet is 155.0 kPa, and the temperature is 89°C. What is the density of the atmosphere?

19. The atmosphere of planet Yaza, from the same star system as Xylo, is made of fluorine gas, F_2. The density of the atmosphere on Yaza is twice the density of the atmosphere on Xylo. The temperature of both planets is the same. What is the atmospheric pressure of Yaza?

Molar Mass of a Gas

You can find the molar mass of a gaseous element or compound in the same way that you find the molar mass of any other element or compound: by adding up the masses of the atoms. You can also find the molar mass by dividing the mass by the number of moles. (See Figure 12.12.)

COURSE CHALLENGE

How can you use what you learn in this Sample Problem to help you identify the gas given off by water plants? You will apply your learning later on, in the Chemistry Course Challenge.

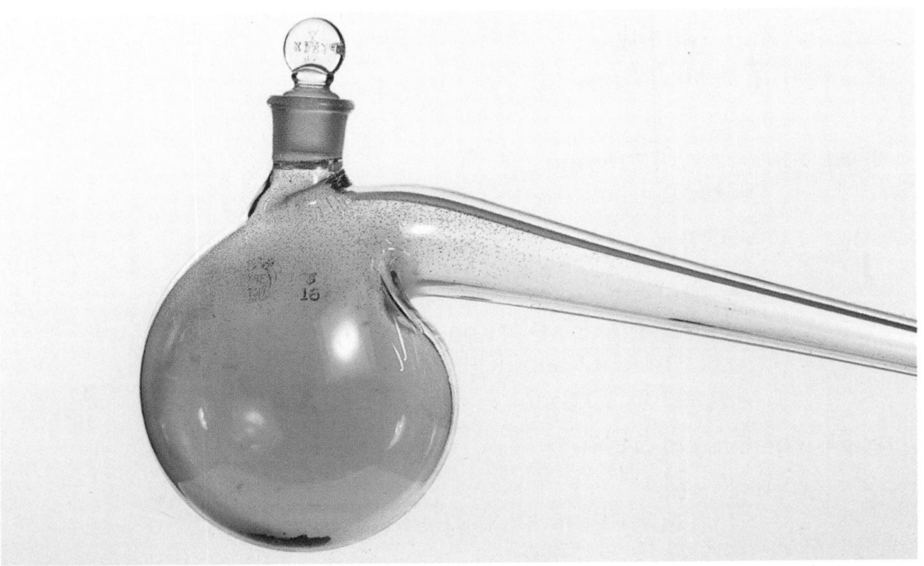

Figure 12.12 How could you determine the molar mass of vaporized iodine gas on paper? How could you determine it in a laboratory?

In the laboratory, calculating the molar mass of an unknown gas can help you identify it. The next Sample Problem will demonstrate this.

Sample Problem

Using Molar Mass to Identify an Unknown Gas

Problem

A scientist isolates 2.366 g of a gas. The sample occupies a volume of 800 mL at 78.0°C and 103 kPa. Use these data to calculate the molar mass of the gas. Is the gas most likely to be bromine, krypton, neon, or fluorine?

What Is Given?

$P = 103$ kPa

$V = 800$ mL

$m = 2.366$ g

$R = 8.314$ kPa·L/mol·K

$T = 78.0$°C

What Is Required?

You need to find the number of moles. Then you need to use the mass and the number of moles to find the molar mass of the gas.

Continued ...

Continued ...
FROM PAGE 494

Plan Your Strategy

Convert the temperature to kelvins. Solve $PV = nRT$ for n. Then substitute in the known values to find the number of moles. Finally, set up a proportion to find the number of grams that would be in one mole, using the equation $M = \frac{m}{n}$, where M is the molar mass, m is the mass, and n is the number of moles.

Act on Your Strategy

$$\begin{aligned} T(\text{K}) &= T(^\circ\text{C}) + 273 \\ &= 78.0^\circ\text{C} + 273 \\ &= 351\ \text{K} \end{aligned}$$

Using the ideal gas law,

$$PV = nRT$$

$$\begin{aligned} \therefore n &= \frac{PV}{RT} \\ &= \frac{103\ \cancel{\text{kPa}} \times 0.800\ \cancel{\text{L}}}{8.314\ \cancel{\text{kPa}\cdot\text{L}}/\text{mol}\cdot\cancel{\text{K}} \times 351\ \cancel{\text{K}}} \\ &= 0.0282\ \text{mol} \end{aligned}$$

$$\begin{aligned} M &= \frac{m}{n} \\ &= \frac{2.366\ \text{g}}{0.0282\ \text{mol}} \\ &= 83.9\ \text{g/mol} \end{aligned}$$

The molar mass of the gas is 83.9 g/mol.

To identify the gas, compare the molar masses of the four gases mentioned.

Bromine, Br_2, has a molar mass of 2×79.9 g/mol = 141.8 g/mol.

Krypton, Kr, has a molar mass of 83.8 g/mol.

Neon, Ne, has a molar mass of 20.2 g/mol.

Fluorine, F_2, has a molar mass of 2×18.9 g/mol = 38.0 g/mol.

Therefore, the gas must be krypton.

Check Your Solution

The units of the answer are g/mol, the correct units for molar mass. The answer has three significant digits, equal to the least number of digits in the question.

The answer is probably correct, since it is so close to the molar mass of one of the given gases.

In the following investigation, you will find the molar mass of an unknown gas. You will use this mass to identify the gas.

The ideal gas law gives great flexibility for solving many different types of problems. After the investigation, you will find another Sample Problem. It illustrates how you can use the ideal gas law with methods you have previously learned, to identify an unknown gas. Practice problems are located at the end of this Sample Problem.

Investigation 12-A

SKILL FOCUS

Performing and recording

Analyzing and interpreting

Communicating results

Calculating the Molar Mass of an Unknown Gas: Teacher Demonstration

Cigarette lighters contain a gaseous fuel that burns quickly. It produces a large amount of heat using only a small amount of gas. Your teacher will measure the volume and mass of a sample of this gas. Then you will use these data to calculate the molar mass of the gas.

Materials

4 L beaker or plastic pail
500 mL graduated cylinder
disposable cigarette lighter
needle nose pliers
plastic wrap
balance or scale
tap water
thermometer
barometer
hair dryer

Your teacher will take the following Safety Precautions and perform the following steps.

Safety Precautions

- Remember that this gas is flammable. Do not try to produce flames with the lighters. Before beginning, check that there are no flames (such as lit Bunsen burners) in the laboratory.
- If water is spilled on the floor, wipe it up immediately so that no one steps in it.
- Release all the gas collected into an operating fume hood after the investigation. Expel the remaining gas in the lighter outside before disposing of the lighter.

Procedure

1. Use pliers to remove the striker, flint, and spring from the disposable lighter.
2. Fill the 4 L beaker (or pail) about two-thirds full of tap water. Determine the temperature of the tap water. Use this measurement to approximate the temperature of the gas.
3. Briefly immerse the lighter in the water, then shake it dry. Use the hair dryer on a low or cool setting to dry the lighter as much as possible. This is to set a standard for drying the empty lighter later. CAUTION Do not overheat the lighter.
4. Determine the mass of the lighter.
5. Fill the graduated cylinder with water. Cover the cylinder tightly with a piece of plastic wrap. With your hand over the plastic wrap, place the cylinder upside down into the beaker. Make sure that no air bubbles are trapped in the cylinder. Slide the plastic wrap away. A water-filled measuring tube to collect gas that has been created. The gas will displace the water as it rises, giving an accurate measurement.

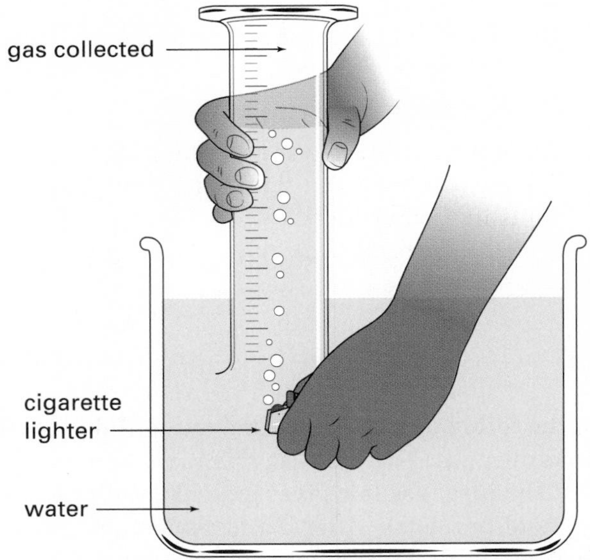

6. As shown in the diagram, hold the lighter underwater, below the graduated cylinder in the beaker. Carefully depress the button on the lighter to release gas into the cylinder. The entire lighter does not need to be emptied. Just gather enough gas for an accurate measurement.
7. Add tap water to the beaker (or pail), or lift the cylinder, so that *the water inside the cylinder is at exactly the same level as the water in the beaker.* This equalizes the pressure in the cylinder with the pressure of the atmosphere. Record the volume of the gas collected when the water levels are equal inside and outside the cylinder.
8. Dry the lighter with the hair dryer. Measure its mass.
9. Record the air pressure in the room.
10. Wash your hands.

Analysis

1. **(a)** Subtract the final mass of the lighter from its initial mass. This will give you the mass of the gas used.
   (b) Use the volume, the mass, the temperature of the water, and the air pressure to calculate the number of moles of gas.
2. Use the mass of the gas and the number of moles to calculate the molar mass.

Conclusions

3. The gas in the lighter has the formula C_nH_{2n+2}.
   (a) Use the periodic table to calculate the molar masses of the compounds with $n = 1$ to $n = 5$.
   (b) Identify the gas in the lighter. You will learn how to name this gas in Chapter 13.
4. How do your results compare with the theoretical molar mass calculated from the periodic table?
5. Calculate the percent error for your results.
6. What were the sources of error?

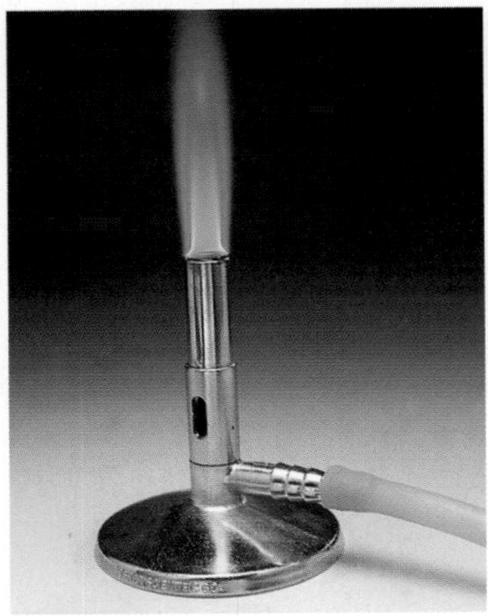

Figure 12.13 Methane gas, CH_4, is used as a fuel for Bunsen burners. Suppose that you are given a container of an unknown gas. What methods could you use to find out if it is methane?

Sample Problem

Identifying a Compound Using Percent Composition and the Ideal Gas Law

Problem

As geologists study the area where an ancient marsh was located, they discover an unknown gas seeping from the ground. They collect a sample of the gas, and take it to a lab for analysis. Lab technicians find that the gas is made up of 80.0% carbon and 20.0% hydrogen. They also find that a 4.60 g sample occupies a volume of 2.50 L at 1.50 atm and 25.0°C. What is the molecular formula of the gas?

What Is Required?

You need to find the molecular formula (and thus the identity) of an unknown gas.

What Is Given?

$P = 1.50$ atm

$V = 2.50$ L

$m = 4.60$ g

$R = 8.314$ kPa·L/mol·K

$T = 25.0$°C

The percentage composition of the gas is 80.0% carbon and 20.0% hydrogen.

Plan Your Strategy

Step 1 Find the empirical formula of the gas, using the molar masses of carbon and hydrogen and the percent compositions.

Step 2 Solve the ideal gas law for the number of moles. First change the temperature to kelvins. Then change the pressure to kilopascals and use $R = 8.314$ kPa·L/mol·K. (You could also use $R = 0.08206$ atm·L/mol·K.)

Step 3 Find the molar mass (M) of the compound by dividing the mass of the sample (m) by the number of moles (n) in the sample.

Step 4 Compare the molar mass of the unknown gas with the molar mass of the empirical formula. To find the molecular formula, multiply the empirical formula by the ratio of the two molar masses.

Act on Your Strategy

Step 1 Find the empirical formula. Assume that the total mass of the sample is 100.0 g. Thus the mass of the carbon in the sample is 80.0% × 100.0 g = 80.0 g. The mass of the hydrogen is 20.0% × 100.0 g = 20.0 g.

Continued ...

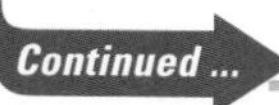

FROM PAGE 498

Now find the number of moles of carbon and hydrogen using the formula $n = \frac{m}{M}$

For carbon,

$$n = \frac{80.0\ \text{g}}{12.01\ \text{g/mol}} = 6.67\ \text{mol}$$

For hydrogen,

$$n = \frac{20.0\ \text{g}}{1.01\ \text{g/mol}} = 20.0\ \text{mol}$$

Finally, find the simplest mole ratio of the two elements in the compound. This will be the empirical, or simplest, formula of the gas.

The ratio of the elements in the compound is

$\frac{6.67}{6.67}$ mol of C to $\frac{20.0}{6.67}$ mol of H

or 1.0 mol of C to 3.0 mol of H

The empirical formula of the unknown gas is CH_3.

Step 2 Use the ideal gas law.

$$T = (25.0^\circ\text{C} + 273) = 298\ \text{K}$$

$$PV = nRT$$

$$\therefore n = \frac{RT}{PV}$$

If you convert the units of R ...	If you convert the pressure to kPa ...
$R = 0.08206\ \text{atm·L/mol·K}$	$P = 1.50\ \text{atm} \times 101.3\ \text{kPa/atm} = 152\ \text{kPa}$
$n = \frac{(1.50\ \text{atm})(2.50\ \text{L})}{(0.08206\ \text{atm·L/mol·K})(298\ \text{K})} = 0.153\ \text{mol}$	$n = \frac{(1.52\ \text{kPa})(2.50\ \text{L})}{(8.314\ \text{kPa·L/mol·K})(298\ \text{K})} = 0.153\ \text{mol}$

Step 3 Find the molar mass.

$$M = \frac{m}{n} = \frac{4.60\ \text{g}}{0.153\ \text{mol}} = 30.1\ \text{g/mol}$$

Step 4 Find the molecular formula.

The molar mass of the compound is 30.1 g/mol.

The molar mass of the empirical formula, CH_3, is 15.04 g/mol.

$$\text{Ratio of molar masses} = \frac{30.07\ \text{g/mol}}{15.04\ \text{g/mol}} = 2.00$$

Thus, the molecular formula of the unknown gas is twice the empirical formula.

$$CH_3 \times 2 = C_2H_6$$

Continued ...

Continued ...
FROM PAGE 499

Check Your Solution

In step 2, the units in the ideal gas equation cancel out to give moles. C_2H_6 is a reasonable answer for the molecular formula of the gas, since it is a simple integer ratio of the two molar masses.

Practice Problems

20. A 1.56 L gas sample has a mass of 3.22 g at 100 kPa and 281 K. What is the molar mass of the gas?

21. 2.0 L of haloethane has a mass of 14.1 g at 344 K and 1.01 atm. What is the molar mass of haloethane?

22. A vapour has a mass of 0.548 g in 237 mL, at 373 K and 755 torr. What is the molar mass of the vapour?

23. The mass of a 5.00 L evacuated container is 125.00 g. When the container is filled with argon gas at 298 K and 105.0 kPa, it has a mass of 133.47 g.

(a) Calculate the density of argon under these conditions.

(b) What is the density of argon at STP?

24. A gaseous compound contains 92.31% carbon and 7.69% hydrogen by mass. 4.35 g of the gas occupies 4.16 L at 22.0°C and 738 torr. Determine the molecular formula of the gas.

Section Wrap-up

In this section, you learned how density and molar mass are related to the ideal gas law. You also learned how to identify an unknown substance by calculating its molar mass, both theoretically and in the laboratory. Before you continue, take the time to complete the following Section Review questions. They will help you remember what you have learned.

Unit Issue Prep

You will be debating a question related to gas pollution in the Unit Issue. How does the density of a gas determine whether the gas pollutes Earth's surface or the atmosphere?

Section Review

1 I What is the density of methane, $CH_{4(g)}$, if 4.5 mol are in 100 L?

2 I 8.1 g of a gas occupy 12.3 L of space at 27°C and 8 atm.

(a) What is the molar mass of the gas?

(b) What might the gas be?

3 I A gas that consists of only nitrogen and oxygen atoms is found to contain 30% nitrogen. A 9.23 g sample of the gas occupies 2.2 L at STP. What is the gas?

4 I You are given a sample of an unknown gas. Describe how you can identify the gas in the laboratory. What measurements will you take? What apparatus might you need?

5 K/U How can the densities of two gases at STP be different, even though their volumes are the same?

Gas Law Stoichiometry

12.3

Many chemical reactions in everyday life involve gases. Figure 12.14 shows a common reaction that has a gas as a reactant. Other reactions, such as the electrolyzation of salt to give chlorine gas, have gases as products. To carry out an accurate and efficient reaction, scientists must know the number of moles of all the reactants. When one or more of the reactants is a gas, this means using the ideal gas law.

Section Preview/ Specific Expectations

In this section, you will

- **perform** stoichiometric calculations involving the number of moles, number of atoms, number of molecules, mass, and volume of substances in a balanced chemical reaction
- **determine** the molar volume of hydrogen in an investigation

Figure 12.14 In this photograph, oxygen gas reacts with calcium to produce calcium oxide.

You have already learned that the ideal gas law can be used to solve for different variables in several different types of situations. As you may recall, *the term "stoichiometry" refers to the relationship between the number of moles of the reactants and the number of moles of the products in a chemical reaction.* In this section, you will learn how to use Gay-Lussac's law of combining volumes and the ideal gas law to solve stoichiometric problems that involve gases.

Volume to Volume Stoichiometry

At the beginning of this chapter, you were introduced to Gay-Lussac's law of combining volumes: *When gases react, the volumes of the reactants and the products, measured at equal temperatures and pressures, are always in whole number ratios.* As well, you learned that the mole ratios from a chemical equation are the same as the ratios of the volumes of the gases.

This information will help you with a certain type of gas stoichiometry problem. When a gas reacts to produce another gas, you can use Gay-Lussac's law of combining volumes to find the volumes of the gases. The following Sample Problem shows you how.

CHECKPOINT

Go back to the beginning of this chapter. Make sure that you understand why the mole ratios are the same as the volume ratios.

Sample Problem

Gay-Lussac's Law of Combining Volumes

Problem

Ammonia is produced by a reaction of nitrogen gas and hydrogen gas. The chemical equation for the reaction is

$$N_{2(g)} + 3H_{2(g)} \rightarrow 2NH_{3(g)}$$

Suppose that 12.0 L of nitrogen gas reacts with hydrogen gas at the same temperature and pressure.

(a) What volume of ammonia gas is produced?

(b) What volume of hydrogen is consumed?

What Is Required?

(a) Calculate the volume of ammonia gas produced when 12.0 L of nitrogen gas reacts.

(b) Calculate the volume of hydrogen gas used up by the reaction.

What Is Given?

From the equation, you know that 12.0 L of nitrogen gas is used. The mole ratios from the equation are

$$\frac{2\text{ mol } NH_{3(g)}}{1\text{ mol } N_{2(g)}}$$

$$\frac{3\text{ mol } H_{2(g)}}{1\text{ mol } N_{2(g)}}$$

Plan Your Strategy

You know that the mole ratios of the volumes of gases are the same as the ratios of the volumes. Therefore, you can use the mole ratios to find the volumes of ammonia gas and hydrogen gas. You do not need to use the temperature and pressure, since they remain the same in this problem.

Act on Your Strategy

(a) Let x be the volume of ammonia gas.

$$\frac{2\text{ mol } NH_{3(g)}}{1\text{ mol } N_{2(g)}} = \frac{x\text{ L } NH_{3(g)}}{12.0\text{ L } N_{2(g)}}$$

$$(12.0\text{ L } N_{2(g)})\frac{2\text{ }\cancel{\text{mol}}\text{ } NH_{3(g)}}{1\text{ }\cancel{\text{mol}}\text{ } N_{2(g)}} = \frac{x\text{ L } NH_{3(g)}}{12.0\text{ L } N_{2(g)}}(12.0\text{ L } N_{2(g)})$$

$$x = 24.0\text{ L } NH_{3(g)}$$

Therefore, 24.0 L of ammonia gas is produced.

Continued ...

Continued ...

FROM PAGE 502

(b) Let y be the volume of hydrogen gas.

$$\frac{3 \text{ mol } H_{2(g)}}{1 \text{ mol } N_{2(g)}} = \frac{y \text{ L } H_{2(g)}}{12.0 \text{ L } N_{2(g)}}$$

$$(12.0 \text{ L } N_{2(g)})\frac{3 \text{ mol } H_{2(g)}}{1 \text{ mol } N_{2(g)}} = \frac{y \text{ } H_{2(g)}}{12.0 \text{ L } N_{2(g)}}(12.0 \text{ L } N_{2(g)})$$

$y = 36.0 \text{ L } H_{2(g)}$

Therefore, 36.0 L of hydrogen gas is consumed.

Check Your Solution

The number of significant digits in the answer is the same as the number of significant digits in the question.

The mole ratio of ammonia to nitrogen is 2:1. Thus, it makes sense that the volume of ammonia gas is twice the volume of nitrogen gas.

The mole ratio of hydrogen to nitrogen is 3:1. It makes sense that the volume of hydrogen gas is three times the volume of nitrogen gas.

Practice Problems

25. Use the following balanced equation to answer the questions below.

$$2H_{2(g)} + O_{2(g)} \rightarrow 2H_2O_{(g)}$$

(a) What is the mole ratio of oxygen gas to water vapour?

(b) What is the volume ratio of oxygen gas to water vapour?

(c) What is the volume ratio of hydrogen gas to oxygen gas?

(d) What is the volume ratio of water vapour to hydrogen gas?

26. 1.5 L of propane gas are burned in a barbecue. The following equation shows the reaction. Assume all gases are at STP.

$$C_3H_{8(g)} + 5O_{2(g)} \rightarrow 3CO_{2(g)} + 4H_2O_{(g)}$$

(a) What volume of carbon dioxide gas is produced?

(b) What volume of oxygen is consumed?

27. Use the following equation to answer the questions below.

$$SO_{2(g)} + O_{2(g)} \rightarrow SO_{3(g)}$$

(a) Balance the equation.

(b) 12.0 L of sulfur trioxide, $SO_{3(g)}$, are produced at 100°C. What volume of oxygen is consumed?

(c) What assumption must you make to answer part (b)?

28. 2.0 L of gas A react with 1.0 L of gas B to produce 1.0 L of gas C. All gases are at STP.

(a) Write the balanced chemical equation for this reaction.

(b) Each molecule of gas A is made of two identical "a" atoms. That is, gas A is really $a_{2(g)}$. In the same way, each molecule of gas B is made of two identical "b" atoms. What is the chemical formula of gas C in terms of "a" and "b" atoms?

Solving Gas Stoichiometry Problems

Earlier in this course, you learned how to do stoichiometry calculations. To solve gas stoichiometry problems, you will incorporate the ideal gas law into what you learned previously. The following steps will help you do this.

How to Solve Gas Stoichiometry Problems

1. Write a balanced equation for the reaction.
2. Write the given information under the appropriate reactants and products. Put a question mark under the reactant or product for which information is needed.
3. Convert all amounts to moles.
4. Compare molar amounts using stoichiometry ratios from the balanced equation. Solve for the unknown molar amount.
5. Convert the new molar amount into the units required. You may multiply by a conversion factor, or use a set of conditions with the ideal gas law, $PV = nRT$.

Using the Ideal Gas Law for the Gaseous Product of a Reaction

The best way to find out how to do a stoichiometry problem using the ideal gas law is to study an example. In the following Sample Problem, you will use a balanced equation and the ideal gas law to find the volume of a gas produced. (Refer to Chapter 4, section 4.1, if you want to review how to write balanced equations.)

Sample Problem

Mass to Volume Stoichiometry

Problem

Ancient alchemists liked to use strong sulfuric acid to produce dramatically dangerous effects. One interesting reaction occurs when sulfuric acid reacts with iron metal to produce gas and an iron(II) compound. What volume of gas is produced when excess sulfuric acid reacts with 40.0 g of iron at 18.0°C and 100.3 kPa?

What Is Required?

Calculate the volume of gas that is produced when sulfuric acid reacts with iron under specific temperature and pressure conditions.

What Is Given?

Reactants: sulfuric acid and iron

Products: an iron(II) compound and a gas

Mass of iron = 40.0 g

Temperature = 18.0°C

Pressure = 100.3 kPa

Continued ...

Continued ...
FROM PAGE 504

Plan Your Strategy

Step 1 Write a balanced equation for the chemical reaction.

Step 2 Find the number of moles of iron present. Use this value, along with the mole ratios from the balanced equation, to find the number of moles of gas produced.

Step 3 Use the ideal gas law. You know the number of moles of gas, the temperature, and the pressure. (Do not forget to change the temperature to kelvins.) Solve for the volume of the gas.

Act on Your Strategy

Step 1 Write the balanced equation. (This reaction is a single displacement reaction.)

$$Fe_{(s)} + H_2SO_{4(aq)} \rightarrow H_{2(g)} + FeSO_{4(aq)}$$
$$40.0\text{ g} \qquad\qquad ?\text{ L}$$

Step 2 Find the number of moles of iron, and the number of moles of gas.

To find the number of moles of iron, divide the mass by the molar mass. You can find the molar mass of iron in the periodic table: 55.85 g/mol. **Note:** If the reactant was a compound, such as $FeCl_2$, you would need to calculate the molar mass by adding the molar masses of all the atoms.

$$\begin{aligned} n &= \frac{m}{M} \\ &= \frac{40.0\text{ g Fe}}{55.85\text{ g/mol}} \\ &= 0.716\text{ mol Fe} \end{aligned}$$

From the balanced equation, the mole ratio is

$$\frac{1\text{ mol H}_2}{1\text{ mol Fe}}$$

Use this ratio to find the number of moles of hydrogen gas formed by the reaction.

$$\frac{n\text{ mol H}_2}{0.716\text{ mol Fe}} = \frac{1\text{ mol H}_2}{1\text{ mol Fe}}$$

$$(0.716\text{ mol Fe})\frac{n\text{ mol H}_2}{0.716\text{ mol Fe}} = \frac{1\text{ mol H}_2}{1\text{ mol Fe}}(0.716\text{ mol Fe})$$

$n = 0.716\text{ mol H}_2$

Step 3 Use the ideal gas law to solve for the volume, since all the other quantities are now known.

First change the temperature to kelvins.

$18.0^\circ\text{C} + 273 = 291\text{ K}$

You now have all the values you need to solve for volume.

$P = 100.3\text{ kPa}$

$n = 0.716\text{ mol}$

$R = 8.314\text{ kPa·L/mol·K}$

$T = 291\text{ K}$

Continued ...

Web LINK

www.school.mcgrawhill/resources

One of the problems with air bags (see Figure 12.15 on the next page) is that they can harm a child or small adult because too much gas is produced. To find out what volume of nitrogen gas in an air bag is safe for a child, go to the web site above. Go to **Science Resources**, then to **Chemistry 11** to find out where to go next. Use the volume you find to calculate the mass of $NaN_{3(s)}$ that is needed to produce this volume at 22.0°C and 105 kPa.

Continued ... FROM PAGE 505

$$PV = nRT$$
$$\therefore V = \frac{nRT}{P}$$
$$= \frac{0.716\ \cancel{mol} \times 8.314\ \cancel{kPa{\cdot}L/mol{\cdot}K} \times 291\ \cancel{K}}{100.3\ \cancel{kPa}}$$
$$= 17.3\ L$$

Therefore, 17.3 L of hydrogen gas is produced by this reaction at 18.0°C and 100.3 kPa.

Check Your Solution

The answer is slightly less than the molar volume of hydrogen gas at STP. Since less than one mole of hydrogen gas was formed, this seems reasonable.

Figure 12.15 Air bags must be tested thoroughly before being manufactured for public use.

Practice Problems

29. Engineers design automobile air bags that deploy almost instantly on impact. To do this, an air bag must provide a large amount of gas in a very short time. Many automobile manufacturers use solid sodium azide, NaN_3, along with suitable catalysts, to provide the gas that is needed to inflate the air bag. The balanced equation for this reaction is

$$2NaN_{3(s)} \rightarrow 2Na_{(s)} + 3N_{2(g)}$$

   (a) What volume of nitrogen gas will be produced if 117.0 g of sodium azide are stored in the steering wheel at 20.2°C and 101.2 kPa?

   (b) How many molecules of nitrogen are present in this volume?

   (c) How many atoms are present in this volume?

30. 0.72 g of hydrogen gas, H_2, reacts with 8.0 L of chlorine gas, Cl_2, at STP. How many litres of hydrogen chloride gas, HCl, are produced?

31. How many grams of baking soda (sodium hydrogen carbonate, $NaHCO_3$), must be used to produce 45 mL of carbon dioxide gas at 190°C and 101.3 kPa in a pan of muffins? (The mole ratio of $NaHCO_3$ to CO_2 is 2:1.)

32. How much zinc (in grams) must react with hydrochloric acid to produce 18 mL of gas at SATP? (**Hint:** Zinc chloride, $ZnC\ell_{2(s)}$ is a product.)

33. 35 g of propane gas burned in a barbecue, according to the following equation:

$$C_3H_{8(g)} + 5O_{2(g)} \rightarrow 3CO_{2(g)} + 4H_2O_{(g)}$$

   All the gases are measured at SATP.

   (a) What volume of water vapour is produced?

   (b) What volume of oxygen is consumed?

34. What mass of oxygen is reacted to produce 0.62 L of water vapour at 100°C and 101.3 KPa? Start by balancing the following equation:

$$H_{2(g)} + O_{2(g)} \rightarrow H_2O_{(g)}$$

CHEM FACT

An automobile air bag fills up with about 65 L of nitrogen gas in approximately 27 ms. This can prevent a driver from being seriously injured. The sodium that is produced is extremely caustic, however. It reacts with iron(III) oxide as follows:

$$6Na_{(s)} + Fe_2O_{3(s)} \rightarrow 3Na_2O_{(s)} + 2Fe_{(s)}$$

The sodium oxide then reacts with carbon dioxide and water vapour.

$$Na_2O(s) + 2CO_{2(g)} + H_2O_{(g)} \rightarrow 2NaHCO_{3(s)}$$

The sodium hydrogen carbonate that is produced is a harmless substance. It is better known as baking soda.

Including Water Vapour Pressure in Gas Calculations

Figure 12.16 This is an efficient and convenient method for collecting hydrogen gas. Unfortunately molecules of water vapour mix with the gas sample.

You can collect many gases by allowing them to bubble up through water into a container that is filled with water. (See Figure 12.16). This is the method your teacher used to collect the gas in Investigation 12-A. Unfortunately, molecules of water vapour mix with the gas sample. To avoid error, the pressure that was contributed by the water vapour must be subtracted when finding the pressure of the gas.

As an example, consider hydrogen gas, which is often collected over water. The hydrogen that is collected is a mixture of hydrogen and water vapour. As you learned from Dalton's law of partial pressures in Chapter 11, the pressure of this mixture is

$$P_{\text{total}} = P_{\text{hydrogen}} + P_{\text{water vapour}}$$

To find the partial pressure of dry hydrogen, subtract the pressure of the water vapour from the total pressure.

$$P_{\text{hydrogen}} = P_{\text{total}} - P_{\text{water vapour}}$$

The pressure of the water vapour is the same for any gas that is collected at a particular temperature. For example, the pressure of water vapour at 25°C is 3.17 kPa. Table 12.3 gives the pressure of water vapour at different temperatures.

When using the ideal gas law for a gas collected over water, you must correct the pressure before you substitute it into the gas law. The following Sample Problem shows you how to do this.

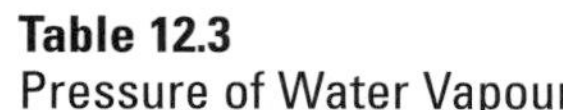

Table 12.3
Pressure of Water Vapour

Temperature (°C)	Pressure (kPa)
17	1.94
18	2.06
19	2.20
20	2.34
21	2.49
22	2.64
23	2.81
24	2.98
25	3.17
26	3.36
27	3.56
28	3.78
29	4.00
30	4.24

Sample Problem

Calculating the Volume of a Gas Collected Over Water

Problem

A student reacts magnesium with excess dilute hydrochloric acid to produce hydrogen gas. She uses 0.15 g of magnesium metal. What volume of dry hydrogen does she collect over water at 28°C and 101.8 kPa?

What Is Required?

You need to find the volume of hydrogen collected over water in this reaction.

What Is Given?

T = 28.0°C

P = 101.8 kPa

Mass of magnesium (m) = 0.15 g

Pressure of water vapour at 28°C = 3.78 kPa

Continued ...

Continued ...

FROM PAGE 507

Plan Your Strategy

Step 1 Write a balanced chemical equation for the reaction.

Step 2 Calculate the number of moles of magnesium by dividing the mass given (m) by the molar mass of magnesium (M). Use the number of moles of magnesium, along with the mole ratio from the equation, to calculate the number of moles of hydrogen gas produced by the reaction.

Step 3 Convert the temperature to kelvins. Since the hydrogen is collected over water, subtract the pressure of the water vapour at 28°C from the atmospheric pressure.

Step 4 Use the ideal gas law to find the unknown volume (V) of hydrogen gas.

Act on Your Strategy

Step 1 The balanced chemical equation is

$$\underset{0.15\text{ g}}{Mg_{(s)}} + 2HCl_{(aq)} \rightarrow MgCl_{2(aq)} + \underset{?\text{ L}}{H_{2(g)}}$$

Step 2 Find the number of moles of magnesium.

From the periodic table, the molar mass of magnesium is 24.31 g/mol.

$$\begin{aligned} n &= \frac{m}{M} \\ &= \frac{0.15\ \cancel{\text{g}}}{24.31\ \cancel{\text{g}}/\text{mol}} \\ &= 6.2 \times 10^{-3}\text{ mol} \end{aligned}$$

The mole ratio of hydrogen gas to magnesium in this reaction is

$$\frac{1\text{ mol }H_2}{1\text{ mol Mg}}$$

Using the mole ratio,

$$\frac{n\text{ mol }H_2}{6.2 \times 10^{-3}\text{ mol}} = \frac{1\text{ mol }H_2}{1\text{ mol Mg}}$$

Cross multiply to get

$$(6.2 \times 10^{-3}\text{ mol Mg})\frac{n\text{ mol }H_2}{6.2 \times 10^{-3}\text{ mol Mg}} = \frac{1\text{ mol }H_2}{1\text{ mol Mg}}(6.2 \times 10^{-3}\text{ mol Mg})$$

$$n = 6.2 \times 10^{-3}\text{ mol}$$

Step 3 Convert the temperature and pressure to kelvins.

$$\begin{aligned} T &= (28^\circ\text{C} + 273) \\ &= 301\text{ K} \end{aligned}$$

The pressure that is exerted by the hydrogen gas is

$$\begin{aligned} P_{\text{hydrogen}} &= 101.8\text{ kPa} - 3.78\text{ kPa} \\ &= 98.0\text{ kPa} \end{aligned}$$

Continued ...

Continued ...
FROM PAGE 508

Step 4 Use the ideal gas law.

$$PV = nRT$$

$$\therefore V = \frac{nRT}{P}$$

$$= \frac{(6.2 \times 10^{-3}\ \cancel{mol})(8.314\ \cancel{kPa}\cdot L/\cancel{mol}\cdot\cancel{K})(301\ \cancel{K})}{(98.0\ \cancel{kPa})}$$

$$= 0.16\ L$$

The student collects 0.16 L of dry hydrogen.

Check Your Solution

The final answer is rounded to two significant digits. This is the least number of significant digits in the question. The mass of the magnesium is a small number. Therefore, the volume of the hydrogen produced is also a small number.

Using the Ideal Gas Law for the Gaseous Reactant of a Reaction

So far, you have used the ideal gas law for reactions with a gas *product*. You can also use the ideal gas law for reactions with a gas as a *reactant*. The following Sample Problem shows you how to do this.

Sample Problem

Space Shuttle Science: Gas as a Reactant

Problem

When astronauts travel in a space shuttle (Figure 12.17), carbon dioxide must be removed from the air they breathe. One method is to bubble the air in the shuttle though a solution of lithium hydroxide. The lithium hydroxide converts any carbon dioxide into lithium carbonate.

$$2LiOH_{(aq)} + CO_{2(g)} \rightarrow Li_2CO_{3(aq)} + H_2O_{(\ell)}$$

(You learned about a different method in Chapter 7 using solid LiOH.) Air containing 25.0 L of carbon dioxide is passed through 1.5 mol/L LiOH solution over a 20 min period. The atmospheric pressure in the shuttle is 0.85 atm, and the temperature is 28.3°C. What mass of lithium carbonate is produced?

Note: Although the air is bubbled through an aqueous solution, you do not need to consider the pressure of the water vapour. This is because you are dealing with the *reactant* not the *product*.

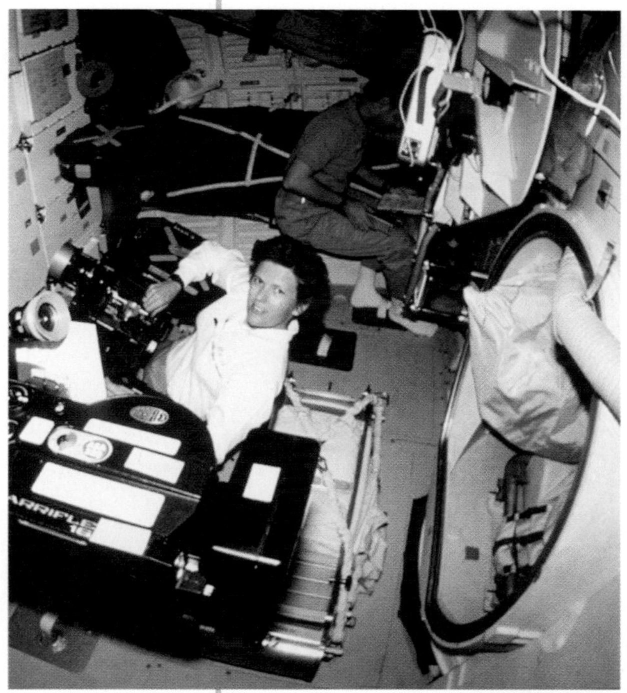

Figure 12.17

Continued ...

Continued ...
FROM PAGE 509

What Is Required?

Find the mass of lithium carbonate that is produced when 25.0 L of carbon dioxide is bubbled through an aqueous solution of lithium hydroxide.

What Is Given?

$2LiOH_{(aq)} + CO_{2(g)} \rightarrow Li_2CO_{3(aq)} + H_2O_{(\ell)}$

1.0 L $V = 25.0$ L
1.5 mol/L $T = 28.3$°C
$P = 0.85$ atm

Plan Your Strategy

Step 1 Convert the temperature to kelvins. Convert the pressure from atm to kPa, or use $R = 0.8206$ atm·L/mol·K.

Step 2 Use the ideal gas law to find the number of moles of carbon dioxide that reacts.

Step 3 Use the stoichiometry of the equation to determine the number of moles of lithium carbonate produced.

Step 4 Use the periodic table to determine the molar mass of lithium carbonate. To find the mass of lithium carbonate produced, multiply the number of moles by the molar mass.

Act on Your Strategy

Step 1 Convert the temperature and pressure.

$$T = 28.3°\text{C} + 273$$
$$= 301 \text{ K}$$

$$0.85 \cancel{\text{atm}} \times \frac{101.3 \text{ kPa}}{1 \cancel{\text{atm}}} = 86 \text{ kPa}$$

Step 2 Find the number of moles of $CO_{2(g)}$.

$$PV = nRT$$

$$\therefore n = \frac{PV}{RT}$$

$$= \frac{(0.85 \cancel{\text{atm}})(25.0 \cancel{\text{L}})}{(0.08206 \cancel{\text{atm}}\cdot\cancel{\text{L}}/\text{mol}\cdot\cancel{\text{K}})(301 \cancel{\text{K}})}$$

$$= 0.86 \text{ mol}$$

Therefore, 0.86 mol of $CO_{2(g)}$ passes through the LiOH solution.

Step 3 Find the number of moles of Li_2CO_3 produced.

From the balanced equation, we know that 1 mol of carbon dioxide produces 1 mol of lithium carbonate.

$$\frac{n \text{ mol } Li_2CO_3}{0.86 \text{ mol } CO_2} = \frac{1 \text{ mol } Li_2CO_3}{1 \text{ mol } CO_2}$$

Continued ...

Continued ...

FROM PAGE 510

$$(0.86 \text{ mol } CO_2)\ \frac{n \text{ mol } Li_2CO_3}{0.86 \text{ mol } CO_2} = \frac{1\text{mol } Li_2CO_3}{1\text{mol } CO_2}\ (0.86 \text{ mol } CO_2)$$

$n = 0.86$ mol of Li_2CO_3

Step 4 Find the molar mass of Li_2CO_3. Then find the mass of Li_2CO_3 produced.

$$\begin{aligned} M_{Li_2CO_3} &= (2 \times 6.94 + 12.01 + 3 \times 16.00) \text{ g/mol} \\ &= 73.89 \text{ g/mol} \end{aligned}$$

Thus, the mass of Li_2CO_3 produced is

$$\begin{aligned} m &= n \times M \\ &= 0.86 \text{ mol} \times 73.89 \text{ g/mol} \\ &= 64 \text{ g} \end{aligned}$$

Therefore, 64 g of Li_2CO_3 is produced in this reaction.

Check Your Solution.

The answer has two significant digits. This is the least number of significant digits in the question.

When the units cancel in the ideal gas equation, mol remains. When the units cancel in the final calculation, g remains.

The answer seems reasonable.

Practice Problems

35. Oxygen, O_2, reacts with magnesium, Mg, to produce 243 g of magnesium oxide, MgO, at 101.3 kPa and 45°C. How many litres of oxygen are consumed? Start by writing the balanced equation.

36. Zinc reacts with nitric acid to produce 34 L of dry hydrogen gas at 900 torr and 20°C. How many grams of zinc are consumed?

37. 0.75 L of hydrogen gas is collected over water at 25.0°C and 101.6 kPa. What volume will the dry hydrogen occupy at 103.3 kPa and 25.0°C?

38. 3070 kg of coal burns to produce carbon dioxide. Assume that the coal is 95% pure carbon and the combustion is 80% efficient. (**Hint:** The mole ratio of $C_{(s)}$ to $CO_{2(g)}$ is 5:4.) How many litres of carbon dioxide are produced at SATP?

39. When 7.48 g of iron reacts with chlorine gas, 21.73 g of product is formed.

(a) How many moles of chlorine are used?

(b) What is the formula for the product?

(c) Write the equation for the reaction that occurs.

Now you have a chance to do an exciting investigation: the reaction of magnesium metal with strong acid. Remember to take appropriate safety precautions and follow your teacher's directions when working with the acid.

SKILL FOCUS

Predicting

Performing and recording

Analyzing and interpreting

Investigation 12-B

The Production of Hydrogen Gas

In this investigation, you will produce hydrogen gas by reacting strong acid with magnesium metal.

$$Mg_{(s)} + 2HCl_{(aq)} \rightarrow MgCl_{2(aq)} + H_{2(g)}$$

You will collect the hydrogen gas over water in a graduated cylinder.

Question

What is the molar volume of dry hydrogen gas at STP? Calculate it using the volume of hydrogen gas that you produce and the mass of one reactant.

Prediction

Predict the volume of hydrogen gas that will be produced. Use the mass of the magnesium and the balanced chemical equation given above. Assume 100% yield, and use regular stoichiometry. Organize your calculations clearly.

Safety Precautions

- Before beginning this investigation, check that there are no open flames (such as lit Bunsen burners) in the laboratory.
- The acid that you are using in this investigation is strong enough to burn. Wear your safety glasses and lab apron at all times. Handle the acid carefully. Wipe up any spills of water or acid immediately. If you accidentally spill any acid on your skin, wash it off immediately with large amounts of cool water.
- When you have finished the investigation, you can safely wash the products down the sink. You must dilute them, however, by running water down afterward.

Materials

scale or balance
100 mL graduated cylinder
stopper with two holes to fit graduated cylinder
1 L beaker or bowl
water at room temperature
6.0 mol/L hydrochloric acid, HCl
6 to 7 cm piece of magnesium ribbon
10 to 15 cm piece of copper wire
steel wool
barometer and thermometer
clamp and ring stand (optional)

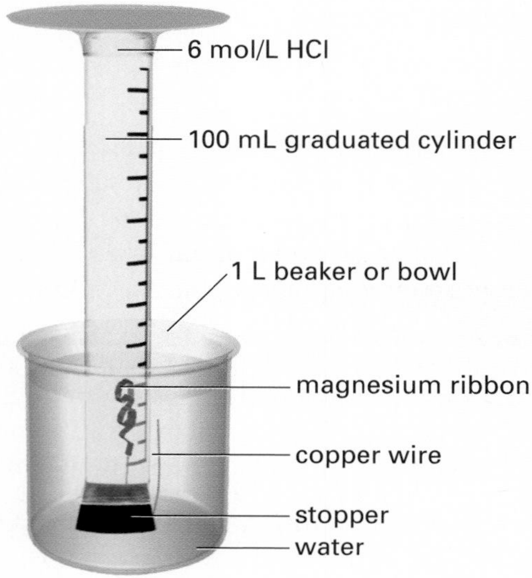

Procedure

1. Prepare a table, in your notebook. Show your calculations.

Observations and Results

Observations	Trial 1	Trial 2 (if time permits)
mass of magnesium ribbon (g)		
temperature of water (°C)		
barometric pressure (kPa)		
volume of hydrogen collected (mL → L)		
volume of pressure of water at this temperature (kPa)*		

Results	Trial 1	Trial 2 (if time permits)
number of moles of magnesium (mol)		
volume of collected dry hydrogen at STP (L)		
molar volume of hydrogen at STP (L/mol)		

*Find the pressure of the water vapour in Table 12.3, or ask your teacher.

2. Obtain a piece of magnesium ribbon that is about 6 to 7 cm long. Use steel wool to clean the outside of the ribbon. Measure the mass of the ribbon.

3. Use the mass of the magnesium and the balanced equation to predict the volume of hydrogen gas that will be produced. Show your calculations in your notebook.

4. Fill the beaker (or bowl) about half full of water at room temperature. Measure the temperature of the water. (You will use this temperature to approximate the temperature of the gas produced.)

5. Measure and record the barometric pressure.

6. Add 15 mL of water to the graduated cylinder. Then, *very carefully*, pour 10 to 15 mL of 6 mol/L HCl into the graduated cylinder. *Very slowly and carefully*, pour water at room temperature down the *sides* of the cylinder until the cylinder is completely filled. Your objective in pouring the water this way is to avoid mixing it with the acid at the bottom of the cylinder. **CAUTION** Normally you should avoid adding water to an acid. Be particularly careful during this step.

7. Attach the magnesium ribbon to the copper wire. Dangle the magnesium in the graduated cylinder. The magnesium should hang 1 to 2 cm below the stopper. Put the stopper in the cylinder. Do not worry if a small amount of water overflows out of the cylinder.

8. Hold your gloved finger over the holes in the stopper. Tip the cylinder upside down into the 1 L beaker. Be careful that no air bubbles get into the cylinder. Hold or clamp the tube into place. Watch the reaction proceed. Record your observations in your notebook.

9. Add water, at room temperature, to the beaker until the level of the water inside the cylinder is exactly the same as the level of the water in the beaker. This equalizes the pressure of the hydrogen gas with the air pressure outside the tube. **Note:** Another way to equalize the pressure is to raise the graduated cylinder slightly to align the water levels.

10. Record the volume of the trapped gas.

11. All the magnesium should be used up by the reaction, since it is the limiting reagent. If any magnesium ribbon does remain after the reaction, rinse it with water, dry it with a paper towel, and measure its mass. To find the mass of the magnesium used up by the reaction, subtract the final mass from the initial mass.

12. Empty and clean all your apparatus. Clean your work space. Wash your hands.

Analysis

1. Calculate the molar volume of hydrogen. Use the volume of the H_2 gas, and the water temperature. Also, use the barometric pressure minus the pressure of the water vapour.

2. Use the combined or ideal gas law to translate the conditions to STP. Redo the calculations.

Conclusions

3. What was the class average for the molar volume of dry hydrogen gas at STP? How close was your molar volume to the class average?

4. How close was your molar volume to the accepted molar volume of a gas at STP? Calculate the percent error for your molar volume.

5. How would your results have been different had you not cleaned the magnesium ribbon before you used it?

6. What were some possible sources of error in your investigation?

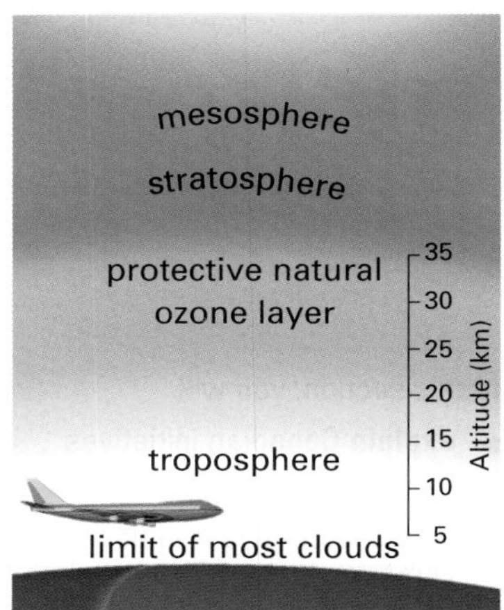

Figure 12.20 Most ozone is located about 25 km above Earth's surface. This is a very thin level, however. If all the ozone in the atmosphere were compressed, it would be only a few millimetres thick.

The Ozone Cycle

High in the atmosphere, a gas called ozone, O_3, absorbs ultraviolet (UV) radiation from the Sun. The radiation separates the ozone into oxygen gas, O_2, and an oxygen atom. After passing through a few more steps, ozone is re-formed when molecules of oxygen gas combine with oxygen atoms. By absorbing energy from the Sun in this way, ozone prevents harmful UV radiation from reaching Earth's surface. Figure 12.19 illustrates the cycle that occurs as ozone is formed, absorbs UV radiation, and breaks up. Figure 12.20 shows where ozone is located in the atmosphere.

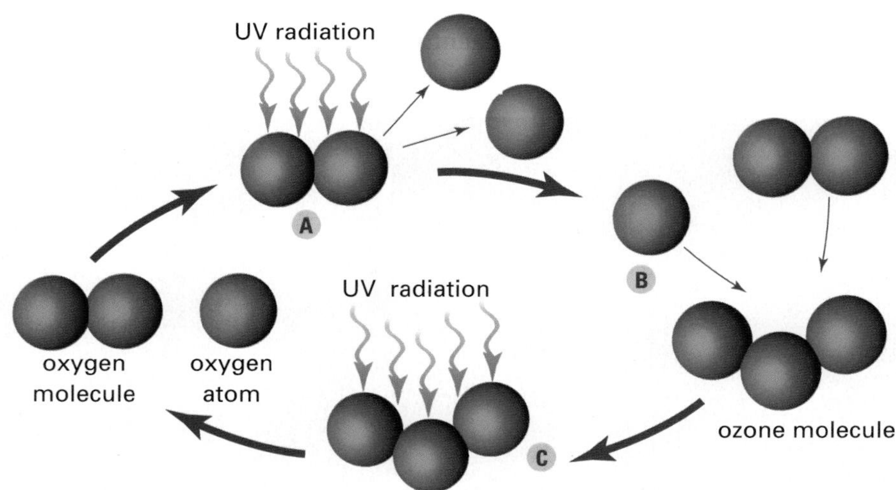

Figure 12.19 The ozone cycle: (A) Oxygen gas in the atmosphere absorbs energy from the Sun. Each oxygen molecule breaks up into two oxygen atoms. (B) An oxygen atom combines with a molecule of oxygen gas to form ozone. Another molecule is needed to absorb extra energy. (C) Ozone absorbs ultraviolet radiation and breaks up again into oxygen gas and an oxygen atom.

Pollutants in the Atmosphere

Gases from living and non-living processes on Earth's surface interact with the gases in the atmosphere. For example, oxygen and carbon dioxide in the air are involved in animal and plant respiration. A forest fire burns plants to produce carbon dioxide gas. On a more damaging level, human technology and industrial processes produce many polluting gases. As you will learn in Chapter 14, burning fossil fuels is a major source of gas pollution in the atmosphere.

Chlorofluorocarbons (CFCs) are an important class of polluting gases that are not usually caused by burning fossil fuels. CFCs are stable and harmless near the ground. When they make their way up into the atmosphere, however, they interact and interfere with atmospheric processes. In particular, these gases interfere with the production and reactions of ozone, O_3. You will learn more about CFCs later in this section.

Figure 12.21 Smog was originally defined as a mixture of smoke and fog. Today it can also be photochemical, caused by sunlight breaking down air pollutants.

Ozone Near the Ground

Ozone does not only exist high up in the stratosphere. It is also present much closer to us, in the *troposphere*: the layer of the atmosphere that lies directly over Earth. Ozone near the ground is largely produced when nitrogen oxide gas from car and truck exhaust fumes reacts with oxygen gas in sunlight. Ozone is a major component of smog in cities. (See Figure 12.21.)

At this level, ozone is a pollutant with a harsh odour. In humans and animals, it causes respiratory problems, including coughing, wheezing, and eye irritation. It retards plant growth, reduces the productivity of crops, and damages forests. Concentrations of ozone as low as 0.1 ppm (parts per million) can decrease photosynthesis by 50%. In addition, it damages plastics, breaks down rubber, and corrodes metals.

Careers  in Chemistry

Environmental Technician

Gases in the atmosphere (such as carbon dioxide and methane) allow heat from the Sun to enter the atmosphere and prevent it from leaving again. This is called the *greenhouse effect*. Thanks to the greenhouse effect, Earth remains fairly warm, with an average temperature of about 15.5°C. Without this effect, Earth's temperature would be about −18°C.

Human activities over the last hundred years have caused the level of carbon dioxide and other gases in the atmosphere to increase. As a result, more heat is trapped in Earth's atmosphere. According to Environment Canada, Canada's average temperature has risen by about 1°C over the last century. This is causing more frequent and more intense winter storms.

Change from the Ground Up

How can we decrease the production of greenhouse gases? One way is to focus on the sources of these gases. As global warming increases, chemists who study atmospheric processes will be more in demand.

Change begins with accurate measurements. Environmental technicians assess and monitor pollution levels in air, water, and soil. To become an environmental technician, you need a high school diploma, with advanced-level credits in mathematics, English, and science (preferably chemistry or physics). You also need a two- or three-year community college program in environmental technology.

Environmental technicians gather gas samples from smokestacks and PCBs from transformers. They also set up equipment in the field to create baseline studies and monitor changes in the environment.

Make Career Connections

Human Resource Development Canada has offices in every province. It also has a web site where you can access descriptions and requirements for many careers, including that of environmental technician. What other environmental career opportunities can you locate? Prepare a brief report of your findings, and present it to the class.

CFCs and Ozone Depletion

As you learned earlier, **chlorofluorocarbons (CFCs)** are chemicals that interfere with the ozone cycle high up in the atmosphere. CFCs are non-toxic, nonflammable compounds that contain atoms of chlorine, carbon, and fluorine. These gases are human-made compounds that were released into the atmosphere primarily from refrigeration and aerosol devices.

In 1928, Thomas Midgley invented the first CFC compound. Because they were useful but safe, they were referred to as "miracle compounds." In particular, dichlorodifluoromethane, CCl_2F_2, also known as Freon, was discovered to be an efficient refrigerant.

INSTANT
CAR
KOOLER
FOR HOT
INTERIORS

Figure 12.22 **CFCs have been used as refrigerants, coolants in home and automobile air conditioners, and propellants in aerosol containers such as hair sprays.**

Because of Freon, household refrigeration became common. In fact, much of our modern-day life style first became possible because of CFCs. By 1974, millions of tonnes of CFCs had been produced (see Figure 12.22). At the University of California, chemists F. Sherwood Rowland and Mario Molina began to wonder where all of these CFCs ended up. They realized that CFCs are chemically very stable. However, they began to calculate what happens when CFCs are exposed to high levels of radiation far up in the atmosphere. As it turned out, their fears were well-founded. In 1985, British scientists in the Antarctic noticed a large decrease in the ozone layer above the Antarctic. A "hole" in the ozone layer was beginning to form. In 1995, Rowland and Molina, along with a third ozone scientist, won the Nobel Prize for their work with CFCs.

How CFCs Attack Ozone

Today we know that CFCs high in the atmosphere break apart under ultraviolet radiation to produce chlorine atoms. These chlorine atoms destroy ozone molecules.

$$Cl + O_3 \rightarrow ClO + O_2$$

The product, ClO, reacts with an oxygen atom and releases the chlorine atom. The chlorine atom attacks another ozone molecule. Over time, one chlorine atom can destroy thousands of ozone molecules. Figure 12.23 illustrates this process for the CFC trichlorofluoromethane, CCl_3F. Eventually the chlorine atom reacts with a different compound in the atmosphere to form a stable, less harmful product.

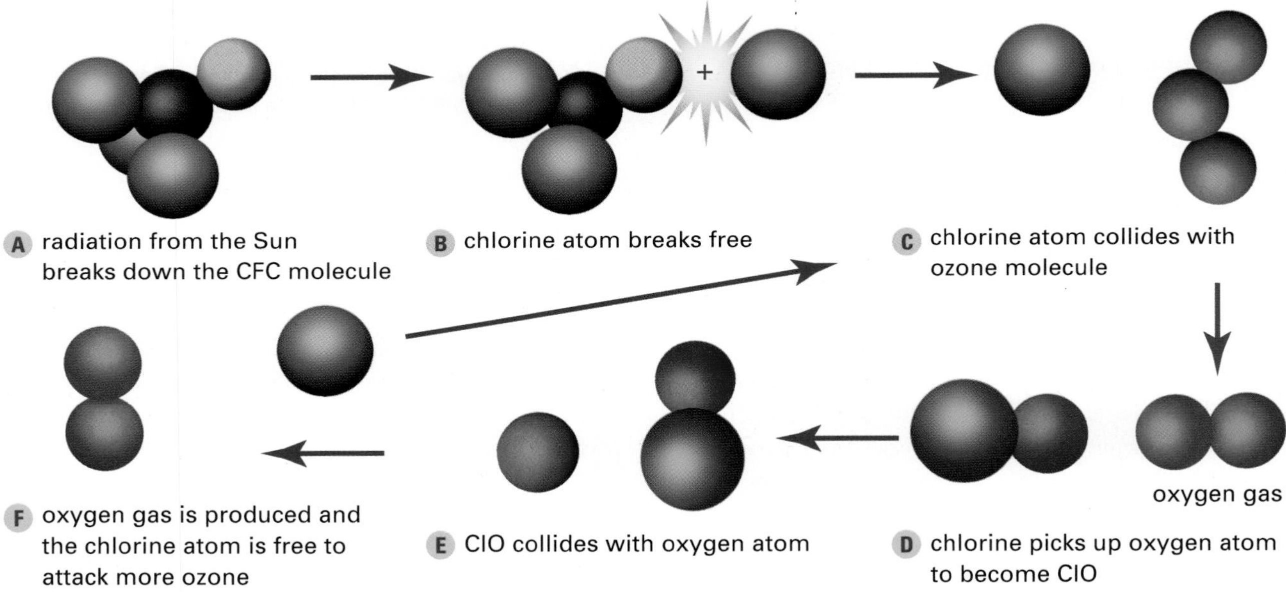
+
A radiation from the Sun breaks down the CFC molecule
B chlorine atom breaks free
C chlorine atom collides with ozone molecule
oxygen gas
D chlorine picks up oxygen atom to become ClO
E ClO collides with oxygen atom
F oxygen gas is produced and the chlorine atom is free to attack more ozone

Figure 12.23 **The breakdown of a CFC by ultraviolet radiation produces single chlorine atoms that attack and destroy ozone. Because the chlorine atoms are released, they attack ozone again and again in a chain reaction.**

Although they are the most abundant ozone-depleting substance, CFCs are not the only culprits. Other chemicals that damage the ozone layer include methyl bromide, CH_3Br, carbon tetrachloride, CCl_4, and halons such as carbon trifluorobromide, CF_3Br.

The Effects of Ozone Depletion

How do we know that CFCs and other ozone-destroying molecules are having an effect on the atmosphere? As Figure 12.24 illustrates, ozone levels over Canada and other parts of the world have decreased significantly since the late 1970s.

With reduced ozone levels, more ultraviolet radiation from the Sun reaches Earth. Among humans, UV-induced skin cancer and eye damage are becoming a serious threat. The increased levels of radiation also damage phytoplankton in fresh and marine ecosystems. Since phytoplankton are the base of the aquatic food chain, this damage affects all other water species. As you learned earlier, the *presence* of ozone close to Earth damages crops and forests. A *lack* of ozone in the atmosphere, however, also reduces the yield of crops, such as barley and canola, and harms forests.

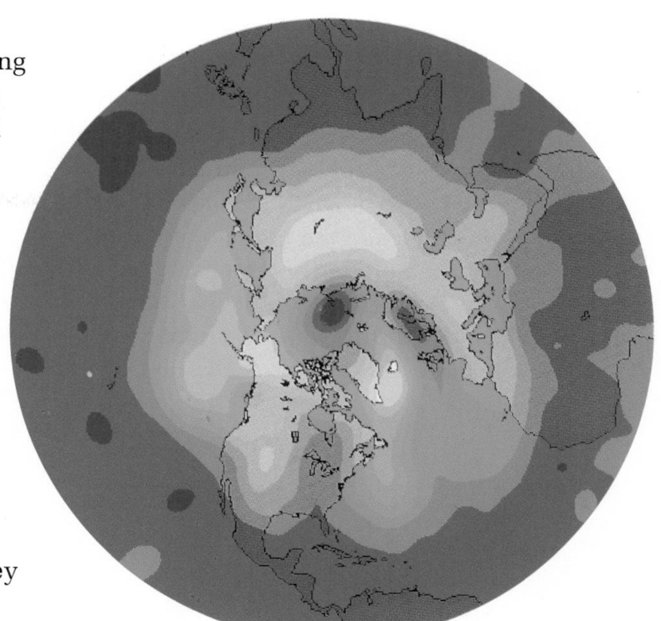

Figure 12.24 Ozone thinning is still occurring over the northern hemisphere, as shown here. The dark blue in the centre indicates the presence of an ozone "hole."

Improving Air Quality

What has happened since the discovery that CFCs and other chemicals were harming the ozone layer? The chemical industry has invented and produced environmentally friendly refrigerants. All new cars produced in North America have air conditioners that contain ozone-friendly refrigerants. Used CFCs from older refrigeration units are being recycled so that they do not escape into the atmosphere. A lot has already been accomplished, but more remains to be done.

Much of the positive activity comes from an international agreement called the **Montréal Protocol**. This agreement was signed in Montréal, Canada, on September 16, 1987. It has been successful in drastically reducing the use of CFCs worldwide. The Montréal Protocol is particularly significant because, for the first time, individual countries put a common planetary goal ahead of their own economic interests.

The Montréal Protocol stated that the production and consumption of all substances that deplete the ozone layer would be phased out by the year 2000 in developed countries. (Methyl chloroform would be phased out by 2005.) The chemicals that are named in the agreement include CFCs, halons, carbon tetrachloride, methyl chloroform, and methyl bromide. Once CFC production and consumption are stopped, scientists hope that the ozone layer will recover within 50 or 60 years. The success of the Montréal Protocol depends however, on the co-operation of both developed and developing countries.

In 1993, a scientific link was established between ozone depletion and increases in ultraviolet radiation. It was found that increased exposure to UV-B radiation causes skin cancer, the formation of cataracts, and the suppression of the human immune system. Research has shown skin cancer to be as common as all other types of cancer combined. Sunscreens can protect humans from the risk of some skin cancers. Unfortunately they do not appear to provide protection against damage to the immune system.

Section Wrap-up

In this section, you learned how ozone high in the atmosphere interacts with UV light from the Sun. In addition, you learned where CFCs come from, and how they damage the ozone layer. Finally, you saw that the Montréal Protocol is striving to prevent further ozone damage.

Canadians  in Chemistry

Parisa Ariya was born in Tehran, the capital of Iran. She chose atmospheric chemistry for a career. Scientists in this field study the transformation of molecules in the atmosphere (the layer of gases surrounding Earth). They also study the atmosphere's interactions with oceans, land, and living things. After studying in several countries, Dr. Ariya became a professor at McGill University in Montréal.

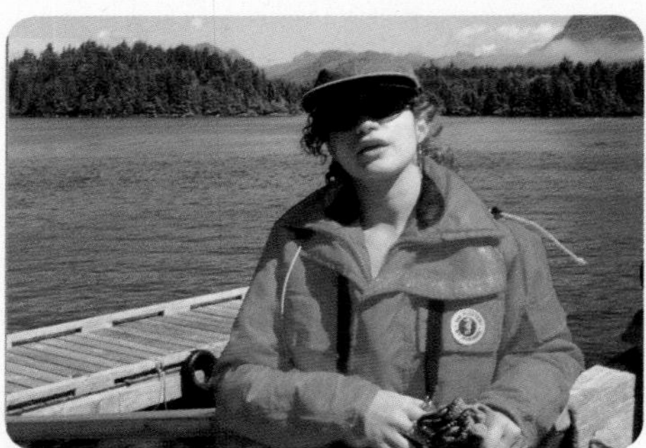

Dr. Parisa Ariya

One of Dr. Ariya's particular areas of interest is halogen chemistry. Halogens such as chlorine, Cl, and bromine, Br, occur naturally in ocean waters. As well, they enter ocean waters as run-off from human activities. The oceans emit these halogens into the atmosphere. There they react with, and destroy, ozone, O_3.

Ariya and her students are trying to determine what kinds of halogens exist in Earth's atmosphere, how quickly they are produced and degraded, and what their major sources are. As they find answers to these questions, they may be able to recommend ways to reduce halogens in the oceans and atmosphere. They may also develop ways to modify halogens' reactions with ozone and other gases so that the halogens do less harm.

Another of Ariya's research interests is sulfur, S. She and her students are studying its atmospheric reactions with ozone and hydrogen peroxide, H_2O_2. Through field studies, laboratory experiments, and modelling, they are trying to determine the impacts of such reactions.

"You can enjoy nature through sports," says Dr. Ariya, who is an avid soccer player and swimmer. "But as a scientist you also enjoy nature intellectually, methodically. Science keeps the mind alive because you're constantly learning. Science is fun!"

Section Review

Unit Issue Prep

Research the Ontario Drive-Clean program. How is pollution from car exhausts being regulated by the government? Search for information in preparation for the Unit Issue.

1. C Describe the cycle that ozone goes through as it absorbs ultraviolet radiation. Use a diagram.
2. C Compare the effects of ozone near the ground with its effects high in the atmosphere.
3. K/U What are chlorofluorocarbons? How do they affect the ozone layer?
4. K/U Describe the Montréal Protocol. Why is it so significant?
5. MC Canadian scientists developed the Brewer Ozone Spectrophotometer, a state-of-the-art ozone-measuring device. It is the most accurate ozone-measuring device in the world. Use the Internet or reference books to find out how the Brewer Ozone Spectrophotometer works. Report your findings to the class.

CHAPTER 12 Review

Reflecting on Chapter 12

Summarize this chapter in the format of your choice. Here are a few ideas to use as guidelines:

- Describe Avogadro's hypothesis, and explain how it relates to the properties of gases.
- Explain how the ideal gas law came about. Describe how to use it to calculate gas properties.
- Explain how to use the ideal gas law for stoichiometric problems involving gases as reactants or products in a chemical reaction.
- Describe how to identify an unknown gas using the ideal gas law.
- Explain the importance of ozone, and describe the action of CFCs on ozone. Explain what the Montréal Protocol has done for the ozone issue.

Reviewing Key Terms

For each of the following terms, write a sentence that shows your understanding of its meaning.

law of combining volumes	molar volume
	ideal gas law
law of multiple proportions	chlorofluorocarbons (CFCs)
Avogadro's hypothesis	Montréal Protocol

Knowledge/Understanding

1. (a) How did the work of John Dalton and Joseph Gay-Lussac lead to Avogadro's hypothesis?
   (b) How does Avogadro's hypothesis help us to understand gas behaviour and gas reactions?
2. (a) What is the law of multiple proportions?
   (b) State Gay-Lussac's law of combining volumes.
   (c) How are the two laws related? What useful problem-solving gas law did they lead to?
3. What is the molar volume of an ideal gas at STP?
4. What are the characteristics of an ideal gas? How is it different from a real gas?
5. Is the density of one mole of hydrogen gas at STP the same as the density of one mole of oxygen gas at STP? Explain your answer.
6. (a) Additional gas is added to a container with a fixed volume. What happens to the pressure and temperature of the gas?
   (b) A balloon filled with gas experiences a drop in pressure. What happens to the volume of the gas?
7. A student reacts a sample of zinc with hydrochloric acid. He collects the gas as it bubbles through water. How will this student need to correct his measurements? Explain your answer.
8. What are the effects of ozone pollution near the ground?
9. Why are high levels of ultraviolet radiation potentially dangerous?
10. How is the Montréal Protocol attempting to improve future air quality?

Inquiry

11. As you work on gas reactions in the laboratory, your barometer shatters. Now it is impossible to measure the pressure of the room.
    (a) Design a simple investigation that allows you to calculate the pressure of the room. Assume you have a thermometer.
    (b) Identify the materials and equipment you will need to carry out the investigation.
    (c) Which variables will you hold constant? Which will you change? Which will you measure?
12. What volume does each amount of helium occupy at STP?
    (a) 1.00 mol
    (b) 12.5 mol
    (c) 100.0 g
13. How many molecules are in 0.250 m^3 of oxygen at STP?
14. What volume does 2.00 mol of oxygen occupy at 750 torr and 30.0°C?
15. What volume does 1.50 g of nitrogen gas occupy at 100.0°C and 5.00 atm?
16. Oxygen that is needed in a school laboratory is stored in a pressurized 2.00 L cylinder. 25.0 g of oxygen is contained in the cylinder at 20.0°C. Under what pressure is the oxygen stored?
17. Propane, C_3H_8, that is needed for a barbecue is stored in a 40.0 dm^3 metal cylinder. The pressure gauge on the cylinder reads 25.0 atm

at 20.0°C. What mass of propane is in the cylinder? (**Hint:** 1 dm^3 = 1 L)

18. Find the density (in g/L) of each atmospheric gas.
(a) oxygen at 1000 torr and 30.0°C
(b) helium at 10.0 atm and 20°C

19. **(a)** Pressurized CO_2 is used in the soft-drink manufacturing industry. How many grams of carbon dioxide are in a 500.0 cm^3 tank at −50.0°C and 2.00 atm?
(b) How many grams of oxygen does this tank hold at the same temperature and pressure?

20. 9.0 g of an unknown gas is stored in a 5.00 L metal tank at 0.0°C and 202.0 kPa. To identify the gas, investigators decide to find out its molecular mass.
(a) What is the molecular mass of the gas?
(b) What is the gas?

21. A 25.0 L tank, stored at −20.0°C, contains 10.0 g of helium and 10.0 g of hydrogen gas.
(a) What is the total number of moles of gas in the tank?
(b) What is the total pressure (in kPa) in the tank?
(c) What is the partial pressure of helium in the tank?

22. A 60.0 g sample of nitrogen gas is stored in a 5.0 L tank at a pressure of 10.0 atm. At what temperature (in °C) is the gas stored?

23. A 13.4 g sample of an unknown liquid is vapourized at 85.0°C and 100.0 kPa. The vapour has a volume of 4.32 L. The percentage composition of the liquid is found to be 68.5% carbon, 8.6% hydrogen, and 22.8% oxygen. What is the molecular formula of the liquid?

24. An unlabelled bottle of an unknown liquid is found on a shelf in a laboratory storeroom. 10.0 g of the liquid is vaporized at 120.0°C and 5.0 atm. The volume of the vapour is found to be 568.0 cm^3. The liquid is found to be made up of 84.2% carbon and 15.8% hydrogen. What is the molecular formula of the liquid?

25. A 4.2 g sample of a volatile liquid contains 1.0 g of carbon and 0.25 g of hydrogen. The rest of the liquid is chlorine. When the sample is vaporized at 101.0 kPa and 60.0°C, it occupies a volume of 2.2 L. What is the molecular formula of the liquid?

26. Methanol has potential to be used as an alternative fuel. It burns in the presence of oxygen to produce carbon dioxide and water.
$CH_3OH_{(\ell)} + O_{2(g)} \rightarrow CO_{2(g)} + 2H_2O_{(g)}$
(a) Balance this equation.
(b) 10 L of oxygen is completely consumed at STP. What volume of CO_2 is produced?
(c) What mass of methanol is consumed in this reaction?

27. A student wants to prepare carbon dioxide using sodium carbonate and dilute hydrochloric acid.
$Na_2CO_{3(s)} + 2HCl_{(aq)} \rightarrow 2NaCl_{(aq)} + CO_{2(g)} + H_2O_{(\ell)}$
How much sodium carbonate should the student react with excess hydrochloric acid to produce 1.00 L of carbon dioxide at 24.0°C and 760 torr?

28. A scientist makes hydrogen gas in the laboratory by reacting calcium metal with an excess of hydrochloric acid.
$Ca_{(s)} + 2HCl_{(aq)} \rightarrow CaCl_{2(aq)} + H_{2(g)}$
The scientist reacts 5.00 g of calcium and collects the hydrogen over water at 25.0°C and 103.0 kPa. What volume of dry hydrogen is produced?

29. A chemist collects oxygen over water at 22.0°C and 105.0 kPa using the following reaction:
$2KClO_{3(s)} \rightarrow 2KCl_{(s)} + 3O_{2(g)}$
What volume of dry oxygen is obtained if the chemist heats 25.0 g of potassium chlorate?

30. Ammonia, a useful fertilizer, is produced by the following reaction:
$CH_{4(g)} + H_2O_{(\ell)} + N_2O_{(g)} \rightarrow 2NH_{3(g)} + CO_{2(g)}$
500.0 g of methane reacts with excess H_2O and N_2O. At 27.0°C and 1.20 atm, what volume of ammonia gas is produced?

31. Hydrochloric acid dissolves limestone, as shown in the following chemical equation:
$CaCO_{3(s)} + 2HCl_{(aq)} \rightarrow CaCl_{2(aq)} + CO_{2(g)} + H_2O_{(\ell)}$
12.0 g of $CaCO_3$ reacts with 110 mL of 1.25 mol/L HCl. At 22.0°C and 99.0 kPa, what volume of carbon dioxide is produced?

32. Butane from a disposable lighter burns according to the following equation:
$C_4H_{10(g)} + O_{2(g)} \rightarrow CO_{2(g)} + H_2O_{(\ell)}$

(a) Balance this chemical equation.

(b) How many grams of butane are needed to produce 300.0 mL of carbon dioxide at 50.0°C and 1.25 atm? How many grams of oxygen are needed?

Communication

33. Draw a concept map showing the connections between the following gas laws:
    - Boyle's law
    - Charles' law
    - Gay-Lussac's law
    - the combined gas law
    - the ideal gas law
    - Gay-Lussac's law of combining volumes
    - Dalton's law of partial pressures

34. Prepare a poster that will explain the relationships between the pressure, volume, and temperature of a gas to students in a younger grade.

35. Express the ideal gas constant in each of the following units.
    (a) kPa·L/mol·K
    (b) atm·L/mol·K
    (c) torr·L/mol·K

Making Connections

36. Complex carbohydrates are starches that your body can convert to glucose, a type of sugar. Simple carbohydrate foods contain glucose, ready for immediate use by the human body. Breathing and burning glucose, $C_6H_{12}O_6$, produces energy in a jogger's muscles, according to the following unbalanced equation:
$C_6H_{12}O_{6(aq)} + O_{2(g)} \rightarrow CO_{2(g)} + H_2O_{(g)}$
Just before running, Myri eats two oranges. The oranges give her body 25 g of glucose to make energy. The temperature outside is 27°C, and the atmospheric pressure is 102.3 kPa. Although 21% of the air Myri breathes in is oxygen, she breathes out about 16% of this oxygen. (In other words, she only uses about 5%.)
    (a) How many litres of air does Myri breathe while running to burn up the glucose she consumed?
    (b) How many litres of water vapour does she produce? How many litres of carbon dioxide gas does she produce?
    (c) Suppose that the water vapour Myri breathes out is condensed to its liquid form. If the density of the water is 1.0 g/mL, what is its volume?

37. Sulfur dioxide reacts with oxygen in air to produce sulfur trioxide. The sulfur trioxide then reacts with water to produce sulfuric acid. Write balanced chemical equations for these reactions.

38. Write a story to describe what your life would be like if you did not participate in any polluting activities (such as riding in petroleum-powered cars) or use any products (such as plastic) that cause pollution. Include as many pollution-causing products as possible.

Answers to Practice Problems and Short Answers to Section Review Questions:
Practice Problems: **1.** 24 L/mol **2.** 42.8 L/mol **3.** 25.9 L/mol **4.(a)** 0.446 mol **(b)** 12.5 g **5.** 0.089 mol **6.** 0.500 mol, 3.01×10^{23} molecules **7.** 77.3 L **8.** 1.0×10^2 L **9.** 78.38 L **10.(a)** 56. 0 L **(b)** 1.51×10^{24} molecules **(c)** 3.01×10^{24} atoms **11.** 74.4 L **12.** 45.7 K **13.** 626 kg **14.** 23.3 L **15.** 25 kPa **16.** 1.71 g/L **17.** 2.35 g/L **18.** 3.65 g/L **19.** 578 kPa **20.** 48.2 g/mol **21.** 2.0×10^2 g/mol **22.** 71.3 g/mol **23.(a)** 1.69 g/L **(b)** 1.78 g/L **24.** C_2H_2 **25.(a)** 1:2 **(b)** 1:2 **(c)** 2:1 **(d)** 1:1 **26.(a)** 4.5 L **(b)** 7.5 L **27.(b)** 6.00 L **28.(b)** a_4b_2 **29.(a)** 65.0 L **(b)** 1.63×10^{24} molecules **(c)** 3.25×10^{24} atoms **30.** 16 L **31.** 0.20 g **32.** 0.047 g **33.(a)** 79 L **(b)** 98 L **34.** 0.32 g **35.** 79 L **36.** 1.1×10^2 g **37.** 0.71 L **38.** 4.8×10^6 L **39.(a)** 0.201 mol **(b)** $FeCl_3$
Section Review: **12.1:** **1.** 104 g **2.** 8.7×10^2 L **3.** 2.5×10^2 kPa **4.** 58 g; 2.2×10^{23} molecules **5.** –19°C **7.** 6.2 atm **12.2:** **1.** 0.72 g/L **2.(a)** 2.0 g/mol **(b)** hydrogen gas **3.** N_2O_4 **12.3:** **5.** 2.0 L **6.(b)** 1.7 L **7.** 41 L **8.(a)** 1.1×10^2 g **(b)** 0.075 atm

An Issue to Analyze

The Costs of Getting Around

Many people blame the government for not enforcing sufficient regulations to protect and clean up the environment. New, cleaner technologies have been developed by science. Why don't we use them? The fact is, Canadians are accustomed to a comfortable lifestyle, and may not be willing to change it. How would you feel about stricter government regulations that affected *your* transportation and recreation?

Background

People have always turned to the cheapest and most efficient technologies. In the last few centuries, fossil fuels such as coal and oil were discovered in large amounts. These fuels were also very efficient, producing large amounts of energy. It is only recently that we have become aware of the environmental consequences of our use of fossil fuels.

Today, we are attempting to reduce environmental harm. Coal- and oil-burning engines have been replaced by cleaner gas-burning engines. Many engines are equipped with "scrubbers" that help remove toxins before releasing gases into the atmosphere. Most environmental problems have not been solved, however. Automobile emissions are one of our most serious environmental problems today.

Automobiles expel an enormous amount of toxins into the air every day. Besides harming the environment, these toxins are also dangerous to human health. It is estimated that many cases of bronchitis, asthma, and even early death could be eliminated, by removing sulfur from fossil fuels before burning them. This process is possible, but it costs money. Ultimately, these additional costs would be paid by the consumer—you!

Often, Canadians do not appreciate or approve of government efforts to reduce automobile emissions. Some people were angry when, decades ago, the government asked them to switch to unleaded gasoline. Even now, many people remove factory-installed, pollution-reducing resonators and catalytic converters from their new cars. They do this in an effort to increase the power of their vehicles.

Environment Canada has copious amounts of research suggesting that reducing automobile emissions would improve the health of Canadians. The restrictions that would make these improvements, however, have not yet been enforced. For example, motorboats can be equipped with motors that are nearly silent and that do not dump oil into the water. Because these cleaner, quieter motors are more expensive, most boaters still use the old cheaper motors.

To what extent should individuals be allowed to make such choices? Should the good of the community and the overall environment have a higher priority than personal choice?

Plan and Present

As a class, you will debate the following proposal:

All vehicles must meet strict environmentally-friendly emissions standards. Any vehicles, new or old, that do not meet these standards must not be used until they are upgraded to the new emission standards. Individual owners must pay for this upgrading. Those individuals who cannot afford to upgrade their vehicles will be obliged to rely on public transportation.

There are many angles to this issue. You may wish to consider motorboat engines, private vehicles, commercial vehicles such as transport trucks, and the development and availability of public transportation.

1 Before starting your research, make a journal entry describing what you think of the proposal.

2 Look at the people listed below. How would each person react to the proposal? Why? For each person, write two comments expressing that persons point of view.
   (a) someone your age who lives in a rural area where there is no public transportation at all
   (b) someone your age who lives downtown in a city with excellent public transportation
   (c) a commercial truck driver who uses thousands of litres of fuel each week
   (d) a used-car salesperson
   (e) an autoworker

3 Write a counterpoint or amendment to the resolution in order to respond to each of the comments in your answer to question 2.

4 Research the facts behind this situation. Before you start, brainstorm some specific questions to research. Some suggestions are given below.
   - What gases do cars and trucks emit?
   - Why are these gases harmful?
   - What emission standards might be set?
   - Who uses the most fuel?
   - What standards do other developed countries have?
   - What standards (if any) are in place in third world countries?

5 Predict two future consequences of not adopting the resolution, and two consequences of adopting it.

6 You will be assigned a viewpoint in a debate of the proposal. Prepare your arguments and participate in the debate.

Evaluate the Results

1 Did your class come to a conclusion about the proposal? Was a compromise reached, or are some parties still unsatisfied?

2 List any solutions or suggestions your class developed. Are these suggestions likely to work? Why or why not?

3 Suppose you were a member of the government who had to decide on this issue.
   (a) Now that you have debated the issue, what is your decision?
   (b) How will you respond to special interest groups who approach you and disagree with your decision?

4 How do you think the issue of vehicle emissions will be addressed in the future?

Assessment

After you have explored this issue,

- Assess the debate based on feedback from your classmates. How could you improve your debating skills?
- Assess your research skills during the course of analyzing this issue, How have your skills improved.

Web LINK

To help you with your research, visit **www.school.mcgrawhill.ca/resources/**; go to Science Resources, then to Chemistry 11 for web site suggestions.

UNIT 4 Review

Knowledge/Understanding

In your notebook, write the letter of the best answer for each of the following questions.

1. Which of the following statements is not consistent with the kinetic molecular theory?
   - **(a)** Gases react with each other in simple whole number ratios of volumes.
   - **(b)** Particles in the gaseous state move slower than particles in the liquid state.
   - **(c)** The particles of a solid are strongly attracted to each other.
   - **(d)** Gas particles move in all directions, in straight lines.
   - **(e)** Elements combine in simple ratios to form compounds.
2. Gases that obey the postulates of the kinetic molecular theory:
   - **(a)** are not affected by pressure
   - **(b)** behave according to Charles' law at very low temperatures only
   - **(c)** are called ideal gases
   - **(d)** are affected by a change in pressure, which causes particles to move faster
   - **(e)** are called real gases
3. After it has been driven for several hours, a car tire heats up. As a result, the tire pressure increases. This is an example of:
   - **(a)** Boyle's law
   - **(b)** Charles' law
   - **(c)** Avogadro's law
   - **(d)** Gay-Lussac's law
   - **(e)** Torricelli's relationship
4. At 20°C, a moving piston reduces the volume of a cylinder by one half. The gas pressure inside the cylinder will:
   - **(a)** decrease by one half
   - **(b)** double
   - **(c)** increase by one half
   - **(d)** remain the same
   - **(e)** cause the gas particles to speed up
5. A weather balloon floats up into the atmosphere. The air pressure decreases as the balloon floats higher. Assuming the temperature remains constant, the balloon:
   - **(a)** increases in volume
   - **(b)** decreases in volume
   - **(c)** maintains a steady height in the moving air currents
   - **(d)** maintains a constant volume
   - **(e)** remains in the stratosphere
6. An engineer experiments with different air mixtures in a car tire. First, she removes the air from a tire and fills it with pure nitrogen. She measures the pressure exerted inside the tire, then adds a measured amount of helium. After measuring the pressure a second time, she adds a certain amount of argon. This engineer is investigating:
   - **(a)** Boyle's law
   - **(b)** Dalton's law of partial pressures
   - **(c)** Charles' law
   - **(d)** Gay-Lussac's law
   - **(e)** The combined gas law
7. Avogadro made an important contribution to the understanding of gases when he concluded that:
   - **(a)** The molar volume of a gas is 24.8 L at STP.
   - **(b)** Any volume of a gas contains 6.02×10^{23} molecules.
   - **(c)** At the same temperature and pressure, equal volumes of gases contain equal numbers of molecules.
   - **(d)** $PV = nRT$
   - **(e)** As the temperature of a gas increases, so does its pressure and volume.
8. At constant temperature, the pressure-volume relationship of an ideal gas is governed by:
   - **(a)** Boyle's law
   - **(b)** Avogadro's law
   - **(c)** The fact that pressure and temperature are directly related
   - **(d)** Gay-Lussac's law
   - **(e)** The fact that temperature and volume are directly related
9. Which if the following sets of coefficients will balance this chemical equation?

   $$NaN_{3(s)} \rightarrow Na_{(s)} + N_{2(g)}$$

   - **(a)** 3, 2, 2
   - **(b)** 2, 2, 3
   - **(c)** 1, 2, 1.5
   - **(d)** 1, 2, 3
   - **(e)** 1, 1, 2

10. Which of the following compounds is not responsible for removing ozone from the stratosphere?
   (a) carbon dioxide
   (b) chlorofluorocarbons
   (c) methyl bromide
   (d) CCl_2F_2
   (e) carbon tetrachloride

11. The main reason why the Montreal Protocol is so important is:
   (a) It cuts down on greenhouse gases in our atmosphere.
   (b) It protects our oceans.
   (c) It brought industrialized countries together to sign an environmental agreement reducing the production of carbon dioxide from fuel emissions from vehicles and industries.
   (d) It preserves our Arctic regions.
   (e) It cuts down on ozone depleting gases in our atmosphere.

Inquiry

12. A group of students measured the effect of pressure on the volume of different gases. One group observed the effect of pressure on pure oxygen gas. The other group observed pure nitrogen gas. The results obtained are shown below.

Pressure (kPa)	Volume O_2 (mL)	Volume N_2 (mL)
60.0	333.0	212.8
80.0	250.0	156.2
120.0	166.5	104.0
160.0	125.0	78.8
200.0	100.0	63.4

   (a) On the same set of axes, plot a graph of *P* versus *V* for oxygen and for nitrogen. Use a different colour for each gas.
   (b) For each of the volumes given, calculate a value for the inverse of the volume (1/*V*).
   (c) On the same set of axes, plot a graph of P versus 1/*V*. Use a different colour for each gas.
   (d) What conclusions can you make from the graphs that you have drawn?

13. What is a real gas? How does it differ in the laboratory from an ideal gas?

14. Boyle's law gives the relationship between the pressure and volume of a gas at constant temperature. Avogadro's law gives the relationship between the number of moles and the volume of a gas.
   (a) Write these two relationships in equation form, using *k* as a constant.
   (b) What is the relationship between pressure and number of moles?

15. **(a)** At constant pressure, the temperature, in kelvins, is doubled. What effect will this have on a gas? Explain.
   (b) At constant pressure, the temperature, in degrees Celcius, is doubled. How is this different from the situation in part (a)? How will the effect on a gas be different? Explain.
   (c) At constant temperature, the pressure on a gas is reduced by a factor of 5. What effect will this have on the volume of a gas? Why?

16. A sample of oxygen gas occupies a volume of 10.0 L at 546 K. At what temperature (in °C) would the gas occupy a volume of 5.0 L?

17. A sample of nitrogen gas occupies 11.20 L at 0°C and 101.3 kPa. How many moles of nitrogen are there in this sample? Explain.

18. The chemical equation below describes what happens when a match is struck against a rough surface to produce light and heat.

$$P_4S_{3(s)} + O_{2(g)} \rightarrow P_4O_{10(g)} + SO_{2(g)}$$

   (a) Balance this chemical equation.
   (b) If 5.3 L of oxygen gas were consumed, how many litres of sulfur dioxide would be produced?

19. A 250.0 mL balloon, full of pure helium at 101.3 kPa, is subjected to a pressure of 125.0 kPa at a constant temperature. What is the final volume of the balloon?

20. A sample of chlorine gas is used as a pool disinfectant. The sample occupies a volume of 500 mL in a cylinder with a moveable piston. The piston forces the gas out into the water when needed. The gas is stored at 25.0 atm at standard temperature. If the temperature remains unchanged and the piston compresses the gas to 220 mL, what is the pressure inside the cylinder?

21. A sample of air occupies a volume of 35.75 L at 25.0°C and 101.3 kPa. What pressure is needed to reduce the volume of air to 9.85 L at 25.0°C?

22. Hydrogen gas occupies a volume of 500 dm^3 at 125°C. If the pressure remains constant, what volume will the hydrogen occupy at 25.0°C?

23. A sample of natural gas occupies a volume of 350 L at 20.0°C. The pressure remains unchanged, and the temperature is increased until the volume of natural gas becomes 385 L. What is the final temperature (in °C) of the gas?

24. A sample of gas is collected at 25°C. If the temperature of the gas is tripled and the pressure on the gas is doubled, what portion of the original volume of gas will remain?

25. For each blank cell in the table below, calculate the missing quantity.

Initial temperature (°C)	Final temperature (°C)	Initial pressure (kPa)	Final pressure (kPa)	Initial volume (L)	Final volume (L)
100	100	101.3	110.0	5.0	
0	0	1.5×10^2		25.0	10.0
35	150	101.3	101.3	750	
85.0		125.0	125.0	35.5	25.5
27.0	45.5	102.5	86.7	1.00	
	85	99.5	88.7	450	500

26. A student collects 55.0 mL of hydrogen gas over water at 23.0°C and 750 torr. What volume will the dry hydrogen occupy at 40°C and 775 torr?

27. A mixture of gases in a cylinder contains 0.85 mol of methane (CH_4), 0.55 mol of oxygen, 1.25 mol of nitrogen, and 0.27 mol of propane (C_3H_8). The pressure on the cylinder gauge reads 2573 kPa. What pressure does each gas exert in the cylinder?

28. What volume will a weather balloon occupy if it contains 10 mol of air at 75.5 kPa and –45°C?

29. A 180 mL light bulb contains approximately 5.8×10^{-3} mol of argon in a 20°C room. What is the pressure of the argon inside the light bulb?

30. A 13.65 L vessel contains 0.750 mol of chlorine at 135 kPa. What is the temperature of the chlorine gas (in °C)?

31. Nitrogen gas is used to produce ammonia fertilizer. A sample of gas occupies a 2500 L tank at 5.5 atm and 27.5°C. What is the mass of nitrogen in the tank?

32. Ethanol vapour burns in air according to the following equation:

$$2C_2H_5OH_{(g)} + 6O_{2(g)} \rightarrow 4CO_{2(g)} + 6H_2O_{(\ell)}$$

If 2.5 L of ethanol burns at STP, what volume of oxygen is required? What volume of carbon dioxide will be produced?

33. The head of a match contains approximately 0.75 g of diphosphorus trisulfide. When the match is struck on a rough surface, it explodes into flame producing diphosphorus pentaoxide and sulfur dioxide (this is a slightly different reaction from that in question 18). What volume of sulfur dioxide will be produced if the temperature is 26.5°C and the pressure is 102.8 kPa?

34. Nitrogen monoxide, NO, is one of the gases that is responsible for smog. It is produced in various ways, one of which is during the combustion of ammonia.

$$4NH_{3(g)} + 5O_{2(g)} \rightarrow 4NO_{(g)} + 6H_2O_{(\ell)}$$

If 25.0 L of ammonia reacts with 27.5 L of oxygen at STP, what mass of nitrogen monoxide will be produced?

35. When baking a tray of blueberry muffins, you need to use baking soda (sodium hydrogen carbonate). The baking soda acts as a leavening agent, causing the muffin dough to rise as it bakes. One of the reactions that occurs during baking is:

$$2NaHCO_{3(s)} \rightarrow Na_2CO_{3(s)} + H_2O_{(\ell)} + CO_{2(g)}$$

Suppose your recipe calls for 5 mL (approximately 3.0 g) of baking soda. What volume of carbon dioxide will be generated at 195°C and 100 kPa to make the dough rise?

36. An anesthetic used in hospitals after World War II was made up of 64.8% carbon, 13.67% hydrogen, and 21.59% oxygen. It was found that a 5.0 L sample of this anesthetic had a mass of 16.7 g at STP. What is the molecular formula of this gas?

37. When you peel an orange, you usually smell a pleasant, tangy odour. This odour is due to the presence of a chemical compound called an ester. The ester in orange peel is made up of

69.8% carbon, 18.6% oxygen, and 11.6% hydrogen. When 5.3 g of this compound is vapourized at 125°C and 102 kPa, it is found to occupy a volume of 1.0 L. What is the molecular formula of this compound?

38. Each time you inhale, you take in about 0.50 L of air. How many molecules of each of the following gases do you inhale in one breath at 22°C and 101.3 kPa?
   (a) nitrogen
   (b) oxygen
   (c) argon
   (d) carbon dioxide

39. The space shuttle orbits Earth at an altitude of approximately 300 km. Canadian astronauts worked outside the shuttle using the Canadarm to repair a satellite. They were working in what is commonly referred to as a vacuum. However, accurate measurements show that the atmospheric pressure at this altitude is 1.33×10^{-9} kPa. Facing the sun, the average temperature is 223°C. How many molecules of gas are there in one litre of this so-called "vacuum" in outer space?

Communication

40. Define each of the following:
   (a) Charles' law
   (b) standard temperature
   (c) standard pressure
   (d) ideal gas
   (e) Gay-Lussac's law

41. Convert each of the following temperatures as indicated:
   (a) 250.1°C to kelvins
   (b) 373 K to degrees Celcius
   (c) absolute zero to degrees Celcius

42. Convert each of the following pressure values as indicated:
   (a) 725 torr to kPa
   (b) 105.1 kPa to torr
   (c) 1.4×10^2 kPa to atm
   (d) 320 torr to atm

43. Prepare a concept map to explain the depletion of ozone in the atmosphere. Include details on the different chemicals responsible. Also, make sure you include information on how this depletion occurs.

Making Connections

44. Nitrogen monoxide is used to produce nitric acid, a widely used industrial compound. Unfortunately, this same gas is also partly responsible for the production of smog and acid rain. Research the amount of nitrogen monoxide produced by an average automobile per hour.
   (a) Estimate the number of cars in your community (assume two cars per family). Calculate the number of grams of nitrous oxide released into the air in your area per hour at rush hour.
   (b) Approximately how many grams of nitrogen monoxide are produced per day in a large city such as Toronto, Montreal, or Vancouver?

45. During World War II, Allied pilots carried lithium hydride tablets taped inside their life belts. Research why this was done. Hint: Look for chemical reactions involving lithium hydride that might be useful for these pilots.

46. The Montreal Protocol is an agreement between industrialized nations to reduce the amount of ozone-depleting chemicals released into the atmosphere. Do research to discover answers to these questions:
   (a) What amount of ozone-depleting chemicals was released in a typical year before the Montreal Protocol was signed?
   (b) What amount of ozone-depleting chemicals was released this year (or in the most recent year you can find)?
   (c) Compare the two values. Have there been any measurable decreases? By how much?
   (d) How does Canada compare to the other countries in the agreement?

COURSE CHALLENGE

Planet Unknown

In the Course Challenge, you will collect and analyze an unknown gas.

- What laboratory techniques have you learned that could help you identify an unknown gas?
- How do you collect a gas?
- What gases are used and produced by plants?

UNIT

5

Hydrocarbons and Energy

UNIT 5 CONTENTS

UNIT 5 OVERALL EXPECTATIONS

- What are the structures of different types of hydrocarbons?
- What are the properties of hydrocarbons?
- What energy changes occur when hydrocarbons are combusted?
- How do hydrocarbons affect our lives? How do they affect the environment?

Unit Project Prep

Look ahead to the end-of-unit project, "Consumer Chemistry." Start thinking now about the consumer item that you would like to investigate. As you study this unit, plan how you will investigate the chemical nature of the product you have chosen.

Think of what the world must have been like 560 million years ago. Temperatures were much warmer than they are now. Lush green vegetation covered much of the North American landscape. Through photosynthesis, these ancient plants (like plants today) stored energy from the Sun as food. Abundant marine life filled the oceans. Other animals thrived on the land. Humans did not yet exist.

Today our society relies heavily on the remains of ancient forests and long-dead marine life. From these remains, we manufacture products such as the gasoline that fuels cars and the plastic wrap that covers the sandwich in your lunch. How does this incredible transformation take place?

In this unit, you will learn about fossil fuels. Fossil fuels are carbon compounds that have been produced over millions of years from the remains of ancient living things. You will examine the structures and properties of many hydrocarbon compounds that we obtain from fossil fuels. As well, you will find out how these compounds can be refined to produce many useful materials.

Later in this unit, you will discover how modern society obtains energy from hydrocarbons.

CHAPTER

13 The Chemistry of Hydrocarbons

Did you know that you have bark from a willow tree in your medicine cabinet at home? The model at the bottom right of the opposite page shows a compound that is found naturally in willow bark. This chemical is called salicin. It is a source of pain relief for moose, deer, and other animals that chew the bark. For thousands of years, Aboriginal people in Canada and around the world have relied on salicin's properties for the same pain-relieving purpose.

The model at the bottom left of the opposite page shows a close relative of salicin. Scientists made, or *synthesized,* this chemical near the end of the nineteenth century. It is called acetyl salicylic acid (ASA). You probably know it better by its brand name, Aspirin™.

Chemists refer to salicin, ASA, and more than ten million other chemicals like them as organic compounds. An **organic compound** is a molecular compound of carbon. Despite the tremendous diversity of organic compounds, nearly all of them share something in common. They are structured from a "backbone" that consists of just two kinds of atoms: carbon and hydrogen.

Compounds that are formed from carbon and hydrogen are called **hydrocarbons**. In this chapter, you will explore the sources, structures, properties, and uses of hydrocarbons—an enormous class of compounds. As well, you will learn how scientists and engineers use the properties of hydrocarbons to produce a seemingly infinite variety of chemicals and products.

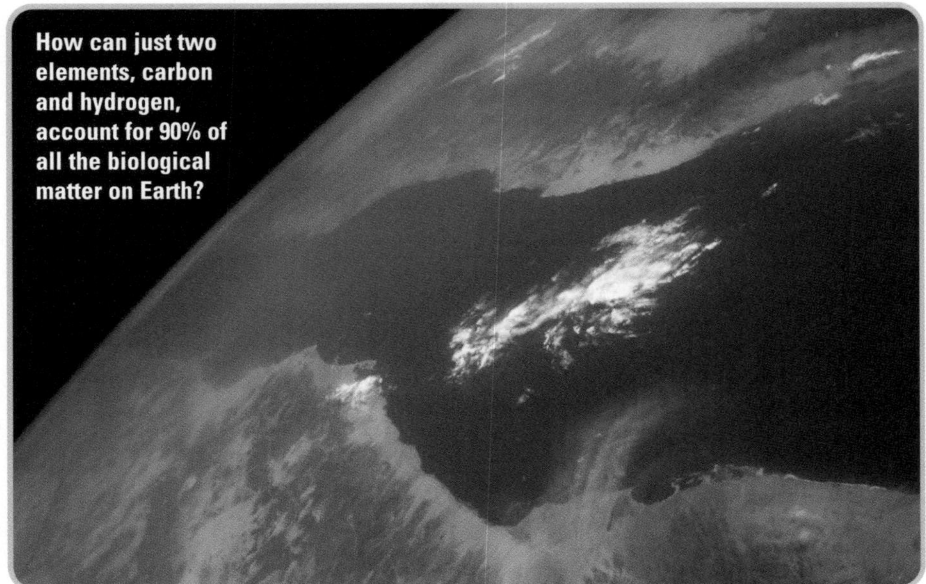
How can just two elements, carbon and hydrogen, account for 90% of all the biological matter on Earth?

Chapter Preview

Concepts and Skills You Will Need

Review the following concepts and skills before you begin this chapter:

- identifying characteristics of covalently bonded compounds (Chapter 3, sections 3.1, 3.3, 3.4.)
- relating physical properties to the polarity of molecules and intermolecular forces (Chapter 3, section 3.3; Chapter 8, section 8.2)
- drawing Lewis structures and structural formulas (Chapter 3, sections 3.2, 3.3)

13.1 Introducing Organic Compounds

Section Preview/ Specific Expectations

In this section, you will

- **identify** the origins and major sources of hydrocarbons and other organic compounds
- **communicate** your understanding of the following terms: *organic compound, hydrocarbons, petroleum*

As stated in the chapter opener, an organic compound is a molecular compound of carbon. There a few exceptions to this definition, however. For example, scientists classify oxides of carbon, such as carbon dioxide and carbon monoxide, as *in*organic. However, the vast majority of carbon-containing compounds are organic.

Organic Compounds: Natural and Synthetic

Organic compounds abound in the natural world. In fact, you probably ate sugar or starch at your last meal. Sugars, starches, and other carbohydrates are natural organic compounds. So are fats, proteins, and the enzymes that help you digest your food. Do you wear clothing made from wool, silk, or cotton? These are natural organic compounds, too. So are the molecules of DNA in the nuclei of your cells.

Until 1828, the only organic compounds on Earth were those that occur naturally. In that year, a German chemist named Friedrich Wohler synthesized urea—an organic compound found in mammal urine—from an inorganic compound, ammonium cyanate. (See Figure 13.1.) This was a startling achievement. Until then, chemists had assumed that only living or once-living organisms could be the source of organic compounds. They believed that living matter had an invisible "vital force." According to these early chemists, this vital (life) force made organic matter fundamentally different from inorganic (non-living) matter.

CHECKPOINT

Cyanides (containing carbon bonded to nitrogen, CN) and carbides, (such as calcium carbide, CaC_2) are compounds that contain carbon, but scientists classify them as inorganic. The same is true of carbonates, such as sodium hydrogen carbonate, $NaHCO_3$ (baking soda). Why are carbonate compounds inorganic, rather than organic, compounds?

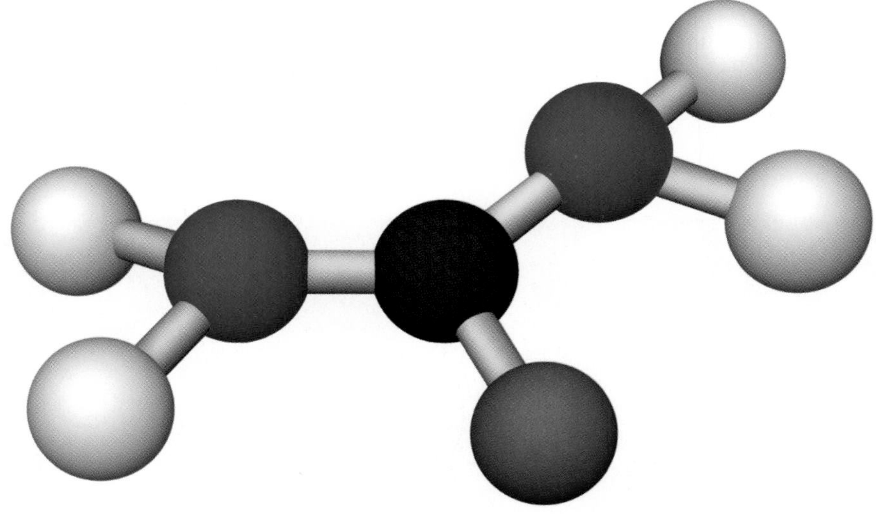

Figure 13.1 When Friedrich Wohler (1800–1882) synthesized urea, he wrote a letter to his teacher. In his letter, he said, "I must tell you that I can make urea without the use of kidneys...Ammonium cyanate is urea." About 20 years earlier, Wohler's teacher, Jons Jakob Berzelius (1779–1848), had invented the system that distinguishes organic substances from inorganic substances.

Language LINK

Acetylsalicylic acid (ASA) was first produced commercially under the brand name Aspirin™ by Frederick Bayer and Company in 1897. The word "aspirin" comes from "a," for acetyl, and "spir," for spirea. *Spirea* is a genus of plants that is another natural source of salicylic acid.

During the mid-1850s, chemists synthesized other organic compounds, such as methane and ethanol, from inorganic chemicals. Eventually chemists abandoned their vital-force ideas. We still use the terms "organic" and "inorganic," however, to distinguish carbon-based compounds from other compounds. For example, sugar is an organic compound since it is carbon based. Salt is inorganic since it contains no carbon.

During the last century, the number of synthetic (human-made) organic compounds has skyrocketed. Chemists invent more than 250 000 new synthetic organic chemicals *each year*. With almost endless variations in properties, chemists can synthesize organic compounds to make products as diverse as life-giving drugs, and toys. Nearly all medicines, such as painkillers, cough syrups, and antidepressants, are based on organic compounds. Perfumes, food flavourings, materials such as rubber and plastic, and fabrics such as nylon, rayon, and polyester are all organic compounds as well.

The next ExpressLab gives you a chance to make one such product. It is up to you to decide what purpose it could have.

ExpressLab  Making Polymer Putty

Polymers are a fundamental part of your life. They also happen to be organic. *Polymers* are long molecular chains that are made up of smaller molecular units called monomers. Natural polymers include cellulose (the "fibre" in your food) and DNA. Synthetic polymers include plastics, polystyrene (see below), and the material you will produce in this activity.

$$—CH_2CHCH_2CHCH_2CH—$$

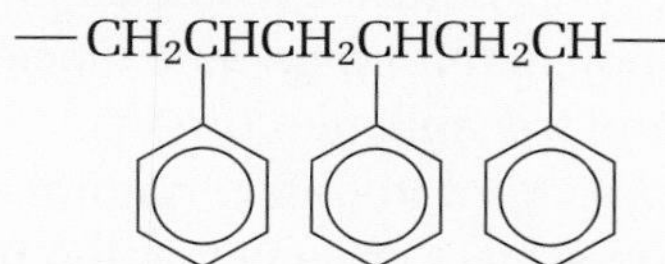

Polystyrene is an example of a polymer.

Safety Precautions

Avoid inhaling the powdered borax. It may cause an allergic reaction. Wash your hands after working with the putty.

Procedure

1. Measure about 45 mL of white glue into a 250 mL beaker. Add an equal amount of warm water and a few drops of food colouring.
2. Measure 15 mL of solid borax into another 250 mL beaker. Add 60 mL of warm water. Stir for about 2 min. **Note:** You may find that not all of the solid borax dissolves. This is all right.
3. Pour 30 mL of the borax solution into the glue mixture. Then quickly and thoroughly stir the mixture.
4. Remove the material from the beaker, and "experiment" with it. Here are a few suggestions. Record your observations.
   - Hold the putty in one hand. Put your other hand below the putty, and let the putty slowly ooze into it.
   - Pull the putty apart slowly.
   - Pull the putty apart quickly.
   - Try bouncing the putty.
5. If time permits, you may wish to try changing the ratio of borax to glue to water. Test your results.
6. Dispose of any excess borax solution as directed by your teacher.

Analysis

1. Compare the properties of the starting materials with the properties of the product. Is the putty a solid or a liquid? Explain.
2. What practical applications for this product can you think of?

COURSE CHALLENGE

Why does society depend on fossil fuels? What are they used for? In the Course Challenge at the end of this book, you will investigate a different type of fossil fuel from an imaginary planet.

The Origins of Hydrocarbons and Other Organic Compounds

Most hydrocarbons and other organic compounds have their origins deep below Earth's present surface. In the past, as now, photosynthetic organisms used energy from the Sun to convert carbon dioxide and water into oxygen and carbohydrates, such as sugars, starches, and cellulose. When these organisms died, they settled to the bottom of lakes, rivers, and ocean beds, along with other organic matter. Bacterial activity removed most of the oxygen and nitrogen from the organic matter, leaving behind mainly hydrogen and carbon.

Over time, the organic matter was covered with layers of mud and sediments. As layer upon layer built up, heat and tremendous pressure transformed the sediments into shale and the organic matter into solid, liquid, and gaseous materials. These materials are the fossil fuels—coal, oil, and natural gas—that society depends on today. (See Figure 13.2.)

Canadians in Chemistry

Dr. Raymond Lemieux

Raymond Lemieux was born into a carpenter's family in 1920 at Lac La Biche, Alberta. He obtained a B.Sc. degree at the University of Alberta and a Ph.D. at McGill University in Montréal. Lemieux then worked briefly at Ohio State University, where he met Dr. Virginia McConaghie. They were married and soon moved to Saskatoon, Saskatchewan. There Lemieux became a senior researcher at a National Research Council (NRC) laboratory.

Raymond Lemieux (1920–2000)

In 1953, while at the NRC lab, Lemieux conquered what some have called "the Mount Everest of organic chemistry." He became the first person to completely synthesize sucrose, or table sugar. Sucrose is a carbohydrate with the chemical formula $C_{12}H_{22}O_{11}$. It is the main sugar in the sap of plants such as sugar beets and sugar cane. Sucrose is related to glucose, $C_6H_{12}O_6$, and other sugars.

Lemieux continued his research at the University of Ottawa, and then at the University of Alberta in Edmonton. He was especially interested in how molecules "recognize" each other and interact in the human body. For example, different blood groups, such as group A and group B, are determined by carbohydrate molecules that differ by only a single sugar. The body is able to recognize these specific sugars and adapt its response to foreign substances, such as bacteria and transplanted organs.

Since it was hard to obtain natural samples of the body sugars that Lemieux wanted to study, he found ways to synthesize them. This was groundbreaking research. Seeing the practical applications of his research, Lemieux was instrumental in starting several chemical companies. Today these companies make products such as antibiotics, blood-group determinants, and immunizing agents that are specific to various human blood groups. They also make complex carbohydrates that absorb antibodies from the blood in order to prevent organ transplant rejection.

Lemieux and his wife had six children and a number of grandchildren. With all that he had accomplished in life, Lemieux said, "My proudest achievement is my family." His autobiography, titled *Explorations with Sugar: How Sweet It Was*, was published in 1990.

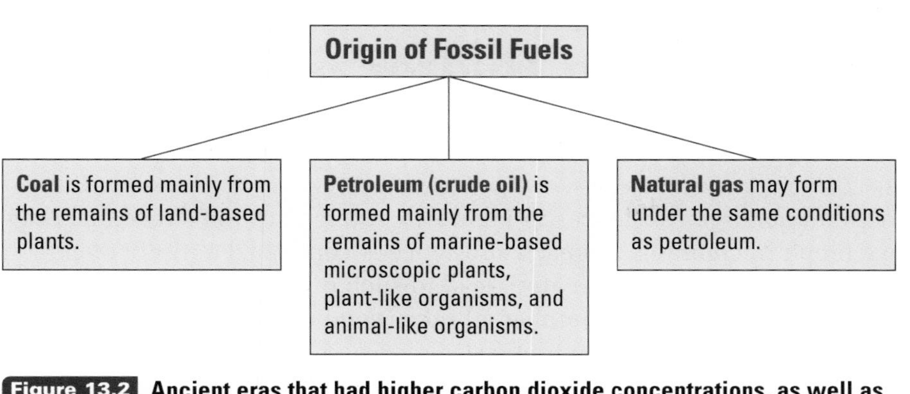

Origin of Fossil Fuels
Coal is formed mainly from the remains of land-based plants.
Petroleum (crude oil) is formed mainly from the remains of marine-based microscopic plants, plant-like organisms, and animal-like organisms.
Natural gas may form under the same conditions as petroleum.

Figure 13.2 **Ancient eras that had higher carbon dioxide concentrations, as well as warmer climates, gave rise to abundant plant and animal life on land and under water. Over time, as these organisms died, the organic substances that made up their bodies were chemically transformed into the materials known today as fossil fuels.**

Sources of Hydrocarbons

Sources of hydrocarbons include wood, the products that result from the fermentation of plants, and fossil fuels. However, one fossil fuel—petroleum—is the main source of the hydrocarbons that are used for fuels and many other products, such as plastics and synthetic fabrics.

Petroleum, sometimes referred to as crude oil, is a complex mixture of solid, liquid, and gaseous hydrocarbons. To understand the importance of petroleum in our society, you need to become better acquainted with hydrocarbons. Your introduction begins, in the next section, with carbon—one of the most versatile elements on Earth.

Geology LINK

The origin of fossil fuels, depicted in Figure 13.2, is based on a theory called the *biogenic theory*. Most geologists accept this theory. A small minority of geologists have proposed an alternative theory, called the *abiogenic theory*. Use print and electronic resources to investigate the following:

- the main points of each theory, and the evidence used to support these points
- the reasons why one theory is favoured over the other

Record your findings in the form of a brief report. Include your own assessment of the two theories.

Section Review

1 (a) K/U Name three compounds that you know are organic.
(b) K/U Name three compounds that you know are inorganic.
(c) K/U Name three compounds that may be organic, but you do not know for sure.

2 K/U What are the origins of hydrocarbons and other organic compounds?

3 K/U Identify at least two sources of hydrocarbons and other organic compounds.

4 C Design a concept map (or another kind of graphic organizer) to show the meanings of the following terms and the relationships among them: organic compound, inorganic compound, hydrocarbon, fossil fuels, petroleum, natural gas.

5 K/U Copy the following compounds into your notebook. Identify each carbon as organic or inorganic, and give reasons for your answer. If you are not sure whether a compound is organic or inorganic, put a question mark beside it.

(a) CH_4 (c) CO_2 (e) C_6H_6 (g) CH_3COOH
(b) CH_3OH (d) HCN (f) NH_4SCN (h) $CaCO_3$

13.2 Representing Hydrocarbon Compounds

Section Preview/ Specific Expectations

In this section, you will

- **demonstrate** an understanding of the bonding characteristics of the carbon atom in hydrocarbons
- **draw** structural representations of aliphatic hydrocarbon molecules
- **demonstrate** the arrangement of atoms in isomers of hydrocarbons using molecular models
- **communicate** your understanding of the following terms: *expanded molecular formula, isomers, structural model, structural diagram*

Examine the three substances in Figure 13.3. It is hard to believe that they have much in common. Yet each substance is composed entirely of carbon atoms. Why does the carbon atom lead to such diversity in structure? Why do carbon compounds outnumber all other compounds so dramatically? The answers lie in carbon's atomic structure and behaviour.

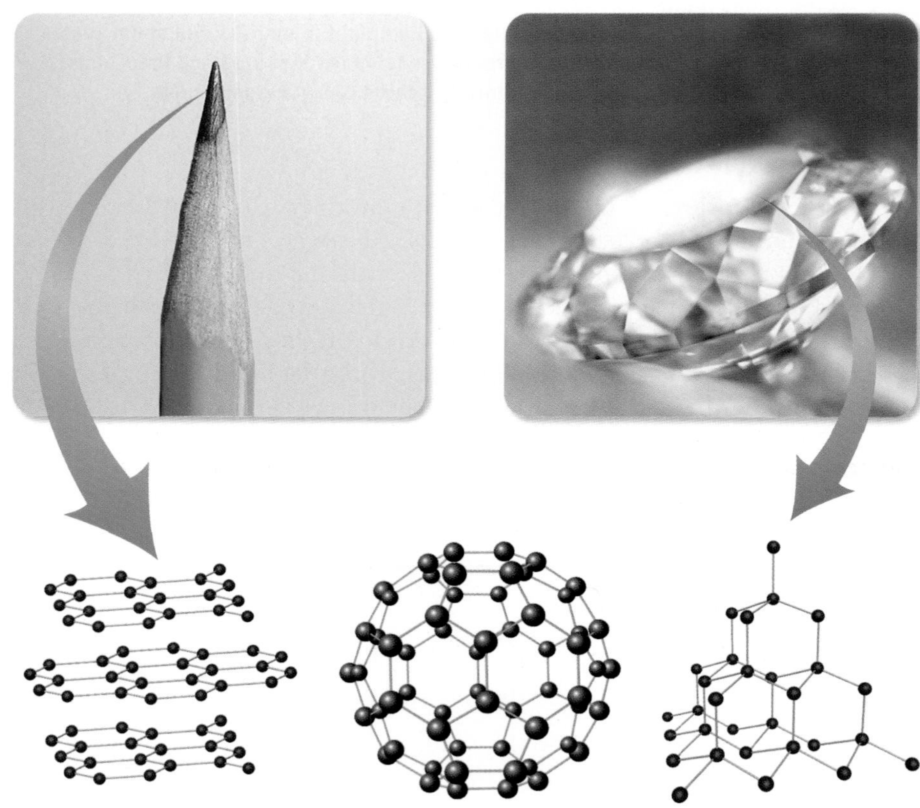

Figure 13.3 Each of these substances is pure carbon. What makes carbon a "chemical chameleon?"

Figure 13.4 outlines three key properties of carbon. Throughout this section and the next, you will explore the consequences of each of these properties.

Properties of Carbon

Carbon has four bonding electrons. This electron structure enables carbon to form four strong covalent bonds. As a result, carbon may bond to itself, as well as to many different elements (mainly hydrogen, oxygen, and nitrogen, but also phosphorus, sulfur, and halogens such as chlorine).

Carbon can form strong single, double, and triple bonds with itself. This allows carbon to form long chains of atoms—something that very few other atoms can do. In addition, the resulting compounds are fairly stable under standard conditions of temperature and pressure.

Carbon atoms can bond together to form a variety of geometrical structures. These structures include straight chains, branched chains, rings, sheets, tubes, and spheres. No other atom can do this.

Figure 13.4 Three key properties of carbon

Representing Structures and Bonding

You have written chemical formulas for inorganic compounds such as ammonia, NH_3, and calcium carbonate, $CaCO_3$. As well, you have represented these compounds using Lewis diagrams, and perhaps other models. Such compounds are fairly small, so they are easy to represent using these methods. Many organic compounds—such as methane, CH_4, and ethanol, C_2H_6O—are also fairly small. With patience, you might even figure out how to draw a Lewis diagram for cholesterol, $C_{27}H_{46}O$! Most hydrocarbons and other organic compounds are quite large, however. They are also structurally complex. Therefore chemists have devised other methods to represent them, as explained below.

Using Expanded Molecular Formulas to Represent Hydrocarbons

One method that chemists use to represent a large molecule is the **expanded molecular formula**. This type of formula shows the groupings of atoms, and it often gives an idea of molecular structure. For example, the chemical formula for propane is C_3H_8. This formula tells you that propane contains three carbon atoms and eight hydrogen atoms. It gives no clue, though, about the way in which the atoms are bonded together.

Propane's expanded molecular formula is $CH_3CH_2CH_3$. As you can see from this formula, the expanded molecular formula gives a clearer idea of atomic arrangement. It implies that a CH_3 group is attached to a CH_2 group, which is attached to another CH_3 group.

Writing expanded molecular formulas becomes more helpful when you are dealing with larger hydrocarbons. For example, C_6H_{14} is a component of gasoline. It is also used as a solvent for extracting oils from soybeans and other edible oil seeds. Depending on how the carbon and hydrogen atoms are bonded together, C_6H_{14} can have any of the five structural arrangements shown in Figure 13.5. Each arrangement has a different name.

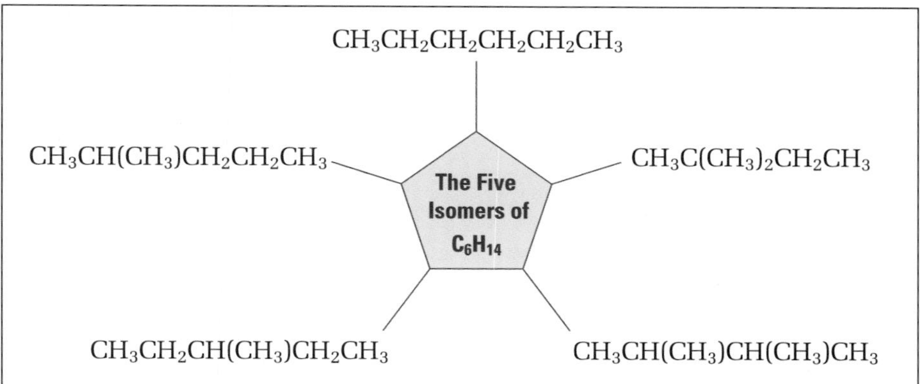

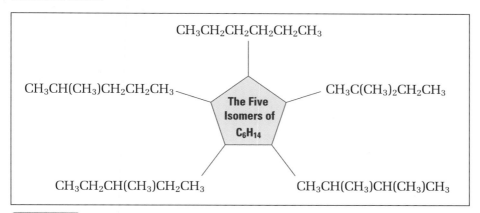

Figure 13.5 Expanded molecular formulas for five structural arrangements of C_6H_{14}

Keep in mind that all five of these arrangements have the same chemical formula: C_6H_{14}. Compounds that have the same formula, but different structural arrangements, are called **isomers**. Hexane, for example, is one isomer of C_6H_{14}. You will learn more about isomers as you study this chapter.

Math LINK

Graph the data in the table below. **Note:** Bond energy is the amount of energy that is needed to break a chemical bond.

Some Average Bond Energies

Bond	Bond energy (kJ/mol)
C—C	346
C=C	610
C≡C	835
Si—Si	226
Si=Si	318 (estimate)
C—H	413
Si—H	318

Infer a relationship between the stability of a compound and bond energy. Then suggest a reason why there are many more carbon-based compounds than silicon-based compounds.

The number of isomers of an organic compound increases greatly as the number of carbon atoms increases. For example, C_5H_{12} has three isomers, C_6H_{14} has five isomers, and C_8H_{18} has 18 isomers. $C_{30}H_{62}$, a large hydrocarbon, has over four billion isomers!

The expanded molecular formulas of C_6H_{14} give you a better idea of the arrangement of the carbon and hydrogen atoms in its five possible isomers. You can use other methods as well, however, to represent hydrocarbons and other organic compounds. These methods are outlined below.

Using Structural Models to Represent Hydrocarbons

A **structural model** is a three-dimensional representation of the structure of a compound. There are two kinds of structural models: *ball-and-stick models* and *space-filling models*. Figure 13.6 shows ball-and-stick models for the five isomers of C_6H_{14}. Notice that they show how the carbon and hydrogen atoms are bonded within the structures.

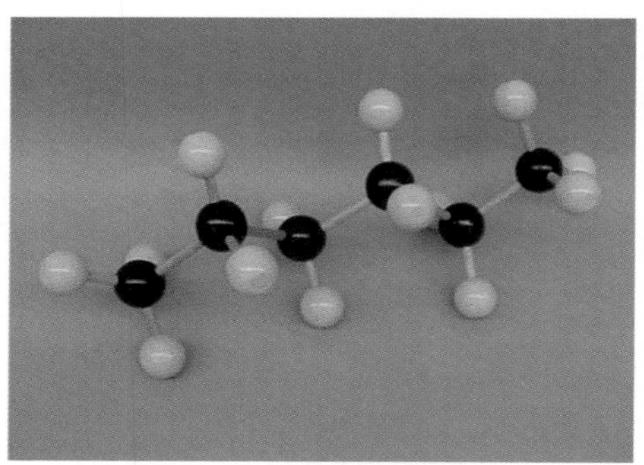

$CH_3CH_2CH_2CH_2CH_2CH_3$

$CH_3CH(CH_3)CH(CH_3)CH_3$

$CH_3CH(CH_3)CH_2CH_2CH_3$

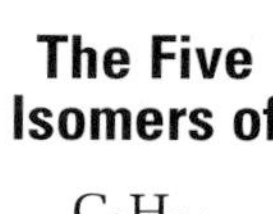

C_6H_{14}

$CH_3C(CH_3)_2CH_2CH_3$

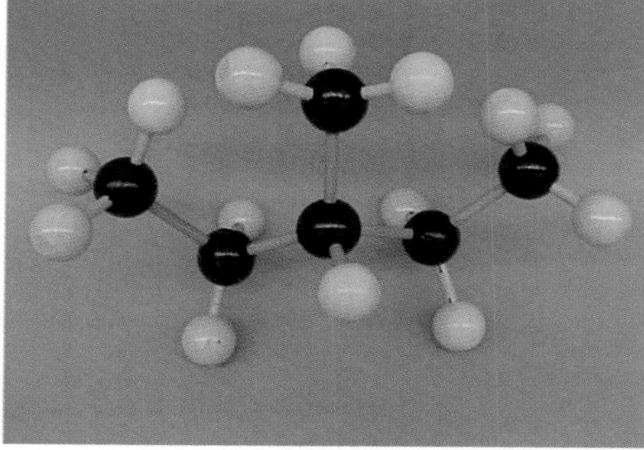

$CH_3CH_2CH(CH_3)CH_2CH_3$

Figure 13.6 C_6H_{14} **can have any one of these five structural arrangements. How do the ball-and-stick models compare with the expanded molecular formulas for** C_6H_{14}**? How does each isomer differ from the rest?**

A space-filling model, such as the one in Figure 13.7, also shows the arrangement of the atoms in a compound. As well, it represents the molecular shape and the amount of space that each atom occupies within the structure.

Electronic Learning Partner

To see an animation comparing structural diagrams with molecular formulas, go to your Chemistry 11 Electronic Partner now.

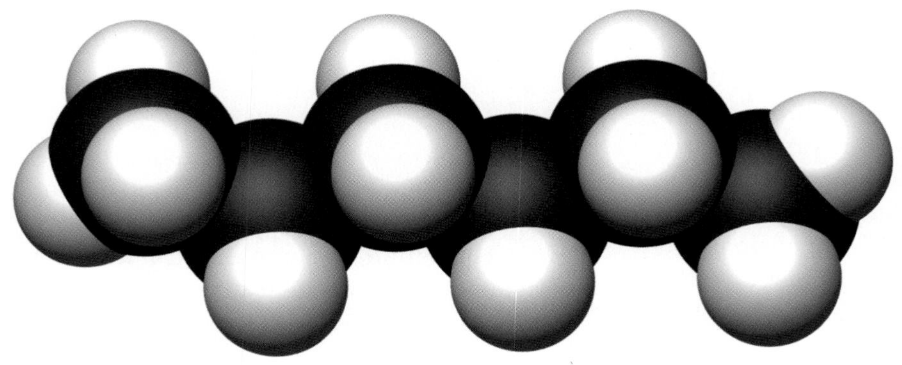

Figure 13.7 A space-filling model for hexane, one of the isomers of C_6H_{14}

Using Structural Diagrams to Represent Hydrocarbons

A **structural diagram** is a two-dimensional representation of the structure of a compound. (In some chemistry textbooks, structural diagrams are called structural formulas.) There are three kinds of structural diagrams: *complete structural diagrams*, *condensed structural diagrams*, and *line structural diagrams*. As you can see in Figure 13.8, each serves a specific purpose in describing the structure of a compound.

In the next investigation, you will have an opportunity to make your own models of the isomers of several organic compounds.

```
        H
        |
     H—C—H
     H  |  H  H  H
     |  |  |  |  |
  H—C—C—C—C—C—H
     |  |  |  |  |
     H  H  H  H  H
```

A **complete structural diagram** shows all the atoms in a structure and the way they are bonded to one another. Straight lines represent the bonds between the atoms.

```
          CH3
          |
  CH3—CH—CH2—CH2—CH3
```

A **condensed structural diagram** simplifies the presentation of the structure. It shows the bonds between the carbon atoms but not the bonds between the carbon and hydrogen atoms. Chemists assume that these bonds are present. Notice how much cleaner and clearer this diagram is, compared with the complete structural diagram.

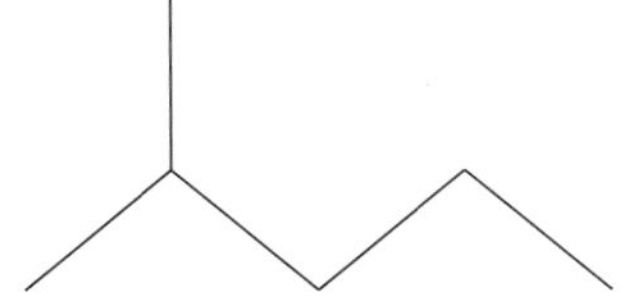

A **line structural diagram** is even simpler than a condensed structural diagram. The end of each line, and the points at which the lines meet, represent carbon atoms. This kind of diagram gives you a sense of the three-dimensional nature of a hydrocarbon. **Note:** Line structural diagrams are used only for hydrocarbons, not for other organic compounds.

Figure 13.8 Comparing complete, condensed, and line structural diagrams

SKILL FOCUS

Predicting

Analyzing and interpreting

Communicating results

Modelling concepts

Investigation 13-A

Modelling Organic Compounds

Figure 13.6 showed you that an organic compound can be arranged in different structural shapes, called isomers. All the isomers of a compound have the same molecular formula. In this investigation, you will make two-dimensional and three-dimensional models of isomers. Your models will help you explore the arrangements of the atoms in organic compounds.

Question

How do models help you visualize the isomers of organic compounds?

Predictions

Predict the complete, condensed, and line structural diagrams for the three isomers of C_5H_{12}. Then predict the complete and condensed structural diagrams for the four isomers of C_3H_9N.

Materials

paper and pencil
molecular modelling kit

These representations of the hydrocarbon ethane, C_2H_6, were made using different modelling kits. Your school may have one or more of these kits available.

Procedure

1. Construct three-dimensional models of the three isomers of C_5H_{12}. Use your predictions to help you. As you complete each model, draw a careful diagram of the structure. Your diagram might be similar to the one below.

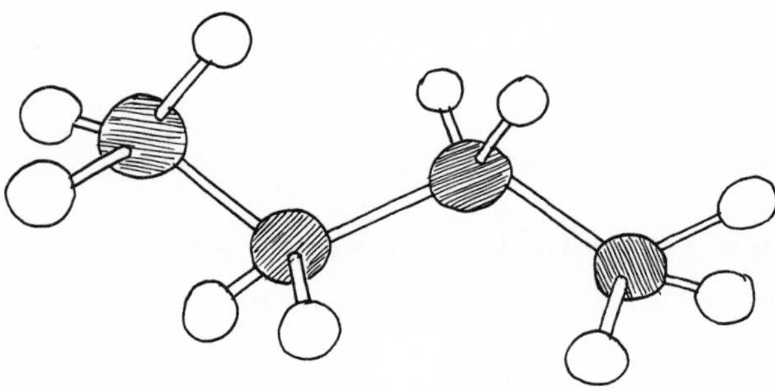

2. Repeat step 1 for each isomer of C_3H_9N.

Analysis

1. In what ways were your completed models similar to your predictions? In what ways were they different?
2. How do the models of each compound help you understand the concept of isomers?

Conclusion

3. How accurately do you think your models represent the real-life structural arrangements of C_5H_{12} and C_3H_9N?

Applications

4. In earlier units, you considered how the structure and polarity of molecules can affect the boiling point of a compound. For each compound you studied in this investigation, predict which isomer has the higher boiling point. Explain your prediction.
5. Construct models for C_7H_{16}. How many isomers are possible?

Section Wrap-up

The names of the isomers of hydrocarbons and other organic compounds are quite different from the names of inorganic compounds. For example, go back to the structural diagrams shown in Figure 13.8. They all represent the same isomer of C_6H_{14}. Its name is 2-methylpentane. Now look at Figure 13.9. It shows the names and condensed structural diagrams for the four other isomers of C_6H_{14}. Why do these names look so different from the names of inorganic compounds? Scientists have devised a systematic way to communicate all the possible atoms and structures for each organic compound—all in its name! In the next section, you will learn the rules for naming hydrocarbons. You will also learn how to interpret the information that hydrocarbon names communicate.

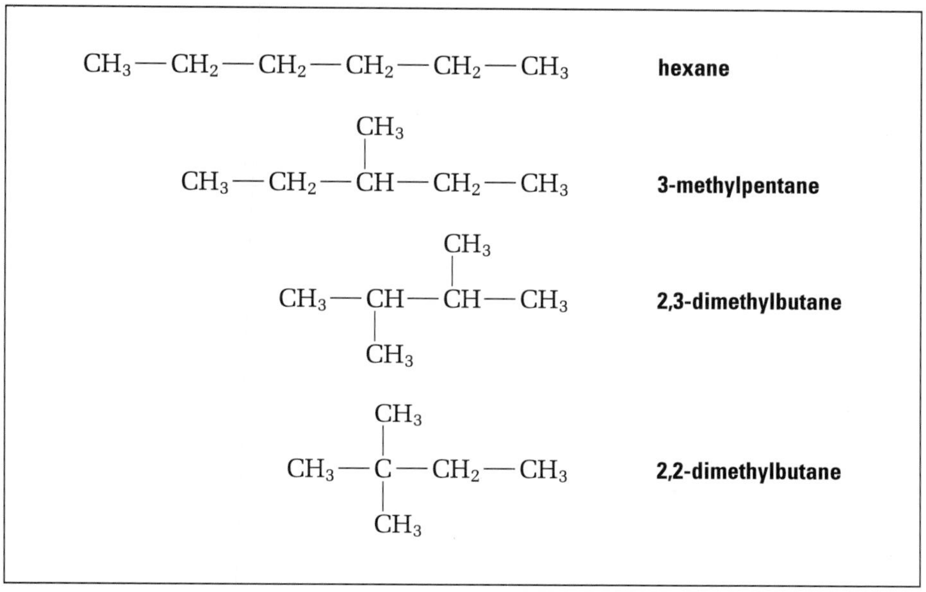

$CH_3—CH_2—CH_2—CH_2—CH_2—CH_3$ **hexane**

$CH_3—CH_2—CH(CH_3)—CH_2—CH_3$ **3-methylpentane**

$CH_3—CH(CH_3)—CH(CH_3)—CH_3$ **2,3-dimethylbutane**

$CH_3—C(CH_3)_2—CH_2—CH_3$ **2,2-dimethylbutane**

Figure 13.9 Four of the five isomers of C_6H_{14}

CHEM FACT

When is an isomer not an isomer? When it is *exactly* the same compound! The hydrocarbon C_6H_{14} has only five isomers. Examine these two representations for one of them, hexane.

$CH_3—CH_2—CH_2—CH_2—CH_2—CH_3$

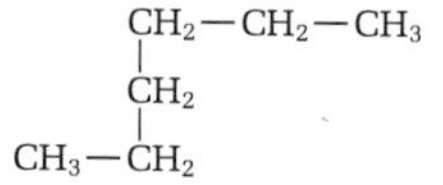

$CH_3—CH_2—CH_2—CH_2—CH_2—CH_3$ (bent: $CH_3—CH_2$, with CH_2 above, then $CH_2—CH_2—CH_3$ on top)

You might think that these are different isomers, but they are actually the same. The structure on the bottom is just a bent version of the structure on the top. Think about a length of chain. You can lay it out straight, or you can bend it. The chain is still a chain in either case. It just looks different. Naming a compound helps you recognize the difference between a true isomer and a structure that just looks like an isomer.

Section Review

1 K/U Identify the three properties of carbon that allow it to form such a great variety of compounds.

2 I Chose one of the hydrocarbon molecules in this section to represent, using as many different kinds of models as you can. Identify each model you used.

3 C You have seen the expanded molecular formulas and condensed structural diagrams for the five isomers of C_6H_{14}. Draw the complete and line structural diagrams for each of these isomers.

4 C Draw the complete, condensed, and line structural diagrams for the isomers of C_5H_{12}. How many *true* isomers did you draw? (See the ChemFact above to help you interpret this question.)

5 C Many organic compounds contain elements such as oxygen, nitrogen, and sulfur, as well as carbon and hydrogen. For example, think about ethanol, CH_3CH_2OH. Draw complete and condensed structural diagrams for ethanol. Can you draw a line structural diagram for ethanol? Explain your answer.

13.3 Classifying Hydrocarbons

Section Preview/ Specific Expectations

In this section, you will

- **demonstrate** an understanding of the carbon atom by classifying hydrocarbons and by analyzing the bonds that carbon forms in aliphatic hydrocarbons
- **name** alkanes, alkenes, and alkynes, and **draw** structural representations for them
- **describe** some of the physical and chemical properties of hydrocarbons
- **determine** through experimentation some of the characteristic properties of saturated and unsaturated hydrocarbons
- **communicate** your understanding of the following terms: *alkanes, aliphatic hydrocarbons, saturated hydrocarbons, homologous series, alkenes, unsaturated hydrocarbons, cis-trans isomer, alkynes, cyclic hydrocarbons*

Language LINK

The name "aliphatic" comes from the Greek word *aleiphatos*, meaning "fat." Early chemists found these compounds to be less dense than water and insoluble in water, like fats. "Aliphatic" now refers to the classes of hydrocarbons called alkanes, alkenes, and alkynes.

Chemists group hydrocarbons and other organic compounds into the categories shown in Figure 13.10. The International Union of Pure and Applied Chemistry (IUPAC) has developed a comprehensive set of rules for naming the compounds within each category. Using these rules, you will be able to classify and name all the hydrocarbon compounds that you will encounter in this unit.

The names that are based on the IUPAC rules are called *systematic names*. During your study of organic chemistry, you will also run across many common names for organic compounds. For example, the systematic name for the organic acid CH_3CO_2H is ethanoic acid. You are probably more familiar with its common name: vinegar.

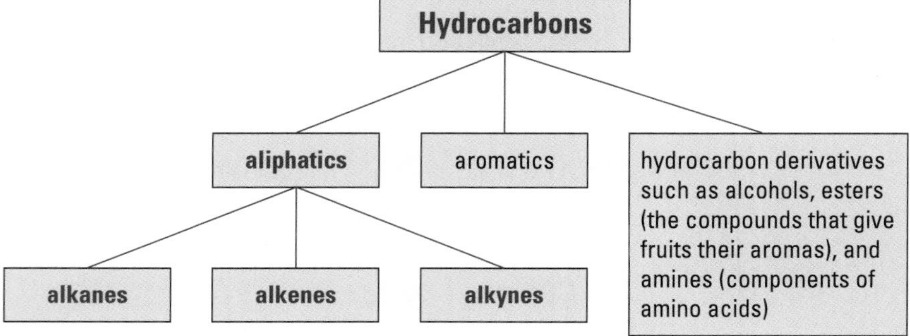

Figure 13.10 This concept map illustrates a system for classifying organic compounds. In this chapter, you will explore only part of the family of organic compounds—the aliphatic hydrocarbons shown by the boldface type.

Alkanes

Alkanes are hydrocarbon molecules that are joined by *single* covalent bonds. They are the simplest hydrocarbons. Methane, CH_4, is the simplest alkane. It is the main component of natural gas. Alkanes are **aliphatic hydrocarbons:** organic compounds in which carbon atoms form chains and non-aromatic rings.

Figure 13.11 on the next page compares the structural formulas of methane and the next three members of the alkane family. Notice three facts about these alkanes:

1. Each carbon atom is bonded to the maximum possible number of atoms (either carbon or hydrogen atoms). As a result, chemists refer to alkanes as **saturated hydrocarbons**.
2. Each molecule differs from the next molecule by the structural unit $-CH_2-$. A series of molecules like this, in which each member increases by the same structural unit, is called a **homologous series**.
3. A mathematical pattern underlies the number of carbon and hydrogen atoms in each alkane. All alkanes have the general formula C_nH_{2n+2}, where n is the number of carbon atoms. For example, propane has 3 carbon atoms. Using the general formula, we find that

$$2n + 2 = 2(3) + 2 = 8$$

Thus propane should have the formula C_3H_8, which it does.

$$
\begin{array}{ccc}
 & H & \\
 & | & \\
H - & C & - H \\
 & | & \\
 & H &
\end{array}
$$

methane

$$
\begin{array}{cccc}
 & H & H & \\
 & | & | & \\
H - & C - & C & - H \\
 & | & | & \\
 & H & H &
\end{array}
$$

ethane

$$
\begin{array}{ccccc}
 & H & H & H & \\
 & | & | & | & \\
H - & C - & C - & C & - H \\
 & | & | & | & \\
 & H & H & H &
\end{array}
$$

propane

$$
\begin{array}{cccccc}
 & H & H & H & H & \\
 & | & | & | & | & \\
H - & C - & C - & C - & C & - H \\
 & | & | & | & | & \\
 & H & H & H & H &
\end{array}
$$

butane

Figure 13.11 Carefully examine these four molecules. They are the first four alkanes. In what ways are they similar? In what ways are they different?

Why is methane the simplest of all the millions of hydrocarbons? **Hint:** Recall what you know about chemical bonding and the common valences of elements.

Figure 13.12 illustrates these three important facts about alkanes. Study the two alkanes, then complete the Practice Problems that follow.

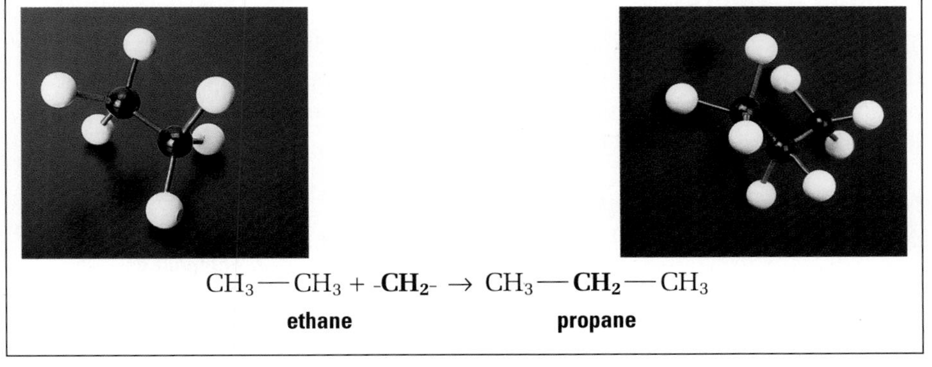

Figure 13.12 How are these two alkanes similar? How are they different? Use the ideas and terms you have just learned to help you answer these questions.

Practice Problems

1. Heptane has 7 carbon atoms. What is the chemical formula of heptane?
2. Nonane has 9 carbon atoms. What is its chemical formula?
3. An alkane has 4 carbon atoms. How many hydrogen atoms does it have?
4. Candle wax contains an alkane with 52 hydrogen atoms. How many carbon atoms does this alkane have?

Properties of Alkanes

Alkanes (and all other aliphatic compounds) have an important physical property. They are non-polar. As you know from Chapter 8, non-polar molecules have fairly weak intermolecular forces. As a result, hydrocarbons such as alkanes have relatively low boiling points. As the number of atoms in the hydrocarbon molecule increases, the boiling point increases. Because of this, alkanes exist in a range of states under standard conditions.

Table 13.1 compares the sizes (number of atoms per molecule) and boiling points of alkanes. Notice how the state changes as the size increases.

Table 13.1 Comparing the Sizes and Boiling Points of Alkanes

Size (number of atoms per molecule)	Boiling point range (°C)	Examples of products
1 to 5	below 30	gases: used for fuels to cook and heat homes
5 to 16	30 to 275	liquids: used for automotive, diesel, and jet engine fuels; also used as raw materials for the petrochemical industry
16 to 22	over 250	heavy liquids: used for oil furnaces and lubricating oils; also used as raw materials to break down more complex hydrocarbons into smaller molecules
over 18	over 400	semi-solids: used for lubricating greases and paraffin waxes to make candles, waxed paper, and cosmetics
over 26	over 500	solid residues: used for asphalts and tars in the paving and roofing industries

CHECKPOINT

Many industries rely on alkane hydrocarbons. The states of these hydrocarbons can affect how they are stored at industrial sites. For example, methane is a gas under standard conditions. In what state would you expect a large quantity of methane to be stored? What safety precautions would be necessary?

Naming Alkanes

The IUPAC system for naming organic compounds is very logical and thorough. The rules for naming alkanes are the basis for naming the other organic compounds that you will study. Therefore it is important that you understand how to name alkanes.

Naming Straight-Chain Alkanes

Recall that carbon can bond to form long, continuous, chain-like structures. Alkanes that bond in this way are called *straight-chain alkanes*. (They are also called *unbranched alkanes*.) Straight-chain alkanes are the simplest alkanes. Table 13.2 lists the names of the first ten straight-chain alkanes.

Table 13.2 The First Ten Straight-Chain Alkanes

Name	Number of carbon atoms	Expanded molecular formula
methane	1	CH_4
ethane	2	CH_3CH_3
propane	3	$CH_3CH_2CH_3$
butane	4	$CH_3(CH_2)_2CH_3$
pentane	5	$CH_3(CH_2)_3CH_3$
hexane	6	$CH_3(CH_2)_4CH_3$
heptane	7	$CH_3(CH_2)_5CH_3$
octane	8	$CH_3(CH_2)_6CH_3$
nonane	9	$CH_3(CH_2)_7CH_3$
decane	10	$CH_3(CH_2)_8CH_3$

The root of each name (highlighted in colour) serves an important function. It tells you the number of carbon atoms in the chain. The suffix -ane tells you that these compounds are alkane hydrocarbons. Thus the root and the suffix of one of these simple names provide the complete structural story of the compound.

Naming Branched-Chain Alkanes

The naming rules for straight-chain alkanes can, with a few additions, help you recognize and name other organic compounds. You now know that the name of a straight-chain alkane is composed of a root (such as meth-) plus a suffix (-ane). Earlier in the chapter, you saw the isomers of C_6H_{14}. Figure 13.13 shows one of them, called 2-methylpentane.

$$
\begin{array}{ccccccc}
 & & CH_3 & & & & CH_3 \\
 & & | & & & & | \\
CH_3 & — & CH & — & CH_2 & — & CH_2
\end{array}
$$

Figure 13.13 **2-methylpentane**

CHECKPOINT

The root and the suffix of an alkane name do not tell you directly about the number of hydrogen atoms in the compound. If you did not know the molecular formula of heptane, for example, how would you still know that heptane contains 16 hydrogen atoms?

Notice four important facts about 2-methylpentane:

1. Its structure is different from the structure of a straight-chain alkane. Like many hydrocarbons, this isomer of C_6H_{14} has a branch-like structure. Alkanes such as 2-methylpentane are called *branched-chain alkanes*. (The branch is sometimes called a *side-chain*.)
2. The name of this alkane has a prefix (2-methyl-) as well as a root and a suffix. Many of the hydrocarbons you will name from now on have a prefix.
3. This alkane has a single CH_3 unit that branches off from the main (parent) chain of the compound.
4. There is another CH_3 unit bonded to a CH_2 unit at the right end of the chain. *This is not another branch.* It is a bend in the parent chain. Before continuing, refer to the ChemFact on page 543. Make sure that you understand why the CH_3 unit is not another branch.

Language LINK

You are probably familiar with the origins of some of the roots that are used for naming organic compounds, especially the roots for naming alkanes with five or more carbons. The roots for the first four alkanes may be unfamiliar, however. Use a comprehensive dictionary (one that gives information about word origins) to find out the meanings of the roots meth- through dec-.

Rules for Naming Alkanes

The names of branched-chain alkanes (and most other aliphatic compounds) have the same general format, as shown in Figure 13.14. This format will become clearer as you learn and practise the rules for naming hydrocarbons. To start, read the steps on the next page to see how 2-methylpentane gets its name.

2-methylpentane is a simple example of a branched hydrocarbon. Later you will see more complicated examples.

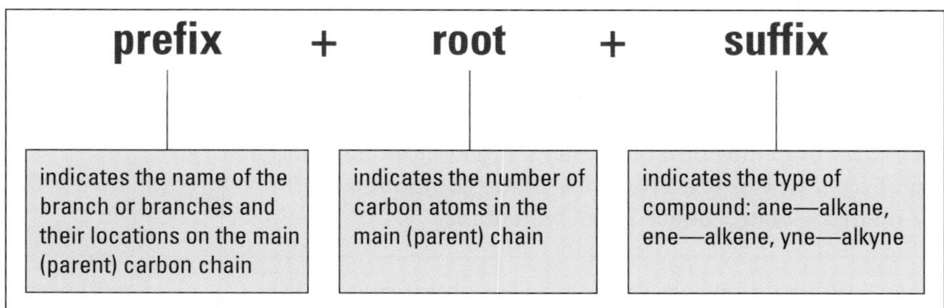

Figure 13.14 **Parts of a hydrocarbon name**

Naming 2-Methylpentane

Step 1

Find the longest continuous chain (the parent chain). It does not have to be straight. The number of carbons in the parent chain forms the root of the name.

$$\begin{array}{ccccccc} & & CH_3 & & & & CH_3 \\ & & | & & & & | \\ CH_3 & — & CH & — & CH_2 & — & CH_2 \end{array}$$

- The parent chain for 2-methylpentane is not straight. It is bent.
- There are five carbons in the longest chain, as highlighted. Therefore the root of the name is -pent-.
- Since the compound contains only single carbon-carbon bonds, the suffix of the name is -ane.
- So far, then, you have pentane as part of the name.

Step 2

Identify any branches that are present. Then number the main chain from the end that gives the lowest number to the first location at which branching occurs.

$$\begin{array}{ccccccc} & & CH_3 & & & & {}_5CH_3 \\ & & | & & & & | \\ {}_1CH_3 & — & {}_2CH & — & {}_3CH_2 & — & {}_4CH_2 \end{array}$$

- There is only one branch here. Since the branch is closer to the left end of the molecule, you number the carbon at the left end "1." Then you number the other main chain carbons consecutively.

Step 3

Identify the location of any branches with numbers. Use the number of carbons in each branch to name it.

$$\begin{array}{ccccccc} & & CH_3 & & & & {}_5CH_3 \\ & & | & & & & | \\ {}_1CH_3 & — & {}_2CH & — & {}_3CH_2 & — & {}_4CH_2 \end{array}$$

- You name a branch based on the appropriate root name for the number of carbons it contains. You change the ending to –yl, however. (IUPAC rules identify branches by using the -yl ending.)
- Here there is only one carbon. Instead of calling the branch methane, you replace the -ane ending with -yl. So the name for this branch is methyl.
- You also give the branch name a number. The number indicates which carbon in the main chain it is bonded with. Here the branch is bonded with the number 2 carbon in the main chain. So the numeral 2 is added to the prefix.
- You now have the full prefix name for the compound: 2-methyl.

Step 4

Put the complete compound name together in this general form: prefix + root + suffix. The name of the compound is 2-methyl + pent + ane = 2-methylpentane.

The structure of an alkane can be much more complex than the structure of 2-methylpentane. For instance, there can be many branches bonded to the main chain, and the branches can be quite long. As a result, you need to know several other IUPAC rules for naming branched-chain alkanes and other aliphatic compounds.

Additional IUPAC Rules for Naming Branched-Chain Alkanes and Other Aliphatic Compounds

1. If there are two or more of the same type of branch, give each branch a position number. Also, use multiplying prefixes such as di- (meaning 2), tri- (meaning 3), and tetra- (meaning 4) to indicate the number of branches.

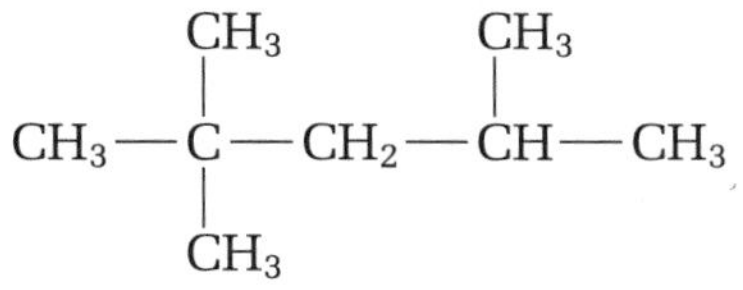

$$\begin{array}{ccccccccc} & & CH_3 & & & & CH_3 & & \\ & & | & & & & | & & \\ CH_3 & - & C & - & CH_2 & - & CH & - & CH_3 \\ & & | & & & & & & \\ & & CH_3 & & & & & & \end{array}$$

Figure 13.15 **This compound is 2,2,4-trimethylpentane, an isomer of C_8H_{18}. It is one of the main ingredients in gasoline.**

2. Put commas between numbers, and hyphens between numbers and letters.

3. When possible, put numbers in ascending order. (For example, the compound in Figure 13.15 compound is 2,2,4-trimethylpentane, not 4,2,2-trimethylpentane.)

4. If there is more than one type of branch, name the branches in alphabetical order. Determine the alphabetical order by using the first letter of the root (for example, -methyl-or-ethyl-), not the multiplying prefix (for example, di- or tri-).

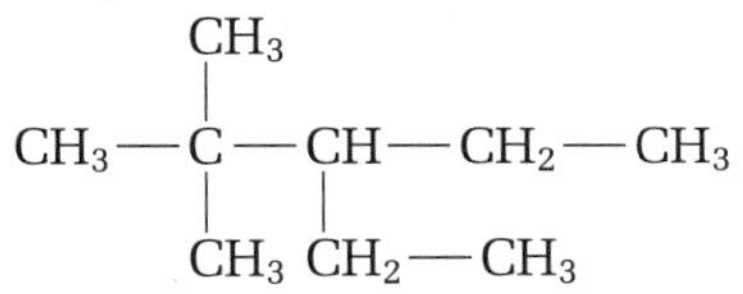

$$\begin{array}{ccccccccc} & & CH_3 & & & & & & \\ & & | & & & & & & \\ CH_3 & - & C & - & CH & - & CH_2 & - & CH_3 \\ & & | & & | & & & & \\ & & CH_3 & & CH_2 & - & CH_3 & & \end{array}$$

Figure 13.16 **3-ethyl-2,2-dimethylpentane**

5. If more than one chain could be the main chain (because they are the same length), choose the chain that has the most branches attached.

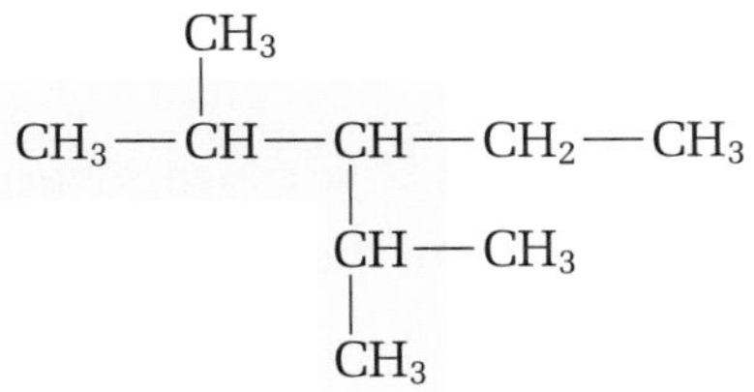

$$\begin{array}{ccccccccc} & & CH_3 & & & & & & \\ & & | & & & & & & \\ CH_3 & - & CH & - & CH & - & CH_2 & - & CH_3 \\ & & & & | & & & & \\ & & & & CH & - & CH_3 & & \\ & & & & | & & & & \\ & & & & CH_3 & & & & \end{array}$$

Figure 13.17 **3-ethyl-2,4-dimethylpentane**

Now practise naming the compounds in the Practice Problem below. Work slowly and patiently. The names become more challenging as you proceed.

Practice Problems

5. Name each compound.

(a)

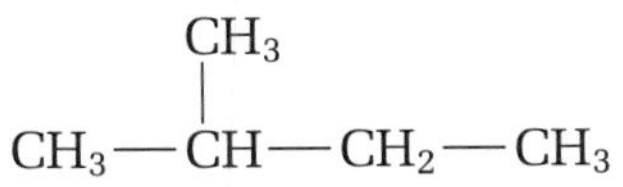

$$\begin{array}{ccccccc} & & CH_3 & & & & \\ & & | & & & & \\ CH_3 & - & CH & - & CH_2 & - & CH_3 \end{array}$$

(b)

$$\begin{array}{ccccc} & & CH_3 & & \\ & & | & & \\ CH_3 & - & C & - & CH_3 \\ & & | & & \\ & & CH_3 & & \end{array}$$

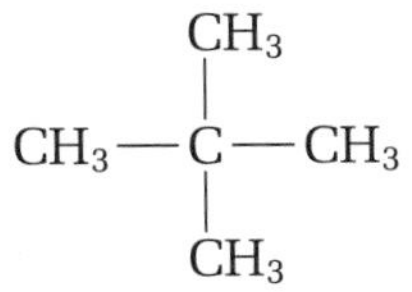

Continued ...

Continued ...
FROM PAGE 549

(c)

$$\begin{array}{ccccccccccc} & & & & & & CH_2 & - & CH_3 & & \\ & & & & & & | & & & & \\ CH_3 & - & CH & - & CH_2 & - & CH & - & CH & - & CH_3 \\ & & | & & & & & & | & & \\ & & CH_2 & & & & & & CH_3 & & \\ & & | & & & & & & & & \\ & & CH_3 & & & & & & & & \end{array}$$

(d)

$$\begin{array}{ccccccccc} & & CH_3 & & & & CH_3 & & \\ & & | & & & & | & & \\ CH_3 & - & C & - & CH_2 & - & C & - & CH_2 - CH_3 \\ & & | & & & & | & & \\ & & CH_3 & & & & CH_3 & & \end{array}$$

(e)

$$\begin{array}{ccccccccccccc} & & & & & & CH_3 & & & & CH_3 & & \\ & & & & & & | & & & & | & & \\ CH_3 & - & CH_2 & - & CH_2 & - & C & - & CH_2 & - & C & - & CH_3 \\ & & & & & & | & & & & | & & \\ & & & & & & CH_2 & & & & CH_3 & & \\ & & & & & & | & & & & & & \\ & & & & & & CH_2 & & & & & & \\ & & & & & & | & & & & & & \\ & & & & & & CH_2 & & & & & & \\ & & & & & & | & & & & & & \\ & & & & & & CH_3 & & & & & & \end{array}$$

Drawing Alkanes

As you learned earlier in this chapter, three kinds of diagrams can be used to represent the structure of a hydrocarbon. The easiest kind is probably the condensed structural diagram. When you are asked to draw a condensed structural diagram for an alkane, such as 2,3-dimethylhexane, you can follow several simple rules. These rules are listed below. After you have studied the rules, use the Practice Problems to practise your alkane-drawing skills.

You have learned how the non-polar nature of alkanes affects their boiling point. This non-polarity also affects another physical property: the solubility of alkanes in water. For example, the solubility of pentane in water is only 5.0×10^{-3} mol/L at 25°C. Hydrocarbon compounds, such as those found in crude oil, do not dissolve in water. Instead they float on the surface. This physical property helps clean-up crews minimize the devastating effects of an oil spill.

Rules for Drawing Condensed Structural Diagrams

1. Identify the root and the suffix of the name. In 2,3-dimethylhexane, for example, the root and suffix are -hexane. The *-hex-* tells you that there are six carbons in the main chain. The *-ane* tells you that the compound is an alkane. Therefore this compound has single carbon-carbon bonds only.
2. Draw the main chain first. Draw it straight, to avoid mistakes caused by a fancy shape. Do not include any hydrogen atoms. You will need to add branches before you finalize the number of hydrogen atoms on each carbon. Leave space beside each carbon on the main chain to write the number of hydrogen atoms later.

$$C\ -C\ -C\ -C\ -C\ -C$$

3. Choose one end of your carbon chain to be carbon number 1. Then locate the carbon atoms to which the branches must be added. Add the appropriate number and size of branches, according to the prefix in the name of the compound. In this example, *2,3-dimethyl* tells you that there is one methyl (single carbon-containing) branch on the second

carbon of the main chain, and another methyl branch on the third carbon of the main chain. It does not matter whether you place both branches above the main chain, both below, or one above and one below. The compound will still be the same.

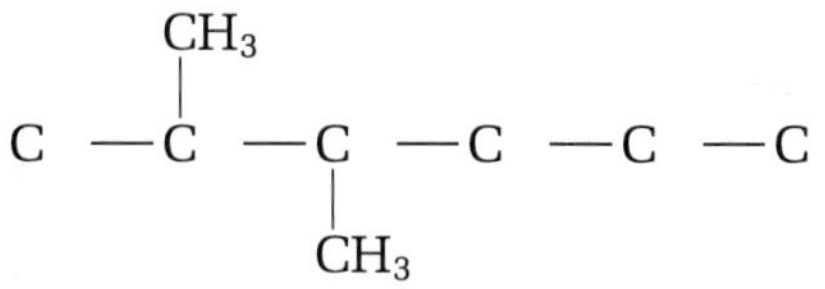

$$
\begin{array}{cccccc}
 & CH_3 & & & & \\
 & | & & & & \\
C\;— & C\;— & C\;— & C\;— & C\;— & C \\
 & & | & & & \\
 & & CH_3 & & &
\end{array}
$$

4. Finish drawing your diagram by adding the appropriate number of hydrogen atoms beside each carbon. Remember that each carbon has a valence of four. So if a carbon atom has one other carbon atom bonded to it, you need to add three hydrogen atoms. If a carbon atom has two other carbon atoms bonded to it, you need to add two hydrogen atoms, and so on.

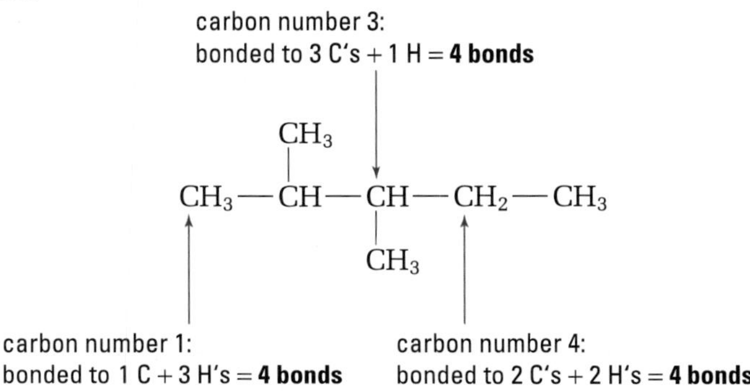

Practice Problems

6. Draw a condensed structural diagram for each alkane.
   (a) 3-ethyl-3,4-dimethylhexane
   (b) 2,3,4-trimethylpentane
   (c) 5-ethyl-3,3-dimethylheptane
   (d) 4-butyl-6-ethyl-2,5-dimethylnonane

7. One way to assess how well you have learned a new skill is to identify mistakes. Examine the following compounds and their names. Identify any mistakes, and correct the names.
   (a) 4-ethyl-2-methylpentane

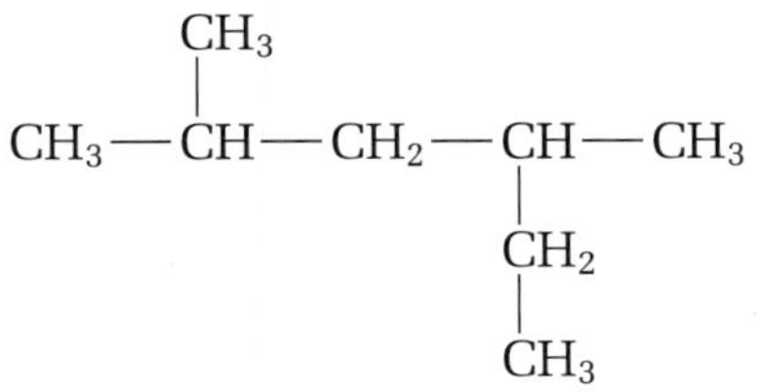

$$
\begin{array}{ccccc}
 & CH_3 & & & \\
 & | & & & \\
CH_3— & CH— & CH_2— & CH— & CH_3 \\
 & & & | & \\
 & & & CH_2 & \\
 & & & | & \\
 & & & CH_3 &
\end{array}
$$

   (b) 4,5-methylhexane

$$
\begin{array}{cccccc}
 & & & CH_3 & & \\
 & & & | & & \\
CH_3— & CH_2— & CH_2— & CH— & CH— & CH_3 \\
 & & & & | & \\
 & & & & CH_3 &
\end{array}
$$

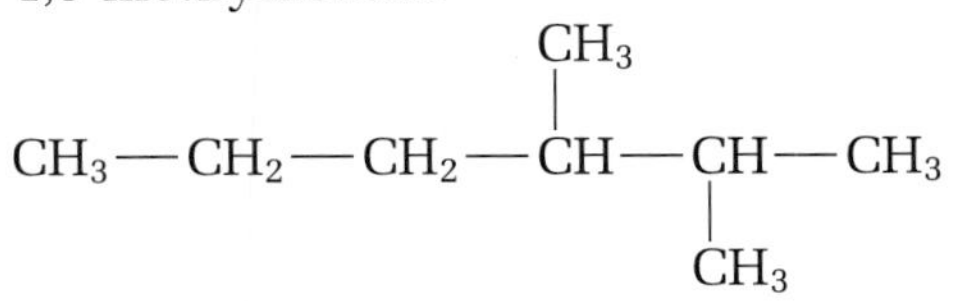

Continued ...

Continued ...
FROM PAGE 551

(c) 3-methyl-3-ethylpentane

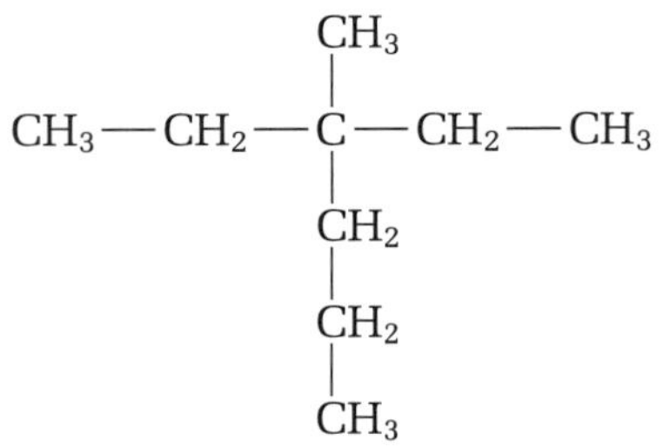

Alkenes

Did you know that bananas are green when they are picked? How do they become yellow and sweet by the time they reach your grocer's produce shelf or your kitchen? Food retailers rely on a hydrocarbon to make the transformation from bitter green fruit to delicious ripe fruit. The hydrocarbon is ethene. It is the simplest member of the second group of aliphatic compounds: the alkenes.

Alkenes are hydrocarbons that contain one or more double bonds. Like alkanes, alkenes can form continuous chain and branched-chain structures. They also form a homologous series. As well, they are non-polar, which gives them physical properties similar to those of alkanes.

Alkenes are different from alkanes, however, in a number of ways. First, their bonds are different, as indicated by their suffixes. As you will recall, the -ane ending tells you that alkane compounds are joined by single bonds. *The -ene ending for alkenes tells you that these compounds have one or more double bonds.* A double bond involves four bonding electrons between two carbon atoms, instead of the two bonding electrons in all alkane bonds. Examine Figure 13.18 to see how the presence of a double bond affects the number of hydrogen atoms in an alkene.

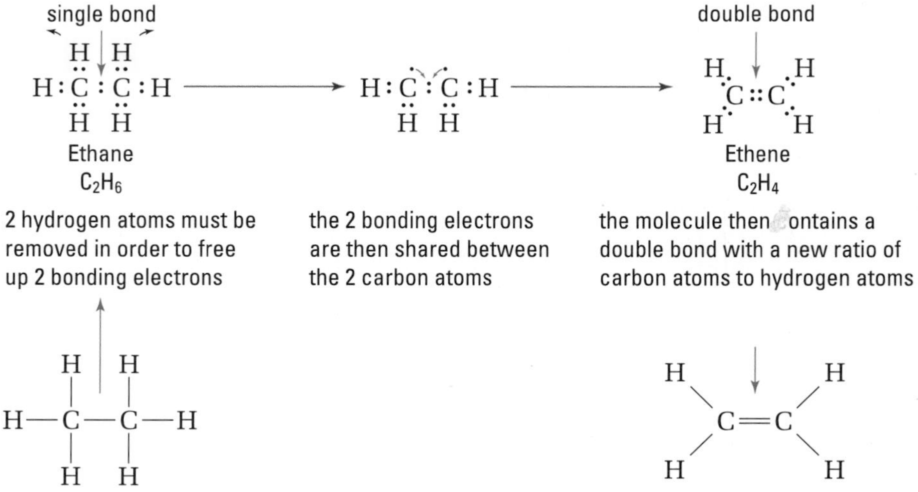

Figure 13.18 **This diagram shows how an ethane molecule can become an ethene molecule. The two hydrogen atoms that are removed from ethane often form hydrogen gas, $H_{2(g)}$.**

The general formula for an alkene is C_nH_{2n}. You can check this against the next two members of the alkene series. Propene has three carbon atoms, so you would expect it to have six hydrogens. Butene has four carbon atoms, so it should have eight hydrogens. The formulas for propene and butene are C_3H_6 and C_4H_8, so the general formula is accurate. Note, however, that the general formula applies only if there is one double bond per molecule. You will learn about alkenes with multiple double bonds in future chemistry courses.

Figure 13.19 **Plastics made from ethene**

Properties of Alkenes

The general formula for alkenes implies that at least two carbon atoms in any alkene compound have fewer than four bonded atoms. As a result, chemists refer to alkenes as unsaturated compounds. Unlike saturated compounds, **unsaturated hydrocarbons** contain carbon atoms that can potentially bond to additional atoms.

Unsaturated compounds have physical and chemical properties that differ from those of saturated compounds. For example, the boiling points of alkenes are usually slightly less than the boiling points of similar-sized alkanes (alkanes with the same number of carbon atoms). This difference reflects the fact that the forces between molecules are slightly less for alkenes than for alkanes. For example, the boiling point of ethane is −89°C, whereas the boiling point of ethene is −104°C. On the other hand, both alkenes and alkanes have a low solubility in water. Alkenes, like all aliphatic compounds, are non-polar.

The double bond in alkenes has important consequences for their chemical properties. Alkenes are much more reactive than alkanes. For example, alkenes react with halogens in the absence of light, but alkanes do not.

PROBEWARE

If you have access to probeware, do the Properties of Hydrocarbons lab, or a similar lab available from a probeware company.

The chemical reactivity of alkenes makes them a popular choice among chemical engineers. For example, nearly half of the ethene used industrially is in the plastics industry. Beverage containers, boil-in-the-bag food pouches, milk bottles, motor oil bottles, many toys, shrink-wrap, and plastic bags are all based on the small ethene molecule. (See Figure 13.19.) The ethene in these products undergoes a process called *polymerization.* In polymerization, hundreds of ethene molecules are reacted and strung together to make long chains of molecules. Another alkene, propene, undergoes a similar process and thus increases the variety of possible polymers.

On the previous page, you learned that ethene is used to ripen fruit. How does this work? When a fruit ripens, enzymes in the fruit begin to produce ethene gas. The ethene is responsible for the colour change, as well as the softening and sweetening that occur as the fruit ripens. Food chemists have learned that they can suppress ethene production (and delay ripening) by keeping fruits at a low temperature as they are transported. Once the fruits reach their final destination, ethene can be pumped into the fruit containers to hasten the ripening process. No wonder you can get such a variety of fresh fruit at the supermarket!

In the next investigation, you will use reactivity to identify saturated and unsaturated compounds in some everyday products. Saturated and unsaturated compounds play an important part in healthy eating, as you will discover.

Investigation 13-B

SKILL FOCUS

- Predicting
- Performing and recording
- Analyzing and interpreting
- Conducting research

Comparing the Reactivity of Alkanes and Alkenes

Because of differences in reactivity, you can use aqueous potassium permanganate, $KMnO_{4(aq)}$, to distinguish alkanes from alkenes. When the permanganate ion comes in contact with unsaturated compounds, such as alkenes, a reaction occurs. The permanganate ion changes to become manganese dioxide. This is shown by a colour change, from purple to brown. When the permanganate ion comes in contact with saturated compounds, such as alkanes, that have only single bonds, no reaction occurs. The colour of the permanganate does not change.

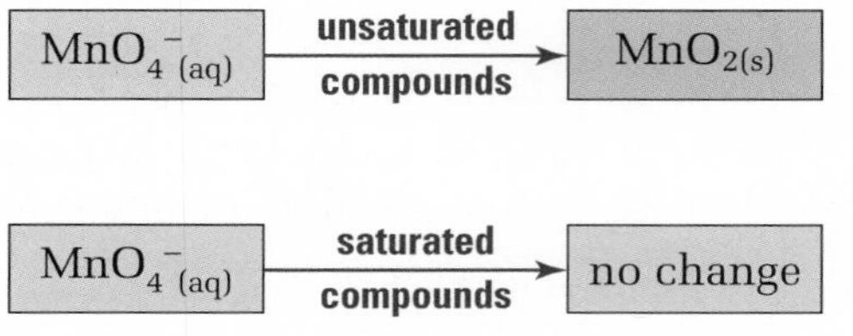

Aqueous potassium permanganate was added to the two test tubes on the left. One test tube contains an alkane compound. The other test tube contains an alkene compound. Which is which?

Question

How can you use aqueous potassium permanganate in a test to identify unsaturated compounds in fats and oils?

Predictions

Predict whether each substance to be tested will react with aqueous potassium permanganate. (In part, you can base your predictions on what you know about saturated and unsaturated fats from the media, as well as from biology and health classes.)

Safety Precautions

Do not spill any $KMnO_{4(aq)}$ on your clothing or skin, because it will stain. If you do accidentally spill $KMnO_{4(aq)}$ on your skin, remove the stain using a solution of sodium bisulfite.

Materials

test tubes (13 × 100 mm)
test tube rack
stoppers
medicine droppers
hot plate
5.0 mmol/L $KMnO_{4(aq)}$
water bath
samples of vegetable oils, such as varieties of margarine, corn oil, and coconut oil
samples of animal fats, such as butter and lard

Procedure

1. Read steps 2 and 3. Design a table to record your predictions, observations, and interpretations. Give your table a title.
2. Place about two full droppers of each test substance in the test tubes. Use a different dropper for each substance, or clean the dropper each time. Melt solids, such as butter, in a warm water bath (40°C to 50°C). Then test them as liquids.
3. Use a clean dropper to add one dropper full of potassium permanganate solution to each substance. Record your observations.

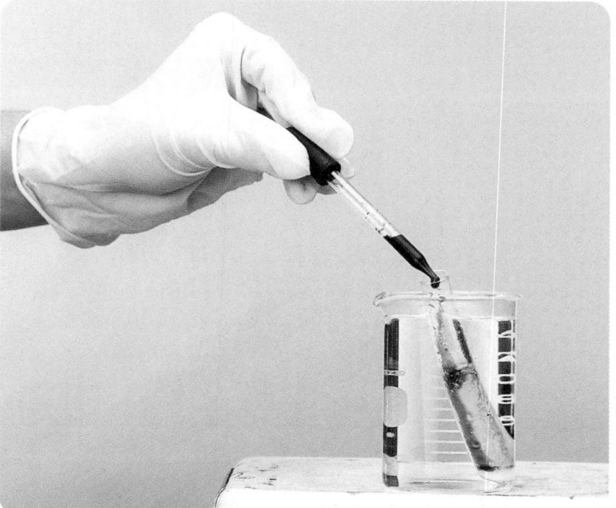

4. Dispose of the reactants as directed by your teacher.

Analysis

1. On what basis did you make your predictions? How accurate were they?
2. Did any of your results surprise you? Explain your answer.

Conclusion

3. Is there a connection between your results and the unsaturated or saturated compounds in the substances you tested? Explain your answer.

Application

4. Investigate the possible structures of some of the compounds (fatty acids) that are contained in various fats and oils. Use both print and electronic resources. See if you can verify a link between the tests you performed and the structures of these fats and oils. Check out compounds such as trans-fats (trans-fatty acids) and cis-fats (cis-fatty acids). Record your findings, and compare them with your classmates' findings.

$$\begin{array}{cccc} & & CH_3 & \\ & & | & \\ & & CH_2 & \\ & & | & \\ CH_3- & CH_2- & C & =CH_2 \end{array}$$

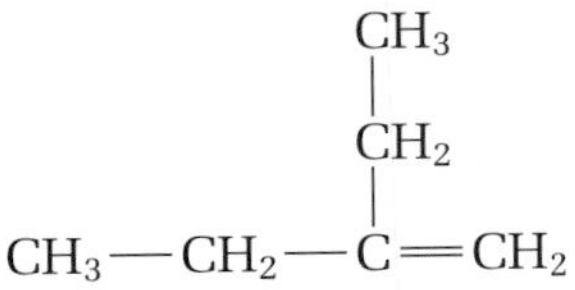

Figure 13.20

Biology LINK

According to many nutrition scientists, foods that contain double or triple bonds (unsaturated compounds) are healthier for us than foods that contain single bonds (saturated compounds). Research the implications of including unsaturated and saturated fats and oils in your diet. Use a library, the Internet, or any other sources of information. Decide on a suitable format in which to present your findings.

Naming Alkenes

The names of alkenes follow the same format as the names of alkanes: **prefix + root + suffix**. The prefixes and the steps for locating and identifying branches are the same, too. The greatest difference involves the double bond. The suffix -ene immediately tells you that a compound has at least one double bond. The rest of the necessary information—the location of the double bond, and the number of carbon atoms in the main chain—is communicated in the root. Follow the steps below to find out how to name the compound in Figure 13.20.

Naming 2-Ethyl-1-Butene

Step 1

Find the longest continuous chain that contains the double bond. (This step represents the main difference from naming alkanes.)

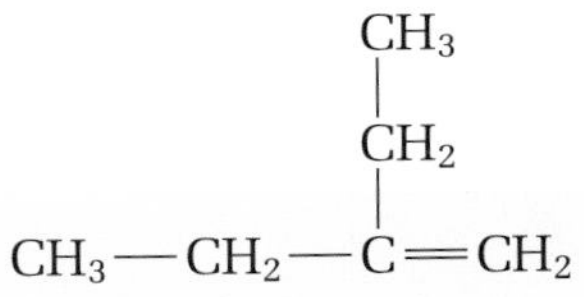

$$\begin{array}{cccc} & & CH_3 & \\ & & | & \\ & & CH_2 & \\ & & | & \\ CH_3- & CH_2- & C & =CH_2 \end{array}$$

- The main chain must contain the double bond, even though it may not be the longest chain.
- Here it is possible to have a main chain of five carbons. However, this chain would not include the double bond. Instead, choose the shorter chain with only four carbons. This chain includes the double bond. We now have part of the root, along with the suffix: -butene.

Step 2

Number the main chain from the end that is closest to the double bond.

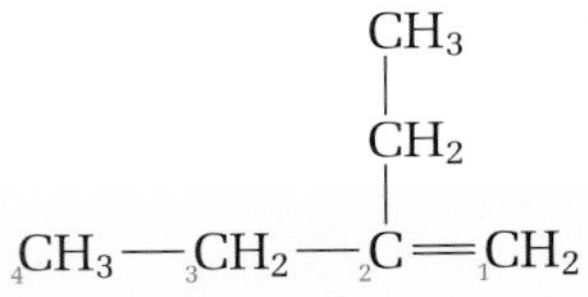

$$\begin{array}{cccc} & & CH_3 & \\ & & | & \\ & & CH_2 & \\ & & | & \\ {}_4CH_3- & {}_3CH_2- & {}_2C & ={}_1CH_2 \end{array}$$

- Here you number the main chain from the right side of the chain. This ensures that the double bond gets the lowest possible position number.

- You need the number that locates the double bond. Here the double bond is between carbon number 1 and carbon number 2. Which should you use? Logic and the IUPAC rules suggest that you should use the lowest possible number. Therefore the double bond is located with the number 1. So the root and suffix of the name are -1-butene.

Step 3

To create the prefix, identify and locate the branches with numbers.

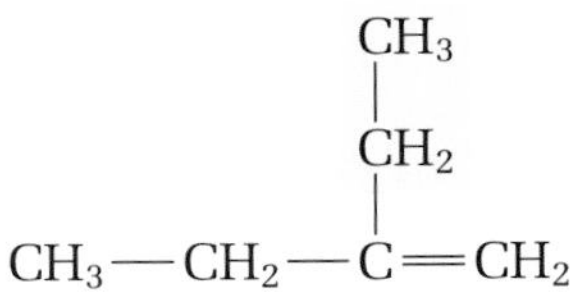

$$\begin{array}{cccc} & & CH_3 & \\ & & | & \\ & & CH_2 & \\ & & | & \\ CH_3- & CH_2- & C & =CH_2 \end{array}$$

- If there is more than one branch, remember to put the branches in alphabetical order.
- There are two carbons in the branch that is attached to the number 2 carbon of the main chain. Therefore the prefix is 2-ethyl-. (Make sure that you understand why -ethyl is used.)

Step 4

Use the formula prefix + root + suffix to put the name together. The full name of this compound is 2-ethyl-1-butene. Remember that a hyphen is placed between numbers and letters, and commas are placed between consecutive numbers.

In 2-ethyl-1-butene, the double bond and the branch are near each other on the carbon chain. What happens if the double bond is close to one end of the carbon chain, but the branches are close to the other end? For example, how would you name the compound in Figure 13.21?

The methyl branch is close to the right end of the main chain. The double bond is close to the left end of the main chain. Therefore you need to follow the rule in step 2: *the double bond has the lowest possible position number.* Now you can combine the name as follows:

- The 5-carbon main chain is numbered from the left. The double bond is given the number 2 (because 2 is lower than 3). The root and the suffix are -2-pentene.
- The prefix is 4-methyl- since there is a 1-carbon branch on the fourth carbon of the main chain.
- The complete name is 4-methyl-2-pentene.

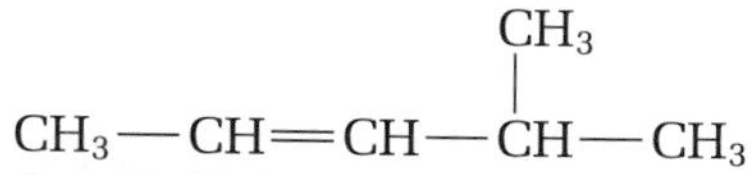

$$\begin{array}{l} \qquad\qquad\qquad\qquad\quad CH_3 \\ \qquad\qquad\qquad\qquad\quad | \\ CH_3—CH{=}CH—CH—CH_3 \end{array}$$

Figure 13.21

Although the compounds you have just named have different structural formulas, they both have the same molecular formula: C_6H_{12}. These compounds are both isomers of C_6H_{12}! You worked with structural isomers earlier in this chapter. You have seen that isomers can be made by rearranging carbon and hydrogen atoms and creating new branches. Now you will learn that isomers can also be made by rearranging double bonds.

Drawing Alkenes

You draw alkenes using the same method you learned for drawing alkanes. There is only one difference: you have to place the double bond in the main chain. Remember the valence of carbon, and be careful to count to four for each carbon atom on the structure. (Figure 13.22 gives an example of another alkene, 2-methyl-2-butene.) Be especially careful with the carbon atoms on each side of the double bond. The double bond is worth two for each carbon! Now complete the Practice Problems to reinforce what you have learned about naming and drawing alkenes.

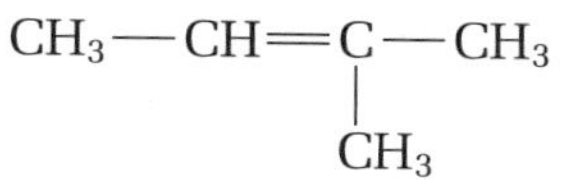

$$\begin{array}{l} CH_3—CH{=}C—CH_3 \\ \qquad\qquad\quad\;\, | \\ \qquad\qquad\quad CH_3 \end{array}$$

Figure 13.22 **Each carbon atom is bonded four times, once for each valence electron.**

Practice Problems

8. Name each hydrocarbon.

(a) $CH_3—CH_2—CH{=}CH—CH_2—CH_3$

(b)

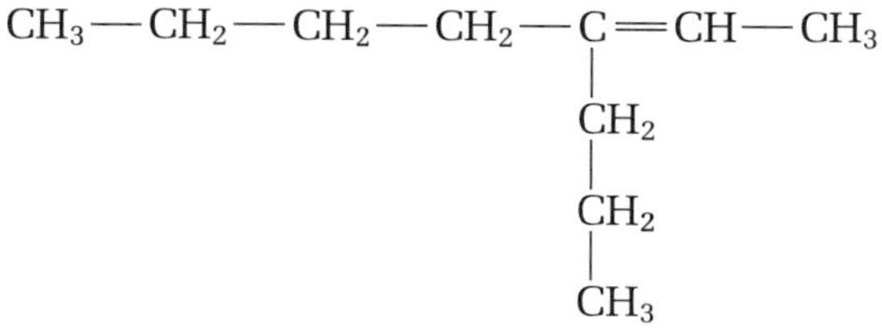

$$\begin{array}{l} CH_3—CH_2—CH_2—CH_2—C{=}CH—CH_3 \\ \qquad\qquad\qquad\qquad\qquad\quad\;\; | \\ \qquad\qquad\qquad\qquad\qquad\quad CH_2 \\ \qquad\qquad\qquad\qquad\qquad\quad\;\; | \\ \qquad\qquad\qquad\qquad\qquad\quad CH_2 \\ \qquad\qquad\qquad\qquad\qquad\quad\;\; | \\ \qquad\qquad\qquad\qquad\qquad\quad CH_3 \end{array}$$

(c)

$$\begin{array}{l} \qquad\qquad\qquad CH_3 \\ \qquad\qquad\qquad\;\; | \\ CH_3—CH—CH—C—CH_2—CH_3 \\ \qquad\quad\;\; | \qquad\qquad \| \\ \qquad\quad CH_3 \qquad CH—CH_2—CH_2—CH_3 \end{array}$$

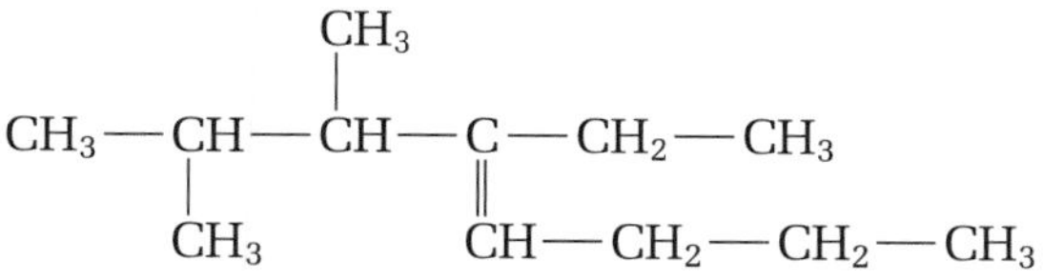

Continued ...

PROBLEM TIP

The easiest way to tell whether or not isomers are true isomers is to name them. Two structures that look different may turn out to have the same name. If this happens, they are not true isomers.

Continued ...

FROM PAGE 557

9. Draw a condensed structural diagram for each compound.

(a) 2-methyl-1-butene

(b) 5-ethyl-3,4,6-trimethyl-2-octene

10. You have seen that alkenes, such as C_6H_{12}, can have isomers. Draw condensed structural formulas for the isomers of C_4H_8. Then name the isomers.

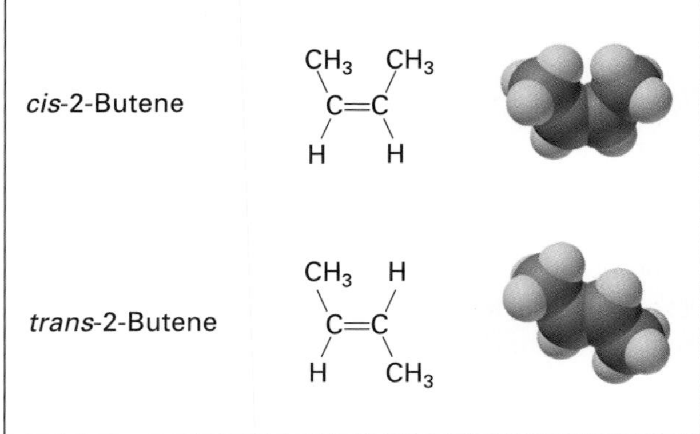

Figure 13.23 These diagrams show the cis and trans isomers of 2-butene. Notice that the larger methyl groups are on the same side of the double bond (both above) in the cis isomer. They are on opposite sides (one above and one below) in the trans isomer.

Cis-Trans (Geometric) Isomers

You have seen that isomers result from rearranging carbon atoms and double bonds in alkenes. Another type of isomer results from the presence of a double bond. It is called a **cis-trans isomer** (or **geometric isomer**). Cis-trans isomers occur when different groups of atoms are arranged around the double bond. Unlike the single carbon-carbon bond, which can rotate, the double carbon-carbon bond remains fixed. Figure 13.23 shows one of the compounds you have worked with already: 2-butene.

Remember these general rules:

- To have a cis-trans (geometric) isomer, each carbon in the C=C double bond must be attached to two different groups.
- In a cis isomer, the two larger groups are attached to each C=C double bond on the same side.
- In a trans isomer, the two larger groups are attached to each C=C double bond on opposite sides.

Like all isomers, cis-trans isomers have different physical and chemical properties. For example, the cis-2-butene isomer has a boiling point of 3.7°C, while the trans-2-butene isomer has a boiling point of 0.9°C.

Practice Problems

11. Draw and name the cis-trans isomers for C_5H_{10}.

12. Why can 1-butene not have cis-trans isomers? Use a structural diagram to explain.

13. Like other isomers, two cis-trans isomers have the same atomic weight. They also yield the same elements when decomposed. How might you distinguish between two such isomers in the lab?

14. C_6H_{12} has four possible pairs of cis-trans isomers. Draw and name all four pairs.

Electronic Learning Partner

Go to the Electronic Learning Partner to find out more about cis-trans isomers.

Chemistry Bulletin

Science | Technology | Society | Environment

Elastomer Technology: Useful or Harmful?

What do tires and chewing gum have in common? They are both made of *elastomers*: any substance you can pull or flex. Elastomer molecules can be *crosslinked*, or chemically linked, to produce rubber that retains its shape. You can find elastomers and/or rubber in many everyday products, such as running shoes, tires, underwear, and bubble gum.

Elastomers are a type of polymer, or giant molecule. To form a polymer such as an elastomer, many small molecules are connected to form a chain with thousands of repeating units.

Elastomer technology is useful in medicine. Muscles and arteries contain giant molecules, called *elastin*, that make muscles and arteries contract. Doctors can give artificial arteries to people with severe heart problems or other diseases. Unfortunately these artificial arteries do not last long enough. Can they be replaced by a new type of elastomer?

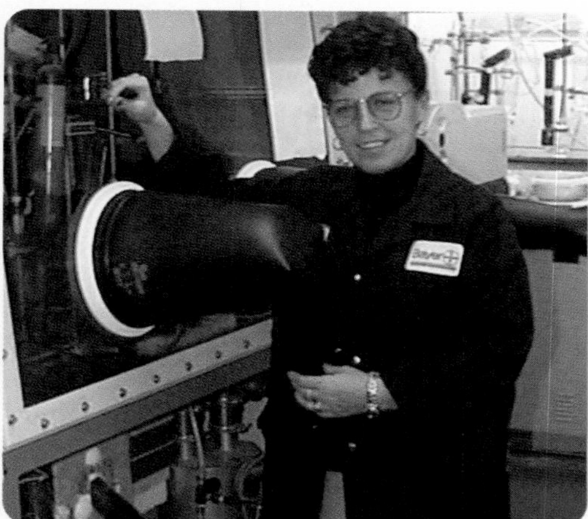

Judit Puskas

Dr. Judit Puskas holds Canada's first Industrial Research Chair in Elastomer Technology at the University of Western Ontario in London, Ontario. She thinks that an elastomer called *polyisobutylene*, along with some of its derivatives, looks promising. In the future, it may be used to make better artificial arteries. It may also be useful for other implants, since it can imitate the rubber-like properties of elastin.

Although elastomers and rubber have many helpful and useful properties, they can also cause problems. For example, rubber is used to make safe tires for cars, trucks, and other vehicles. These tires wear out and must be changed regularly. As a result, Canada and the United States accumulate 275 million used tires every year! What can be done with these used tires?

Scrap tires that are left in piles may catch on fire. Since tires usually burn with incomplete combustion, many dangerous and polluting gases are emitted into the environment. In addition, burning tires can leak oil and aromatic hydrocarbons into the soil. Thus they can contaminate drinking water in the area.

Tire manufacturers try to reduce waste by making tires that last longer. Unfortunately, when these durable radial tires break down, they disintegrate into small airborne particles, instead of large pieces as older tires do. This "tire dust" contains latex rubber, which is an allergen. As well, tire dust is small enough to be breathed deeply into the lungs. Tire dust may be one reason why asthma is becoming more common in North America.

Since tires are made of hydrocarbons, scrap tires can be used as a fuel source. Tire derived fuel (TDF) is used in power plants, paper mills, and cement kilns. Environmentalists are concerned, however, that incomplete combustion may result in the release of toxins.

Making Connections

A large company is proposing to build a cement kiln to produce cement. It will use scrap tires as fuel. Divide the class into various stakeholders, such as

- local citizens
- an environmental group
- the company
- the provincial ministry of the environment
- the owners of a scrap tire storage facility

Debate the proposal.

Alkynes

Carbon and hydrogen atoms can be arranged in many ways to produce a great variety of compounds. Yet another way involves triple bonds in the structure of compounds. This bond structure creates a class of aliphatic compounds called alkynes. **Alkynes** are aliphatic compounds that contain one or more triple bonds.

Naming and Drawing Alkynes

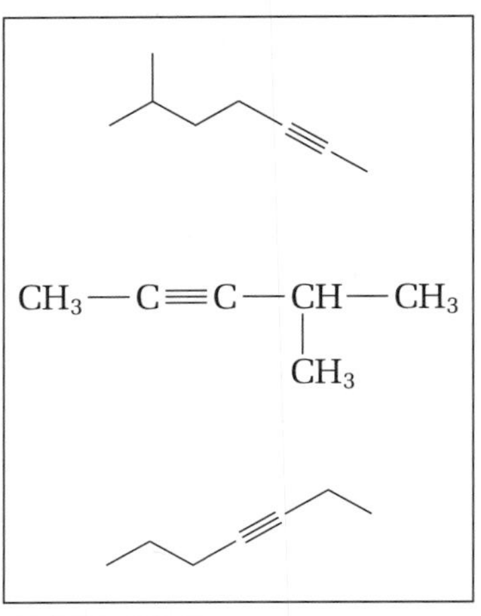

Figure 13.24 **Alkynes are linear around the triple bond, as these examples show.**

Both double and triple bonds are multiple bonds. Therefore alkynes are unsaturated hydrocarbons, just as alkenes are. To name alkynes and draw their structures, you follow the same rules that you used for alkenes. The only difference is the suffix -yne, which you need to use when naming alkyne compounds. Also, remember to count the number of bonds for each carbon. An alkyne bond counts as three bonds.

As you might expect, the presence of a triple bond in alkynes makes their physical and chemical properties different from those of alkanes and alkenes. A structure with a triple bond must be linear around the bond. (See Figure 13.24.) This means that the shapes of alkynes are different from the shapes of alkanes and alkenes. As well, the triple bond makes the molecule much more reactive—even more so than the double bond. In fact, few alkynes occur naturally because they are so reactive.

Alkynes are similar to both alkanes and alkenes because they form a homologous series. Alkynes have the general formula of C_nH_{2n-2}. So, for example, the first member of the alkyne series, ethyne, has the formula C_2H_2. (You may know this compound by its common name: acetylene.) The next member, propyne, has the formula C_3H_4.

Practice Problems

15. Name each alkyne.

(a)

```
        CH3
        |
CH3—C—C≡C—CH3
        |
        CH3
```

(b)

```
                        CH3
                        |
        CH3             CH2
        |               |
CH3—CH—CH2—C—C≡CH
                        |
                        CH2
                        |
                        CH2
                        |
                        CH3
```

16. Draw a condensed structural diagram for each compound.

(a) 2-pentyne

(b) 4,5-dimethyl-2-heptyne

(c) 3-ethyl-4-methyl-1-hexyne

(d) 2,5,7-trimethyl-3-octyne

Cyclic Hydrocarbons

You have probably heard the term "steroid" used in the context of athletics. (See Figure 13.25.) Our bodies contain steroids, such as testosterone (a male sex hormone) and estrone (a female sex hormone). Steroids also have important medicinal uses. For example, budesonide is a steroid that is used to treat asthma. One of the most common steroids is cholesterol. This compound is essential to your normal body functions, but it has been linked to blocked artery walls and heart disease, as well.

Steroids have also been associated with misuse, especially at the Olympics and other sporting events. Some athletes have tried to gain an advantage by using steroids to increase their muscle mass.

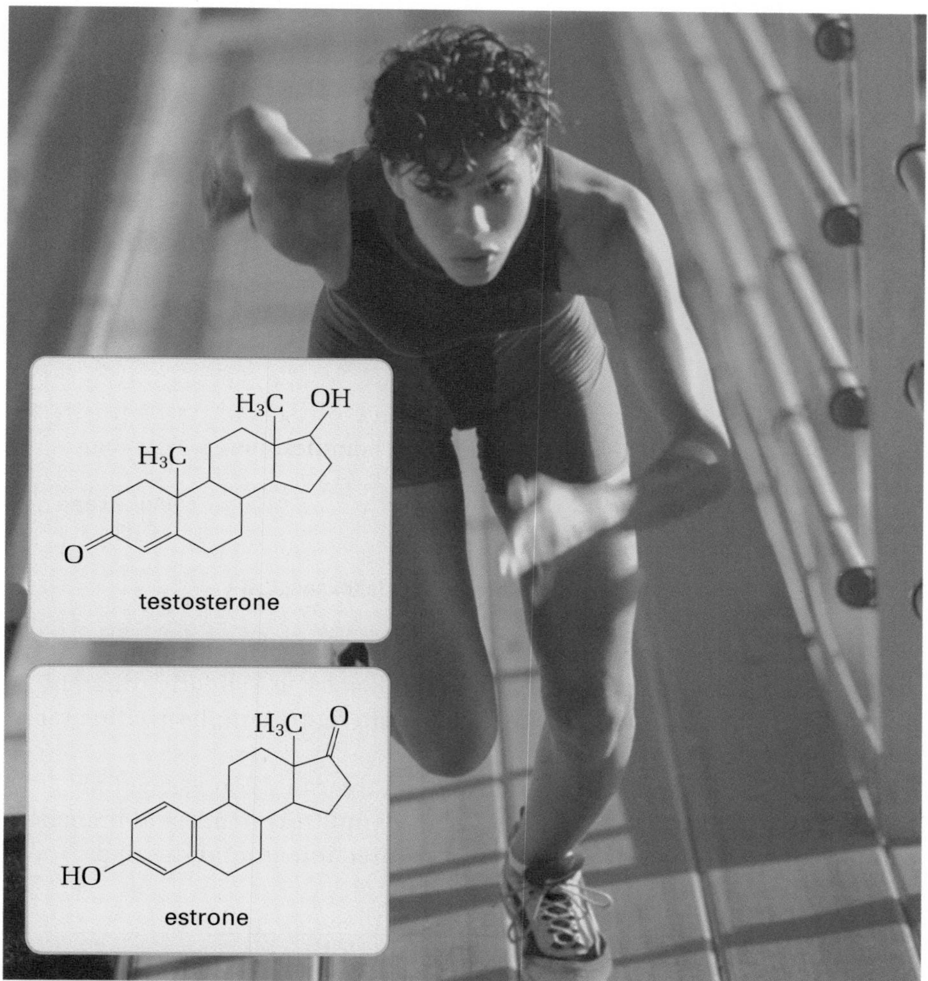

Figure 13.25 **Steroids are organic compounds. Our bodies make steroids naturally. Steroids may also be synthesized in chemical laboratories. What do the structures of these steroids have in common?**

What do steroids have to do with hydrocarbons? Steroids are unsaturated compounds. Although they are complex organic molecules, their basic structure centres on four rings of carbon atoms. In other words, steroids are built around ring structures of alkanes and alkenes.

Hydrocarbon ring structures are called **cyclic hydrocarbons**. They occur when the two ends of a hydrocarbon chain join together. In order to do this, a hydrogen atom from each end carbon must be removed, just as in the formation of a multiple bond. (See Figure 13.26 on the next page.)

You should have no trouble recognizing cyclohexane as a member of the alkane family. Notice, however, that cycloalkanes, such as cyclohexane, have two fewer hydrogen atoms compared with other alkanes. Thus they have the general formula C_nH_{2n}. (This is the same as the general formula for alkenes.)

Naming and Drawing Cyclic Hydrocarbons

To draw the structure of a cyclic hydrocarbon, use a line diagram in a ring-like shape, such as the one shown in Figure 13.26. Each carbon-carbon bond is shown as a straight line. Each corner of the ring represents a carbon atom. Hydrogen atoms are not shown, but they are assumed to be present in the correct numbers.

hexane
C_6H_{14}
lose
2 H's
bond
formation
+ 2H
cyclohexane
C_6H_{12}

Figure 13.26 **How hexane, C_6H_{14}, can become cyclohexane, C_6H_{12}**

Because of the ring structure, the naming rules for cyclic hydrocarbons, including cycloalkanes and cycloalkenes, are slightly different from those for alkanes and alkenes. Below are four examples to illustrate the naming rules.

To draw cyclic hydrocarbons, start with the rules you learned for drawing other types of compounds. To place multiple bonds and branches, you have the option of counting in either direction around the ring.

Naming Cyclic Hydrocarbons

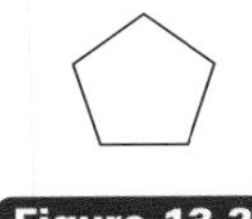

Figure 13.27

Example 1: You can still use the general formula: prefix + root + suffix. In Figure 13.27, there are only single carbon-carbon bonds. There are also five corners (carbon atoms) in the ring, which is the main chain. Since there are no branches, the name of this compound is cyclopentane. Notice the addition of cyclo- to indicate the ring structure.

CH_3

Figure 13.28

Example 2: When naming cyclic compounds, all carbon atoms in the ring are treated as equal. This means that any carbon can be carbon number 1. In Figure 13.28, only one branch is attached to the ring. Therefore the carbon that the branch is attached to is carbon number 1. Because this branch automatically gets the lowest possible position number, no position number is required in the name. Thus the name of this compound is methylcyclohexane.

Example 3: When two or more branches are on a ring structure, each must have the lowest possible position number. Which way do you count the carbons around the ring? You can count in either direction around the ring. In Figure 13.29, a good choice is to make the ethyl branch carbon number 1 and then count counterclockwise. This allows you to sequence the branches in alphabetical order and to add the position numbers in ascending order. So the name of this structure is 1-ethyl-3-methylcyclohexane.

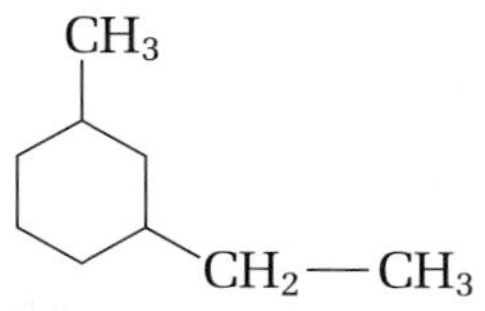

CH_3

$CH_2—CH_3$

Figure 13.29

Example 4: In Figure 13.30, there is a double bond, represented by the extra vertical line inside the ring structure. You must follow the same rules as for alkenes. That is, the double bond gets priority for the lowest number. This means that one of the carbon atoms, on either end of the double bond, must be carbon number 1. The carbon atom at the other end must be carbon number 2. Next you have to decide in which direction to count so that the branch gets the lowest possible position number. In this compound, the carbon atom on the bottom end of the double bond is carbon number 1. Then you can count clockwise so that the methyl group on the top carbon of the ring has position number 3. (Counting in the other direction would give a higher locating number for the branch.) The name of this structure is 3-methyl-1-cyclohexene.

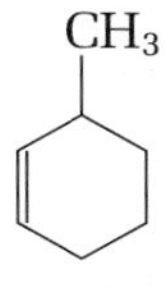

CH_3

Figure 13.30

Practice Problems

17. Name each compound.

(a) 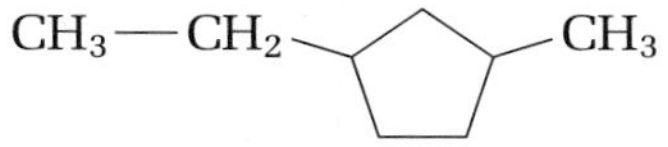 $CH_3—CH_2$ — (cyclopentane) — CH_3

(b) 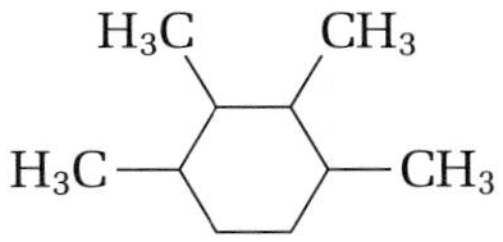 H_3C CH_3 H_3C— (cyclohexane) —CH_3

(c) CH_3 (cyclobutane)

(d) $CH_3—CH_2—CH_2$ — (cyclopentene) — CH_3

(e) CH_3 (cyclooctene)

(f) CH_2CH_3 —CH_3 CH_3 (cyclooctene)

(g) CH_3CH_2 (cyclohexene) —CH_2CH_3

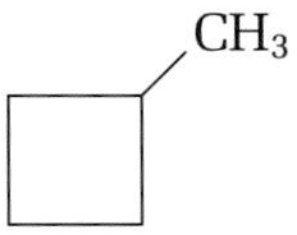

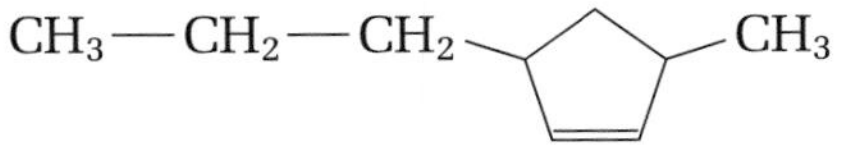

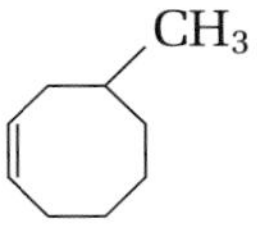

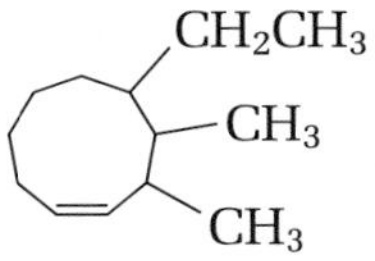

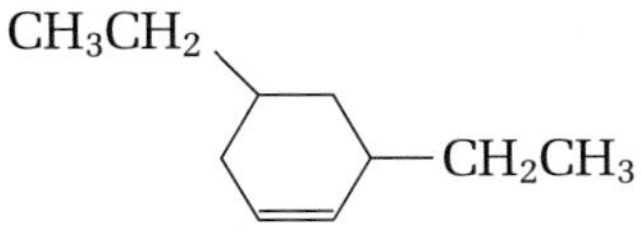

(h) CH_3 (cyclopentane) $CH_2CH_2CH_2CH_2CH_3$

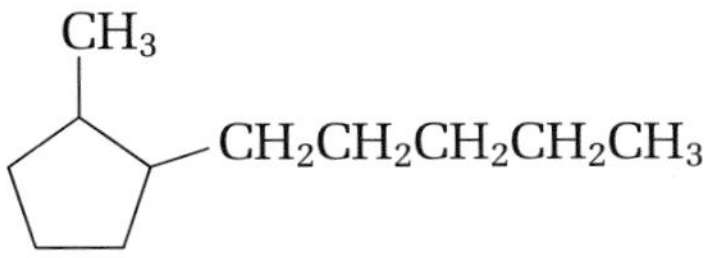

18. Draw a condensed structural diagram for each compound.

(a) 1,2,4-trimethylcycloheptane

(b) 2-ethyl-3-propyl-1-cyclobutene

(c) 3-methyl-2-cyclopentene

(d) cyclopentene

(e) 1,3-ethyl-2-methylcyclopentane

(f) 4-butyl-3-methyl-1-cyclohexene

(g) 1,1-dimethylcyclopentane

(h) 1,2,3,4,5,6-hexamethyl cyclohexane

You have now had some experience naming and drawing aliphatic compounds. In the next investigation, you will develop a more thorough understanding of them by examining their structural and physical properties.

Investigation 13-C

SKILL FOCUS

Predicting

Modelling concepts

Analyzing and interpreting

Structures and Properties of Aliphatic Compounds

To compare the properties of alkanes, alkenes, and alkynes, you will be working with compounds that have the same number of carbon atoms. First, you will construct and compare butane, trans-2-butene, 2-butyne, and cyclobutane. You will use a graph to compare the boiling points of each compound. Next, you will use what you have just observed to predict the relative boiling points of pentane, trans-2-pentene, 2-pentyne, and cyclopentane. You will construct and compare these structures and graph their boiling points.

Question

How can constructing models of butane, trans-2-butene, 2-butyne, and cyclobutane help you understand and compare their physical properties?

Predictions

Predict the structural formula for each compound. After completing steps 1 to 4, predict what the graph of the boiling points of pentane, trans-2-pentene, 2-pentyne, and cyclopentane will look like.

Materials

molecular model kits
reference books

Procedure

1. Construct models of butane, trans-2-butene, 2-butyne, and cyclobutane.
2. Examine the structure of each model. Draw a diagram of it in your notebook.
3. If possible, rotate the molecule around each carbon-carbon bond to see if this changes the appearance of the structure.
4. Look up the boiling points of these four compounds in the table below. Draw a bar graph to compare the boiling points.
5. Repeat steps 1 to 3 for pentane, trans-2-pentene, 2-pentyne, and cyclopentane.
6. Predict the relative boiling points for these four compounds. Use a reference book to find and graph the actual boiling points.

Comparing the Boiling Points of Four-Carbon Compounds

Compound	Boiling point (°C)
butane	–0.5
trans-2-butene	0.9
2-butyne	27
cyclobutane	12

Analysis

1. What are the differences between the multiple-bond compounds and the alkanes?
2. What type of compound has the highest boiling point? What type has the lowest boiling point?

Conclusion

3. Identify possible reasons for the differences in boiling points between compounds.

Application

4. Compare the boiling points of cyclopentane and cyclobutane. Use this information to put the following compounds in order from highest to lowest boiling point: cyclohexane, cyclobutane, cyclopropane, cyclopentane. Use a reference book to check your order.

Summary: Rules for Naming and Drawing Aliphatic Compounds

Naming Alkanes

- Find the longest continuous chain. This is the parent (main) chain. Number the parent chain so that the branches have the lowest possible position numbers.
- Locate any branches. Use the number of carbons in each branch to name it (for example, ethyl). Give it a position number.
- When writing the parts of the name, separate two numbers by a comma. Separate a number from a word by a hyphen.
- Write the prefix. Each branch has a position number. Write the branches in alphabetical order. More than one of each type of branch is shown by di-, tri-, and so on.
- Write the root. This depends on the number of carbons in the main chain.
- Write the suffix -ane. (See Figure 13.31 for practice.)

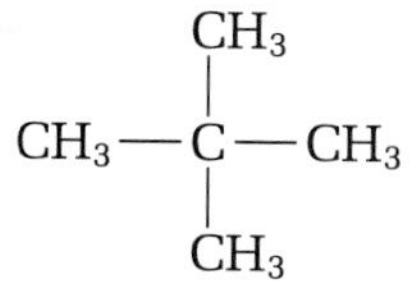

Drawing Alkanes

- In a straight line, draw the carbon atoms in the parent (main) chain. The number of carbon atoms is indicated by the root of the name.
- Give position numbers to the carbon atoms in the parent chain. Attach all the branches to their appropriate parent chain carbon atoms.
- Add enough hydrogen atoms and bonds for each carbon atom to have four bonds.

$$\begin{array}{c} CH_3 \\ | \\ CH_3—C—CH_3 \\ | \\ CH_3 \end{array}$$

Figure 13.31 Can you name this alkane?

Naming Alkenes and Alkynes

- Find the longest continuous chain containing the multiple bond. This is the parent chain. Number the parent chain so that the multiple bond has the lowest possible position number.
- Locate any branches. Identify the position number and number of carbons for each branch.
- When writing the parts of the name, separate two numbers by a comma. Separate a number from a word by a hyphen.
- Write the prefix. Each branch has a position number. Write the branches in alphabetical order. More than one of each type of branch is shown by di-, tri-, and so on.
- Write the root. This includes the position number of the multiple bond and the name for the number of carbon atoms in the parent chain.
- Write the suffix -ene for alkenes and -yne for alkynes. (See Figure 13.32 for practice.)

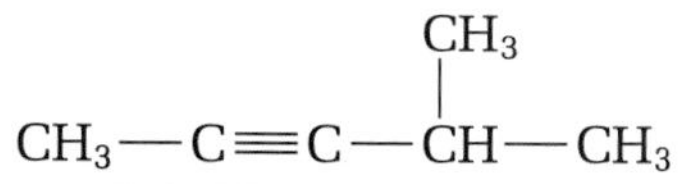

Drawing Alkenes and Alkynes

- In a straight line, draw the carbon atoms in the parent chain. The number of carbon atoms is indicated by the root of the name. Place the multiple bond between two carbon atoms in the chain, as indicated by the number in the name.
- Give position numbers to the carbon atoms in the parent chain. Attach all the branches to their appropriate parent chain carbon atoms.
- Add enough hydrogen atoms and bonds for each carbon atom to have four bonds.

$$\begin{array}{r} CH_3 \\ | \\ CH_3—C\equiv C—CH—CH_3 \end{array}$$

Figure 13.32 Can you name this alkyne?

Naming Cycloalkanes

- Identify the branches.
- Number the carbons in the ring, in either direction, so that the branches have the lowest possible position numbers.
- Write the prefix, as for naming alkanes.
- Write the root -cyclo- plus the name for the number of carbon atoms in the ring. For example, a five-carbon ring would have the root -cyclopent-.
- Write the suffix -ane.

Drawing Cycloalkanes

- Draw the ring, according to the root of the name.
- Choose one of the ring carbon atoms as carbon number 1. Place each branch accordingly.

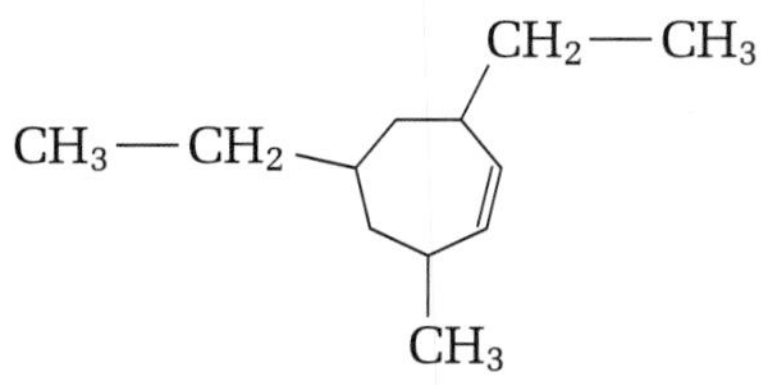

$CH_2—CH_3$

$CH_3—CH_2$

CH_3

Figure 13.33 Can you name this cycloalkene?

Naming Cycloalkenes and Cycloalkynes

- Identify the branches.
- Number the carbons in the ring, in either direction, so that the multiple bond is between the two lowest numbers, and the branches get the lowest possible position numbers.
- Write the prefix and root, as for cycloalkanes.
- Write the suffix -ene for cycloalkenes or -yne for cycloalkynes. (See Figure 13.33 for practice.)

Drawing Cycloalkenes and Cycloalkynes

- Draw the ring, according to the root of the name.
- Choose one of the ring carbon atoms as carbon number 1. Place the multiple bond accordingly.
- Add each branch according to its type and position number.

Section Wrap-up

Now you know how to name and draw alkanes, alkenes, alkynes, and ring compounds. In the next section, you will discover how petroleum, the source of most hydrocarbons, can be separated into its components.

Section Review

1 (a) K/U What are the names of the three types of aliphatic compounds that you studied in this section?

(b) K/U Which of these are saturated compounds, and which are unsaturated compounds? How does this difference affect their properties?

2 K/U List the roots used to name the first ten members of the alkane homologous series. Indicate the number of carbon atoms that each represents.

3 C If water and octane are mixed, does the octane dissolve in the water? Explain.

4 **(a)** K/U Name each compound.

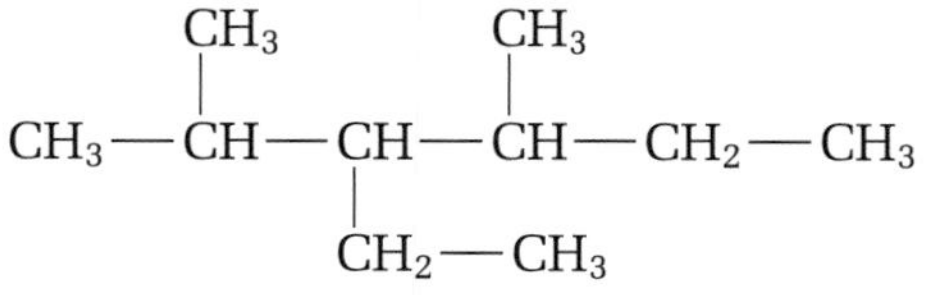

```
        CH3          CH3
        |            |
CH3—CH—CH—CH—CH2—CH3
             |
             CH2—CH3
```

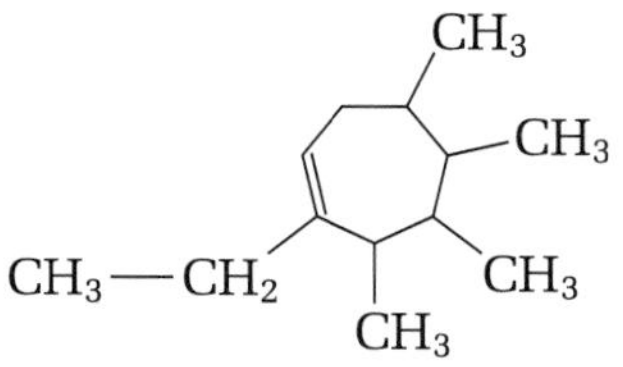

(b)

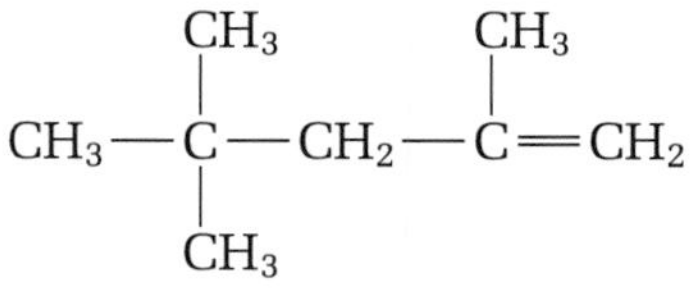

```
      CH3          CH3
      |            |
CH3—C—CH2—C=CH2
      |
      CH3
```

(d)

```
                 CH3
     (ring)      —CH3
CH3—CH2     CH3     CH3
```

(c)

```
                  CH3
                  |
                  CH2
                  |
CH≡C—CH—CH—CH2—CH3
          |
          CH2
          |
          CH2
          |
          CH3
```

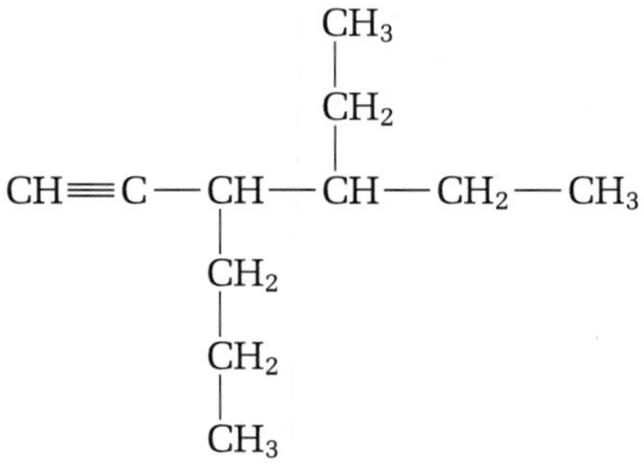

5 C Draw a condensed structural diagram for each compound.

(a) 2,4-dimethyl-3-hexene

(b) 5-ethyl-4-propyl-2-heptyne

(c) 3,5-diethyl-2,4,7,8-tetramethyl-5-propyldecane

(d) trans-4-methyl-3-heptene

6 C Draw and name all the isomers that are represented by each molecular formula.

(a) C_3H_4 **(b)** C_5H_{12} **(c)** C_5H_{10}

7 K/U Identify any mistakes in the name and/or structure of each compound.

(a) 3-methyl-2-butene

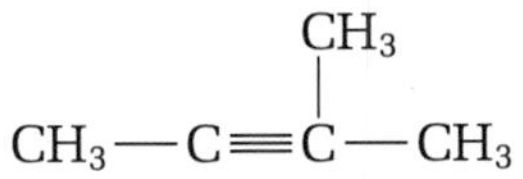

```
           CH3
           |
CH3—C≡C—CH3
```

(b) 2-ethyl-4,5-methyl-1-hexene

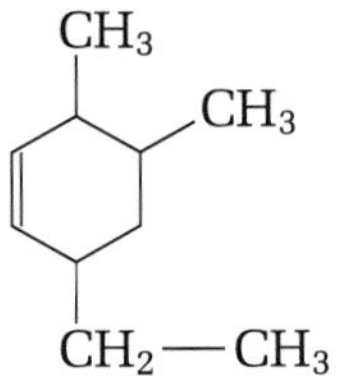

```
CH3
        CH3
 (ring)

CH2—CH3
```

8 C What happens when there is more than one double bond or more than one ring? Try to draw a condensed structural diagram for each of the following compounds.

(a) propadiene

(b) 2-methyl-1,3-butadiene

(c) 1,3,5-cycloheptatriene

(d) cyclopentylcyclohexane

13.4 Refining and Using Hydrocarbons

Section Preview/ Specific Expectations

In this section, you will

- **describe** the steps involved in refining petroleum to obtain gasoline and other useful fractions
- **explain** the importance of hydrocarbons in the petrochemical industry
- **communicate** your understanding of the following terms: *petrochemicals, fractional distillation, cracking, reforming*

Is it possible to take hydrocarbons straight out of the ground and use them as they are? Since the earliest recorded history, people have done just that. In the past, people used crude oil that seeped up to Earth's surface to waterproof boats and buildings. They also used it to grease wheels and even dress the wounds of animals and humans. As well, people burned natural gas, mainly to supply lighting for temples and palaces.

Today hydrocarbons are extracted from the ground at well sites, then processed further at refineries. (See Figure 13.34.) The first commercial oil well in North America was located in southern Ontario's Enniskillen Township. It began production in 1858. At that time, kerosene (which was used to fuel lamps) was the principal focus of the young petroleum industry. Paraffin (for making candles) and lubricating oils were also produced, but there was little demand for other hydrocarbon materials, such as gasoline.

Figure 13.34 **This oil refinery is in Clarkson, Ontario.**

Reliance on hydrocarbons has increased substantially since the nineteenth century. Our society now requires these compounds for fuel, as well as for the raw materials that are used to synthesize petrochemicals. **Petrochemicals** are basic hydrocarbons, such as ethene and propene, that are converted into plastics and other synthetic materials. Petroleum is the chief source of the petrochemicals that drive our cars and our economy. Petroleum is not a pure substance, however. Rather, petroleum is a complex mixture of hydrocarbons—mainly alkanes and alkenes—of varying molecular sizes and states. Because petroleum is a mixture, its composition varies widely from region to region in the world. An efficient process is essential for separating and collecting the individual, pure hydrocarbon components. Read on to learn about this process in greater detail.

Using Properties to Separate Petroleum Components

Fractional distillation is a process for separating petroleum into its hydrocarbon components. This process relies on a physical property—boiling point. Each of the hydrocarbon components, called *fractions*, has its own range of boiling points. At an oil refinery, the separating (refining) of petroleum begins in a large furnace. The furnace vaporizes the liquid components. The fluid mixture then enters a large fractionation tower. Figure 13.35 outlines how the various hydrocarbon fractions are separated in the tower. (**Note:** The temperatures shown are approximate boiling points for the hydrocarbon fractions.)

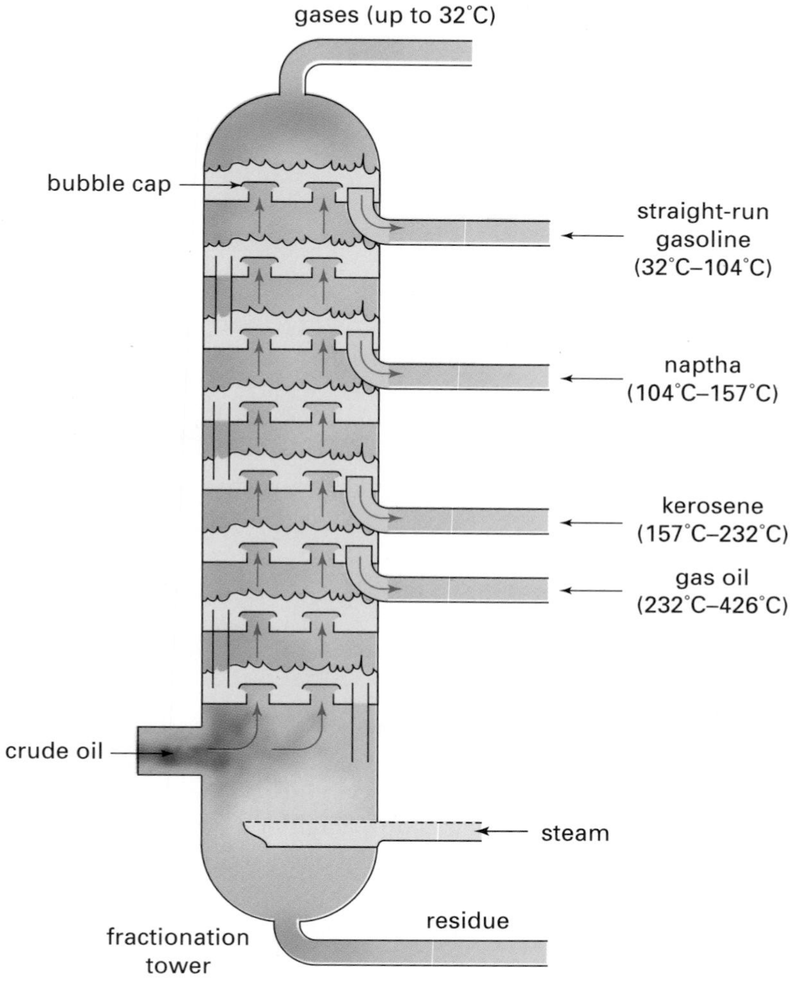

Figure 13.35 This diagram shows how petroleum is separated into its hydrocarbon fractions. Each fraction has a different range of boiling points. The tower separates the fractions by a repeated process of heating, evaporating, cooling, and condensing.

Perforated plates, which are fitted with bubble caps, are placed at various levels in the tower. As each fraction reaches a plate where the temperature is just below its boiling point, it condenses and liquefies. The liquid fractions are taken from the tower by pipes. Other fractions that are still vapours continue to pass up through the plates to higher levels.

Several plates are needed to collect each fraction. Heavier hydrocarbons (larger molecules) have higher boiling points. They condense first and are removed in the lower sections of the tower. The lighter hydrocarbons, with lower boiling points, reach the higher levels of the tower before they are separated.

Web LINK

www.school.mcgrawhill.ca/resources/

Canada's land mass has experienced dynamic changes over tens of millions of years. Climatic conditions, along with the deposition of countless remains of organisms, created the areas in which petroleum is found today. These areas are called *sedimentary basins*. Where are the sedimentary basins? Are they all active sites for oil and gas extraction? How much oil and gas do scientists estimate there is? How do Canada's reserves compare with those of other petroleum-producing nations? Go to the web site above to "tap" into Canada's and the world's petroleum resources. Go to **Science Resources**, then to **Chemistry 11** to find out where to go next. Decide on a suitable format in which to record your findings.

Cracking and Reforming

Once the fractions are removed from the distillation tower, they may be chemically processed or purified further to make them marketable. (See Figure 13.36.) There has been a tremendous increase in the demand for a variety of petroleum products in the early twentieth century. This demand has forced the oil industry to develop new techniques to increase the yield from each barrel of oil. These techniques are called cracking and reforming.

Petroleum Fractionation Products

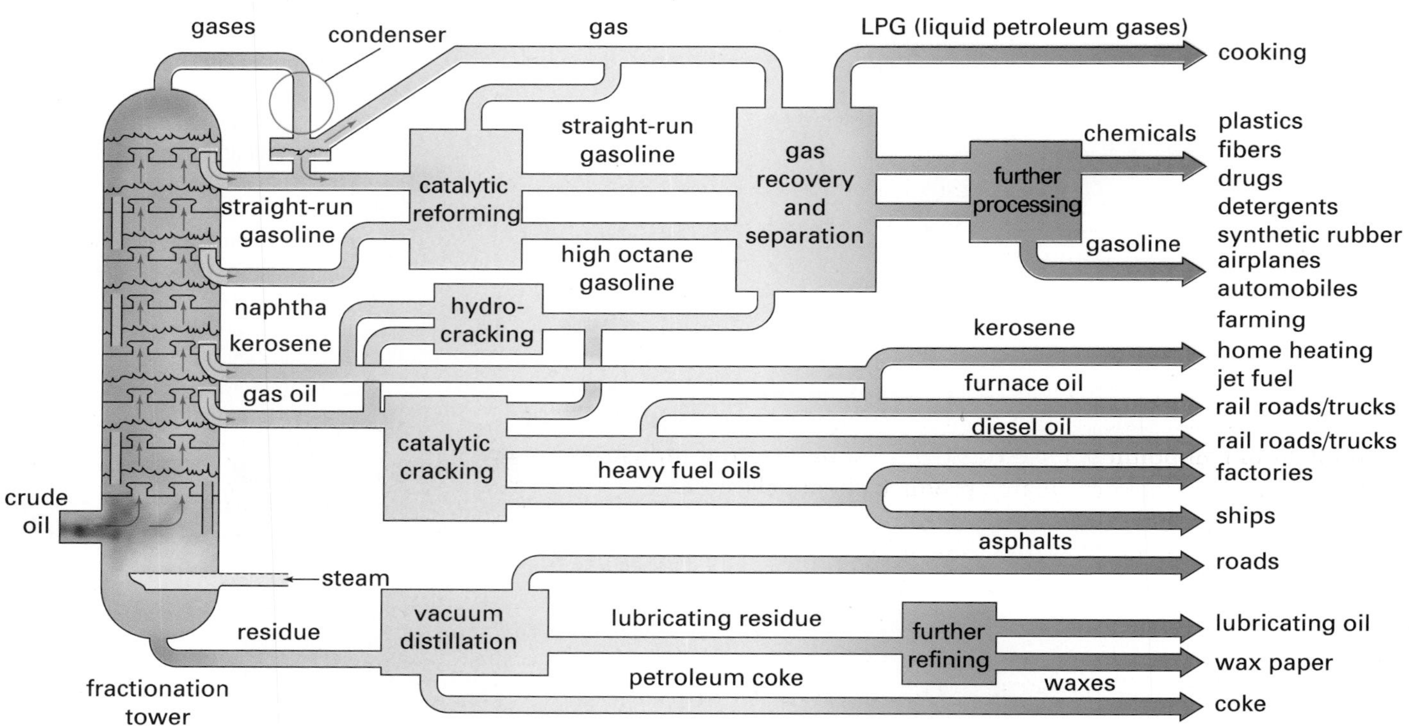

Figure 13.36 **Selected uses of petroleum fractions**

Cracking was first introduced in Sarnia, Ontario. This process uses heat to break larger hydrocarbon molecules into smaller gasoline molecules. Cracking is done in the absence of air and can produce different types of hydrocarbons. For example, the cracking of propane can produce methane and ethene, as well as propene and hydrogen. (See Figure 13.37.)

500°C–700°C

$$2\ CH_3CH_2CH_3 \rightarrow CH_4 + CH_2{=}CH_2 + CH_3CH{=}CH_2 + H_2$$

propane **methane** **ethene** **propene** **hydrogen**

Figure 13.37 **This chemical equation summarizes the cracking of propane.**

The Chemistry 11 Electronic Learning Partner has an animation that illustrates how a fractionation tower works.

Reforming is another technique that uses heat, pressure, and catalysts to convert large hydrocarbons into other compounds. Reforming can produce larger hydrocarbons or a different type of hydrocarbon, called aromatic compounds. (See Figure 13.38.) Aromatic compounds contain special ring structures. You will learn about them in a later chemistry course.

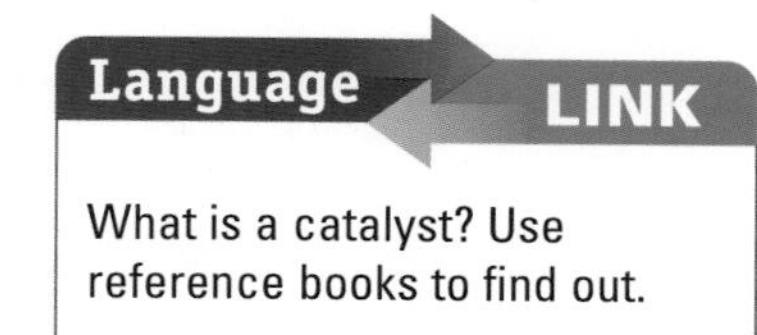

What is a catalyst? Use reference books to find out.

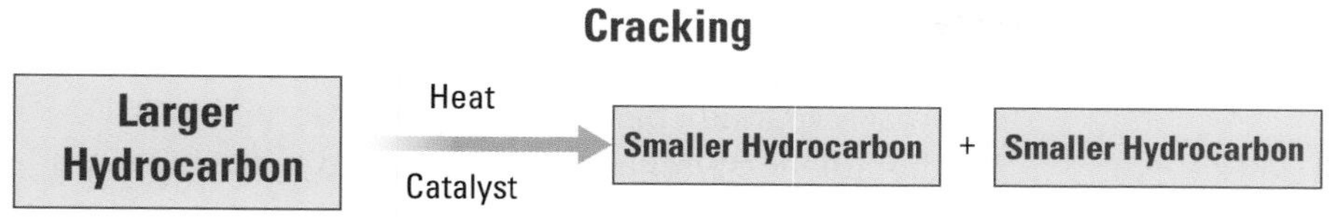

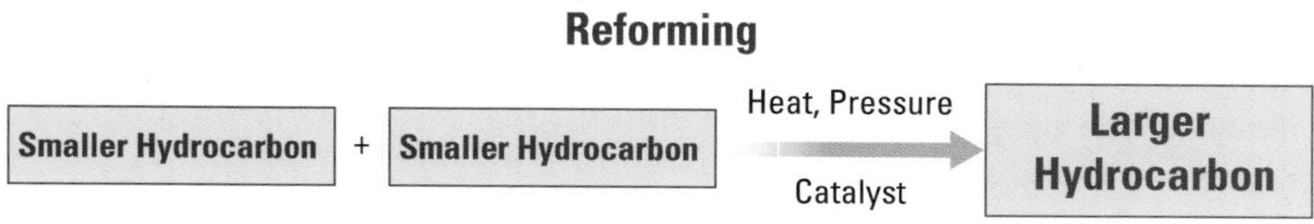

Figure 13.38 **Comparing cracking and reforming**

Some of the products of the refining process are transported to petrochemical plants. These plants convert complex hydrocarbons, such as naphtha, into simple chemical compounds, or a small number of compounds, for further processing by other industries. Canadian petrochemical plants produce chemicals such as methanol, ethylene, propylene, styrene, butadiene, butylene, toluene, and xylene. These chemicals are used as building blocks in the production of other finished products. Nearly every room in your school, your home, and your favourite shopping centre contains, or is made from, at least one petrochemical product. Yet, as you can see in Figure 13.39, petrochemicals represent only a small fraction of society's uses of petroleum.

In the next chapter, you will discover why hydrocarbons are so highly valued as fuels. As well, you will unravel the mystery of how so much energy can be locked up inside a hydrocarbon molecule.

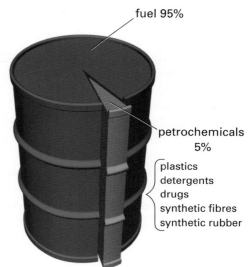

fuel 95%

petrochemicals 5%

plastics
detergents
drugs
synthetic fibres
synthetic rubber

Petroleum

Figure 13.39 **A tremendous number of petrochemical products enrich your life. Yet they still make up only about 5% of what is produced from a barrel of crude oil. A staggering 95% of all petroleum is used as fuel.**

Section Wrap-up

In this section, you learned that petrochemicals, an essential part of our society and technology, are obtained from petroleum. You discovered how petroleum is separated into its components by fractional distillation, cracking, and reforming.

In this chapter, you learned a great deal about hydrocarbons. You learned how hydrocarbons are formed in section 13.1. In sections 13.2 and 13.3, you learned to draw, classify, and name different hydrocarbons. Finally, you learned about the practical side of hydrocarbons—how they are produced and used in everyday life.

Careers in Chemistry

Polymer Chemist

Bulletproof vests used to be heavy, bulky, and uncomfortable. Stephanie Kwolek changed that. Now police officers, police dogs, and soldiers, can wear light, strong bulletproof vests made of a synthetic fibre called Kevlar™. Kwolek (shown above) developed Kevlar™ while working for the DuPont chemical company. Kevlar™ first came into general use in 1971. It is five times stronger than steel, gram for gram, but almost as light as nylon. It is flame resistant, resists wear and tear, and does not conduct electricity. This versatile material is used not only in bulletproof vests, but also in other manufactured items, including hockey helmets, firefighters' suits, spacecraft shells, and surgeons' gloves.

Kwoluk's branch of organic chemistry—polymer chemistry—specializes in creating synthetic materials that are cheaper, faster, and stronger than natural materials. Polymer chemists often work by stringing together thousands of atoms to form long molecules called polymers. Then they manipulate these polymers in various ways. Polymer chemists have invented an amazing array of materials. These include polyvinyl chloride (used to make garden hose and duct tape), polyurethane (used to stuff teddy bears and make spandex bicycle pants), and acrylonitrile-butadiene-styrene, or ABS (used to make brakes and other auto body parts).

Make Career Connections

What kind of education do polymer chemists need for their jobs?

- Find out about polymer chemistry programs that are offered by a university near you.
- Research different companies that employ polymer chemists. If possible, interview a polymer chemist who is employed by one of these companies.

Section Review

Unit Project Prep

Before you choose a product to investigate for the end-of-unit Project, consider what you have learned about petrochemical products. What compound is used to produce most plastics, and thus almost all products' packaging?

1 (a) K/U What physical property allows various fractions of crude oil to be separated in a fractionation tower?

(b) C What is this process of separation called? Briefly describe it. Include a diagram in your answer if you wish.

2 MC Our society's demand for petroleum products has increased dramatically over the last century. Describe two techniques that can be used to transform the petroleum fractions from a fractionation tower, in order to meet this demand.

3 MC List five types of petrochemicals that can be produced from refined hydrocarbons. Describe briefly how each type has affected your life.

4 MC Petrochemicals are not always helpful. Do research and prepare a presentation illustrating the effects of petrochemicals on the environment.

CHAPTER 13 Review

Reflecting on Chapter 13

Summarize this chapter in the format of your choice. Here are a few ideas to use as guidelines:

- Identify the origins and sources of hydrocarbons and other organic compounds.
- Describe the characteristics that enable carbon to form so many, varied compounds.
- Distinguish among complete, condensed, and line structural diagrams.
- Identify, draw, and name at least two examples of each kind of hydrocarbon you studied: alkanes, alkenes, alkynes, and cyclic hydrocarbons.
- Demonstrate, using suitable examples, true isomers of a hydrocarbon.
- Compare physical properties, such as boiling point, of aliphatic compounds.
- Describe the processes and techniques that the petrochemical industry depends on. Identify at least ten products of this industry.

Reviewing Key Terms

For each of the following terms, write a sentence that shows your understanding of its meaning.

aliphatic hydrocarbons	alkanes
alkenes	alkynes
cis-trans isomer	cracking
cyclic hydrocarbons	expanded molecular formula
fractional distillation	geometric isomer
homologous series	isomers
hydrocarbons	organic compound
petrochemicals	petroleum
reforming	saturated hydrocarbons
structural diagram	structural model
unsaturated hydrocarbons	

Knowledge/Understanding

1. **(a)** What are the origins of most hydrocarbons and other organic compounds?
   (b) List three sources of hydrocarbons.
   (c) What is the main source of hydrocarbons used for fuels?
2. List three factors that are required to change biological matter into petroleum.
3. What are the key properties of the carbon atom that allow it to form such diverse compounds?
4. Define the following terms: fractional distillation, petrochemical.
5. What are isomers? Give one example of a set of isomers for a molecular formula.
6. Describe how the boiling point changes as the chain length of an aliphatic compound increases. Explain why this happens.
7. Briefly compare alkanes, alkenes, and alkynes. (Give both similarities and differences.)
8. Describe the difference between structural isomers and cis-trans isomers.
9. Name each compound.

   (a) $CH_3—CH_2—CH_2—CH_3$

   (b) $CH_3—CH_3$

   (c)
   ```
           CH3         CH3
            |           |
   CH3 — CH — CH — CH — CH3
                |
               CH3
   ```

   (d)
   ```
                      CH2 — CH3
                       |
   CH3 — CH2 — CH — CH — CH3
                |
               CH2 — CH3
   ```

10. For each structural diagram, write the IUPAC name and identify the type of aliphatic compound.

   (a)
   ```
          CH3
           |
   CH2 ═ C — CH3
   ```

   (b) (cyclohexane ring bearing CH_2CH_3 and CH_3 on adjacent carbons)

   (c)
   ```
                CH3
                 |
   CH ≡ C — CH — CH2 — CH3
   ```

   (d)
   ```
   CH3           CH3
       \        /
        C ═ C
       /        \
   H             CH2 — CH3
   ```

   (e)
   ```
                  CH3
                   |
                  CH2
                   |
   CH3 — CH2 — C — CH3
                   |
   CH3 — CH2 — CH — CH3
   ```

11. Name the following isomers of 2-heptene.

```
H           CH2—CH2—CH2—CH3
  \        /
   C == C
  /        \
CH3          H
```

```
CH3         CH2—CH2—CH2—CH3
  \        /
   C == C
  /        \
H            H
```

12. Match the following names and structural diagrams. Note that only four of the six names match.

(a) cis-3-methyl-3-hexene
(b) trans-3-hexene
(c) trans-3-methyl-2-hexene
(d) trans-3-methyl-3-hexene
(e) cis-3-methyl-2-hexene
(f) cis-3-hexene

(A)
```
         CH3          CH2—CH3
            \        /
             C == C
            /        \
  CH3—CH2             H
```

(B)
```
          H            H
            \        /
             C == C
            /        \
  CH3—CH2             CH2—CH3
```

(C)
```
         CH3          CH3
            \        /
             C == C
            /        \
          H           CH2—CH2—CH3
```

(D)
```
         CH3          H
            \        /
             C == C
            /        \
  CH3—CH2             CH2—CH3
```

13. Is 1-methyl-2-cyclobutene the correct name for a compound? Draw the structural diagram for the compound, and rename it if necessary.

14. Explain why the rule "like dissolves like" is very useful when cleaning up an oil spill on a body of water.

15. Why is it impossible for the correct name of a linear alkane to begin with 1-methyl?

16. Which combinations of the following structural diagrams represent true isomers of C_6H_{14}? Which structural diagrams represent the same isomer?

(a)
```
        CH3
        |
  CH3—CH—CH—CH3
           |
           CH3
```

(b)
```
        CH3
        |
  CH3—CH—CH2—CH2—CH3
```

(c)
```
  CH3  CH3
  |    |
  CH—CH
  |    |
  CH3  CH3
```

(d)
```
  CH3          CH3
  |            |
  CH—CH2—CH2
  |
  CH3
```

(e)
```
        CH3
        |
  CH3—C—CH2—CH3
        |
        CH3
```

Inquiry

17. Someone has left two colourless liquids, each in an unlabelled beaker, on the lab bench. You know that one liquid is an alkane, and one is an alkene. Describe a simple test that you can use to determine which liquid is which.

18. Suppose that you were given a liquid in a beaker labelled "C_6H_{12}." Discuss how you would determine whether the substance was an alkane, a cycloalkane, an alkene, or an alkyne.

19.In Investigation 13-C, you constructed cyclobutane. In order to construct this isomer, you had to use springs for bonds.

(a) Why did you have to use springs?
(b) What do you think the bulging springs tell you about the stability of cyclobutane in real life?

Communication

20. Draw a complete structural diagram for each of these compounds.

(a) 3-ethylhexane
(b) 1-butene
(c) 2,3-dimethylpentane

21. Draw a condensed structural diagram for each of these compounds.
(a) methylpropane
(b) 1-ethyl-3-propylcyclopentane
(c) cis-3-methyl-3-heptene
(d) 3-butyl-4-methyl-1-octyne
(e) 4-ethyl-1-cyclooctene

22. (a) Draw four structural isomers of C_5H_{12}.
(b) Draw four cis-trans isomers of C_8H_{16}.

23. In a fractionation tower, the gaseous fractions are removed from the top and the solid residues are removed from the bottom.
(a) Explain why the fractions separate like this.
(b) A sample of crude oil was tested and found to contain mostly smaller hydrocarbon molecules, less than 15 carbon atoms long. From what part of the tower would most of the fractions of this sample be removed? Why?

24. Use a diagram to describe the steps involved in refining petroleum to obtain gasoline.

25. Draw a concept map to illustrate how hydrocarbon molecules can start as crude oil and end up as synthetic rubber.

26. Imagine that you are the owner of an oil refinery. Your main supply of crude oil contains a high percent of longer-chain hydrocarbons (greater than 15 carbon atoms per molecule). A new customer is looking for a large supply of gasoline, which contains 7- and 8-carbon molecules. How will you meet your customer's needs?

27. Place the following alkanes in order from lowest to highest boiling point. Explain your reasoning.
(a) $CH_3—CH_2—CH_2—CH_2—CH_3$
(b) $CH_3—CH_2—CH_3$
(c) $CH_3—CH(CH_3)—CH_3$
(d) $CH_3—CH_2—CH_2—CH_3$

28. Draw the complete, condensed, and line structural diagrams for 2,3,4-trimethylpentane. Discuss the main advantange of each type of structural diagram.

Making Connections

29. List two ways in which ethene is important in everyday life.

30. Research the similarities and differences in drilling for oil offshore versus onshore. Write a report of your findings.

31. The National Energy Board has estimated that Canada's original petroleum resources included 4.3×10^{11} m^3 (430 billion cubic metres) of oil and bitumen (a tar-like substance) and 1.7×10^{14} m^3 (170 trillion cubic metres) of natural gas. Canada's petroleum-producing companies have used only a small fraction of these resources. Why, then, do you think many people are so concerned about exhausting them?

Answers to Practice Problems and Short Answers to Section Review Questions:
Practice Problems: **1.** C_7H_{16} **2.** C_9H_{20} **3.** 10 **4.** 25 **5.(a)** 2-methylbutane **(b)** 2,2-dimethylpropane **(c)** 3-ethyl-2,5-dimethylheptane **(d)** 2,2,4,4-tetramethylhexane **(e)** 2,2,4-trimethyl-4-propyloctane **7.(a)** 2,4-dimethylhexane **(b)** 2,3-dimethylhexane **(c)** 3-ethyl-3-methylhexane **8.(a)** 3-hexene **(b)** 3-propyl-2-heptene **(c)** 4-ethyl-2,3-dimethyl-4-octene **10.** 2-butene, 1-butene, 2-methyl-1-propene **11.** cis-2-pentene, trans-2-pentene **14.** cis-2-hexene, trans-2-hexene; cis-3-hexene, trans-3-hexene; 3-methyl-cis-2-pentene, 3-methyl-trans-2-pentene; 4-methyl-cis-2-pentene, 4-methyl-trans-2-pentene **15.(a)** 4,4-dimethyl-2-pentyne **(b)** 3-ethyl-5-methyl-3 -propyl-1-hexyne **17.(a)** 1-ethyl-3-methylcyclopentane **(b)** 1,2,3,4-tetramethylcyclohexane **(c)** methylcyclobutane **(d)** 3-methyl-5-propyl-1-cyclopentene **(e)** 4-methyl-1-cyclooctene **(f)** 5-ethyl-3,4-dimethyl-1-cyclononene **(g)** 3,5-diethyl-1-cyclohexene **(h)** 1-methyl-2-pentylcyclopentane **13.1:** **5.(a)** organic **(b)** organic **(c)** inorganic **(d)** inorganic **(e)** organic **(f)** inorganic **(g)** organic **(h)** inorganic **13.3:** **1.(a)** alkane, alkene, alkyne **2.** meth (1), eth (2), prop (3), but (4), pent (5), hex (6), hept (7), oct (8), non (9), dec (10) **4.(a)** 3-ethyl-2,4-dimethylhexane **(b)** 2,4,4-trimethyl-1-pentene **(c)** 4-ethyl-3-propyl-1-hexyne **(d)** 2-ethyl-3,4,5,6-tetramethyl-1-cycloheptene **6.(a)** propyne, cyclopropene **(b)** pentane, 2-methylbutane, 2,2-dimethylpropane **(c)** 1-pentene, cis-2-pentene, trans-2-pentene, 2-methyl-2-butene, 2-methyl-1-butene, cyclopentane **13.4:** **1.(a)** boiling point **(b)** fractional distillation

CHAPTER 14

Energy Trapped in Hydrocarbons

The whole world runs on energy—and so do you! Fossil fuels provide energy to power cars and heat buildings. Food provides energy to keep your body alive and warm. Both sources of energy come from organic compounds, such as hydrocarbons, sugars, and proteins.

Green plants, algae, and plankton trap the Sun's energy through the process of photosynthesis. After these organisms die, they are broken down by natural processes. Their remains accumulate on Earth's surface. In some areas, these remains build up in thick layers, which are eventually covered by rock and soil. Under certain conditions, over billions of years, pressure changes these layers into something new: fossil fuels. **Fossil fuels** (such as coal, natural gas, and petroleum) are fuels that are made from fossilized organic materials. The trapped energy from the Sun is still present in fossil fuels. To use this energy, we need to extract it. **Combustion**, or burning, is the most common way to extract energy from fossil fuels.

In this chapter, you will explore the ways in which our society obtains energy from fossil fuels. You will get a chance to measure exactly how much energy is obtained from an organic substance by doing your own combustion reaction. As well, you will learn how dangerous incomplete combustion reactions can be.

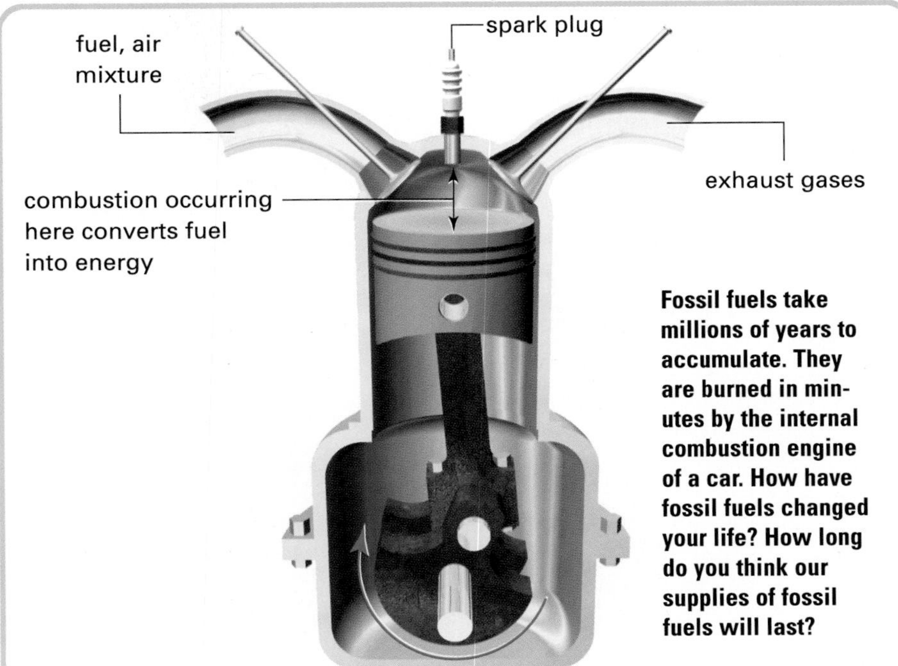

Fossil fuels take millions of years to accumulate. They are burned in minutes by the internal combustion engine of a car. How have fossil fuels changed your life? How long do you think our supplies of fossil fuels will last?

Chapter Preview

Concepts and Skills You Will Need

Before you begin this chapter, review the following concepts and skills:

- explaining bonding in molecular compounds (Chapter 3, section 3.3)
- writing and balancing chemical equations for different reactions (Chapter 4, section 4.1)
- significant digits (Chapter 1, section 1.2)
- problem solving in gas laws (Chapter 12, section 12.3)
- naming and drawing aliphatic compounds (Chapter 13, section 13.2)

14.1 Formation and Combustion Reactions

Section Preview/ Specific Expectations

In this section, you will

- **write** balanced chemical equations for the complete and incomplete combustion of hydrocarbons
- **perform** an experiment to produce and burn a hydrocarbon
- **recognize** the importance of hydrocarbons as fuels and as precursors for the production of petrochemicals
- **identify** the risks and benefits of the uses of hydrocarbons, for society and the environment
- **communicate** your understanding of the following terms: *complete combustion, incomplete combustion*

One of our most common uses of hydrocarbons is as fuel. (See Figure 14.1.) The combustion of fossil fuels gives us the energy we need to travel and to keep warm in cold climates. Fossil fuel combustion is also an important source of energy in the construction and manufacturing industries. As well, many power plants burn natural gas when generating electricity. Even goldsmiths use hydrocarbons, such as butane, as a heat source when crafting gold jewellery. At home, we often burn fossil fuels, such as natural gas, to cook our food.

How do we get energy from these compounds? In this section, you will learn how complete and incomplete combustion can be expressed as chemical equations. Combustion in the presence of oxygen is a *chemical* property of all hydrocarbons. (In Chapter 13, you learned about some *physical* properties of hydrocarbons, such as boiling point and solubility.)

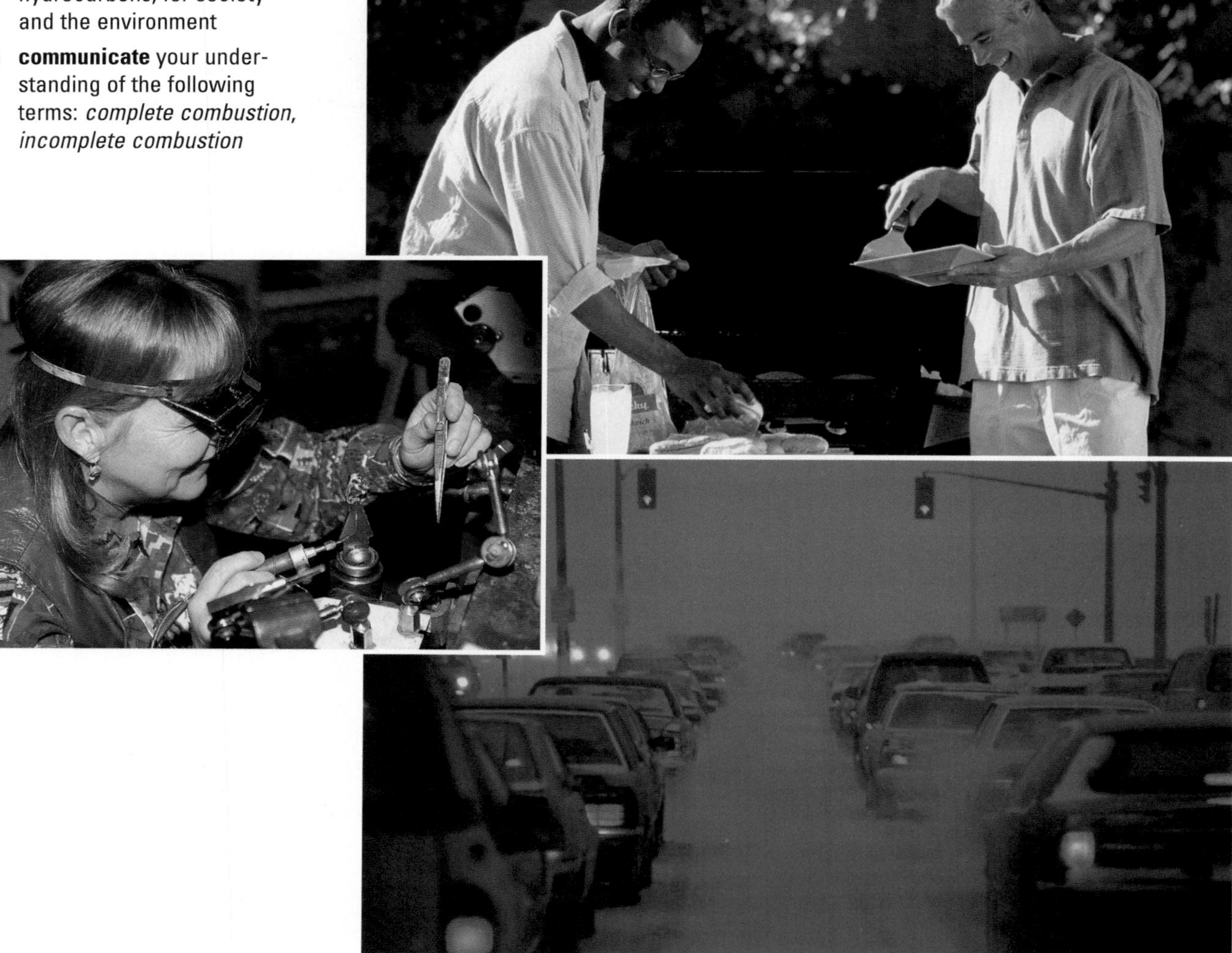

Figure 14.1 How are fossil fuels being used in these photographs?

Chemistry Bulletin

Science | Technology | Society | Environment

Lamp Oil and the Petroleum Age

Abraham Gesner was born in 1797 near Cornwallis, Nova Scotia. Although Gesner became a medical doctor, he was much more interested in fossils. Gesner was fascinated by hydrocarbon substances, such as coal, asphaltum (asphalt), and bitumen. These substances were formed long ago from fossilized plants, algae, fish, and animals.

When Gesner was a young man, the main light sources available were fire, candles, and whale oil lamps. Gesner had made several trips to Trinidad. He began to experiment with asphaltum, a semisolid hydrocarbon from Trinidad's famous "pitch lake." In 1846, while giving a lecture in Prince Edward Island, he startled his audience by lighting a lamp that was filled with a fuel he had distilled from asphaltum. Gesner's lamp fuel gave more light and produced less smoke than any other lamp fuel the audience had ever seen used.

Gesner needed a more easily obtainable raw material to make his new lamp fuel. He tried a solid, black, coal-like bitumen from Albert County, New Brunswick. This substance, called albertite, worked better than any other substance that Gesner had tested.

Making Kerosene

One residue from Gesner's distillation process was a type of wax. Therefore, he called his lamp fuel *kerosolain*, from the Greek word for "wax oil." He soon shortened the name to *kerosene*. To produce kerosene, Gesner heated chunks of albertite in a retort (a distilling vessel with a long downward-bending neck). As the albertite was heated, it gave off vapours. The vapours passed into the neck of the retort, condensed into liquids, and trickled down into a holding tank. Once Gesner had finished the first distillation, he let the tank's contents stand for several hours. This allowed water and solid to settle to the bottom. Then he drew off the oil that remained on top.

Gesner distilled this oil again, and then treated it with sulfuric acid and calcium oxide. Finally he distilled the oil once more.

By 1853, Gesner had perfected his process. In New York, he helped to start the North American Kerosene Gas Light Company. Gesner distinguished between three grades of kerosene: grades A, B, and C. Grade C, he said, was the best lamp oil. Grades A and B could also be burned in lamps, but they were dangerous because they could cause explosions and fires.

Although Gesner never knew, his grades A and B kerosene became even more useful than the purer grade C. These grades were later produced from crude oil, or petroleum, and given a new name: gasoline!

Gesner laid the groundwork for the entire petroleum industry. All the basics of later petroleum refining can be found in his technology.

Making Connections

1. In the early nineteenth century, whales were hunted extensively for their oil, which was used mainly as lamp fuel. When kerosene became widely available, the demand for whale oil decreased. Find out what effect this had on whalers and whales.
2. How do you think the introduction of kerosene as a lamp oil changed people's lives at the time? What conclusions can you draw about the possible impact of technology?

Figure 14.2 Sour gas, $H_2S_{(g)}$, is sometimes "flared off" (burned) from an oil well. This combustion produces sulfur dioxide gas, which reacts with the water in the atmosphere to produce acid rain. Oil companies are now making an effort to reduce this type of pollution.

Complete and Incomplete Combustion

During a typical combustion reaction, an element or a compound reacts with oxygen to produce oxides of the element (or elements) found in the compound. Figure 14.2 shows an example of a combustion reactions.

Hydrocarbon compounds will burn in the presence of air to produce oxides. This is a chemical property of all hydrocarbons. **Complete combustion** occurs if enough oxygen is present. A hydrocarbon that undergoes complete combustion produces carbon dioxide and water vapour. The following equation shows the complete combustion of propane. (See also Figure 14.3.)

$$C_3H_{8(g)} + 5O_{2(g)} \rightarrow 3CO_{2(g)} + 4H_2O_{(g)}$$

If you burn a fuel, such as propane, in a barbecue, you want complete combustion to occur. Complete combustion ensures that you are getting maximum efficiency from the barbecue. More importantly, toxic gases can result from **incomplete combustion**: combustion that occurs when not enough oxygen is present. During incomplete combustion, other products (besides carbon dioxide and water) can form. The equation below shows the incomplete combustion of propane. Note that unburned carbon, $C_{(s)}$, and carbon monoxide, $CO_{(g)}$, are produced as well as carbon dioxide and water.

$$2C_3H_{8(g)} + 7O_{2(g)} \rightarrow 2C_{(s)} + 2CO_{(g)} + 2CO_{2(g)} + 8H_2O_{(g)}$$

Figure 14.4 shows another example of incomplete combustion. Go back to the equation for the complete combustion of propane. Notice that the mole ratio of oxygen to propane for the complete combustion (5 mol oxygen to 1 mol propane) is higher than the mole ratio for the incomplete combustion (7 mol oxygen to 2 mol propane, or 3.5 mol oxygen to 1 mol propane). These ratios show that the complete combustion of propane used up more oxygen than the incomplete combustion. In fact, the incomplete combustion probably occurred because not enough oxygen was present. You just learned that incomplete combustion produces poisonous carbon monoxide. This is why you should never operate a gas barbecue or gas heater indoors, where there is less oxygen available. This is also why you should make sure that any natural gas or oil-burning furnaces and appliances in your home are working at peak efficiency, to reduce the risk of incomplete combustion. Carbon monoxide detectors are a good safeguard. They warn you if there is dangerous carbon monoxide in your home, due to incomplete combustion.

Figure 14.4 The yellow flame of this candle indicates that incomplete combustion is occurring. Carbon, $C_{(s)}$, emits light energy in the yellow wavelength region of the visible spectrum.

Balancing Combustion Equations

Have you ever seen a construction worker using an oxyacetylene torch? (See Figure 14.6.) A brilliant white light comes from the torch as it cuts through steel. The intense heat that is associated with this flame comes from the combustion of ethyne, a very common alkyne. Ethyne is also known as *acetylene*.

Figure 14.3 Propane burning in a propane torch: A blue flame indicates that complete combustion is occurring.

How do you write the balanced equation for the complete combustion of acetylene (ethyne)? Complete hydrocarbon combustion reactions follow a general format:

hydrocarbon + oxygen → carbon dioxide + water vapour

You can use this general format for the complete combustion of any hydrocarbon, no matter how large or how small. For example, both acetylene and propane burn completely to give carbon dioxide and water vapour. Each hydrocarbon, however, produces different amounts, or mole ratios, of carbon dioxide and water.

You have seen, written, and balanced several types of reaction equations so far in this textbook. In the following sample problem, you will learn an easy way to write and balance hydrocarbon combustion equations.

Sample Problem

Complete Combustion of Acetylene

Problem

Write the balanced equation for the complete combustion of acetylene (ethyne).

What Is Required?

You need to write the equation. Then you need to balance the atoms of the reactants and the products.

What Is Given?

You know that acetylene (ethyne) and oxygen are the reactants. Since the reaction is a complete combustion reaction, carbon dioxide and water vapour are the products.

Plan Your Strategy

Step 1 Write the equation.

Step 2 Balance the carbon atoms first.

Step 3 Balance the hydrogen atoms next.

Step 4 Balance the oxygen atoms last.

Act on Your Strategy

Step 1 Write the chemical formulas and states for the reactants and products.

ethyne	+	oxygen	→	carbon dioxide	+	water vapour
$C_2H_{2(g)}$	+	$O_{2(g)}$	→	$CO_{2(g)}$	+	$H_2O_{(g)}$

Step 2 Balance the carbon atoms first.

$$C_2H_{2(g)} + O_{2(g)} \rightarrow 2CO_{2(g)} + H_2O_{(g)}$$

(2 carbons) (2 carbons)

Continued ...

FROM PAGE 581

Step 3 Balance the hydrogen atoms next.

$$C_2H_{2(g)} + O_{2(g)} \rightarrow 2CO_{2(g)} + H_2O_{(g)}$$

(2 hydrogens) (2 hydrogens)

Step 4 Balance the oxygen atoms last.

The product coefficients are now set. Therefore, count the total number of oxygen atoms on the product side. Then place an appropriate coefficient in front of the reactant oxygen.

$$C_2H_{2(g)} + ?O_{2(g)} \rightarrow 2CO_{2(g)} + H_2O_{(g)}$$

(4 + 1 oxygens)

You end up with an *odd* number of oxygen atoms on the product side of the equation. When this happens, use a fractional coefficient so that the reactant oxygen balances. Here you have

$$C_2H_{2(g)} + \frac{5}{2}O_{2(g)} \rightarrow 2CO_{2(g)} + H_2O_{(g)}$$

You may prefer to balance the equation with whole numbers. If so, multiply everything by a factor that is equivalent to the denominator of the fraction. Since the fractional coefficient of $O_{2(g)}$ has a 2 in the denominator, multiply *all* the coefficients by 2 to get whole number coefficients.

$$2C_2H_{2(g)} + 5O_{2(g)} \rightarrow 4CO_{2(g)} + 2H_2O_{(g)}$$

Check Your Solution

The same number of carbon atoms appear on both sides of the equation.

The same number of hydrogen atoms appear on both sides of the equation.

The same number of oxygen atoms appear on both sides of the equation.

Sample Problem

Incomplete Combustion of 2,2,4-Trimethylpentane

Problem

2,2,4-trimethylpentane is a major component of gasoline.
Write one possible equation for the incomplete combustion of 2,2,4-trimethylpentane.

What Is Required?

You need to write the equation for the incomplete combustion of 2,2,4-trimethylpentane. Then you need to balance the atoms of the reactants and the products. For an incomplete combustion reaction, more than one balanced equation is possible.

Continued ...

FROM PAGE 582

What is given?

You know that 2,2,4-trimethylpentane and oxygen are the reactants. Since the reaction is an incomplete combustion reaction, the products are unburned carbon, carbon monoxide, carbon dioxide, and water vapour.

Plan Your Strategy

Draw the structural diagram for 2,2,4-trimethylpentane to find out how many hydrogen and oxygen atoms it has. Then write the equation and balance the atoms. There are many carbon-containing products but only one hydrogen-containing product, water. Therefore, you need to balance the hydrogen atoms first. Next balance the carbon atoms, and finally the oxygen atoms.

Act on Your Strategy

$$\begin{array}{ccccccccc} & & CH_3 & & & & CH_3 & & \\ & & | & & & & | & & \\ CH_3 & — & C & — & CH_2 & — & CH & — & CH_3 + O_2 \rightarrow C + CO + CO_2 + H_2O \\ & & | & & & & & & \\ & & CH_3 & & & & & & \\ & & (C_8H_{18}) & & & & & & \end{array}$$

Count the carbon and hydrogen atoms, and balance the equation. Different coefficients are possible for the carbon-containing product molecules. Follow the steps you learned in the previous sample problem to obtain the balanced equation shown below.

$$\begin{array}{ccccccccc} & & CH_3 & & & & CH_3 & & \\ & & | & & & & | & & \\ CH_3 & — & C & — & CH_2 & — & CH & — & CH_3 + \frac{15}{2}O_2 \rightarrow 4C + 2CO + 2CO_2 + 9H_2O \\ & & | & & & & & & \\ & & CH_3 & & & & & & \end{array}$$

or

$$2\left(\begin{array}{ccccccccc} & & CH_3 & & & & CH_3 & & \\ & & | & & & & | & & \\ CH_3 & — & C & — & CH_2 & — & CH & — & CH_3 \\ & & | & & & & & & \\ & & CH_3 & & & & & & \end{array}\right) + 15O_2 \rightarrow 8C + 4CO + 4CO_2 + 18H_2O$$

Check Your Solution

The same number of carbon atoms appear on both sides of the equation.

The same number of hydrogen atoms appear on both sides of the equation.

The same number of oxygen atoms appear on both sides of the equation.

Continued ...

Continued ...

FROM PAGE 583

Practice Problems

1. The following equation shows the combustion of 3-ethyl-2,5-dimethylheptane:

$$C_{11}H_{24} + 17O_2 \rightarrow 11CO_2 + 12H_2O$$

   (a) Does this equation show *complete* or *incomplete* combustion?

   (b) Draw the structural formula for 3-ethyl-2,5-dimethylheptane.

2. (a) Write a balanced equation for the complete combustion of pentane, C_5H_{12}.

   (b) Write a balanced equation for the complete combustion of octane, C_8H_{18}.

   (c) Write two possible balanced equations for the incomplete combustion of ethane, C_2H_6

3. (a) The flame of a butane lighter is usually yellow, indicating incomplete combustion of the gas. Write a balanced chemical equation to represent the incomplete combustion of butane in a butane lighter. Use the condensed structural formula for butane.

   (b) If you supplied enough oxygen, the butane would burn with a blue flame. Write a balanced chemical equation for the complete combustion of butane.

4. The paraffin wax in a candle burns with a yellow flame. If it had sufficient oxygen to burn with a blue flame, it would burn rapidly and release a lot of energy. It might even be dangerous! Write the balanced chemical equation for the complete combustion of candle wax, $C_{25}H_{52(s)}$.

5. 4-propyldecane burns to give solid carbon, water vapour, carbon monoxide, and carbon dioxide.

   (a) Draw the structural formula for 4-propyldecane.

   (b) Write two different balanced equations for the reaction described in this problem.

   (c) Name the type of combustion. Explain.

Large quantities of acetylene are produced each year by an inexpensive process that combines calcium carbide and water. In the next investigation, you will use this process to produce your own acetylene.

Investigation 14-A

SKILL FOCUS

Predicting

Performing and recording

Analyzing and interpreting

Communicating results

The Formation and Combustion of Acetylene

In this investigation, you will produce acetylene (ethyne) gas by mixing solid calcium carbide with water.

$$CaC_{2(s)} + 2H_2O_{(\ell)} \rightarrow C_2H_{2(g)} + Ca(OH)_{2(s)}$$

Then you will combine the acetylene with different quantities of air to determine the best reaction ratio for complete combustion. Make sure that you follow all the safety precautions given in this investigation and by your teacher.

Question

What is the ideal ratio of fuel to air for the complete combustion of acetylene (ethyne) gas?

Prediction

Air contains 20% oxygen, $O_{2(g)}$. How much air do you think is needed for the complete combustion of acetylene gas? Predict which proportion will react best:

- $\frac{1}{2}$ acetylene to $\frac{1}{2}$ air
- $\frac{1}{3}$ acetylene to $\frac{2}{3}$ air
- $\frac{1}{5}$ acetylene to $\frac{4}{5}$ air
- $\frac{1}{10}$ acetylene to $\frac{9}{10}$ air.

Safety Precautions

- Be careful of the flames from Bunsen burners. Check that there are no flammable solvents close by. If your hair is long, tie it back. Confine loose clothing.

Materials

4 test tubes (100 mL)
4 rubber stoppers
grease pencil
ruler
400 mL beaker
tweezers
matches (or Bunsen burner and splints)
1 or 2 calcium carbide chips
phenolphthalein indicator
limewater
medicine dropper
distilled water
test tube tongs

Procedure

1. Make a table to record your observations. Give your table a title.
2. Mark each test tube with a grease pencil to indicate one of the following volumes: $\frac{1}{2}$, $\frac{1}{3}$, $\frac{1}{5}$, and $\frac{1}{10}$. To find out where to mark the test tube, measure the total length of the test tube with a ruler. Then multiply the length by the appropriate fraction. Measure the fraction from the *bottom* of the test tube.
3. Fill the four test tubes completely with distilled water.
4. Invert the four test tubes in a 400 mL beaker, half full of distilled water. Make sure that the test tubes stay completely full.
5. Add three to five drops of phenolphthalein to the water in the beaker.
6. Using tweezers, drop a small chip of calcium carbide into the water. **CAUTION** Do not touch calcium carbide with your hands!
7. Capture the gas that is produced by holding the test tube marked $\frac{1}{2}$ over the calcium carbide chip. Fill the test tube to the $\frac{1}{2}$ mark with the gas. Remove the test tube from the beaker, still inverted. Let the water drain out. Air will replace the water and mix with the gas in the test tube. Insert a rubber stopper. Invert the test tube a few times to mix the acetylene gas with the air in the test tube.

8. Repeat step 7 with the other three test tubes. Fill each test tube to the volume that you marked on it: $\frac{1}{3}$, $\frac{1}{5}$, or $\frac{1}{10}$. After filling each test tube, remove it from the water and insert a rubber stopper.

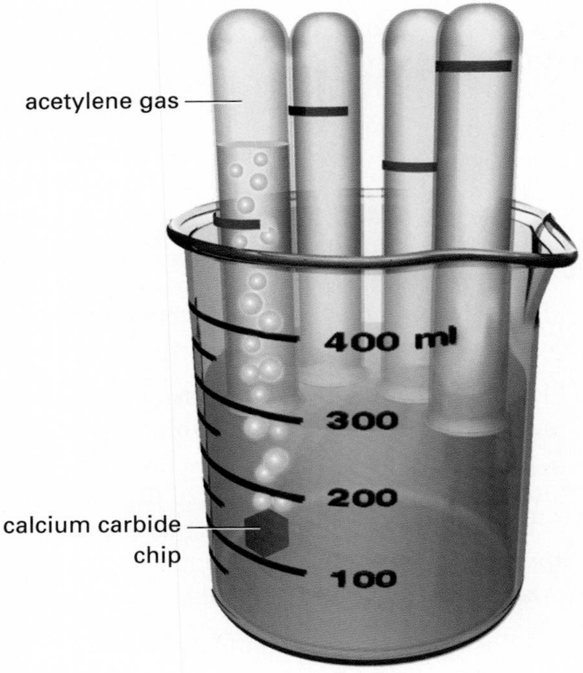

Figure 14.5

9. Invert the first test tube ($\frac{1}{2}$ full). Light a match or a splint. Using test tube tongs, hold the test tube inverted, and take out the stopper. Ignite the gas in the test tube by holding the lighted match or splint near the mouth of the test tube. CAUTION If you are using a Bunsen burner and wooden splints, take appropriate safety precautions. Extinguish burning splints by immersing them in water. Be aware of the Bunsen burner flame. Make sure that long hair is tied back and loose clothing is confined.

10. *Immediately* after the reaction in the test tube occurs, use the medicine dropper to add about 1 mL of limewater to the test tube. Stopper the test tube and shake it. CAUTION The mouth of the test tube may be hot.

11. Record your observations of the gas when it was ignited. Record your observations of what happened when you added the limewater. Describe any residue left on the test tube.

12. Repeat steps 9 to 11 with the other test tubes in order: $\frac{1}{3}$, $\frac{1}{5}$, and then $\frac{1}{10}$ full.

13. Dispose of all chemical materials as instructed by your teacher.

Analysis

1. What happened to the phenolphthalein indicator during the production of the gas? Explain your observation. **Note**: phenolphthalein is an acid/base indicator.

2. What products may have formed during the combustion of the gas? Support your answers with experimental evidence. **Note**: Limewater reacts with carbon dioxide to produce a milky white solid.

Conclusions

3. (a) Write a balanced chemical equation for the incomplete combustion of acetylene gas.

   (b) Write a balanced chemical equation for the complete combustion of acetylene gas.

4. The air that we breathe is approximately 20% oxygen. Think about the reaction you just wrote for the complete combustion of acetylene. Which ratio in this investigation ($\frac{1}{2}$, $\frac{1}{3}$, $\frac{1}{5}$, or $\frac{1}{10}$) allowed the closest amount of oxygen needed for complete combustion? Support your answer with calculations. Do the observations you made support your answer? Explain.

Applications

5. An automobile engine requires a carburetor or fuel injector to mix the fuel with air. The fuel and air must be mixed in a particular ratio to achieve maximum efficiency in the combustion of the fuel. What might happen if the fuel and air mixture is too rich (if there is too much fuel)?

6. What does the limewater test indicate? Write the balanced chemical equation for the limewater test. **Hint**: Limewater is a dilute solution of calcium hydroxide. A carbonate forms.

In this section, you were introduced to the complete and incomplete combustion of hydrocarbons. You learned that the complete combustion of a hydrocarbon produces water and carbon dioxide. You also learned that the incomplete combustion of a hydrocarbon produces additional products, such as unburned carbon and dangerous carbon monoxide. In the investigation, you had the chance to make and combust a hydrocarbon.

In the next section, you will learn about an important factor of combustion reactions: energy. The combustion of hydrocarbons produces a large amount of energy. This is why they are so useful as fuels. How can you include energy as part of a combustion or other equation? How can you calculate the energy released by fossil fuels? You will learn the answers to these questions in the rest of this chapter.

Section Review

1. K/U What is the difference between incomplete and complete hydrocarbon combustion reactions?

2. K/U Explain why you would usually write (g) for the state of the product water in these combustion reactions. When might you identify the state of water as liquid?

3. (a) K/U Write the balanced equation for the complete combustion of heptane, C_7H_{16}.

   (b) K/U Write a balanced equation for the incomplete combustion of 1-pentene, C_5H_{10}.

4. K/U Natural gas is mainly methane gas. If you have a natural gas furnace, stove, or water heater in your home, you must ensure that these appliances are always running at peak efficiency. In other words, the methane gas should undergo complete combustion so that carbon monoxide is not produced. Write a balanced chemical equation to show the complete combustion of methane gas. Write a second balanced equation to show the incomplete combustion of methane gas.

5. C All hydrocarbons have the chemical property of combustion in the presence of oxygen. How do the complete combustion reactions of methane, ethane, and propane differ? How are they similar? **Hint:** Compare the balanced equations for the complete combustion of each.

6. I The complete combustion of ethane is given by the following unbalanced equation:

$$C_2H_{6(g)} + O_{2(g)} \rightarrow CO_{2(g)} + H_2O_{(g)}$$

   (a) Balance this equation.

   (b) If one mole of ethane is combusted, how many grams of water vapour are produced?

   (c) Assume that air contains 20% oxygen gas. What volume of air at STP is needed for the complete combustion of one mole of ethane?

14.2 Thermochemical Equations

Section Preview/ Specific Expectations

In this section, you will

- **write** thermochemical equations
- **relate** bond breaking and bond making to endothermic and exothermic energy changes
- **communicate** your understanding of the following terms: *exothermic, thermochemical equation, endothermic, bond energy*

In the last section, you learned about acetylene, an important fuel in our society (Figure 14.6). You balanced the equation for the complete combustion of acetylene. You then produced acetylene in an investigation. What you did not consider, however, was the most useful product that our society gets from acetylene: heat energy. How can you represent the heat that is released during combustion as part of a chemical equation?

For now, you will use the word "energy" to represent the heat in an equation. In the next section, you will calculate numerical values for this energy and use these energy values in chemical equations.

In previous science courses, you studied reactions that involve a change in energy. Figure 14.7 reviews some important terms.

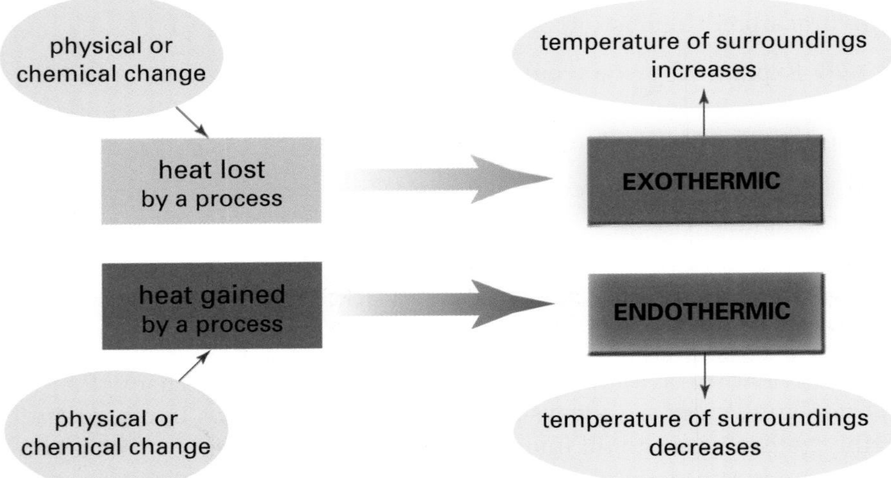

Figure 14.7

Figure 14.6 **The combustion of acetylene (ethyne) in an oxyacetylene torch produces the highest temperature (about 3300°C) of any known mixture of combustible gases. Metal workers can use the heat from this combustion to cut through most metal alloys.**

An **exothermic** reaction gives off heat. Combustion reactions are exothermic reactions because they produce heat. *Since the energy is a product of the reaction, it is shown on the product side of the equation.*

The combustion of acetylene is a good example of an exothermic reaction:

$$2C_2H_{2(g)} + 5O_{2(g)} \rightarrow 4CO_{2(g)} + 2H_2O_{(g)} + \text{energy}$$

When the energy is written as part of the chemical equation, the equation is called a **thermochemical equation.**

An **endothermic** reaction absorbs heat energy. Energy must be added to the reactants for an endothermic reaction to occur. *Since the energy is needed as a reactant, it is shown on the reactant side of the equation.*

Have you ever noticed that perfume or rubbing alcohol feels cool on your skin as it evaporates? The physical process of evaporation is endothermic. Energy is taken away from the surface of your skin, so you feel cool. The energy is added to the liquid alcohol or perfume solvent to make it a gas. The following equation shows the endothermic evaporation of isopropanol (a type of rubbing alcohol) from a liquid to a gas:

$$\text{energy} + CH_3CHOHCH_{3(\ell)} \rightarrow CH_3CHOHCH_{3(g)}$$

How can you tell that this process is not a chemical reaction?

Bond Breaking and Bond Making

The formation of acetylene (ethyne) gas from its elements is an endothermic reaction. The combustion of acetylene, however, is exothermic. In fact, it releases enough heat energy to cut steel! How can you explain the formation and combustion of acetylene in terms of bonds being broken and made? The answer to this question is fundamental to your understanding of the energy changes that occur in chemical reactions.

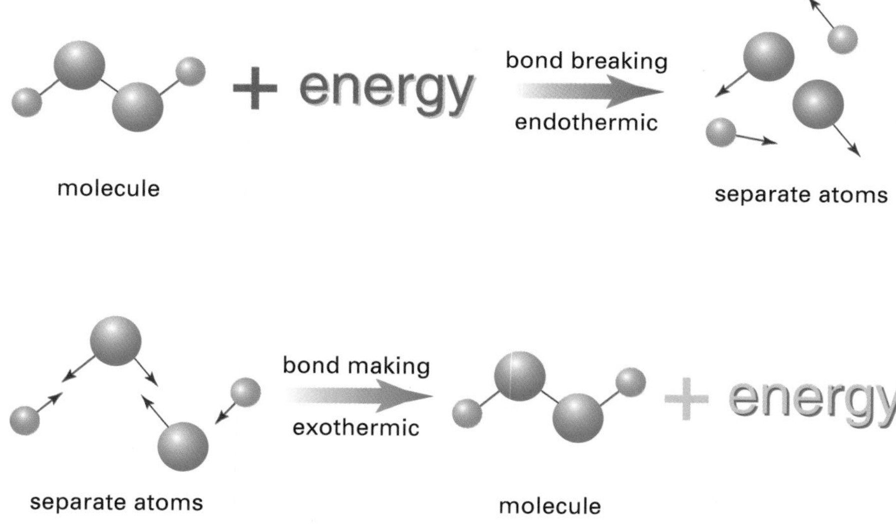

Figure 14.8 **This illustration shows bonds being broken and made during a chemical reaction. If the bonds are strong, there is a large change in energy. If the bonds are weak, there is a small change in energy.**

A chemical bond is caused by the attraction between the electrons and nuclei of two atoms. Energy is needed to break a chemical bond, just like energy is needed to break a link in a chain. On the other hand, making a chemical bond releases energy. The strength of a bond depends on how much energy is needed to break the bond. (see Figure 14.8.)

A specific amount of energy is needed to break each type of bond. When the same type of bond is formed, the same specific amount of energy is released. The energy that is absorbed or released when breaking or making a bond is called the **bond energy**. Bond energy is usually measured in kilojoules (kJ). Table 14.1 shows some average bond energies.

From the table, you can see that 347 kJ of energy is needed to break one mole of C—C bonds in a sample of propane or any other carbon. Similarly, 347 kJ of energy is released if one mole of C—C bonds forms in a sample of butane.

Every chemical reaction involves both bond breaking (reactant bonds are broken) *and* bond making (product bonds are formed). Since there are different types of bonds inside the reactant and product molecules, the bond breaking and bond making energies are different. This results in a net amount of energy for each reaction.

The next sample problem compares bond breaking and bond making in a combustion reaction.

mind STRETCH

(a) Use a molecular model kit to model the formation of butane from its elements: $4C_{(s)} + 5H_{2(g)} \rightarrow C_4H_{10(g)}$ Compare the bonds you break with the bonds you form.

(b) Using Table 14.1, estimate the energy needed to break the bonds of 5 mol of hydrogen gas. Compare this energy with the energy produced by making the bonds of 1 mol of butane gas. Predict whether the formation of butane is exothermic or endothermic.

(c) Look up "heat of formation" in a reference book such as *The CRC Handbook of Chemistry and Physics* to find the actual energy of this reaction. Is it exothermic (negative) or endothermic (positive)?

Note: Since bond energies are only a way of *estimating* the energy produced by the formation of a compound, your answer will not agree with the recognized value. Look at the Thinking Critically question in the Section Review to find out how to estimate the energy of a formation reaction more accurately.

Table 14.1 Average Bond Energies

Type of bond	Average energy (kJ/mol)
H—H	432
C—C	347
C═C	614
C≡C	839
C—H	413
C—O	358
C═O	745
O—H	467
O═O (in O_2)	498

Sample Problem

The Combustion of Acetylene

Problem

Consider bond breaking and bond making to explain why the combustion of acetylene is exothermic. Then write a thermochemical equation for the reaction, using the following balanced equation:

$$2C_2H_{2(g)} + 5O_{2(g)} \rightarrow 4CO_{2(g)} + 2H_2O_{(g)}$$

What Is Required?

You need to describe what happens when the reactant bonds are broken and what happens when the product bonds are formed. You need to compare the energy that is absorbed when the reactant bonds are broken with the energy that is released when the product bonds are formed. Then you need to write a thermochemical equation, using the word "energy."

What Is Given?

You know that energy is absorbed when bonds are broken and energy is released when bonds are formed. You also know that the equation is exothermic. (Overall, energy is released in this reaction.)

Plan Your Strategy

Step 1 Describe what happens when the reactant bonds are broken.

Step 2 Describe what happens when the product bonds are formed.

Step 3 Compare the energy that is absorbed when the reactant bonds are broken with the energy that is released when the product bonds are formed.

Step 4 Write the thermochemical equation, using the word "energy."

Act on Your Strategy

Step 1 The reactant bonds are broken. This process absorbs enough energy to split the reactants into separate atoms.

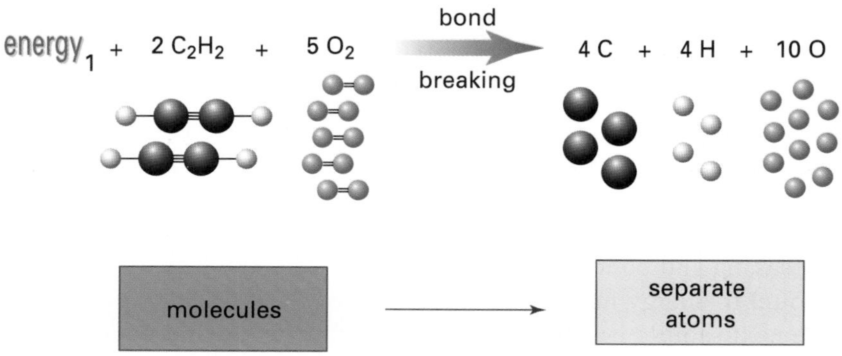

PROBLEM TIP

- If the energy that is needed to break the reactant bonds is greater than the energy that is released when the product bonds form, the reaction is endothermic.
- If the energy that is needed to break the reactant bonds is less than the energy that is released when the product bonds form, the reaction is exothermic.
- If the reaction is endothermic, the energy is on the left side of the equation.
- If the reaction is exothermic, the energy is on the right side of the equation.

Continued ...

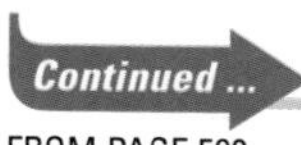

FROM PAGE 590

Step 2 The product bonds are made. This process releases energy as the product molecules are formed.

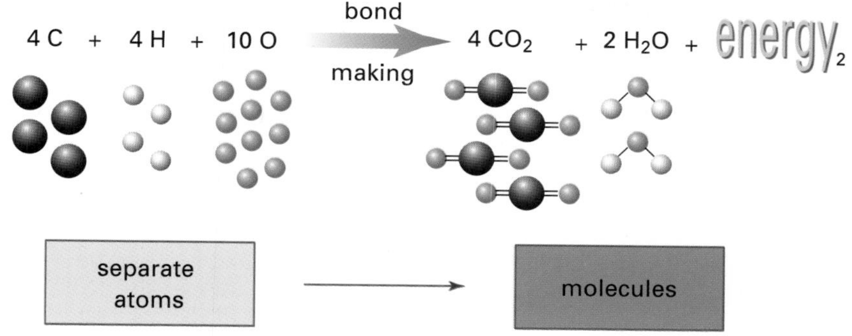

Step 3 The combustion of acetylene is exothermic. Therefore, the energy that is used when the reactant bonds are broken must be less than the energy that is released when the product bonds are formed.

$$\text{energy}_1 < \text{energy}_2$$

Step 4 The thermochemical equation is

$$2C_2H_{2(g)} + 5O_{2(g)} \rightarrow 4CO_{2(g)} + 2H_2O_{(g)} + \text{energy}$$

Check Your Solution

The equation is exothermic, so the energy is on the product side of the equation.

Practice Problems

6. The formation of propane from its elements is an exothermic reaction. The combustion of propane is also exothermic.

(a) Write the balanced thermochemical equation for the formation of propane.

(b) Write the balanced thermochemical equation for the combustion of propane. (The balanced equation is on page 580.)

(c) Consider the combustion of propane. Compare the energies of bond breaking and bond making to explain why the reaction is exothermic.

7. (a) Explain why the formation of ethene, $C_2H_{4(g)}$, from its elements is endothermic, while its combustion is exothermic.

(b) Write the balanced thermochemical equations for the formation and the combustion of ethene.

PROBLEM TIP

Remember that different types of bonds have different bond energies. The bonds that are broken in the reactants are different from the bonds that are formed in the products. The net energy for the entire reaction is the difference between energy_1 and energy_2.

So far, you have been using the word "energy" in thermochemical equations to represent the net energy. It is preferable to have a numerical value for the amount of energy when talking about endothermic and exothermic processes. In the next section, you will see how the net energy in a process can be measured. You will even measure some energy values yourself!

Section Review

1 K/U Choose the correct term in each box to describe the given thermochemical equation. Name the organic compounds.

(a)

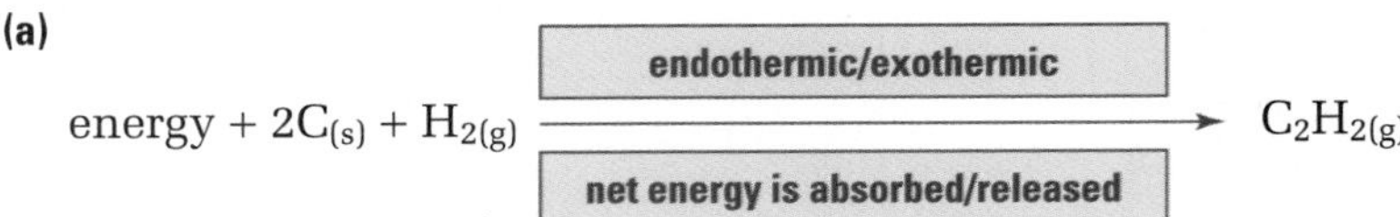

(b)

endothermic/exothermic

$$CH_{4(g)} + 2O_{2(g)} \longrightarrow CO_{2(g)} + 2H_2O_{(g)} + \text{energy}$$

net energy is absorbed/released

2 K/U If an overall reaction is endothermic, which process involves more energy: breaking the reactant bonds or making the product bonds?

3 **(a)** C When energy is absorbed in a reaction to break the bonds in the reactants, where does the energy go?

(b) C When energy is released to form the bonds in the products, where does the energy go?

4 MC Flour, sugar, eggs, and milk are combined and baked to produce cookies.

(a) Is this reaction exothermic or endothermic? Explain.

(b) Write a word equation that includes the word "energy" to describe this reaction.

5 K/U In the MindStretch on page 589, you used bond energies to estimate the energy of a reaction. Your estimated value was very different from the actual value, however. This is because solid carbon must become gaseous *before* reacting to form a hydrocarbon. The change of state takes additional energy.

$$C_{(s)} + 715 \text{ kJ} \rightarrow C_{(g)}$$

Use the information above, along with the bond energies in Table 14.1, to estimate the energy needed to form acetylene, $C_2H_{2(g)}$, from its elements.

$$2C_{(s)} + H_{2(g)} \rightarrow C_2H_{2(g)}$$

Remember to consider the breaking of the H—H bonds, and the forming of the C—H and C≡C bonds. Compare your answer with the recognized value, 226.7 kJ/mol.

Measuring Energy Changes

14.3

You have probably had plenty of experience with heating and cooling materials around you. You know how to cool yourself by taking a cold shower. You know how to heat hot chocolate on a stove. However, have you ever stopped to think about what energy changes are occuring? Where does the energy come from, and where does it go?

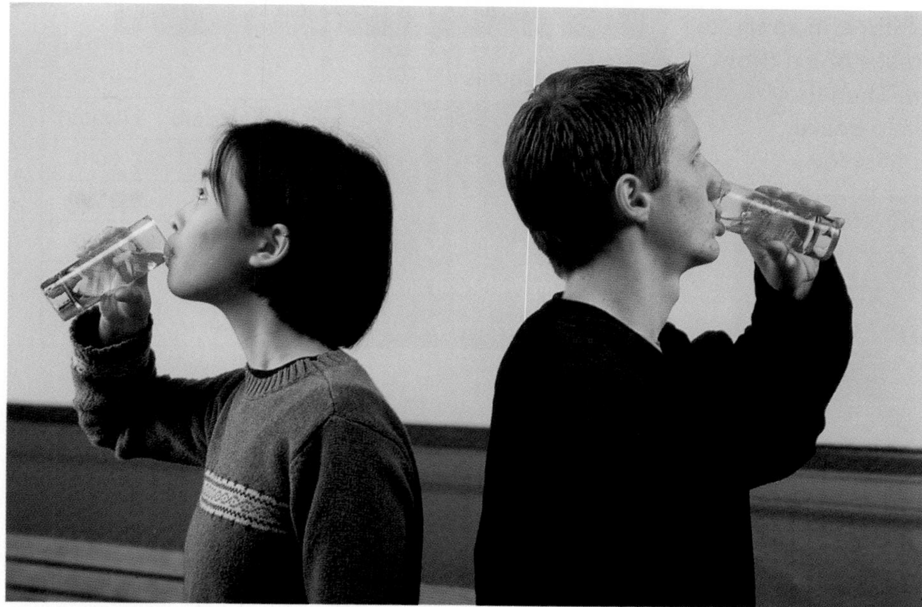

Figure 14.9 **Suppose that you are very thirsty, but the tap water is not cold enough. How can you get the water as cold as you want it to be? Your experience tells you to add ice to the water. How does the ice cool the water?**

Any time that something is cooled or heated, a change in thermal energy occurs. **Thermal energy** is the kinetic energy of particles of matter. In a glass of tap water with ice (as shown in Figure 14.9), the water started out warmer than the ice. Energy was transferred from the water to the ice. As the ice absorbed the energy, the ice melted. As the water lost energy, the water cooled.

These examples involve heat. **Heat** is the transfer of thermal energy between objects with different temperatures. In this section, you will study how to measure the amount of thermal energy that is transferred during a process involving an energy change.

Important Factors in Energy Measurement

How can you measure the change in energy when you add ice to water? **Temperature** is a measure of the average kinetic energy of a system. You have already had experience using a thermometer to measure temperature. By placing a thermometer in the water, you can monitor the drop in temperature when the ice is added. Temperature is an important factor in heat energy changes. What other factors are important? Work through the following ThoughtLab to find out.

Section Preview/ Specific Expectations

In this section, you will

- **explain** how mass, specific heat capacity, and change in temperature of an object determine how much heat the object gains or loses
- **solve** problems using the equation $Q = mc\Delta T$
- **communicate** your understanding of the following terms: *thermal energy, heat, temperature, ΔT, specific heat capacity*

CHECKPOINT

What other forms of energy can you think of? Make a list, starting with thermal energy.

mind STRETCH

How does drinking hot chocolate warm you? Describe the energy transfer that takes place.

ThoughtLab  Factors in Heat Transfer

Two students performed an experiment to determine what factors need to be considered when determining the amount of heat lost or gained by a substance undergoing an energy change. They set up their experiment as follows.

Part A

The students placed two different masses of water, at the same initial temperature, in separate beakers. They placed an equal mass of ice (from the same freezer) in each beaker. Then they monitored the temperature of each beaker. Their results are listed in the table below.

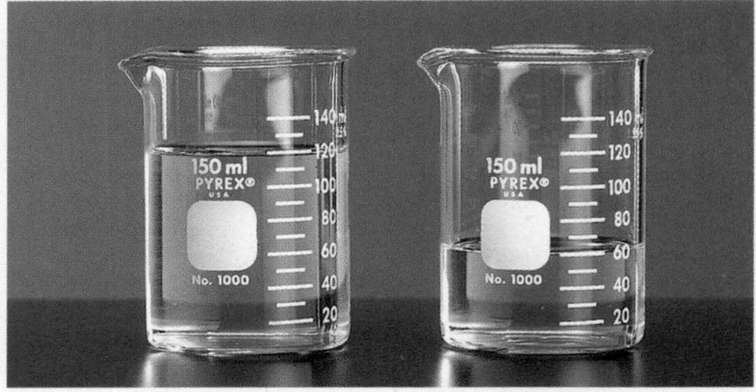

Different Masses of Water

Beaker	1	2
Mass of water (g)	60.0	120.0
Initial temperature of water (°C)	26.5	26.5
Mass of ice added (g)	10.0	10.0
Final temperature of mixture (°C)	9.7	17.4
Temperature change (°C)	16.8	9.1

Part B

The students placed equal masses of canola oil and water, at the same initial temperature, in separate beakers. They placed equal masses of ice (from the same freezer) in the two beakers. Then they monitored the temperature of each beaker. Their results are listed in the following table.

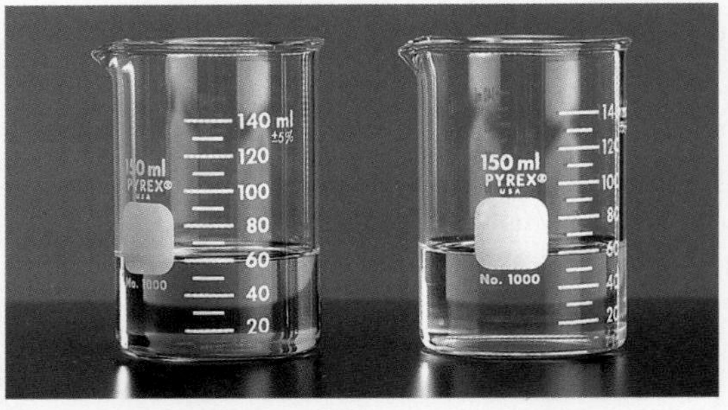

Different Liquids

Beaker	1 (canola oil)	2 (water)
Mass of liquid (g)	60.0	60.0
Initial temperature of liquid (°C)	35.0	35.0
Mass of ice added (g)	10.0	10.0
Final temperature of mixture (°C)	5.2	16.9
Temperature change (°C)	29.8	18.1

Procedure

1. For each part of the experiment, identify
   (a) the variable that was changed by the students (the manipulated variable)
   (b) the variable that changed as a result of changing the manipulated variable (the responding variable)
   (c) the variables that were kept constant to ensure a fair test (the controlled variables)
2. Interpret the students' results by answering the following questions.
   (a) If ice is added to two different masses of water, how does the temperature change?
   (b) If ice is added to two different liquids, how does the temperature change?

Analysis

Think about your interpretation of the students' experiment and the discussion prior to this ThoughtLab. What are three important factors to consider when measuring the thermal energy change of a substance?

The ThoughtLab gave you some insight into the factors that are important when measuring energy changes. How can you use these factors to calculate the amount of heat that is transferred? First you must examine each factor and determine its relationship to heat transfer. You will begin with the most obvious factor: temperature. Then you will look at how the mass of a substance affects the amount of heat it can store. Finally, you will look at the type of substance and how it affects heat transfer.

Temperature

Figure 14.10 How do you control the temperature of your body?

In the summer, your body temperature is fairly close to the temperature of your surroundings. You do not need to wear extra clothing to keep warm. When winter hits with fierce winds and cold temperatures, however, dressing warmly becomes a necessity. The temperature of your surroundings is now much colder than your body temperature. Heat is transferred from your body to your surroundings, making you feel cold. The extra clothing helps to minimize heat loss from your body. (See Figure 14.10.)

Temperature is directly related to heat transfer. A large change in temperature indicates a large energy change. A small change in temperature indicates a small energy change. Therefore, temperature is an important factor when calculating heat transfer. The temperature variable that is used is the *change in temperature*. This is symbolized by ΔT.

Mass of Substance

Did you know that 70% of Earth's surface is covered with water? This enormous mass of water absorbs and releases tremendous amounts of heat energy. Water makes our climate more moderate by absorbing heat in hot weather and releasing heat in cold weather. The greater the mass of the water, the greater the amount of heat it can absorb and release. (Areas without much water, such as deserts, experience huge variations in temperature.) Therefore, mass is directly related to heat transfer. Mass is a variable in the calculation of heat energy. It is symbolized by a lower-case m.

Type of Substance

In the ThoughtLab on, you probably noticed that the amount of heat being transferred depends on the type of substance. When you added equal masses of ice to the same mass of oil and water, the temperature change of the oil was almost double the temperature change of the water. "Type of substance" cannot be used as a variable, however, when calculating energy changes. Instead, we use a variable that reflects the individual nature of different substances: specific heat capacity. The **specific heat capacity** of a substance is the amount of energy, in joules (J), that is required to change one gram (g) of the substance by one degree Celsius (°C). The specific heat capacity of a substance reflects how well the substance can store energy. A substance with a large specific heat capacity can absorb and release more energy than a substance with a smaller specific heat capacity. The symbol that is used for specific heat capacity is a lower-case c. The units are J/g·°C.

The specific heat capacity of water is relatively large: 4.184 J/g·°C. This value helps to explain how water can absorb and release enough energy to moderate Earth's temperature. Examine the values in Table 14.2. Notice that the specific heat capacities of most substances are much lower than the specific heat capacity of water.

Table 14.2 Specific Heat Capacities of Various Substances

Substance	Specific heat capacity (J/g·°C at 25°C)
Elements	
aluminum	0.900
carbon (graphite)	0.711
copper	0.385
gold	0.129
hydrogen	14.267
iron	0.444
Compounds	
ammonia (liquid)	4.70
ethanol	2.46
water (solid)	2.01
water (liquid)	4.184
water (gas)	2.01
Other materials	
air	1.02
concrete	0.88
glass	0.84
granite	0.79
wood	1.76

You have just considered three variables: change in temperature (ΔT), mass (m), and type of substance, which is characterized by specific heat capacity (c). How can you combine these variables into a formula to calculate heat transfer (Figure 14.11)?

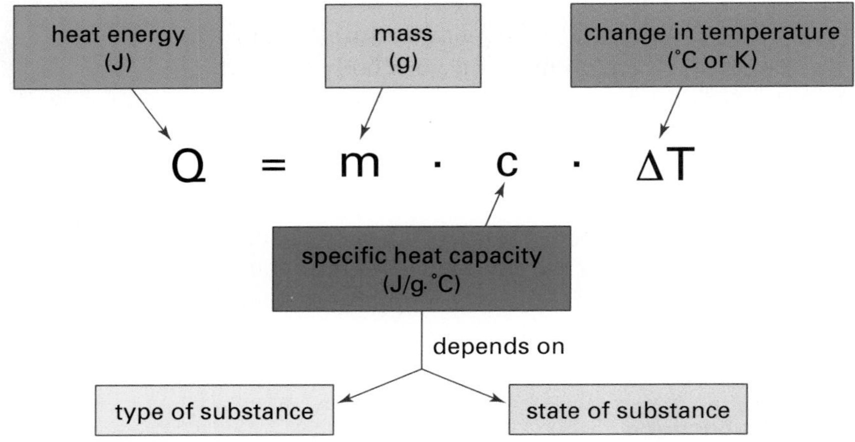

Figure 14.11 **Use this formula to calculate heat (Q) transfer.**

How do you solve heat energy problems using $Q = mc\Delta T$? Go back to the ThoughtLab. Some of the data in this ThoughtLab can be used to illustrate the calculation of heat transfer, as shown below.

Sample Problem

Heat Transferred From Water to Ice

Problem

In the ThoughtLab on page 594, 10.0 g of ice was added to 60.0 g of water. The initial temperature of the water was 26.5°C. The final temperature of the mixture was 9.7°C. How much heat was lost by the water?

What Is Required?

You need to calculate the amount of heat (Q) that was lost by the water.

What Is Given?

You know the mass of the water. You also know the initial and final temperatures of the water.

Mass of water (m) = 60.0 g

Initial temperature (T_i) = 26.5°C

Final temperature (T_f) = 9.7°C

Plan Your Strategy

You have enough information to solve this problem using $Q = mc\Delta T$. Use the initial and final temperatures to calculate ΔT. You need the specific heat capacity (c) of liquid water. This is given in Table 14.2 (4.184 J/g·°C). Because you are only concerned with the water, you will not use the mass of the ice.

Continued ...

mind STRETCH

Why does water have such a high specific heat capacity? Do some research to find out. **Hint**: Water's specific heat capacity has something to do with bonding.

Continued ...
FROM PAGE 596

Act on Your Strategy

Substitute the values into the following heat formula, and solve. Remember that $\Delta T = T_f - T_i$

$$
\begin{aligned}
Q &= mc\Delta T \\
&= (60.0\ \text{g})(4.184\frac{\text{J}}{\text{g}\cdot{}^\circ\text{C}})(9.7^\circ\text{C} - 26.5^\circ\text{C}) \\
&= -4217.472\ (\cancel{\text{g}})(\frac{\text{J}}{\cancel{\text{g}}\cdot\cancel{{}^\circ\text{C}}})(\cancel{{}^\circ\text{C}}) \\
&= -4217.472\ \text{J} \\
&= -4.22 \times 10^3\ \text{J (or } -4.22\ \text{kJ)}
\end{aligned}
$$

The water lost 4.22×10^3 J of heat.

Check Your Solution

The water lost heat, so the heat value should be negative.
Heat is measured in joules or kilojoules. Make sure that the units cancel out to give the appropriate unit for your answer.

Practice Problems

8. 100 g of ethanol at 25°C is heated until it reaches 50°C. How much heat does the ethanol gain? **Hint**: Find the specific heat capacity of ethanol in Table 14.2.

9. In Part A of the ThoughtLab on page 594, the students added ice to 120.0 g of water in beaker 2. Calculate the heat lost by the water. Use the information given for beaker 2, as well as specific heat capacities in Table 14.2.

10. A beaker contains 50 g of liquid at room temperature. The beaker is heated until the liquid gains 10°C. A second beaker contains 100 g of the same liquid at room temperature. This beaker is also heated until the liquid gains 10°C. In which beaker does the liquid gain the most thermal energy? Explain.

11. As the diagram on the next page illustrates, the sign of the heat value tells you whether a substance has lost or gained heat energy. Consider the following descriptions. Write each heat value, and give it the appropriate sign to indicate whether heat was lost or gained.

(a) In Part A of the ThoughtLab on page 594, the ice gained the heat that was lost by the water. When ice was added to 60.0 g of water, it gained 4.22 kJ of energy. When ice was added to 120.0 g of water, it gained 4.6 kJ of energy.

Continued ...

Continued ...
FROM PAGE 597

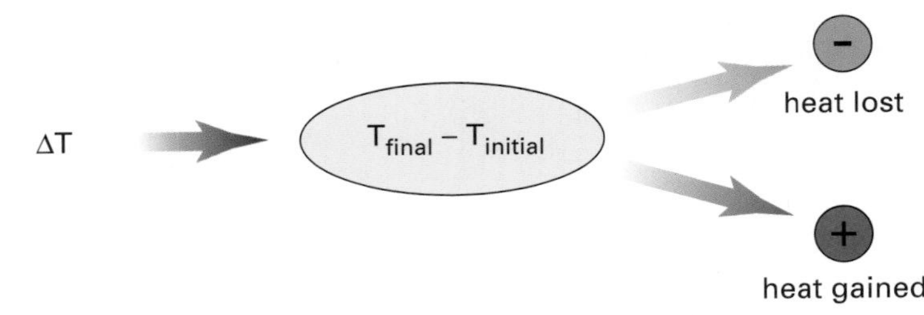

(b) When 2.0 L of water was heated over a campfire, the water gained 487 kJ of energy.

(c) A student baked a cherry pie and put it outside on a cold winter day. There was a change of 290 kJ of heat energy in the pie.

Math LINK

Heat values are often very large. Therefore it is convenient to use kilojoules (kJ) to calculate heat. How does this affect the units of the other variables in the heat equation? Does the specific heat capacity have to change? The following diagram shows how units must be modified in order to end up with kilojoules.

$$Q = m \cdot c \cdot \Delta T$$

Units: → $kJ = kg \cdot \frac{4.184\ kJ}{kg \cdot ^\circ C}\ ^\circ C$

mass-must be kg

Specific heat capacity
- must have kJ (top) and kg (bottom)
- Since "k" is on top andbottom, the number stays the same

In the Sample Problem, heat was *lost* by the water. Therefore the value of Q was *negative*. If the value of Q is *positive*, this indicates that heat is *gained* by a substance.

The heat equation $Q = mc\Delta T$ can be rearranged to solve for any of the variables. For example, in Part B of the ThoughtLab on page 594, ice was added to both canola oil and water. How can you use the information given in Part B to calculate the specific heat capacity of the canola oil?

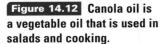

Figure 14.12 Canola oil is a vegetable oil that is used in salads and cooking.

Sample Problem

Calculating Specific Heat Capacity

Problem

Calculate the specific heat capacity of canola oil, using the information given in Part B of the ThoughtLab on page 594. Note that the ice gained 4.0×10^3 J of energy when it came in contact with the canola oil.

What Is Required?

You need to calculate the specific heat capacity (c) of the canola oil.

What Is Given?

From the ThoughtLab, you know the mass (m) and the initial and final temperatures of the canola oil.

Mass of oil (m) = 60.0 g

Initial temperature (T_i) = 35.0°C

Final temperature (T_f) = 5.2°C

Continued ...

Continued ...
FROM PAGE 598

You also know the amount of heat gained by the ice. This must be the same as the heat lost by the oil.
Heat gained by the ice = Heat lost by the canola oil = 4.0×10^3 J

Plan Your Strategy

Rearrange the equation $Q = mc\Delta T$ to solve for c. Then substitute the values for Q, m, and ΔT ($T_f - T_i$) into the equation.

Act on Your Strategy

$$c = \frac{Q}{m\Delta T}$$
$$= \frac{-4.0 \times 10^3 \text{ J}}{(60.0 \text{ g})(5.2^\circ\text{C} - 35.0^\circ\text{C})}$$
$$= 2.2437 \frac{\text{J}}{\text{g}\cdot^\circ\text{C}}$$
$$= 2.24 \frac{\text{J}}{\text{g}\cdot^\circ\text{C}}$$

Check Your Solution

The specific heat capacity should be positive, and it is. It should have the units $\frac{\text{J}}{\text{g}\cdot^\circ\text{C}}$.

Practice Problems

12. Solve the equation $Q = mc\Delta T$ for the following quantities.

(a) m

(b) c

(c) ΔT

13. You know that $\Delta T = T_f - T_i$. Combine this equation with the heat equation, $Q = mc\Delta T$, to solve for the following quantities.

(a) T_i (in terms of Q, m, c, and T_f)

(b) T_f (in terms of Q, m, c, and T_i)

14. How much heat is required to raise the temperature of 789 g of liquid ammonia, from 25.0°C to 82.7°C?

15. A solid substance has a mass of 250.00 g. It is cooled by 25.00°C and loses 4937.50 J of heat. What is its specific heat capacity? Look at Table 14.2 to identify the substance.

16. A piece of metal with a mass of 14.9 g is heated to 98.0°C. When the metal is placed in 75.0 g of water at 20.0°C, the temperature of the water rises by 28.5°C. What is the specific heat capacity of the metal?

17. A piece of gold ($c = 0.129$ J/g°C) with mass of 45.5 g and a temperature of 80.5°C is dropped into 192 g of water at 15.0°C. What is the final temperature of the system? (**Hint:** Use the equation $Q_w = -Q_g$.)

In this section, you learned that temperature, mass, and type of substance are all important factors to consider when measuring heat change. You saw how these factors can be combined to produce the heat equation: $Q = mc\Delta T$

In the next section, you will learn how to use a calorimeter to measure heat change. You will perform a heat transfer investigation of your own. Using the knowledge you have gained in this section, you will then calculate the specific heat capacity. As well, you will see why the Procedure that was used in the ThoughtLab could be improved.

Section Review

1. K/U Define the term "heat."
2. I What are three important factors to consider when measuring heat energy?
3. I In Part B of the ThoughtLab, 60.0 g of water was in beaker 2. The initial temperature of the water was 35.0°C, and the final temperature was 16.9°C.
   - **(a)** Calculate the heat that was lost by the water in beaker 2.
   - **(b)** Where did the heat go?
4. I When iron nails are hammered into wood, friction causes the nails to heat up.
   - **(a)** Calculate the heat that is gained by a 5.2 g iron nail as it changes from 22.0°C to 38.5°C. (See Table 14.2.)
   - **(b)** Calculate the heat that is gained by a 10.4 g iron nail as it changes from 22.0°C to 38.5°C.
   - **(c)** Calculate the heat that is gained by the 5.2 g nail if its temperature changes from 22.0°C to 55.0°C.
5. **(a)** I A 23.9 g silver spoon is put in a cup of hot chocolate. It takes 0.343 kJ of energy to change the temperature of the spoon from 24.5°C to 85.0°C. What is the specific heat capacity of solid silver?
   - **(b)** I The same amount of heat energy, 0.343 kJ, is gained by 23.9 g of liquid water. What is the temperature change of the water?
6. C The specific heat capacity of aluminum is 0.902 J/g°C. The specific heat capacity of copper is 0.389 J/g°C. The same amount of heat is applied to equal masses of these two metals. Which metal increases more in temperature? Explain.
7. K/U Explain why there is an energy difference between the following reactions.

$$CH_{4(g)} + 2O_{2(g)} \rightarrow CO_{2(g)} + 2H_2O_{(g)} + 802 \text{ kJ}$$
$$CH_{4(g)} + 2O_{2(g)} \rightarrow CO_{2(g)} + 2H_2O_{(\ell)} + 890 \text{ kJ}$$

The Technology of Heat Measurement

14.4

Figure 14.13 **How can you measure the energy in a substance?**

Section Preview/ Specific Expectations

In this section, you will

- **describe** some of the physical and chemical properties of hydrocarbons
- **apply** calorimetric techniques to the calculation of energy changes
- **gather** and **interpret** experimental data, and **solve** problems involving calorimetry and the combustion of hydrocarbons
- **communicate** your understanding of the following terms: *isolated system, calorimeter, calorimetry, bomb calorimeter, thermal equilibrium, heat of combustion, heat of solution, heat capacity, Calorie, potential energy*

In this unit, you have learned about the importance of hydrocarbons as fuels. Hydrocarbons are useful because of the energy that is released when they burn. It is often necessary, however, to know the amount of energy that is released. For example, engineers need to know how much energy is released from different fuels when they design an engine and choose an appropriate fuel. Firefighters need to know how much heat can be given off by different materials so they can decide on the best way to fight a specific fire. (See Figure 14.13.)

What about food—the fuel for your body? In order to choose an appropriate and balanced diet, you need to know how much energy each type of food releases when it is digested. Food energy is measured in Calories. (You will learn more about Calories later in this section.)

How do you measure the amount of energy that is produced? In this section, you will focus on measuring heat changes. You will learn about some technology and techniques to measure heat. You will then apply what you have learned by performing your own heat experiments.

Calorimetry

In the ThoughtLab in section 14.3, two students used beakers with no lids when they measured change in temperature.
The students assumed that energy was being exchanged only between the ice and the water. In fact, energy was also being exchanged with the surroundings. As a result, the data that the students obtained had a large experimental error. How could the students have prevented this error?

Much of the technology in our lives is designed to stop the flow of heat. Your home is insulated to prevent heat loss in the winter and heat gain in the summer. If you take hot soup to school for your lunch, you probably use a Thermos™ to prevent heat loss to the environment. Whenever there is a temperature difference between two objects, thermal energy flows from the hotter object to the colder object. When you measure the heat being transferred in a reaction or other process, you must minimize any heat that is exchanged with the surroundings.

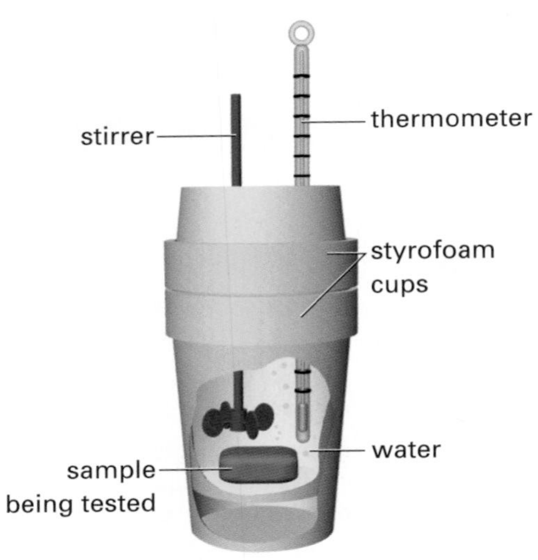

Figure 14.14 **A polystyrene (coffee cup) calorimeter usually consists of two nested polystyrene cups with a polystyrene lid, to provide insulation from the surroundings.**

To measure the heat flow in a process, you need an isolated system, like a Thermos™. An **isolated system** stops matter and energy from flowing in or out of the system. You also need a known amount of a substance, usually water. The water absorbs the heat that is produced by the process, or it releases heat if the process is endothermic. To determine the heat flow, you can measure the temperature change of the water. With its large specific heat capacity (4.184 J/ g·°C) and its broad temperature range (0°C to 100°C), liquid water can absorb and release a lot of energy.

Water, a thermometer, and an isolated system are the basic components of a calorimeter. A **calorimeter** is a device that is used to measure changes in thermal energy. (Figures 14.14 and 14.15 show two types of calorimeters.) The technological process of measuring changes in thermal energy is called **calorimetry**.

How Calorimeters Work

In a polystyrene calorimeter, a known mass of water is inside the polystyrene cup. The water surrounds, and is in direct contact with, the process that produces the energy change. The initial temperature of the water is measured. Then the process takes place and the final temperature of the water is measured. The water is stirred to maintain even energy distribution, and the system is kept at a constant pressure. This type of calorimeter can measure heat changes during processes such as dissolving, neutralization, heating, and cooling.

A **bomb calorimeter** is used for the measurement of heat changes during combustion reactions at a constant volume. It works on the same general principle as the polystyrene calorimeter. The reaction, however, takes place inside an inner metal chamber, called a "bomb." This "bomb" contains pure oxygen. The reactants are ignited using an electric coil. A known quantity of water surrounds the bomb and absorbs the energy that is released by the reaction. You will learn more about bomb calorimeters later in this section.

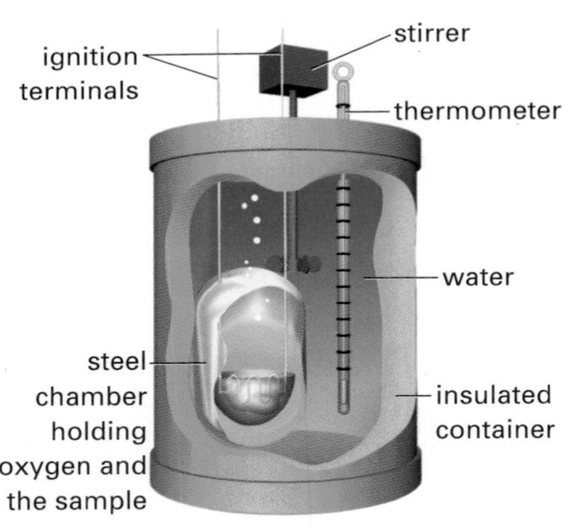

Figure 14.15 **A bomb calorimeter is more sophisticated than a polystyrene calorimeter.**

The law of conservation of energy states that energy can be changed into different forms, but it cannot be created or destroyed. This law allows you to calculate the energy change in a calorimetry experiment. However, you need to make the following assumptions:

- The system is isolated. (No heat is exchanged with the surroundings outside the calorimeter.)
- The amount of heat energy that is exchanged with the calorimeter itself is small enough to be ignored.
- If something dissolves or reacts in the calorimeter water, the solution still retains the properties of water. (For example, density and specific heat capacity remain the same.)

The First Ice Calorimeter

A calorimeter measures the thermal energy, that is absorbed or released by a material. Today we measure heat using joules (J) or calories (cal). Early scientists accepted one unit of heat as the amount of heat required to melt 1 kg of ice. Thus two units of heat could melt 2 kg of ice.

The earliest measurements of heat energy were taken around 1760, by a Scottish chemist named Joseph Black. He hollowed out a chamber in a block of ice. Then he wiped the chamber dry and placed a piece of platinum, heated to 38°C, inside. He used another slab of ice as a lid. As the platinum cooled, it gave up its heat to the ice. The ice melted, and water collected in the chamber. When the platinum reached the temperature of the ice, Black removed the water and weighed it to find out how much ice had melted. In this way, he measured the amount of heat that was released by the platinum.

In 1780, two French scientists, Antoine Lavoisier and Pierre Laplace, developed the first apparatus formally called a calorimeter. Like Black, they used the amount of melted ice to measure the heat released by a material. Their calorimeter consisted of three concentric chambers. The object to be tested was placed in the innermost chamber. Broken chunks of ice were placed in the middle chamber. Ice was also placed in the outer chamber to prevent any heat reaching the apparatus from outside. As the object in the inner chamber released heat, the ice in the middle chamber melted. Water was drawn from the middle chamber by a tube, and then measured.

Lavoisier made many important contributions to the science of chemistry. Unfortunately his interest in political reform led to his arrest during the French Revolution. He was beheaded after a trial that lasted less than a day.

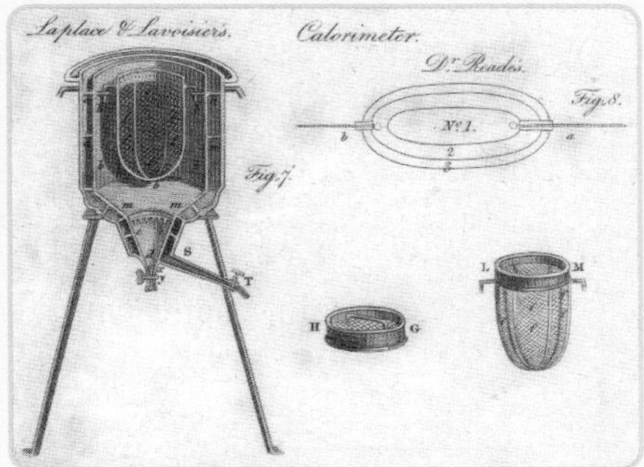

The original calorimeter used by Lavoisier and Laplace

When a process causes an energy change in a calorimeter, the change in temperature is measured by a thermometer in the water. If you know the mass of the water and its specific heat capacity, you can calculate the change in thermal energy caused by the process. See Figures 14.16 and 14.17 for examples.

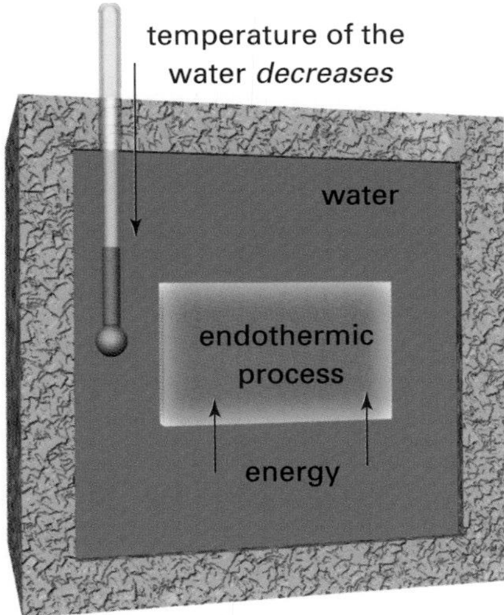

Figure 14.16 An endothermic process, such as ice melting

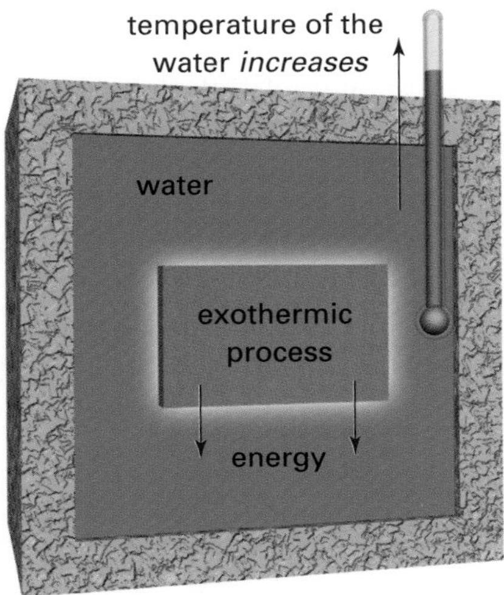

Figure 14.17 An exothermic process, such as the combustion of propane

The energy change in a calorimetry experiment can be summarized as follows:

Heat lost by the process = Heat gained by the water

or

Heat gained by the process = Heat lost by the water

In the next Sample Problem, you will use what you have just learned to calculate the specific heat capacity of a metal. This problem is similar to the calculation of the specific heat capacity of canola oil in section 14.3. Here, however, a calorimeter is used to reduce the heat exchange to the environment.

Sample Problem

Determining a Metal's Specific Heat Capacity

Problem

A 70.0 g sample of a metal was heated to 95.0°C in a hot water bath. Then it was quickly transferred to a polystyrene calorimeter. The calorimeter contained 100.0 g of water at an initial temperature of 19.8°C. The final temperature of the contents of the calorimeter was 22.6°C.

(a) How much heat did the metal lose? How much heat did the water gain?

(b) What is the specific heat capacity of the metal?

What Is Required?

(a) You need to calculate the heat lost by the metal (Q_m) and the heat gained by the water (Q_w).

(b) You need to calculate the specific heat capacity of the metal.

What Is Given?

You know the mass of the metal, and its initial and final temperatures.
Mass of metal (m_m) = 70.0 g

Initial temperature of metal (T_i) = 95.0°C

Final temperature of metal (T_f) = 22.6°C

You also know the mass of the water, and its initial and final temperatures.

Mass of water (m_w) = 100.0 g

Initial temperature of water (T_i) = 19.8°C

Final temperature of water (T_f) = 22.6°C

As well, you know the specific heat capacity of water: 4.184 J/g·°C.

Continued ...

Continued ...
FROM PAGE 604

Plan Your Strategy

(a) You have all the information that you need to find the heat gained by the water. Use the heat equation $Q = mc\Delta T$. To find the heat lost by the metal, assume that $Q_m = -Q_w$.

(b) Calculate the specific heat capacity of the metal by rearranging the heat equation and solving for c.

It is very important that you do not mix up the given information. For example, when solving for the thermal energy change of the water, Q_w, make sure that you only use variables for the water. You must use the initial temperature of the water, 19.8°C, *not* the initial temperature of the metal, 95.0°C. Also, remember that $\Delta T = T_f - T_i$.

Act on Your Strategy

(a) Solve for Q_w.

$$\begin{aligned} Q_w &= m_w c_w \Delta T_w \\ &= (100.0\ \text{g})(4.184\ \frac{\text{J}}{\text{g}\cdot{}^\circ\text{C}})(22.6^\circ\text{C} - 19.8^\circ\text{C}) \\ &= 1171.52\ (\cancel{\text{g}})(\frac{\text{J}}{\cancel{\text{g}}\cdot\cancel{{}^\circ\text{C}}})(\cancel{{}^\circ\text{C}}) \\ &= 1.2 \times 10^3\ \text{J} \end{aligned}$$

The water gained 1.2×10^3 J of thermal energy.

Solve for Q_m.

$$\begin{aligned} Q_m &= -Q_w \\ &= -1.2 \times 10^3\ \text{J} \end{aligned}$$

The metal lost 1.2×10^3 J of thermal energy.

(b) Solve for c_m.

$$\begin{aligned} c_m &= \frac{Q_m}{m_m \Delta T_m} \\ &= \frac{-1.2 \times 10^3\ \text{J}}{(70.0\ \text{g})(22.6^\circ\text{C} - 95.0^\circ\text{C})} \\ &= 0.24\ \text{J/g}\cdot{}^\circ\text{C} \end{aligned}$$

The specific heat capacity of the metal is 0.24 J/g·°C.

Check Your Solution

The heat gained by the water is a positive value. The heat lost by the metal is a negative value. The heat is expressed in joules. (Kilojoules are also acceptable.)

The specific heat capacity of the metal is positive, and it has the correct units.

Notice that all the materials in the calorimeter in the Sample Problem had the same final temperature. This is called **thermal equilibrium**.

Figure 14.18

Heat of Combustion

How much energy is needed for a natural gas water heater, like the one shown in Figure 14.18, to heat the hot water in your home? It is probably much more than you think! In the next Sample Problem, you will find out. You will also calculate the **heat of combustion** of a hydrocarbon: the heat that is released when combustion occurs.

Sample Problem

Calculating Thermal Energy

Problem

Many homes in North America use natural gas for general heating and for water heating. Like calorimeters, natural gas water heaters have an insulated container that is filled with water. A gas flame at the bottom heats the water. A typical water heater might hold 151 L of water.

(a) How much thermal energy is needed to raise the temperature of 151 L of water from 20.5°C to 65.0°C? Note: Make the same three assumptions that you made for calorimeters.

(b) If it takes 506 g of methane to heat this water, what is the heat of combustion of methane per gram?

What Is Required?

(a) You need to calculate the amount of thermal energy (Q) needed to heat 151 L of water.

(b) You need to calculate the heat released per gram of methane burned.

What Is Given?

(a) You know the initial and final temperatures of the water. You also know the volume of the water.
Initial temperature (T_i) = 20.5°C
Final temperature (T_f) = 65.0°C
Volume = 151 L
As well, you know the specific heat capacity of water (c):

4.184 J/g·°C or 4.184 kJ/kg·°C.

(b) You know the mass of the methane. Mass of methane (m) = 506 g

Plan Your Strategy

(a) This problem involves thermal energy and a change in temperature. You can use the heat equation $Q = mc\Delta T$. First calculate the mass of 151 L of water. (Remember that the density of water at room temperature is 1 g/mL, or 1 kg/L.) If you express the mass of the water in kilograms, you must also use the appropriate specific heat capacity of water: 4.184 kJ/kg·°C.

Note: To keep the calculation simple, assume that the density of the water remains the same when it is heated. (This is not strictly true.)

Continued ...

FROM PAGE 606

(b) Use the concept of heat lost = heat gained. Since a loss of heat gives a negative value, use the following equation.

$$Q_m = -Q_w$$

To find the heat per gram, divide the amount of heat by the mass of methane.

Act on Your Strategy

(a) Mass of water = Volume × Density

$$= (151 \text{ L})(1 \text{ kg/L})$$
$$= 151 \text{ kg}$$

Substitute into $Q = mc\Delta T$, and solve.

$$\begin{aligned} Q &= mc\Delta T \\ &= (151 \text{ kg})(4.184 \text{ kJ/kg}\cdot{}^\circ\text{C})(65.0^\circ\text{C} - 20.5^\circ\text{C}) \\ &= 28\ 114\ (\cancel{\text{kg}})(\text{kJ}/\cancel{\text{kg}}\cdot\cancel{^\circ\text{C}})(\cancel{^\circ\text{C}}) \\ &= 2.81 \times 10^4 \text{ kJ} \end{aligned}$$

Therefore, 2.81×10^4 kJ, or 28.1 MJ (megajoules), of energy is needed to heat the water. (This is a great deal of energy!)

(b) $Q_m = -Q_w$

$$= -2.81 \times 10^4 \text{ kJ}$$

Divide the amount of heat by the mass of methane to find the heat per gram.

$$\begin{aligned} Q_m \text{ (per gram)} &= \frac{Q_m}{m_m} \\ &= \frac{-2.81 \times 10^4 \text{ kJ}}{506 \text{ g}} \\ &= -55.5 \text{ kJ/g} \end{aligned}$$

This means that 55.5 kJ of thermal (heat) energy is *released* for each gram of methane that burns.

Check Your Solution

(a) The water gains heat, so the heat value is positive.
Heat is expressed in kilojoules. (Joules are also acceptable.)

(b) The methane loses energy, so the heat value is negative. Since this value is the heat per gram, the unit is kJ/g.

Practice Problems

18. A reaction lowers the temperature of 500.0 g of water in a calorimeter by 1.10°C. How much heat is absorbed by the reaction?

19. Aluminum reacts with iron(III) oxide to yield aluminum oxide and iron. The temperature of 1.00 kg of water in a calorimeter increases by 3.00°C during the reaction. Calculate the heat that is released in the reaction.

Continued ...
FROM PAGE 607

20. 5.0 g of an unknown solid was dissolved in 100 g water in a polystyrene calorimeter. The initial temperature of the water was 21.7°C, and the final temperature of the solution was 29.6°C

(a) Calculate the heat change caused by the solid dissolving.

(b) What is the heat of solution per gram of solid dissolved?

21. A 92.0 g sample of a substance, with a temperature of 55.0°C, is placed in a polystyrene calorimeter. The calorimeter contains 1.00 kg of water at 20.0°C. The final temperature of the system is 25.2°C.

(a) How much heat did the substance lose? How much heat did the water gain?

(b) What is the specific heat capacity of the substance?

Heat of Solution

In Practice Problem 20, you calculated the thermal energy change as a solid dissolved in water. This value is called the **heat of solution**: the energy change caused by a substance dissolving. The following ExpressLab deals with the heat of solution of a solid.

ExpressLab  The Energy of Dissolving

In this lab you will measure the heat of solution of two solids.

Safety Precautions

- NaOH and KOH can burn skin. If you accidentally spill NaOH or KOH on your skin, wash immediately with copious amounts of cold water.

Materials

balance and beakers or weigh boats
polystyrene calorimeter
thermometer and stirring rod
distilled water
2 pairs of solid compounds:
- ammonium nitrate and potassium hydroxide
- potassium nitrate and sodium hydroxide

Procedure

1. Choose *one* pair of chemicals from the list.
2. For each of the two chemicals, calculate the mass required to make 100.0 mL of a 1.00 mol/L aqueous solution.
3. Measure the required mass of one of the chemicals in a beaker or a weigh boat.
4. Measure exactly 100 g of distilled water directly into your calorimeter.
5. Measure the initial temperature of the water.
6. Pour one of the chemicals into the calorimeter. Put the lid on the calorimeter.
7. Stir the solution. Record the temperature until there is a maximum temperature change.
8. Dispose of the chemical as directed by your teacher. Clean your apparatus.
9. Repeat steps 3 to 8, using the other chemical.

Analysis

1. For each chemical you used, calculate the heat change per gram and the heat change per mole of substance dissolved.
2. Which chemical dissolved endothermically? Which chemical dissolved exothermically?
3. One type of cold pack contains a compartment of powder and a compartment of water. When the barrier between the two compartments is broken, the solid dissolves in the water and causes an energy change. What chemical could be used in this type of cold pack? Why?

A Closer Look at Bomb Calorimetry

Polystyrene calorimeters are reasonably efficient for measuring heat changes during physical processes, such as dissolving and phase changes. They can also be used to measure heat changes during chemical processes, such as neutralization. A stronger and more precise type of calorimeter is needed, however, to measure the heat of combustion of foods, fuels, and other materials. As you learned earlier, bomb calorimeters are used for this purpose. (See Figure 14.19.)

Figure 14.19 **Bomb calorimeters give more accurate measurements than polystyrene calorimeters.**

A bomb calorimeter has many more parts than a polystyrene calorimeter. All of these parts can absorb or release small quantities of energy. Therefore, you cannot assume that the heat lost to the calorimeter is small enough to be negligible. To obtain precise heat measurements, you must know or find out the heat capacity of the bomb calorimeter. **Heat capacity** is the ratio of the heat gained or lost by a system to the change in temperature caused by this heat. It is usually expressed in kJ/°C. Unlike specific heat capacity, which refers to a single substance, heat capacity refers to a system. Thus, the heat capacity of a calorimeter takes into account the heat that *all* parts of the calorimeter can lose or gain. (See Figure 14.20.)

$$Q_{total} = Q_{water} + Q_{thermometer} + Q_{stirrer} + Q_{container}$$

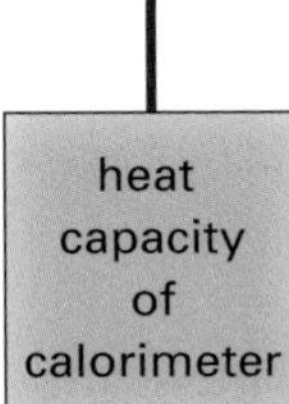

Figure 14.20 **Heat capacity is symbolized by an upper-case *C*. It is usually expressed in the unit J/°C.**

Web LINK

www.school.mcgrawhill.ca/resources

Why do some solids dissolve exothermically, while other solids dissolve endothermically? What factors may be involved? Research two factors: *lattice energies* and *solvation energies*. Go to the web site above. Go to **Science Resources**, then to **Chemistry 11** to find out where to go next. Report your results to the class.

CHEM FACT

Have you ever strained a muscle or sprained a joint? You may obtain temporary relief by applying the right heat of solution. A hot pack or cold pack consists of a thick outer pouch that contains water and a thin inner pouch that contains a salt. A squeeze on the outer pouch breaks the inner pouch, and the salt dissolves. Most hot packs use anhydrous $CaCl_2$. Most cold packs use NH_4NO_3. The change in temperature can be quite large. A cold pack, for instance, can bring the solution from room temperature down to 0°C. The usable time, however, is limited to around half an hour.

A bomb calorimeter is calibrated for a constant mass of water. Since the mass of the other parts remain constant, there is no need for mass units in the heat capacity value. The manufacturer usually includes the heat capacity value(s) in the instructions for the calorimeter.

Heat calculations must be done differently when the heat capacity of a calorimeter is included. The next Sample Problem illustrates this.

Sample Problem

Calculating Heat Change in a Bomb Calorimeter

Problem

A laboratory decided to test the energy content of peanut butter. A technician placed a 16.0 g sample of peanut butter in the steel bomb of a calorimeter, along with sufficient oxygen to burn the sample completely. She ignited the mixture and took heat measurements. The heat capacity of the calorimeter was calibrated at 8.28 kJ/°C. During the experiment, the temperature increased by 50.5°C.

(a) What was the thermal energy released by the sample of peanut butter?

(b) What is the heat of combustion of the peanut butter per gram of sample?

What Is Required?

(a) You need to calculate the heat (Q) lost by the peanut butter.

(b) You need to calculate the heat lost per gram of peanut butter.

What Is Given?

You know the mass of the peanut butter, the heat capacity of the calorimeter, and the change in temperature of the system.

Mass of peanut butter (m) = 16.0 g

Heat capacity of calorimeter (C) = 8.28 kJ/°C

Change in temperature (ΔT) = 50.5°C

Plan Your Strategy

(a) The heat capacity of the calorimeter takes into account the specific heat capacities and masses of all the parts of the calorimeter. Calculate the heat change of the calorimeter, Q_{cal}, using the equation

$$Q_{cal} = C\Delta T$$

Note: C is the heat capacity of the calorimeter in J/°C or kJ/°C. It replaces the m and c in other calculations involving specific heat capacity.

First calculate the heat gained by the calorimeter. When the peanut butter burns, the heat lost by the peanut butter sample equals the heat gained by the calorimeter.

$$Q_{sample} = -Q_{cal}$$

Continued ...

Continued ...

FROM PAGE 610

(b) To find the heat of combustion per gram, divide the heat by the mass of the sample.

Act on Your Strategy

(a) $Q_{cal} = C\Delta T$
$= (8.28\text{kJ}/^\circ\text{C})(50.5^\circ\text{C})$
$= 418.14(\text{kJ}/\cancel{^\circ\text{C}})(\cancel{^\circ\text{C}})$
$= 418 \text{ kJ}$

The calorimeter gained 418 kJ of thermal energy.

$$Q_{sample} = -Q_{cal} = -418 \text{ kJ}$$

The sample of peanut butter released 418 kJ of thermal energy.

(b) $$\text{Heat of combustion per gram} = \frac{\text{Heat released}}{\text{Mass of sample}} = \frac{-418 \text{ kJ}}{16.0 \text{ g}} = -26.2 \text{ kJ/g}$$

The heat of combustion per gram of peanut butter is –26.2 kJ/g.

Check Your Solution

Heat was lost by the peanut butter, so the heat value is negative.

Practice Problems

22. Use the heat equation for a calibrated calorimeter, $Q_{cal} = C\Delta T$. Recall that $\Delta T = T_f = T_i$. Solve for the following quantities.

(a) C

(b) ΔT

(c) T_f (in terms of C, ΔT, and T_i)

(d) T_i (in terms of C, ΔT, and T_f)

23. A lab technician places a 5.00 g food sample into a bomb calorimeter that is calibrated at 9.23 kJ/°C. The initial temperature of the calorimeter system is 21.0°C. After burning the food, the final temperature of the system is 32.0°C. What is the heat of combustion of the food in kJ/g?

24. A scientist places a small block of ice in an uncalibrated bomb calorimeter. The ice melts, gains 10.5 kJ (10.5×10^3 J) of heat and undergoes a temperature change of 25.0°C. The calorimeter undergoes a temperature change of 1.2°C.

(a) What mass of ice was added to the calorimeter? (Use the heat capacity of liquid water.)

(b) What is the calibration of the bomb calorimeter in kJ/°C?

CHECKPOINT

It takes one calorie (small c) to heat 1 g of water by 1°C. What mass of water can one Calorie (large C) heat by 1°C?

Food as a Fuel

Food is the fuel for your body. It provides you with the energy you need to function every day. Unlike the peanut butter in the previous problem, the food you digest is not burned. The process of digestion, however, is very similar to burning. In fact, people often talk about "burning off Calories." When food is digested, it undergoes slow combustion (without flames!) as it reacts with the oxygen you breathe. Eventually this combustion produces the materials that your body needs. It also releases carbon dioxide and water vapour as waste.

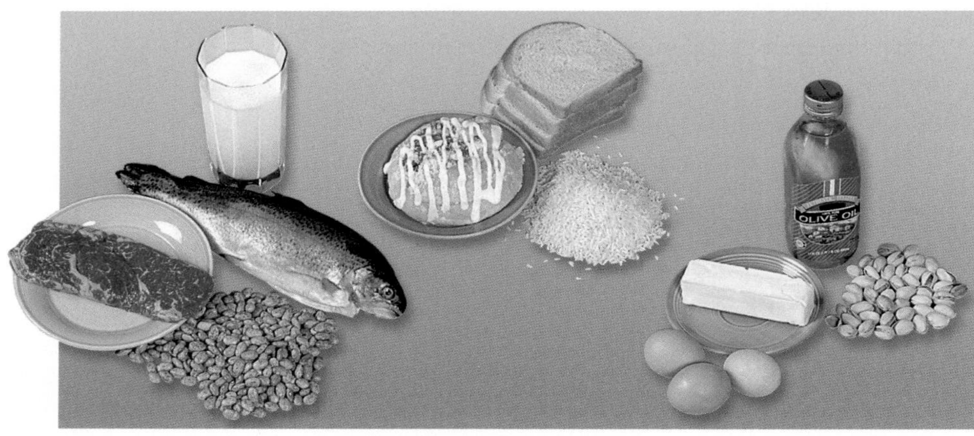

Figure 14.21 Food contains energy, which is usually measured in Calories.

You have probably noticed that Calories are used more often than kilojoules when discussing the energy in food (Figure 14.21). How are these terms related?

For years, chemists used the calorie as a unit of energy. One calorie is equal to 4.184 J. This is the amount of energy that is required to heat 1 g of liquid water by 1°C. The food **Calorie** (notice the upper-case C) is equal to one thousand calories, or one kilocalorie. *Therefore one food Calorie is equal to 4.184 kJ.*

You will recall that in the last sample problem, 16.0 g of peanut butter released 418 kJ of energy. To translate a value in kilojoules into Calories, multiply it by the fraction $\frac{1\text{ Cal}}{4.184\text{ kJ}}$. For the Sample Problem,

$$418\text{ kJ} \times \frac{1\text{ Cal}}{4.184\text{ kJ}} = 100\text{ Cal}$$

Therefore, 16.0 g of peanut butter released 100 Cal of energy. Table 14.3 gives energy values in Calories for some foods.

Table 14.3 Energy Values for Some Common Foods

Food	Quantity	Energy (kJ)	Energy (Cal)
almonds (shelled, whole)	75 g	1880	449
apple	100 g	283	68
beef (broiled)	90 g	1330	318
chicken (breast, broiled)	84 g	502	120
tuna (canned)	90 g	740	177
carrots (raw)	50 g	80	19
bread (white, enriched)	30 g	340	81
spaghetti (cooked)	148 g	690	165
olive oil	232 g (1 cup)	8580	2051
caramels (plain)	30 g (3 caramels)	480	115

When you take in food energy, your body stores excess food energy in the form of fat. If your body needs energy later, it will use up some of this fat. This is the secret behind hibernation. Why is excess energy in your body stored as fat? Why is it not stored as protein or carbohydrates? The next ThoughtLab will examine these questions. You will do calculations to find out which substance releases the most heat when burned.

Figure 14.22 **Many mammals in the animal kingdom rely on fat that is stored in their bodies. By surviving on fat reserves that are stored during the autumn, bears can hibernate throughout the winter without eating.**

ThoughtLab  Energy Content in Fat and Carbohydrates

By comparing the thermal energy that is released when fats, proteins, and carbohydrates are burned in a bomb calorimeter, you can compare the energy that is stored in these compounds. Natural fats are made up of various types of *fatty acids*. Fatty acids are long chain organic acids. The most common fatty acid in nature is oleic acid.

Glucose is a common sugar in the body. Most sugars that you ingest are broken down into glucose before they are digested further. Collagen is one of the most common proteins in your body.

Which compound releases more energy per gram: a fat (assume oleic acid, $C_{17}H_{33}COOH_{(\ell)}$), a carbohydrate such as sugar (assume glucose, $C_6H_{12}O_{6(s)}$), or a protein (assume collagen, molar mass 300 000 g/mol)?

A sample of glucose is placed in a bomb calorimeter, along with oxygen. The glucose is completely burned. The process is repeated with a sample of oleic acid and a sample of collagen. The following results are recorded:

Compound	sugar (glucose)	fat (oleic acid)	protein (collagen)
mass of compound (g)	1.35	1.23	1.31
initial temperature (°C)	25.20	25.00	25.10
final temperature (°C)	27.65	30.56	28.74
heat capacity of calorimeter (kJ/°C)	8.28	8.28	8.28

Procedure

1. For each substance, calculate the heat energy released per gram of substance burned.
2. For each substance, calculate the heat of combustion per mole of substance burned.

Analysis

1. Based on your calculations, which substance stores more energy? How do you know?

Application

2. Investigate why fat produces more energy than sugar when combusted. You may want to research and compare the bond structures of glucose and oleic acid.

Comparing Fats and Hydrocarbons

One of the substances that you considered in the last ThoughtLab was oleic acid. Oleic acid has a long hydrocarbon chain in each molecule. This hydrocarbon chain is similar to the hydrocarbon chains of fossil fuels, such as octane. (See Figure 14.23.)

oleic acid

$CH_3—CH_2—CH_2—CH_2—CH_2—CH_2—CH_2—CH_2—CH=CH—CH_2—CH_2—CH_2—CH_2—CH_2—CH_2—CH_2—COOH$

hydrocarbon chain

$CH_3—CH_2—CH_2—CH_2—CH_2—CH_2—CH_2—CH_3$

octane

Figure 14.23 By comparing octane and oleic acid, you can see that the heat of combustion of fossil fuels must be similar to the heat of combustion of fats.

mind STRETCH

Which stores more energy: *saturated* fat or *unsaturated* fat? Do research to find out. Then compare the structures of these two types of fat to explain why.

What is the relationship between the chemical bonds in a compound and the amount of energy that the compound can store? You already know that the net energy of a reaction equals the difference between *the energy absorbed when the reactant bonds are broken* and *the energy released when the product bonds are formed.* The size of the difference reflects the strength of the bonds in the reactant molecules compared with the strength of the bonds in the product molecules.

Figure 14.24 shows an exothermic reaction: the combustion of methane. The stored energy of the products is less than the stored energy of the reactants. Therefore, a net amount of energy is released by the reaction. In an endothermic reaction, the stored energy of the products is greater than the stored energy of the reactants.

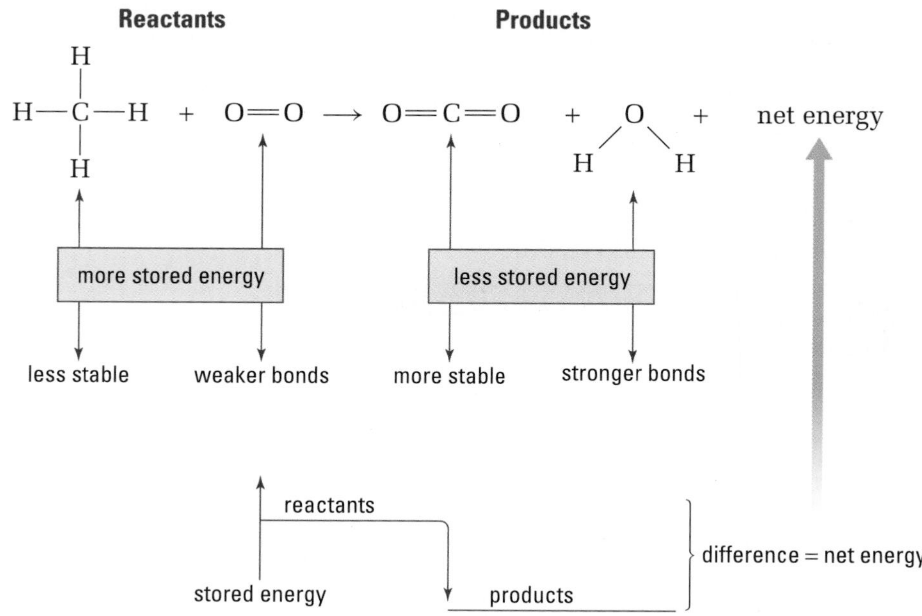

Figure 14.24 Energy stored in bonds

Most of the energy change in a reaction is due to a change in the **potential energy** (stored energy) of the bonds. All hydrocarbon fuels are made of the same elements: carbon and hydrogen. They react to give the same combustion products. As a result, the difference in the amount of heat energy that is released by any two hydrocarbons is directly related to the stored energy in the bonds of these compounds. The next ThoughtLab compares the heat energy that is released by the combustion of two fuels: propane and butane.

ThoughtLab  Heat Combustion of Propane and Butane

Propane and butane are very common hydrocarbon fuels. As successive members of the alkane family, they are closely related, but have a different number of bonds. Does this difference in the number of bonds affect the amount of heat released during combustion?

Samples of propane and butane were completely burned in a bomb calorimeter. The calorimeter was calibrated with a heat capacity of 8.28 kJ/°C. The observations are given in the table below.

Substance	Mass of sample (g)	Initial temperature (°C)	Final temperature (°C)
propane	1.50	25.00	34.03
butane	1.50	25.00	33.87

Procedure

1. Predict which compound will release the greater amount of energy per mole of gas.
2. Calculate the heat of combustion per mole for each substance.
3. (a) Which substance has the higher heat of combustion per mole?
   (b) Draw complete structural diagrams of propane and butane. Use these diagrams to explain your answer to part (a).

Analysis

1. Was your prediction correct?
2. Find and compare the boiling points of propane and butane. Which fuel would be better for winter camping? Why?

The Combustion of Candles

So far in this chapter, you have focussed on gaseous hydrocarbons, such as methane and acetylene. You have examined their heats of combustion and various processes in which they are used. In your everyday life, you may have encountered another type of hydrocarbon: paraffins. Paraffins are long chain hydrocarbons. They are semisolid or solid at room temperature. One type of paraffin has been a household item for centuries—paraffin wax, $C_{25}H_{52(s)}$, better known as candle wax. (See Figure 14.25.)

Like other hydrocarbons, the paraffin wax in candles undergoes combustion when burned. It releases thermal energy in the process. In the following investigation, you will measure this thermal energy.

Figure 14.25 Paraffin wax candles have been an important light source for hundreds of years.

SKILL FOCUS

Predicting

Performing and recording

Analyzing and interpreting

Investigation 14-B

The Heat of Combustion of a Candle

You have probably gazed into the flame of a candle without thinking about chemistry! Now, however, you will use the combustion of candle wax to gain insight into the measurement of heat changes. You will also evaluate the design of this investigation and make suggestions for improvement.

Question

What is the heat of combustion of candle wax?

Prediction

Will the heat of combustion of candle wax be greater or less than the heat of combustion of other fuels, such as propane and butane? Record your prediction, and give reasons.

Safety Precautions

- Tie back long hair and confine any loose clothing. Before you light the candle, check that there are no flammable solvents near by.

Materials

balance
calorimeter apparatus (see the diagram to the right)
thermometer
stirring rod
matches
water
candle

Procedure

1. Burn the candle to melt some wax. Use the wax to attach the candle to the smaller can lid. Blow out the candle.
2. Set up the apparatus as shown in the diagram, but do not include the large can yet. Adjust the ring stand so that the small can is about 5 cm above the wick of the candle. The tip of the flame should just touch the bottom of the small can.
3. Measure the mass of the candle and the lid.
4. Measure the mass of the small can and the hanger.
5. Place the candle inside the large can on the retort stand.
6. Fill the small can about two-thirds full of cold water (10°C to 15°C). You will measure the mass of the water later.

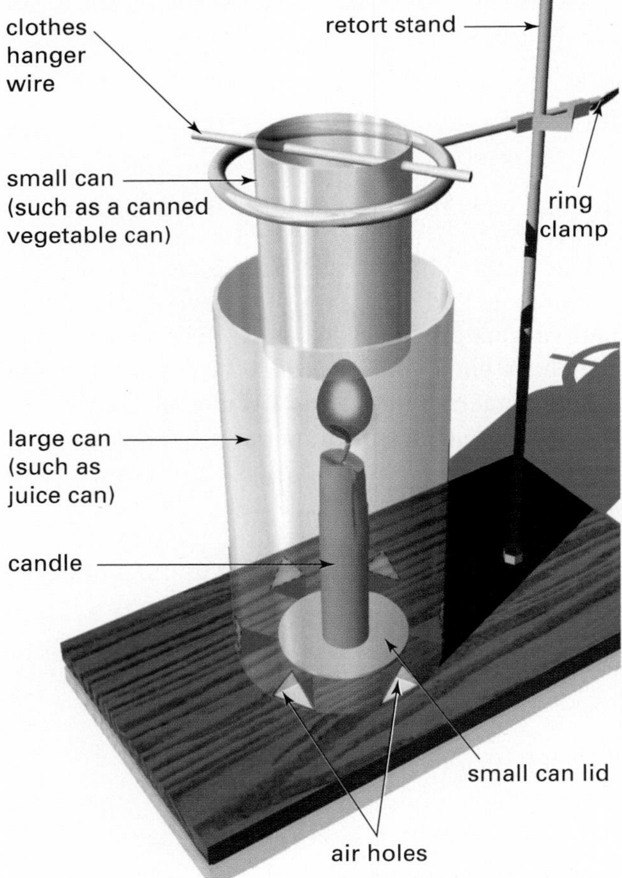

7. Stir the water in the can. Measure the temperature of the water.

8. Light the candle. Quickly place the small can in position over the candle. **CAUTION** Be careful of the open flame.

9. Continue stirring. Monitor the temperature of the water until it has reached 10°C to 15°C above room temperature.

10. Blow out the candle. Continue to stir. Monitor the temperature until you observe no further change.

11. Record the final temperature of the water. Examine the bottom of the small can, and record your observations.

12. Measure the mass of the small can and the water.

13. Measure the mass of the candle, lid, and any drops of candle wax.

Analysis

1. **(a)** Calculate the mass of the water.

   (b) Calculate the mass of candle wax that burned.

2. Calculate the thermal energy that was absorbed by the water.

3. Calculate the heat of combustion of the candle wax per gram.

4. **(a)** Assume that the candle wax is pure paraffin wax, $C_{25}H_{52(s)}$. Calculate the heat of combustion per mole of paraffin wax.

   (b) Write a balanced thermochemical equation for the complete combustion of paraffin wax. **Hint**: You calculated the heat of combustion per mole of paraffin wax in part (a). The chemical equation is balanced for one mole of $C_{25}H_{52(s)}$. Therefore, you can use an actual value for the energy in the equation.

Conclusions

5. **(a)** List some possible sources of error that may have affected the results you obtained.

   (b) Evaluate the design and the procedure of this investigation. Consider the apparatus, the combustion, and anything else you can think of. Make suggestions for possible improvements.

6. What if soot (unburned carbon) accumulated on the bottom of the small can? Would this produce a greater or a lower heat value than the value you expected? Explain.

CHECKPOINT

Hydrocarbons, such as fossil fuels, carry both risks and benefits. In a group, brainstorm to identify some risks and benefits.

Risks and Benefits

A **risk** is a chance of possible negative or dangerous results. Riding a bicycle carries the risk of falling off. Driving a car carries the risk of an accident. Almost everything you do has some kind of risk attached. Fortunately most risks are relatively small, and they may never happen. Many of the activities that carry risks also carry benefits. A **benefit** is an advantage, or positive result. For example, riding a bicycle provides the benefits of exercise, transportation, and enjoyment. When deciding to do an activity, it may be a good idea to compare the risks and benefits involved. (See Figure 14.29.)

Figure 14.29 Would you like to have a coal-burning power plant near your home? Some people might be upset by this idea because there is a health risk caused by pollution from the plant. Other people might think that a coal-burning power plant poses no threat at all. Who is right? How do you decide?

Risk-Benefit Analysis

Knowing more about an issue helps you assess its risks and benefits more accurately. How can you make the most informed decision possible? Follow these steps to do your own assessment of risks and benefits, called a **risk-benefit analysis**.

Step 1 Identify possible risks and benefits of the activity. Decide how to research these risks and benefits.

Step 2 Research the risks and benefits. You need information from reliable sources to make an accurate analysis.

Step 3 Weigh the effects of the risks and benefits. You may find that the risks are too great and decide not to do the activity. On the other hand, you may find that the benefits are greater than the risks.

Step 4 Compare your method for doing the activity with other possible methods. Do you use the safest method to do the activity? One method may be much safer than another.

In the next Sample Problem, you will see how a risk-benefit analysis can help you make informed decisions.

Sample Problem

Smoking

Problem

Many of your friends smoke, including some people you respect. Lately you have been thinking about taking up smoking too. Should you smoke? Perform a risk-benefit analysis to help you decide.

What Is Required?

You need to perform a risk-benefit analysis of smoking. This includes identifying, researching, and weighing the risks and benefits of smoking. It also includes looking at different methods of smoking, which might reduce the risk.

What Is Given?

The problem mentions that many of your friends smoke. This indicates a possible benefit of smoking—you will be imitating people

Continued ...

Continued ...
FROM PAGE 620

you like and respect. You need to do some research to identify more benefits and risks.

Plan Your Strategy

Use these four steps to identify and assess the risks and benefits of smoking.

Step 1 Identify possible risks and benefits.

Step 2 Research the risks and benefits. Use the Internet, reference books such as encyclopedias, and other sources to help you. It is important to choose reliable sources from which to obtain information.

Step 3 Weigh the risks and benefits.

Step 4 Compare different methods of smoking. Is one method less risky than another?

Act on Your Strategy

Step 1 Identify possible risks and benefits.

Possible risks	Possible benefits
Smoking is hazardous to your health.	Your friends smoke.
	Smoking may help you relax.

At this point, the benefits appear to outweigh the risks. You need to do more research, however, to make an informed decision.

Step 2 Research the risks and benefits.

Further research on the Internet provides more information on smoking risks:

Risks	Benefits
The tobacco smoke inhaled when smoking contains many toxic substances, such as carbon monoxide and ammonia gas.	Your friends smoke.
Smoking is the number one cause of lung cancer, heart disease, and emphysema.	Smoking may help you relax.
Smoking is addictive.	
About 50% of smokers end up dying from a tobacco-related disease.	
Second-hand smoke harms the people around you.	

Step 3 Weigh the risks and benefits of smoking.

The risks of smoking heavily outweigh the benefits.

Step 4 Compare different methods of smoking. Is one method less risky?

Smoking only one cigarette a day is less risky than smoking a pack a day. You cannot count on this method, however. Any kind of smoking is risky.

Continued ...

Continued ...
FROM PAGE 621

Practice Problems

25. Your town is considering dumping its plastic waste in a nearby lake. Identify possible risks and benefits for this plan. Explain where you could find more information to help your town make a decision.
26. Earth's reservoir of fossil fuels, including natural gas, will not last forever. As well, burning fossil fuels releases carbon dioxide (a greenhouse gas) and other pollutants that can cause acid rain. On the other hand, alternate energy sources, such as solar panels, are more expensive. They can also be less reliable than using fossil fuels. Perform a risk-benefit analysis to decide if you should heat your home using natural gas or solar panels. You will need to do more research to make an informed decision. (See the Internet Link on this page.)

Web LINK

www.school.mcgrawhill.ca/resources
To access more information on solar panels, natural gas heating, and the environmental effects of burning fossil fuels, go to the web site above. Go to **Science Resources**, then to **Chemistry 11** to find out where to go next. Use this information to help you answer Practice Problem 26.

Hydrocarbons: A Risky Business?

How do hydrocarbons benefit our society? How do they affect the environment? What are the benefits and risks of using hydrocarbon fuels and petrochemicals? These are important questions for our global community. See Figure 14.30 for some ideas.

Figure 14.30

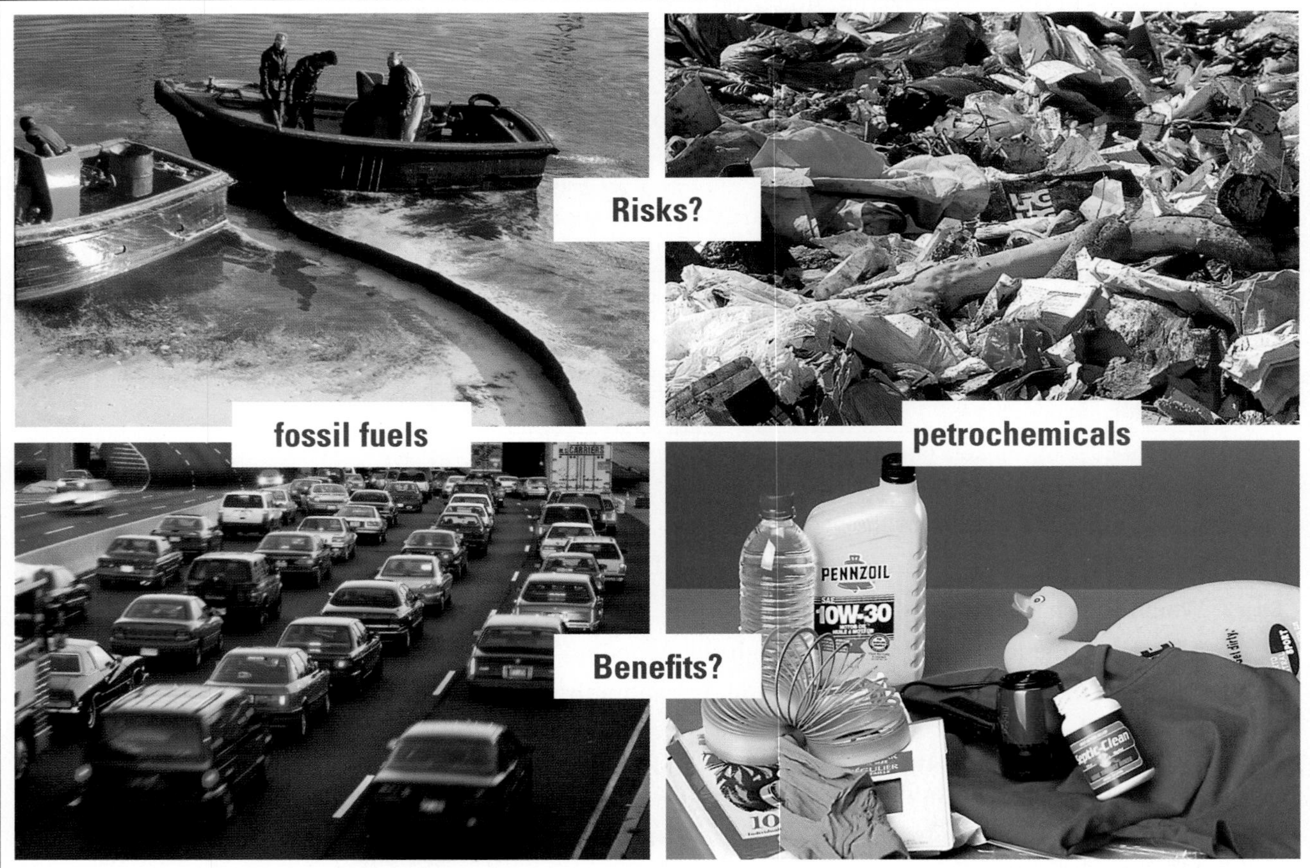

Hydrocarbon fuels have changed the way we live. Our dependence on them, however, has affected the world around us. The greenhouse effect, global warming, acid rain, and pollution are familiar topics on the news today. Our use of petroleum products, such as oil and gasoline, is linked directly to these problems.

Web LINK

www.school.mcgrawhill.ca/resources/
To learn more about greenhouse gases, global warming, and acid rain, go to the web site above. Go to **Science Resources**, then to **Chemistry 11** to find out where to go next.

The Greenhouse Effect and Global Warming

Roads, expressways, service stations, and parking lots occupy almost 40% of Toronto. They are the result of our demand for fast and efficient transportation. Every day, Toronto's vehicles produce nearly 16 000 t of carbon dioxide by the combustion of fossil fuels. Carbon dioxide is an important greenhouse gas. **Greenhouse gases** trap heat in Earth's atmosphere and prevent the heat from escaping into outer space. Scientists think that a build-up of carbon dioxide in the atmosphere may lead to an increase in global temperature, known as **global warming**. The diagram below shows how these concepts are connected to fossil fuels.

Concept Organizer Hydrocarbons and the Enviroment

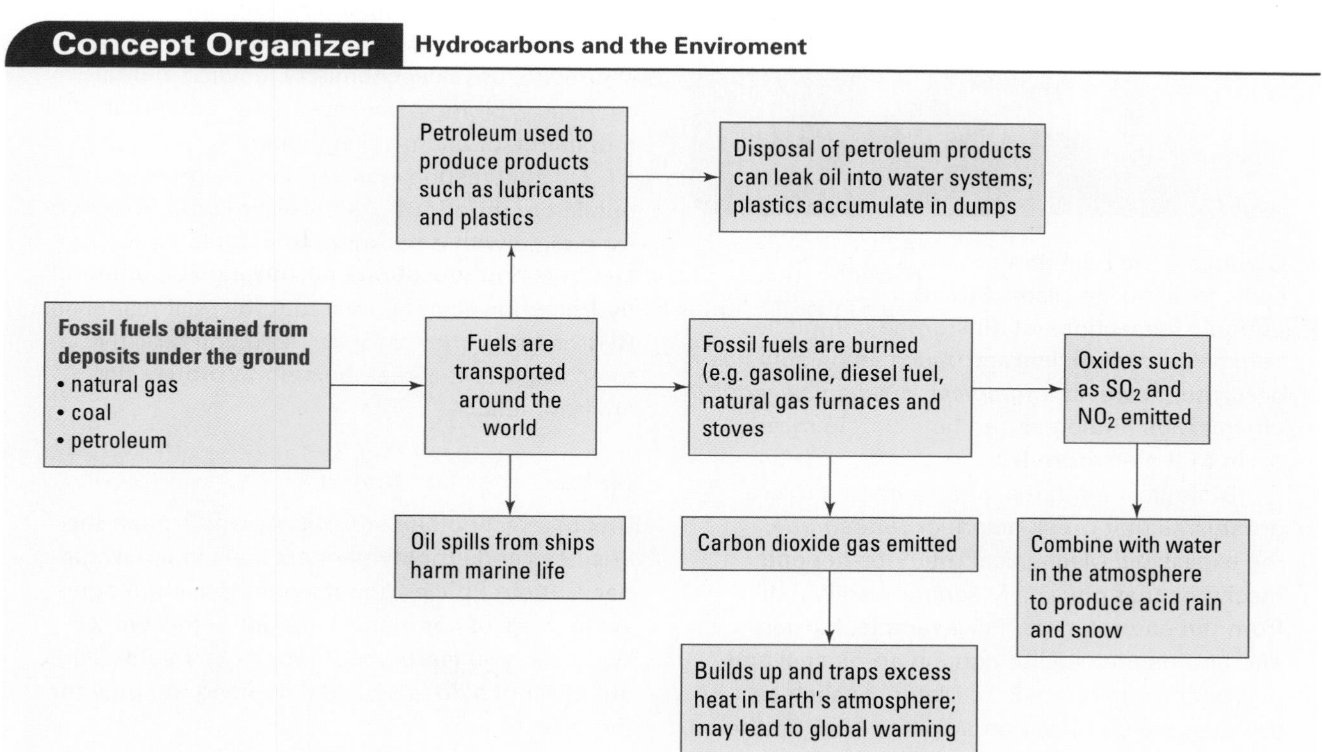

Acid Rain

The combustion of fossil fuels releases sulfur and nitrogen oxides. These oxides react with water vapour in the atmosphere to produce acid rain. Some lakes in northern Canada are "dead" because acid rain has killed the plants, algae, and fish that used to live in them. Forests in Québec and other parts of Canada have also suffered from acid rain.

Oil Spill Pollution

Our society demands a regular supply of fossil fuels. Petroleum is transported from oil-rich countries to the rest of the world. If an oil tanker carrying petroleum has an accident, the resulting oil spill can be disastrous to the environment.

Careers  in Chemistry

Oil Spill Advisor

Developed nations, such as Canada, depend heavily on petroleum. Our dependence affects the environment in many ways. Oil spills are a dramatic example of environmental harm caused by petrochemicals. In the news, you may have seen oceans on fire and wildlife choked with tar. What can we do?

Obviously the best thing to do is to prevent oil spills from taking place. Stricter regulations and periodic inspections of oil storage companies help to prevent oil leakage. Once an oil spill has occurred, however, *biological*, *mechanical*, and *chemical* technologies can help to minimize harm to the environment.

Biological methods involve helpful micro-organisms that break down, or *biodegrade*, the excess oil. Mechanical methods depend on machines that physically separate spilled oil from the environment. For example, barriers and booms are used to contain an oil spill and prevent it from spreading. Materials such as sawdust are sprinkled on a spill to soak up the oil.

Two main chemical strategies are also used to clean up oil spills. In the first strategy, *gelling agents* are added to react with the oil. The reaction results in a bulky product that is easier to collect using mechanical methods. In the second strategy, *dispersing agents* break up oil into small droplets that mix with the water. This prevents the oil from reaching nearby shorelines. Dispersing agents work in much the same way as a bar of soap!

The scientific advisor for an oil spill response unit assesses a spill and determines the appropriate clean-up methods. She or he acts as part of a team of advisors. Most advisors have an M.Sc. or Ph.D. in an area of expertise such as organic chemistry, physical chemistry, environmental chemistry, biology, oceanography, computer modelling, or chemical engineering.

Oil spill response is handled by private and public organizations. All these organizations look for people with a background in chemistry. In fact, much of what you are learning about hydrocarbons can be related to oil spill response. Hydrocarbon chemistry can lead you directly to an important career, helping to protect the environment.

Make Career Connections

Create a technology scrapbook. Go through the business and employment sections in a newspaper. Cut out articles about clean-up technologies. What kinds of companies are doing this work? What can you learn about jobs in this field? What qualifications does a candidate need to apply for this type of job?

Web LINK

www.school.mcgrawhill.ca/resources/
Go to the web site above to learn about the famous oil spill caused by the *Exxon Valdez*. Go to **Science Resources**, then to **Chemistry 11** to find out where to go next. When and where did this oil spill occur? How did it affect the environment? What was done to clean it up?

Everyday Oil Pollution

The biggest source of oil pollution comes from the everyday use of oil by ordinary people. Oil that is dumped into water in urban areas adds to oil pollution from ships and tankers. In total, *three million tonnes* of oil reach the ocean each year. This is equivalent to having an oil spill disaster every day!

A student from Thornhill, Ontario, did a home experiment to discover how much oil remains in "empty" motor oil containers that are thrown out. He collected 100 empty oil containers from a local gas station. Then he measured the amount of oil that was left in each container. He found an average of 36 mL per container. Over 130 million oil containers are sold and thrown out in Canada each year. Using these figures, he calculated that nearly five million litres of oil are dumped into landfill sites every year, just in "empty" oil containers!

Once oil reaches the environment, it is almost impossible to clean up. Oil leaking from a landfill site can contaminate drinking water in the area. Because oil can dissolve similar substances, pollutants such as chlorine and pesticides, and other organic toxins, mix with the oil. They are carried with it into the water system, increasing the problem.

Solutions to Environmental Problems

All of the problems described above hinge on our use of fossil fuels. Thus, cutting back on our use of fossil fuels will help to reduce environmental damage. Cutting back on fossil fuels, however, depends on the consumers who buy petrochemicals and use fossil fuels. In other words, it depends on you and the people you know.

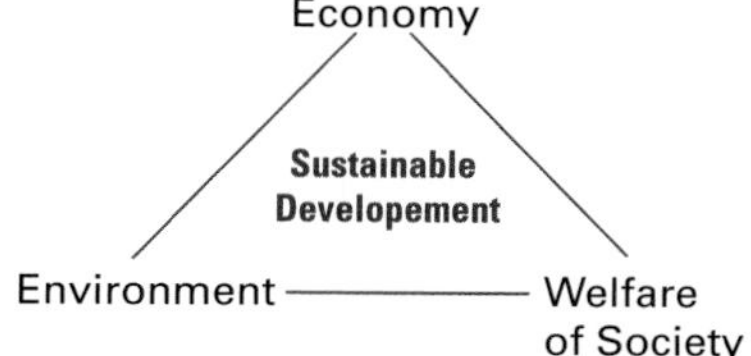

Figure 14.31 **Canada and other members of the United Nations endorse the principle of sustainable development. This principle states that the world must find ways to meet our current needs, without compromising the needs of future generations.**

Corporations that are looking for profit have little incentive to change their use of fossil fuels. For example, the technology is available to build cars that can drive about 32 km on a single litre of fuel. Because this technology is not financially profitable, cars are still being produced that drive about 8 km per litre of fuel. If consumers demand and purchase more fuel-efficient cars, however, car manufacturers will have an incentive to produce such cars. Tougher government standards may also help to push the vehicle industry towards greater fuel efficiency.

Governments can also bring about change by endorsing the principle of sustainable development. This principle was introduced at the 1992 Earth Summit Conference. **Sustainable development** takes into account *the environment, the economy, and the health and needs of society.* (See Figure 14.31.)

Hydrocarbon fuels and products can benefit our society if they are managed well. They can cause great environmental damage, however, if they are managed irresponsibly. With enough knowledge, you can learn to make informed decisions on these important issues. Here are some suggestions of way you can reduce your consumption of petroleum products. Why not choose one or more methods to practice? Or, brainstorm with your classmates to think of other ways to reduce consumption.

- Contact your local government and local power companies. Suggest using alternative fuels, such as solar energy and wind power.
- Ride a bicycle or walk more.
- Express your concerns by writing letters to the government or to newspapers.
- Become more informed by researching issues that concern you.
- Fix oil leaks in vehicles, and avoid dumping oil down the sink.
- If you are cold at home, put on an extra sweater instead of turning up the heat.
- Recycle and re-use petrochemical products, such as plastic shopping bags.
- Repair a broken item rather than buying a new one.

Section Wrap-up

In this section, you learned about some of the risks and benefits resulting from our use of fossil fuels. We obtain gasoline, heating oil, jet fuel, diesel fuel, fertilizers, and plastics from the oil and petroleum industry. Burning fossil fuels, however, produces carbon dioxide (a greenhouse gas) and other pollutants that lead to acid rain. Transporting oil also carries the risk of oil spills. Do the benefits of fossil fuels outweigh the risks? Complete these Section Review questions to help you decide.

Section Review

1 K/U You are about to try water-skiing for the first time. Should you try it? Will your activity affect the environment? How will you make your decision?

(a) Describe the steps you would follow to do a risk-benefit analysis.

(b) What are some possible risks and benefits?

(c) Where might you find the information you need to make a decision?

2 C Identify three benefits and three risks associated with the use of petroleum products and petrochemicals.

3 C Identify some steps that you can take to reduce your dependence on petroleum.

4 MC A construction company is planning to level a forest near your home to build a strip mall with a large parking lot. Many people enjoy walking in the forest, and many children play there. You have also observed wildlife, such as rabbits, snakes, frogs, and many kinds of birds living in the forest. In a group, brainstorm ways that the company could consider the environment and human welfare, as well as the economy. (Think of the economy as the owner of the property and the stores that will be built.)

5 MC Perform a risk-benefit analysis of the petroleum industry. Use the information in this section to help you identify possible risks and benefits. Use the Internet, or reference books to do more research on these risks and benefits.

6 MC Impure coal and gasoline contain nitrogen and sulfur compounds ($NO_{(g)}$ and $S_{(s)}$). The combustion of nitrogen and sulfur produces oxides that lead to acid rain.

(a) Balance the following six equations.

$S_{8(s)} + O_{2(g)} \rightarrow SO_{2(g)}$

$S_{8(s)} + O_{2(g)} \rightarrow SO_{3(g)}$

$NO_{(g)} + O_{2(g)} \rightarrow NO_{2(g)}$

$SO_{2(g)} + H_2O_{(\ell)} \rightarrow H_2SO_{3(aq)}$

$SO_{3(g)} + H_2O_{(\ell)} \rightarrow H_2SO_{4(aq)}$

$NO_{2(g)} + H_2O_{(\ell)} \rightarrow HNO_{3(aq)} + HNO_{2(aq)}$

(b) Suggest possible sources for the reactants $O_{2(g)}$ and $H_2O_{(\ell)}$.

(c) Explain how these equations show the production of acid rain.

Review

Reflecting on Chapter 14

Summarize this chapter in the format of your choice. Here are a few ideas to use as guidelines:

- You learned about fossil fuels as an important source of energy in our society.
- You developed skills in writing combustion equations, and you learned to measure the heat changes caused by physical processes such as dissolving.
- While investigating calorimetry, you recognized the importance of isolating a system to reduce heat flow when measuring heat.
- You have become more informed about our society's use of fossil fuels, and the resulting effects on the environment.
- You learned to weigh the risks and benefits of an activity, and perform a risk-benefit analysis. This skill will help you make more informed decisions on issues that affect society, the economy, and the environment.

Reviewing Key Terms

For each of the following terms, write a sentence that shows your understanding of its meaning.

benefit	bomb calorimeter
bond energy	Calorie
calorimeter	calorimetry
ΔT	combustion
complete combustion	endothermic
exothermic	fossil fuels
global warming	greenhouse gases
heat	heat capacity
heat of combustion	incomplete combustion
isolated system	potential energy
risk	risk-benefit analysis
specific heat capacity	sustainable development
temperature	thermal energy
thermal equilibrium	thermochemical equation

Knowledge/Understanding

1. **(a)** What are the products of the complete combustion of a hydrocarbon?
   (b) What products form if the combustion is incomplete?
2. **(a)** Why can incomplete combustion be dangerous if it occurs in your home?
   (b) How can you tell, by looking at a flame, that incomplete combustion is taking place?
3. Indicate whether each process is endothermic or exothermic.
   (a) water evaporating
   (b) a piece of paper burning
   (c) rubbing your hands together
   (d) clouds forming
4. How can a balanced thermochemical equation tell you whether a chemical reaction is exothermic or endothermic?
5. Describe the relationship between the amount of thermal energy that is released by water and
   (a) the mass of water
   (b) the temperature change of the water
6. Why is energy needed to sustain an endothermic reaction?
7. **(a)** The combustion of paraffin, $C_{25}H_{52(s)}$, is exothermic. Explain why by comparing the energy changes observed when chemical bonds are broken and formed.
   (b) The formation of 1-pentyne, $C_5H_{8(\ell)}$, is endothermic. Explain why by comparing the energy changes observed when chemical bonds are broken and formed.
8. Hydrogen is used as a fuel for the space shuttle because it provides more energy per gram than many other fuels. The combustion of hydrogen is described by the following equation.
   $$2H_{2(g)} + O_{2(g)} \rightarrow 2H_2O_{(g)} + 484 \text{ kJ}$$
   (a) Is this reaction exothermic or endothermic?
   (b) How much energy does the complete combustion of 1 g of hydrogen provide?
9. Write a balanced chemical equation for each reaction.
   (a) complete combustion of 4-methyl-1-pentene, $C_6H_{12(\ell)}$
   (b) incomplete combustion of benzene, $C_6H_{6(\ell)}$
   (c) incomplete combustion of propene, $C_3H_{6(g)}$
   (d) complete combustion of 3-ethyhexane, $C_8H_{18(\ell)}$
10. List three assumptions that you make when using a polystyrene calorimeter.

11. The same amount of heat is added to aluminum (c_{Al} = 0.900 J/g·°C) and nickel (c_{Ni} = 0.444 J/g·°C). Which metal will have a greater temperature increase? Explain.

12. Does propane burning in an outdoor barbecue have a negative or positive heat of combustion? Explain.

Inquiry

13. To make four cups of tea, 1.00 kg of water is heated from 22.0°C to 99.0°C. How much energy is needed?

14. Two different foods are burned in a calorimeter. Sample 1 has a mass of 6.0 g and releases 25 Cal of heat. Sample 2 has a mass of 2.1 g and releases 9.0 Cal of heat. Which food releases more heat per gram?

15. A 3.00 g sample of a new snack food is burned in a calorimeter. The 2.00 kg of surrounding water change in temperature from 25.0°C to 32.4°C. What is the food value in Calories per gram?

16. A substance is burned completely in a bomb calorimeter. The temperature of the 2000 g of water in the calorimeter rises from 25.0°C to 43.9°C. How much energy is released?

17. A horseshoe can be shaped from an iron bar when the iron is heated to temperatures near 1500°C. The hot iron is then dropped into a bucket of water and cooled. An iron bar is heated from 1500°C and then cooled in 1000 g of water that was initially at 20.0°C. How much heat energy does the water absorb if its final temperature is 65.0°C?

18. A group of students decide to measure the energy content of certain foods. They heat 50.0 g of water in an aluminum can by burning a sample of the food beneath the can. When they use 1.00 g of popcorn as their test food, the temperature of the water rises by 24°C.
    (a) Calculate the heat energy that is released by the popcorn. Express your answer in both kilojoules and Calories per gram of popcorn.
    (b) Another student tells the group that she has read the label on the popcorn bag. The label states that 30 g of popcorn yields 110 Cal. What is this value in Calories per gram? How can you account for the difference between the two values?

19. In Chapters 13 and 14, you have examined many properties of hydrocarbons. Describe one physical property and one chemical property of hydrocarbons. Explain how these two properties vary from one hydrocarbon to another. Describe how you might measure each property in a lab.

20. A reaction in a calorimeter causes 250.0 g of water to decrease in temperature by 2.40°C. How much heat did the reaction absorb?

21. A chemist wants to calibrate a new bomb calorimeter. He completely burns a mass of 0.930 g of carbon in a calorimeter. The temperature of the calorimeter changes from 25.00°C to 28.15°C. If the thermal energy change is 32.8 kJ/g of carbon burned, what is the heat capacity of the new calorimeter? What evidence shows that the reaction was exothermic?

22. 200 g of iron at 350°C is added to 225 g of water at 10.0°C. What is the final temperature of the iron-water mixture?

23. In this chapter, you learned that fats have long hydrocarbon sections in their molecular structure. Therefore, they have many C—C and C—H bonds. Sugars have fewer C—C and C—H bonds but more C—O bonds. Use Table 14.1 in this chapters. Explain why you can obtain more energy from burning a fat than from burning a sugar.

24. 2,2,4-trimethylpentane, an isomer of C_8H_{18}, is a major component of gasoline.
    (a) Write the balanced thermochemical equation, using the word "energy," for the complete combustion of this compound. Use $C_8H_{18(\ell)}$ as the formula for the compound.
    (b) What is the ideal ratio of fuel to air for this fuel?
    (c) In the previous unit, you learned how to solve problems involving gases. Calculate the volume of carbon dioxide, at 20.0°C and 105 kPa, that is produced from the combustion of 1.00 L of $C_8H_{18(\ell)}$. **Note:** The density of $C_8H_{18(\ell)}$ is 0.69 g/mL.

25. 100 g of calcium carbide is used to produce acetylene in a laboratory, as you did in Investigation 14-A.

(a) What volume of water (density 1.00 g/mL) is needed to react completely with the calcium carbide?

(b) What volume of acetylene gas will be produced at STP when the calcium carbide and water are mixed?

Communication

26. Earlier in the chapter, you learned that poisonous carbon monoxide can form during incomplete hydrocarbon combustion. The use of carbon monoxide detectors in homes and businesses has reduced the number of deaths due to carbon monoxide poisoning. Are all carbon monoxide detectors the same? Telephone your local fire department, go to a library, or search the Internet to find out about carbon monoxide detectors.

27. Design a poster or a brochure to explain the concept of sustainable development to a student in a much younger grade.

28. Prepare a concept map to illustrate the effects of hydrocarbons on our society and the environment.

Making Connections

29. When energy is wasted during an industrial process, what actually happens to this energy?

30. Look at Table 14.3. Compare caramels with raw carrots. Which food gives more Calories per gram?

31. Petrochemical products, such as plastics, have affected your life. Identify one benefit and one possible risk associated with the use of petrochemical products.

32. Define sustainable development. Suggest a condition that you feel society must agree on to achieve sustainable development.

33. The Hibernia Oil Field is located off the Grand Banks of Newfoundland, on Canada's east coast. It started oil production in the fall of 1997. Research and write a report on some of the risks and benefits of this massive oil operation. Consider ecological, economic, and social issues.

34. On an episode of *The Nature of Things*, Dr. David Suzuki made the following comment: "As a society and as individuals, we're hooked on it [oil]." Discuss his comment. Explain how our society has benefitted from hydrocarbons. Describe some of the problems that are associated with the use of hydrocarbons. Also describe some possible alternatives for the future.

Answers to Practice Problems and Short Section Review Questions:

Practice Problems **1.(a)** complete **2.(a)** $C_5H_{12} + 8O_2 \rightarrow 5Co_2 + 6H_2O$ **(b)** $2C_8H_{18} + 25O_2 \rightarrow 16CO_2 + 18H_2O$ **(c)** $6C_2H_6 + 15O_2 \rightarrow 4CO_2 + 4CO + 4C + 18H_2O$, $3C_2H_6 + 6O_2 \rightarrow CO_2 + CO + 4C + 9H_2O$, **3.(a)** $C_4H_{10} + 4O_2 \rightarrow CO_2 + CO + 2C + 5H_2O$ **(b)** $2C_4H_{10} + 13O_2 \rightarrow 8CO_2 + 10H_2O$ **4.** $C_{25}H52 + 38O_2 \rightarrow 25CO_2 + 26H_2O$ **5.(a)** $C_{13}H_{28} + 9O_2 \rightarrow CO_2 + 2CO + 10C + 14H_2O$, $C_{13}H_{28} + 10O_2 \rightarrow 2CO_2 + 2CO + 8C + 14H_2O$ **(c)** incomplete **6.(a)** $3C_{(s)} + 4H_{2(g)} \rightarrow C_3H_{8(g)} +$ energy **(b)** $C_3H_{8(g)} + 5O_{2(g)} \rightarrow 3CO_{2(g)} + 4H_2O_{(g)} +$ energy **7.(b)** energy $+ 2C_{(s)} + 2H_{2(g)} \rightarrow C_2H_{4(g)}$, $C_2H_{4(g)} + 3O_{2(g)} \rightarrow 2CO_{2(g)} + 2H_2O_{(g)} +$ energy **8.** 6.2×10^3J **9.** -4.6×10^3J **11.(a)** +4.22 kJ, +4.6 kJ **12.(a)** $m = Q/c\Delta T$ **(b)** $c = Q/m\Delta T$ **(c)** $\Delta T = Q/mc$ **13.(a)** $T_I = T_f - Q/mc$ **(b)** $T_f = T_I + Q/mc$ **14.** 2.14×10^5 J **15.** 0.7900 J/g•°C, granite **16.** 1.21 J/g•°C **17.** 15.5•°C **18.** -2.30×10^3 J **19.** 1.26×10^4 J **20.** 661 J/g **21.(a)** -2.2×10^4 J, 2.2×10^4 J **(b)** 8.0 J/g•°C **22.(a)** $C = Q/\Delta T$ **(b)** $\Delta T = Q/C$ **(c)** $T_f = T_I + Q/C$ **(d)** $T_i = T_f - Q/C$ **23.** −20.3 kJ/g **24.(a)** 100 g **(b)** 8.75 kJ/°C

Section Review: 14.1: **3.(a)** $C_7H_{16} + 11O_2 \rightarrow 7CO_2 + 8H_2O$ **(b)** $C_5H_{10} + 6O_2 \rightarrow 3CO_2 + CO + C + 5H_2O$ **4.** $CH_4 + 2O_2 \rightarrow CO_2 + 2H_2O$, $4CH_4 + 6O_2 \rightarrow CO_2 + 2CO + C + 8H_2O$ **6.(a)** $2C_2H_6 + 7O_2 \rightarrow 4CO_2 + 6H_2O$ **(b)** 54 g **(c)** 392 L **14.2:** **1.(a)** endothermic, net energy is absorbed, formation of ethyne (acetylene) **(b)** exothermic, net energy is released, combustion of methane **4.(a)** endothermic **5.** 197 kJ/mol **14.3:** **3.(a)** -4.54×10^3 J **4.(a)** 38.1 J **(b)** 76.2 J **(c)** 76.2 J **5.(a)** 0.237 J/g•°C **(b)** 3.43°C **6.** copper **14.4:** **3.** −39.8 kJ **4.** -3.1×10^3 J **5.** 9.63 Cal, yes **14.5:** **6.(a)** $S_8 + 8O_2 \rightarrow 8SO_2$, $S_8 + 12O_2 \rightarrow 8SO_3$, $2NO + O_2 \rightarrow 2NO_2$, $SO_2 + H2O \rightarrow H_2SO_3$, $SO_3 + H_2O \rightarrow H_2SO_4$, $2NO_2 + H_2O \rightarrow HNO_3 + HNO_2$

Consumer Chemistry

Background

As a consumer, you are continually being exposed to new products. Some of these products can be controversial. For example, foods such as tomatoes and soybeans are genetically modified organisms (GMOs). As a citizen of the world, it is your responsibility to become informed. You should know what is in the products you use. You should understand how to use them appropriately and safely, and how to dispose of them in a responsible manner.

Challenge

Choose a common consumer item from the food, drug, or hardware department of a local supermarket. Your teacher may suggest a range of suitable products. If you choose an item that is not from one of these departments, your teacher must approve your choice. Your task is to communicate (using text and graphics) the chemical nature of the consumer product, and its container or packaging.

Materials

If you choose to make a poster or pamphlet, you will need art supplies and construction paper. You may need other materials if you choose a different presentation format.

Design Criteria

Your project can take the form of a pamphlet, poster, multimedia presentation, or other format of your choice.

Your project should include as many of these parts as possible. Alternatively, assign one part to each team working on a common product.

- **A** A creative and visually pleasing format (Use original artwork/graphics, and/or images taken from web sites, CD ROMS, or reference books.)
- **B** A description of all the ingredients in the product (Give the common or IUPAC names, molecular or ionic formulae, and structural diagrams for each ingredient.)
- **C** A description of the packaging material (Give the common or IUPAC names, molecular or ionic formulae, and structural diagrams for the packaging materials. Include recycling information.)
- **D** Any interesting historical information related to the project
- **E** Environmental or health/safety concerns related to specific ingredients or components of the product and/or packaging
- **F** A description of how the product is manufactured
- **G** A list of the references you used to obtain information

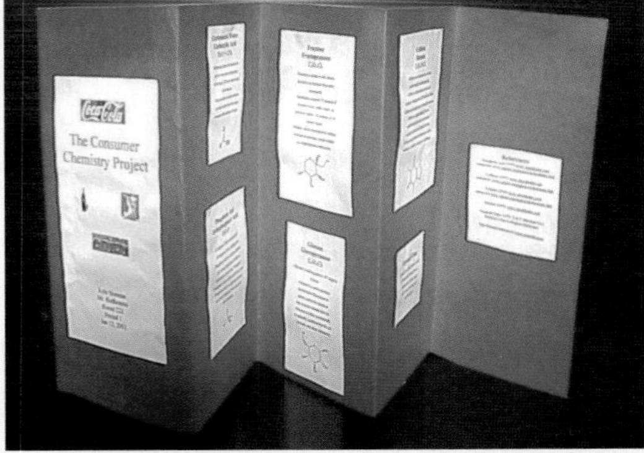

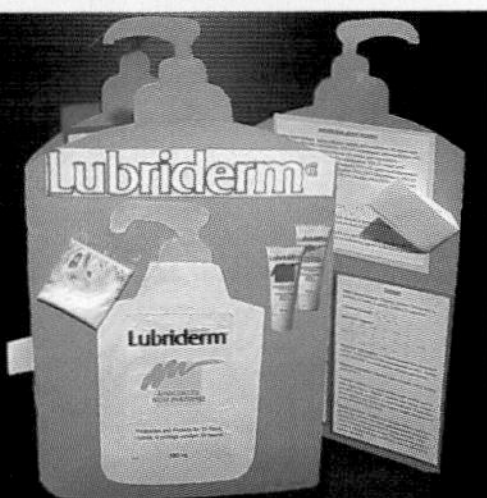

Action Plan

1. Decide whether you will work individually, or in a small group of two or three people.

Decide which consumer product you would like to research. If you are not sure of the suitability of your choice, check with your teacher.

3 Prepare a brief design proposal to answer these questions:
- Will you create a pamphlet, poster, or other display?
- What information do you want to include?
- How large will your final product be?

4 If you are working in a group, outline the duties to be performed. Assign duties to each group member.

5 Research your project. Look for chemical structures of ingredients, recycling codes, safety and WHMIS considerations, additives to foods, etc. You may wish to research in any of these areas:
- The Internet, at http://www.school.mcgrawhill.ca/resources/ (Go to this web site. Go to Science Resources, then to Chemistry 11 to find out where to go next.)
- Your local pharmacy (The pharmacist is a great source of information on medicinal ingredients.)
- Your local library (Your librarian is a valuable resource and can direct you to many sources of information and reference materials.)
- The 1-800 number or web site address given on the product's packaging

6 Put your project together. Be creative in your layout. Make sure that all the topics are covered in a logical fashion. Be sure to include input from all members of your group.

Evaluate

1 Is your project scientifically accurate? Is it complete?

2 Did you include interesting or unusual information?

3 Is your presentation colourful and visually attractive?

4 Are your graphics clear and informative? Did you pay close attention to detail?

5 Are all your references included in an approved format?

Assessment

After you complete this project,

- Assess the success of your project based on how well your project meets the design criteria.
- Assess your project based on how clearly the chemistry concepts are conveyed.
- Assess your project based on how interesting it is to others in your class.

UNIT 5 Review

Knowledge/Understanding

Multiple Choice

In your notebook, write the letter beside the best answer for each question.

1. An alkane contains 5 carbon atoms. How many hydrogen atoms does it contain?
   (a) 5 hydrogen atoms
   (b) 8 hydrogen atoms
   (c) 10 hydrogen atoms
   (d) 12 hydrogen atoms
   (e) 14 hydrogen atoms
2. An alkyne contains 12 hydrogen atoms. What is its chemical formula?
   (a) C_5H_{12}
   (b) C_6H_{12}
   (c) C_7H_{12}
   (d) C_8H_{12}
   (e) C_9H_{12}
3. Which set of properties best describes a small alkane, such as ethane?
   (a) non-polar, high boiling point, insoluble in water, extremely reactive
   (b) non-polar, low boiling point, insoluble in water, not very reactive
   (c) polar, low boiling point, soluble in water, not very reactive
   (d) polar, high boiling point, insoluble in water, not very reactive
   (e) non-polar, low boiling point, soluble in water, extremely reactive
4. Which equation shows the complete combustion of propane?
   (a) $2C_2H_6 + 7O_2 \rightarrow 4CO_2 + 6H_2O$
   (b) $2C_2H_6 + 5O_2 \rightarrow C + 2CO + CO_2 + 6H_2O$
   (c) $2C_3H_6 + 9O_2 \rightarrow 6CO_2 + 6H_2O$
   (d) $C_3H_8 + 5O_2 \rightarrow 3CO_2 + 4H_2O$
   (e) $2C_3H_8 + 7O_2 \rightarrow 2C + 2CO + 2CO_2 + 8H_2O$
5. Which situation describes an exothermic reaction?
   (a) The energy that is released to form the product bonds is greater than the energy that is used to break the reactant bonds.
   (b) The energy that is released to form the product bonds is less than the energy that is used to break the reactant bonds.
   (c) The energy that is released to form the product bonds is equal to the energy that is used to break the reactant bonds.
   (d) The energy that is used to break the product bonds is greater than the energy that is released to form the reactant bonds.
   (e) The energy that is used to break the product bonds is less than the energy that is released to form the reactant bonds.
6. Beaker A contains 500 mL of a liquid. Beaker B contains 1000 mL of the same liquid. What happens when 200 kJ of thermal energy is added to the liquid in each beaker?
   (a) The temperature of the liquid increases the same amount in both beakers.
   (b) The temperature of the liquid decreases the same amount in both beakers.
   (c) The temperature of the liquid remains the same in both beakers.
   (d) The temperature change of the liquid in Beaker B is twice as large as the temperature change of the liquid in Beaker A.
   (e) The temperature change of the liquid in Beaker B is one half as large as the temperature change of the liquid in Beaker A.

Short Answer

7. Copy the table below into your notebook. Fill in the blanks.

Name	Structural diagram	Molecular formula	Classification
ethane	CH_3-CH_3	C_2H_6	alkane
		C_3H_4	
1-ethyl-3-methy-cyclopentane			
	CH_3 H C=C CH_3-CH_2 CH_2-CH_3		
1, 2-dimethyl-cyclobutane			
	CH_3 $CH_3-CH_2-C-CH_3$ $CH_3-CH-CH_2-CH_3$		
		C_4H_6 (isomer 1)	
		C_4H_6 (isomer 2)	

Name	Structural diagram	Molecular formula	Classification
trans-4-ethyl-6-methyl-3-heptene			
	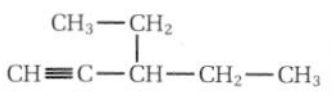 CH_3-CH_2 (on the CH) $CH \equiv C-CH-CH_2-CH_3$		
	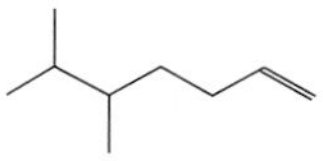		
4-ethyl-3,6-dimethyl-5-propyl-decane			

8. Identify the error(s) in each structure.

(a) $CH_3 - CH_3 - CH_3$

(b)

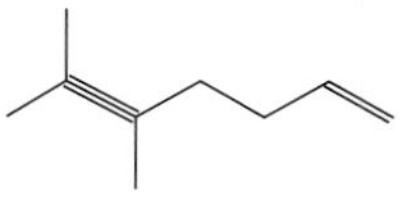

(c) CH_3 (on the CH_2)
$CH_3-CH_2-CH_3$

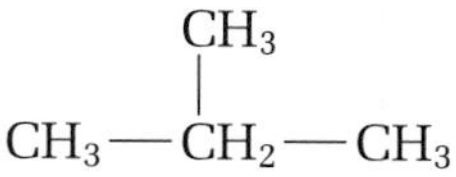

(d) H–C=C–H (ring), $CH-CH_2-CH \equiv CH$

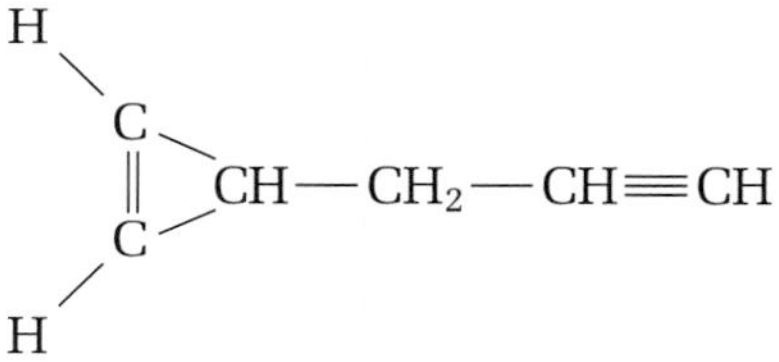

9. Identify the error(s) in each name.

(a) 2-ethene
(b) 1,2,2-trimethylpropane
(c) 3, 5-dimethyl-4-ethyl-4-hexene
(d) 4-ethyl-2-pentene

10. Copy the table below into your notebook. Complete it as follows:

- Write the general formulas for the alkane, alkene, alkyne, and cycloalkane homologous series.
- Draw the structural diagram for the third member of each homologous series. Hint: The first member of the alkene series is ethene, C_2H_4.
- Name the compound you drew.
- Indicate whether the compound you drew is saturated or unsaturated.

Homologous series	General formula	Structural diagram	Name	Saturated or unsaturated
alkane				
alkene				
alkyne				
cylcoalkane				

11. Four fuels are listed in the table below. Equal masses of these fuels are burned completely.

Fuel	Heat of combustion at SATP (kJ/mol)
octane, $C_8H_{18(\ell)}$	5513
methane, $CH_{4(g)}$	890
ethanol, $C_2H_5OH_{(\ell)}$	1367
hydrogen $H_{2(g)}$	285

(a) Put the fuels in order, from greatest to least amount of energy provided.
(b) Write a balanced chemical equation for the complete combustion of each fuel.
(c) Suggest one benefit and one risk of using hydrocarbon fuels.

12. You are given a 70 g sample of each of the following metals, all at 25°C. You heat each metal under identical conditions. Which metal will be first to reach 30°C? Which will be last? Explain your reasoning.

Metal	Specific Heat Capacity (J/g•°C)
Platinum	0.133
Titanium	0.528
Zinc	0.388

Inquiry

13. Design an investigation to differentiate between these two compounds:

$$CH_3 - CH_2 - CH_2 - CH_2 - CH_2 - CH_3 \text{ and}$$
$$CH_3 - CH = CH - CH_2 - CH_2 - CH_3$$

14. When kerosene, $C_{12}H_{26(\ell)}$, is burned in a device such a space heater, energy is released. Design an investigation to determine the amount of energy that is released, per gram of kerosene. Discuss potential problems with using a kerosene heater in a confined area, such as a camper trailer, where the supply of air may be limited. Support your discussion with balanced chemical equations.

15. A group of students tested two white, crystalline solids, A and B, to determine their heats of solution. The students dissolved 10.00 g of each solid in 100.0 mL of water in a polystyrene calorimeter and collected the temperature data. They obtained the following data:

Time	Temperature (A)	Temperature (B)
0.0	15	25
0.5	20.1	18.8
1.0	25	16.7
1.5	29.8	15.8
2.0	31.9	15.2
2.5	32.8	15
3.0	33	15
3.5	33	15.2
4.0	32.8	15.5
4.5	32.5	15.8
5.0	32.2	16.1
5.5	31.9	16.4

(a) Graph the data in the table above, placing time on the *x*-axis and temperature on the *y*-axis. Use your graph to answer (b) to (d).
(b) Classify the heat of solution of each solid as exothermic or endothermic.
(c) From the data given, calculate the heat of solution for each solid. Your answer should be in kJ/g.
(d) Is there any evidence from the data that the students' calorimeter could have been more efficient? Explain your answer.

16. An unknown hydrocarbon gas was collected in test tubes. It was mixed with pure oxygen gas in the following ratios:

Test tube	% gas	% oxygen
1	75	25
2	50	50
3	33	67
4	25	75

A match was held up to the mouth of each test tube to ignite the contents. Afterwards, limewater was added to each test tube to test for the presence of carbon dioxide. (**Note:** The presence of large amounts of carbon dioxide indicates more complete combustion.) The following results were recorded.

Test tube	Result of combustion	Carbon dioxide test
1	no sound; smoke and black solid formed	remained almost clear, so very little $CO_{2(g)}$ present
2	small pop; less smoke; less black solid	turned slightly milky so some $CO_{2(g)}$ present
3	loud pop; no smoke or black solid	turned very milky so alot of $CO_{2(g)}$ present
4	small pop; no smoke or black solid	turned slightly milky so some $CO_{2(g)}$ present

The gas may be methane, ethane, or propane.

(a) Write balanced chemical equations for the complete combustion of methane, ethane, and propane.
(b) Which test tube indicates the most complete combustion? Why?
(c) Use your answer to question (b) and your balanced chemical equations to determine the most likely identity of the unknown gas. Support your answer with a clear explanation.
(d) Write one possible balanced equation for the incomplete combustion of the gas you chose in question (c).
(e) Both methane and propane are used in Canada to heat homes. Discuss the possible consequences if these gases are burned in furnaces or appliances that do not have adequate ventilation.

Communication

17. Use a concept map to trace the path of a six-carbon hydrocarbon molecule as it goes from its source in the ground, through processing, into a can of paint. Describe each step in its processing. Explain any changes that the molecule undergoes.

18. What do "saturated" and "unsaturated" mean when applied to hydrocarbons? Give an example of a saturated hydrocarbon and an example of an unsaturated hydrocarbon.

19. Why do alkanes and alkynes, unlike alkenes, have no geometric isomers?

20. Explain why carbon is able to form so many more compounds than any other element.

21. Discuss how you can determine whether the following compounds are alkanes, cycloalkanes, alkenes, or alkynes, without drawing their structural diagrams: C_6H_{12}, C_4H_6, C_5H_{12}, C_7H_{14}, and C_3H_4.

22. Describe the information that is included in the following thermochemical equation:

$$N_{2(g)} + 3H_{2(g)} \rightarrow 2HN_{3(g)} + 92.2 \text{ kJ}$$

23. Define calorimetry. Describe two commonly used calorimeters.

24. When taking a calorimetric measurement, why do you need to know the heat capacity of the calorimeter?

25. Describe two exothermic processes and two endothermic processes.

26. Compare the following terms: specific heat capacity and heat capacity.
   (a) Write their symbols and their units.
   (b) Write a mathematical formula in which each term would be used.

Making Connections

27. You will often see numbers on gas pumps that indicate how efficiently the fuel burns. A higher number indicates a more efficient fuel. These numbers refer to the "octane rating" of the fuel. There is a direct relationship between the octane rating and the degree of branching in fuel hydrocarbons. The presence of branches increases the efficiency of a fuel. Arrange the following hydrocarbon compounds in order of increasing octane rating: 2-methylhexane, heptane, and 2,2,4-trimethylpentane. Explain.

28. When walking briskly, you use about 20 kJ of energy per minute. One serving of a whole wheat cereal (37.5 g) provides about 126 Cal (food calories) of energy. Hint: 1 Cal = 4.184 kJ
   (a) How long could you walk after eating one serving of cereal?
   (b) How far could you walk at 60 km/h?

29. As you learned in previous chapters, research has been focused on using hydrogen gas as a fuel. Hydrogen is being researched as a fuel. It burns according to the following equation:

$$2H_{2(g)} + O_{2(g)} \rightarrow 2H_2O_{(\ell)}$$

The heat of combustion for hydrogen is -141.5 kJ/g.
   (a) Calculate the energy produced by the combustion of one litre (0.702 kg) of octane, $C_8H_{8(\ell)}$, in gasoline. The heat of combustion of octane is -41.3 kJ/g.
   (b) Calculate the volume of hydrogen gas at SATP that is required to produce the same amount of energy as the octane in part (a).
   (c) How could hydrogen gas be stored in order to carry a practical amount in a passenger vehicle? What possible danger might this create?

30. Government leaders often use the perspectives of human health and well-being, the environment, and the economy to study the risks and benefits of various fuels and alternatives. Suppose you are a leader in a developing country. You need to choose one of these energy sources to meet your country's needs: hydrocarbon fuels, hydroelectric power, or nuclear energy. Write a small report, analyzing each energy source from the three perspectives given above. Which energy source would you choose? Why?

COURSE CHALLENGE

Planet Unknown

Consider the following as you continue to plan for your Chemistry Course Challenge:

- How can you measure the energy content of a fossil fuel in the laboratory?
- How can you calculate heat using mass, temperature change, and specific heat capacity?
- What conditions are necessary in order for a large amount of fungus to be converted into a fossil fuel?

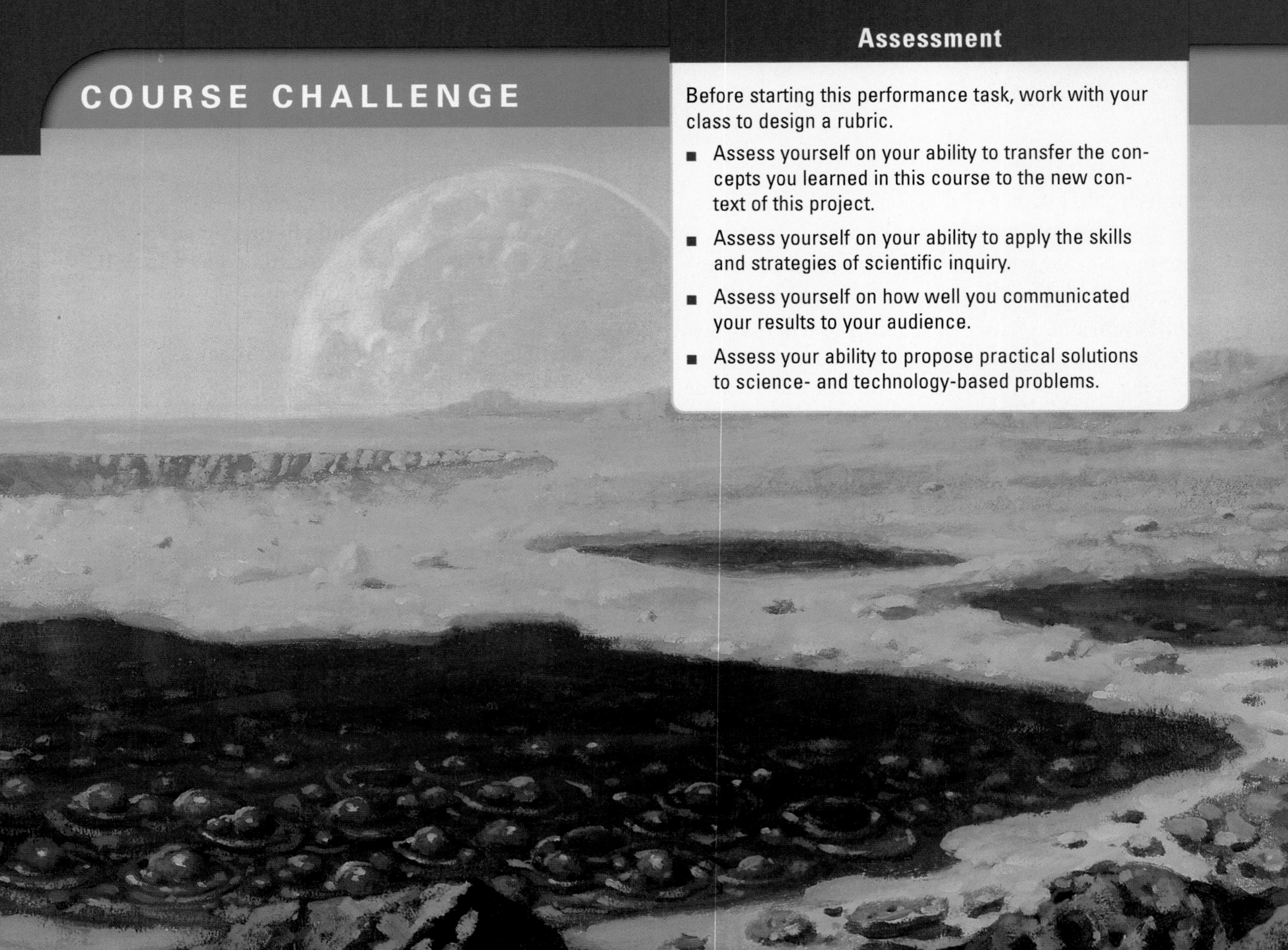

COURSE CHALLENGE

Assessment

Before starting this performance task, work with your class to design a rubric.

- Assess yourself on your ability to transfer the concepts you learned in this course to the new context of this project.
- Assess yourself on your ability to apply the skills and strategies of scientific inquiry.
- Assess yourself on how well you communicated your results to your audience.
- Assess your ability to propose practical solutions to science- and technology-based problems.

Planet Unknown

It is the year 3000, and space travel is commonplace. Nearly one trillion human beings inhabit the Milky Way galaxy now. They reside on thousands of planets that have been made habitable. You are part of a scientific team that has been sent to a newly discovered planet. Your team must analyse the resources present on the planet. You will also decide on the technological processes needed to make the planet a safe place to live. This will involve removing dangerous toxins from the environment, and identifying possible sources of fuel, oxygen, water, and food.

The newly discovered planet has an insulating atmosphere and temperature range similar to Earth's. Each day on the planet is 25 Earth hours long. Every day contains 8 hours of bright sunlight. However, the planet has no ocean, no surface water, and no atmospheric water at all. There is evidence that plant life once existed on this planet. Without surface water, however, there are no plants living on the planet's surface at the present time. A small population of bacteria may be present.

One side of the planet contains a dry and rocky desert, covered in part by a large crater. A thick layer of rock covers the other side of the planet. Under this rocky surface runs a wide network of lava-tube tunnels. Scientists hypothesize that a very large meteor may have struck the planet at one time. The meteor knocked off a chunk of the planet. This chunk now orbits the planet as a moon.

CHECKPOINT

Choose a name for the planet you will explore.

Scientists believe that the enormous heat of this collision decomposed all surface water into hydrogen and oxygen gases. This explains the lack of surface water on the planet. The heat also triggered massive volcanic eruptions. These eruptions resulted in the rocky surface and lava tunnels that cover half the planet.

Fortunately, organic dirt deposits still exist underground, in pockets beneath the lava rock. These dirt deposits would be useful for growing food. A sufficient air supply, plant nutrients, and light will also be necessary in order to grow food.

This challenge comprises five parts, each featuring a unique challenge of its own. Each part reflects the content of a unit from the *Chemistry 11* course you have completed. As you work through the parts, you will apply your knowledge from this chemistry course to make this planet habitable. CAUTION **In any hands-on investigation, carefully consider and carry out all necessary safety precautions**.

Part 1: Trends in Elements

Preview

In Part 1, you will use the skills and content you learned in Unit 1, Matter and Chemical Bonding. In particular, you will be applying your knowledge of elements and the periodic table.

Your science team has discovered some elemental metals on the planet. After testing each element, you have obtained the information given in the data chart below. You have not yet been able to obtain the atomic masses. In addition, these are preliminary tests, so the degree of accuracy is not high. However, you now have enough information to analyse the periodic trends of these new elements.

	Atomic radius (nm)	Observations	Electronegativity (Pauling)	Ionization energy (kJ/mol)	Melting point (K)
A	0.164	silvery white; turns yellow or pinkish in air	1.37	632	1801
B	0.180	lustrous, silver, metallic	1.20	614	1784
C	0.188	silvery white; can be cut with a knife	1.11	537	1188
D	0.189	silver metal; radioactive	1.10	492	1330
E	0.183	soft, silver, malleable, and ductile; gives an intense yellow colour when used as a component in glass	1.12	523	1211
F	0.182	bright silver, metallic; gives a violet, wine red, or grey colour when used as a component	1.13	532	1295
G	0.181	the salts of the metal glow pale blue or green in the dark; radioactive	1.15	537	1321
H	0.180	bright silver; stable in air; absorbs infrared light when used as a component in glass	1.17	545	1352
I	0.199	silver metal	1.19	548	1099
J	0.180	silver metal	1.20	596	1580

mind STRETCH

One of the geologists on your team finds some fossilized wood on the planet. She attempts to find the age of the wood using carbon dating. In carbon dating, the relative amounts of carbon isotopes are compared. Since carbon isotopes decay at a regular rate, this information can be used to find the age of carbon compounds. However, the isotopic ratios of carbon on this planet are different from those on Earth. The isotopic ratios of carbon on this planet are 80.2 % ^{12}C, 19.3% ^{13}C, and 0.5% ^{14}C.

(a) What are the isotopic ratios of carbon on Earth?

(b) What is the average molar mass of carbon on the unknown planet?

Challenge

Analyse any periodic trends that appear in the data. Trends will be similar to those in Earth's periodic table. (**Hint:** Part of the list belongs to a group. Another part of the list belongs to a period.)

1. I Explain the periodic trends seen in the chart. How does your explanation correlate with Earth's periodic table? Which trends on Earth's periodic table increase from left to right? Which trends generally increase from top to bottom? Which data do not seem to fit the trend? Can you suggest a possible explanation?

2. K/U Which metal is more reactive, A or C? Which metal is the least reactive, F or G? How do you know?

3. C Assume that these metals are also present on Earth. Use Earth's periodic table to identify a possible element on Earth that matches each unknown element. Explain your reasoning.

Part 2: Growing Food

In your explorations, your team discovers a particularly large underground cavern. The cavern has a steady water source seeping out of phosphate rock. This trickling water picks up a weak concentration of phosphate ions as it filters through the rock. There is plenty of soil in this cavern, although extra plant nutrients will need to be added on a regular basis as crops are grown and harvested. This cavern may be an excellent site for growing vegetables and other plants. A source of light will be needed.

Preview

In Part 2, you will apply the skills you learned in Unit 2, Quantities in Chemical Reactions.

Challenge

To grow sufficient crops, you will need large amounts of calcium phosphate fertilizer. You can make this compound by combining the phosphate solution in the stream with calcium cations. Simulate this reaction in the laboratory. Design a procedure to find the mass of dry calcium phosphate precipitate. Mix 50 mL of a 0.10 mol/L aqueous solution of sodium phosphate with sufficient calcium nitrate to obtain a precipitate. Filter the precipitate, dry it, and determine its mass.

1. K/U Write the balanced chemical equation and net ionic equation for the reaction described in the Challenge.
2. I Write up a procedure for your investigation, and have your teacher approve it before you begin. Before starting, discuss appropriate safety precautions with your teacher. Wear eye protection. Wash your hands after any contact with chemicals.
3. C Prepare a detailed report to share your findings. Include all of your observations and answers to the questions in Part 2.
4. I In your report, compare the theoretical yield with the actual yield of the product. Identify possible sources of error in your procedure.
5. I If time permits, modify your procedure to improve your results. Keep a detailed record of the changes you made to the procedure, the observed benefits, and the new percent yields obtained.

Think and Research

1. I Using the reactant amounts from your experiment, what mass of sodium nitrate would you expect to get from the reaction? How could you collect this product to test your calculations?
2. I Show the calculations you used to find the amount of calcium nitrate necessary to react with the phosphate solution.
3. I If 600 L of 0.10 mol/L solution of sodium phosphate were reacted with 10 kg of calcium nitrate, what would the limiting reagent be?

Extension

4. MC What other modifications will you need to make the cavern suitable for growing plants? For example, would you need to add other nutrients as fertilizer? Do sources of these nutrients exist on the planet already? Write the chemical equations for the reactions.

Part 3: What's in the Water?

Preview

This part will call upon what you learned in Unit 3, Solutions and Solubility. In particular, you will review solutions, titrations, and acid-base chemistry.

Life as we know it is impossible without water. Thus it is essential to establish a good water source. Although there is no surface water of any kind, the planet does have water—underground! Since this water is constantly filtered through underground rocks and soil, it contains many dissolved substances. Some of these substances may be dangerous to human health. These materials will have to be removed in order to obtain clean water. How can you find out what substances are present in the water?

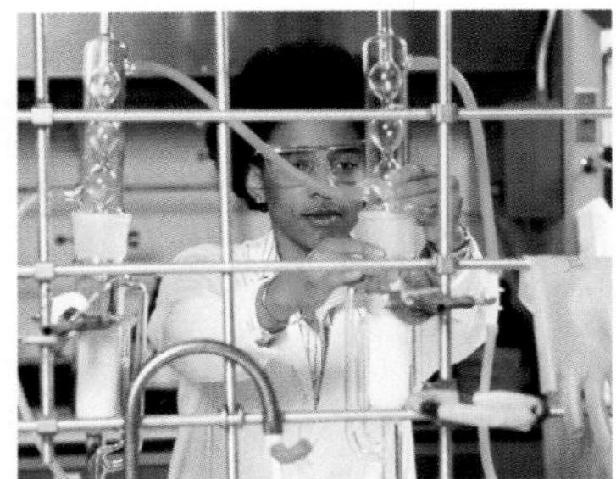

Challenge

Your teacher will give you a sample of water representing the planet's underground water. Analyse the chemical composition of the water sample. (Note that the sample you have may not be the same as the samples of your classmates.)

There are several methods you can use to analyse specific components of your sample. These may involve:

- titration
- precipitation
- acid-base tests
- evaporation
- distillation

1 I Choose one or more methods to analyse your sample. Your teacher will give you further guidance on the methods that are most useful for your specific sample.

2 I Write up a procedure for your investigation, and have your teacher approve it before you begin. Before starting, discuss appropriate safety precautions with your teacher. Wear eye protection, gloves, and an apron.

3 C After you have analysed your sample, prepare a detailed report.

Think and Research

1 I What else might you need to know about your water sample? How could you find out?

2 C Write a brief report that lists the different methods used to clean water on Earth. Explain how each method works. Remember to include diagrams and any relevant balanced chemical equations.

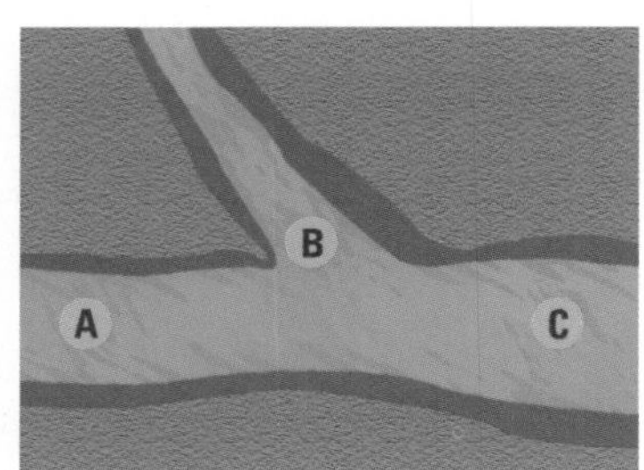

At point A, the river is clean. At point C, however, the river is contaminated with the toxic nitrite ion. How can you contain the pollution at point B?

Extension

3 I A large river of clean, fresh water flows through the largest lava tunnel. At point B, a small river flows into the larger river. This smaller river contains dangerous levels of toxic nitrite ions (NO_2^-). To prevent the smaller river from contaminating the larger one, you have decided to limit the flow of the small river. Safety standards permit a maximum of 3.2 mg of nitrite per litre of water. The flow of the larger river is 1.2×10^4 L/s. The concentration of nitrite in the smaller river is 51.2 ppm. What amount of water (in L/s) can be allowed to enter the larger river from the smaller river?

Part 4: Planetary Gases

The atmosphere of the planet is composed of the following gases:

argon (49.2%)	carbon dioxide (1.3%)
oxygen (30.8%)	ammonia (0.9%)
nitrogen (14.6%)	dinitrogen oxide (0.5%)
hydrogen cyanide (2.3%)	hydrogen sulfide (0.4%)

Although the atmosphere contains oxygen, it also contains gases that are toxic to humans. Eventually, you hope to build a settlement inside a huge synthetic bubble with its own safe atmospheric mixture of gases. First, your team must find an alternate source of oxygen gas. On Earth, plants supply oxygen through the process of photosynthesis. This may be the best method of providing a constant supply of breathable air for the settlement bubble on the new planet. But do plants give off pure oxygen, or a mixture of gases? How much oxygen does a certain mass of plants give off?

Challenge

Design an investigation to collect the gas given off by a certain mass of Rotela water plants. Find the mass and volume of the gas. Use this data to calculate the molar mass and molar volume of the gas. (**Note:** This experiment will take several days to do correctly. If possible, do multiple trials.)

1. K/U Write the balanced equations for photosynthesis and for cellular respiration.
2. I Write up a procedure for your investigation, and have your teacher approve it before you begin. Before starting, discuss appropriate safety precautions with your teacher. Wash your hands after handling the *Rotela* plants.
3. I Carry out your procedure. Use the volume and the mass of the gas, along with the temperature of the water and pressure of the room, to calculate the average molar mass and the molar volume of the gas.
4. C Prepare a detailed report to share your findings. In your report, compare the molar mass of the gas you calculated to the molar mass of oxygen. Explain the difference between these values. Identify possible sources of error in your procedure.

Think and Research

1. I Your team sends a balloon 50 m up into the atmosphere of the planet. The balloon contains 20.3 mol of helium gas. At a temperature of 313 K, the balloon has a volume of 614 L. Assume that the universal gas constant, R, has the same value as it would on Earth. What is the atmospheric pressure on this planet at 313 K? What is it at 0°C?
2. I Use the pressure at 0°C from the previous question. What is the partial pressure of each of the components of the atmosphere?
3. I What volume will 1.44 kg of oxygen gas occupy at the planet's atmospheric pressure? (Assume a temperature of 0°C.)
4. I Rearrange the ideal gas equation to calculate the molar volume for any ideal gas on this planet at 0°C.

Preview

In this part, you will use the skills you learned in Unit 4, Gases and Atmospheric Chemistry.

mind STRETCH

During cellular respiration, plant cells use oxygen to break down food into energy. Thus, plants take in oxygen and give off carbon dioxide. When in the presence of light, plants also take in carbon dioxide and give off oxygen during photosynthesis. Does cellular respiration use up the same amount of oxygen that is given off by photosynthesis? Do plants give off a net amount of oxygen, carbon dioxide, or a combination of the two? Do research to find out.

PROBLEM TIP

(a) Photosynthesis only occurs during the hours of sunlight. At night, only respiration is taking place. How might this affect your investigation?

(b) Remember to include the vapour pressure of water in your calculations.

(c) Did the temperature or pressure change over the course of the experiment? Design your investigation to include some checks on this data.

Part 5: Finding Fuel

Preview

This part draws upon the skills and concepts you learned in Unit 5. Your main focus will be hydrocarbon fuels and the heat equation.

While exploring the lava tunnels, your team discovers a coal-like substance. This substance may have originated as a fungus growing in underground caves. When the meteor hit the planet, the caves collapsed. Over millions of years, the pressure of the rock converted the organic material into a fossil fuel.

Challenge

Your teacher will give you a sample of the coal-like substance discovered on the planet. Design a calorimeter to measure the heat of combustion of this substance. Decide whether this substance will be useful as a fuel.

1. I Write up a procedure for your investigation, and have your teacher approve it before you begin. Before starting, discuss appropriate safety precautions with your teacher. Wear eye protection, and use care with hot surfaces.
2. I Carry out your procedure. Calculate the heat of combustion per gram of your sample.
3. C Prepare a detailed report to share your findings.

Think and Research

Your team has also discovered large pools of hydrocarbon oils on the surface of the planet. The hydrocarbons seem similar to hydrocarbons on Earth, and may have been formed in the same way. When analysed, the oil proves to be composed entirely of compounds with the molecular formula C_7H_{16}.

1. K/U How might these hydrocarbon oils have originated?
2. K/U Draw and name all the possible isomers that have this molecular formula.
3. MC What might be done with the open pools of oil? Give two or three possibilities.
4. I On another area of the planet, your team discovers a vent with gas seeping from it. The gas is found to be 81.7% carbon and 18.3% hydrogen by mass. The molecular mass of the gas is 44.11 g/mol. What is the formula of the compound? Name the gas. Draw the structural formula of the compound.
5. MC Suppose you decide to use the coal-like substance and the pools of oil as fuel. How might your use of these fuels affect the environment on the planet? What precautions will you take to prevent environmental damage?

Part 6: Final Presentation

What other challenges might humans face while developing this planet? You may want to investigate a different problem. For example:

Unit 1

- Suppose matter on the new planet were not composed of electrons, protons, and neutrons. Devise a new atomic theory of your own. Explain how it works in "real life" on the planet. Make up a small, functional periodic table with new, imaginary elements.
- Develop rules that explain the bonding between your new atoms. How do these rules affect the reactivity of your elements and structure of your compounds?

Unit 2

- Suggest reactions that may make life easier on the planet. Research and perform these reactions in the laboratory. Write balanced equations, and calculate percent yields in moles and in grams. Explain the purpose of these reactions in the settlers lives.

Unit 3

- Suggest one or more substances that may already be dissolved in the water on the newly-discovered planet. Test the solubility of the substance and plot solubility curves for the solute in water. Explain why the substance is in the water, and how the substance will affect the life of the settlers. For example, is it safe to drink? How can it be removed from the water? Once extracted, can it be used for anything else?

Unit 4

- Choose a reaction producing a gas that may be useful to settlers on the new planet. Research the reaction, then carry it out in the laboratory. Collect the gas produced and determine its molar volume.

Unit 5

- Choose three of these compounds: NaOH, anhydrous $CaCl_2$, LiBr, HCl, anhydrous Na_2SO_4, $AgNO_3$, KI, NH_4NO_3, $Na_2CO_3 \cdot 10H_2O$, and $KMnO_4$. Design an investigation to measure and compare the heats of solution per mole of these chemicals. Choose one chemical based on its heat of solution, and its safety for everyday use. Come up with a plan or device that uses the heating or cooling effect produced by this process to help the planet's settlers remain comfortable.

Your presentation After preparing lab reports on one or more sections of this Course Challenge, you will share your results in a class presentation. Assume this presentation is being made to the Canadian Minister of Emigration, the Canadian Minister of the Economy, and a group of business people who may wish to invest in the development of the planet.

- Provide your audience with solutions to the different scenarios and problems you investigated. Be prepared to respond to other possible questions and problems your audience (your class and teacher) may raise.
- To enhance your presentation, try to include two or more of the following: posters, brochures, drama, music, lab demonstrations, video or audio clips, charts or graphs, surveys, models, multimedia, or web pages with links.

Web LINK

If you have time, work as a class to create a web site based on this Course Challenge. Create a virtual tour of your planet's resources, projected living spaces, and scientific adaptations. Share your findings with other grade eleven classes in your province over the Internet.

Appendix A

Answers to Numerical Chapter and Unit Review Questions

Chapter 1

5. (a) 4.0×10^2 mL; 10.0 mL; 1.0×10^2 mL
   (b) 1.7×10^2 mL
6. (a) 1.0×10^4 g (b) 2.23×10^{-1} m
   (c) 52 cm^3 (d) 1.0×10^3 cm^3
7. (a) 1 (b) 4 (c) 1 (d) 2
   (e) 5 (f) 4 (g) 5
8. (b) 5.7×10^3 km (c) 5.700×10^3 km
9. (a) 8.73 mL (b) 1.1×10^5 m^2
   (c) 2.2×10^2 kg/L (d) 0.7
   (e) 1.225×10^4 L (f) 1.8×10^1 g/mL
10. (a) 6.21×10^3 (b) 3×10^1
    (c) 6×10^2 (d) 1.72×10^1
11. 1.9×10^4 cm^3
12. (a) 2.4×10^1 °C
    (b) the tenths digit, to the right of the decimal

Chapter 2

6. (a) 7 (b) 7 (c) 10 (d) 3–
   (e) ${}^{79}_{34}Se^{2-}$ (f) 2– (g) Cr (h) 24
   (i) 28 (j) 21 (k) 3+ (l) 19
   (m) 9 (n) 9 (o) 9 (p) 0
7. 32 neutrons; 27 electrons
8. (a) 2.40×10^6 (b) 2.13×10^6

Chapter 3

2. (a) 1.79 (b) 1.35 (c) 1.28 (d) 0.40
14. (a) 1.55; 1.49; 2.43
    (b) 2.41; 1.59; 1.39
    (c) 1.62; 1.33; 1.26; 1.23; 1.30
16. (a) Ag, 1; Cl, 1 (b) Mn, 2; P, 3
    (c) P, 5; Cl, 1 (d) C, 4; H, 1
    (e) Ti, 4; O, 2 (f) Hg, 2; F, 1
    (g) Ca, 2; O, 2 (h) Fe, 2; S, 2

Unit 1

42. (a) 21.5 (b) 58 cm^3
    (c) 19.3 kg/dm^3 (d) 17.5 g
    (e) 298°C
44. 7, 7, 10, ${}^{32}S^{2-}$, 16, ${}^{4}He$, 2, 4, 2, ${}^{38}Ca$, 18

Chapter 5

9. 40
10. 69.8 u
11. 72.71 u
12. K-39, 95.0%; K-40, 5.0%
13. (a) 2.84×10^{-3} mol (b) 5.17×10^{-1} mol
    (c) 8.16×10^{-5} mol (d) 4.38 mol
    (e) 0.126 mol
14. (a) 7.09×10^{-3} mol (b) 1.23 mol
    (c) 8.93×10^{-2} mol (d) 4.83×10^{-1} mol
    (e) 2.69 mol (f) 9.27×10^{-4} mol
15. NH_3: 17.0, 8.77×10^{23}, 1.46, 5.84; H_2O: 18.0, 1.58, 8.77×10^{-2}, 2.63×10^{-1}; Mn_2O_3: 158, 1.05×10^1, 4.00×10^{22}, 6.64×10^{-2}; K_2CrO_4: 246, 2.37×10^{21}, 3.92×10^{-3}, 2.75×10^{-2}; $C_8H_8O_3$: 152, 2.00×10^3, 1.31×10^1, 2.49×10^2; $Al(OH)_3$: 78.0, 6.66×10^4, 5.14×10^{26}, 5.98×10^3
16. (a) 355 (b) 74.1
    (c) 142 (d) 252
    (e) 310 (f) 183
17. (a) 6.66×10^3 (b) 3.35×10^2
    (c) 3.75×10^3 (d) 1.45
    (e) 5.74×10^1 (f) 2.05×10^{-2}
18. 2.11×10^{24}
19. 1.3×10^{24}
20. (a) 131 (b) 131
    (c) 2.18×10^{-22} (d) 7.89×10^{25}
    (e) 6.02×10^{23}
21. 3.01×10^{24}
22. 6.57×10^{24}
23. 4.53×10^{23}
24. 192 g
25. Br-79, 55.0%; Br-81, 45.0%
26. (a) 1.00 (b) 84.9
27. (a) 14 (b) 16
    (c) 4.2×10^{-3}
32. (a) 2.65×10^{-4} (b) 1.19×10^{-1}
33. 101 mg

Chapter 6

5. 2.64 g
6. (a) 9.9% C; 58.6% Cl; 31.4% F
   (b) 80.1% Pb; 16.5% O; 0.3% H; 3.1% C
7. (a) 6.86 g (b) 1.74 g
8. (a) 63.5% (b) 127 kg
9. 26.9 g
10. 168 g
11. $C_4H_8O_4$
12. $C_{12}H_4O_2Cl_4$
13. (b) C_3H_8
14. $C_{21}H_{30}O_2$
15. $Na_2Cr_2O_7$
16. $HgSO_4$
17. (a) $Ca_3P_2O_8$ (b) $Ca_3(PO_4)_2$
18. (a) $C_{18}H_{26}O_3N$ (b) $C_{18}H_{26}O_3N$
19. V

20. C_3H_6O

21. 7

22. (a) 37.5% C; 4.2% H; 58.3% O
(b) $C_6H_8O_7$ (c) $C_6H_8O_7$

23. CO_2, 1.37 g; H_2O, 1.13 g

28. (a) 0.87 g (b) 0.56 g

29. (b) 60.3% (c) 1.0×10^2 kg

30. (b) $C_9H_8O_4$

Chapter 7

6. 9.60 g

7. 8.0×10^{22}

8. 292 g

9. (b) 8.94 g (c) 0.407 g S

10. 36.1 g

11. 2.04 g

12. 22.6 g

13. 1.65 g

14. 2.58 g

15. (a) 7.32 g (b) 34.2%
(c) 7.23 g

16. 53.7%

17. (a) 15.1 g (b) 0.414 g
(c) 14.2 g (d) 94.3%

19. (a) 3.24 g (b) 270 g

23. 5.9×10^5 L air

24. (d) 7×10^{-2}

Unit 2

29. (a) 6.02×10^{23} molecules; 1.21×10^{24} atoms
(b) 3.0×10^{24} ions (c) 2.3×10^{24} atoms

33. (a) 0.167 mol (b) 1.00×10^{23} mol
(c) 3.01×10^{23} atoms

34. 1.94×10^{23}

35. 1.20×10^{22}

37. 68.1% C; 13.7% H; 18.1% O

38. 2.84 g

42. (a) 0.015 mm (b) 2.7 nm

Chapter 8

11. 6.25 g

12. (a) 25 g (b) 225 g
(c) 2.50 mol/L

13. 96 mL

14. 1.2 ppm

15. 5.67 mol/L

16. 0.427 mol/L

17. (a) 9.89 g (b) 83 g

18. (a) 1.7 mol/L (b) 1.44 mol/L

19. (a) 0.381 mol/L (b) 0.25 mol/L

20. 10.0 g

23. (b) approximately 90 g (c) approximately 145 g
(d) approximately 76°C

24. approximately 380 g

25. 40°C

29. 0.25 ppm; 250 ppb

Chapter 9

13. 1.5×10^{-2} mol/L

14. (a) 2.69×10^{-3} mol/L (b) 1.88×10^{-2} mol/L
(c) 0.999 mol/L

15. 0.104 mol/L

16. $K^+ = 6.67 \times 10^{-2}$ mol/L; $Ca^{2+} = 6.67 \times 10^{-2}$ mol/L; $NO_3^- = 0.200$ mol/L

17. iron(III) chloride

18. (a) $Ca^{2+} = 4.5 \times 10^{-2}$ mol/L; $Mg^{2+} = 2.4 \times 10^{-2}$ mol/L
(b) $Ca^{2+} = 1.8 \times 10^3$ ppm; $Mg^{2+} = 5.8 \times 10^2$ ppm

19. 1.4×10^{-2} mol

20. 12.2 g

22. (b) 0.01057 mol (c) 120.4 g/mol

24. (b) 0.09600%

Chapter 10

10. (b) half

13. (a) 1.0×10^3 g (b) 55 mol

14. 0.800 mol/L

15. 0.020 mol/L

16. 1.0×10^{-5} mol

18. (a) 2.399 mol/L (b) no (because 9.317%)

Unit 3

22. 3.6×10^{-3} g

23. 1.25×10^{-5} g

Chapter 11

16. 0.50 L

17. 37.5 atm

18. 1.31×10^3 torr

19. 21 L

20. 2.8 L

21. −233

22. (a) 16.2 L (b) 138 cm^3
(c) 109 mL

23. 2.2 L

24. 50.2 kPa

25. 3.7×10^2 °C

26. 62.5 kPa

27. 273°C

28. 262 kPa

29. 68 days

30. –68°C

31. 24 atm

32. 2.3×10^3 L

33. 21 mL

34. 752 mm

35. (a) 1.5×10^2 (b) 1.017×10^3 (c) 750

36. 1.666×10^1, 1.066×10^1

37. (a) 87.5 (b) −148

Chapter 12

12. (a) 22.4 L (b) 2.80×10^2 L (c) 5.60×10^2 L

13. 6.72×10^{24}

14. 50.4 L

15. 3.28×10^{-1}

16. 9.68×10^2 kPa

17. 1.80×10^3 g

18. (a) 1.69 g/L (b) 1.66 g/L

19. (a) 2.40 g (b) 1.75 g

20. (a) 20 g/mol

21. (a) 7.46 mol (b) 628 kPa (c) 210 kPa

22. 12°C

23. C_2H_6O; $C_4H_{12}O_2$

24. C_8H_{18}

25. CH_3Cl

26. (b) 10 L (c) 14.3 g

27. 4.35 g

28. 3.00 L

29. 7.33 L

30. 1.28×10^3 L

31. 1.70 L

32. (b) 0.0514 g (c) 0.735 g

35. (a) 8.314 kPa•L/mol•K (b) 0.08206 atm•L/mol•K (c) 62.36 torr•L/mol•K

36. (a) 68 L (b) 3.4 L; 3.4 L (c) 2.5 mL

Unit 4

16. 0°C

17. 0.0005 mol

18. 2.0 L

19. 202.6 mL

20. 56.8 atm

21. 368 kPa

22. 374 dm^3

23. 49.3°C

24. 1.5

25. 4.6; 6.2×10^2; 1.03×10^3; −15.8; 15.1; 88

26. 54.7 mL

27. 7.5×10^2 kPa, 4.8×10^2 kPa, 1.10×10^3 kPa, 2.4×10^2 kPa

28. 2.5×10^2 L

29. 78 kPa

30. 22.4°C

31. 16 kg

32. 7.5 L, 5.0 L

33. 0.34 L

34. 29.5 g

35. 0.69 L

36. $C_4H_{10}O$

37. $C_{10}O_2H_{20}$

38. (a) 9.7×10^{21} molecules (b) 2.6×10^{21} molecules (c) 1.2×10^{20} molecules (d) 3.7×10^{18} molecules

39. 1.94×10^{11} mol

Chapter 14

8. (b) 121 kJ

13. 322 kJ

14. Sample 2

15. 4.93 Cal/g

16. 158 kJ

17. 188 kJ

18. (a) 5.0 kJ/g, 1.2 Cal/g (b) 3.6 Cal/g

20. 2.51 kJ

21. 9.68 kJ/°C

22. 39.3°C

24. 1.1×10^3 L

25. (a) 56.2 mL (b) 34.9 L

Unit 5

15. (c) –0.75 kJ/g, 0.42 kJ/g

28. (a) 26 min (b) 2.6 km

29. (a) 2.90×10^4 kJ (b) 2.52×10^3 L

Appendix B

Supplemental Practice Problems

UNIT 1

Chapter 1

1. A student measures the mass of five different ingots of aluminum to be 28.6 g, 28.72 g, 28.5 g, 29.0 g, 28.6 g. What is the average mass of these ingots?
2. Label each as either a physical or a chemical property.
   (a) The boiling point of water is 100°C.
   (b) Chlorine gas reacts violently with sodium metal.
   (c) Bromine has a brown colour.
   (d) Sulfuric acid causes burns when it comes in contact with skin.
3. The population of Canada is about 30 million. Express this amount to the correct number of significant digits.
4. How many significant digits are in the following quantities?
   (a) 624 students
   (b) 22.40 mL of water
   (c) 0.00786 g of platinum
5. Round off the following measured quantities to the number of significant digits specified.
   (a) 9.276×10^3 m (2 significant digits)
   (b) 87.45 g (3 significant digits)
   (c) 93.951 kg (3 significant digits)
6. The masses of several samples of titanium were measured to be: 193.67g; 28.9 g; 78 g; 4.946×10^{-1} kg. These samples were all put into an overflow can together. The water displaced had a volume of 176.1 mL. What is the average density of the titanium pieces?
7. Characterize each of the following occurrences as a physical or as a chemical change.
   (a) sugar is heated over a flame and caramelises (turns black)
   (b) blood clots
   (c) a rubber band is stretched until it snaps
   (d) a match burns
   (e) a grape is crushed
   (f) salt is put on the roads in the winter, melting the ice.

Chapter 2

8. Draw Lewis structures to represent each of the following atoms.
   (a) Mg **(b)** K **(c)** Ne
   (d) B **(e)** C **(f)** Al
9. Draw Lewis structures to represent each of the following ions.
   (a) H^+ **(b)** K^+ **(c)** F^-
   (d) S^{2-} **(e)** Al^{3+} **(f)** Br^-
10. Using only a periodic table (not the values for atomic radius), rank the following sets of atoms in order of increasing size.
    (a) W, Cr, Mo **(b)** As, Ca, K
    (c) I, Cl, F, Br **(d)** Cl, P, Mg
    (e) Zn, Cd, Hg
11. Using only a periodic table (not the values for ionization energy), rank the following sets of atoms in order of increasing ionization energy.
    (a) Ar, Xe, Ne **(b)** P, Al, Cl
    (c) Rb, Li, K **(d)** Mg, Be, Ca

Chapter 3

12. Predict whether each of the following bonds has a primarily ionic or covalent character.
    (a) B–F **(b)** C–H
    (c) Na–Cl **(d)** Si–O
13. Draw Lewis structures representing the following ionic compounds.
    (a) KBr **(b)** CaO
    (c) $MgCl_2$ **(d)** Mg_3N_2
14. Draw Lewis structures to represent the following covalent compounds.
    (a) F_2 **(b)** CH_4
    (c) CO_2 **(d)** CO
    (e) NO **(f)** N_2
15. For the previous problem, indicate any polar covalent bonds with a partial negative or positive charge on the appropriate atom.
16. Name each of the following ionic compounds.
    (a) $MgCl_2$ **(b)** Na_2O
    (c) $FeCl_3$ **(d)** CuO
    (e) $AlBr_3$
17. Write the chemical formula for each of the following compounds.
    (a) aluminum bromide
    (b) magnesium oxide
    (c) sodium sulfide
    (d) iron(II) oxide
    (e) copper(II) chloride
18. Write the formula for each of the following.
    (a) sodium hydrogen carbonate
    (b) potassium dichromate
    (c) sodium hypochlorite
    (d) lithium hydroxide
    (e) potassium permanganate
19. Name each of the following compounds.
    (a) K_2CrO_4 **(b)** NH_4NO_3
    (c) Na_2SO_4 **(d)** KH_2PO_4
    (e) $Sr_3(PO_4)_2$

20. Name each of the following covalent compounds.
    (a) Cl_2O_7 **(b)** H_2O
    (c) BF_3 **(d)** N_2O_4
    (e) N_2O

21. Write the formula for each of the following compounds.
    (a) tetraphosphorus decoxide
    (b) nitrogen trichloride
    (c) sulfur tetrafluoride
    (d) xenon hexafluoride

Chapter 4

22. Balance each of the following skeleton equations. Classify each chemical reaction.
    (a) $Fe + Cl_2 \rightarrow FeCl_2$
    (b) $FeCl_2 + Cl_2 \rightarrow FeCl_3$
    (c) $C_4H_{10}O + O_2 \rightarrow CO_2 + H_2O$
    (d) $Al + H_2SO_4 \rightarrow Al_2(SO_4)_3 + H_2$
    (e) $N_2O_5 + H_2O \rightarrow HNO_3$
    (f) $(NH_4)_2CO_3 \rightarrow NH_3 + CO_2 + H_2O$

23. Write the product(s) of each of the following chemical reactions. Also, identify the reaction type. In the case of no reaction, state NR.
    (a) $MgCO_{3(s)} \rightarrow$
    (The magnesium carbonate is heated.)
    (b) $Ca_{(s)} + Cl_{2(g)} \rightarrow$
    (c) $NH_4CO_{3(aq)} + KOH_{(aq)} \rightarrow$
    (Group I ions, hydrogen ions, and ammonium ions always form soluble ionic compounds.)
    (d) $I_{2(aq)} + KBr_{(aq)} \rightarrow$
    (e) $Na_2CO_{3(aq)} + MgCl_{2(aq)} \rightarrow$
    (Carbonate compounds form precipitates except when they contain ions from Group I, hydrogen, or ammonium. Group I ions form soluble ionic compounds.)
    (f) $K_{(s)} + O_{2(g)} \rightarrow$

24. Balance the following chemical equation.
    $BiCl_3 + NH_3 + H_2O \rightarrow Bi(OH)_3 + NH_4Cl$

25. Balance the equation.
    $NiSO_4 + NH_3 + H_2O \rightarrow Ni(NH_3)_6(OH)_2 + (NH_4)_2SO_4$

26. Complete and balance if necessary, each of the following nuclear equations.
    (a) ${}^{237}_{92}U \rightarrow {}^{0}_{-1}e +$
    (b) ${}^{231}_{90}Th \rightarrow {}^{231}_{91}Pa +$
    (c) ${}^{215}_{84}Po \rightarrow {}^{4}_{2}He +$

27. Write a balanced nuclear equation for each of the following.
    (a) Radon-233 decays with the emission of an alpha particle.
    (b) Actinium-228 decays with the emission of a beta particle.

28. Complete and balance each of the following nuclear equations.
    (a) ${}^{23}_{11}Na + \qquad \rightarrow {}^{23}_{12}Mg + {}^{1}_{0}n$
    (b) ${}^{96}_{42}Mo + {}^{4}_{2}He \rightarrow {}^{100}_{43}Tc +$
    (c) $\qquad + {}^{1}_{1}H \rightarrow {}^{29}_{14}Si + {}^{0}_{0}?$
    (d) ${}^{209}_{83}Bi + \qquad \rightarrow {}^{210}_{84}Po + {}^{1}_{0}n$

UNIT 2

Chapter 5

29. Gallium exists as two isotopes, Ga-69 and Ga-71.
    (a) How many protons and neutrons are in each isotope?
    (b) If Ga-69 exists in 60.0% relative abundance, estimate the average atomic mass of gallium using the mass numbers of the isotopes.

30. Rubidium exists as two isotopes: Rb-85 has a mass of 84.9117 u and Rb-87 has a mass of 86.9085 u. If the average atomic mass of rubidium is 85.4678, determine the relative abundance of each isotope.

31. You have 10 mL of isotopically labelled water, ${}^{3}H_2O$. That is, the water is made with the radioactive isotope of hydrogen, tritium, ${}^{3}H$. You pour the 10 mL of tritium-labelled water into an ocean and allow it to thoroughly mix with all the bodies of water on the earth. After the tritium-labelled water mixes thoroughly with the earth's ocean water, you remove 100 mL of ocean water. Estimate how many molecules of ${}^{3}H_2O$ will be in this 100 mL sample. (Assume that the average depth of the ocean is 5 km. The earth's surface is covered roughly two-thirds with water. The radius of the earth is about 6400 km.)

32. Calculate the molar mass of each of the following compounds.
    (a) $Al_2(CrO_4)_3$
    (b) $C_4H_9SiCl_3$ (n-butyltrichlorosilane, an intermediate in the synthesis of silicones)
    (c) $Cd(ClO_3)_2 \cdot 2H_2O$ (cadmium chlorate dihydrate, an oxidizing agent)

33. How many atoms are contained in 3.49 moles of manganese?

34. How many atoms are there in 8.56 g of sodium?

35. What is the mass of 5.67×10^{23} molecules of pentane, C_5H_{12}?

36. Consider a 23.9 g sample of ammonium carbonate, $(NH_4)_2CO_3$.
    (a) How many moles are in this sample?
    (b) How many formula units are in this sample?
    (c) How many atoms are in this sample?

Chapter 6

37. Pyridine, C_5H_5N, is a slightly yellow liquid with a nauseating odour. It is flammable and toxic by ingestion and inhalation. Pyridine is used in the synthesis of vitamins and drugs, and has many other uses in industrial chemistry. Determine the percentage composition of pyridine.

38. Bromine azide is an explosive compound that is composed of bromine and nitrogen. A sample of bromine azide was found to contain 2.35 g Br and 1.24 g N.
 (a) Calculate the percentage by mass of Br and N in bromine azide.
 (b) Calculate the empirical formula of bromine azide.
 (c) The molar mass of bromine azide is 122 g/mol. Determine its molecular formula.

39. Progesterone is a female hormone. It is 80.2% C, 9.62% H and 10.2% O by mass.
 (a) Determine the empirical formula of progesterone.
 (b) From the given data, is it possible to determine the molecular formula of progesterone? Explain your answer.

40. Potassium tartrate is a colourless, crystalline solid. It is 34.6% K, 21.1% C, 1.78% H, 42.4% O by mass.
 (a) Calculate the empirical formula of potassium tartrate.
 (b) If the molar mass of potassium tartrate is 226 g/mol, what is the molecular formula of potassium tartrate?

41. Menthol is a compound that contains C, H and O. It is derived from peppermint oil and is used in cough drops and chest rubs. When 0.2393 g of menthol is subjected to carbon-hydrogen combustion analysis, 0.6735 g of CO_2 and 0.2760 g of H_2O are obtained.
 (a) Determine the empirical formula of menthol.
 (b) If each menthol molecule contains one oxygen atom, what is the molecular formula of menthol?

42. Glycerol, $C_3H_8O_3$, also known as glycerin, is used in products that claim to protect and soften skin. Glycerol can be purchased at the drug store. If 0.784 g of glycerol is placed in a carbon-hydrogen combustion analyzer, what mass of CO_2 and H_2O will be expected?

43. Calculate the percentage by mass of water in potassium sulfite dihydrate, $K_2SO_3 \cdot 2H_2O$.

44. What mass of water is present in 24.7 g of cobaltous nitrate hexahydrate, $Co(NO_3)_2 \cdot 6H_2O$?

45. A chemist requires 1.28 g of sodium hypochlorite, NaOCl, to carry out an experiment, but only has sodium hypochlorite pentahydrate, $NaOCl \cdot 5H_2O$ in the lab. How many grams of the hydrate should the chemist use?

Chapter 7

46. Consider the equation corresponding to the decomposition of mercuric oxide.

 $2HgO_{(s)} \rightarrow 2Hg_{(l)} + O_{2(g)}$

 What mass of liquid mercury is produced when 5.79 g of mercuric oxide decomposes?

47. Examine the following equation.

 $C_3H_{8(g)} + 5\ O_{2(g)} \rightarrow 3\ CO_{2(g)} + 4\ H_2O_{(g)}$

 (a) What mass of propane, C_3H_8, reacting with excess oxygen, is required to produce 26.7 g of carbon dioxide gas?
 (b) How many oxygen molecules are required to react with 26.7 g of propane?

48. Metal hydrides, such as strontium hydride, SrH_2, react with water to form hydrogen gas and the corresponding metal hydroxide.

 $SrH_{2(s)} + 2H_2O_{(l)} \rightarrow Sr(OH)_{2(s)} + 2H_{2(g)}$

 (a) When 2.50 g of SrH_2 is reacted with 8.03×10^{22} molecules of water, what is the limiting reagent?
 (b) What mass of strontium hydroxide will be produced?

49. Consider the following successive reactions.
 reaction (1): A → B
 reaction (2): B → C
 If reaction (1) proceeds with a 45% yield and reaction (2) has a 70% yield, what is the overall yield for the reactions that convert A to C?

50. Disposable cigarette lighters contain liquid butane, C_4H_{10}. Butane undergoes complete combustion to carbon dioxide gas and water vapour according to the skeleton equation below:

 $C_4H_{10(l)} + O_{2(g)} \rightarrow CO_{2(g)} + H_2O_{(l)}$

 A particular lighter contains 5.00 mL of butane, which has a density of 0.579 g/mL.
 (a) How many grams of O_2 are required to combust all of the butane?
 (b) How many molecules of water will be produced?
 (c) Air contains 21.0% O_2 by volume. What mass of air is required to combust 5.00 mL of butane?

51. If the following reaction proceeds with a 75% yield, how much diborane, B_2H_6, will be produced when 23.5 g of sodium borohydride, $NaBH_4$ reacts with 50.0 g of boron trifluoride, BF_3?

 $NaBH_{4(s)} + BF_{3(g)} \rightarrow B_2H_{6(g)} + NaBH_{4(s)}$

UNIT 3

Chapter 8

52. What is the molar concentration of the solution made by dissolving 1.00 g of solid sodium nitrate, $NaNO_3$, in enough water to make 315 mL of solution?

53. What volume of 4.00×10^{-2} mol/L calcium nitrate solution, $Ca(NO_3)_{2(aq)}$ will contain 5.0×10^{-2} mol of nitrate ions?

54. By the addition of water, 80.0 mL of 4.00 mol/L sulfuric acid, H_2SO_4, is diluted to 400.0 mL. What is the molar concentration of the sulfuric acid after dilution?

55. How many moles of NaOH are in 100.0 mL of 0.00100 mol/L NaOH solution?

56. If a burette delivers 20 drops of solution per 1.0 mL, how many moles of $HCl_{(aq)}$ are in one drop of a 0.20 mol/L HCl solution?

57. What is the mass percent concentration of nicotine in the body of a 70 kg person smokes a pack of cigarettes (20 cigarettes) in one day? Assume that there is 1.0 mg of nicotine per cigarette, and that all the nicotine is absorbed into the person's body.

58. The blood of an average adult contains about 2.0 L of red blood cells. The hemoglobin present in these cells contains approximately 0.33% iron by mass. Make a reasonable guess about the density of red blood cells, and use this value to estimate the mass of iron present in the red blood cells of an average adult.

59. Ozone is a highly irritating gas that reduces the lung capability of healthy people in concentrations as low as 0.12 ppm. Older photocopy machines could generate ozone gas and they were often placed in closed rooms with little air circulation. Calculate the volume of ozone gas that would result in a concentration of 0.12 ppm in a room with dimensions of 5.0 m × 4.0 m × 3.0 m.

60. Human blood serum contains about 3.4 g/L of sodium ions. What is the molar concentration of Na^+ in blood serum?

Chapter 9

61. Write the net ionic equation for the reaction between aqueous solutions of barium chloride and sodium sulfate. Be sure to include the state of each reactant and product.

62. Write the net ionic equation for the reaction between aqueous sodium hydroxide and aqueous nitric acid. Be sure to include the state of each reactant and product.

63. What are the spectator ions when solutions of Na_2SO_4 and $Pb(NO_3)_2$ are mixed?

64. Iron(II) sulfate reacts with potassium hydroxide in aqueous solution to form a precipitate.
    (a) What is the net ionic equation for this reaction?
    (b) Which ions are spectator ions?

65. Write the balanced molecular and net-ionic equations for the following reactions:
    (a) $Na_3PO_{4(aq)} + Ca(OH)_{2(aq)} \rightarrow NaOH_{(aq)} + Ca_3(PO_4)_{2(s)}$
    (b) $Zn_{(s)} + Fe_2(SO_4)_{3(aq)} \rightarrow ZnSO_{4(aq)} + Fe_{(s)}$

Chapter 10

66. Name each of the following acids. Indicate whether each one is a strong or weak acid.
    (a) $H_2SO_{4(aq)}$
    (b) $HNO_{3(aq)}$
    (c) $HBr_{(aq)}$
    (d) $HCl_{(aq)}$
    (e) $HF_{(aq)}$

67. A sample of lemon juice was found to have a pH of 2.50. What is the concentration of hydronium ions in the lemon juice?

68. How many millilitres of sodium hydroxide solution are required to neutralize 20 mL of 1.0 mol/L acetic acid if 32 mL of the same sodium hydroxide solution neutralized 20 mL of 1.0 mol/L hydrochloric acid?

69. What are the concentrations of hydrogen ions and hydroxide ions in a solution that has a pH of 5?

70. What is the pH of a 1.0×10^{-5} mol/L $Ca(OH)_2$ (calcium hydroxide) solution?

71. How many moles of calcium hydroxide will be neutralized by one mole of hydrochloric acid, according to the following equation?
    $Ca(OH)_{2(s)} + 2HCl_{(aq)} \rightarrow CaCl_{2(aq)} + 2H_2O(l)$

72. In an experiment, 50.0 mL of 0.0800 mol/L NaOH is titrated by the addition of 0.0500 mol/L HNO_3. What is the hydroxide ion concentration after 30.0 mL of HNO_3 solution has been added?

73. A100 mL volume of 0.200 mol/L HCl was placed in a flask. What volume of 0.400 mol/L NaOH solution must be added to bring the solution to a pH of 7.0?

74. What is the pH of a solution in which 2.0×10^{-4} mol of HCl is dissolved in enough distilled water to make 300 mL of solution?

75. What is the pH of a solution containing 2.5 g of NaOH dissolved in 100 mL of water?

76. For each of the following reactions, identify the acid, the base, the conjugate base, and the conjugate acid:
    (a) $HF_{(aq)} + NH_{3(aq)} \rightarrow NH4^+_{(aq)} + F^-_{(aq)}$
    (b) $Fe(H_2O)_6{}^{3+}{}_{(aq)} + H_2O_{(l)} \rightarrow Fe(H_2O)_5(OH)_2{}^+{}_{(aq)} + H_3O^+_{(aq)}$
    (c) $NH_4{}^+{}_{(aq)} + CN^-_{(aq)} \rightarrow HCN_{(aq)} + NH_{3(aq)}$
    (d) $(CH_3)_3N_{(aq)} + H_2O_{(l)} \rightarrow (CH_3)_3NH^+_{(aq)} + OH^-_{(aq)}$

77. A solution was prepared by mixing 70.0 mL of 4.00 mo/L $HCl_{(aq)}$ and 30.0 mL of 8.00 mol/L $HNO_{3(aq)}$. Water was then added until the final volume was 500 mL. Calculate $[H^+]$ and find the pH of the solution.

UNIT 4

Chapter 11

78. The gas in a large balloon occupies 30.0 L at a pressure of 300 kPa. If the temperature is kept constant at 300 K, what volume will the balloon be at a pressure of 1200 kPa?

79. A gas occupies a 2.0 L container at 25°C and 300 kPa pressure. If the gas is transferred to a 3.0 L container at the same temperature, what will be the new pressure?

80. If the volume of a given amount of gas is tripled while the temperature remains constant, what will be the new pressure of the gas, relative to the initial pressure?

81. To what temperature must an ideal gas at 27°C be cooled to reduce its volume by one third? In other words, the new volume will be $\frac{2}{3}$ the original volume.

82. If 2.0 L of gas in a piston at 400 K is expanded to 3.0 L while keeping the pressure constant, what is the final temperature of the gas in kelvins?

83. If a certain mass of gas occupies 55 cm^3 at 303 K and 780 mm Hg, what is its volume in L at SATP?

84. If 1.00 L of helium gas at 20°C and 100 kPa is forced into a 250 mL container and subjected to a pressure of 400.0 kPa, what will be the new temperature of the gas?

85. A mixture of gases contains equal masses of H_2, O_2 and CH_4. If the partial pressure of CH_4 is 80 kPa, what is the partial pressure of H_2?

86. The volume of an automobile tire does not change appreciably when the car is driven. Before starting on a journey, a tire contains air at 220 kPa and 20°C. After being driven for an hour, the tire and the air in it become warmer and the pressure increases to 240 kPa. What is the temperature of the air inside the tire?

87. The gases inside a balloon exerted a total pressure of 150.0 kPa on the walls of the balloon. Seventy-five percent of the gas was nitrogen and twelve percent was oxygen. There was also some water vapour, which exerted a pressure of 2.4 kPa and some carbon dioxide. Calculate the pressure exerted by the CO_2 gas.

Chapter 12

88. A container holds one mole of gaseous neon at a certain temperature and pressure. A second, identical container holds gaseous nitrogen at three times the pressure and twice the temperature (in kelvins). How many moles of neon are in the second container?

89. If the mass of a gas is tripled and the pressure is quadrupled while the temperature is constant, by what factor will the volume of the gas change?

90. A cylinder with a volume of 25.0 L contains carbon dioxide at a pressure of 120 kPa and a temperature of 25°C. How much carbon dioxide is in the cylinder?

91. A closed vessel contains the following gases: 6.0 g of hydrogen, 14 g of nitrogen, and 44 g of carbon dioxide. If the total pressure in the vessel is 400 kPa, what is the partial pressure (in kPa) exerted by the nitrogen in the mixture?

92. One litre of a certain gas has a mass of 2.05 g at SATP. What is the molar mass of this gas?

93. What amount of gas is contained in a 10.0 L flask at a pressure of 180 kPa and a temperature of 300 K?

94. When a spark ignites a mixture of hydrogen gas and oxygen gas, water vapour is formed. What mass of oxygen gas would be required to react completely with 1.00 g of hydrogen?

95. Drinking a solution of baking soda (sodium hydrogen carbonate, $NaHCO_3$) can neutralize excess hydrochloric acid in the stomach in water. A student stirred 5.0 g of baking soda in water and drank the solution, then calculated the size of "burp" expected from the carbon dioxide generated in the following reaction.

$$NaHCO_{3(aq)} + HCl_{(aq)} \rightarrow NaCl_{(aq)} + H_2O_{(l)} + CO_{2(g)}$$

Assuming the gas will be at a pressure of 101 kPa, and body temperature is 37°C, what volume of carbon dioxide will be generated?

96. Rutherford proved that alpha particles were helium nuclei. In an experiment 1.82×10^{17} alpha particles were counted by use of a Geiger counter. The resulting helium gas occupied a volume of 7.34×10^{-3} mL at 19°C and 99.3 kPa. Use this information to calculate Avogadro's number.

97. A compound was found to contain 54.5% carbon, 9.10% hydrogen, and 36.4% oxygen. The vapour from 0.082 g of the compound occupied 21.8 mL at 104 kPa and 20°C.
   - **(a)** Calculate the empirical formula of the compound.
   - **(b)** Calculate the molecular mass of the compound.
   - **(c)** Calculate the molecular formula of the compound.

98. A student collected 375 mL of oxygen gas from the decomposition of hydrogen peroxide. The gas was collected over water at 19°C and 100.2 kPa. What mass of oxygen was collected? The vapour pressure of water at 19°C is 2.2 kPa.

99. Examine the reaction below and answer the following questions.

$$C_7H_{16(g)} + 11O_{2(g)} \rightarrow 7CO_{2(g)} + 8H_2O_{(g)}$$

   - **(a)** if 10.0 L of $C_7H_16_{(g)}$ are burned, what volume of oxygen gas, measured at the same temperature and pressure, is required?
   - **(b)** if 200 g of CO_2 are formed, what mass of $C_7H_{16(g)}$ was burned?
   - **(c)** if 200 L of CO_2 are formed, measured at STP, what mass of oxygen was consumed?

UNIT 5

Chapter 13

100. Name each of the following hydrocarbons.
   - **(a)** $CH_3—CH—CH_2—CH_2—CH_3$ with CH_3 attached to the second carbon (CH)

(b)

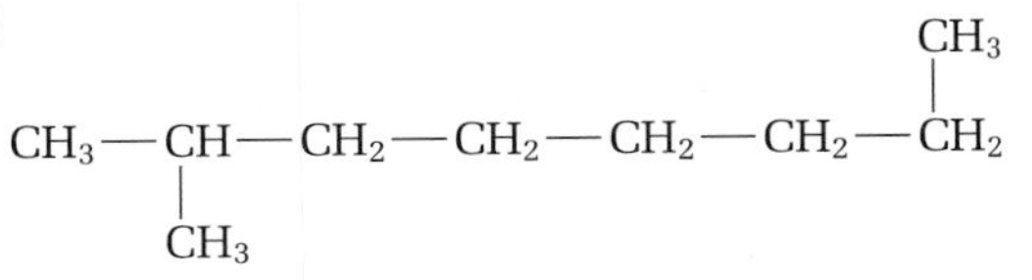

```
                                           CH3
                                            |
CH3—CH—CH2—CH2—CH2—CH2—CH2
     |
    CH3
```

(c)

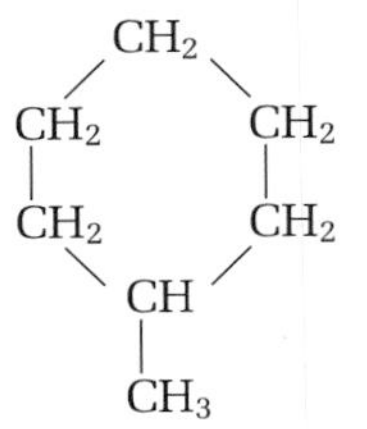

```
      CH2
     /   \
  CH2     CH2
   |       |
  CH2     CH2
     \   /
      CH
      |
     CH3
```

(d)

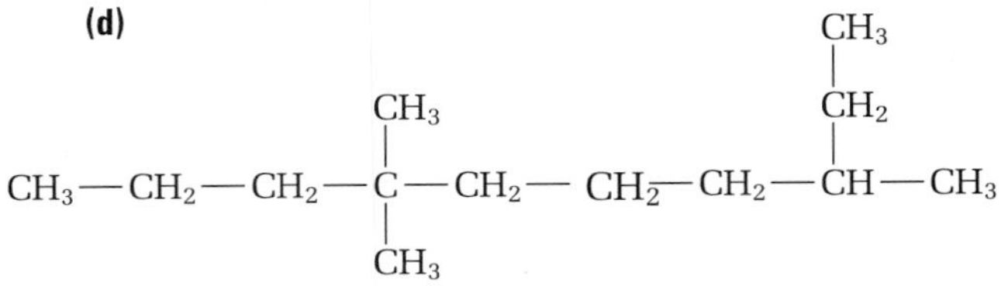

```
                                           CH3
                                            |
                 CH3                       CH2
                  |                         |
CH3—CH2—CH2—C—CH2—CH2—CH2—CH—CH3
                  |
                 CH3
```

(e)

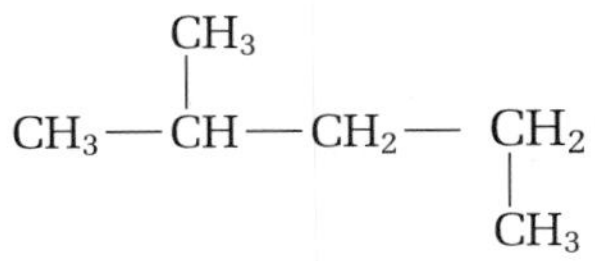

```
       CH3
        |
CH3—CH—CH2—CH2
                  |
                 CH3
```

(f)

```
        CH3
         |
CH3—CH
         |
        CH2
         |
        CH
       /   \
   CH2     CH2
    |       |
   CH2     CH2
      \   /
       CH2
```

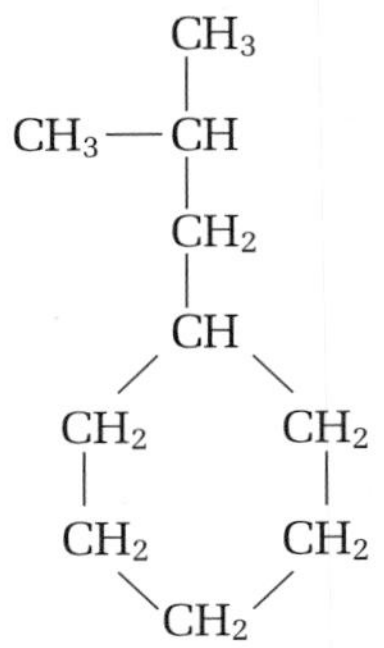

101. Draw condensed structural diagrams for each of the following compounds.

(a) 2-ethyl-1-pentene
(b) 2,6-octadien-4-yne
(c) 2,5-dimethyl-3-hexyne
(d) 2-methyl-3-butene
(e) 2-butyl-3-ethyl-1-cyclobutene
(f) 1,4-dimethylcycloheptane

102. Write IUPAC names for each of the following compounds.

(a) C_3H_6 **(b)** C_5H_8 **(c)** C_5H_{10}
(d) C_5H_{12} **(e)** C_5H_{14} **(f)** C_5H_{12}

103. Draw condensed structural diagrams for the following compounds:

(a) 2,5-dimethylhexane
(b) 2-ethyl-5-methyl-hexane
(c) 2-methyl-5-ethyl-hexane
(d) 2,5-dimethylheptane
(e) 3,6-dimethylheptane
(f) 2,3,4-trimethylpentane

104. Draw and name three structural isomers of C_6H_{14}.

105. Name the following alkenes and alkynes.

(a) $CH_2{=}CH{-}CH_3$

(b)

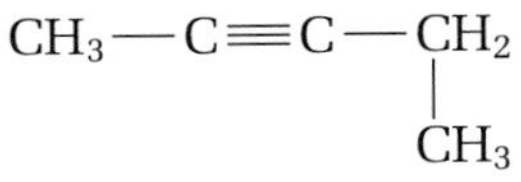

```
CH3—C≡C—CH2
             |
            CH3
```

(c)

```
       CH2
      /   \
   CH      CH2
   ||       |
   CH      CH
      \   /   \
       CH2     CH3
```

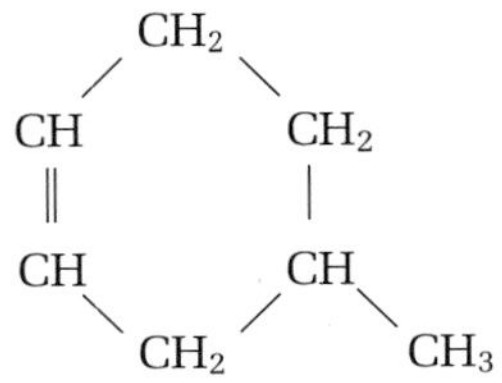

(d) $CH_3{-}CH{=}CH{-}CH_2{-}CH_2{-}C{\equiv}C{-}CH_3$

(e)

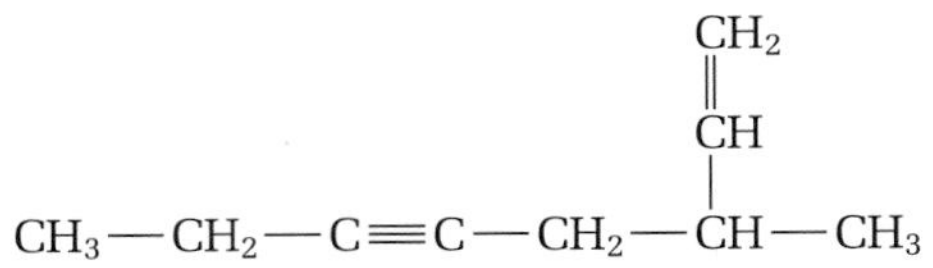

```
                               CH2
                               ||
                               CH
                               |
CH3—CH2—C≡C—CH2—CH—CH3
```

(f)

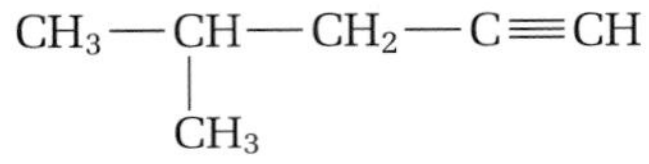

```
CH3—CH—CH2—C≡CH
     |
    CH3
```

(g)

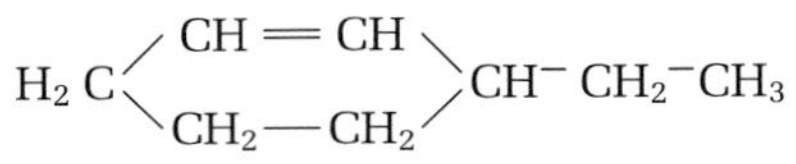

```
     / CH = CH \
H2 C            CH—CH2—CH3
     \ CH2—CH2 /
```

(h)

```
CH3—CH=CH—CH2—C=CH2—CH3
                   |
                   CH—CH3
                   |
                   CH3
```

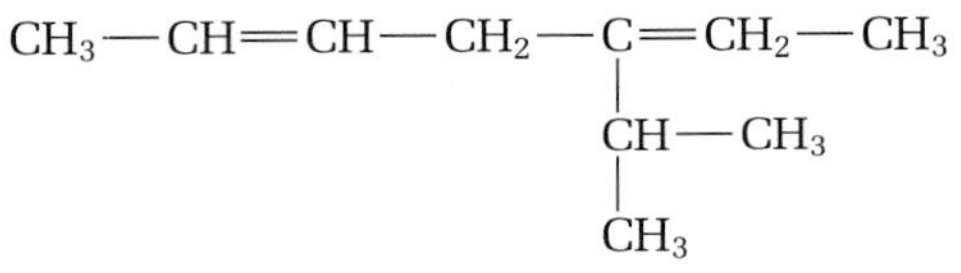

106. Examine the following compounds. Correct any flaws that you see in the structural diagrams.

(a) Is this compound 4-ethyl-2-methylpentane?

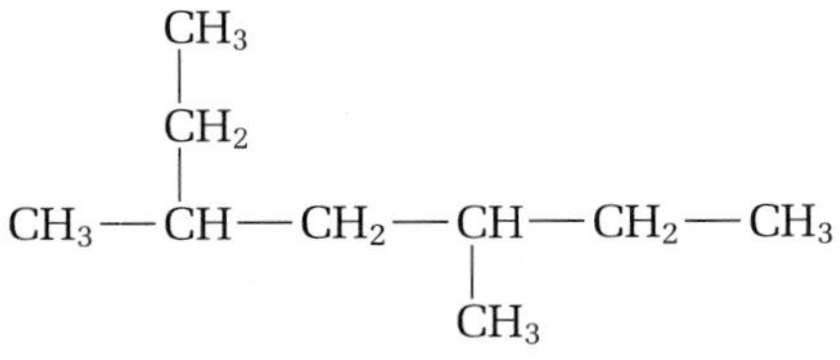

```
       CH3
        |
       CH2
        |
CH3—CH—CH2—CH—CH2—CH3
                |
               CH3
```

(b) Is this compound 4,5-dimethylhexane?

```
              CH2
             /   \
        CH2         CH2
         |           |
        CH          CH2
       /   \        /
   CH3        CH
              |
             CH3
```

(c) Is this compound 2-methyl-3-ethylpentane?

```
         CH3
          |
         CH2
          |
  CH3 —  CH — CH — CH2 — CH3
               |
              CH2
               |
              CH3
```

(d) Is this compound 1-methyl-3-cyclobutene?

```
  CH ═ CH
  |     |
  CH2 — CH — CH3
```

Chapter 14

107. 100 g of ethanol at 25°C is heated until it reaches 75°C. How much thermal energy did the ethanol gain? Hint: Use the information in Table 14.2.

108. An unknown material with a mass of 18 g was heated from 22°C until it reached 232°C. During this process, the material gained 751 J of energy. What is the heat capacity of the unknown material per gram?

109. A beaker containing 25 g of liquid at room temperature is heated until it gains 5°C. A second beaker containing 50 g of the same liquid at room temperature is heated until it also gains 5°C. Which beaker has gained the most thermal energy? Explain.

110. An unknown solid was dissolved in the water of a polystyrene calorimeter in order to find its heat of solution. The following data was recorded:
mass of solid (g) = 5.5
mass of calorimeter water (g) = 120.0
initial temperature of water (°C) = 21.7
final temperature of solution (°C) = 32.6

(a) Calculate the heat change of the water.

(b) Calculate the heat change caused by the solid dissolving.

(c) What is the heat of solution per gram of solid dissolved.

111. A 100 -g sample of food is placed in a bomb calorimeter calibrated at 7.23 kJ/°C. When the food is burned, the calorimeter gains 512 kJ of heat. If the initial temperature of the calorimeter was 19°C, what is the final temperature of the calorimeter and its contents?

Periodic Table of the Elements

Key (carbon): 6 = Atomic number; 12.01 = Average atomic mass*; 2.5 = Electronegativity; +4 = Common ion charge; 1086 = First ionization energy (kJ/mol); +2 = Other ion charges; 4765 = Melting point (K); 4098 = Boiling point (K); C; carbon

Legend: Gases; Liquids; Synthetics; metals (main group); metals (transition); metals (inner transition); metalloids; nonmetals

MAIN-GROUP ELEMENTS

Period	Group	Atomic number	Symbol	Name	Average atomic mass*	Electronegativity	First ionization energy (kJ/mol)	Melting point (K)	Boiling point (K)	Common ion charge	Other ion charges
1	1 (IA)	1	H	hydrogen	1.01	2.20	1312	13.81	20.28	1+	1-
2	1 (IA)	3	Li	lithium	6.94	0.98	520	453.7	1615	1+	
2	2 (IIA)	4	Be	beryllium	9.01	1.57	899	1560	2744	2+	
3	1 (IA)	11	Na	sodium	22.99	0.93	496	371	1156	1+	
3	2 (IIA)	12	Mg	magnesium	24.31	1.31	738	923.2	1363	2+	
4	1 (IA)	19	K	potassium	39.10	0.82	419	336.7	1032	1+	
4	2 (IIA)	20	Ca	calcium	40.08	1.00	590	1115	1757	2+	
5	1 (IA)	37	Rb	rubidium	85.47	0.82	403	312.5	941.2	1+	
5	2 (IIA)	38	Sr	strontium	87.62	0.95	549	1050	1655	2+	
6	1 (IA)	55	Cs	cesium	132.91	0.79	376	301.7	944	1+	
6	2 (IIA)	56	Ba	barium	137.33	0.89	503	1000	2170	2+	
7	1 (IA)	87	Fr	francium	(223)	0.7	~375	300.2	–	1+	
7	2 (IIA)	88	Ra	radium	(226)	0.9	509	973.2	–	2+	

TRANSITION ELEMENTS

Period	Group	Atomic number	Symbol	Name	Average atomic mass*	Electronegativity	First ionization energy (kJ/mol)	Melting point (K)	Boiling point (K)	Common ion charge	Other ion charges
4	3 (IIIB)	21	Sc	scandium	44.96	1.36	631	1814	3109	3+	
4	4 (IVB)	22	Ti	titanium	47.87	1.54	658	1941	3560	4+	2+, 3+
4	5 (VB)	23	V	vanadium	50.94	1.63	650	2183	3680	5+	2+, 3+, 4+
4	6 (VIB)	24	Cr	chromium	52.00	1.66	653	2180	2944	3+	2+, 6+
4	7 (VIIB)	25	Mn	manganese	54.94	1.55	717	1519	2334	4+	2+, 3+, 7+
4	8 (VIIIB)	26	Fe	iron	55.85	1.83	759	1811	3134	3+	2+
4	9 (VIIIB)	27	Co	cobalt	58.93	1.88	760	1768	3200	2+	3+
5	3 (IIIB)	39	Y	yttrium	88.91	1.22	616	1795	3618	3+	
5	4 (IVB)	40	Zr	zirconium	91.22	1.33	660	2128	4682	4+	
5	5 (VB)	41	Nb	niobium	92.91	1.6	664	2750	5017	5+	3+
5	6 (VIB)	42	Mo	molybdenum	95.94	2.16	685	2896	4912	6+	
5	7 (VIIB)	43	Tc	technetium	(98)	2.10	702	2430	4538	7+	4+, 6+
5	8 (VIIIB)	44	Ru	ruthenium	101.07	2.2	711	2607	4423	3+	
5	9 (VIIIB)	45	Rh	rhodium	102.91	2.28	720	2237	3968	3+	
6	3 (IIIB)	57	La	lanthanum	138.91	1.10	538	1191	3737	3+	
6	4 (IVB)	72	Hf	hafnium	178.49	1.3	642	2506	4876	4+	
6	5 (VB)	73	Ta	tantalum	180.95	1.5	761	3290	5731	5+	
6	6 (VIB)	74	W	tungsten	183.84	1.7	770	3695	5828	6+	
6	7 (VIIB)	75	Re	rhenium	186.21	1.9	760	3459	5869	4+	6+, 7+
6	8 (VIIIB)	76	Os	osmium	190.23	2.2	840	3306	5285	4+	3+
6	9 (VIIIB)	77	Ir	iridium	192.22	2.2	880	2719	4701	4+	3+
7	3 (IIIB)	89	Ac	actinium	(227)	1.1	499	1324	3471	3+	
7	4 (IVB)	104	Rf	rutherfordium	(261)	–	–	–	–	4+	
7	5 (VB)	105	Db	dubnium	(262)	–	–	–	–	–	
7	6 (VIB)	106	Sg	seaborgium	(266)	–	–	–	–	–	
7	7 (VIIB)	107	Bh	bohrium	(264)	–	–	–	–	–	
7	8 (VIIIB)	108	Hs	hassium	(265)	–	–	–	–	–	
7	9 (VIIIB)	109	Mt	meitnerium	(268)	–	–	–	–	–	

INNER TRANSITION ELEMENTS

Period	Series	Atomic number	Symbol	Name	Average atomic mass*	Electronegativity	First ionization energy (kJ/mol)	Melting point (K)	Boiling point (K)	Common ion charge	Other ion charges
6	Lanthanoids	58	Ce	cerium	140.12	1.12	527	1071	3716	3+	4+
6	Lanthanoids	59	Pr	praseodymium	140.91	1.13	523	1204	3793	3+	
6	Lanthanoids	60	Nd	neodymium	144.24	1.14	530	1294	3347	3+	
6	Lanthanoids	61	Pm	promethium	(145)	–	536	1315	3273	3+	
6	Lanthanoids	62	Sm	samarium	150.36	1.17	543	1347	2067	3+	2+
6	Lanthanoids	63	Eu	europium	151.96	–	547	1095	1802	3+	2+
6	Lanthanoids	64	Gd	gadolinium	157.25	1.20	593	1586	3546	3+	
7	Actinoids	90	Th	thorium	232.04	1.3	587	2023	5061	4+	
7	Actinoids	91	Pa	protactinium	231.04	1.5	568	1845	–	5+	4+
7	Actinoids	92	U	uranium	238.03	1.7	584	1408	4404	6+	3+, 4+, 5+
7	Actinoids	93	Np	neptunium	237.05	1.3	597	917	–	5+	3+, 4+, 6+
7	Actinoids	94	Pu	plutonium	(244)	1.3	585	913.2	3501	4+	3+, 5+, 6+
7	Actinoids	95	Am	americium	(243)	–	578	1449	2284	3+	4+, 5+, 6+
7	Actinoids	96	Cm	curium	(247)	–	581	1618	3373	3+	

*Average atomic mass data in brackets indicate atomic mass of most stable isotope of the element.
Data obtained from *The CRC Handbook of Chemistry and Physics*, 81st Edition

MAIN-GROUP ELEMENTS

10	11 (IB)	12 (IIB)	13 (IIIA)	14 (IVA)	15 (VA)	16 (VIA)	17 (VIIA)	18 (VIIIA)
								2 4.00 – – 2372 5.19 5.02 **He** helium
			5 10.81 2.04 – 800 2348 4273 **B** boron	6 12.01 2.55 – 1086 4765 4098 **C** carbon	7 14.01 3.04 3– 1402 63.15 77.36 **N** nitrogen	8 16.00 3.44 2– 1314 54.36 90.2 **O** oxygen	9 19.00 3.98 1– 1681 53.48 84.88 **F** fluorine	10 20.18 – – 2080 24.56 27.07 **Ne** neon
			13 26.98 1.61 – 577 933.5 2792 **Al** aluminum	14 28.09 1.90 – 786 1687 3538 **Si** silicon	15 30.97 2.19 – 1012 317.3 553.7 **P** phosphorus	16 32.07 2.58 2– 999 392.8 717.8 **S** sulfur	17 35.45 3.16 1– 1256 171.7 239.1 **Cl** chlorine	18 39.95 – – 1520 83.8 87.3 **Ar** argon
28 58.69 1.91 2+ 737 3+ 1728 3186 **Ni** nickel	29 63.55 1.90 2+ 745 1+ 1358 2835 **Cu** copper	30 65.39 1.65 2+ 906 692.7 1180 **Zn** zinc	31 69.72 1.81 3+ 579 302.9 2477 **Ga** gallium	32 72.61 2.01 – 761 1211 3106 **Ge** germanium	33 74.92 2.18 – 947 1090 876.2 **As** arsenic	34 78.96 2.55 2– 941 493.7 958.2 **Se** selenium	35 79.90 2.96 1– 1143 266 332 **Br** bromine	36 83.80 – – 1351 115.8 119.9 **Kr** krypton
46 106.42 2.20 2+ 805 3+ 1828 3236 **Pd** palladium	47 107.87 1.93 1+ 731 1235 2435 **Ag** silver	48 112.41 1.69 2+ 868 594.2 1040 **Cd** cadmium	49 114.82 1.78 3+ 558 429.8 3345 **In** indium	50 118.71 1.96 4+ 708 2+ 505 2875 **Sn** tin	51 121.76 2.05 – 834 903.8 1860 **Sb** antimony	52 127.60 2.1 – 869 722.7 1261 **Te** tellurium	53 126.90 2.66 1– 1009 386.9 457.4 **I** iodine	54 131.29 – – 1170 161.4 165 **Xe** xenon
78 195.08 2.2 4+ 870 2+ 2042 4098 **Pt** platinum	79 196.97 2.4 3+ 890 1+ 1337 3129 **Au** gold	80 200.59 1.9 2+ 1107 1+ 234.3 629.9 **Hg** mercury	81 204.38 1.8 1+ 589 3+ 577.2 1746 **Tl** thallium	82 207.20 1.8 2+ 715 4+ 600.6 2022 **Pb** lead	83 208.98 1.9 3+ 703 5+ 544.6 1837 **Bi** bismuth	84 (209) 2.0 4+ 813 2+ 527.2 1235 **Po** polonium	85 (210) 2.2 1– (926) 575 – **At** astatine	86 (222) – – 1037 202.2 211.5 **Rn** radon
110 (269) – – – – – **Uun** ununnilium	111 (272) – – – – – **Uuu** unununium	112 (277) – – – – – **Uub** ununbium		114 (285) – – – – – **Uuq** ununquadium		116 (289) – – – – – **Uuh** ununhexium		118 (293) – – – – – **Uuo** ununoctium

65 158.93 – 3+ 565 1629 3503 **Tb** terbium	66 162.50 1.22 3+ 572 1685 2840 **Dy** dysprosium	67 164.93 1.23 3+ 581 1747 2973 **Ho** holmium	68 167.26 1.24 3+ 589 1802 3141 **Er** erbium	69 168.93 1.25 3+ 597 1818 2223 **Tm** thulium	70 173.04 – 3+ 603 2+ 1092 1469 **Yb** ytterbium	71 174.97 1.0 3+ 524 1936 3675 **Lu** lutetium
97 (247) – 3+ 601 4+ 1323 – **Bk** berkelium	98 (251) – 3+ 608 1173 – **Cf** californium	99 (252) – 3+ 619 1133 – **Es** einsteinium	100 (257) – 3+ 627 1800 – **Fm** fermium	101 (258) – 3+ 635 2+ 1100 – **Md** mendelevium	102 (259) – 3+ 642 2+ 1100 – **No** nobelium	103 (262) – 3+ – 1900 – **Lr** lawrencium

Expanding the Model of the Atom

In science, theoretical models and experimental evidence are always linked. For example, early scientists believed that rotting meat generated maggots. This seemed reasonable, since exposed meat becomes infested with maggots. Experiments proved, however, that if the meat was protected from flies, no maggots appeared. Thus, rotting meat didn't generate maggots; flies did. When a model fails to explain experimental evidence, scientists must discard or modify that model. The model of the atom is no exception.

Emission Spectra

In Chapter 2, you used a diffraction grating to observe the spectra of various elements. These elements were sealed in gas discharge tubes. The spectra resulted from electrons in atoms moving from higher energy levels to lower energy levels and releasing energy as light.

Each wavelength of visible light is associated with a colour. When white light is shone through a gas discharge tube, it produces a line spectrum. A line spectrum is a series of narrow lines having specific colours (energy), separated by colourless spaces. Each element produces a different and characteristic emission spectrum. You can see several examples of these spectra in Figure D.1. Scientists who were thinking about how to describe the atom needed to take the emission spectra into account.

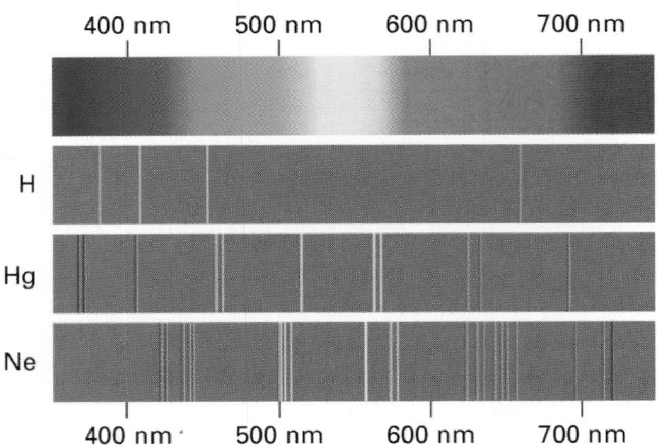

Figure D.1 The top spectrum is the continuous spectrum of white light. The others are emission spectra for hydrogen, mercury, and neon. Each element has its own distinct spectrum, which is like a "fingerprint" for that element.

Bohr's Model of the Atom

In 1913, the Danish physicist Niels Bohr developed a model of the atom that explained the hydrogen emission spectrum. In Bohr's model, electrons orbit the nucleus in the same way that Earth orbits the Sun, as shown in Figure D.2. The following three points of Bohr's theory help to explain hydrogen's emission spectrum.

1. In Bohr's model, atoms have specific allowable energy levels. He called these energy levels *stationary states*. Each of these levels corresponds to a fixed, circular orbit around the nucleus.
2. An atom does not give off energy when its electrons are in a stationary state.
3. An atom changes stationary states by giving off or absorbing a quantity of light energy exactly equal to the difference in energy between the two stationary states.

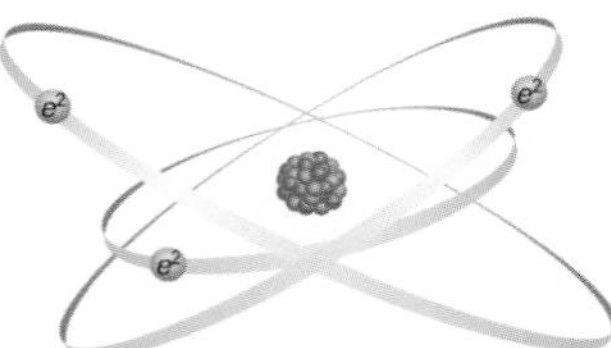

Figure D.2 In Bohr's atomic model, electrons move around the nucleus with fixed, circular orbits.

Bohr's model was revolutionary, because he proposed that the energy absorbed or emitted by an atom needed to have specific values. The energy change was quantized, rather than continuous. When something is *quantized*, it means that it is limited to discrete amounts or multiples of discrete amounts. Two great scientists paved the way for this surprising idea. German physicist Max Planck had already suggested that energy in general was quantized, meaning that it exists in "packets." Building on this idea, Einstein proposed that light could behave as particles, which he called *photons*.

The energy associated with the light in a line spectrum corresponds with the change in energy of an electron as it moves up or down an energy level. For example, when electrons in hydrogen atoms that have been excited to the third energy level subsequently drop to the second energy level, they emit light that has a specific energy. They emit photons of red light that have a wavelength of 656.3 nm. These photons cause the red line on the line spectrum for hydrogen, which you can see in Figure D.1.

Why a New Model?

So what was the problem? Bohr's model worked beautifully, correctly predicting the line spectrum for hydrogen. It also worked fine for ions with one electron, such as He^{+}, Li^{2+}, Be^{3+}, etc. The model failed, however, when it was applied to the emission spectra of atoms that had more than one electron. Bohr's model needed to be modified, because it was too simple to explain the experimental evidence.

Sublevels

The spectra of many-electron atoms suggested that a more complex structure was needed. Notice in Figure D.1 that spectra for these more complex atoms have groups of lines close together. The groups are separated by spaces. The large spaces represent the energy differences between energy levels, while the smaller spaces represent energy differences within the levels.

If the electrons are changing energy within the levels, this suggests that there are *sublevels* within each level, each with its own slightly different energy. The idea of energy levels 1, 2, 3, 4, etc. remains, but each energy level is split up into sublevels called *s*, *p*, *d*, and *f*. Examine Table D.1 to see how electrons are arranged in sublevels for energy levels 1 to 4.

Table D.1 Distribution of Electrons in Energy Levels and Sublevels

Energy level	Number of electrons in energy level	Sublevel	Number of electrons in sublevel
1	2	1*s*	2
2	8	2*s*	2
		2*p*	6
3	18	3*s*	2
		3*p*	6
		3*d*	10
4	32	4*s*	2
		4*p*	6
		4*d*	10
		4*f*	14

Each sublevel has its own energy. Examine Figure D.3 to see the relative energies of the levels and sublevels.

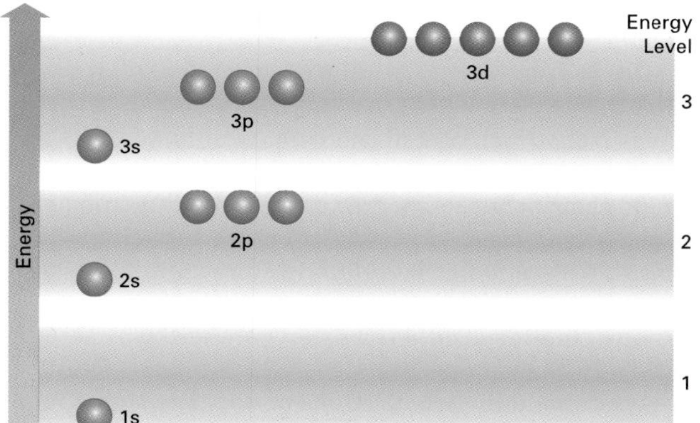

Figure D.3 This figure shows the relative energies of sublevels in energy levels 1, 2, and 3.

Visualizing the New Model

But where do you find the electrons in this new model? How are the sublevels oriented in space? Several theories enabled scientists to describe what the newly conceived atom "looked like."

Particles with Wave-Like Properties

In 1924, Louis de Broglie, a young physics student, suggested that all matter had wave-like properties. This seemed to follow from Planck and Einstein's idea that electromagnetic radiation has matter-like properties. De Broglie developed an equation that allowed him to calculate the wavelength associated with any object, from a bowling ball to an electron. Objects that we can see and can interact with have calculated wavelengths that are smaller than electrons. Their wavelengths are so tiny compared to their size that they do not have any measurable effect on the motion of the objects.

For very tiny moving particles, however, such as electrons, the wavelength becomes very significant. In fact, an electron moving at an average speed has a wavelength even larger than the size of the entire hydrogen atom!

De Broglie's theory was proven by experiment when streams of electrons produced diffraction patterns similar to those produced by electromagnetic radiation, which was already known to travel in waves.

Orbitals

In 1926, Erwin Schrödinger used de Broglie's idea that matter has wavelike properties. Schrödinger proposed what is now known as the *quantum mechanical model of the atom*. In this new model, he abandoned the notion of the electron as a small particle orbiting the nucleus. Instead, he took into account the particle's wavelike properties, and described the behaviour of electrons in terms of wave functions.

The imprecise nature of Schrödinger's model was supported shortly afterwards by a principle proposed by Werner Heisenberg, in 1927. Heisenberg demonstrated that it is impossible to know both an electron's pathway and its exact location. *Heisenberg's uncertainty principle* is a mathematical relationship that shows that you can never know both the position and the momentum of an object beyond a certain measure of precision.

The momentum of an object is its mass multiplied by its velocity. An object's momentum is directly related to the amount of energy it has.

Heisenberg also showed that if you could know either velocity or position precisely, then the other property would be uncertain. Therefore, when talking about where electrons are found in an atom, you cannot talk in terms of certainties, but only in terms of probabilities.

Schrödinger used a mathematical wave equation to define the probability of finding an electron within an atom. There are multiple solutions to this wave equation and Schrödinger called these solutions wave functions, or *orbitals*. Each solution provides information about the energy and location of an electron within an atom. Each orbital has a specific energy associated with it, and each contains information about where, inside the atom, the electrons would spend most of their time. The actual paths of the moving electrons cannot be determined. However, the solutions can be used to show where, for each orbital, there is a high probability of finding an electron.

Imagine you were able to take many exact measurements of the position of an electron in the 1*s* sublevel at fixed intervals. At each second, you would mark the electron's position on a graph. After a while, your graph might look something like the first diagram in Figure D.4. If someone tried to use your graph to determine where the electron was, they would not be able to state its position exactly. They would, however, be able to see that the probability of finding the electron would be greater near the nucleus and would decrease farther away from the nucleus.

The probability graph is "fuzzy," because the probability of finding an electron anywhere in a 1*s* sublevel is never zero. The probability becomes extremely small when it is far away from the nucleus, but it never quite reaches zero. Therefore, to obtain an exact shape for an orbital, you need to choose a level of probability. For example, drawing a contour that encompasses 95 percent of the probability graph results in a spherical shape, like the third diagram in Figure D.4. In other words, at any time, there is a 95 percent chance of finding the electron within the spherical contour.

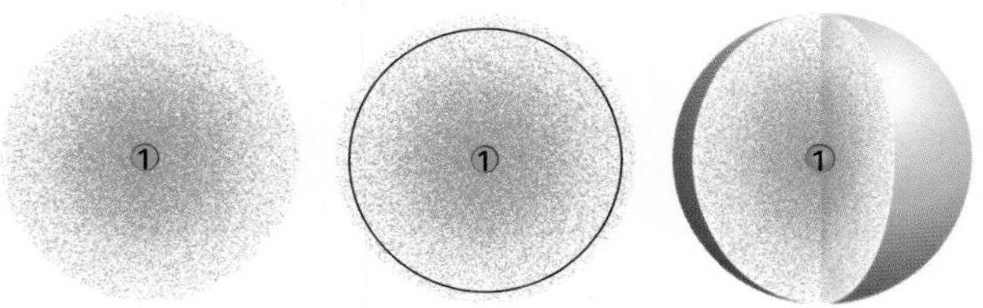

Figure D.4 These three figures represent the probability of finding an electron at any point in a hydrogen atom. The two-dimensional drawing on the left shows that the electron spends most of its time fairly close to the nucleus. The circle around the cloud in the centre encompasses 95 percent of the two-dimensional cloud. The diagram on the right shows the 95 percent probability contour in three dimensions.

Shapes of Orbitals

The shapes of the probability graphs from Schrödinger's wave functions are the shapes of the orbitals in which electrons reside in atoms. You can visualize orbitals as electron clouds. The shape of each cloud is based on probability—it tells you where the electron spends most of its time.

Examine Figure D.5. The *s* orbitals are spherical in shape, as described above. Each of these spherical shells contains two electrons. There are three *p* orbitals for each sublevel, each with a capacity for two electrons. Each orbital is shaped something like a dumbbell. For each sublevel, the *p* orbitals are oriented along the *x*, *y*, and *z* axes. Therefore, the three *p* orbitals in each sublevel are sometimes designated with subscripts to show this. For example, the 2*p* orbitals may be designated $2p_x$, $2p_y$, $2p_z$.

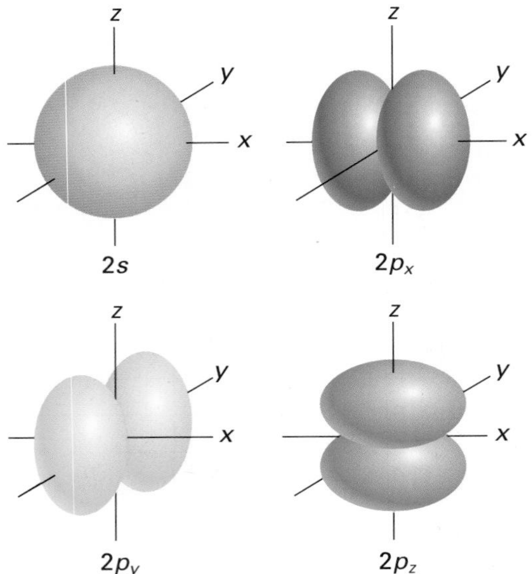

Figure D.5 This figure shows the shapes of the 2*s* and 2*p* orbitals.

The *d* and *f* orbitals are quite complex in shape. Each *d* sublevel contains five *d* orbitals, while each *f* sublevel contains seven *f* orbitals.

Filling the Orbitals

Why is it that each orbital can contain only two electrons? A hypothesis suggests that electrons spin around their own axes as they move around the nucleus, generating magnetic fields. They can spin either in a positive direction or in a negative direction. Electrons with opposite spins attract each other. This attraction partially counteracts the repulsion between two negatively charged electrons. In 1925, Wolfgang Pauli proposed that only two electrons of opposite spin could occupy an orbital. This idea became known as the *Pauli exclusion principle*.

How do electrons fill orbitals within atoms? They do so in such a way as to minimize the potential energy of the atom.

1. They will fill orbitals with the lowest energy first. The 1*s* orbital will fill before the 2*s* orbital, which will fill before the 2*p* orbitals.
2. When occupying two or more orbitals with the same energy (for example, any of the three 2*p* orbitals), electrons will half fill each orbital until all are half filled before adding a second electron to each one. This is called *Hund's rule.*

You can show how electrons fill orbitals, using superscripts. For example, a boron atom contains five electrons. Here's how you show the electron configuration:

$$1s^2, 2s^2, 2p_x^{\,1}$$

How would you show the electron configuration of a nitrogen atom? Remember Hund's rule:

$$1s^2, 2s^2, 2p_x^{\,1}, 2p_y^{\,1}, 2p_z^{\,1}$$

Using the Model

The quantum mechanical model of the atom is useful for explaining phenomena in addition to emission spectra. For example, in Chapter 2, you learned about trends in ionization energy. You learned that, in general, ionization energy increases across a period. Figure D.6 shows the ionization energy as a function of atomic number for the first 18 elements. In most cases, the ionization energy follows the trend. There are some exceptions, however, such as oxygen, boron, and sulfur. You can use the quantum model of the atom to explain these discrepancies.

First Ionization Energy versus Atomic Number

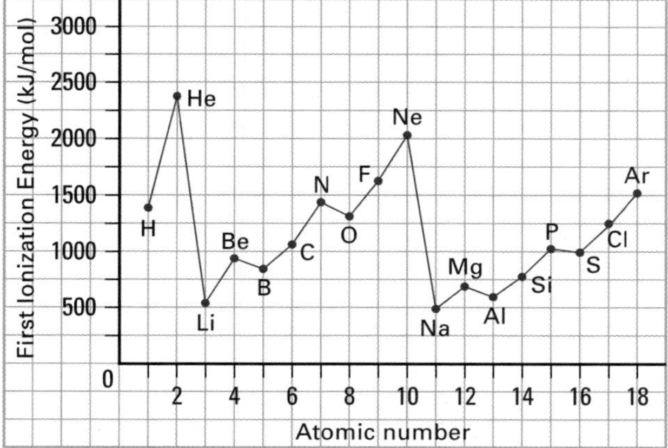

Figure D.6 **This graph shows the relationship between first ionization energy and atomic number for the first 18 elements.**

Practice Problems

1. Write the electron configurations for the following atoms:
   (a) carbon
   (b) oxygen
   (c) fluorine
   (d) sodium
   (e) silicon
   (f) hydrogen
2. Explain why the ionization energy of the following elements is less than the element that precedes it, even though its nucleus has a greater positive charge.
   (a) oxygen has a lower ionization energy than nitrogen
   (b) boron has a lower ionization energy than beryllium
   (c) sulfur has a lower ionization energy than phosphorus

PROBLEM TIP

Keep in mind two things when answering question 2:

(a) A completely filled sublevel is favourable in terms of energy.

(b) When removing electrons from the same sublevel, it is easier to remove an electron from a filled orbital than from a half-filled orbital. In spite of their opposite spins, the negative charges of electrons together in a filled orbital repel each other more than those in half-filled orbitals.

3. Explain why Bohr's model of the atom explained emission spectra for one-electron atoms, but failed to explain emission spectra for many-electron atoms.

Math and Chemistry

In chemistry, quantities and values are often very, very large or very, very small. One mole of H_2O contains a huge number (602 214 199 000 000 000 000 000) of molecules. The mass of each molecule is minuscule (0.000 000 000 000 000 000 000 029 9 g). To calculate the mass of one mole of H_2O using the above data, you multiply the number of molecules in a mole by the mass of an individual molecule. Imagine doing this by hand:

$$\begin{array}{r} 602\ 214\ 199\ 000\ 000\ 000\ 000\ 000 \\ \times\ \underline{0.000\ 000\ 000\ 000\ 000\ 000\ 000\ 029\ 9}\ \text{g} \\ 18.0062045501\ \text{g} \end{array}$$

There is another problem with writing all these zeros: It implies that they are all significant digits. Yet the non-final zeros to the right of the decimal point in the mass of an H_2O molecule are used as placeholders. They are not significant digits. There are only three significant digits in

0.000 000 000 000 000 000 000 **029 9** g

Scientific Notation

To simplify reporting large numbers and doing calculations, you can use scientific notation. As you learned in an earlier mathematics course, scientific notation works by using *powers of 10* as multipliers.

The number of particles in one mole of a substance (the *Avogadro constant*) is often rounded down to 602 000 000 000 000 000 000 000. When performing calculations, however, this number is still too unwieldy. To simplify it, you can express it as a number between 1 and 10, multiplied by factors of 10. To do this, move the decimal point behind the left-most non-zero digit, counting the number of places that the decimal point moves. The number of places moved is the exponent for the base 10.

6.02 000 000 000 000 000 000 000.

23 21 18 15 12 9 6 3

6.02×10^{23}

Figure E.1 **The decimal point moves to the left.**

Numbers less than 1 are converted into scientific notation by moving the decimal point to the right of the left-most non-zero digit. The exponent is expressed as a negative power of 10.

0.000 000 000 000 000 000 000 02.9 9 g

3 6 9 12 15 18 21

2.99×10^{-23}

Figure E.2 **The decimal point moves to the right.**

Once you have numbers in scientific notation, calculations become much simpler. To find the mass of one mole of H_2O, you can multiply $(6.02 \times 10^{23}/\text{mol}) \times (2.99 \times 10^{-23}\ \text{g})$. To do this without a calculator, you first multiply.

$6.02 \times 2.99 = 17.9998 \rightarrow 18.0$ (rounded up to three significant digits)

When multiplying, exponents are added to find the new exponent: $-23 + 23 = 0$.

This gives 18.0×10^0 g/mol. In scientific notation, however, there must be only one digit before the decimal place. Here the decimal place must be moved one place to the left: 1.80×10^1 g/mol.

Figure E.3 shows how to do this calculation with a scientific calculator. When you enter an exponent on a scientific calculator, you do not have to enter (× 10) at any time.

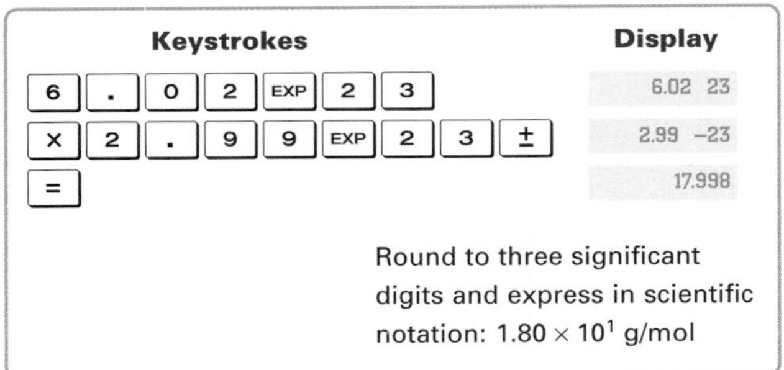

Figure E.3 **On some scientific calculators, the EXP key is labelled EE. Key in negative exponents by entering the exponent, then striking the ± key.**

Rules for Scientific Notation

- When multiplying, exponents are added algebraically.

$$\begin{aligned}&(2.37 \times 10^4\ \text{g}) \times (1.89 \times 10^{-1}\ \text{g}) \\ &= (2.37 \times 1.89) \times 10^{(4+(-1))} \\ &= 4.4793 \times 10^3\ \text{g} \rightarrow 4.48 \times 10^3\ \text{g}\end{aligned}$$

- When dividing, exponents are subtracted algebraically.

$$\begin{aligned}&(9.23 \times 10^6\ \text{L}) \div (4.11 \times 10^4\ \text{L}) \\ &= (9.23 \div 4.11) \times 10^{(6-4)} \\ &= 2.245742 \times 10^2\ \text{L} \rightarrow 2.25 \times 10^2\ \text{L}\end{aligned}$$

- When adding or subtracting numbers in scientific notation, the numbers must be converted to the same power of 10 as the measurement with the greatest power of 10. Once the numbers are all expressed to the same power of 10, the power of 10 is neither added nor subtracted in the calculation.

$$\begin{aligned}&(8.22 \times 10^4\ \text{cm}) + (5.66 \times 10^3\ \text{cm}) \\ &= (8.22 \times 10^4\ \text{cm}) + (.566 \times 10^4\ \text{cm}) \\ &= 8.786 \times 10^4\ \text{cm} \rightarrow 8.79 \times 10^4\ \text{cm}\end{aligned}$$

$(9.73 \times 10^1\ dm^3) - (8.11 \times 10^{-1}\ dm^3)$
$= (9.73 \times 10^1\ dm^3) - (.0811 \times 10^1\ dm^3)$
$= 9.6489 \times 10^1\ dm^3 \rightarrow 9.65 \times 10^1\ dm^3$

Practice Problems

Scientific Notation

1. Convert each number into scientific notation.
   (a) 0.000 945
   (b) 39 230 000 000 000
   (c) 0.000 000 000 000 003 497
   (d) 879×10^5
   (e) 0.00142×10^3
   (f) 31271×10^{-6}
2. Add, subtract, multiply, or divide. Round your answer, and express it in scientific notation to the correct number of significant digits.
   (a) $(3.00 \times 10^3) + (8.50 \times 10^1)$
   (b) $(6.99 \times 10^3) + (8.13 \times 10^2)$
   (c) $(1.01 \times 10^1) - (9.34 \times 10^{-2})$
   (d) $(12.01 \times 10^6) \times (8.32 \times 10^7)$
   (e) $(4.75 \times 10^{-3}) \div (3.21 \times 10^{-2})$

Logarithms

Calculators have made multiplication and division as easy as adding and subtracting. Before calculators were invented, however, what did people do? In 1614, John Napier (1550–1617) invented *logarithms* to help people multiply and divide large numbers.

Why do logarithms make multiplication and division easier? Recall, from the power laws, how you multiply and divide powers. To multiply powers with the same base, you add the exponents. To divide powers with the same base, you subtract the exponents. Logarithms are exponents. Therefore, to multiply large numbers using logarithms, you simply add the logarithms. To divide large numbers using logarithms, you subtract the logarithms.

Find the log (for "logarithm") key on your calculator. Enter 1000 or 10^3 into your calculator, then press the log key. The number 3 will be in the display. Now enter 10 000 or any other power of 10. Press the log key. What is displayed for the log of the number? (The logarithm of any power of 10 is just the exponent.)

How do you multiply one million by one billion, or $10^6 \times 10^9$, using logarithms? You add the exponents, or logarithms, of these two numbers: $6 + 9 = 15$. The answer is 10^{15}, a power of 10. This power is called the *antilogarithm* of the logarithm 15.

Logarithms may not seem like a major discovery, but John Napier also evaluated logarithms for numbers such as 2, 3, 5, and 764. (Table E.1 has more examples of logarithms.) Enter 2 into your calculator, and press the log key. The display will show 0.3010, to four significant digits. This means that $10^{0.3010} = 2$. Repeat to find the log of 3. The display will show 0.4771. This means that $10^{0.4771} = 3$.

You know that $2 \times 3 = 6$. If you add 0.3010 to 0.4771, you get 0.7781. What is $10^{0.7781}$? Enter 0.7781 into your calculator. Find the 10^x key. (Usually it is above the log key, as a second function.) Press the 10^x key. If you round the answer in the display, you will get 6.

Find the log of 8 and the log of 4. You know that $8 \div 4 = 2$. Subtract the log of 4 from the log of 8: $0.9031 - 0.6021 = 0.3010$. This answer is the log of 2: $10^{0.9031} \div 10^{0.6021} = 10^{0.3010} = 2$. Evaluate $10^{0.3010}$ on your calculator by typing in 0.3010 and using the 10^x key.

Technology LINK

Logarithms were used for 350 years to do many complex calculations. Even today, calculators are programmed to do mathematical manipulations with logarithms. So, indirectly, we are still using John Napier's invention.

Table E.1 Some Numbers and Their Logarithms

Number	Scientific notation	As a power of 10	Logarithim
1 000 000	1×10^6	10^6	6
7 895 900	7.8590×10^5	$10^{5.8954}$	5.8954
1	1×10^0	10^0	0
0.000 001	1×10^{-6}	10^{-6}	−6
0.004 276	4.276×10^{-3}	$10^{-2.3690}$	−2.3690

Notice that numbers above 1 correspond to logarithms whose value is positive; numbers below 1 correspond to logarithms whose value is negative. Over the years, scientists have found logarithms convenient for more than calculations. Logarithms can be used to express values that span a range of powers of 10, such as pH. In Chapter 10, section 10.2, concentrations of acids are discussed. The pH of an acid solution is defined as $-\log[H_3O^+]$. (The square brackets mean "concentration.") For example, suppose that the hydronium ion concentration in a solution is 0.0001 mol/L (10^{-4} mol/L). The pH is −log(0.0001). To calculate this, enter 0.0001 into your calculator. Then press the log key. Since the logarithm is negative, press the [±] key. The answer in the display is 4. Therefore the pH of the solution is 4.

As stated above, there are logarithms for all numbers, not just whole multiples of 10. What is the pH of a solution if $[H_3O^+] = 0.00476$ mol/L? Enter 0.00476. Press the log key and then the [±] key. The answer is 2.322. Remember that the concentration was expressed to three significant digits. The pH to three significant digits is 2.32.

What if you want to find the $[H_3O^+]$ from the pH? You would need to find 10^{-pH}. For example, what is the $[H_3O^+]$ if the pH is 5.78? Enter 5.78, and press the [±] key. Then use the 10^x function. The answer is $10^{-5.78}$. Therefore, the $[H_3O^+]$ is 1.66×10^{-6} mol/L.

Remember that the pH scale is a negative log scale. Thus, a decrease in pH from pH 7 to pH 4 is an increase of 10^3, or 1000, in the acidity of a solution. An increase from pH 3 to pH 6 is a decrease of 10^3, or 1000, in acidity.

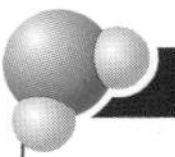

CHEM FACT

You may be familiar with two other logarithmic scales: the decibel scale for sound, and the Richter scale for earthquake strength.

Practice Problems

Logarithms

1. Calculate the logarithm of each number. Note the trend in your answers.
   (a) 1 **(b)** 5 **(c)** 10 **(d)** 50 **(e)** 100 **(f)** 500 **(g)** 1000 **(h)** 5000 **(i)** 10 000 **(j)** 50 000 **(k)** 100 000 **(l)** 500 000 **(m)** 1 000 000
2. Calculate the antilogarithm for each number.
   (a) 0 **(b)** 1 **(c)** –1 **(d)** 2 **(e)** –2 **(f)** 3 **(g)** –3
3. **(a)** How are your answers for question 2, parts (b) and (c) related?
   (b) How are your answers for question 2, parts (d) and (e) related?
   (c) How are your answers for question 2, parts (f) and (g) related?
   (d) Calculate the antilogarithm of 3.5.
   (e) Calculate the antilogarithm of –3.5.
   (f) Take the reciprocal of your answer for part (d).
   (g) How are your answers for parts (e) and (f) related?
4. **(a)** Calculate log 76 and log 55.
   (b) Add your answers for part (a).
   (c) Find the antilogarithm of your answer for part (b).
   (d) Multiply 76 and 55.
   (e) How are your answers in parts (c) and (d) related?

Linear Graphs

Suppose that you conduct a Boyle's law experiment. (See Chapter 11, section 11.2.) You obtain data that are similar to the data in Table E.2.

Table E.2 Boyle's Law Data

1/Pressure (kPa^{-1})	Volume (mL)
1.00×10^{-2}	50.0
8.93×10^{-3}	44.6
7.87×10^{-3}	39.4
7.04×10^{-3}	35.2
6.13×10^{-3}	30.7

Looking at the data does not tell you much about the relationship between the two variables: the volume (V) and the inverse pressure ($\frac{1}{P}$). What if you graph the data? Put $\frac{1}{P}$ (the independent variable) on the x-axis. Put V (the dependent variable) on the y-axis. The scale for $\frac{1}{P}$ should go from 0 to at least 0.01. The scale for V should go from 0 to at least 50.0. To plot each point, go along the x-axis until you reach the value of $\frac{1}{P}$ (such as 0.01). Then go up from this point to the corresponding value of V (50.0). After you have plotted all the points, draw the line of best fit through the points. (See Figures E.4A and E.4B).

Finding a Point on a Graph

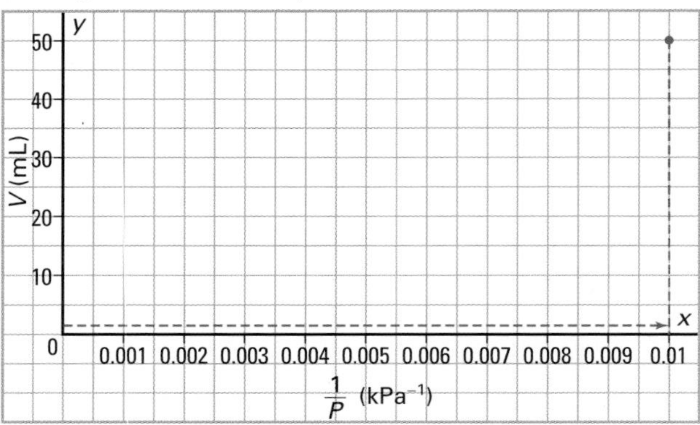

Figure E.4A

Drawing a Line of Best Fit

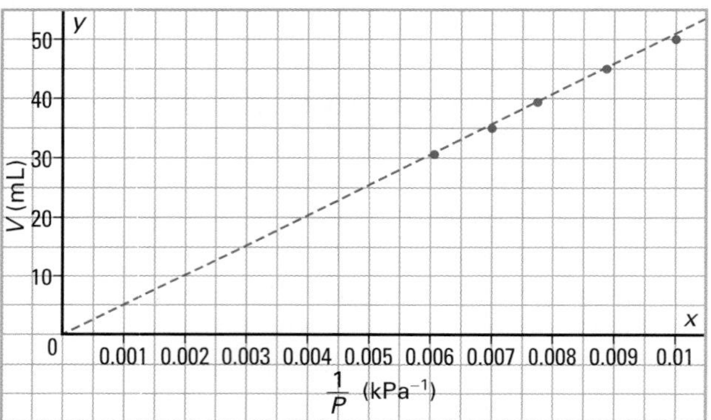

Figure E.4B

Examine the graph in Figure E.4B. You can see that the points form a straight line that goes through the origin, (0,0). You can also see that there is a linear relationship between the volume and the inverse pressure.

Mathematically, this means that the volume is directly proportional to the inverse pressure, or $V \propto \frac{1}{P}$. To remove the proportionality sign and replace it with an equal sign, V must be multiplied by a proportionality constant, k.

$$\therefore V = k(\frac{1}{P})$$
$$V = \frac{k}{P}$$
$$PV = k$$

How can you determine the proportionality constant? Recall, from mathematics courses, that the equation of a straight line that goes through the origin is given by $y = mx$, where m is the slope of the line. Also recall the equation for the slope of a straight line.

$$\text{Slope} = \frac{\text{Rise}}{\text{Run}}$$

This equation is illustrated in Figure E.4C.

Finding the Rise and Run on a Graph

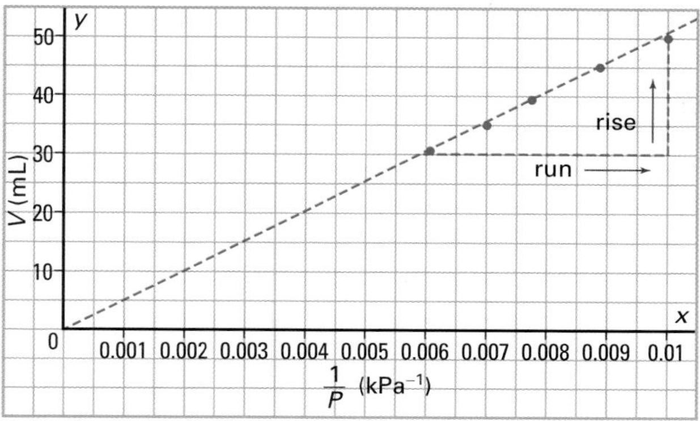

Figure E.4C

$$m = \frac{\Delta y}{\Delta x}$$
$$= \frac{(y_2 - y_1)}{(x_2 - x_1)}$$

For this graph, the change in the y values corresponds to a change in V, from 50.0 mL to 30.7 mL. The change in the x values corresponds to a change in $\frac{1}{P}$, from 1.00×10^{-2} kPa^{-1} to 6.13×10^{-3} kPa^{-1}.

$$m = \frac{(y_2 - y_1)}{(x_2 - x_1)}$$
$$= \frac{(50.0 - 30.7)\text{ mL}}{(0.010 - 0.00613)\text{ kPa}^{-1}}$$
$$= 4987.08\text{ (kPa mL)}$$

The equation of the line is therefore $V = 4.99 \times \frac{10^3}{P}$.

What happens if the data do not give a line that goes through the origin? The data in Table E.3 came from an experiment to determine the density of an unknown liquid. The student forgot to measure the mass of the container that was used to hold the various volumes of the liquid.

Table E.3 Mass and Volume of a Liquid

Volume of liquid (mL)	Mass of liquid and container (g)
22	58.4
41	71.7
65	87.2
83	99.2
100	110.6

Graph of Mass and Volume of a Liquid

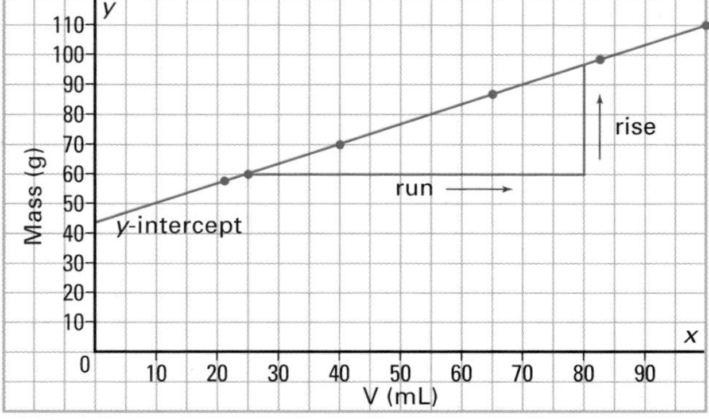

Figure E.5

These data are plotted, as shown in Figure E.5, with the volume on the x-axis and the mass on the y-axis. The equation of this line is in the form $y = mx + b$, where m is the slope of the line and b is the y-intercept. (The y-intercept is the value of y where the line crosses the y-axis. This value occurs when $x = 0$.) Thus, the relationship here is $m = kV + b$, where the k is the slope of the line. As before,

$$k = \frac{(y_2 - y_1)}{(x_2 - x_1)}$$
$$= \frac{(97.2 - 63.8)\text{ g}}{(80 - 30)\text{ mL}}$$
$$= 0.668\text{ g/mL}$$

Therefore, the equation of the line is $m = 0.668V + b$. If there is no liquid, both the mass and the volume of the liquid are zero. Notice, however, that there is still a value for mass on the graph. This is the value of b, the y-intercept, which is 43.8 g. It represents the mass of the container. Therefore, the final equation is $m = 0.668V + 43.8$.

Consider one more example. Two different sets of data for Charles' law experiments are given in Table E.4.

Table E.4 Charles' Law Data

Set 1

Temperature (°C)	Volume (mL)
20	150
34	157
42	161
55	168
68	175

Set 2

Temperature (°C)	Volume (mL)
15	240
26	249
38	259
51	270
80	294

The linear graphs of both sets of data are given in Figure E.6, on the same pair of axes. Since two different samples of gas were used, there are two different lines in the form $y = mx + b$. Notice that the lines intersect at −273°C, or 0 K, the theoretical lowest temperature possible. (The linear graphs for the temperature and volume of all gases go through this point.) If the y-axis was moved to this point and a new origin was established here, then the equations for the lines would be in the form $y = mx$. This is what happens when you convert temperatures to the Kelvin scale.

Charles' Law Graph

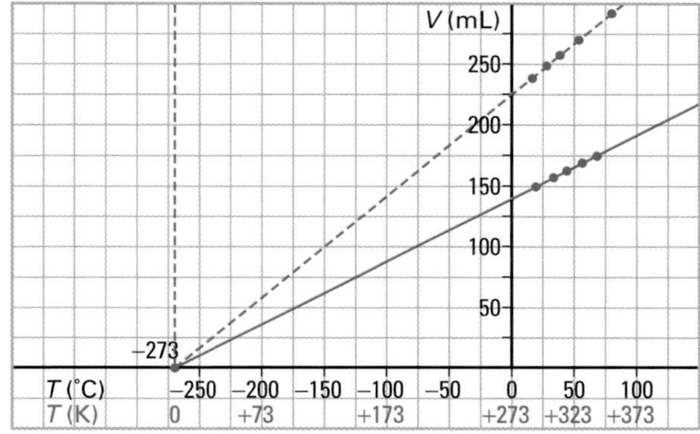

Figure E.6

Practice Problems

Linear Graphs

1. A spring, 20 cm long, has hooks on each end. One end is attached to the ceiling. The other has small, metal blocks, with hooks hanging from it. The following data are generated.

Mass attached (g)	Spring length (cm)
50	20.1
200	20.4
300	20.7
500	21.0
600	21.3

(a) Which variable is the independent variable?
(b) Which variable is the dependent variable?
(c) Graph these data. Is the graph linear?
(d) Determine the slope and the y-intercept for your graph. Do not forget to use appropriate units.
(e) Write an equation to represent your graph.

2. An object is dropped off the roof of a building that is 100 m high. Its position at various times is given below.

Time (s)	Distance fallen (m)
0	0
0.5	1.3
1	4.9
1.5	11
2	20

(a) Which variable is the independent variable?
(b) Which variable is the dependent variable?
(c) Graph these data. Is the graph linear?
(d) Try graphing *distance* versus some function of *time* (such as $1\sqrt{\text{time}}$ or time^2) until a linear graph appears to be generated.
(e) Write an equation from your linear graph in part (d).
(f) How long will it take for the object to reach the ground?

The Unit Analysis Method of Problem Solving

The unit analysis method of problem solving is extremely versatile. You can use it to convert between units or to solve simple formula problems. If you forget a formula during a test, you may still be able to solve the problem using unit analysis.

The unit analysis method involves analyzing the units and setting up conversion factors. You match and arrange the units so that they divide out to give the desired unit in the answer. Then you multiply and divide the numbers that correspond with the units.

Steps for Solving Problems Using Unit Analysis

Step 1 Determine which data you have and which conversion factors you need to use. (A conversion factor is usually a ratio of two numbers with units, such as 1000 g/1 kg. You multiply the given data by the conversion factor to get the desired units in the answer.) It is often convenient to use the following three categories to set up your solution: Have, Need, and Conversion Factors.

Step 2 Arrange the data and conversion factors so that you can cross out the undesired units. Decide whether you need any additional conversion factors to get the desired units in the answer.

Step 3 Multiply all the numbers on the top of the ratio. Then multiply all the numbers on the bottom of the ratio. Divide the top result by the bottom result.

Step 4 Check that the units have cancelled correctly. Also check that the answer seems reasonable, and that the significant digits are correct.

Sample Problem
Active ASA

Problem

In the past, pharmacists measured the active ingredients in many medications in a unit called grains (gr). A grain is equal to 64.8 mg. If one headache tablet contains 5.0 gr of active acetylsalicylic acid (ASA), how many grams of ASA are in two tablets?

What Is Required?

You need to find the mass in grams of ASA in two tablets.

What Is Given?

There are 5.0 gr of ASA in one tablet. A conversion factor for grains to milligrams is given.

Plan Your Strategy

Multiply the given quantity by conversion factors until all the unwanted units cancel out and only the desired units remain.

Have	Need	Conversion factors
5.0 gr	? g	64.8 mg/1 gr and 1 g/1000 mg

Act on Your Strategy

$$\frac{5.0\ \cancel{\text{gr}}}{1\ \cancel{\text{tablet}}} \times \frac{64.8\ \cancel{\text{mg}}}{1\ \cancel{\text{gr}}} \times \frac{1\ \text{g}}{1000\ \cancel{\text{mg}}} \times 2\ \cancel{\text{tablets}}$$

$$= \frac{5.0 \times 64.8 \times 1 \times 2\ \text{g}}{1000}$$

$$= 0.648\ \text{g}$$

$$= 0.65\ \text{g}$$

Continued ...

Continued ...

There are 0.65 g of active ASA in two headache tablets.

Check Your Solution

There are two significant digits in the answer. This is the least number of significant digits in the given data.

Notice how conversion factors are multiplied until all the unwanted units are cancelled out, leaving only the desired unit in the answer.

The next Sample Problem will show you how to solve a simple stoichiometric problem.

Sample Problem
Stoichiometry and Unit Analysis

Problem

What mass of oxygen, O_2, can be obtained by the decomposition of 5.0 g of potassium chlorate, $KClO_3$? The balanced equation is given below.

$$2KClO_3 \rightarrow 2KCl + 3O_2$$

What Is Required?

You need to calculate the amount of oxygen, in grams, that is produced by the decomposition of 5.0 g of potassium chlorate.

What Is Given?

You know the mass of potassium chlorate that decomposes.

$$\text{Mass} = 5.0 \text{ g}$$

From the balanced equation, you can obtain the molar ratio of the reactant and the product.

$$\frac{3 \text{ mol } O_2}{2 \text{ mol } KClO_3}$$

Plan Your Strategy

Calculate the molar masses of potassium chlorate and oxygen. Use the molar mass of potassium chlorate to find the number of moles in the sample.

Use the molar ratio to find the number of moles of oxygen produced. Use the molar mass of oxygen to convert this value to grams.

Act on Your Strategy

The molar mass of potassium chlorate is

$$\begin{aligned} 1 \times M_K &= 39.10 \\ 1 \times M_{Cl} &= 35.45 \\ 3 \times M_O &= 48.00 \\ \hline &\quad 122.55 \text{ g/mol} \end{aligned}$$

The molar mass of oxygen is

$$2 \times M_O = 36.00 \text{ g/mol}$$

Continued ...

Find the number of moles of potassium chlorate.

$$\begin{aligned} \text{mol } KClO_3 &= 5.0 \cancel{\text{g}} \times \left(\frac{1 \text{ mol}}{122.55 \cancel{\text{g}}\, KClO_3}\right) \\ &= 0.0408 \text{ mol} \end{aligned}$$

Find the number of moles of oxygen produced.

$$\frac{\text{mol } O_2}{0.0408 \text{ mol } KClO_3} = \frac{3 \text{ mol } O_2}{2 \text{ mol } KClO_3}$$

$$\begin{aligned} \text{mol } O_2 &= 0.0408 \text{ mol } KClO_3 \times \frac{3 \text{ mol } O_2}{2 \text{ mol } KClO_3} \\ &= 0.0612 \text{ mol} \end{aligned}$$

Convert this value to grams.

$$\begin{aligned} \text{mass } O_2 &= 0.0612 \cancel{\text{mol}} \times \frac{32.00 \text{ g}}{1 \cancel{\text{mol}}\, O_2} \\ &= 1.96 \text{ g} \\ &= 2.0 \text{ g} \end{aligned}$$

Therefore, 2.0 g of oxygen are produced by the decomposition of 5.0 g of potassium chlorate. As you become more familiar with this type of question, you will be able to complete more than one step at once. Below, you can see how the conversion factors we used in each step above can be combined. Set these conversion ratios so that the units cancel out correctly.

$$\begin{aligned} \text{mass } O_2 &= 5.0 \cancel{\text{g}\, KClO_3} \times \left(\frac{1 \cancel{\text{mol}}}{122.6 \cancel{\text{g}\, KClO_3}}\right) \times \\ &\left(\frac{3 \cancel{\text{mol}}\, O_2}{2 \cancel{\text{mol}}\, KClO_3}\right) \times \left(\frac{32.0 \text{ g}}{1 \cancel{\text{mol}}\, O_2}\right) \\ &= 1.96 \text{ g} \\ &= 2.0 \text{ g} \end{aligned}$$

Check Your Solution

The oxygen makes up only part of the potassium chlorate. Thus, we would expect less than 5.0 g of oxygen, as was calculated.

The smallest number of significant digits in the question is two. Thus, the answer must also have two significant digits.

Practice Problems

Unit Analysis

Use the unit analysis method to solve each problem.

1. The molecular mass of nitric acid is 63.02 g/mol. What is the mass, in g, of 7.00 mol of nitric acid?
2. To make a salt solution, 0.50 mol of NaCl are dissolved in 750 mL of water. What is the concentration, in g/L, of the salt solution?
3. The density of solid sulfur is 2.07 g/cm^3. What is the mass, in kg, of a 345 cm^3 sample?
4. How many grams of dissolved potassium iodide are in 550 mL of a 1.10 mol/L solution?

Chemistry Data Tables

Table F.1 Ionic Charges of Representative Elements

IA 1	IIA 2	IIIA 13	IVA 14	VA 15	VIA 16	VIIA 17	VIIIA 18
H^+						H^-	noble
Li^+	Be^{2+}			N^{3-}	O^{2-}	F^-	gases
Na^+	Mg^{2+}	Al^{3+}		P^{3-}	S^{2-}	Cl^-	do not
K^+	Ca^{2+}				Se^{2-}	Br^-	ionize
Rb^+	Sr^{2+}					I^-	
Cs^+	Ba^{2+}						

Table F.2 Charges of Some Transition Metal Ions

1+	2+	3+
silver, Ag^+	cadmium, Cd^{2+} nickel, Ni^{2+} zinc, Zn^{2+}	scandium, Sc^{3+}

Table F.3 Common Metal Ions with More Than One Ionic Charge

Formula	Stock Name	Classical Name
Cu^+	copper(I) ion	cuprous ion
Cu^{2+}	copper(II) ion	cupric ion
Fe^{2+}	iron(II) ion	ferrous ion
Fe^{3+}	iron(III) ion	ferric ion
$Hg_2^{2+}(Hg^+)$	mercury(I) ion	mercurous ion
Hg^{2+}	mercury(II) ion	mercuric ion
Pb^{2+}	lead(II) ion	plumbous ion
Pb^{4+}	lead(IV) ion	plumbic ion
Sn^{2+}	tin(II) ion	stannous ion
Sn^{4+}	tin(IV) ion	stannic ion
Cr^{2+}	chromium(II) ion	chromous ion
Cr^{3+}	chromium(III) ion	chromic ion
Mn^{2+}	manganese(II) ion	
Mn^{3+}	manganese(III) ion	
Mn^{4+}	manganese(IV) ion	
Co^{2+}	cobalt(II) ion	cobaltous ion
Co^{3+}	cobalt(III) ion	cobaltic ion

Table F.4 Common Polyatomic Ions

Formula	Name
PO_4^{3-}	phosphate
PO_3^{3-}	phosphite
SO_4^{2-}	sulfate
SO_3^{2-}	sulfite
CO_3^{2-}	carbonate
NO_3^-	nitrate
NO_2^-	nitrite
ClO_4^-	perchlorate
ClO_3^-	chlorate
ClO_2^-	chlorite
ClO^-	hypochlorite
CrO_4^{2-}	chromate
$Cr_2O_7^{2-}$	dichromate
$C_2H_3O_2^-$	acetate or ethanoate

Formula	Name
CN^-	cyanide
OH^-	hydroxide
MnO_4^-	permanganate
$C_2O_4^{2-}$	oxalate
SiO_3^{2-}	silicate
NH_4^+	ammonium
HPO_4^{2-}	hydrogen phosphate or biphosphate
$H_2PO_4^-$	dihydrogen phosphate
HPO_3^{2-}	hydrogen phosphite
$H_2PO_3^-$	dihydrogen phosphite
HSO_4^-	hydrogen sulfate
HSO_3^-	hydrogen sulfite
HCO_3^-	hydrogen carbonate or bicarbonate

Table F.5 Solubility of Compounds at SATP

	aluminum	ammonium	barium	calcium	copper(II)	iron(II)	iron(III)	lithium	magnesium	potassium	silver	sodium	strontium	zinc
acetate	S	S	S	S	S	S	S	S	S	S	ss	S	S	S
bromide	S	S	S	S	S	S	S	S	S	S	I	S	S	S
carbonate	–	S	I	I	–	I	–	ss	I	S	I	S	I	I
chlorate	S	S	S	S	S	S	S	S	S	S	S	S	S	S
chloride	S	S	S	S	S	S	S	S	S	S	I	S	S	S
chromate	I	S	I	S	I	–	I	S	S	S	I	S	ss	S
hydroxide	I	S	S	S	I	I	I	S	I	S	–	S	S	I
iodide	S	S	S	S	S	S	S	S	S	S	I	S	S	S
nitrate	S	S	S	S	S	S	S	S	S	S	S	S	S	S
oxide	I	–	ss	ss	I	I	I	S	I	S	I	S	S	I
perchlorate	S	S	S	S	S	S	S	S	S	S	S	S	S	S
phosphate	I	S	I	I	I	I	I	ss	I	S	I	S	I	I
sulfate	S	S	I	ss	S	S	ss	S	S	S	ss	S	I	S
sulfide	d	S	d	I	I	I	d	S	d	S	I	S	I	I

Legend

S = soluble
ss = slightly soluble
I = insoluble
– = no compound
d = decomposes in water

Table F.6 Chemicals in Everyday Life

Common name	Chemical formula and name (other names)	Physical properties	Safety concerns	Comments
acetone	CH_3COCH_3 2-propanone	clear; evaporates quickly	flammable; toxic by ingestion and inhalation	solvent; contained in some nail polish removers
acetylene	C_2H_2 ethyne	smells sweet	highly explosive	burns very hot, with oxygen, in oxyacetylene welding torches; used to produce a wide range of synthetic products
ASA	$CH_3COOC_6H_4COOH$ o-acetoxy benzoic acid (acetylsalicylic acid)	white crystals with a slightly bitter taste	excessive use may cause hearing loss or Reye's syndrome, especially in young people	used in Aspirin™ and related medicines for pain, fever, and inflammation
baking soda	$NaHCO_3$ sodium hydrogen carbonate (sodium bicarbonate)	tiny white crystals	none	used for baking and cleaning, as an antacid and mouthwash, and in fire extinguishers
battery acid	H_2SO_4 sulfuric acid	clear and odourless	corrosive	used in lead-acid storage batteries (automobile batteries)
bleach	$NaOCl_{(aq)}$ sodium hypochlorite solution	yellowish solution with a chlorine smell	toxic, strong oxidizing agent	household chlorine bleach; used for bleaching clothes and for cleaning
bluestone	$CuSO_4 \cdot 5H_2O$ copper(II) sulfate pentahydrate (cupric sulfate pentahydrate)	blue crystals or blue crystalline granules	toxic by ingestion; strong irritant	used in agriculture and industry, as a germicide, and for wood preservation
borax	$Na_2B_4O_7 \cdot 10H_2O$ sodium borate decahydrate	white crystals	none	main source is mining; used in the glass and ceramics industries; used for making Silly Putty® and for washing clothes
carborundum	SiC silicon carbide	hard, black solid	none	used as an abrasive
citric acid	$(HOOCCH_2)_2C(OH)(COOH)$ 2-hydroxy-1,2,3-propane (tricarboxylic acid)	translucent crystals with a strongly acidic taste	none	used in foods and soft drinks as an acidifying agent and an antioxidant
CFCs	CCl_2F_2, CCl_3F, $CClF_3$ chlorofluorocarbons (freons, Freon-12)	colourless, odourless gas	CFCs are now banned by the Montréal Protocol	in the past, were used as refrigerants and aerosols
charcoal/ graphite	$C_{(s)}$ pure carbon, in a less structured form than diamond	soft grey or black solid that rubs easily onto other substances	none	used as pencil "lead"and artists' charcoal, as a de-colourizing and filtering agent, in gunpowder, and for barbeque briquettes
cream of tartar	HOOC $(CHOH)_2$ COOK potassium hydrogen tartrate	white, crystalline solid	none	used as a leavening agent in baking powder

dry ice	CO_2 solid carbon dioxide	cold white solid that sublimates	damaging to skin and tissue after prolonged exposure	used as a refrigerant in laboratories when cold temperatures (as low as –79°C) are required
Epsom salts	$MgSO_4 \cdot 7H_2O$ magnesium sulfate heptahydrate	colourless crystals	can cause abdominal cramps and diarrhea	used as a bath salt and in cosmetics and dietary supplements; has industrial uses
ethylene	C_2H_4 ethene	colourless gas with sweet odour and taste	flammable	used to accelerate fruit ripening and to synthesize polymers such as polystyrene; occurs naturally in plants
ethylene glycol	CH_2OHCH_2OH glycol	clear, colourless, syrupy liquid	toxic by ingestion and inhalation	used in antifreeze and cosmetics, and as a de-icing fluid for airport runways
Glauber's salt	$Na_2SO_4 \cdot 10H_2O$ sodium sulfate decahydrate	large, transparent crystals, needles, or granular powder	none	a laxative; used for paper and glass making, and in solar heat storage and air conditioning; energy storage capacity more than seven times that of water
glucose	$C_6H_{12}O_6$ dextrose, grape sugar, corn sugar	white crystals with a sweet taste	none	source of energy for most organisms
grain alcohol	C_2H_5OH ethanol (ethyl alcohol)	clear, volatile liquid with distinctive odour	flammable	beverage alcohol, antiseptic, laboratory/industrial solvent; produced by the fermentation of grains or fruits
gyp rock	$CaSO_4 \cdot 2H_2O$ gypsum	hard, beige mineral	none	used in plaster of Paris and as a core for drywall
hydrogen peroxide	H_2O_2	clear, colourless liquid	damaging to skin in high concentrations	sold as 3% solution in drugstores; non-chlorine bleach often 6% H_2O_2
ibuprofen	$C_{13}H_{18}O_2$ p-isobutyl-hydratropic acid	white crystals	can conflict with other medications	ingredient in over-the-counter pain relievers
laughing gas	N_2O nitrous oxide, dinitrogen oxide	colourless, mainly odourless, soluble gas	prolonged exposure causes brain damage and infertility	used as a dental anesthetic, an aerosol propellant, and to increase fuel performance in racing cars
lime	CaO calcium oxide (hydrated lime, hydraulic lime, quicklime)	white powder	reacts with water to produce caustic calcium hydroxide, $Ca(OH)_2$, with liberation of heat	used to make cement and to clean and nullify odours in stables
limestone	$CaCO_3$ calcium carbonate	soft white mineral	none	used for making lime and for building; has industrial uses
lye	$NaOH$ sodium hydroxide (caustic soda)	white solid, found mainly in form of beads or pellets; quickly absorbs water and CO_2 from the air	corrosive, strong irritant	produced by the electrolysis of brine or the reaction of calcium hydroxide and sodium carbonate; has many laboratory and industrial uses; used to manufacture chemicals and make soap
malachite	$CuCO_3 \cdot Cu(OH)_2$ basic coper(II) carbonate	clear, hard, bright green mineral	none	ornamental and gem stone; copper found in the ore

milk of magnesia	$Mg(OH)_2$ magnesium hydroxide (magnesia magma)	white powder	harmless if used in small amounts	antacid, laxative
moth balls	$C_{10}H_8$ naphthalene	white, volatile solid with an unpleasant odour	toxic by ingestion and inhalation	used to repel insects in homes and gardens, and to make synthetic resins; obtained from crude oil
MSG	$COOH(CH_2)_2CH(NH_2)$-COONa monosodium glutamate	white, crystalline powder	may cause headaches in some people	flavour enhancer for foods in concentrations of about 0.3%
muriatic acid	$HCl_{(aq)}$ hydrochloric acid	colourless or slightly yellow aqueous solution	toxic by ingestion and inhalation; strong irritant	has many industrial and laboratory uses; used for processing food, cleaning, and pickling
natural gas	about 85% methane, CH_4, 10% ethane, C_2H_6, and some propane, C_3H_8, butane, C_4H_{10}, and pentane, C_5H_{12}	odourless, colourless gas	flammable and explosive; a warning odour is added to household gas as a safety precaution	used for heating, energy, and cooking; about 3% is used as a feedstock for the chemical industry
oxalic acid	HO_2CCO_2H ethanedoic acid	strongly flavoured acid; white crystals	toxic by inhalation and ingestion; strong irritant in high concentrations	occurs naturally in rhubarb, wood sorrel, and spinach; used as wood and textile bleach, rust remover, and deck cleaner; has many industrial and laboratory uses
Pepto-Bismol™	bismuth subsalicylate calcium carbonate	pink solid or solution	may cause stomach upset if taken in excess of recommended dose	relieves digestive difficulties by coating the digestive tract and reducing acidity
PCBs	polychlorinated biphenyls: class of compounds with two benzene rings and two or more substituted chlorine atoms	colourless liquids	highly toxic, unreactive, and persistent; cause ecological damage	used as coolants in electrical transformers
potash	K_2CO_3 potassium carbonate	white, granular, translucent powder	solutions irritating to tissue	laboratory and industrial uses; used in special glasses, in soaps, and as a dehydrating agent
PVCs	$(C_2H_3Cl)_n$ polyvinyl chloride, polychloroethene	tough, white, unreactive solid	none	used extensively as a building material
road salt	$CaCl_2$ calcium chloride	white crystalline compound	none	by-product of the Solvay process
rotten-egg gas	H_2S hydrogen sulfide	colourless gas with an offensive odour	highly flammable, therefore high fire risk; explosive; toxic by inhalation; strong irritant to eyes and mucous membranes	obtained from sour gas during natural gas production

rubbing alcohol	$(CH_3)_2CHOH$ isopropanol (isopropyl alcohol)	colourless liquid with a pleasant odour	flammable, therefore high fire risk; explosive; toxic by inhalation and ingestion	has industrial and medical uses
salicylic acid	HOC_6H_4COOH 2-hydroxybenzoic acid	white crystalline solid	damages skin in high concentrations	can be used in different amounts in foods and dyes, and in wart treatment
sand	SiO_2 silica	large, glassy cubic crystals	toxic by inhalation; chronic exposure to dust may cause silicosis	occurs widely in nature as sand, quartz, flint, and diatomite
slaked lime	$Ca(OH)_2$ calcium hydroxide	white powder that is insoluble in water	none	used to neutralize acidity in soils and to make whitewash, bleaching powder, and glass
soda ash	Na_2CO_3 sodium carbonate	white powdery crystals	none	used to manufacture glass, soaps, and detergents
sugar	$C_{12}H_{22}O_{11}$ sucrose (cane or beet sugar)	cubic white crystals	none	used in foods as a sweetener; source of metabolic energy
table salt	NaCl sodium chloride (rock salt, halite)	cubic white crystals	none	produced by the evaporation of natural brines and by the solar evaporation of sea water; also mined from underground sources; used in foods and for de-icing roads
Tylenol™	$CH_3CONHC_6H_4OH$ N-acetyl-p-aminophenol (acetaminophen, APAP)	colourless, slightly bitter crystals	can be toxic if an overdose is taken	pain reliever (analgesic)
TSP	Na_3PO_4 trisodium phosphate (sodium phosphate, sodium orthophosphate)	white crystals	toxic by ingestion; irritant to tissue; pH of 1% solution is 11.8 to 12	used as a water softener and cleaner (for example, to clean metals and to clean walls before painting); has many industrial uses
vinegar	5% acetic acid, CH_3COOH, in water	clear solution with a distinctive smell	none	used for cooking and household cleaning
vitamin C	$C_6H_8O_6$ ascorbic acid	white crystals or powder with a tart, acidic taste	none	required in diet to prevent scurvy; found in citrus fruits, tomatoes, potatoes, and green leafy vegetables
washing soda	$Na_2CO_3{\cdot}H_2O$ sodium carbonate monohydrate (soda ash)	white powdery crystals	may be irritating to skin	used for cleaning and photography, and as a food additive; has many industrial and laboratory uses
wood alcohol	CH_3OH methanol (methyl alcohol)	clear, colourless liquid with faint alcoholic odour	flammable; toxic by ingestion, skin absorption, and inhalation; causes blindness and death	has many industrial and household uses; used in gasoline antifreeze and as a thinner for shellac and paint; can be mixed with vegetable oil and lye to make diesel

Alphabetical List of Elements

Element	Symbol	Atomic Number
Actinium	Ac	89
Aluminum	Al	13
Americium	Am	95
Antimony	Sb	51
Argon	Ar	18
Arsenic	As	33
Astatine	At	85
Barium	Ba	56
Berkelium	Bk	97
Beryllium	Be	4
Bismuth	Bi	83
Bohrium	Bh	107
Boron	B	5
Bromine	Br	35
Cadmium	Cd	48
Calcium	Ca	20
Californium	Cf	98
Carbon	C	6
Cerium	Ce	58
Cesium	Cs	55
Chlorine	Cl	17
Chromium	Cr	24
Cobalt	Co	27
Copper	Cu	29
Curium	Cm	96
Dubnium	Db	105
Dysprosium	Dy	66
Einsteinium	Es	99
Erbium	Er	68
Europium	Eu	63
Fermium	Fm	100
Fluorine	F	9
Francium	Fr	87
Gadolinium	Gd	64
Gallium	Ga	31
Germanium	Ge	32
Gold	Au	79
Hafnium	Hf	72
Hassium	Hs	108
Helium	He	2
Holmium	Ho	67
Hydrogen	H	1
Indium	In	49
Iodine	I	53
Iridium	Ir	77
Iron	Fe	26
Krypton	Kr	36
Lanthanum	La	57
Lawrencium	Lr	103
Lead	Pb	82
Lithium	Li	3
Lutetium	Lu	71
Magnesium	Mg	12
Manganese	Mn	25
Meitnerium	Mt	109
Mendelevium	Md	101
Mercury	Hg	80
Molybdenum	Mo	42
Neodymium	Nd	60
Neon	Ne	10
Neptunium	Np	93
Nickel	Ni	28
Niobium	Nb	41
Nitrogen	N	7
Nobelium	No	102
Osmium	Os	76
Oxygen	O	8
Palladium	Pd	46
Phosphorus	P	15
Platinum	Pt	78
Plutonium	Pu	94
Polonium	Po	84
Potassium	K	19
Praseodymium	Pr	59
Promethium	Pm	61
Protactinium	Pa	91
Radium	Ra	88
Radon	Rn	86
Rhenium	Re	75
Rhodium	Rh	45
Rubidium	Rb	37
Ruthenium	Ru	44
Rutherfordium	Rf	104
Samarium	Sm	62
Scandium	Sc	21
Seaborgium	Sg	106
Selenium	Se	34
Silicon	Si	14
Silver	Ag	47
Sodium	Na	11
Strontium	Sr	38
Sulfur	S	16
Tantalum	Ta	73
Technetium	Tc	43
Tellurium	Te	52
Terbium	Tb	65
Thallium	Tl	81
Thorium	Th	90
Thulium	Tm	69
Tin	Sn	50
Titanium	Ti	22
Tungsten	W	74
Ununbium	Uub	112
Ununhexium	Uuh	116
Ununnilium	Uun	110**
Ununoctium	Uuo	118
Ununquadium	Uuq	114
Unununium	Uuu	111
Uranium	U	92
Vanadium	V	23
Xenon	Xe	54
Ytterbium	Yb	70
Yttrium	Y	39
Zinc	Zn	30
Zirconium	Zr	40

**The names and symbols for elements 110 through 118 have not yet been chosen

Glossary

Section numbers are provided in parentheses.

A

accuracy: the closeness of a measurement to an accepted value (1.2)

acid: a substance that produces hydrogen ions in aqueous solutions (10.1)

acid-base indicator: a substance, usually a weak monoprotic acid, that changes colour in acidic and basic solutions (10.3)

activity series: a ranking of the relative reactivity of metals or halogens in aqueous reactions (4.3)

actual yield: the measured quantity of product obtained in a chemical reaction (7.3)

aliphatic hydrocarbons: hydrocarbons that consist of chains or non-aromatic rings; the carbon atoms are bonded to the maximum number of hydrogen or carbon atoms (13.3)

alkanes: hydrocarbons that contain only carbon-carbon single bonds; general formula C_nH_{2n+2} (13.3)

alkenes: hydrocarbons that contain one or more carbon-carbon double bond; general formula C_nH_{2n} (13.3)

alkynes: hydrocarbons that contain one or more carbon-carbon triple bond; general formula C_nH_{2n-2} (13.3)

alloys: solid metallic solutions (8.1)

alpha particle emission: a radioactive process that involves the loss of one alpha ($\propto$) particle (a helium nucleus, ${}^{4}_{2}He$); also called *alpha decay* (4.4)

anhydrous: a term used to describe a compound that does not have any water molecules bonded to it; applies to compounds that can be hydrated (6.4)

aqueous solution: a solution in which water is the solvent (8.1)

Arrhenius theory of acids and bases: the theory stating that an acid is a substance that produces hydrogen ions in water and a base is a substance that produces hydroxide ions in water (10.1)

atmosphere (atm): a unit of pressure; equal to 101.325 kPa (11.2)

atom: the basic unit of an element, which still retains the element's properties (2.1)

atomic mass unit (u): a unit of mass that is 1/12 of the mass of a carbon-12 atom; equal to 1.66×10^{-24} g (2.1)

atomic number (*Z*): the unique number of protons in the nucleus of a particular element (2.1)

atomic symbol: a one- or two-letter abbreviation of the name of an element; also called *element symbol* (2.1)

average atomic mass: the average of the masses of all the isotopes of an element; given in atomic mass units (u) (5.1)

Avogadro constant (N_A): the experimentally-determined number of particles in 1 mol of a substance; the currently accepted value is $6.022\ 141\ 99 \times 10^{23}$ (5.2)

Avogadro's hypothesis: equal volumes of gases, at the same temperature and pressure, contain the same number of particles (12.1)

B

balanced chemical equation: a statement that uses chemical formulas and coefficients to show the identity and quantity of the reactants and products involved in a chemical reaction (4.1)

base: a substance that produces hydroxide ions in aqueous solutions (10.1)

benefit: a desirable result of an action (14.4)

beta decay: a nuclear reaction that results in the emission of a beta (β) particle (electron) from a nucleus (4.4)

binary acid: an acid that is composed of two elements: hydrogen and a non-metal (10.2)

binary compound: a compound that is composed of two elements (3.4)

bomb calorimeter: a device that combusts a substance in pure oxygen in order to measure the heat of combustion of that substance (14.4)

bond energy: the amount of energy that is produced or absorbed when a specific bond in a molecule is broken or formed; measured in kJ/mol (14.2)

Boyle's law: the law stating that the volume of a given amount of gas varies inversely with the applied pressure, if the temperature is constant: $V \propto \frac{1}{P}$ (11.2)

branched-chain alkane: an alkane with one or more side-chains that branch off the parent chain (13.3)

Brønsted-Lowry theory of acids and bases: the theory defining an acid as a substance from which a hydrogen ion can be removed and a base as a substance that can remove a hydrogen ion from an acid (10.1)

C

calorie: the amount of energy that is needed to raise the temperature of 1 g of liquid water by 1°C; equal to 4.184 J (14.4)

Calorie: a unit of energy equal to one thousand calories (1 kcal = 1000 cal) or 4.184 kJ (14.4)

calorimeter: a device that burns compounds containing carbon, hydrogen, and other elements in a stream of oxygen, O_2, to determine their composition (14.4)

calorimetry: the process of measuring changes in thermal energy (14.4)

carbon-hydrogen combustion analyzer: a device that uses the combustion-enabling properties of O_2 to determine the composition of compounds containing carbon, hydrogen, and oxygen (6.4)

cation: a positively charged ion (2.3)

Charles' law: the law stating that the volume of a fixed mass of gas is directly proportional to its kelvin temperature, if the pressure remains constant: $V \propto T$ (11.3)

chemical bond: the force that holds atoms together in compounds (3.1)

chemical change: the type of change that occurs when elements and/or compounds interact with each other to form different substances with different properties; involves the rearrangement of atoms (1.3)

chemical equation: a statement of what occurs in a chemical reaction; can be a word equation, a skeleton equation, or a balanced chemical equation (4.1)

chemical formula: a representation, in atomic symbols and numerical subscripts, of the type and number of atoms that are present in a compound (3.4)

chemical nomenclature: the system that is used to name chemical compounds (3.4)

chemical property: a property of a substance that can only be observed as the substance changes into another substance (1.2)

chemical reaction: a process in which a substance (or substances) changes, forming one or more different substances (4.1)

chemistry: the study of matter, its composition, and its interactions (1.1)

chlorofluorocarbon (CFC): a compound containing carbon, fluorine, and chlorine atoms that is chemically inert in the troposphere, but that is broken down by solar radiation in the stratosphere (12.4)

cis-trans isomers: compounds that have the same formula but different arrangements of atoms around a fixed carbon-carbon double bond; also called *geometric isomers* (13.3)

closed system: a system in which the total amount of matter remains constant; matter can neither enter nor leave this system (11.2)

coefficient: in a balanced chemical equation, a positive number that is placed in front of a formula to show how many units of the substance are involved (4.1)

combined gas law: a combination of Boyle's law and Charles' law, which states that the pressure and volume of a given amount of gas are inversely proportional, and directly proportional to the kelvin temperature of the gas: $V \propto \frac{T}{P}$ (11.4)

combustion: the reaction of a substance with oxygen, producing oxides, heat, and light; burning (14)

common name: a name for a compound that does not necessarily suggest anything about the chemical composition of the compound (e.g., water, baking soda); also called the *trivial name* (3.4)

competing reaction: a reaction that occurs at the same time as a principal reaction and consumes the reactants and/or products of the principal reaction (7.3)

complete combustion: combustion in which a hydrocarbon fuel is completely reacted in the presence of sufficient oxygen, producing only carbon dioxide gas and water vapour (14.1)

complete combustion reaction: a synthesis reaction in which a compound burns in the presence of oxygen gas, forming the most common oxides of the elements in the compound (4.2)

complete structural diagram: a symbolic representation of all the atoms in a molecule, showing how they are bonded (13.2)

compound: a pure substance that is composed of two or more elements chemically combined in fixed proportions (1.3)

concentration: the ratio of the amount of solute per quantity of solvent (8)

condensation: a physical change from the gaseous state to the liquid state (11.1)

condensed structural diagram: a symbolic representation of a compound, showing most atoms present, and the bonds between carbon atoms (13.2)

conjugate acid: the particle that results when a base receives a proton (10.1)

conjugate acid-base pair: two molecules or ions that are linked by the transfer of a proton (10.1)

conjugate base: the particle that remains when a proton is removed from an acid (10.1)

covalent bond: a chemical bond in which two electrons are shared by two atoms (3.1)

cracking: the use of heat or catalysts, in the absence of air, to break down or rearrange large hydrocarbon molecules (13.4)

cyclic hydrocarbon: a hydrocarbon that consists of one or more rings; can be a cycloalkane, a cycloalkene, or a cycloalkyne (13.3)

Dalton's law of partial pressures: the law stating that the total pressure of a mixture of gases is the sum of the pressures of each of the individual gases (11.4)

decomposition reaction: a chemical reaction in which a compound breaks down into elements or simpler compounds (4.2)

ΔT: a symbol used to indicate change in temperature (14.3)

diatomic element: an atom of this element tends to bond with another atom of the same element, forming a molecule that contains two atoms (3.2)

diffraction grating: a device that separates light into a spectrum (2.2)

dipole: dipole moment; a distribution of molecular charge consisting of two opposite charges that are separated by a short distance (8.2)

dipole-dipole attraction: the intermolecular force between oppositely charged ends of two polar molecules (molecules with dipoles) (8.2)

diprotic acid: an acid that contains two hydrogen ions that can dissociate (10.2)

double bond: a covalent bond in which two atoms share two pairs of electrons (3.2)

double displacement reaction: a chemical reaction in which the cations of two ionic compounds exchange places, resulting in the formation of two new compounds (4.3)

E

elastomer: a polymer that can be bent or twisted by an outside force; it will return to its previous shape once the force is removed (13.3)

electrolyte: a solute that conducts a current in an aqueous solution (8.2)

electron: a negatively charged subatomic particle that occupies the space around the nucleus of an atom (2.1)

electron affinity: the change in energy that accompanies the addition of an electron to an atom in the gaseous state (2.3)

electronegativity: a relative measure of an atom's ability to attract shared electrons in a chemical bond (3.1)

element: a pure substance that cannot be broken down into smaller particles and retain the same properties (1.3)

empirical formula: shows the lowest whole number ratio of atoms of each element in a compound (6.2)

endothermic process: a process that absorbs thermal energy (14.2)

end-point: the point in a titration when the acid-base indicator changes colour (10.3)

energy level: fixed, three-dimensional volume in which electrons travel around the nucleus (2.2)

equivalence point: the point in a titration when the number of moles of added solution is stoichiometrically equal to the number of moles of standard solution (10.3)

exothermic process: a process that produces thermal energy (14.2)

expanded molecular formula: a symbolic representation that shows the arrangement of atoms in a molecule (e.g., $CH_3CH_2CH_3$ for propane, C_3H_8) (13.2)

F

forensic scientist: a scientist who uses specialized knowledge to analyze evidence in legal cases (6.3)

fossil fuel: a fuel that is formed over geologic time by the action of pressure and heat on organic materials (e.g., petroleum, coal) (14)

fractional distillation: a process that uses the specific boiling points of substances to refine a mixture into separate components (13.4)

fraction: one component of a substance that has been refined by fractional distillation (13.4)

fuel cell: a technology that produces energy by the reaction of hydrogen and oxygen, leaving water as a by-product (11.5)

fusible plug: a safety device that melts at a high temperature to relieve gas pressure inside a container (11.3)

G

gamma radiation: a type of high-energy electromagnetic radiation in which gamma (γ) photons are emitted from a nucleus (4.4)

Gay-Lussac's law: the law stating that the pressure of a fixed amount of gas is directly proportional to its Kelvin temperature, if the volume is constant: $P \propto T$ (11.3)

general solubility guidelines: a set of guidelines that characterize the solubility of substances in water (9.1)

geometric isomers: compounds that have the same formula but different arrangements of atoms around a fixed carbon-carbon double bond; also called cis-trans isomers (13.3)

global warming: a gradual increase in the average temperature of Earth's atmosphere (14.5)

greenhouse gas: a gas that prevents some of the heat produced by solar radiation from leaving the atmosphere (e.g., carbon dioxide) (14.5)

H

hard water: water with a high concentration of dissolved ions (9.4)

heat: the transfer of thermal energy between objects with different temperatures (14.3)

heat capacity: the amount of energy that is needed to change the temperature of a particular substance or system by 1°C; measured in kJ/°C (14.4)

heat of combustion: the heat that is released by a combustion reaction; usually measured in kJ/mol (14.4)

heat of solution: the change in the thermal energy when a solute dissolves in a solvent (14.4)

heterogeneous mixture: a mixture in which the different components can be distinctly seen (1.3)

homogeneous mixture: a mixture in which the different components are mixed so that they appear to be a single substance; a solution (1.3)

homologous series: a series of molecules in which each member differs from the next by an additional specific structural unit (e.g. $-CH_2-$) (13.3)

hydrate: a compound that has a specific number of water molecules bonded to each formula unit (6.4)

hydrated: a term used to describe ions in aqueous solutions, surrounded by and attached to water molecules (8.2)

hydrocarbon: a molecular compound that contains only hydrogen and carbon atoms (13)

hydrogen bonding: the strong intermolecular attraction between molecules containing a hydrogen atom bonded to an atom of a highly electronegative element, especially oxygen (8.2)

hydronium ion: a proton that is bonded to a water molecule; chemical formula H_3O^+ (10.1)

I

ideal gas: a hypothetical gas with particles that have mass but no volume or attractive forces between them (11.1)

ideal gas law: the law stating that the pressure times the volume of a gas equals the number of moles of the gas times the universal gas constant and the temperature of the gas; $PV = nRT$ (12.1)

immiscible: a term used to describe substances that are not able to combine with each other in a solution (8.1)

incomplete combustion: combustion in which insufficient oxygen prevents a hydrocarbon fuel from reacting completely, leaving products other than carbon dioxide gas and water vapour (4.2, 14.1)

insoluble: a term used to describe a substance that has a solubility of less than 0.1 g per 100 mL in a particular solvent (8.1)

intermolecular forces: the forces that exist between molecules (3.2)

International System of Units (SI): the international system of measurement units, including units, including units such as the metre, the kilogram, and the mole; from the French *Système international d'unités* (1.2)

intramolecular forces: the forces that bond atoms together within a molecule (e.g., covalent bonds) (3.2)

ion: a positively or negatively charged particle that results from a neutral atom or group of atoms giving up or gaining electrons (2.2)

ion-dipole attractions: the intermolecular forces between ions and polar molecules (8.2)

ion exchange: a process for softening water by exchanging one type of ion with another (9.4)

ionic bond: a bond between oppositely charged ions that arises from electron transfer; usually involves a metal atom and a non-metal atom (3.1)

ionization energy: the energy that is needed to remove an electron from a neutral atom (2.3)

isolated system: a system in which the total amount of matter and energy remains constant (14.4)

isomers: compounds that have the same chemical formula but different molecular arrangements and properties (13.2)

isotopes: atoms of an element that are chemically similar but have different numbers of neutrons and thus, different mass numbers (2.1)

isotopic abundance: the relative amount of an isotope of an element; expressed as a percent or a decimal fraction (5.1)

IUPAC: the acronym for *International Union of Pure and Applied Chemistry*, an organization that specifies rules for chemical names and symbols (3.4)

K

Kelvin scale: a temperature scale that begins at the theoretical point of absolute zero kinetic energy (0 K = −273.15°C); each unit (a kelvin) is equal to 1°C (11.3)

kilopascal: a unit of pressure equal to 1000 Pa (11.2)

kinetic molecular theory: the theory explaining gas behaviour in terms of the random motion of particles with negligible volume and negligible intermolecular forces (11.1)

L

law of combining volumes: the law stating that when gases react, the volumes of the gaseous reactants and products, at constant temperatures and pressures, are always in whole number ratios (12.1)

law of conservation of mass: the law stating that matter can be neither created nor destroyed; in any chemical reaction, the mass of the products is always equal to the mass of the reactants (4.1)

law of definite proportions: the law stating that the elements in a chemical compound are always present in the same proportions by mass (6.1)

law of multiple proportions: the law stating that the masses of two or more elements that combine to form a compound can be expressed in small whole number ratios (12.1)

Lewis structure: a symbolic representation of the arrangement of the valence electrons of an element (2.2)

limiting reactant: the reactant that is completely consumed during a chemical reaction, limiting the amount of product produced (7.2)

line structural diagram: a graphical representation of the bonds between carbon atoms in a hydrocarbon (13.2)

lone pairs: pairs of electrons in an atom's outer valence shell that are not involved in covalent bonding (3.3)

M

mass/mass percent: the mass of a solute divided by the mass of the solution, expressed as a percent (8.3)

mass number (*A*): the sum of the protons and neutrons in the nucleus of one atom of a particular element (2.1)

mass percent: the mass of an element in a compound, expressed as a percent of the compound's total mass (6.1)

mass spectrometer: an instrument that uses magnetic fields to separate the isotopes of an element and measure the mass and abundance of each isotope (5.1)

mass/volume percent: the mass of a solute divided by the volume of the solution, expressed as a percent (8.3)

matter: anything that has mass and occupies space (1.2)

metathesis reaction: a double displacement reaction (9.2)

millimetre of mercury (mm Hg): a unit of pressure that is based on the height of a column of mercury in a barometer or manometer; equal to 1 torr (11.2)

miscible: a term used to describe substances that are able to combine with each other in any proportion (8.1)

mixture: a combination of two or more kinds of matter, in which each component retains its own characteristics (1.3)

molar concentration (*C*): a unit of concentration expressed as the number of moles of solute present in one litre of solution; also called *molarity* (8.3)

molar mass (*M*): the mass of 1 mol of a substance, numerically equal to the element's average atomic mass; expressed in g/mol (5.2)

molar volume: the amount of space that is occupied by 1 mol of a substance; equal to 22.4 L for a gas at standard temperature and pressure (STP) (12.1)

molecular compound: a non-conducting compound whose intramolecular bonds are not broken when the compound changes state (3.2)

molecular formula: a formula that gives the actual number of atoms of each element in a molecule or formula unit (6.2)

mole (mol): the SI base unit for amount of substance; contains the same number of atoms, molecules, or formula units as exactly 12 g of carbon-12 (5.2)

mole ratio: a ratio that compares the number of moles of different substances in a balanced chemical equation (7.1)

monomer: a small, repeating molecular unit in a polymer chain (13.1)

monoprotic acid: an acid that contains only one hydrogen ion that can dissociate (10.2)

Montréal Protocol: an international agreement that limits the global use of CFCs and other ozone-destroying chemicals (12.4)

N

net ionic equation: a representation of a chemical reaction in a solution that shows only the ions involved in the chemical change (9.2)

neutralization reaction: a double displacement reaction in which an acid and a base combine to form water and a salt (4.3, 10.3)

neutron: an uncharged subatomic particle in the nucleus of an atom (2.1)

non-electrolyte: a solute that does not conduct a current in an aqueous solution (8.2)

non-polar molecule: a covalently bonded molecule that does not possess a dipole moment, because of the arrangement of its molecules (3.3)

nuclear equation: a symbolic representation of a nuclear reaction, showing how a nucleus gains or loses subatomic particles (4.4)

nuclear fission: the process in which an unstable, heavy isotope splits into smaller, lighter nuclei (4.4)

nuclear fusion: the process by which a nucleus absorbs lighter, accelerated nuclei (4.4)

nuclear reaction: a reaction that involves changes in the nuclei of atoms (4.4)

nucleus: the central core of an atom, composed of protons and neutrons (2.1)

O

octet: an arrangement of eight electrons in the valence shell of an atom (2.2)

octet rule: the rule stating that atoms bond in such a way as to attain eight electrons in their valence shells (3.2)

organic compound: a molecular compound based on carbon, almost always containing carbon-carbon and carbon-hydrogen bonds (13.1)

oxoacid: an acid formed from a polyatomic ion that contains oxygen, hydrogen, and one other acid (10.2)

P

parts per million/parts per billion: units of concentration used to express very small quantities of solute (8.3)

pascal: the SI unit of pressure; equal to 1 N/m^2 (11.2)

percentage composition: the relative mass of each element in a compound (6.1)

percentage purity: the percent of a sample that is composed of a specific compound or element (7.3)

percentage yield: the actual yield of a reaction, expressed as a percent of the theoretical yield (7.3)

periodic table: a system for organizing the elements by atomic number into groups (columns) and periods (rows), so that elements with similar properties are in the same column (2.2)

periodic trend: a pattern that is evident when elements are organized by their atomic numbers (2.2)

petrochemical: a product that is derived from petroleum (13.4)

petroleum: a complex mixture of solid, liquid, and gaseous hydrocarbons (13.1)

pH: the negative logarithm of the concentration of hydronium ions, $-\log [H_3O^+]$, measured in mol/L (10.2)

pH scale: a mathematical scale that is used to express the concentration of hydronium ions in a solution as a number from 0 to 14 (10.2)

physical change: a change, such as change of state, that does not alter the composition of matter (1.3)

physical property: a property of a substance that can be observed without the substance changing into or interacting with another substance (1.2)

polar covalent bond: a covalent bond between atoms that have significantly different electronegativities, in which the electron pair is unevenly shared (3.3)

polar molecule: a molecule that has an uneven distribution of charge; one end has a partial positive charge and one end has a partial negative charge (3.3)

polyatomic ion: an ion that is made up of two or more atoms; it has a positive or negative charge (3.4)

polymerization: a process, common in the plastics industry, in which polymers are formed by reacting monomers (13.3)

polymer: a very long molecule that is formed by the covalent bonding of many smaller, identical molecular units (monomers) (13.1)

potential energy: stored energy; the energy of an object due to its position (14.4)

precipitate: an insoluble solid that is formed by a chemical reaction between two soluble compounds (4.3) (9.1)

precipitation reaction: a double displacement reaction that forms a precipitate (9.2)

precision: the closeness of a measurement to other measurements of the same object or phenomena (1.2)

pressure: the force that is exerted on an object, per unit of surface area (11.2)

pressure relief valve: a device that regulates the pressure of a gas inside a container (11.3)

product: a substance that is formed by a chemical reaction (4.1)

property: a characteristic that distinguishes different types of matter; (e.g., colour, melting or boiling point, conductivity, density) (1.2)

proton: a positively charged subatomic particle in the nucleus of an atom (2.1)

pure covalent bond: a chemical bond between two atoms with identical or nearly identical electronegativities (3.2)

pure substance: a material that is composed of only one type of particle (e.g., iron, water, sodium chloride) (1.3)

Q

qualitative analysis: the process of separating and identifying ions in an aqueous solution (9.2)

qualitative property: a property of matter that can be observed but cannot be precisely measured or expressed numerically (e.g., colour, odour) (1.2)

quantitative property: a property of matter that can be measured and expressed numerically (e.g., density, boiling point) (1.2)

R

radioactivity: the process in which unstable nuclei spontaneously decay, releasing energy and subatomic particles (2.1)

radioisotope: an unstable isotope of an element, which undergoes radioactive decay (2.1)

rate of dissolving: the speed at which a solute dissolves in a solvent (8.2)

reactant: a substance that undergoes a chemical change in a chemical reaction (4.1)

reforming: the use of heat, pressure, and catalysts to convert a large hydrocarbon molecule into other compounds (13.4)

risk: a potential danger; a chance of an undesirable consequence (14.4)

risk-benefit analysis: a thoughtful assessment of both the positive and negative results that may be caused by a particular course of action (14.4)

rotational motion: the motion of particles around other particles; characteristic of liquids (11.1)

S

salt: any ionic compound that is formed in a neutralization reaction from the anion of an acid and the cation of a base (10.3)

saturated hydrocarbon: a hydrocarbon that consists of chains or non-aromatic rings, whose carbon atoms are bonded to the maximum number of hydrogen or carbon atoms (13.3)

saturated solution: a solution in which no more of a particular solute can be dissolved at a specific temperature (8.1)

SI: the international system of measurement units, including units such as the metre, the kilogram, and the mole; from the French *Système international d'unités*) (1.2)

significant digits: the number of meaningful digits, including a final uncertain digit, that is obtained by measurement or used in calculations (1.2)

single displacement reaction: a chemical reaction in which one element in a compound is replaced (displaced) by another element (4.3)

skeleton equation: an equation that identifies the reactants and products in a chemical reaction by their chemical formulas but does not quantify them (4.1)

soft water: water with a low concentration of dissolved ions (9.4)

solubility: the amount of solute that dissolves in a given quantity of solvent at a specific temperature (8.1)

soluble: a term used to describe a substance that has a solubility greater than 1 g per 100 mL of a particular solvent (8.1)

solute: a substance that is dissolved in a solution (8.1)

solution: a homogeneous mixture of a solvent and one or more solutes (8.1)

solvent: a substance that has other substances dissolved in it (8.1)

specific heat capacity (*c*): the amount of energy (in J) required to change the temperature of 1 g of a substance by 1°C; measured in J/g•°C (14.3)

spectator ions: ions that are present in a solution but are not involved in the chemical reaction (9.2)

stable octet: an arrangement of eight electrons in the valence shell of an atom (2.2)

standard ambient temperature and pressure (SATP): 25°C and 100 kPa (11.4)

standard atmospheric pressure: 101.325 kPa at sea level and 0°C; the pressure that supports a column of mercury exactly 760 mm in height (11.2)

standard solution: a solution of known concentration (8.4)

standard temperature: 0°C, the freezing point of water (11.4)

standard temperature and pressure (STP): 0°C and 101.325 kPa (11.4)

Stock system: the current system for naming compounds that have elements that can have more than one valence; the valence of the first element name (usually a metal) in roman numerals in parentheses (e.g., copper(II)) (3.4)

stoichiometric amount: the exact molar amounts of a reactant or a product, as predicted by a balanced chemical equation (7.2)

stoichiometric coefficient: a number that is placed in front of the formula of the formula of a product or a reactant of a chemical equation to indicate how many moles are involved in the reaction (7.2)

stoichiometry: the study of the mass-mole-number relationships in chemical reactions and formulas (7.1)

straight-chain alkane: a hydrocarbon whose carbon atoms form a continuous chain of single carbon-carbon bonds (13.3)

strong acid: an acid that completely dissociates into ions in aqueous solutions (10.2)

strong base: a base that completely dissociates into ions in aqueous solutions (10.2)

structural diagram: a two-dimensional representation of the structure of a compound; can be a complete diagram, a condensed diagram, or a line diagram (13.2)

structural model: a three-dimensional representation of the structure of a compound (13.2)

STSE: an abbreviation for the interactions between science, technology, society, and the environment (1.1)

subatomic particle: one of the small particles (protons, neutrons, and electrons) that make up an atom (2.1)

sustainable development: the use of resources in a way that meets our current needs, without jeopardizing the ability of other people, or future generations, to meet their needs (14.5)

synthesis reaction: a chemical reaction in which two or more reactants combine to produce a single, different substance (4.2)

systematic name: a name that is based on the IUPAC rules for naming compounds (13.3)

T

temperature: a measure of the average kinetic energy of a substance or a system (14.3)

theoretical yield: the amount of product that is produced by a chemical reaction as predicted by the stoichiometry of the chemical equation (7.3)

thermal energy: the kinetic energy of particles; the energy possessed by vibrating particles (14.3)

thermal equilibrium: the state that is achieved when all the substances in a system have the same final temperature (14.4)

thermochemical equation: an equation that shows the energy produced or absorbed in a reaction (14.2)

titration: a laboratory process that is used to determine the concentration of a acidic or basic solution by reacting it with a solution of known concentration (10.3)

torr: a unit of pressure; equal to 1 mm of mercury in the column of a barometer or manometer (11.2)

total ionic equation: a form of chemical equation that shows dissociated ions of soluble ionic compounds (9.2)

translational motion: the independent motion of particles from one point in space to another; characteristic of gases (11.1)

triple bond: a covalent bond in which two atoms share three pairs of electrons (3.2)

trivial name: a name for a compound that does not necessarily suggest anything about the chemical composition of the compound (e.g., water, baking soda); also called the *common name* (3.4)

troposphere: the layer of the atmosphere that is closest to the surface of Earth (12.4)

U

universal gas constant (R): a proportionality constant that relates pressure, temperature, volume, and amount of gas; equal to 8.31 kPa·L/mol·K (12.1)

unsaturated hydrocarbon: a hydrocarbon that contains carbon-carbon double or triple bonds; the carbon atoms can potentially bond to additional atoms (13.3)

unsaturated solution: a solution in which more of a particular solute can be dissolved at a specific temperature (8.1)

V

valence: a number, positive or negative, that describes the bonding capacity of an element or ion (3.4)

valence electron: an electron that occupies the outermost energy level of an atom (2.2)

variable composition: a term used to describe a solution; capable of having different ratios of solutes to solvent (8.1)

vibrational motion: the motion of particles that are fixed in position; a characteristic of solids (11.1)

volumetric flask: a flat-bottomed, tapered glass vessel that is used to prepare standard solutions; accurate to ± 0.1 mL (8.4)

volume/volume percent: the volume of a liquid solute divided by the volume of the solution, expressed as a percent (8.3)

W

waste-water treatment: the cleaning of used water by physical, chemical, and biological processes (9.4)

water treatment: the process of removing chemical, biological, and physical contaminants to make water suitable for consumption (9.4)

weak acid: an acid that only slightly dissociates into ions in aqueous solutions (10.2)

weak base: a base that only slightly dissociates into ions in aqueous solutions (10.2)

weighted average: an average that takes into account the abundance or importance of each value (5.3)

word equation: an equation that identifies the reactants and products of a chemical reaction by name, but does not specify their amounts (4.1)

Z

zero sum rule: the rule stating that for chemical formulas of neutral compounds involving ions, the sum of positive valences and negative valences must equal zero (3.4)

Index

The page numbers in **boldface** type inidcate the pages where the terms are defined. Terms that occur in Sample Problems (*SP*), Investigations (*inv*), ExpressLabs (*EL*), and ThoughtLabs (*TL*) are also indicated.

Photo Credits

2-3 (background), © R. Kord/Firstlight.ca; **2-3** (background), Artbase Inc.; **2-3** (background), Ludek Pesek/Science Photo Library/Photo Researchers Inc; **4** (background), NASA/NSSCD; **4** (bottom right), Artbase Inc.; **5** (bottom left), Artbase Inc.; **6** (bottom right), © Ron Sherman/Stone; **6** (centre right), © Ron Fehling/Masterfile; **7** (centre left), Artbase Inc.; **7** (centre right), © Tom Stewart/The Stock Market/Firstlight.ca; **9** (centre left), © Sun Media Corporation; **16** (bottom centre), from *Chemistry: The Molecular Nature of Matter and Change*, Second Edition, by Martin S. Silberberg © 2000, The McGraw-Hill Companies Inc.; **23** (top left), Artbase Inc.; **32** (background), Darius Kochli/Firstlight.ca; **32** (bottom left), Armstong Roberts/Comstock; **32** (background), Artbase Inc.; **40** (bottom), © Stephen Frish Photography; **42** (top), from *Chemistry: The Molecular Nature of Matter and Change*, Second Edition, by Martin S. Silberberg © 2000, The McGraw-Hill Companies Inc.; **56** (top left), © Patrick Bennett/Corbis/Magma; **58** (top left), © Magma Photo News; **64** (background), © Underwood & Underwood/Corbis/Magma; **64** (top right), Anthony Howard/Masterfile; **65** (bottom left), © Catherine Ursillo/Photo Researchers Inc.; **66** (centre left), © George Bernard/Science Photo Library/Photo Researchers, Inc.; **66** (centre right), © Geoff Tompkinson/Science Photo Library/Photo Researchers, Inc.; **69** (top right), © Larry Stepanowicz/Visuals Unlimited, Inc.; **69** (centre left), © Gary Griffen/Animals Animals/Earth Scenes; **69** (bottom right), Artbase Inc.; **70** (top left), © Dick Hemingway; **72** (top right), from *Chemistry: The Molecular Nature of Matter and Change*, Second Edition, by Martin S. Silberberg © 2000, The McGraw-Hill Companies Inc.; **83** (bottom right), from *Chemistry: Concepts and Applications* © 2000, The McGraw-Hill Companies Inc.; **85** (centre left), © Jean Marc Barey/Agence Vandystadt/Photo Researchers, Inc.; **85** (centre left), Artbase Inc.; **90** (centre right), © Stephen Dalton/Photo Researchers, Inc.; **93** (background), Artbase Inc.; **110** (background), © Mike Timo/STONE; **110** (bottom left), © Mike & Carol Werner/Stock, Boston Inc./PictureQuest; **111** (bottom left), Trevor Bonderud/Firstlight.ca; **114** (top left), Artbase Inc.; **119** (bottom left), Artbase Inc.; **119** (bottom right), © Dale Sarver/Animals Animals; **120** (centre right), Bill Ivy/Ivy Images; **121** (centre left), © Bruce Forster/STONE; **123** (top right), © Charles D. Winters/Photo Researchers; **127** (top right), from *Sciencepower 10* © 2000, McGraw-Hill Ryerson Limited; **127** (centre left), from *Chemistry: Concepts and Applications* © 2000, The McGraw-Hill Companies Inc.; **141** (centre right), © Dave G. Houser/Corbis/Magma; **142** (centre left), © David Joel/STONE; **143** (centre left), VU/National Archives/Visuals Unlimited; **145** (top left), Notman Photographic Archives, McCord Museum of Canadian History, Montreal; **147** (top right), © AFP/CORBIS/MAGMA; **148-149** (background), © Notts/Firstlight.ca; **152** (centre left), COMSTOCK IMAGES/B. Pizaro; **152** (bottom left), © Russell D. Curtis/Photo Researchers, Inc.; **172** (centre right), from *Chemistry: Concepts and Applications* © 2000, The McGraw-Hill Companies Inc.; **173** (bottom left), from *Chemistry: Concepts and Applications* © 2000, The McGraw-Hill Companies Inc.; **174** (centre left), © Walter Schmid/Stone Images; **177** (bottom right), Artbase Inc; **192** (top left), © Nino Mascardi/Image Bank; **196** (background), Inga Spence/Index Stock Imagery; **196** (bottom right), Jackson Vereen/Food Pix; **198** (bottom right), Index Stock Imagery; **198** (bottom left), © Larry Williams/Masterfile; **198** (bottom right), Didier Givois/Allsport USA; **205** (centre), from *Sciencepower 8* © 2000, McGraw-Hill Ryerson Limited; **211** (centre right), © Will & Deni McIntyre/Photo Researchers, Inc.; **216** (top left), Chris Mikula/The Ottawa Citizen; **222** (centre left), © Mark Stephenson/Firstlight.ca; **223** (centre right), Angle Wynn/Native Stock; **232** (bottom left), NASA/Science Photo Library/Photo Researchers; **233** (bottom left), NASA; **236** (top right), © Thomas Dressler/First Light Associated Photographers; **242** (centre right), NASA; **244** (centre left), Neil Bartlett/B. C. Jennings Photographer; **248** (centre left), © Crown/Health & Safety Laboratory/Science Photo Library/Photo Researchers, Inc.; **252** (top left), © Nancy Richmond/Ivy Images; **260** (centre right), Artbase Inc; **265** (top left), V.C.L/Spencer Rowell/FPG International; **268** (top left), © Ed Gifford/Masterfile; **274** (centre left), © 1991, Richard T. Nowitz/Photo Researchers, Inc.; **280-281** (background), © Carolyn A. McKeone/Photo Researchers, Inc.; **282** (background), © C. Dass/Firstlight.ca; **282** (bottom centre), © Chris Vincent/Firstlight.ca; **283** (bottom left), © Christian Michaels/FPG International; **285** (top left), © Larry Stepanowicz/Visuals Unlimited, Inc.; **285** (top left), © Larry Stepanowicz/Visuals Unlimited, Inc.; **285** (top centre), © Chris Vincent/Firstlight.ca; **285** (top centre), © David Burch/Index Stock; **285** (top right), Dick Hemingway Photographs; **285** (background), © NMR/Firstlight.ca; **285** (top right), © IPS/Index Stock; **286** (bottom left), © Sovfoto/Eastfoto/PictureQuest; **290** (bottom left), © Lawrence Migdale/Science Source/Photo Researchers, Inc.; **298** (top left), © Terry C. Hazen/Visuals Unlimited, Inc.; **299** (top right), Dick Hemingway Photographs; **300** (bottom left), © Minden/Firstlight.ca; **307** (top left), Artbase Inc; **307** (top right), Artbase Inc; **309** (top right), © Charles D. Winters/Photo Researchers, Inc.; **310** (centre left), Dick Hemingway Photographs; **313** (centre right), © Matthias Kulka/Firstlight.ca; **319** (centre right), © Larry Stepanowicz/VISUALS UNLIMITED, INC.; **324** (top left), Artbase Inc.; **328** (background), Artbase Inc.; **328** (bottom left), Artbase Inc.; **350** (centre right), © Hans Reinhard/Okapia/Photo Researchers, Inc.; **358** (bottom right), © Geoff Tompkinson/Science Photo Library/Photo Researchers, Inc.; **360** (bottom right), © Andrew Syred/Science Photo Library/Photo Researchers, Inc.; **368** (bottom right), © McGraw Hill Higher Education/Stephen Frisch, Photographer; **368** (bottom right), © McGraw Hill Higher Education/Stephen Frisch, Photographer; **369** (bottom left), Kathleen Campbell/Stone; **373** (bottom right), © E.F. Smith Collection University of Pennsylania Library; **375** (bottom left), © Science Photo Library/Photo Researchers, Inc.; **375** (bottom centre), © E.F. Smith Collection University of Pennsylania Library; **385** (top left), © VU/Jeff Greenberg/MR/Visuals Unlimited; **408** (centre left), © Garry Black/Masterfile; **408** (top right), © Brian Sytnyk/Masterfile; **408** (centre right), © Gilbert&Brant/Photo Researchers, Inc.; **408** (top right), Artbase Inc.; **416** (background), © Pat Morrow/Firstlight.ca; **416** (bottom right), Comstock Photofile Limited; **417** (bottom left), © Firstlight.ca; **425** (centre left), © John Lemker/Earth Scenes; **427** (bottom left), © Damien Lovegrove/Science Photo Library/Photo Researchers, Inc.; **428** (centre right), Westerkov/Earth Scenes; **436** (bottom centre), © Bruce Davidson/Earth Scenes; **436** (bottom left), © Bruce Davidson/Earth Scenes; **436** (bottom right), © Bruce Davidson/Earth Scenes; **437** (top right), © Roy Ooms/Masterfile; **450** (bottom left), Artbase Inc.; **451** (top right), Michelle Del Guercio/Photo Researchers Inc.; **452** (centre left), Paul Rocheleau/Index Stock; **458** (centre left), Science Photo Library/Photo Researchers; **458** (bottom left), D&J Heaton/Stock Boston; **462** (bottom left), © James King-Holmes/Science Photo Library/Photo Researchers, Inc.; **462** (bottom right), Artbase Inc.; **463** (centre left), © Michael Creagen Photography; **464** (bottom left), © Bob Peterson/FPG International; **464** (bottom right), Artbase Inc.; **465** (bottom right), © Kevin Aitken/Firstlight; **468** (background), © Benelux Press/Masterfile; **468** (bottom left), North Wind Picture Archives; **469** (bottom left), COMSTOCK IMAGES; **470** (centre left), © Science Photo Library/Photo Researchers, Inc.; **470** (centre), © Science Photo Library/Photo Researchers, Inc.; **470** (centre right), © Science Photo Library/Photo Researchers, Inc.; **487** (centre left), © Bart Vallecoccia/Masterfile; **488** (bottom left), © Jeff Perkell/Masterfile; **489** (centre left), © David & Peter Turnley/CORBIS/MAGMA; **490** (top right), © Ralph Reinhold/Earth Scenes; **492** (top right), © E.R. Degginger/Earth Scenes; **496** (top left), © Charles D. Winters/Science Photo Library/Photo Researchers, Inc.; **504** (top left), Samuel Ashfield/FPG International; **505** (top right), © Charles D. Winters/Science Photo Library/Photo Researchers, Inc.; **507** (bottom right), Visuals Unlimited, Inc.; **513** (centre left), © Ned Therrien/Visuals Unlimited, Inc.; **514** (bottom left), © Peter Christopher/Masterfile; **515** (centre right), Artbase Inc.; **516** (top left), © Charles D. Winters/Science Photo Library/Photo Researchers, Inc.; **517** (top right), © NOAA/Science Photo Library/Photo Researchers, Inc.; **518** (top left), P.A. Ariya; **526** (bottom left), Comstock Images; **526** (top right), VU/© Bernd Wittich/Visuals Unlimited, Inc.; **532** Richard Dunoff/First Light Associates; **532** (background), © Richard Dunoff/Firstlight.ca; **533** (bottom left), Artbase Inc.; **536** (bottom left), © University of Alberta Imaging Centre 1999; **538** (top right), © 2000, Rosemary Weller/Stone; **561** (centre left), © David Hall/Masterfile; **568** (centre right), © Mike Dobel, MCMXCI/Masterfile; **572** (top left), © Michael Branscom Photography; **572** (centre right), © K9 Storm Inc.; **576** (background), Artbase Inc.; **576** (bottom right), Artbase Inc.; **578** (bottom right), © Peter Christopher/Masterfile; **578** (centre right), © Firstlight.ca; **578** (centre left), © Marilyn "Angel" Wynn/Native Stock; **580** (top left), © Lowell Georgia/Photo Researchers, Inc.; **580** (bottom right), © Phil Degginger/Color-Pic, Inc.; **580** (bottom left), © Alan Marsh/First Light; **588** (bottom left), © Charles D. Winters/Photo Researchers, Inc.; **595** (top right), Artbase Inc.; **601** (top left), Artbase Inc.; **603** (centre right), Holton Getty/Designed by Laplace and Lavoisier; **606** (top left), Comstock Photofile Limited; **609** (centre left), © Stephen Frish/Stock, Boston/PictureQuest; **612** (centre), from *Chemistry: Concepts and Applications* © 2000, The McGraw-Hill Companies Inc.; **613** (top left), © Eastcott/Momatiuk/Animals Animals; **613** (bottom left), © Stephen Frish/Stock, Boston/PictureQuest; **615** (bottom right), Artbase Inc.; **619** (centre right), Image Courtesy GORE-TEX; **619** (centre left), © 1988, Michael Ventura/Bruce Coleman, Inc.; **619** (bottom left), Artbase Inc.; **619** (bottom right), Artbase Inc.; **620** (top left), Artbase Inc.; **622** (bottom left), © Richard Berenholtz/Firstlight.ca; **622** (centre left), © S.L.I.D.E./Phototake; **622** (centre right), © Tim Davis/Photo Researchers, Inc.; **624** (top left), © Vanessa Vick/Photo Researchers, Inc.; **629** (top right), © George Hunter/Comstock; **630** (centre left), © SIU/Visuals Unlimited; **639** (centre right), COMSTOCK IMAGES/Grant Heilman Photography; **640** (centre left), © Ken Lax/Photo Researchers, Inc.; **641** (centre right), Holt Studios International (Nigel Cattlin)/Photo Researchers Inc.; **641** (centre right), Bernd Wittich/Visuals Unlimited; **642** (centre left), © Jonathan Blair/CORBIS/MAGMA; **643** (bottom right), COMSTOCK IMAGES

Illustration Credits

16 (centre), from *Chemistry: Concepts and Applications* © 2000, The McGraw-Hill Companies Inc.; **29** (centre right), from *Chemistry: The Molecular Nature of Matter and Change*, Second Edition, by Martin S. Silberberg © 2000, The McGraw-Hill Companies Inc.; **35** (bottom right), from *Chemistry: The Molecular Nature of Matter and Change*, Second Edition, by Martin S. Silberberg © 2000, The McGraw-Hill Companies Inc.; **54** (top left), **42** (top), from *Chemistry: The Molecular Nature of Matter and Change*, Second Edition, by Martin S. Silberberg © 2000, The McGraw-Hill Companies Inc.; **54** (centre right), **42** (top), from *Chemistry: The Molecular Nature of Matter and Change*, Second Edition, by Martin S. Silberberg © 2000, The McGraw-Hill Companies Inc.; **57** (bottom left), from *Chemistry: The Molecular Nature of Matter and Change*, Second Edition, by Martin S. Silberberg © 2000, The McGraw-Hill Companies Inc.; **141** (bottom), from *Chemistry: The Molecular Nature of Matter and Change*, Second Edition, by Martin S. Silberberg © 2000, The McGraw-Hill Companies Inc.; **166** (bottom), from *Chemistry: The Molecular Nature of Matter and Change*, Second Edition, by Martin S. Silberberg © 2000, The McGraw-Hill Companies Inc.; **215** (centre), from *Chemistry: The Molecular Nature of Matter and Change*, Second Edition, by Martin S. Silberberg © 2000, The McGraw-Hill Companies Inc.; **219** (bottom), from *Chemistry: The Molecular Nature of Matter and Change*, Second Edition, by Martin S. Silberberg © 2000, The McGraw-Hill Companies Inc.; **294** (centre right), from *Chemistry: Concepts and Applications* © 2000, The McGraw-Hill Companies Inc.; **301** (centre right), from *Chemistry: Concepts and Applications* © 2000, The McGraw-Hill Companies Inc.; **344** (bottom centre), from *Chemistry: The Molecular Nature of Matter and Change*, Second Edition, by Martin S. Silberberg © 2000, The McGraw-Hill Companies Inc.; **359** (centre), from *Chemistry: The Molecular Nature of Matter and Change*, Second Edition, by Martin S. Silberberg © 2000, The McGraw-Hill Companies Inc.; **418** (centre left), from *Chemistry: The Molecular Nature of Matter and Change*, Second Edition, by Martin S. Silberberg © 2000, The McGraw-Hill Companies Inc.; **419** (top right), from *Chemistry: The Molecular Nature of Matter and Change*, Second Edition, by Martin S. Silberberg © 2000, The McGraw-Hill Companies Inc.; **419** (centre right), from *Chemistry: The Molecular Nature of Matter and Change*, Second Edition, by Martin S. Silberberg © 2000, The McGraw-Hill Companies Inc.; **432** (centre right), from *Chemistry: The Molecular Nature of Matter and Change*, Second Edition, by Martin S. Silberberg © 2000, The McGraw-Hill Companies Inc.; **442** (top), from *Chemistry: The Molecular Nature of Matter and Change*, Second Edition, by Martin S. Silberberg © 2000, The McGraw-Hill Companies Inc.; **476** (bottom centre), from *Chemistry: The Molecular Nature of Matter and Change*, Second Edition, by Martin S. Silberberg © 2000, The McGraw-Hill Companies Inc.; **480** (bottom left), from *Chemistry: The Molecular Nature of Matter and Change*, Second Edition, by Martin S. Silberberg © 2000, The McGraw-Hill Companies Inc.

Text Credits

632-633 A portion of this material adapted from *Essential Chemistry* by Raymond Chang © 1996, McGraw-Hill

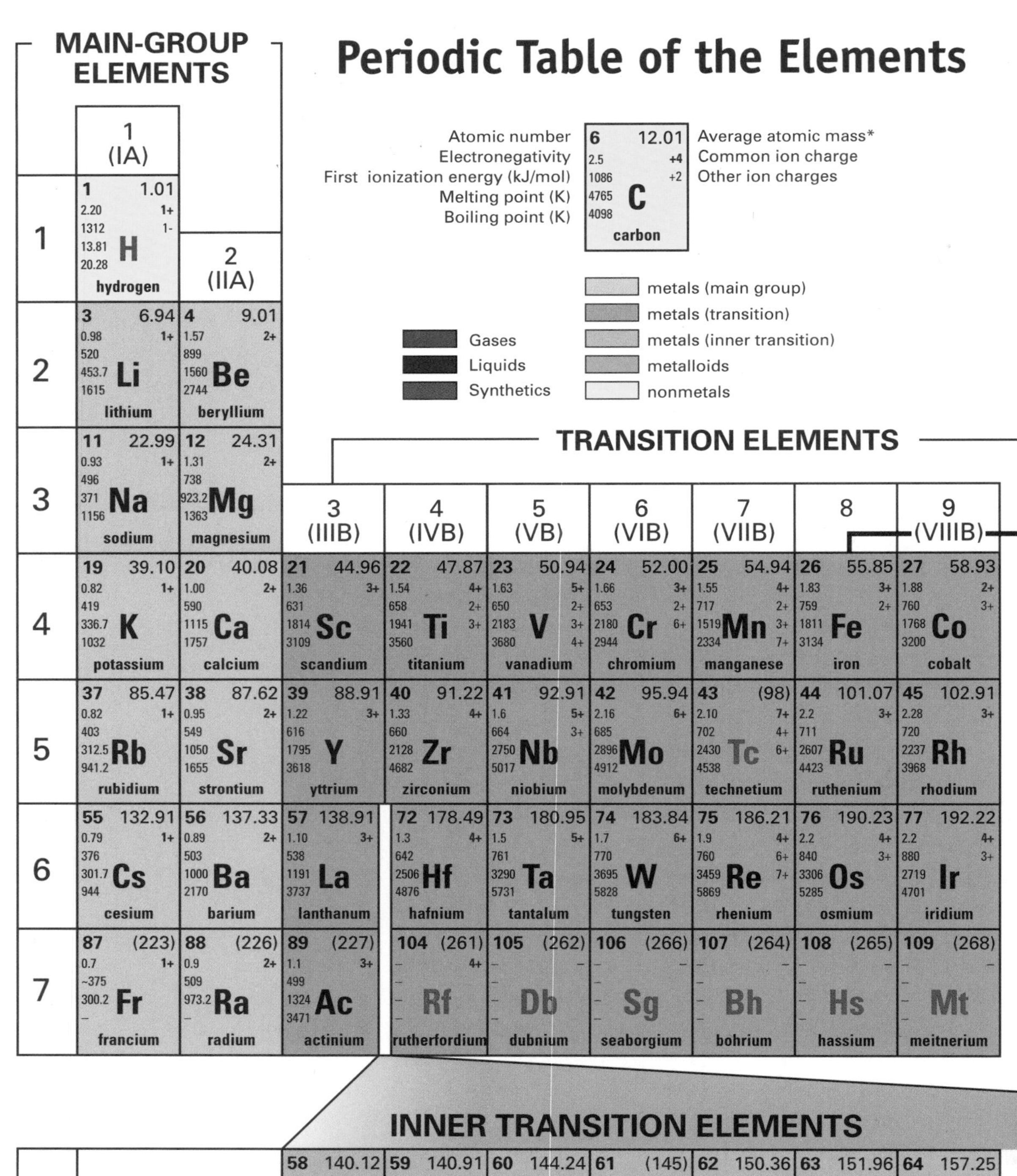

Periodic Table of the Elements

MAIN-GROUP ELEMENTS

TRANSITION ELEMENTS

INNER TRANSITION ELEMENTS

Key (carbon): Atomic number 6; Average atomic mass* 12.01; Electronegativity 2.5; Common ion charge +4; First ionization energy (kJ/mol) 1086; Other ion charges +2; Melting point (K) 4765; Boiling point (K) 4098; C; carbon

Legend: metals (main group); metals (transition); metals (inner transition); metalloids; nonmetals; Gases; Liquids; Synthetics

Period	Group	Atomic number	Symbol	Name	Average atomic mass*	Electronegativity	First ionization energy (kJ/mol)	Melting point (K)	Boiling point (K)	Ion charges
1	1 (IA)	1	H	hydrogen	1.01	2.20	1312	13.81	20.28	1+, 1-
2	1 (IA)	3	Li	lithium	6.94	0.98	520	453.7	1615	1+
2	2 (IIA)	4	Be	beryllium	9.01	1.57	899	1560	2744	2+
3	1 (IA)	11	Na	sodium	22.99	0.93	496	371	1156	1+
3	2 (IIA)	12	Mg	magnesium	24.31	1.31	738	923.2	1363	2+
4	1 (IA)	19	K	potassium	39.10	0.82	419	336.7	1032	1+
4	2 (IIA)	20	Ca	calcium	40.08	1.00	590	1115	1757	2+
4	3 (IIIB)	21	Sc	scandium	44.96	1.36	631	1814	3109	3+
4	4 (IVB)	22	Ti	titanium	47.87	1.54	658	1941	3560	4+, 2+, 3+
4	5 (VB)	23	V	vanadium	50.94	1.63	650	2183	3680	5+, 2+, 3+, 4+
4	6 (VIB)	24	Cr	chromium	52.00	1.66	653	2180	2944	3+, 2+, 6+
4	7 (VIIB)	25	Mn	manganese	54.94	1.55	717	1519	2334	4+, 2+, 3+, 7+
4	8	26	Fe	iron	55.85	1.83	759	1811	3134	3+, 2+
4	9 (VIIIB)	27	Co	cobalt	58.93	1.88	760	1768	3200	2+, 3+
5	1 (IA)	37	Rb	rubidium	85.47	0.82	403	312.5	941.2	1+
5	2 (IIA)	38	Sr	strontium	87.62	0.95	549	1050	1655	2+
5	3 (IIIB)	39	Y	yttrium	88.91	1.22	616	1795	3618	3+
5	4 (IVB)	40	Zr	zirconium	91.22	1.33	660	2128	4682	4+
5	5 (VB)	41	Nb	niobium	92.91	1.6	664	2750	5017	5+, 3+
5	6 (VIB)	42	Mo	molybdenum	95.94	2.16	685	2896	4912	6+
5	7 (VIIB)	43	Tc	technetium	(98)	2.10	702	2430	4538	7+, 4+, 6+
5	8	44	Ru	ruthenium	101.07	2.2	711	2607	4423	3+
5	9 (VIIIB)	45	Rh	rhodium	102.91	2.28	720	2237	3968	3+
6	1 (IA)	55	Cs	cesium	132.91	0.79	376	301.7	944	1+
6	2 (IIA)	56	Ba	barium	137.33	0.89	503	1000	2170	2+
6	3 (IIIB)	57	La	lanthanum	138.91	1.10	538	1191	3737	3+
6	4 (IVB)	72	Hf	hafnium	178.49	1.3	642	2506	4876	4+
6	5 (VB)	73	Ta	tantalum	180.95	1.5	761	3290	5731	5+
6	6 (VIB)	74	W	tungsten	183.84	1.7	770	3695	5828	6+
6	7 (VIIB)	75	Re	rhenium	186.21	1.9	760	3459	5869	4+, 6+, 7+
6	8	76	Os	osmium	190.23	2.2	840	3306	5285	4+, 3+
6	9 (VIIIB)	77	Ir	iridium	192.22	2.2	880	2719	4701	4+, 3+
7	1 (IA)	87	Fr	francium	(223)	0.7	~375	300.2	–	1+
7	2 (IIA)	88	Ra	radium	(226)	0.9	509	973.2	–	2+
7	3 (IIIB)	89	Ac	actinium	(227)	1.1	499	1324	3471	3+
7	4 (IVB)	104	Rf	rutherfordium	(261)	–	–	–	–	4+
7	5 (VB)	105	Db	dubnium	(262)	–	–	–	–	–
7	6 (VIB)	106	Sg	seaborgium	(266)	–	–	–	–	–
7	7 (VIIB)	107	Bh	bohrium	(264)	–	–	–	–	–
7	8	108	Hs	hassium	(265)	–	–	–	–	–
7	9 (VIIIB)	109	Mt	meitnerium	(268)	–	–	–	–	–

6 Lanthanoids

Atomic number	Symbol	Name	Average atomic mass*	Electronegativity	First ionization energy (kJ/mol)	Melting point (K)	Boiling point (K)	Ion charges
58	Ce	cerium	140.12	1.12	527	1071	3716	3+, 4+
59	Pr	praseodymium	140.91	1.13	523	1204	3793	3+
60	Nd	neodymium	144.24	1.14	530	1294	3347	3+
61	Pm	promethium	(145)	–	536	1315	3273	3+
62	Sm	samarium	150.36	1.17	543	1347	2067	3+, 2+
63	Eu	europium	151.96	–	547	1095	1802	3+, 2+
64	Gd	gadolinium	157.25	1.20	593	1586	3546	3+

7 Actinoids

Atomic number	Symbol	Name	Average atomic mass*	Electronegativity	First ionization energy (kJ/mol)	Melting point (K)	Boiling point (K)	Ion charges
90	Th	thorium	232.04	1.3	587	2023	5061	4+
91	Pa	protactinium	231.04	1.5	568	1845	–	5+, 4+
92	U	uranium	238.03	1.7	584	1408	4404	6+, 3+, 4+, 5+
93	Np	neptunium	237.05	1.3	597	917	–	5+, 3+, 4+, 6+
94	Pu	plutonium	(244)	1.3	585	913.2	3501	4+, 3+, 5+, 6+
95	Am	americium	(243)	–	578	1449	2284	3+, 4+, 5+, 6+
96	Cm	curium	(247)	–	581	1618	3373	3+

*Average atomic mass data in brackets indicate atomic mass of most stable isotope of the element.
Data obtained from *The CRC Handbook of Chemistry and Physics*, 81st Edition